THERMODYNAMICS

THERMODYNAMICS

FOURTH EDITION

J. P. Holman
Professor of Mechanical Engineering
Southern Methodist University

McGraw-Hill Book Company
New York St. Louis San Francisco Auckland Bogotá Hamburg
London Madrid Mexico Milan Montreal New Delhi Panama
Paris São Paulo Singapore Sydney Tokyo Toronto

THERMODYNAMICS

2 3 4 5 6 7 8 9 0 D O C D O C 8 9 2 1 0 9 8

ISBN 0-07-029633-2

Library of Congress Cataloging-in-Publication Data

Holman, J. P. (Jack Philip)
Thermodynamics.

Bibliography: p.
Includes index.
1. Thermodynamics. I. Title.
QC311.H76 1988 536′.7 87-3513
ISBN 0-07-029633-2

This book was set in Times Roman by Bi-Comp, Inc.
The editors were Anne Duffy and David A. Damstra;
the designer was Cheryl Hanna;
the production supervisor was Salvador Gonzales.
Drawings were done by J&R Services, Inc.
R. R. Donnelley & Sons Company was printer and binder.

ABOUT THE AUTHOR

Dr. JACK P. HOLMAN received his Ph.D. degree in mechanical engineering from Oklahoma State University in 1958. After two years active duty as a research scientist in the Air Force Aerospace Research Laboratory, he joined the faculty of Southern Methodist University, where he is presently Professor of Mechanical Engineering.

During his tenure at Southern Methodist University he has eight times been voted the Outstanding Engineering Faculty Member by the student body in a poll conducted annually. In addition, he has held administrative positions as Director of Thermal and Fluid Sciences Center, Head of the Civil and Mechanical Engineering Department, and Assistant Provost for Instructional Media.

As a principal investigator for research sponsored by the Atomic Energy Commission, National Science Foundation, NASA, Environmental Protection Agency, and the Strategic Defense Initiative, he has published extensively in such journals as *Industrial and Engineering Chemistry, International Journal of Heat and Mass Transfer, Journal of the Aerospace Sciences,* and others.

He is also the author of three widely used textbooks: *Heat Transfer,* 1963 (6th edition 1986), *Experimental Methods for Engineers,* 1966 (4th edition 1984), and *Thermodynamics,* 1969 (4th edition 1988), all published by the McGraw-Hill Book Company. These books have been translated into Spanish, Chinese, Japanese, Korean, and Portuguese and have received worldwide distribution through International Student Editions printed in Japan. Dr. Holman is the consulting editor for the McGraw-Hill Series in Mechanical Engineering. Dr. Holman also consults widely for industry in the fields of energy conservation and energy systems.

A member of the American Society of Engineering Education, he is past Chairman of the National Mechanical Engineering Division and past Chairman of the A.S.M.E. Region X Mechanical Engineering Department Heads. Dr. Holman is a registered professional engineer in the state of Texas, and received the Mechanical Engineer of the Year award from the North Texas Section of the American Society of Mechanical Engineers in 1971.

In 1972 Dr. Holman was recipient of the George Westinghouse Award from the American Society of Engineering Education for distinguished contributions to engineering education and in 1986 the American Society of Mechanical Engineers awarded him the James Henry Potter Gold Medal for eminent contributions to the science of thermodynamics in mechanical engineering.

CONTENTS

APPENDIX B
PROPERTIES FOR HEAT-TRANSFER CALCULATIONS 771

PREFACE

This book is intended for use in the first course in thermodynamics taken by students in engineering. Naturally, not all the material presented can be covered in one semester, and different courses may demand different emphases. For those programs which require it, the text contains sufficient material for a two-semester sequence.

In this fourth edition a new chapter on the elements of heat transfer has been added. This material may be of interest in courses taught primarily for electrical and computer engineers. Cooling problems of increased severity are now being encountered in electronics applications, so inclusion of this material may find strong interest among electrical engineers.

Some discussion of the chapter sequence is in order at this point. Chapters 1 to 4 present a classical development of the first law of thermodynamics, properties of pure substances, and energy analysis of open systems. Chapter 5 presents the second law of thermodynamics and the concept of available energy from a macroscopic viewpoint. Chapter 6 gives a classical presentation of equations of state and generalized compressibility relations.

Chapter 7 considers gaseous mixtures from the macroscopic viewpoint, with substantial emphasis placed on air-conditioning applications. The calculation of properties of real-gas mixtures is also discussed but can be omitted if sufficient time is not available to cover this material. Chapter 8 gives a classical development of chemical thermodynamics and equilibrium, including energy analyses of such systems. Chapter 9 discusses a broad variety of power cycles, with strong emphasis given to the limitations imposed on efficiency by the second law. Chapter 10 gives an abbreviated treatment of the thermodynamics of compressible flow.

Chapter 11 gives an introduction to the subject of heat transfer. In connection with this material some new tables have been added in the appendix to furnish such property values as thermal conductivity and viscosity, which are necessary for convection heat-transfer calculations.

Chapters 12 and 13 give an introduction to statistical thermodynamics and its application in the calculation of properties of gaseous and solid materials. These topics would normally be covered in a second course but could be included in a first course if preceded by the first- and second-law materials of Chapters 1 to 5 and the Maxwell relations treated in Chapter 6 (Sec. 6-5).

While the text allows for considerable flexibility in topic coverage, one must insist upon the inclusion of materials in Chapters 1 to 5 in order to provide a proper basis for further study. The course, with these materials as a base, can then move in the direction of applications most appropriate to those students enrolled.

It is appropriate to offer remarks regarding the progress of conversion to SI units. We may expect that the most rapid progress, albeit slow, will be in those industries with more exotic thermodynamic applications like aerospace, direct energy conversion, etc. Established industries in the power-generation and the heating, ventilating, and air-conditioning fields move much more slowly. In fact, it is difficult to find practicing engineers in these fields who use SI units at all. So the approach to units in this text has a very pragmatic objective: to prepare students for whatever practice they may encounter. Energy balance and power cycle analyses are presented in both SI and English unit formats. Compressible flow, irreversibility analysis, and heat transfer have more SI emphasis. It is an injustice to educate the student only in SI units, because industry simply does not operate this way. But it is equally unfair to ignore the obvious calculation advantages of SI units and the input which new graduates will have in promulgating their use. Some further comments in this regard are presented in Chapter 1. My personal preference would be for a very rapid changeover to SI units in all applications; but that seems unlikely to happen. For those who wish to conduct a course entirely in SI units there are ample examples and problems to do so.

The author is grateful for the comments and suggestions received from many individuals who have used the first three editions of this text. Of especial importance have been those penetrating questions of students which strike at the heart of a subject and demand clarification and explanation. As new materials have been added and clarifying remarks expanded, it is hoped that many of the student questions will be answered and, most importantly, that further inquiry will be stimulated.

J. P. Holman

THERMODYNAMICS

1

INTRODUCTION

1-1 THE NATURE OF THERMODYNAMICS

Energy propels society. The matchless economic and technological advances of the civilized world are traceable directly to an increasing amount of energy available to perform the tasks previously performed through human muscular effort. The availability of goods and services, and industrial productivity in general, are directly related to per capita energy consumption.

Thermodynamics is the study of energy and its transformation. This statement may seem rather aspiring, because it could be interpreted to mean that thermodynamics is the one science that is most strongly related to societal needs because of our increasing consumption of energy to produce goods and services. There are many different types of energy: the frictional work of a block sliding on a plane, electric energy, magnetic energy, nuclear energy, the energy stored in a quantum of light, the chemical energy of a petroleum fuel, and others. All these types of energy can fall in the province of thermodynamic analysis and we shall examine a variety of applications as the subject develops. As we shall see later, the laws of thermodynamics limit the amount of energy which is available to us for performing useful work.

Various sources of energy are becoming increasingly expensive to produce and one must be concerned with environmental factors involved in all conversion schemes. For these reasons, more and more attention must be focused on the efficiency of energy utilization. Thermodynamics furnishes the scientific basis for analysis of energy-conversion schemes and thus is central to an understanding of future energy-consumption trends and their social and economic impact.

Generally speaking, most studies of thermodynamics are primarily concerned with two forms of energy: heat and work. The principal objectives of the studies are to develop basic principles describing these types of energy and to become conversant with the language surrounding these basic principles. As in all such studies, the first step is to build a vocabulary of definitions and terms which may be used to conserve thought as the development becomes more complex. To define rigorously the various concepts of thermodynamics requires considerable space and effort, and will, of course, form a large part of the discussion in this book. In this introductory chapter we seek to give a brief qualitative picture of the broad subject of thermodynamics to achieve a perspective for detailed studies in subsequent chapters. In this respect it is well to note that many of the qualitative discussions are offered on the basis of physical reasonableness and should be accepted with the view that more rigorous definitions and developments will be presented later. The objective of this chapter is to achieve an overall picture of the scope of thermodynamics.

1-2 THE RELATION BETWEEN CLASSICAL MECHANICS AND THERMODYNAMICS

The study of classical mechanics involves concepts of force, mass, distance, and time. A force has a physical meaning of a "push or pull," which may be represented mathematically as a vector with a point of application. Mechanics is developed through the application of Newton's laws of motion and, particularly, the second law, which states that the summation of forces acting on a particle is proportional to the time rate of change of momentum:

$$\sum F = \frac{d}{d\tau}(mv)$$

For purposes of analyzing mechanical systems a *free body* is used, whereby a definite portion of a mechanism is broken away and all forces acting on this mechanism are specified for use with Newton's second law. It is important to realize that the mechanical *system* is specified in terms of its coordinates of space and velocity. The behavior of the mechanical system is further described in terms of its interaction with its surroundings through the application of various forces. We say that the *state* of the system may be specified with its space and velocity coordinates and its *behavior*; i.e., its change from one state to another is described in terms of its interactions with adjoining mechanisms or surroundings. It may be observed that the mechanical system will not change its state, i.e., its position in space and/or its velocity, unless it is acted upon by some net external force. The important point of this brief reference to classical mechanics is that the concept of a system (free body) and specification of the *state* of a system through the use of space or velocity coordinates are already familiar to those readers with experience in classical mechanics.

A System, Its Surroundings, and Its Boundary

While we are concerned with dynamical quantities in mechanics, the analysis of thermodynamic systems is concerned with energy quantities. A *system* is described in thermodynamics by breaking away a certain quantity of matter similar to the free-body technique in mechanics. The matter outside this system is termed the *surroundings,* and the separation between the system and surroundings is called the *boundary* of the system. As an example of a thermodynamic system consider a mass of air contained under pressure in a steel tank. The boundary of the system would be the inside surface of the tank and the surroundings would consist of the tank and the medium outside the tank. It is well to mention that the boundary of a system may be either a real or an imaginary surface. The air-tank system is shown in Fig. 1-1.

Thermodynamic State, Properties, and Processes

In mechanics the state of a system is specified by its space and velocity coordinates. The *state* of a thermodynamic system is described by specifying its thermodynamic coordinates. We cannot describe all thermodynamic coordinates at this point but may note that temperature, pressure, chemical energy content, etc., are typical examples. These coordinates are usually denoted as *properties* of the system. In mechanics we note that a system will not change its

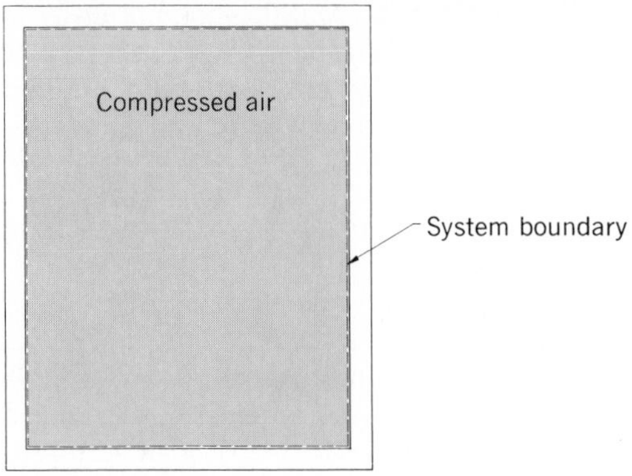

FIG. 1-1 Compressed air in a tank as a thermodynamic system.

state unless there is some interaction with its surroundings to change its spatial position and/or velocity. This interaction usually takes the form of an energy transfer into or out of the system.

When a thermodynamic system changes from one state to another, it is said to execute a *process*.

In the study of thermodynamics we are interested in the changes which a system may undergo as it executes various processes. Clearly, we must be able to define the state of the system of its thermodynamic coordinates if we are to meet with success in describing processes which the system may undergo, for, if we are to describe the process, we must say what is happening each step of the way. The state of the system must be described at each point in the process. To do this, a very fundamental concept must be introduced—the concept of equilibrium.

Equilibrium and the Quasistatic Process

A system is said to be in *equilibrium* when its pressure, temperature, and density are uniform (we shall employ a more rigorous definition of equilibrium later on). The system consisting of a mass of air in a tank would be in equilibrium if its pressure, temperature, and density were uniform throughout; however, if heat is applied to the tank such that the temperature at one end is higher than at the other end, the system will not be in equilibrium. When a system is in equilibrium, its thermodynamic coordinates are related in a definite way. Notice the importance of the concept of equilibrium. The system must be in equilibrium if we are to define its thermodynamic coordinates and their interrelationships.

Obviously we are interested in the process that is a succession of equilibrium states, because in this type of process we can define the state of the system each step along the way. We call this type of process *reversible*. The meaning of this term will be demonstrated later; nevertheless, this type of process is physically described as a succession of equilibrium states and is sometimes called a *quasistatic* process because it is composed of a series of nearly static equilibrium states.

The preceding paragraphs have linked the analysis methods of mechanics with those of thermodynamics; in particular, it has been implied that the free body and thermodynamic system represent similar analysis techniques. They are similar in that in both cases certain quantities of matter are broken away and a study is made of the interactions with the surroundings. There is one important difference, however. In the free body of mechanics we are always careful to study the forces which act on the free body. In thermodynamic system analysis we study the effect of the forces as they interact with the system and its surroundings and focus our attention on energy quantities. The two kinds of interaction energies, i.e., energies which leave or enter the system, are work and heat. We shall see that there are forms of energy "contained" within the system which must also be considered.

1-3 TEMPERATURE, HEAT, AND THE ZEROTH LAW OF THERMODYNAMICS

The terms *temperature* and *heat* are normally used with the implication that the reader understands the meaning of these concepts. The contrary is more likely, however, because a precise understanding of heat and temperature is one of the objectives of the study of thermodynamics. Intuitively, the physical meaning of temperature is that it describes whether a body is "hot" or "cold." For example, we touch a block of metal at 120°F and conclude that it is hotter than a block of ice. The reason for this conclusion is that the hot block of metal gives up heat energy to the hand whereas the cold block of ice extracts energy. Notice that this intuitive concept of temperature is based upon an energy-transfer process which we might simply describe as *heat exchange*. It might therefore be possible to conclude that if two bodies at the same temperature are brought into contact no heat will be exchanged between the two. This serves to define *equality of temperature* but cannot establish an absolute scale of temperature, which may only be accomplished with the second law of thermodynamics, to be discussed in subsequent chapters. The concept of equality of temperature may be stated in the form:

Two bodies, each in thermal equilibrium with a third body, are in thermal equilibrium with each other.

This statement is sometimes referred to as the *zeroth law of thermodynamics*.

It may be noted that for two bodies to be in complete equilibrium it is necessary to specify that more than the temperature of the two bodies must be equal; i.e., the pressure must also be equal and the two bodies must be so constituted that chemical change does not take place when they are brought into contact.

Insofar as everyday observations are concerned, we usually take as an indication of the temperature of a system the effect on some easily observable property of a measuring device. Thus a mercury-in-glass thermometer identifies temperature level with the relative expansion of the mercury and glass, an electrical resistance thermometer identifies temperature level with the electrical resistance of some particular metal (usually platinum), a thermocouple identifies temperature with the electromotive force generated at the junction of two dissimilar metals, and so on. In all these cases temperature, a thermodynamic property (or thermodynamic coordinate), is expressed in terms of other properties of either a liquid or metallic substance.

In accordance with this brief discussion we may tentatively conclude that heat is a form of energy which flows from one body to another as a result of a temperature difference. Note the almost circular definitions: Equality of temperature is related to a condition of zero heat flow—heat is related to temperature difference. We shall clarify this matter later on.

1-4 TEMPERATURE SCALES

The two temperature scales normally employed for measurement purposes are the Fahrenheit and Celsius scales. These scales are based on a specification of the number of increments between the freezing point and boiling point of water at standard atmospheric pressure. The Celsius scale has 100 units between these points, whereas the Fahrenheit scale has 180 units. The zero points on the scales are arbitrary.

It will be shown that the second law of thermodynamics serves to define an absolute *thermodynamic* temperature scale having only positive values. The absolute Celsius scale is called the *Kelvin scale,* and the absolute Fahrenheit scale is termed the *Rankine scale*. The zero points on both *absolute* scales represent the same physical state, and the ratio of two values is the same regardless of the scale used; i.e.,

$$\left(\frac{T_2}{T_1}\right)_{\text{Rankine}} = \left(\frac{T_2}{T_1}\right)_{\text{Kelvin}} \tag{1-1}$$

The boiling point of water is arbitrarily taken as 100° on the Celsius scale and 212° on the Fahrenheit scale. The relationship between the scales is indicated in Fig. 1-2, and it is evident that the following relations apply:

$$°\text{F} = 32.0 + \frac{9}{5}\,°\text{C} \tag{1-2}$$

$$°\text{R} = \frac{9}{5}\,\text{K}$$

$$°\text{R} = °\text{F} + 459.67 \tag{1-3}$$

$$\text{K} = °\text{C} + 273.15$$

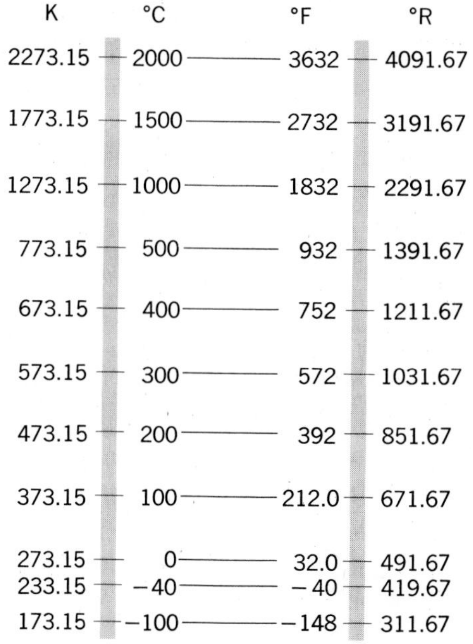

FIG. 1-2 Relationship between Fahrenheit and Celsius temperature scales.

To perform a measurement of temperature it is necessary to set up standards which may be employed for calibration of various thermometer devices. The boiling and freezing points of water are two such "standards," but they certainly do not encompass the whole range of temperatures of interest in experimental measurements. The International Practical Temperature Scale of 1968 [5]† serves to set up standards checkpoints over a wide range of tempera-

† Numbers in brackets pertain to references at end of each chapter.

TABLE 1-1
Primary Points for the International Practical Temperature Scale of 1968

Point	**Temperature**	
Normal pressure = 14.6959 psia = 1.0132 × 10⁵ Pa	**°C**	**°F**
Triple point of equilibrium hydrogen	−259.34	−434.81
Boiling point of equilibrium hydrogen at 25/76 normal pressure	−256.108	−428.99
Normal boiling point (1 atm) of equilibrium hydrogen	−252.87	−423.17
Normal boiling point of neon	−246.048	−410.89
Triple point of oxygen	−218.789	−361.82
Normal boiling point of oxygen	−182.962	−297.33
Triple point of water	0.01	32.018
Normal boiling point of water	100	212.00
Normal freezing point of zinc	419.58	787.24
Normal freezing point of silver	961.93	1763.47
Normal freezing point of gold	1064.43	1947.97

Source: Barber [5].

TABLE 1-2 Secondary Fixed Points for the International Practical Temperature Scale of 1968

Point	Temperature, °C
Triple point, normal H_2	−259.194
Boiling point, normal H_2	−252.753
Triple point, Ne	−248.595
Triple point, N_2	−210.002
Boiling point, N_2	−195.802
Sublimation point, CO_2 (normal)	−78.476
Freezing point, Hg	−38.862
Ice point	0
Triple point, phenoxybenzene	26.87
Triple point, benzoic acid	122.37
Freezing point, In	156.634
Freezing point, Bi	271.442
Freezing point, Cd	321.108
Freezing point, Pb	327.502
Boiling point, Hg	356.66
Boiling point, S	444.674
Freezing point, Cu-Al eutectic	548.23
Freezing point, Sb	630.74
Freezing point, Al	660.74
Freezing point, Cu	1084.5
Freezing point, Ni	1455
Freezing point, Co	1494
Freezing point, Pd	1554
Freezing point, Pt	1772
Freezing point, Rh	1963
Freezing point, Ir	2447
Freezing point, W	3387

Source: Barber [5].

tures. The primary and secondary fixed points for this scale are given in Tables 1-1 and 1-2. In addition to these fixed points the international temperature scale also establishes precise procedures for interpolating between these values.

The reader should realize that we still have not given a precise *definition* of temperature. For a satisfactory analytical definition we must enter into a full discussion of the second law of thermodynamics, whereas experimental measurements (or definitions) of temperature require a set of standards like the ones described previously. Detailed techniques for measurement of temperature are described in [2]. Precise experimental measurements of temperature require considerable laboratory experience, and new techniques are introduced periodically.

1-5 THE STATE PRINCIPLE

From the brief preceding discussion we may easily see that certain properties of matter are functionally related: Thermal expansion is related to temperature, the pressure of a gas is related to temperature and volume, and so on. Mention has been made of the fact that our main interest is with equilibrium states because it is in these states that we are able to define system properties. The primary problem is to be able to define the state of the system, and the question immediately arises as to how many properties (or thermodynamic coordinates) will be necessary to accomplish this. Once again we may draw upon experience with mechanics to illustrate the nature of the problem. In planar motion we need to specify only two coordinates, whereas in spatial motion three coordinates are needed to specify the position of a particle (the particle is the system in this case). Once the particle position is specified in cartesian coordinates, no additional information is required to know the position. Furthermore, any additional coordinates which may be supplied do not furnish us with any additional information about the particle position. For example, we may specify the particle position in cartesian coordinates and also give its cylindrical and spherical coordinates. These additional coordinates convey no new information; the *state* or position of the particle is completely specified with the cartesian coordinates, and we might consider the other coordinates or properties as functionally dependent on the cartesian coordinates, as we know very well they are. It may be noted, of course, that a complete specification of the *dynamical state* of the particle will require a specification of the velocity coordinates as well as the position coordinates.

In thermodynamic systems we would expect to have certain primary properties which are necessary to define the state of a system, whereas other properties would be functionally dependent on these primary state variables. Although the simple particle system previously considered is quite easy to analyze insofar as a specification of position coordinates is concerned, even the simplest thermodynamic system involves quite a bit more thought. Our problem is to discover the thermodynamic properties *necessary* to define the state of the system; they are known as the *primary properties*. The other thermodynamic properties will then be functionally related to the primary properties. As in mechanics, the primary properties are treated as independent variables, while the remaining properties are the dependent variables for use in mathematical analysis. In general, the number of primary variables for each thermodynamic system may be determined only from experiment and experience. As an example, an ideal gas requires only two properties to define the state of the system. Any two of pressure, volume, or temperature may be used.

The state principle unites the experimental axiom that the properties of matter are functionally related and the intuitive concepts of primary and dependent properties. Note that the principle is one we accept without proof; i.e., it is based on a large number of experimental observations and has no other justification. We shall have more to say about the specification of the state of a system later; however, we can clearly see that variables such as chemical composition and the number of components present in the system can have a strong influence on the problem. For now, the reader should consider the state principle as a definition of the meaning and concept of primary and dependent properties of a system. We shall give a more precise statement of the principle in Chap. 3.

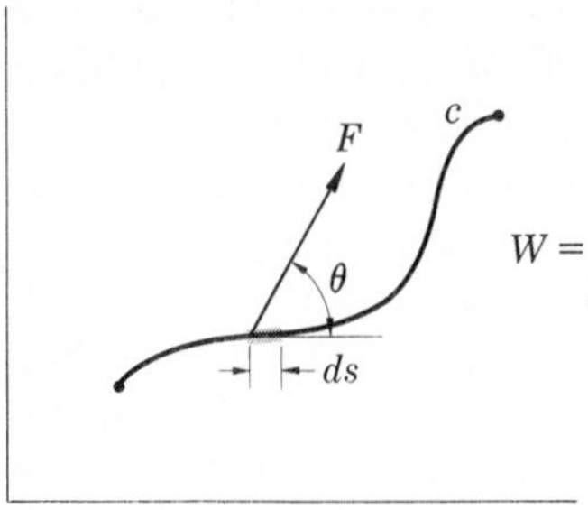

FIG. 1-3 Work expressed as a line integral or path function.

1-6 WORK

Work is defined as the energy expended by a force acting through a displacement and is expressed mathematically as

$$\text{Work} = \int_c \mathbf{F} \cdot d\mathbf{s} \tag{1-4}$$

where the line integral indicates that we consider only the product of force and displacement *in the direction of the force,* as illustrated in Fig. 1-3. To calculate work we only need to be able to specify the force and its movement; if this information is available we need to know nothing about thermal properties of the system. Why then should work be considered in the study of thermodynamics? The answer to this question is that work is an energy quantity and as such falls in the province of thermodynamic analysis. We shall see later how work and heat are related.

1-7 MICROSCOPIC AND MACROSCOPIC THERMODYNAMICS

The discussion so far has centered around consideration of thermodynamic systems composed of quantities of matter of a finite size, such as the compressed air in the steel tank or a hot block of metal. Both of these systems fall into the subject of *macroscopic* thermodynamics. In the analysis of such systems only the bulk nature and properties of matter are considered, and the detailed molecular and atomic structures of the substance are ignored. This type of analysis is sometimes called *classical* thermodynamics.

When we consider the detailed molecular and atomic nature of matter, the analysis is termed *microscopic* thermodynamics. Microscopic thermodynamics is considerably more complex than classical thermodynamics.

Microscopic thermodynamic analysis could proceed by first analyzing the behavior of gas molecules on the basis of classical mechanics. This development is called *kinetic theory*. In this analysis the laws of mechanics are accepted as axiomatic, and certain special macroscopic thermodynamic phenomena are shown to follow from the detailed microscopic analysis.

It is well known that the laws of classical mechanics must be supplemented by quantum theory when operating on an atomic scale of matter. The number of particles which are considered in a microscopic analysis is very large (of the order of 10^{20} cm^{-3}); hence the application of statistical techniques to describe the most probable distribution of the particles among certain energy and momentum states would be expected to meet with good success. This is precisely the case. When statistical techniques are coupled with the limitations imposed by the quantum theory, the general method of analysis is called *statistical mechanics* or *statistical thermodynamics*.

Microscopic thermodynamics has two objectives:

1 To show how energy is distributed at the molecular and atomic level and to indicate calculations of macroscopic energy properties from this distribution.
2 To show how a knowledge of energy distribution at the microscopic level may be employed for a calculation of transport rates. These transport rates may involve energy, momentum, or mass quantities and are important in a variety of engineering applications.

To illustrate further the nature of the difference between classical macroscopic thermodynamic analyses and those which deal with the microscopic nature of matter, consider the case of a volume of gas, say 1 liter, at room temperature and pressure. We know that this volume contains something on the order of 10^{23} molecules which are moving randomly in all directions, hitting each other and the walls of the vessel in which they are contained. Let us suppose that these molecules behave like perfectly elastic spheres in their collision processes. Presumably, we could apply newtonian dynamics to these particles in order to determine their motions in space. For this analysis we would need to write Newton's second law of motion

$$\sum F = \frac{d}{d\tau}(mv)$$

for *each* particle. All of the equations thus obtained would have to be solved simultaneously in order to determine the various particle velocities. What an enormous task that would be! The solution of such a large set of equations is far beyond the reach of even the largest computers. If we complicate the problem by considering intermolecular force fields and the quantization of particle energies, the solution appears even more hopeless. Why, then, have we said that some thermodynamic analyses are performed on a microscopic basis, i.e., by considering the detailed molecular behavior? The answer is simple: We do not solve this formidable set of equations, but instead rely on other methods to infer the molecular behavior. These other methods are statistical in nature.

1-8 THERMODYNAMIC PROPERTIES

Intensive and Extensive Properties

A *thermodynamic property* may be described as any observable characteristic of a system. The general classification of thermodynamic properties may be divided into the specific classes of intensive and extensive properties. *Intensive* properties are those which are independent of the quantity of matter enclosed by the boundary of the system. *Extensive* properties are those whose values are directly proportional to the mass of the system. Thus the total volume of a system would be an extensive property, whereas the specific volume (volume per unit mass) would be an intensive property. Temperature and pressure are other examples of intensive properties. In general, an extensive property when divided by the mass of the system becomes an intensive property.

Phases and Components

A thermodynamic system may contain more than one *phase,* such as a mixture of water and water vapor in which the liquid and vapor phases are present. A mixture of air, water droplets, and ice particles would be a three-phase mixture. A thermodynamic system may also consist of more than one *component.* The mixture of water and water vapor contains only one component, whereas the air-water-ice mixture contains two components: water and air. Note the distinction between phase and component. A mixture of nitrogen and carbon dioxide at atmospheric pressure and temperature would contain two components but only one phase—the gaseous phase. If the components and phases are uniformly distributed throughout the volume of the system, the system is said to be *homogeneous*. If the components and phases are not distributed uniformly, the system is *heterogeneous*. A water and water-vapor mixture may be homogeneous if the water is in the form of tiny liquid droplets uniformly distributed throughout the system volume or heterogeneous if the liquid is contained as a pool in the bottom of a pan with the vapor above it. The mixture of nitrogen and carbon dioxide would be another example of a homogeneous system.

The thermodynamic properties of a system obviously depend on its number of phases and components and whether the system is homogeneous or not.

1-9 FUNDAMENTAL THERMODYNAMIC LAWS

Without exploring any details, we may at least state the fundamental principles of thermodynamics at this point so that the reader can grasp in a qualitative way an overall perspective of the subject. These principles are accepted as axioms, as fundamental physical laws because they explain such a large body of experimental observations.

The first law of thermodynamic states:

The energy of an isolated system remains constant.

It is seen that the first law is nothing more than the law of conservation of energy which is familiar even to the neophyte scientist or engineer. The term *isolated system* describes one which does not exchange energy with its surroundings. Notice that this law is accepted as being axiomatic because it explains such a wide variety of experimental observations in physical science.

The second law of thermodynamics accepts as axiomatic several equivalent experimental observations, a few of which are stated in the following:

1 Heat flows from a high temperature to a low temperature in the absence of other effects. This means that a hot body will cool down when brought into contact with a body at a lower temperature, and not the opposite.
2 Two gases, when placed in an isolated chamber, will mix uniformly throughout the chamber but will not separate spontaneously once mixed.
3 A battery will discharge through a resistor, releasing a certain amount of energy, but it is not possible to make the reverse of this happen, i.e., to add energy to the resistor by heating and thus cause the battery to charge itself.
4 It is not possible to construct a machine or device which will operate continuously while receiving heat from a single reservoir and producing an equivalent amount of work.

We see that the second law of thermodynamics relates to the *direction* in which an energy transfer or conversion may take place; some transformations are allowed, whereas others are not. All of the phenomena described in the foregoing are consequences of the second law, although the most widely used statements are those of observations 1 and 4. Note that observation 4 states that we cannot build an engine (or cycle) which is 100 percent efficient in converting heat from a single reservoir to mechanical energy.†

The third law of thermodynamics pertains to the properties of substances at absolute zero temperature and must be developed after an analytical exposition of the first and second laws.

1-10 APPLICATIONS OF THERMODYNAMICS

Because thermodynamics is a study of energy and its transformations, it is not surprising that the most widely used applications of the science are those involving beneficial energy-conversion schemes. The large steam power plant

† A pithy expression with appeal to most students is: The first law says you cannot get something for nothing; the second law says you cannot even break even.

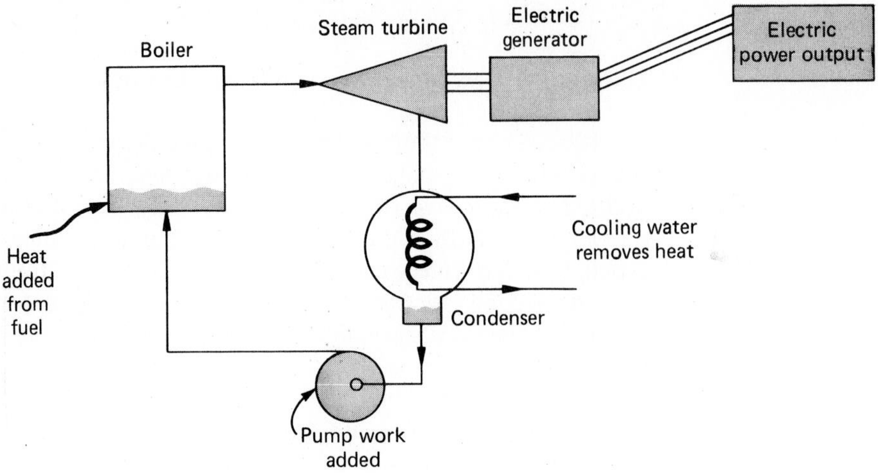

FIG. 1-4 Schematic of a steam power plant.

is a typical example and is illustrated in Fig. 1-4. High-pressure water is supplied to a boiler where heat is added to produce high-pressure, high-temperature steam. The heat may be furnished by converting the chemical energy of a fossil fuel (natural gas, oil, coal) in a combustion process, or as the heat release from a nuclear reaction. The steam is then forced through a turbine which drives an electric generator. During this flow process through the turbine, energy is removed from the steam and, as a result, it leaves the turbine at a much lower temperature and pressure. The low-pressure steam is then fed to a large vessel which contains hundreds of tubes carrying cooling water. The steam condenses on the outside of the tubes and collects at the bottom of the vessel where it is picked up by the pump and returned to the boiler at high pressure.

In the steam power plant we see the characteristic elements of a *power cycle* or *heat engine:*

1. Heat addition at high temperature (boiler)
2. Heat rejection at low temperature (condenser)
3. Net work output (work out of turbine less work into pump)

The *thermal efficiency* of a heat engine is defined as

$$\text{Thermal efficiency} = \frac{\text{net work output}}{\text{heat input at high temperature}} \tag{1-5}$$

This efficiency has definite economic significance because the heat input at the high temperature represents the energy which must be purchased (fuel, uranium, etc.) and the net work output represents what we get for the purchase. Large steam power plants can achieve efficiencies on the order of 40 percent.

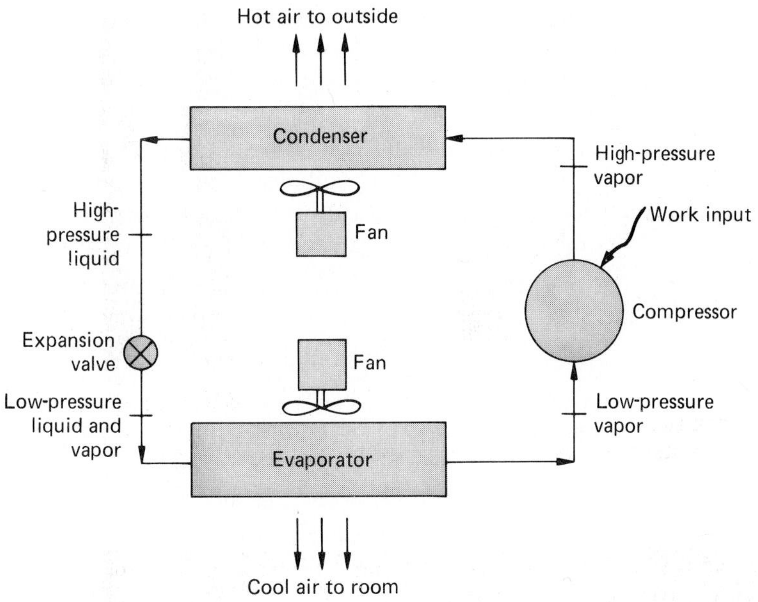

FIG. 1-5 Schematic of an air-conditioning or refrigeration system.

In Fig. 1-5 we have shown a schematic for a typical air conditioner for a home. In this application a fluid, typically Freon, at low pressure and temperature is allowed to enter the finned tubes of the evaporator. Air from the room is blown across the tubes, thereby adding energy to the Freon. As a result of this heat addition the Freon vaporizes (or boils) and leaves the evaporator in a vapor or gaseous state. It is then compressed to high pressure and temperature by adding work in the compressor. This high-temperature vapor is now at a higher temperature than the outside air so it is next passed through another finned tube device where the outside air removes sufficient heat to condense the vapor into a liquid at high pressure. The temperature at which a liquid will boil or condense depends on its pressure; the lower the pressure, the lower the evaporation temperature. Thus, the next step in this cycle is to allow the high-pressure liquid to flow through an expansion valve which lowers the pressure, thereby producing the low temperature entering the evaporator.

The air-conditioner or refrigeration cycle has an entirely different energy-conversion objective from the steam power plant. In the power cycle the objective is to obtain a work output from a heat input. For the refrigeration cycle our objective is the cooling effect for the room, while the energy we must purchase is the work input to the compressor. Thus, we measure our objective in this case by a *coefficient of performance:*

$$\text{Coefficient of performance (COP)} = \frac{\text{cooling effect}}{\text{work input}} \tag{1-6}$$

Refrigeration devices can have a COP on the order of 3.0. Sometimes the refrigeration cycle is called a *heat pump* because it takes the heat at the low room temperature and "pumps" it up to the high temperature outside.

There are many other energy-conversion devices (fuel cells, thermoelectric generators, internal combustion engines) which make use of thermodynamic principles, and we shall discuss these applications after the basic principles have been developed. For now, our purpose is to alert the reader to the fact that these applications exist and furnish motivation for careful study of the analysis techniques which will be employed in later chapters.

1-11 UNIT SYSTEMS

Despite strong efforts in the professional scientific and engineering community to standardize units with a single international system, a multitude of systems are still employed in practice. Because of its simplicity and adaptability, we shall stress the international system as the most desirable one to use; however, full recognition must be given to the fact that the reader of this text will encounter other unit systems. It would be a disservice, then, not to offer explanations of the other systems in wide use and display proper conversion factors which must be employed to obtain correct answers in technical calculations. The main difficulties arise in mechanical and thermal units because electrical units have been standardized for some time. The SI (Système Internationale d'Unités) set of units will eventually prevail, and we shall point toward use of this system wherever appropriate.

The purpose of a unit system is to affix specific numerical values to observable physical phenomena so that these phenomena may be described analytically. A *dimension* is a physical variable used to specify the behavior or nature of a particular system, and a *unit* is the term used to measure the dimension. A length dimension, for example, may be measured in units of inches, feet, centimeters, etc.

The choice and specification of a particular unit system are usually made so that they fit nicely with some basic physical principle. The basic physical dimensions we shall use to describe all physical phenomena are

L = length $\qquad \tau$ = time

M = mass $\qquad T$ = temperature

F = force $\qquad q$ = electric charge

The symbol q will also be used to designate heat, a form of energy, in subsequent chapters. For now we use it to designate charge. Mechanical systems of units are based on two physical principles: Newton's second law of motion.

Force $\sim$ time rate of change of momentum

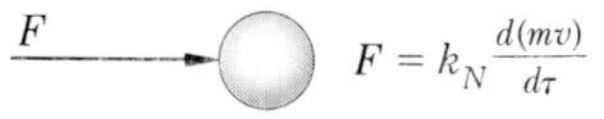

FIG. 1-6 Newton's second law of motion.

and Newton's gravitational attraction principle,

$$\text{Force} \sim \frac{m_1 m_2}{r^2}$$

Rewriting these principles in equation form,

$$F = k_N \frac{d(mv)}{d\tau} \tag{1-7}$$

$$F = k_G \frac{m_1 m_2}{r^2} \tag{1-8}$$

where the k_N and k_G are proportionality constants. The newtonian laws are illustrated in Figs. 1-6 and 1-7. Consider first Eq. (1-7) as applied to a particle of constant mass so that

$$F = k_N ma \tag{1-9}$$

where now a is the acceleration $dv/d\tau$. We may choose any system of units we like for this equation, just so the value of k_N is properly chosen. It is worthwhile to mention, however, that something must be said to define the size of each unit. The most commonly used systems are specified with the following statements:

1 One pound-force (lbf) will accelerate one pound-mass (lbm) 32.174 feet per second squared.
2 One pound-force will accelerate one slug-mass one foot per second squared.
3 One dyne-force will accelerate one gram-mass one centimeter per second squared.
4 One newton (N)-force will accelerate one kilogram-mass one meter per second squared.
5 One kilogram-force will accelerate one kilogram-mass 9.806 65 meters per second squared.

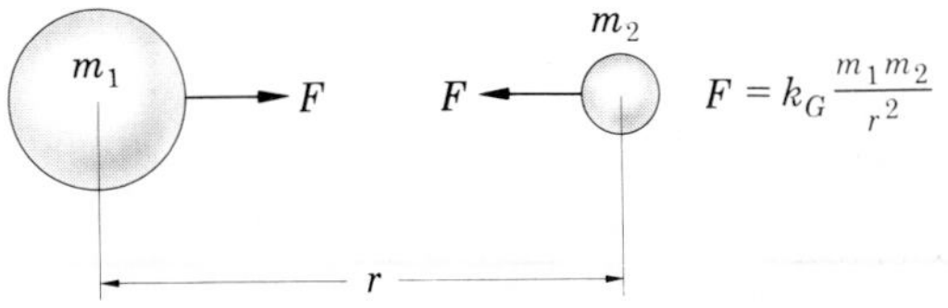

FIG. 1-7 Newton's gravitational law.

These statements serve to relate the quantities $FLM\tau$ through the dynamical Eq. (1-9). We might ask the question: How big is a newton? The answer is that it is big enough to accelerate 1 kilogram 1 meter per second squared. The next questions are: How big is a kilogram, a meter, a second? The answer is that these quantities are maintained as standards at various national laboratories throughout the world—in the United States at the National Bureau of Standards. Other standard units are similarly maintained. The standard unit of time, the second, is defined as the duration of 9 192 631 770 periods of the radiation corresponding to the transition between the two hyperfine levels of the fundamental state of the atom of cesium 133. Eventually all unit systems must rely on these standards, and others, but the subject of standards is beyond the scope of our discussion.

For convenience the constant k_N is sometimes written as

$$k_N = \frac{1}{g_c} \tag{1-10}$$

where g_c is now a new constant having the dimensions of $MLF^{-1}\tau^{-2}$.

The weight of a body is defined as the force exerted on the body as a result of the acceleration of gravity. Thus

$$W = \frac{g}{g_c} m \tag{1-11}$$

where g is the acceleration of gravity. Note that the weight of a body has the dimensions of a force.

Unfortunately, the concept of g_c is confusing to most students. In the first system of units discussed in the foregoing it has the value

1 $g_c = 32.174$ lbm·ft/lbf·s^2

In this system g_c has the same numerical value as the acceleration of gravity at sea level, but *it is not the acceleration of gravity*. Rather, it is a dimensional constant employed to facilitate the use of Newton's second law of motion with this particular system of units. In the other unit systems noted previously, g_c has the following values:

2 $g_c = 1$ slug·ft/lbf·s^2
3 $g_c = 1$ kg-mass·m/N·s^2
4 $g_c = 1$ g-mass·cm/dyn·s^2
5 $g_c = 9.8066$ kg-mass·m/kg force·s^2

The *kilopond-force* (kp) is an alternative nomenclature for kilogram-force and is employed widely in Europe.

The astute reader can see why systems 1 and 5 were devised. When one buys meat or produce at the grocery store, a *weight* or *force* measurement is performed. If a lifting crane is purchased, its capacity is specified in terms of its lifting *force*. The grocery scale or crane is calibrated in terms of force† such that

1 lbm will weigh 1 lbf at sea level

1 kg-mass will weigh 1 kg-force at sea level

In various equations used throughout this book the reader should be careful to ensure that the units of various terms are consistent. If the term involves a dynamical quantity based on Newton's second law, then the g_c constant is understood to be included in the proper value for the unit system under consideration. We shall adopt the practice of omitting it from many equations in order to be consistent with the nomenclature followed in mechanics. Thus the equation

$$F = ma$$

is understood to include the $1/g_c$ term even though it is not written. In another example, kinetic energy will be written as

$$\text{KE} = \frac{1}{2} mV^2$$

but it is understood to mean that

$$\text{KE} = \frac{1}{2g_c} mV^2$$

The reader should note that in three of the five unit systems previously mentioned the numerical value of g_c is unity. This fact accounts for their wide usage for problems in mechanics. The pound-mass (lbm) system is used extensively in many thermodynamic problems and, for this reason, due caution must be exercised when working with g_c.

Energy Units

Work and *energy*, of course, have the dimensions of a force times a distance and would be expressed as pound-force–foot, dyne-centimeter, or newton-meter with the definitions,

1 dyne-centimeter (dyn·cm) = 1 erg

1 newton-meter (N·m) = 1 joule (J)

† In practice the device is probably calibrated against a standard mass, but basically, a force measurement is performed.

Power is the time rate of doing work and some appropriate units are

$$1 \text{ watt (W)} = 1 \text{ J/s}$$

$$1 \text{ horsepower (hp)} = 550 \text{ ft·lbf/s} = 745.7 \text{ W}$$

Some units of energy peculiar to the study of thermodynamics are the so-called *thermal units*—the British thermal units (Btu) and the calorie (cal):

$$1 \text{ Btu} = \text{energy required to raise 1 lbm of water 1°F at 68°F}$$

$$1 \text{ cal} = \text{energy required to raise 1 g of water 1°C at 20°C}$$

$$1 \text{ kcal} = \text{energy required to raise 1 kg of water 1°C at 20°C}$$

Some useful conversion factors are

$$1 \text{ W} = 3.413 \text{ Btu/h}$$

$$1 \text{ hp} = 2545 \text{ Btu/h}$$

$$1 \text{ Btu} = 252.16 \text{ cal} = 1055.04 \text{ J}$$

The electronvolt (eV) is a unit of energy defined as the work required to move an electron through a potential difference of 1 volt (V). This unit may be converted to joules by observing that the charge on the electron, q_e, is

$$q_e = 1.602\ 189 \times 10^{-19} \text{ coulomb (C)}$$

But the volt is defined as the work to displace a unit charge, so we obtain immediately that

$$1 \text{ eV} = 1.602\ 189 \times 10^{-19} \text{ J}$$

The SI System

We can see that unit systems 3, 4, and 5 applied to Newton's second law of motion are "metric," but they employ different units for force, mass, and distance. The SI system is essentially that of number 4 and does not admit use of the thermal units of energy. In other words, the joule is the *only* energy unit which is allowed. Note that the calorie and kilocalorie are "metric" but they are not SI. In addition, the concept of g_c is not normally used in the SI system and the newton is *defined* as

$$1 \text{ newton (N)} \equiv 1 \text{ kilogram-meter per second squared(kg·m/s}^2) \qquad (1\text{-}12)$$

Basic and derived SI units are given in Tables A-3 and A-4 (Appendix). Standardized multiplier prefixes are also specified as shown in Table 1-3. Thus, the centimeter is cm, the kilojoule is kJ, and a meganewton would be MN. Wher-

TABLE 1-3 Standard Prefixes and Multiples in SI Units

Multiples and Submultiples	Prefixes	Symbols	Pronunciations
10^{12}	tera	T	tĕr'ȧ
10^{9}	giga	G	jĭ'gȧ
10^{6}	mega	M	mĕg'ȧ
10^{3}	kilo	k	kĭl'ô
10^{2}	hecto	h	hek'tô
10	deka	da	dĕk'ȧ
10^{-1}	deci	d	dĕs'ĭ
10^{-2}	centi	c	sĕn'tĭ
10^{-3}	milli	m	mĭl ĭ
10^{-6}	micro	μ	mī' krô
10^{-9}	nano	n	năn'ô
10^{-12}	pico	p	pē' cô
10^{-15}	femto	f	fĕm' tô
10^{-18}	atto	a	ăt' tô

ever possible, numerical results should use the standard prefixes. For example, we would write 2.6×10^5 J as either 260 kJ or 0.26 MJ.

Conversion factors for various units are given in the Appendix in Table A-2.

EXAMPLE 1-1

A 10-kilopond (kp) force is applied to a mass of 3 lbm. Calculate the acceleration which results.

SOLUTION We can perform the calculation in whatever unit system we choose, as long as we are consistent. Using the N-m-kg-s system, we observe

$$10 \text{ kp} = 98.066 \text{ N}$$

$$3 \text{ lbm} = (3)(0.454) \text{ kg} = 1.362 \text{ kg}$$

Then, from Eq. (1-9)

$$a = \frac{Fg_c}{m} = \frac{(98.066 \text{ N})(1.0 \text{ kg}\cdot\text{m/N}\cdot\text{s}^2)}{1.362 \text{ kg}}$$

$$= 72.00 \text{ m/s}^2 = 236.3 \text{ ft/s}^2$$

If we choose the lbm-lbf system,

$$10 \text{ kp} = (10)(2.2046) = 22.046 \text{ lbf}$$

$$a = \frac{Fg_c}{m} = \frac{(22.046 \text{ lbf})(32.174 \text{ lbm}\cdot\text{ft/lbf}\cdot\text{s}^2)}{3 \text{ lbm}}$$

$$= 236.3 \text{ ft/s}^2 = 72.00 \text{ m/s}^2$$

EXAMPLE 1-2
A large power plant like that shown in Fig. 1-4 has a net power output of 1000 MW and a thermal efficiency of 40 percent. The heat dissipated in the condenser is achieved by heating water 10°F at a temperature of about 68°F. Calculate the amount of water which must be heated per second.

SOLUTION The basic energy balance of the power plant is

$$\text{Heat added to boiler} = \text{net work output} + \text{heat dissipated in the condenser} \tag{a}$$

From the definition of thermal efficiency [Eq. (1-5)], we have

$$\text{Heat added in boiler} = \frac{1000\ \text{MW}}{0.4} = 2500\ \text{MW}$$

Then, from Eq. (a),

$$\begin{aligned}\text{Heat dissipated in condenser} &= 2500\ \text{MW} - 1000\ \text{MW}\\ &= 1500\ \text{MW}\\ &= 5.12 \times 10^9\ \text{Btu/h}\\ &= 1.427 \times 10^6\ \text{Btu/s}\end{aligned}$$

From the definition of the Btu, it will take 10 Btu to raise 1 lbm of water 10°F, so

$$\begin{aligned}\text{Mass of water required} &= \frac{1.422 \times 10^6\ \text{Btu/s}}{10\ \text{Btu/lbm}}\\ &= 1.411 \times 10^5\ \text{lbm/s}\end{aligned}$$

The gravitation law [Eq. (1-8)] expresses the gravitational attraction force between two masses in terms of their separation distance r and k_G, *the universal gravitational constant*. In the SI system the value of k_G is

$$k_G = 6.672 \times 10^{-11}\ \text{N}\cdot\text{m}^2/\text{kg}^2 \tag{1-13}$$

This value is determined experimentally and is subject to slight modification as better experimental results become available.

Electrostatic units are based on Coulomb's law, which states

$$F = k_c \frac{qq'}{r^2} \tag{1-14}$$

where F is the force exerted between the two charges q and q' and r is the separation distance, as shown in Fig. 1-8. The value of the proportionality

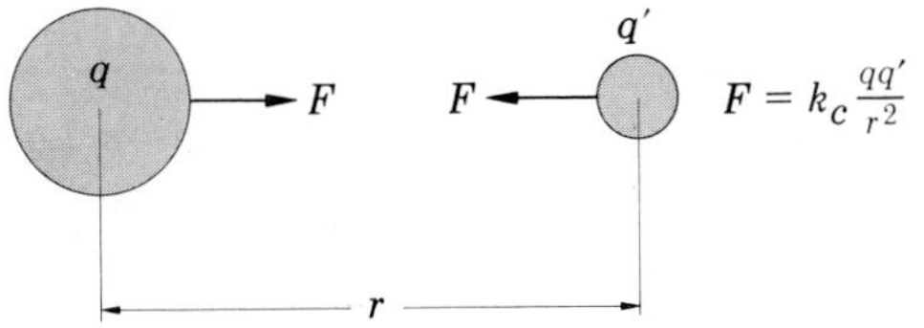

FIG. 1-8 Coulomb's electrostatic force law.

constant k_c is 9×10^9 N·m²/C², when the charge is expressed in coulombs and the force in newtons. As a matter of convention it is standard practice to let

$$k_c = \frac{1}{4\pi\epsilon_0} \tag{1-15}$$

where now $\epsilon_0 = 8.854\ 187\ 818 \times 10^{-12}$ is a new constant called the *permittivity of free space*.

Ampere's law states that the force between two parallel conductors, as shown in Fig. 1-9, is proportional to the currents in the conductors according to

$$F = k_A II' \frac{LL'}{r^2} \tag{1-16}$$

where I and I' are the currents in the conductors, L and L' are the lengths of the conductors, and r is the separation distance. The force is attractive if the currents are in the same direction. The proportionality constant k_A has the value

$$k_A = 10^{-7}\ \text{N/A}^2 \tag{1-17}$$

when the current is in amperes and the force in newtons. The ampere is defined by

$$1\ \text{ampere (A)} = 1\ \text{C/s} \tag{1-18}$$

FIG. 1-9 Ampere's law.

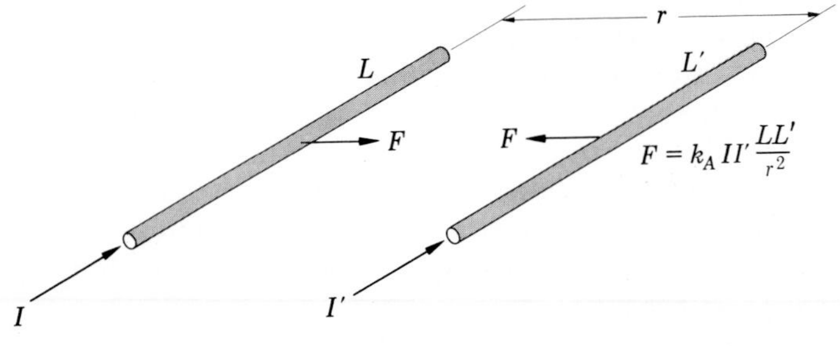

For convenience the following substitution is sometimes made:

$$k_A = \frac{\mu_0}{4\pi} \tag{1-19}$$

where now μ_0 is a new constant called the *magnetic permittivity of free space*.

Ampere's law clearly relates a force to a movement of charge (current). If a charge of q' moving with a velocity v experiences a force in the absence of any electrostatic field, the force is said to be due to a magnetic field B. The force is related by

$$\mathbf{F} = q'\mathbf{v} \times \mathbf{B} \tag{1-20}$$

The vector equation (1-20) serves to define the *magnetic induction* or magnetic flux density **B**. We see that a consistent set of units for **B** would be newton-second per coulomb-meter. It is common practice to introduce the new unit defined by

$$1 \text{ weber (Wb)} = 1 \text{ N·s·m/C} \tag{1-21}$$

Thus the units for **B** could be expressed as webers per square meter.

The *electric intensity* **E** is defined as the force exerted per unit charge resulting from an electric field. Thus

$$\mathbf{E} = \frac{\mathbf{F}}{q'} \tag{1-22}$$

The *electrical potential* **V** is defined as the work necessary to displace a charge per unit charge. A typical set of units for the electric potential would be joules per coulomb (newton-meter per coulomb). This is defined as

$$1 \text{ volt (V)} = 1 \text{ J/C} \tag{1-23}$$

1-12 MOVEMENT TOWARD THE SI UNIT SYSTEM

There is a strong movement to adopt the SI Unit system as the standard for the United States, since it has become widely accepted throughout the world. Considerable benefit will be derived from standardization of units, but we may expect workers in industry, and some research workers as well, to continue to employ a multitude of units for an extended period of time. In other words, units like the Btu, calorie, and horsepower will still be used in technical discussions for the indefinite future. For this reason, we shall use them freely in this text. To give the reader a feel for the magnitudes of the various units we shall express the answers to example problems in several alternative ways when it appears beneficial to do so.

TABLE 1-4
Typical Energy Quantities

	English (USCS)	SI
Heating value of gallon of fuel oil	149 000 Btu	157 MJ
Heating value of bituminous coal	12 900 Btu/lbm	30 MJ/kg
Solar energy on earth, clear day	330 Btu/h·ft^2	1040 W/m^2
Heating value, 1000 ft^3 natural gas	10^6 Btu	1.05 GJ
Nuclear energy, uranium	3.2×10^{10} Btu/lbm	74×10^6 MJ/kg
Air-conditioning cooling load for a typical dwelling, hot summer day	50 000 Btu/h	14.6 kW
Power output, typical large steam power plant	3.41×10^9 Btu/h	1000 MW
Typical body heat dissipation	450 Btu/h	132 W
Energy to vaporize water	1000 Btu/lbm	2326 kJ/kg

Undoubtedly, considerable confusion can arise when working with different unit systems, and different people prefer to use different systems. Furthermore, there can be vexing arguments over what represents a "fundamental" physical quantity. For example, our senses respond to forces and we sense mass through the inertial force exerted by the mass. On the other hand, mass represents a "quantity of matter," whereas force has no such quantitative interpretation, except through an indirect measurement process based on a physical principle relating force to "quantitative" properties of matter such as Eqs. (1-7) and (1-8). The purpose of this brief discussion of units and dimensions has been to show that the selection of unit systems is made on the basis of convenience so that they will be easy to use when applied to fundamental physical phenomena.

We have stated that thermodynamics is the study of energy and its transformation. Since we have now defined the various units of energy, it is worthwhile to give some typical values which may be of interest to the reader. Table 1-4 gives a few such quantities in both English and SI units. (The English system is also called the United States Customary System, or USCS.)

EXAMPLE 1-3

Two large conductors 3 ft long supplying current to an electric arc facility each carry a current of 1500 A and are separated by a distance of 3 in. Calculate the force acting on the conductors.

SOLUTION Ampere's law is employed for this calculation:

$$F = k_A I I' \frac{LL'}{r^2}$$

with

$$L = L' = 3\ \text{ft} = 0.9144\ \text{m}$$

$$I = I' = 1500\ \text{A}$$

$$r = 3\ \text{in} = 0.0762\ \text{m}$$

Thus

$$F = \frac{(10^{-7}\ \text{N/A}^2)(1500\ \text{A})^2(0.9144)^2}{(0.0762\ \text{m})^2}$$

$$= 32.4\ \text{N} = 7.28\ \text{lbf}$$

EXAMPLE 1-4

A man circling the earth in a spaceship is weighed with a device which indicates his weight as 275 N. From navigational measurements it is known that the gravitational acceleration at this particular location is 11.0 ft/s². Calculate the mass of the man and his weight at sea level on the earth.

SOLUTION The value of g is 11.0 ft/s² = 3.3528 m/s² and we use Eq. (1-11):

$$W = m\frac{g}{g_c}$$

$$275\ \text{N} = \frac{m(3.3528\ \text{m/s}^2)}{1.0\ \text{kg}\cdot\text{m/N}\cdot\text{s}^2}$$

which yields

$$m = 82.021\ \text{kg} = 180.83\ \text{lbm}$$

At sea level we have $g = 32.174$ ft/s² = 9.8066 m/s² so that the weight would be

$$W = \frac{mg}{g_c} = \frac{(82.021\ \text{kg})(9.8066\ \text{m/s}^2)}{1.0\ \text{N}\cdot\text{m/kg}\cdot\text{s}^2}$$

$$= 804.34\ \text{N} = 180.83\ \text{lbf}$$

We notice, of course, that 1 lbm weighs 1 lbf at sea level.

1-13 PRESSURE AND THE CONTINUUM

Most readers are familiar with the atomic and molecular nature of matter and know that a finite volume of liquid or gas, large enough to be visible to the eye, contains a very large number of particles. The density of such a finite volume ρ

is defined as the mass per unit volume or

$$\rho = \frac{m}{V} \tag{1-24}$$

This definition assumes that the volume is sufficiently large that the total number of particles in the volume does not experience significant fluctuations with time. In this case we say that the system can be treated as a *continuum*.

Pressure is defined as the normal component of force per unit area exerted by a fluid on a boundary. The pressure is defined only for an area element sufficiently large so that the fluid may be treated as a continuum. Thus

$$p = \lim_{\Delta A \to \Delta A'} \frac{\Delta F_n}{\Delta A} \tag{1-25}$$

where $\Delta A'$ is the minimum area for which the fluid behaves as a continuum. In general, continuum behavior is experienced so long as the average distance that a fluid molecule travels between collisions is small compared with a boundary dimension—in this case, to a side of the area element $\Delta A'$.

The total pressure exerted on a boundary wall is called the *absolute pressure*. The pressure exerted on a wall by the atmosphere is called the *atmospheric pressure* and varies with location and elevation on the earth's surface. The atmospheric pressure is the result of the weight of air at the particular location. *Gage pressure* designates the difference between absolute and atmospheric pressure in a particular system and is normally measured by an instrument which has atmospheric pressure as a reference. *Vacuum* represents the amount by which the atmospheric pressure exceeds the absolute pressure of a system. The term *vacuum* is synonymous with the term *negative gage pressure*. From these definitions we see that the absolute pressure may not be negative and the vacuum may not be greater than the local atmospheric pressure. The various pressure terms are illustrated in Fig. 1-10.

In the U.S. engineering system of units, pressure is usually expressed in pound-force per square inch absolute (lb/in^2 abs, or psia). Gage pressure carry-

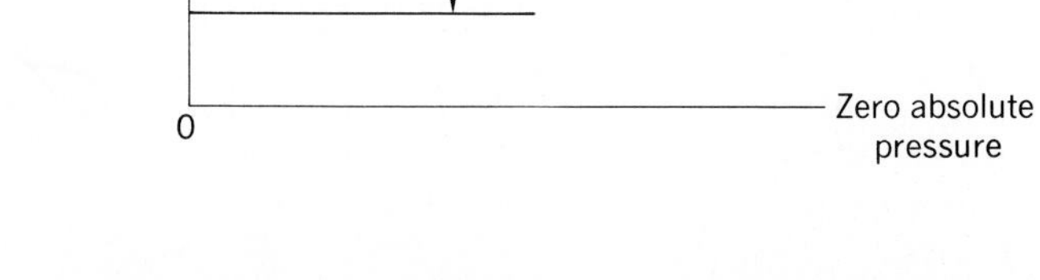

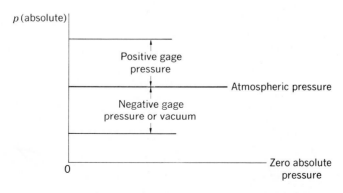

FIG. 1-10 Relationship between pressure terms.

ing the same units is expressed in pounds-force per square inch gage (lb/in^2 gage, or psig). Pressure is frequently expressed in terms of the height of a column of mercury (Hg) which it will support at a temperature of 20°C (68°F). At standard atmospheric pressure of 14.696 psia this height is 760 mm with the density of mercury taken as 13.5951 g/cm^3. The SI unit for pressure is the newton per square meter, or pascal (Pa). Conversion factors for pressure are given in Appendix A, Table A-2. Some other units commonly used are

$$1 \text{ microbar } (\mu\text{bar}) = 1 \text{ dyn/cm}^2 = 10^{-6} \text{ bar} = 0.1 \text{ Pa}$$

$$1 \text{ millimeter} = 1 \text{ mmHg} = 1333.22 \text{ microbars}$$

$$1 \text{ micrometer} = 1\ \mu\text{m} = 10^{-6} \text{ mHg} = 10^{-3} \text{ mmHg}$$

$$1 \text{ torr} = 1 \text{ mmHg}$$

Some of the common units for standard atmospheric pressure are

$$1 \text{ standard atmosphere} = \begin{cases} 14.696 \text{ lbf/in}^2 \text{ abs (psia)} \\ 29.92 \text{ inHg at } 32°\text{F} \\ 1.01325 \times 10^5 \text{ Pa} \\ 1.0332 \text{ kgf/cm}^2 \text{ (kp/cm}^2\text{)} \end{cases}$$

We should note that one bar (10^5 Pa) carries the connotation that it is "one *bar*ometric pressure," but it is in fact about 1 percent less than one standard atmosphere, as noted above.

1-14 THE IDEAL GAS

A *mole* (mol) is a quantity of a substance having a mass numerically equal to its molecular weight. One kilogram mole of oxygen, for example, would be 32.000 kg and 1 lbm mol would be 32.000 lbm. We shall designate the molecular weight with M and the number of moles with η. The mass of the substance m is therefore

$$m = \eta M \tag{1-26}$$

The mass per molecule is designated by M^* so that the mass per mole is

$$\frac{m}{\eta} = M^* N_0 = M \tag{1-27}$$

where N_0 is the number of molecules per mole. Thus

$$N_0 = \frac{M}{M^*} \tag{1-28}$$

This ratio is a constant called *Avogadro's number* and has the value

$$N_0 = 6.022\ 045 \times 10^{23} \text{ molecules/g mol} \tag{1-29}$$

The total volume is designated by the symbol V and the *specific* volume (volume per unit mass) by v. The molal specific volume (volume per mole) is designated by $\bar{v}$.

Now suppose a series of experiments is conducted with different gases. Pressure, volume, and temperature are measured for 1 mol of each gas at various pressures and temperatures. The results of the experiments might be plotted as shown in Fig. 1-11 for a single gas. Regardless of the gas, it is found experimentally that lines of constant temperature on this chart all converge to a single point for the limit of zero pressure. This value is defined as the *universal gas constant,*

$$\mathcal{R} = \lim_{p \to 0} \frac{p\bar{v}}{T} \tag{1-30}$$

and has the numerical value of 1545.35 ft·lbf/lbm·mol·°R or 8314.41 J/kg·mol·K.

To a good approximation many gases behave according to the simple equation

$$p\bar{v} = \mathcal{R}T \tag{1-31}$$

over a rather wide range of temperatures and pressures. Equation (1-31) is called the *equation of state* of an ideal gas. The term *equation of state* means that it establishes the relationships between the thermodynamic properties necessary to define the state of the system.

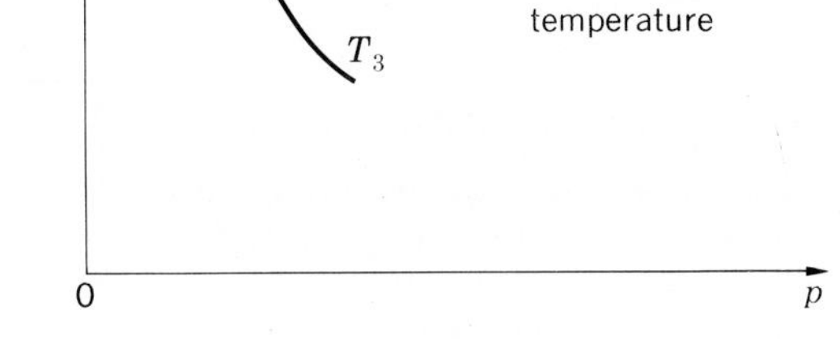

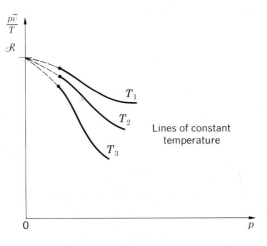

FIG. 1-11 Determination of the universal gas constant.

It is to be noted that the temperature in Eq. (1-31) must be expressed in absolute units, i.e., in either Rankine or Kelvin units.

We have used the indefinite phrase "rather wide range" to describe the applicability of the ideal-gas equation. Hopefully, this will not cause confusion. It turns out that gases like oxygen, nitrogen, air, helium, carbon dioxide, etc., at room temperature do obey Eq. (1-31) up to pressures of tens of atmospheres. In Chap. 6 we shall examine some explicit techniques for handling nonideal gases. In the meantime, when working the problems, the reader should assume that the foregoing gases do exhibit ideal behavior.

We can write Eq. (1-31) in several different ways. Since $\bar{v} = V/\eta$, we have

$$pV = \eta \mathcal{R} T \tag{1-32}$$

Also, since $V = mv$, we have

$$pv = \frac{\eta}{m} \mathcal{R} T \tag{1-33}$$

But $m/\eta = M$, so that

$$pv = \frac{\mathcal{R}}{M} T = RT$$

The ratio $\mathcal{R}/M = R$ is called the *gas constant* for the particular gas. We may also write

$$pV = mRT \tag{1-34}$$

In Eq. (1-24) we defined the density of a finite volume as the mass per unit volume, or

$$\rho = \frac{m}{V} = \frac{1}{v} \tag{1-35}$$

Thus the ideal-gas equation could also be written in the alternative form:

$$p = \rho RT \tag{1-36}$$

It is purely a matter of convenience as to which of the foregoing relations shall be employed for calculation purposes.

EXAMPLE 1-5

A certain automobile tire may be approximated as a torus 20 cm in diameter with an inside radius of 20 cm. Calculate the mass of air contained in the tire at 20°C and 2.0 atm *gage* pressure. The equivalent molecular weight of air is 28.97.

SOLUTION The mass of air is calculated by inserting appropriate values in Eq. (1-34). The volume is calculated as

$$V = \frac{\pi(20)^2}{4}(2\pi)(20 + 10) = 59\ 218\ \text{cm}^3 = 0.0592\ \text{m}^3$$

The gas constant for air is calculated as

$$R = \frac{\mathcal{R}}{M} = \frac{8314\ \text{J/kg·mol·K}}{28.97\ \text{kg/kg mol}} = 287\ \text{J/kg·K}$$

The term *gage pressure* refers to the differential pressure above atmospheric. Thus the absolute pressure is

$$p = 2\ \text{atm} + 1\ \text{atm} = 3\ \text{atm} = 3.04 \times 10^5\ \text{Pa}$$

From Eq. (1-34) the mass is

$$m = \frac{pV}{RT} = \frac{(3.04 \times 10^5)(0.0592)}{(287)(20 + 273)} = 0.214\ \text{kg}$$

EXAMPLE 1-6

Calculate the number of oxygen (O_2) molecules in a 170-cm^3 container at 2 atm and 250°C.

SOLUTION For this problem we first calculate the number of kilogram-moles and then make use of Avogardro's number in Eq. (1-29):

$$V = 170\ \text{cm}^3 = 1.7 \times 10^{-4}\ \text{m}^3$$

$$p = 2\ \text{atm} = 2.0264 \times 10^5\ \text{Pa}\ (\text{N/m}^2)$$

$$T = 250°\text{C} = 523\ \text{K}$$

The number of kilogram-moles may now be calculated with Eq. (1-32) using $\mathcal{R} = 8314$ J/kg·mol·K:

$$\eta = \frac{pV}{\mathcal{R}T} = \frac{(2.0264 \times 10^5)(1.7 \times 10^{-4})}{(8314)(523)} = 7.92 \times 10^{-6}\ \text{kg·mol}$$

The number of molecules is thus

$$N = \eta N_0 = (7.92 \times 10^{-6})(6.022 \times 10^{26}\ \text{molecules/kg·mol})$$
$$= 4.77 \times 10^{21}\ \text{molecules}$$

Note that the type of gas (oxygen) does not influence the answer. If we wish to calculate the mass of oxygen we may use

$$m = \eta M = (7.92 \times 10^{-6}\ \text{kg·mol})(32.00\ \text{kg/kg·mol}) = 2.53 \times 10^{-4}\ \text{kg}$$

1-15 THE IDEAL-GAS THERMOMETER

The foregoing discussion furnishes the basic principles for a temperature-measurement device which may serve as a secondary experimental standard. A fixed volume is filled with gas and exposed to the temperature to be measured, as shown in Fig. 1-12. At the temperature T the gas system pressure is measured. Next the volume is exposed to a standard reference temperature (as discussed in Sec. 1-4) and the pressure is measured under these conditions. According to Eq. (1-31), at constant volume,

$$T = T_{\text{ref}} \left(\frac{p}{p_{\text{ref}}} \right)_{\text{const. vol.}} \tag{1-37}$$

Now suppose that some of the gas is removed from the volume and the pressure measurements are repeated. In general, there will be a slight difference in the pressure ratio in Eq. (1-37) as the quantity of gas is varied. However, regardless of the gas used, the series of measurements may be repeated and the results plotted, as shown in Fig. 1-13. When the curve is extrapolated to zero pressure, the true temperature as defined by the ideal-gas equation of state will be obtained. A gas thermometer may be used to measure temperatures as low as 1 K by extrapolation.

It is most important to realize that a reference temperature must be established in order to utilize the gas-thermometer principle (or any other thermometer for that matter). This reference temperature may be established with the international temperature scale, as discussed in Sec. 1-4.

FIG. 1-12 Schematic of an ideal-gas thermometer.

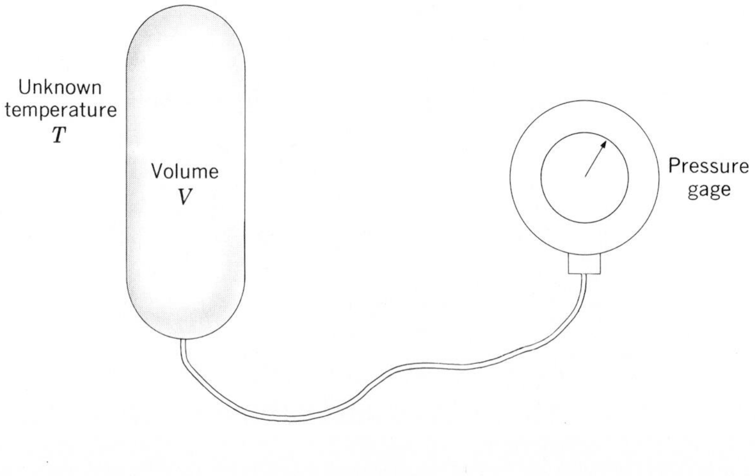

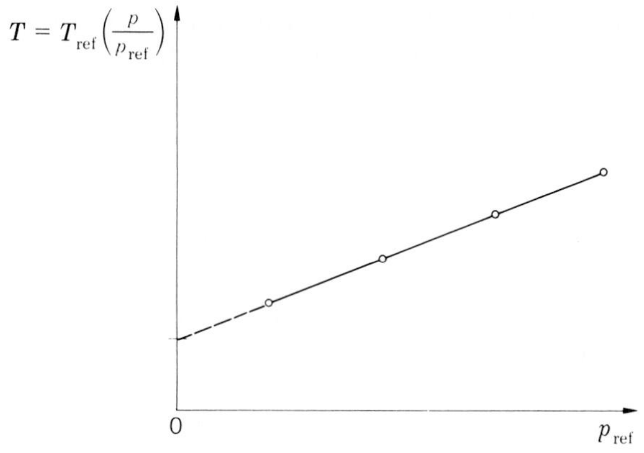

FIG. 1-13 Results of measurements with an ideal-gas thermometer.

1-16 SIMPLE KINETIC THEORY OF AN IDEAL GAS

To give an analytical indication of the relation between macroscopic and microscopic thermodynamics, let us consider a very simple kinetic analysis of gas molecules. We assume that all molecules are spherical in shape and perfectly resilient in their collision processes.

The pressure of the gas is identified with the force resulting from molecular collisions exerted on a containing wall, as shown in Fig. 1-14. The impulse imparted to the wall by each particle is given by

$$F\,d\tau = M^*v_z - M^*(-v_z) = 2M^*v_z \tag{1-38}$$

where M^* is the mass of the particle and v_z is the component of velocity in the z direction. For a perfectly elastic collision this component will have the same

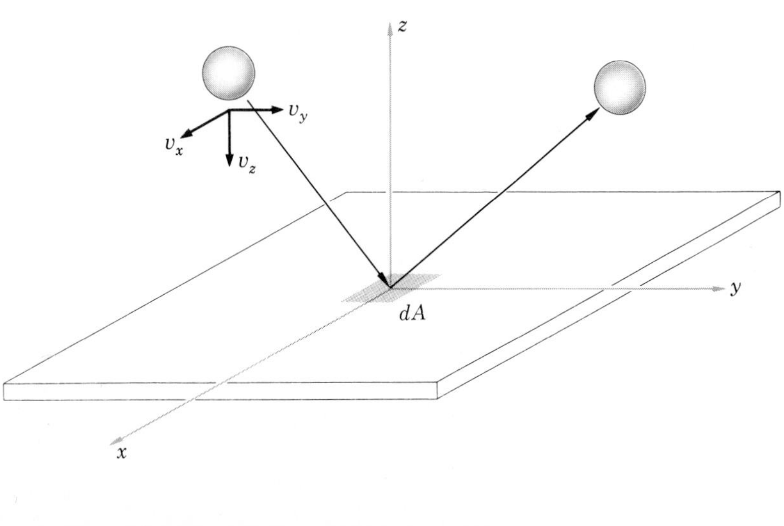

FIG. 1-14 Kinetic theory interpretation of pressure of an ideal gas.

magnitude before and after the collision. Now consider the small area dA. The rate at which molecules strike this area is the *flow rate* of molecules in the z direction, or

$$\dot{n} = n\,dAv_z \qquad \text{molecules per second} \tag{1-39}$$

where n is the molecular density (molecules/unit volume). If we assume random molecular motion, only *half* of the molecules will be headed downward, so that the molecular flow rate of interest is

$$\dot{n}_{-z} = \frac{1}{2}n\,dAv_z \tag{1-40}$$

The total impulse imparted to the surface in time $d\tau$ is therefore given as a product of Eqs. (1-38) and (1-40):

$$F\,d\tau = (2M^*v_z)\left(\frac{1}{2}nv_z\,dA\,d\tau\right)$$

The force per unit area, or pressure, is therefore

$$\frac{F}{dA} = p = M^*nv_z^2 \tag{1-41}$$

In reality, the molecules do not all move with the same speed, and the velocity factor in Eq. (1-41) must be an average value. It is reasonable to assume that the molecules move randomly in all directions so that one component is just as likely as another. We have

$$v^2 = v_x^2 + v_y^2 + v_z^2$$

The "randomness" assumption means that

$$\overline{v_x^2} = \overline{v_y^2} = \overline{v_z^2} = \frac{1}{3}\overline{v^2} \tag{1-42}$$

where the bar sign denotes an average value. Rewriting Eq. (1-41) in accordance with Eq. (1-42) gives

$$p = \frac{1}{3}nM^*\overline{v^2} \tag{1-43}$$

The product nM^* represents the mass per unit volume, which, of course, is the reciprocal of the specific volume (volume per unit mass). Accordingly, Eq. (1-

43) may be written as

$$pv = \frac{1}{3}\overline{v^2}$$

or

$$pV = m\frac{1}{3}\overline{v^2} \qquad \text{(1-44)}$$

Equation (1-44) is the same as Eq. (1-34) if we identify

$$\frac{1}{3}\overline{v^2} = RT \qquad \text{(1-45)}$$

Thus, in this kinetic interpretation of an ideal gas, temperature is found to be a measure of the average kinetic energy of the gas molecules.

EXAMPLE 1-7

Calculate the root-mean-square (rms) velocity of oxygen molecules at 14.7 psia and 70°F. The rms velocity is defined as

$$v_{\text{rms}} = (\overline{v^2})^{1/2}$$

SOLUTION We use Eq. (1-45) to calculate the mean-square velocity and thence the value of v_{rms}:

$$\overline{v^2} = 3RT \qquad (a)$$

The gas constant for oxygen is calculated as

$$R_{0_2} = \frac{\mathcal{R}}{M_{0_2}} = \frac{1545}{32} = 48.3 \text{ ft·lbf/lbm·°R} = 259.9 \text{ J/kg·K} \qquad (b)$$

At this point we remind the reader that g_c must be properly used when dynamical quantities are involved. Since we are working in the lbm-lbf system of units, $g_c = 32.174$ lbm·ft/lbf·s^2 and the kinetic energy is understood to be written as

$$\text{KE} = \frac{1}{2g_c} mv^2$$

This means that relation (*a*) is understood to contain g_c since $\overline{v^2}$ is representative of kinetic energy. We have

$$\overline{v^2} = 3RT = (3)(48.3)(460 + 70) = 7.67 \times 10^4 \text{ ft·lbf/lbm}$$

Note that a simple multiplication by g_c gives the proper units for $\overline{v^2}$:

$$\overline{v^2} = (7.67 \times 10^4 \text{ ft·lbf/lbm})(32.174 \text{ lbm·ft/lbf·s}^2)$$

$$\overline{v^2} = 2.47 \times 10^6 \text{ ft}^2/\text{s}^2$$

The rms velocity is thus

$$v_{\text{rms}} = (\overline{v^2})^{1/2} = (2.47 \times 10^6)^{1/2} = 1570 \text{ ft/s} = 478.5 \text{ m/s}$$

There are several restrictions that must be imposed on the foregoing analysis. Fow now, the reader should glean the fact that microscopic analysis may be employed to yield valuable physical interpretations of macroscopic phenomena. In even this simple case it has (1) identified temperature with molecular kinetic energy, (2) modeled an "ideal gas" as one having perfectly elastic collisions, and (3) modeled an ideal gas as one having perfectly random molecular motion. The last two results were initial assumptions that were made to formulate the model. The fact that these assumptions lead to a prediction of the ideal-gas law lends credence to their validity.

REVIEW QUESTIONS

1 Discuss the following terms:

(*a*) System	(*e*) Surroundings	(*i*) Phase
(*b*) State	(*f*) Quasistatic	(*j*) Component
(*c*) Property	(*g*) Work and heat	(*k*) Dimensions and units
(*d*) Process	(*h*) Boundary	

2 How are temperature scales defined for experimental work?
3 How is work defined?
4 What is the first law of thermodynamics?
5 Describe a physical phenomenon which is a result of the second law of thermodynamics.
6 What are the fundamental units in the SI system?
7 How does a pound-mass differ from a pound-force?
8 Which is larger, a kilopond or a newton?
9 Describe the so-called thermal units of energy.
10 Why isn't g_c the acceleration of gravity?
11 What is a continuum?
12 What is an ideal gas?
13 What is meant by the term *pressure?*
14 What is a mole?
15 What do the terms *gage pressure* and *vacuum* mean?
16 Why is the concept of equilibrium important in describing thermodynamic systems?

17 Discuss the similarity between a free body and the concept of a thermodynamic system.
18 What is a quasistatic process?
19 How are heat and temperature related?
20 Why are properties needed?
21 What determines the state of a thermodynamic system?
22 What is meant by the term *specific volume?*
23 How is thermal efficiency of a heat engine defined?
24 How does one characterize the performance of a refrigeration device?
25 What is Newton's second law of motion?
26 What is a Btu?
27 What is a kilopond-force?

PROBLEMS (MIXED UNITS)

1-1 A force of 1 N is applied to a mass of 3 lbm. Calculate the acceleration of the mass in ft/s^2. Also determine a value of g_c which would be applicable for this particular set of units.

1-2 Two 1-lbm masses are separated by a distance of 4 in. Calculate the gravitational attraction force between the masses. What acceleration would this force produce in ft/s^2?

1-3 Assuming that the diameter of the earth is 8000 mi, calculate an equivalent mass of the earth by using the acceleration of gravity at sea level (32.174 ft/s^2) and Newton's law of gravitation.

1-4 What is the weight of a 1 slug-mass at a location where the acceleration of gravity is 10 ft/s^2?

1-5 Two 1-ft parallel conductors each have a current of 1000 A and are separated by a distance of 1.5 in. Calculate the attraction force between the conductors in units of lbf.

1-6 A mass having a weight of 1 lbf at a location where $g = 11$ ft/s^2 is acted on by a force of 1 N. Calculate the acceleration.

1-7 What force in newtons is necessary to accelerate 1 slug 1 ft/s^2?

1-8 5 lbm of nitrogen are contained in a vessel at 10 psia and 500°F. Calculate the volume of the container and the number of molecules enclosed. Also calculate the rms velocity for the molecules.

1-9 Air ($M = 28.97$) is compressed from standard atmospheric temperature and pressure (14.7 psia, 70°F) to 300 psia and 300°F. The initial volume of the air is 1.0 ft^3. Calculate the volume under the high-pressure conditions.

1-10 Nitrogen is contained in a 3.5-m^3 tank at a pressure of 10 MPa and a temperature of 40°C. Calculate the mass in kilograms.

1-11 A small block of metal is weighed in a satellite where the acceleration of gravity is 3 m/s^2. The observed weight is 10 N. Calculate the mass in kilograms and also in some other units.

1-12 Calculate the rms velocity of nitrogen molecules at a temperature of 100°C.

1-13 A quantity of oxygen is compressed from 100 kPa and 50°C to 200 kPa and 100°C. Calculate the percentage change in volume which results from this compression.

1-14 How many molecules are contained in 1 lbm·mol of air?

1-15 Convert the following temperatures to °C:
(*a*) −40°F (*b*) 0°F (*c*) 1000°F (*d*) 2500°R (*e*) 1300 K

1-16 Calculate the density of the following gases at the specified conditions. Express in units of lbm/ft^3 and kg/m^3.
(*a*) Air at 1 atm and 68°F
(*b*) Oxygen at 500 psia and 300°F
(*c*) Oxygen at 3 MPa and 200°C
(*d*) Helium ($M = 4$) at 50 psia and 400°F
(*e*) Helium at 100 kPa and 20°C
(*f*) Carbon dioxide ($M = 44$) at 3 psia and 100°F
(*g*) Carbon dioxide at 10 kPa and 50°C

1-17 A certain container is connected to a vacuum pump and evacuated to an absolute pressure of 10^{-9} atm. Calculate the number of air molecules per unit volume at this low pressure and 70°F.

1-18 The boundary force on a submerged object is equal to the weight of displaced fluid. Consider a 10-ft-diameter balloon filled with helium at 3-psig pressure and 70°F. Calculate the net lifting force of the balloon when immersed in standard atmospheric air at 14.7 psia and 70°F. Neglect the weight of the balloon.

1-19 Calculate the rms velocity of hydrogen molecules in a gas at 1000 K.

1-20 Calculate the number of air molecules in a 20-by-20-by-10-ft room.

1-21 Calculate the pressure exerted by a 1-ft column of mercury.

1-22 A special manometer fluid has a specific gravity of 2.95 and is used to measure a pressure of 17.5 psia at a location where the barometric pressure is 28.9 inHg. What height will the manometer fluid register?

1-23 A mass of 3 kg is accelerated to a velocity of 20 ft/s. Calculate the work required to accomplish this acceleration and express in units of (*a*) ft·lbf, (*b*) joules, and (*c*) Btu.

1-24 Assuming that two large masses of 1000 kg each obey Newton's gravitational attraction principle in free space, calculate the velocity the two masses would assume when separated by a distance of 50 ft if they are released from rest when separated by a distance of 300 ft.

1-25 Calculate the attractive force between two parallel 6-ft conductors carrying currents of 10 000 A and separated by a distance of 2 in. Express the answer in lbf.

1-26 Calculate the number of O_2 molecules striking the wall per unit area and per unit time in a container where the gas is maintained at a pressure of 80 psia and a temperature of 500°F.

1-27 Repeat Prob. 1-26 for a pressure of 10^{-5} atm and a temperature of −100°F.

1-28 The solar constant is the solar energy flux which arrives at the outer edge of the atmosphere and has a value of 444.2 Btu/h·ft^2. Express this constant in units of W/cm^2, cal/in^2, and J/h·in^2.

1-29 In the economic evaluation of energy systems the cost per million Btu of energy is a figure commonly used to measure alternatives. Express the cost of energy in this unit for each of the following sources:

Source	Unit Cost	Energy Content
Natural gas	\$4.00/1000 ft^3	1000 Btu/ft^3
Gasoline	\$1.00/gal	140 000 Btu/gal
Electricity	\$0.08/kWh	

1-30 Suppose an automobile consumes gasoline at the rate of 12 mi/gal when moving at a speed of 60 mi/h. Using the data of Prob. 1-29, calculate the rate of expenditure of energy and the fuel cost per mile.

1-31 Nitrogen is maintained in a vessel at a pressure of 25 atm and a temperature of 75°C. Calculate the mass of gas in a 1000-cm^3 vessel.

1-32 How much work is required to accelerate a 4800-lbm automobile to a speed of 60 mi/h?

1-33 Approximately 80 cal are required to melt 1 g of ice. A commonly used unit to describe refrigeration capacity is the "ton," defined as the energy required to freeze 1 ton of water during a 24-h period. Express the value of the ton in Btu/h and also in kW.

1-34 In a certain large power plant the heat which must be dissipated in a nearby lake is equivalent to 2800 MW (1 MW = 10^6 watts). Water from the lake is used to accomplish the cooling, and the water is allowed to rise 12°F as it circulates through the power plant. If each gallon of water has a mass of 8.33 lbm and 1 Btu is required to raise each lbm 1°F, calculate the number of gallons of cooling water required each minute.

1-35 An alternative system to that of Prob. 1-34 is proposed whereby a large cooling tower is used to remove the heat by evaporating water. If 1040 Btu are required to vaporize each lbm of water, how many gallons would have to be evaporated each minute to dissipate the 2800 MW?

1-36 An average person consumes about 3000 kcal of food energy each day. Suppose solar energy falls on the earth's surface at the rate of 200 Btu/h·ft^2 after attenuation and absorption in the atmosphere and that this figure is averaged over an 8-h period. How many ft^2 would be necessary to capture the solar energy equivalent to a person's daily consumption of food energy?

1-37 A steam power cycle is designed to produce a power output of 300 000 kW, and the energy supplied to the cycle is 13 500 Btu/kWh of power output. Consider two fuels: (*a*) coal with an energy content of 12 500 Btu/lbm and a cost of \$35.00/ton and (*b*) natural gas with an energy content of 1000 Btu/ft^3 at a cost of \$2.50/1000 ft^3. Calculate the operating cost per hour for each fuel.

REFERENCES

1 Stimson, H. F.: The International Temperature Scale of 1948, *J. Res. Natl. Bur. Std.* (paper, 1962), vol. 42, p. 211, March 1949.
2 Holman, J. P.: "Experimental Methods for Engineers," 4th ed., McGraw-Hill Book Company, New York, 1984.
3 Hatsopoulos, G. N., and J. H. Keenan: "Principles of General Thermodynamics," John Wiley & Sons, Inc., New York, 1965.
4 Sears, F. W.: "Thermodynamics," 2d ed., Addison-Wesley Publishing Company, Inc., Reading, Mass., 1953.
5 Barber, C. R.: The International Practice Temperature Scale of 1968, *Metrologia,* vol. 5, no. 2, p. 35, 1969.
6 Cohen, E. R., and B. N. Taylor: The 1973 Least-Squares Adjustment of the Fundamental Constants, *J. Phys. Chem. Ref. Data.,* vol. 2, no. 4, 1973.
7 The Metric System of Measurement: Interpretation and Modification of the International System of Units for the United States, *Federal Register,* vol. 41, no. 239, pp. 54018–54019, Dec. 10, 1976.
8 Mechtly, E. A.: "The International System of Units, Physical Constants, and Conversion Factors, Revised," NASA SP-7012, 1973.

2

THE FIRST LAW OF THERMO-DYNAMICS

2-1 INTRODUCTION

The first law of thermodynamics has been stated as the conservation-of-energy principle. This statement is a simple one, but a general application of the principle can be quite complex in practice because we must be careful to identify all types of energy involved in a process. Even with all the types of energy identified, the analysis of the problem is not necessarily simple.

In this chapter we seek to amplify the conservation-of-energy principle so that it may be expressed in explicit analytical form. An analysis of systems on the basis of the first law of thermodynamics takes the form of calculation of various energy quantities and a balance of these quantities in accordance with the conservation idea. At times we are tempted to say that the first law is given by this equation or that equation, and the tendency is to fall into an analytical vertigo while the simple physical principle is at least momentarily forgotten. At the risk of seeming repetitious, it is well to remind the reader that all of the analytical formalisms are only mathematical statements of a physical principle. The conservation idea is the basic underlying notion, and should uncertainties about an analysis procedure occur, the procedure should be examined to be sure that it is consistent with the physical principle.

Once the first law is expressed in analytical form, application to specific engineering systems of interest may begin immediately. As the application of the energy conservation principle progresses, it will be necessary to introduce information regarding thermodynamic properties and their interrelationship in order to study a variety of processes used in engineering applications.

2-2 WORK

Work is defined as the energy expended by a force acting through a distance. It is expressed mathematically as

$$W = \int_c \mathbf{F} \cdot d\mathbf{s} = \int F \cos \theta \, ds \tag{2-1}$$

where the dot product of the force vector and elemental displacement vector indicates that the work is computed by considering only the component of force in the direction of the displacement. This calculation is indicated schematically in Fig. 2-1, where a force moves along a specified path. The total work done is given by the line integral along the path. We immediately sense the notion that work is dependent on the path, for it is a path integral which must be evaluated in order to calculate the work.

Sign Convention for Work

It is important to realize the sign convention for work which is implied by Eq. (2-1). A *positive* work is obtained when both the force and displacement vectors

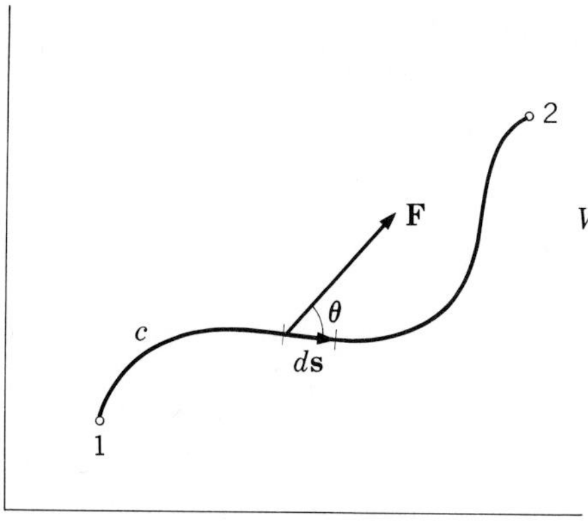

FIG. 2-1 Definition of work.

are in the same directions. A *negative* work is obtained when they are in opposite directions. In other words, a force acting to the right, and displaced to the right, does positive work, while a force acting to the right, and displaced to the *left,* does negative work.

Equation (2-1) is the fundamental analytical definition of work and may always be used as a basis for calculating work. To clarify this calculation, we now consider some applications of Eq. (2-1) to several simple processes.

2-3 EXPANSION OR COMPRESSION WORK IN A CYLINDER

First consider the expansion of a gas behind a piston in a cylinder, as shown in Fig. 2-2. The pressure in the cylinder is p and the volume of the gas is denoted by V. The force exerted on the piston is

$$F = pA$$

where A is the surface area of the piston exposed to the gas. The motion of the piston is in the direction of the applied force, and a differential displacement ds may be expressed in terms of the change in volume of gas dV as

$$ds = \frac{dV}{A}$$

FIG. 2-2 Expansion work in a cylinder.

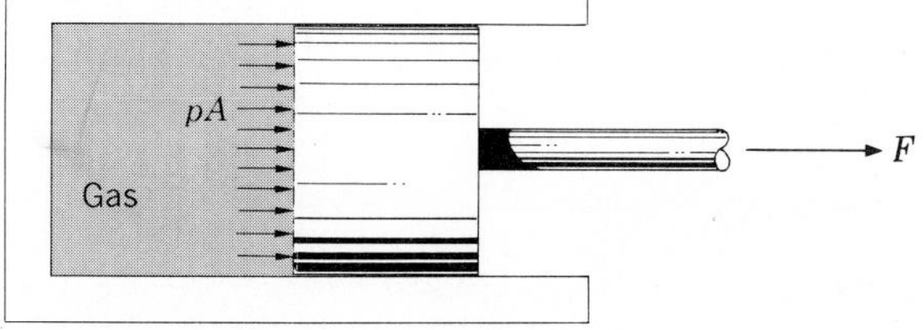

Combining the force and displacement functions, the amount of work done by the force acting *on the piston* is

$$dW = pA\,\frac{dV}{A}$$

or

$$W = \int_{V_1}^{V_2} p\,dV \qquad \text{work done on face of piston} \tag{2-2}$$

Note that the work done on the face of the piston is positive when the piston moves in the direction of the force, i.e., when dV is positive. This is the work done *on* the piston *by* the pressure force of the gas and may be illustrated graphically as the area under the curve in Fig. 2-3. Now let us think in terms of thermodynamic systems for a moment. If we take the piston as a thermodynamic system (or free body), the work delivered to the boundary of that system by the pressure forces of the gas is given by Eq. (2-2). On the other hand, if we took the *gas in the cylinder* as the thermodynamic system (or free body), the work delivered to the boundary of that system would be negative (assuming dV positive) because the force and displacement are in opposite directions. Considering the gas as a thermodynamic system we could write Eq. (2-2) as

$$W = -\int_{V_1}^{V_2} p\,dV \qquad \text{work done on gas by piston} \tag{2-2a}$$

So we see from this simple example that the sign one attaches to the work term depends on the system selected. Consistent with the sign conventions of mechanics we conclude that the work done on a system is positive because that is what we experience when the force acting on the system and the displacement are in the same direction. Conversely, the work done by the system is negative.

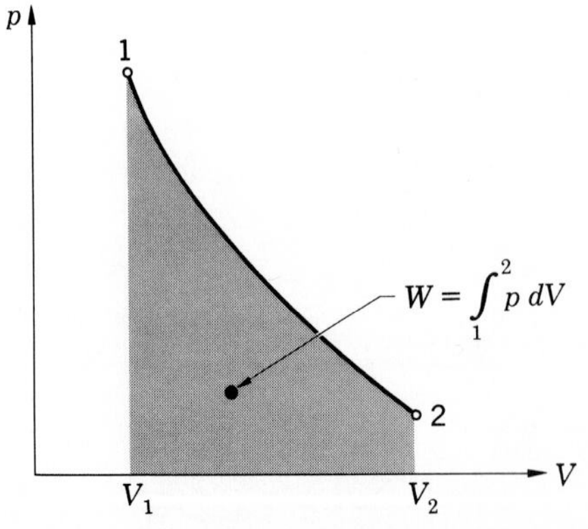

FIG. 2-3 Graphical representation of expansion work.

As we have mentioned previously, we shall tend to focus our discussion in thermodynamics on *work delivered to the boundary of a system,* rather than on forces and displacements. Nevertheless, one should keep in mind that it is the notion of a force acting through a displacement that represents the fundamental concept of work.

Open and Closed Systems

In the piston-cylinder arrangement the fluid is taken as the thermodynamic system. We may note that this is a *closed system,* or one which does not admit the transfer of mass across the system boundary. Analysis techniques for *open systems,* where mass may cross the boundary, will be developed in Chap. 4.

Nonequilibrium Effects

We have already described the quasistatic process in Chap. 1 as a series of equilibrium states. For the gas in the cylinder such a process would be experienced when the expansion occurs slowly and the pressure and temperature remain uniform during each step of the process. If, however, the expansion (or compression) occurs very rapidly the pressure may be nonuniform in the cylinder and the pressure force at the face of the piston may not be indicative of the pressure at other locations in the confined space. The *work at the boundary* is still calculated from an integration of the forces and displacements on the boundary, but for the nonequilibrium case there is no simple relation to the thermodynamic state of the gas in the cylinder. From this discussion we see once again that work is an effect which occurs at the boundary of a system and that it is strongly dependent on the particular way the thermodynamic process is executed.

Now consider the effect friction might have on the calculation of the expansion work in the cylinder. Friction can occur at the contact surface between the piston and cylinder and acts in a direction opposite to the motion of the piston. Thus, friction acts to reduce the work delivered to other things outside the piston-cylinder arrangement even if the process involving the gas is a nice quasistatic one. Sometimes one may speak of a thermodynamic system consisting of the gas *and* piston because it is the work output of this combination which is of frequent interest. For such a combined system a quasistatic process would only be experienced when the piston is slow-moving and frictionless. To evaluate the work, we need only to establish a free body and calculate the work done by the various forces acting on the free body. The calculation is made with Eq. (2-1).

Our discussion of the expansion work in the piston cylinder assumed a gas as the medium contained in the cylinder. We should note at this point that Eq. (2-2*a*) would apply equally well to a liquid or even to a solid; however, in these cases very large pressures may be required to effect any appreciable change in volume.

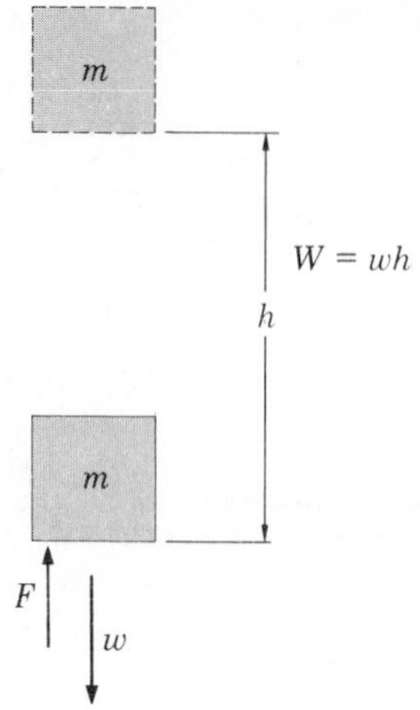

FIG. 2-4 Work to displace weight in a gravitational field.

2-4 WORK IN A GRAVITATIONAL FIELD

Now consider a system consisting of a weight. We wish to compute the work expended in raising the weight through a height h in a gravitational field. Again, the force and displacement are specified in order to calculate the work. The force is equal to the weight, as shown in Fig. 2-4, and is given by

$$F = w = m \frac{g}{g_c}$$

The work done by the force is given by Eq. (2-1) as

$$W = \int \mathbf{F} \cdot d\mathbf{s} = \int_0^h w \cdot ds = wh \tag{2-3}$$

Note the sign convention for the force and displacement, as shown in Fig. 2-4, and the fact that the work done on the system is positive.

2-5 ELECTRICAL AND MAGNETIC WORK

Consider a simple electric system as shown in Fig. 2-5. We take the battery as a thermodynamic system and wish to compute the work done as it discharges through the resistance R. The physical meaning of work in this case relates to the fact that an electric force is necessary to drive the charges through the resistor. We may recall that the electric potential V is the potential for doing work per unit charge. The electric current i is the time rate of change of charge or

$$q = \int i \, d\tau = i \, \Delta\tau$$

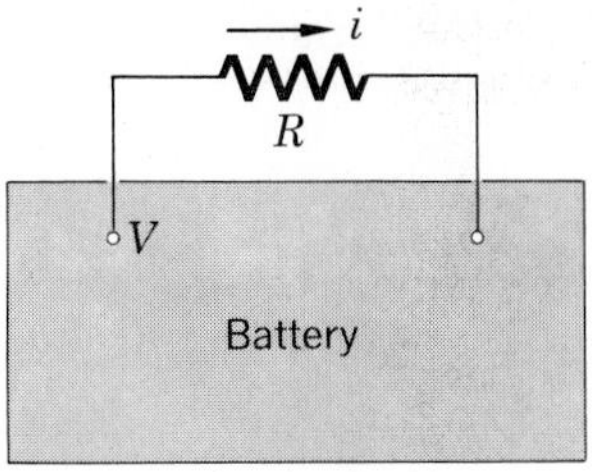

FIG. 2-5 Electrical work.

where $\Delta\tau$ is some specified time increment. Thus the electrical work expended during a time increment $\Delta\tau$ is

$$W = Vi\,\Delta\tau \tag{2-4}$$

In Eq. (2-4) the work W is the work that the thermodynamic system (the battery) does on the resistor since the calculation is performed by analyzing the electric forces acting on the charges external to the battery. The sign convention adopted previously would require that this work carry a negative sign since it is done *by* the system (the battery). If we were considering the resistor as a thermodynamic system, the work would be positive. In simple terms, the battery does work on the resistor through the action of electric forces.

Finally, we consider a very simple system where magnetic work is involved. We wish to calculate the work done by moving a conductor through a magnetic field **B** with a velocity **v**, as shown in Fig. 2-6. The force exerted on a moving charge placed in a magnetic field is

$$\mathbf{F} = q\mathbf{v} \times \mathbf{B} \tag{2-5}$$

Of course, this force tends to displace the charges in the conductor and create a current flow in the direction shown. The movement of the charges in the direction of the current then produces the restraining force exerted on the conduc-

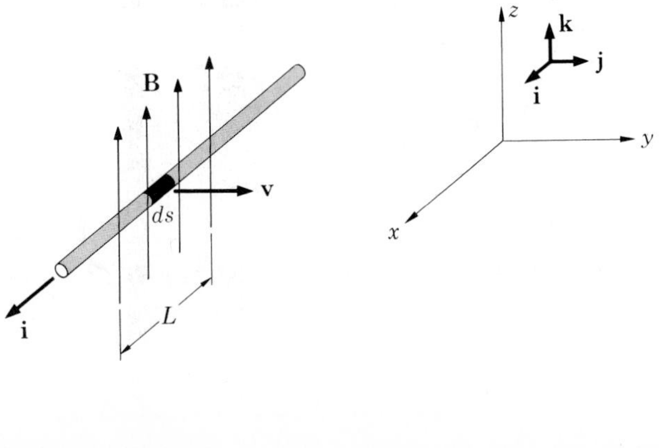

FIG. 2-6 Work performed by conductor moving in a magnetic field.

tor. For an element of the conductor of length $d\mathbf{s}$ we have

$$q\mathbf{v} = \int i \, d\tau \frac{d\mathbf{s}}{d\tau} = \int i \, d\mathbf{s} \tag{2-6}$$

Inserting Eq. (2-6) into Eq. (2-5) gives

$$\mathbf{F} = \int_0^L i \, d\mathbf{s} \times \mathbf{B} = -iBL\mathbf{j} \tag{2-7}$$

The rate of doing work is

$$W = \mathbf{F} \cdot \mathbf{v} = \left(\int_0^L i \, d\mathbf{s} \times \mathbf{B} \right) \cdot \mathbf{v}$$

For the coordinate orientation shown in Fig. 2-6 this becomes

$$W = -(iBL\mathbf{j}) \cdot (v\mathbf{j}) = -iLBv \tag{2-8}$$

In this example we are considering the conductor (set of charges) as the thermodynamic system and computing the work done on the conductor by the magnetic field. The force shown in Eq. (2-7) is the restraining force exerted on the conductor by the field. To move the conductor in the field, we would have to apply an external force equal in magnitude and opposite in direction. The work done on the conductor by this external force would then be

$$W = +iLBv \tag{2-8a}$$

This example is analogous to the problem of calculating the work necessary to raise a weight in a gravitational field. In this example the work of the external force is stored in the magnetic field. In the gravitational problem the work of the external force is stored in the gravitational field.

Another system where magnetic work is involved is shown in Fig. 2-7. The cross-sectional area of the toroidal ring is A and its circumferential length is L.

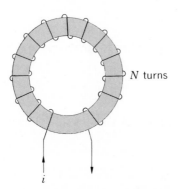

FIG. 2-7 Magnetic work on a toroidal ring.

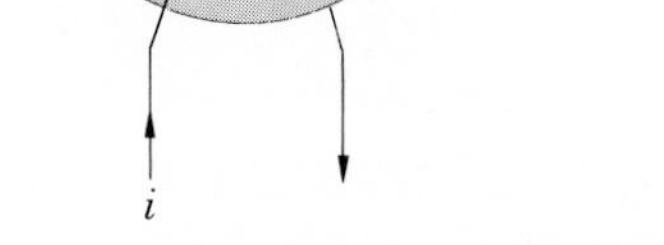

The magnetic flux density is denoted by **B**, the number of turns is N, and the current is i. We wish to calculate the work done as the current is changed in the windings. From Faraday's law of electromagnetic induction the induced electromotive force (emf) E is given by

$$E = N\frac{d(BA)}{d\tau} \tag{2-9}$$

and the work done in the time $d\tau$ is

$$dW = Ei\,d\tau = NAi\,dB$$

The magnetic intensity H is calculated from

$$H = \frac{Ni}{L}$$

so that

$$dW = ALH\,dB \tag{2-10}$$

We note that AL is the total volume of the coil, so the work can be written as

$$dW = VH\,dB \tag{2-11}$$

In this case we observe that work producing force is that required to overcome the back emf of the coil. The calculation has been performed for a toroidal coil, but it would also be valid for long, slender coils of uniform cross section.

EXAMPLE 2-1 Expansion of Gas in a Cylinder
A gas expands in a cylinder according to the relation $pV^{1.3}$ = const. The initial volume of the gas is 1 ft³ and the initial pressure is 200 psia. The final pressure is 15 psia. Calculate the work done on the face of the piston by the pressure forces of the gas.

SOLUTION Equation (2-2) applies to this problem so that

$$W = \int_{V_1}^{V_2} p\,dV$$

From the given process relation, $pV^{1.3} = C$ or $p = CV^{-1.3}$. Thus

$$W = \int_{V_1}^{V_2} CV^{-1.3}\,dV = \frac{-C}{0.3}\left(V^{-0.3}\right)_{V_1}^{V_2} \tag{a}$$

But $C = p_1V_1^{1.3} = p_2V_2^{1.3}$ so that

$$W = \frac{p_2V_2 - p_1V_1}{-0.3} \tag{b}$$

The final volume is

$$V_2 = V_1\left(\frac{p_1}{p_2}\right)^{1/1.3} = 1\left(\frac{200}{15}\right)^{1/1.3} = 7.32\ \text{ft}^3 \qquad (0.2073\ \text{m}^3)$$

Inserting this value in Eq. (*b*) gives

$$W = -\frac{(144)[(15)(7.32) - (200)(1)]}{0.3}$$
$$= 4.33 \times 10^4\ \text{ft·lbf} \qquad (5.87 \times 10^4\ \text{J})$$

EXAMPLE 2-2 Magnetic Work
A 1-ft conductor carrying a current of 10 A is moved through a magnetic field having a magnetic flux density of 1 Wb/m^2. The field and conductor are orthogonal, as shown in Fig. 2-6. Calculate the force necessary to move the conductor and the rate of doing work when the velocity is 15 ft/s.

SOLUTION The first step in the solution of this problem is to get all the physical quantities into a consistent set of units. We choose the metric system since it is simplest to use:

$$L = 1\ \text{ft} = 0.305\ \text{m}$$
$$v = 15\ \text{ft/s} = 4.58\ \text{m/s}$$

From Eq. (2-7) the force is

$$\mathbf{F} = ibL\mathbf{j}$$
$$= (10)(1)(0.305) = 3.05\ \text{N}$$

The work per unit time is then calculated from Eq. (2-8*a*) as

$$W = iBLv = (10)(1)(0.305)(4.58)$$
$$= 13.98\ \text{N·m/s} = 13.98\ \text{J/s}$$
$$= 13.98\ \text{W}$$

2-6 INTERNAL ENERGY

In the foregoing we have shown how the work done on a system may be evaluated by calculating the energy expended by a force acting through a distance. Work, then, is one form of energy which may cross the boundary of a system. It may be remarked again that there must be a force interaction between the system and its surroundings in order for work to be present. We thus sense once again that work is dependent on the process because it is only observed as the system changes from one state to another.

The concept of potential energy is a familiar one which arises in mechanics, electromagnetics, and other subjects. A review of this concept is appropriate because it bears on a more general type of energy which is treated in thermodynamics. It will be recalled that gravitational potential energy is defined as the work necessary to raise a weight to a height above a particular reference level. Kinetic energy is evaluated by determining the work required to accelerate a body from rest to a certain velocity. Thus,

$$F = \frac{1}{g_c} ma = \frac{1}{g_c} m \frac{dv}{d\tau} = \frac{m}{g_c} \frac{dv}{ds} \frac{ds}{d\tau} = \frac{m}{g_c} v \frac{dv}{ds}$$

and

$$W = \int \mathbf{F} \cdot d\mathbf{s} = \int \frac{m}{g_c} v \, dv = \frac{1}{2g_c} mv^2 = \text{KE} \tag{2-12}$$

When the system changes from some initial velocity v_1 to a final velocity v_2, the corresponding *change* in kinetic energy is

$$\text{KE}_2 - \text{KE}_1 = \int_{v_1}^{v_2} \frac{m}{g_c} v \, dv = \frac{1}{2g_c} m(v_2^2 - v_1^2)$$

Thus, we imagine the acceleration work as being stored as kinetic energy. If the body were decelerated through the action of some restraining force this energy would be recovered as a work delivered to the restraining mechanism (assuming no losses due to friction).

In a similar fashion electric potential is defined as the work to displace a charge in an electric field. In all of these cases the concept of potential energy is identified with work. In all of the cases mentioned so far the potential energy is a function of the *state* of the system and does not depend on the particular process under consideration. Thus the kinetic energy of a body does not depend on the manner in which it is brought to rest or even on the direction of the velocity vector; the gravitational potential energy does not depend on the manner in which a weight is raised to a given elevation, but only on the value of the elevation, and so on. Note very carefully once again that the familiar concept of potential energy is defined in terms of work. Observe, furthermore, that the work done by the forces acting on the system may be expressed as

$$\text{PE}_{\text{final}} - \text{PE}_{\text{initial}} = \Delta\text{PE} = W \tag{2-13}$$

for the special class of systems discussed so far in which the only interaction of the system with its surroundings is in the form of work. We may remark that if the only types of work terms were those in which a potential-energy function could be defined, energy analysis of systems would be a simple matter indeed because the work would always be independent of the process. We note also that the technique of defining potential-energy terms automatically forces us to *leave out* of the work terms in Eq. (2-13) those values which may be expressed in terms of a potential function. It will be recalled from mechanics that if the only work done by a system is that which is independent of path, then the system is called *conservative*. The forces acting on such a system are called *conservative forces*. We may thus interpret Eq. (2-13) as

Work done by conservative forces = work done by nonconservative forces

General Internal-Energy Function for an Adiabatic Process

Equation (2-13) is restrictive in that the only kinds of potential energy which it admits are those expressed through the conservative-force concept. The first law of thermodynamics permits a generalization of the concept of potential energy. To clarify this principle, imagine a system which changes from some state I to some state II, as shown in Fig. 2-8. Let us assume that the system is perfectly insulated so that no heat may be transferred to or from the system. We specify that the only interaction with the surroundings is in the form of

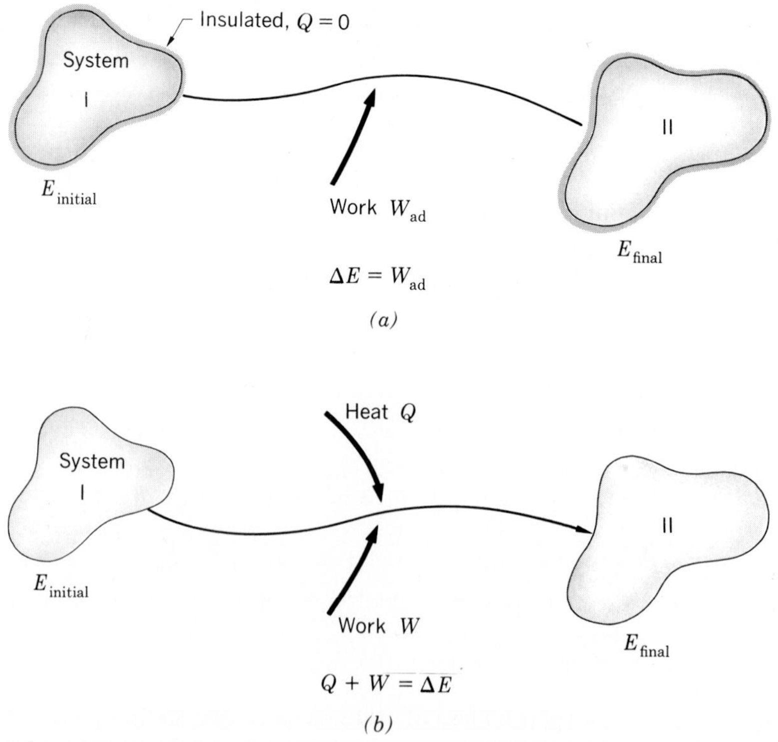

FIG. 2-8 Conservation-of-energy principle for a system changing from state I to II in (*a*) an adiabatic process and (*b*) a general process involving heat transfer.

work, and we exclude from this work all those forms that can be expressed as changes in potential functions. In this way we speak of the work that is, in fact, *delivered* to the system by the surroundings or some adjoining mechanism. As a result of this process, the principle of conservation of energy permits the definition of a *general internal energy function E,* such that

$$E_{\text{final}} - E_{\text{initial}} = \Delta E = W_{\text{ad}} \tag{2-14}$$

where W_{ad} is used to designate the work. The subscript "ad" indicates that the work is done in an *adiabatic process,* i.e., where no heat exchange occurs between the system and the surroundings.

Heat cannot be expressed as a force acting through a distance, and thus we are compelled to omit it from consideration for now. A verbal statement of Eq. (2-14) is

Gain in internal energy = adiabatic work delivered to the system by all forces acting on the system

It is found *by experiment* that the adiabatic work between two end states is always the same, regardless of the process. We thus conclude that the quantity E is a property, and is a function of only the state of the system. The quantity E is called the *internal energy* of the system.

In Chap. 1 we defined a property of a system as a function of state, a point function, or a quantity whose change between two end states does not depend on the process path used to effect the change. The quantity E fits this definition, so we call it a property of the system. Notice that the definition of E in terms of adiabatic work does not serve to establish an absolute scale for internal energy or a reference level for tabulation of properties of substances. Reference levels may be assigned arbitrarily, as in the case of reference points for the international temperature scale, but absolute values of the internal-energy function may only be obtained from detailed microscopic considerations. The important point is that a property E exists and that its existence is a matter of experimental evidence.

The Influence of Heat

We shall have more to say about the constitution of the internal-energy function shortly, but now let us consider the effect of heat transfer in a process which the system undergoes. Suppose we let the system undergo a process between the same end states as before, but now the system is not insulated and it may exchange heat with the surroundings. The change in internal energy is the same as before, since the end states are the same, but the work is altered according to the conservation of energy by the amount of heat energy added, Q. Thus,

$$W + Q = W_{\text{ad}} \tag{2-15}$$

where we define the heat energy as positive when it is added to the system and negative when it is given up by the system. Rewriting Eq. (2-15) according to Eq. (2-14),

$$Q + W = \Delta E \tag{2-16}$$

Equation (2-16) has a very simple physical interpretation. It states that the energy added at the boundary of the system as a result of forces acting on the boundary (work) plus all other energies added at the boundary (heat) must be experienced as an increase in the internal energy of the system. Or,

Energy added to system = accumulation of energy in the system

or, alternatively,

Initial internal energy
+ energy added at boundary in the form of heat and work
= final internal energy

That is,

$$E_1 + W + Q = E_2$$

At this point we remind the reader that our discussion has focused on systems where mass does not cross the boundary, or *closed systems*. When mass is allowed to cross the boundary, there can be a corresponding transport of energy carried by the mass. In Chap. 4 we shall see how such energy transport is analyzed.

2-7 THE FIRST LAW OF THERMODYNAMICS

Equation (2-16) is frequently given as an analytical statement of the first law of thermodynamics, but it must be remembered that it is simply a statement of the conservation-of-energy principle where we recognize heat, work, and internal energy as different forms of energy. We might think of Eq. (2-16) as a definition of heat in terms of work and internal energy. In a physical sense

heat is thought of as an energy exchange with the surroundings which does not take the form of a force acting through a distance.

Notice that the conservation idea is abundantly clear when we consider an *isolated* system, i.e., one where no energy crosses the boundary of the system. Thus

$$\Delta E_{\text{isolated}} = 0 \tag{2-17}$$

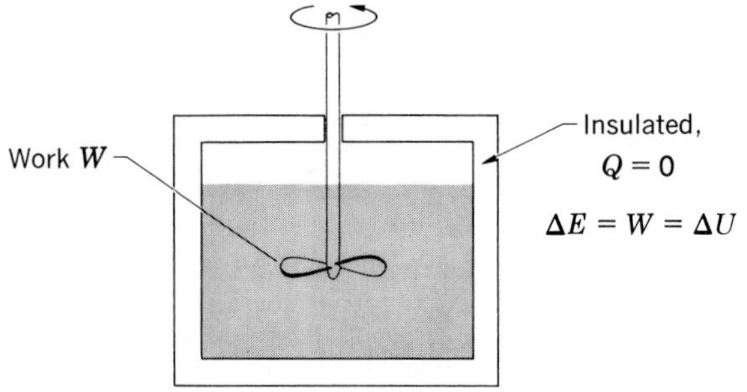

FIG. 2-9 Process in which work goes to raise the internal thermal energy of the system.

or the total internal energy of an isolated system does not change—it may only be changed from one form to another. The preceding reasoning is illustrated in Fig. 2-8.

Let us now consider the nature of the internal-energy function. Clearly it incorporates the potential-energy functions such as gravitational and kinetic energy. But it also includes other types of internal energy. To illustrate this, consider the process indicated in Fig. 2-9. We do work by stirring the water in a container which is insulated with respect to the surroundings. After the stirring, the water is allowed to come to rest again. Obviously the gravitational and kinetic energies do not change in this process, but the internal energy must increase as a consequence of the stirring action because we have added energy to the water. A similar example is the compression of a gas in a cylinder. Once again, if the cylinder is insulated, the internal energy of the gas is increased. In both of the aforementioned systems the increase in total internal energy of the system is the result of a change in the *internal thermal energy* of the system. In a physical sense, this internal thermal energy results from the kinetic energy of the molecules of the substance which compose the system and is usually evidenced by a lower or higher temperature of the system. The internal thermal energy is denoted by the symbol U. We might also consider the inherent chemical energy of the substance which could produce external work, as in the case of a storage battery delivering electrical work to some external circuit. We could carry the consideration of total internal energy further by noting that the relativistic mass-energy relationship

$$E = mc^2$$

must be employed whenever nuclear reactions are involved.

Our discussion of the internal-energy function may now be summarized with the following expression:

$$E = U + \mathrm{KE} + \mathrm{PE} + \mathrm{ChE} + \cdots \tag{2-18}$$

where it is understood that electric, magnetic, and nuclear energies may also be included in this expression when appropriate. Capital letters (E and U) are employed to designate the *total* properties for an entire system. Lowercase characters (e and u) will be used to indicate the properties per unit mass of the system. In some circumstances we shall use $\bar{e}$ and $\bar{u}$ to designate properties per mole of substance. Typical units for e and u are Btu/lbm and kJ/kg.

We remark again that Eq. (2-16) represents the energy-transfer mechanisms in a process wherein the system changes from one state to another. Now suppose we allow the system to undergo a *cyclic* change, i.e., a continuous series of processes such that the system periodically returns to its initial state. If we consider a small change in the system, then the first law may be written in differential form as

$$dQ + dW = dE$$

Clearly, since the internal energy E is a function only of state, its change must be zero over any complete cycle, and we have

$$\oint dQ + \oint dW = 0 \qquad (2\text{-}19)$$

where the integral sign denotes a cyclic process.

If must be emphasized, in general, that both heat and work are path functions and, thus, to evaluate their magnitude we must take into consideration the process involved. On the other hand, internal energy is a state function and is characterized mathematically by

$$\oint dE = 0$$

Thus, in mathematical language, dE may be described as an exact differential, whereas dQ and dW would be called *inexact differentials*. Hereafter we shall use the modified symbols $d'Q$ and $d'W$ to designate the fact that the work and heat differentials are, in general, functions of path. In some writings the differentials are denoted as δW and δQ to give emphasis to the fact that work and heat depend on the process.

2-8 HEAT AND SPECIFIC HEAT

We have seen that heat is an energy interaction between the thermodynamic system and its surroundings; however, this energy interaction may not be calculated by the application of a force through a distance as in the case of work. Heat and work, therefore, are fundamentally different types of energy. Intuitively, we associate heat with the temperature of a system because the temperature of the system usually rises when heat is added to it. A study of a historical development of the subject of thermodynamics will show that at one time scientists thought of heat as a substance called *caloric* which could be poured from one body into another, in the same way that hot water may be poured from one container into another. They therefore reasoned that heat was contained in a body. We now know that this idea is false and that *heat is not contained in a system* but is manifest only as an interaction of the system with its surroundings as the system changes from one state to another.

Hot water and cold water "contain" the same amount of heat: none at all.

A holdover concept from the old calorie theory is that of heat capacity or, as it is now called, *specific heat*. If a quantity of heat $d'Q$ is added to a system, resulting in a change in temperature dT, then the specific heat C is defined by

$$d'Q = C\,dT \tag{2-20}$$

Suppose we consider a thermodynamic system where the only changes in internal energy are present in the thermal energy function U. Furthermore, let us assume that the only work interactions with the surroundings are those which result from a change in volume of the system, as shown in Fig. 2-2. In this process we are assuming that other forms of work are not present and that the system volume increases. For this type of system we may write Eq. (2-16) as

$$d'Q + d'W = dU \tag{2-21}$$

$$d'Q - p\,dV = dU$$

where p is the system pressure. Recall the sign convention for work when developing Eq. (2-2*a*). We may then write for the heat added

$$d'Q = dU + p\,dV \tag{2-21a}$$

Now let this system undergo a process at constant volume so that $dV = 0$. Thus we could write for the specific heat at constant volume, and per unit mass, recalling that uppercase letters denote total properties, while lowercase letters denote values per unit mass,

$$dQ_v = c_v\,dT_v$$

or

$$du_v = c_v\,dT_v$$

where the subscripts indicate a change at constant system volume. The specific heat per unit mass is thus written as

$$c_v = \left(\frac{\partial u}{\partial T}\right)_v \tag{2-22}$$

Although the foregoing relation has been developed for a special system, we shall take Eq. (2-22) as the *defining* relation for specific heat at constant volume. At this point we note a departure from the old caloric theory. The modern concept is to treat specific heat as a *property* of the system which may be expressed in terms of other properties of the system, as in Eq. (2-22) for the specific heat at constant volume. Although we now recognize the terms "specific heat" and "heat capacity" as misnomers, we cannot divorce these concepts from the historical treatment of thermodynamics where they are related to heat transfer in certain specific processes.

Now suppose we allow the simple system discussed previously to undergo a constant-pressure process. The heat transfer per unit mass is expressed as

$$d'Q_p = du_p + p\, dv_p \tag{2-23}$$

Enthalpy

Let us define a new property called the *enthalpy* by

$$h = u + pv \tag{2-24}$$

The differential of the enthalpy is

$$dh = du + p\, dv + v\, dp$$

and, at constant pressure,

$$dh_p = du_p + p\, dv_p \tag{2-25}$$

Thus

$$dQ_p = dh_p = c_p\, dT_p$$

and

$$c_p = \left(\frac{\partial h}{\partial T}\right)_p \tag{2-26}$$

The specific heat at constant pressure is thus a property of the system and may be defined in terms of other properties, as shown in Eq. (2-26).

Typical units for specific heat are Btu per pound-mass–degree Fahrenheit, kilocalories per kilogram-degree Celsius, and joules per kilogram-degree Celsius. We shall use the symbols $\bar{c}_v$ and $\bar{c}_p$ to designate specific heats expressed on a molal basis.

We see that, in general, the heat added in a constant-pressure process will be equal to the change in enthalpy because

$$Q_p = \int dh_p = h_2 - h_1 \tag{2-27}$$

on a unit mass basis.

TABLE 2-1
Typical Values of Specific Heats

Substance	c_p kJ/kg·°C	c_p Btu/lbm·°F	c_v kJ/kg·°C	c_v Btu/lbm·°F
Solids (20°C)				
Aluminum	0.896	0.214		
Copper	0.383	0.091		
Iron	0.452	0.108		
Silver	0.234	0.056		
Tungsten	0.134	0.032		
Brick	0.84	0.201		
Glass wool	0.7	0.167		
Wood (pine)	2.8	0.669		
Liquids (20°C)				
Water	4.18	1.00		
Freon 12	0.966	0.231		
Ammonia	4.8	1.15		
Engine oil	1.9	0.45		
Mercury	0.14	0.033		
Ethylene glycol	2.38	0.57		
Gases (1 atm, 20°C)				
Air	1.005	0.24	0.718	0.1715
Hydrogen	14.32	3.42	10.17	2.43
Carbon dioxide	0.846	0.202	0.653	0.156

Specific heats, like other thermodynamic properties, can vary strongly with temperature and pressure of the substance, and experimental data must be employed to obtain reliable values. For modest temperature ranges the values may frequently be assumed constant for calculation purposes. For liquids and solids c_p and c_v are almost equal, while there is an appreciable difference for gases. Table 2-1 gives some typical values to illustrate the range which may be experienced.

EXAMPLE 2-3 Paddle-Wheel Work
A 1-hp stirring motor is applied to a tank of water, as shown in Fig. 2-9. The tank contains 50 lbm of water, and the stirring action is applied for 1 h. Assuming that the tank is perfectly insulated, calculate the change in internal energy of the water. Also calculate the rise in temperature of the water, assuming that the process occurs at constant volume and that c_v for water may be taken as 1.0 Btu/lbm·°F.

SOLUTION Some appropriate conversion factors for use in this problem are

$$1 \text{ hp} = 550 \text{ ft·lbf/s} = 2545 \text{ Btu/h}$$

$$778.16 \text{ ft·lbf} = 1 \text{ Btu} = 1055 \text{ J}$$

We take the water as the thermodynamic system and write the first law as

$$W + Q = \Delta U$$

neglecting any changes in PE or KE. The increase in the internal energy of the water is clearly the work added during the 1-h time period, since $Q = 0$ for an insulated tank:

$$\Delta U = W = (550)(3600) = 1.98 \times 10^6 \text{ ft·lbf}$$
$$= 2545 \text{ Btu} \qquad (2.685 \times 10^6 \text{ J})$$

For this calculation we assume that the kinetic energies of the water in the tank are negligible. For the low velocities in the stirring process this is a reasonable expectation. From Eq. (2-22) we have

$$c_v = \left(\frac{\partial u}{\partial T}\right)_v$$

For a constant specific heat we may approximate the derivative by

$$c_v \approx \left(\frac{\Delta u}{\Delta T}\right)_v$$

or

$$\Delta u = c_v \, \Delta T$$

Multiplying by the mass, we obtain

$$m \, \Delta u = \Delta U = mc_v \, \Delta T$$

The increase in temperature is now calculated as

$$\Delta T = \frac{\Delta U}{mc_v} = \frac{2545}{(50)(1.0)} = 50.9°\text{F} \qquad (28.28°\text{C})$$

2-9 SPECIFIC HEATS FOR IDEAL GASES

It is worthwhile to examine the specific-heat behavior of gases at this time to provide a basis for calculations in the following chapters.

In Chap. 6 it will be shown that, if a gas obeys the equation of state given in Eq. (1-34), i.e.,

$$pV = mRT$$

then the internal energy and enthalpy are functions of temperature alone, and we may integrate Eqs. (2-22) and (2-26) to give

$$\Delta u = u_2 - u_1 = \int_{T_1}^{T_2} c_v \, dT \tag{2-28}$$

$$\Delta h = h_2 - h_1 = \int_{T_1}^{T_2} c_p \, dT \tag{2-29}$$

Furthermore, if the specific heats are constant, we have the simple results

$$u_2 - u_1 = c_v(T_2 - T_1) \tag{2-30}$$

$$h_2 - h_1 = c_p(T_2 - T_1) \tag{2-31}$$

In a microscopic sense an ideal gas is one in which the molecules are sufficiently far apart that the internal energy of the gas may be calculated on the basis of the average kinetic energy of the molecules. It turns out, as illustrated in the simple analysis of Sec. 1-16, that this energy is a function of the macroscopic temperature of the gas. If a gas is compressed to a high pressure so that the molecules are forced into close proximity to one another and molecular-force field interactions become important, then the internal energy becomes a function of both pressure and temperature.

A casual reader may miss the important point regarding internal energy of an ideal gas, so we reiterate it here in boldface type to provide an increased emphasis:

Internal energy and enthalpy for an ideal gas are functions of temperature alone.

This important fact will be used over and over again in energy-balance calculations.

Some differences in terminology for gases exist in the literature. A gas which obeys Eq. (1-34) is sometimes said to be *thermally perfect*. If it has constant specific heats, it is said to be *calorically perfect*. The terms *ideal gas* and *perfect gas* are used by various authors to indicate that the gas obeys the $pv = RT$ equation of state, and we shall employ this terminology in subsequent sections. For developments in this book the reader may consider the terms *ideal gas* and *perfect gas* as synonyms. The matter of specific heats will be taken into account with an appropriate statement such as "ideal gas with constant specific heats."

For gases at low pressures the specific heats are fairly constant and do not vary over modest temperature ranges. If calculations are to be performed on the basis of an assumption of constant specific heat, then the appropriate values to be used are as given in Table 2-2.

One useful relation between c_v and c_p for an ideal gas may be derived as follows. We have

$$dh = c_p \, dT$$

$$du = c_v \, dT$$

TABLE 2-2

Properties of Ideal Gases at Low Pressures and Normal Room Temperature (20°C)

Gas	Molecular Weight	c_p		c_v		R		γ
		Btu/lbm·°F	kJ/kg·°C	Btu/lbm·°F	kJ/kg·°C	ft·lbf/lbm·°R	J/kg·K	
Air	28.97	0.240	1.005	0.1715	0.718	53.35	287.1	1.40
Hydrogen, H_2	2.016	3.42	14.32	2.43	10.17	767.0	4127	1.41
Helium, He	4.003	1.25	5.234	0.75	3.14	386.3	2078	1.66
Methane, CH_4	16.04	0.532	2.227	0.403	1.687	96.4	518.7	1.32
Water vapor, H_2O	18.02	0.446	1.867	0.336	1.407	85.6	460.6	1.33
Acetylene, C_2H_2	26.04	0.409	1.712	0.333	1.394	59.4	319.6	1.23
Carbon monoxide, CO	28.01	0.249	1.043	0.178	0.745	55.13	296.6	1.40
Nitrogen, N_2	28.02	0.248	1.038	0.177	0.741	55.12	296.6	1.40
Ethane, C_2H_6	30.07	0.422	1.767	0.357	1.495	51.3	276	1.18
Oxygen, O_2	32.00	0.219	0.917	0.156	0.653	48.24	259.6	1.40
Argon, A	39.94	0.123	0.515	0.074	0.310	38.65	208	1.67
Carbon dioxide, CO_2	44.01	0.202	0.846	0.156	0.653	35.1	188.9	1.30
Propane, C_3H_8	44.09	0.404	1.692	0.360	1.507	35.0	188.3	1.12
Isobutane, C_4H_{10}	58.12	0.420	1.758	0.387	1.62	26.6	143.1	1.09

Subtracting these expressions,

$$dh - du = (c_p - c_v)\, dT$$

But

$$dh = du + d(pv) = du + R\, dT$$

so that

$$R\, dT = (c_p - c_v)\, dT$$

and

$$R = c_p - c_v \tag{2-32}$$

We thus have the result that the gas constant may be calculated as the difference between the constant-pressure and constant-volume specific heats. Recip-

TABLE 2-3 Molal Constant-Pressure Specific Heats for Gases at Very Low Pressures (English units)

Gas or Vapor	$\bar{c}_{po}$, Btu/lb·mol·°R; T, °R	Range, °R	Max. Error, %
O_2	$\bar{c}_{po} = 11.515 - \frac{172}{\sqrt{T}} + \frac{1530}{T}$	540–5000	1.1
	$= 11.515 - \frac{172}{\sqrt{T}} + \frac{1530}{T} + \frac{0.05}{1000}(T - 4000)$	5000–9000	0.3
N_2	$\bar{c}_{po} = 9.47 - \frac{3.47 \times 10^3}{T} + \frac{1.16 \times 10^6}{T^2}$	540–9000	1.7
CO	$\bar{c}_{po} = 9.46 - \frac{3.29 \times 10^3}{T} + \frac{1.07 \times 10^6}{T^2}$	540–9000	1.1
H_2	$\bar{c}_{po} = 5.76 + \frac{0.578}{1000} T + \frac{20}{\sqrt{T}}$	540–4000	0.8
	$= 5.76 + \frac{0.578}{1000} T + \frac{20}{\sqrt{T}} - \frac{0.33}{1000}(T - 4000)$	4000–9000	1.4
H_2O	$\bar{c}_{po} = 19.86 - \frac{597}{\sqrt{T}} + \frac{7500}{T}$	540–5400	1.8
CO_2	$\bar{c}_{po} = 16.2 - \frac{6.53 \times 10^3}{T} + \frac{1.41 \times 10^6}{T^2}$	540–6300	0.8
CH_4	$\bar{c}_{po} = 4.52 + 0.007\,37T$	540–1500	1.2
C_2H_4	$\bar{c}_{po} = 4.23 + 0.011\,77T$	350–1100	1.5
C_2H_6	$\bar{c}_{po} = 4.01 + 0.016\,36T$	400–1100	1.5
C_8H_{18}	$\bar{c}_{po} = 7.92 + 0.0601T$	400–1100	4 (est.)
$C_{12}H_{26}$	$\bar{c}_{po} = 8.68 + 0.0889T$	400–1100	4 (est.)

Source: Callen [1].

rocally, if one of the specific heats is known along with the gas constant, then the other specific heat is easily calculated. Finally, if specific heats are expressed on a molal basis, we have

$$\mathcal{R} = \bar{c}_p - \bar{c}_v \tag{2-33}$$

or the difference in molal specific heats for *all* ideal gases is a constant.

When to Treat a Gas as Ideal

The reader may logically ask: When is a gas ideal? In a microscopic sense, it is ideal when the molecules are so widely spaced that molecular force fields do not play a significant role in the pressure exerted by the gas on a containing vessel. This means that ideal-gas behavior is usually observed when the gas density is low (a relatively small number of molecules per unit volume), but we have not yet developed a suitable means to predict the pressure and temperature ranges for which the ideal-gas law is applicable. Section 6-13 will furnish us with this information. For now, the reader can be assured that such gases as

TABLE 2-3M Molal Constant-Pressure Specific Heats for Gases at Low Pressures (SI units)

Gas or Vapor	$\bar{c}_{po}$, kJ/kg·mol·K T, K	Range, K
O_2	$\bar{c}_{po} = 48.212 - \frac{536.8}{\sqrt{T}} + \frac{3559}{T}$	300–2800
	$= 48.212 - \frac{536.8}{\sqrt{T}} + \frac{3559}{T} + 3.768 \times 10^{-4}(T - 2222)$	2800–5000
N_2	$\bar{c}_{po} = 39.65 - \frac{8071}{T} + \frac{1.5 \times 10^6}{T^2}$	300–5000
CO	$\bar{c}_{po} = 39.61 - \frac{7652}{T} + \frac{1.38 \times 10^6}{T^2}$	300–5000
H_2	$\bar{c}_{po} = 24.12 + 4.356 \times 10^{-3}T + \frac{62.41}{\sqrt{T}}$	300–2200
	$= 24.12 + 4.356 \times 10^{-3}T + \frac{62.41}{\sqrt{T}} - 5.94 \times 10^{-4}(T - 2222)$	2200–5000
H_2O	$\bar{c}_{po} = 83.15 - \frac{1863}{\sqrt{T}} + \frac{17445}{T}$	300–3000
CO_2	$\bar{c}_{po} = 67.83 - \frac{15\,189}{T} + \frac{1.82 \times 10^6}{T^2}$	300–3500
CH_4	$\bar{c}_{po} = 18.92 + 0.055T$	300–830
C_2H_4	$\bar{c}_{po} = 17.71 + 0.0887T$	200–650
C_2H_6	$\bar{c}_{po} = 16.79 + 0.123T$	220–600
C_8H_{18}	$\bar{c}_{po} = 33.16 + 0.453T$	220–600
$C_{12}H_{26}$	$\bar{c}_{po} = 36.34 + 0.670T$	220–600

air, nitrogen, oxygen, helium, etc., do obey the ideal-gas equation of state at normal atmospheric temperature and pressure (20°C and 1 atm).

When the pressure of a gas is sufficiently low that the ideal-gas equation of state applies, but one wishes to take account of specific heat variation with temperature, the empirical relations of Table 2-3 are applicable. Due caution should be exercised to see that these expressions are used only when the gas is at low pressures. Water vapor, for example, is *not* an ideal gas at normal atmospheric pressure, and the relations of Tables 2-3 and 2-3M could not be used to calculate the specific heat at this pressure.

For calculation purposes in this and following chapters we shall assume, unless some statement is made to the contrary, that the gases in Table 2-2 *except water vapor* behave as ideal gases.

EXAMPLE 2-4 Constant-Volume Heat Addition
A 1-ft^3 container is filled with air at 20 psia and 100°F. Calculate the final pressure in the container if 10 Btu of heat are added. Assume ideal-gas behavior, with constant specific heats.

SOLUTION We choose air as the thermodynamic system, with the tank as the boundary, and write the first law as

$$Q + W = \Delta U$$

where ΔKE, etc., are assumed to be negligible.

Since the container is rigid, there is no work because the pressure force cannot act through a displacement, and the first-law statement becomes

$$Q = \Delta U = m(u_2 - u_1)$$

For an ideal gas this becomes

$$Q = mc_v(T_2 - T_1) \qquad (a)$$

The mass of air is calculated with

$$m = \frac{pV}{RT} = \frac{(20)(144)(1)}{(53.35)(560)} = 0.0963 \text{ lbm} \qquad (0.043\ 68 \text{ kg})$$

For air, $c_v = 0.1715$ Btu/lbm·°F (from Table 2-2) so that Eq. (*a*) becomes

$$10 = (0.0963)(0.1715)(T_2 - 100)$$

and

$$T_2 = 706°\text{F} = 1166°\text{R} \qquad (374°\text{C}, 647 \text{ K})$$

The final pressure may now be calculated with

$$p = \frac{mRT}{V} \tag{b}$$

$$= \frac{(0.0963)(53.35)(1166)}{1}$$

$$= 5980 \text{ lbf/ft}^2 = 41 \text{ psia} \qquad (2.827 \times 10^5 \text{ Pa})$$

EXAMPLE 2-5

Nitrogen is to be heated at constant pressure from 100 to 3000°F. Calculate the percent error which would result from assuming a constant specific heat taken from Table 2-2 in a computation of the heat transfer.

SOLUTION The "true" heat transfer per mole is given by

$$Q_p = \int_{T_1}^{T_2} \bar{c}_p \, dT \tag{a}$$

where the relation for $\bar{c}_p$ is taken from Table 2-3. This result is then to be compared with

$$Q_p = \bar{c}_p(T_2 - T_1) \tag{b}$$

where now $\bar{c}_p$ is a value calculated from the information in Table 2-2. From Table 2-3 we have, for nitrogen,

$$\bar{c}_p = 9.47 - \frac{3.47 \times 10^3}{T} + \frac{1.16 \times 10^6}{T_2} \text{ Btu/lbm·mol·°R}$$

where T is in degrees Rankine. Thus the "true" heat transfer is

$$Q_p = \int_{T_1}^{T_2} \left(9.47 - \frac{3.47 \times 10^3}{T} + \frac{1.16 \times 10^6}{T^2}\right) dT$$

$$= \left(9.47T - 3.47 \times 10^3 \ln T - \frac{1.16 \times 10^6}{T}\right)_{T_1}^{T_2} \tag{c}$$

where $T_1 = 560$°R and $T_2 = 3460$°R. Evaluating Eq. (c) gives

$$Q_p = 22\,920 \text{ Btu/lbm·mol}$$

Using the data of Table 2-2, we obtain

$$Q_p = (28.02)(0.248)(3460 - 560)$$

$$= 20\,100 \text{ Btu/lbm·mol}$$

The error in assuming a constant specific heat is therefore

$$\text{Error} = \frac{22\,920 - 20\,100}{22\,920} = 0.123 = 12.3 \text{ percent}$$

This example illustrates the necessity of taking specific heat variations into account when wide temperature ranges are encountered.

EXAMPLE 2-6

Air is contained in a piston-cylinder arrangement, as shown, with a cross-sectional area of 4 cm^2 and an initial volume of 20 cm^3. The air is initially at 1 atm and 20°C. Connected to the piston is a spring having a deformation constant of $k_s = 100$ N/cm and the spring is initially undeformed. How much heat must be added to the air to increase the pressure to 3 atm? The apparatus is surrounded by atmospheric air.

SOLUTION The system is shown in the accompanying sketch. The force required to compress the spring is

$$F_s = k_s x \tag{a}$$

where x is the displacement from the equilibrium position. The *work* to compress the spring is therefore

$$W_s = \int F_s\,dx = \int_0^x k_s x\,dx = \frac{1}{2} k_s x^2 = \Delta \text{PE}_s \tag{b}$$

Now consider the air as a thermodynamic system. The work consists of two parts: (1) that to compress the spring W_s and (2) that to displace the surrounding air at 1 atm; or $W_{\text{surr}} = -p\,\Delta V$ for constant-pressure surroundings:

$$Q + W_s + W_{\text{surr}} = \Delta U_{\text{air}} \tag{c}$$

FIG. EXAMPLE 2-6

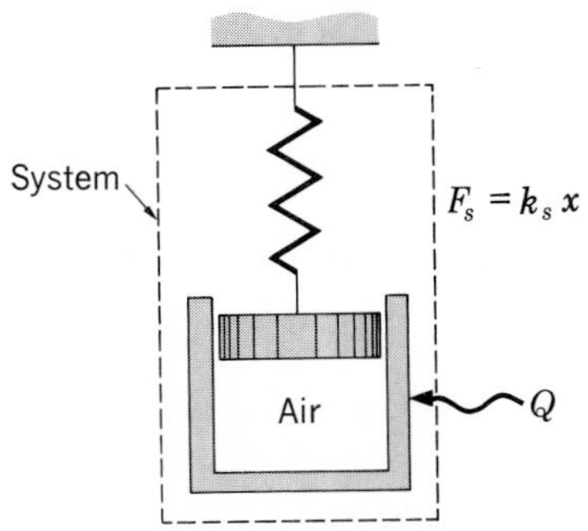

For air we assume $c_v \approx$ const. and

$$\Delta U_{\text{air}} = m_a c_v \, \Delta T_a \qquad (d)$$

From Table 2-2 we have $T_1 = 20°\text{C} = 293$ K, and

$$c_v = 0.1715 \text{ Btu/lbm}\cdot°\text{F} = 718 \text{ J/kg}\cdot°\text{C}$$

$$m_a = \frac{p_1 V_1 M_a}{\mathcal{R} T_1} = \frac{(1.0132 \times 10^5)(20 \times 10^{-6})(28.97)}{(8314)(293)}$$
$$= 2.41 \times 10^{-5} \text{ kg}$$

If we consider the piston as a free body, we see that the initial pressure of the air in the cylinder is just balanced by the surrounding atmospheric pressure, leaving a net force of zero exerted by the spring. At the final condition of 3 atm in the cylinder we still have a balancing pressure of 1 atm from the surroundings, leaving the spring to provide 2 atm pressure over the 4 cm^2 face. Therefore

$$F_{s2} = (2)(1.0132 \times 10^5 \text{ N/m}^2)(4.0 \times 10^{-4} \text{ m}^2)$$
$$= 81 \text{ N}$$

The final displacement of the spring is obtained from Eq. (*a*):

$$x = F_s/k_s = (81 \text{ N})/(100 \text{ N/cm}) = 0.81 \text{ cm}$$

The additional volume swept out by the piston is $A\,\Delta x = (4 \text{ cm}^2)(0.81 \text{ cm}) = 3.24 \text{ cm}^3$ so the final volume of the air is

$$V_2 = V_1 + 3.24 = 20 + 3.24 = 23.24 \text{ cm}^3$$

The work to displace the surroundings is thus

$$W_{\text{surr}} = -p_{\text{atm}} \, \Delta V = -(1.0132 \times 10^5)(3.24 \times 10^{-6}) = -0.3283 \, J$$

The final temperature is obtained from the ideal-gas law expressed as

$$m = \frac{p_2 V_2}{R T_2} = \frac{p_1 V_1}{R T_1}$$

so that

$$T_2 = \frac{(293)(3/1)(23.24)}{20}$$
$$= 1021 \text{ K} = 748°\text{C}$$

From Eq. (*b*)

$$W_s = -\frac{1}{2} k_s x^2 = \frac{1}{2} (100 \text{ N/cm})(0.81 \text{ cm})^2$$
$$= -32.81 \text{ N·cm} = -0.3281 \text{ J}$$

The negative sign applies because this is the work done on the air (the thermodynamic system) *by* the spring. The air does a positive work *on* the spring. Also, from Eq. (*d*),

$$\Delta U_a = (2.41 \times 10^{-5} \text{ kg})(718 \text{ J/kg·°C})(748 - 20)$$
$$= 12.60 \text{ J}$$

The total heat added is now obtained from the energy relation of Eq. (*c*):

$$Q = 0.3283 + 2.3546 + 12.60 = 13.26 \text{ J} \qquad (0.0126 \text{ Btu})$$

In this example we see that most of the heat added goes to increase the temperature of the air enough to raise the pressure.

EXAMPLE 2-7 Constant-Pressure Heat Addition
Air at 1 atm and 20°C occupies an initial volume of 1000 cm^3 in a piston-cylinder arrangement. The air is confined by the piston, which has a constant restraining force, so the gas pressure always remains constant. Heat is added to the air until its temperature reaches 260°C. Calculate the heat added, the work the gas does on the piston, and the change in internal energy of the gas.

SOLUTION The heat transfer is calculated with

$$Q_p = mc_p \, \Delta T_p \tag{a}$$

assuming constant specific heat behavior, as we have been instructed to do. Recalling that the molecular weight of air is 28.97, we find the gas constant for air is

$$R = \frac{\mathcal{R}}{M} = \frac{8314 \text{ J/kg·mol·K}}{28.97 \text{ kg/kg mol}} = 287 \text{ J/kg·K}$$

(which can be obtained directly from Table 2-2), and the mass is calculated as

$$m = \frac{pV}{RT} = \frac{(1.0132 \times 10^5 \text{ N/m}^2)(1000 \times 10^{-6} \text{ m}^3)}{(287 \text{ J/kg·K})(293 \text{ K})} = 1.205 \times 10^{-3} \text{ kg}$$

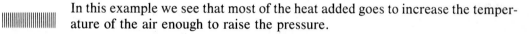

The heat transfer is then

$$Q_p = (1.205 \times 10^{-3}\ \text{kg})(1.005 \times 10^3\ \text{J/kg}\cdot{}^\circ\text{C})(260 - 20)$$
$$= 290.6\ \text{J} \qquad (0.275\ \text{Btu})$$

The work done on the piston is given by Eq. (2-2):

$$W = \int p\, dV$$

For a constant-pressure process this becomes

$$W = p(V_2 - V_1) \tag{b}$$

The final volume is calculated from

$$\frac{p_1V_1}{T_1} = mR = \frac{p_2V_2}{T_2}$$

or

$$V_2 = V_1 \frac{p_1}{p_2}\frac{T_2}{T_1} = (1000)(1)\left(\frac{533}{293}\right) = 1819\ \text{cm}^3$$

Now the work is calculated from Eq. (*b*) as

$$W = (1.0132 \times 10^5\ \text{Pa})(1819 - 1000)(10^{-6})$$
$$= 82.98\ \text{J} \qquad (0.079\ \text{Btu})$$

The work done *on the gas* is the negative of this quantity, or

$$W\ (\text{on gas}) = -82.98\ \text{J}$$

If we take the air as the thermodynamic system, the change in internal energy may now be calculated by applying Eq. (2-16) (neglecting all quantities in the E function except U):

$$Q + W = \Delta U$$
$$290.6 - 82.98 = \Delta U = 207.6\ \text{J} \qquad (0.197\ \text{Btu})$$

Because the air is assumed to be an ideal gas with constant specific heat, we could also calculate the change in internal energy directly from

$$\Delta U = mc_v(T_2 - T_1) \tag{c}$$

From Table 2-2, $c_v = 718$ J/kg·°C so that

$$\Delta U = (1.205 \times 10^{-3}\ \text{kg})(718\ \text{J/kg·°C})(260 - 20)$$
$$= 207.6\ \text{J}$$

in agreement with our previous calculation. Please note that ΔU is calculated with c_v even though this is a process at constant pressure. This results from the fact that U is a function of temperature alone for an ideal gas. Once the temperatures are specified in Eq. (*c*), the value of ΔU is fixed, regardless of the process because U is a property or function of state.

To summarize, the 290.6 J of heat which is added to the air goes to raise the internal energy by 207.6 J and to deliver 82.98 J of work to the surroundings (the piston).

EXAMPLE 2-8 Constant-Temperature Compression
An ideal gas which obeys the equation $pV = mRT$ is compressed in a piston-cylinder arrangement such that the temperature remains constant (isothermal process). Derive an expression for the work done on the gas and calculate the quantity of work when 2 kg of helium is compressed from 1 atm, 20°C, to 1 MPa, holding the temperature constant. Also, calculate the heat added and the change in internal energy.

SOLUTION We apply Eq. (2-2*a*), observing that

$$p = \frac{mRT}{V} \tag{a}$$

so that

$$W = -\int_{V_1}^{V_2} \frac{mRT}{V}\, dV \tag{b}$$

Because $T =$ const., this reduces to

$$W = -mRT \ln \frac{V_2}{V_1} \tag{c}$$

Again, because $T =$ const.,

$$\frac{V_2}{V_1} = \frac{p_1}{p_2}$$

and we have the alternative form

$$W = -mRT \ln \frac{p_1}{p_2} \tag{d}$$

We can now calculate the work for the helium. The gas constant is

$$R = \frac{\mathcal{R}}{M} = \frac{8314 \text{ J/kg·mol·K}}{4 \text{ kg/kg·mol}} = 2078.5 \text{ J/kg·K}$$

or obtained directly from Table 2-2, and

$$W = -(2 \text{ kg})(2078.5 \text{ J/kg·K})(293 \text{ K}) \ln \frac{1.0132 \times 10^5}{1.0 \times 10^6}$$

$$= 2.789 \text{ MJ}$$

If we take the helium as the thermodynamic system, the first law is written as

$$Q + W = \Delta U$$

But we have been told that

$$U(\text{ideal gas}) = \text{function of temperature alone}$$

so if there is no change in temperature (isothermal process), there is no change in internal energy. Thus $\Delta U = 0$ and

$$Q = -W = -2.789 \text{ MJ}$$

EXAMPLE 2-9 Heat Addition to a Room
Twenty people attend a cocktail party in a small room which measures 30 by 25 ft and has an 8-ft ceiling. Each person gives up about 450 Btu of heat per hour. Assuming that the room is completely sealed off and insulated, calculate the air-temperature rise occurring within 15 min. Assume that each person occupies a volume of 2.5 ft^3.

SOLUTION In this problem we take the air in the room as a thermodynamic system and assume that the people are adding heat to the air in a constant-volume process. The room volume is

$$V_{\text{room}} = (30)(25)(8) = 6000 \text{ ft}^3 \qquad (169.9 \text{ m}^3)$$

The volume of air is obtained by subtracting the volume occupied by the people:

$$V_{\text{air}} = 6000 - (20)(2.5) = 5950 \text{ ft}^3 \qquad (168.5 \text{ m}^3)$$

Assuming that the constant-volume heat-addition process starts at standard atmospheric conditions of 14.7 psia and 70°F, the mass of air is

$$m = \frac{pV}{RT} = \frac{(14.7)(144)(5950)}{(53.35)(530)} = 445.4 \text{ lbm} \qquad (202 \text{ kg})$$

For the constant-volume process,

$$c_v \approx \left(\frac{\Delta u}{\Delta T}\right)_v$$

and

$$\Delta U_v = mc_v \, \Delta T_v \tag{a}$$

The change in internal energy of the air is equal to the heat added by the people, since

$$Q + W = \Delta U \tag{b}$$

and there is presumably no work done on the air because the pressure force at the boundary does not act through a displacement. Thus

$$\Delta U = (20)(450) = 9000 \text{ Btu/h} \qquad (2637 \text{ W})$$

or, for a 15-min period,

$$\Delta U = \frac{15}{60}(9000) = 2250 \text{ Btu} \qquad (2133 \text{ kJ})$$

The temperature increase is now calculated from Eq. (*a*) as

$$\Delta T = \frac{\Delta U}{mc_v} = \frac{2250}{(445.4)(0.1715)} = 29.46°\text{F} \qquad (16.37°\text{C})$$

This problem illustrates the rather obvious need for adequate ventilation when a large number of people are confined in a small space.

2-10 PROCESSES FOR IDEAL GASES

In the foregoing examples we have already made considerable use of ideal-gas relations for a number of processes. Energy balances were made for the special cases of constant temperature (isothermal), constant volume, etc., to illustrate system behavior. We remind the reader at this point that we have only been examining *closed* systems, where mass does not cross the boundary. We now wish to look at ideal-gas processes in a more general way. The analytical relationships which will be developed may be used for analysis of both closed systems, as we have done in this chapter, or open systems, to be discussed in Chap. 4.

First let us summarize the terms which apply to special processes:

1 Constant temperature, or *isothermal*
2 Constant pressure, or *isobaric*
3 Constant volume, *isometric* or *isochoric*
4 Zero heat transfer, or *adiabatic* (i.e., perfectly insulated at boundary)

A *polytropic* process is one which may be represented by the relation

$$pv^n = \text{const.} = C \tag{2-34}$$

or, alternatively, in the logarithmic form,

$$\ln p = -n \ln v + \ln C \tag{2-35}$$

where n is appropriately called the *polytropic exponent*. For an ideal gas

$$pv = RT \tag{2-36}$$

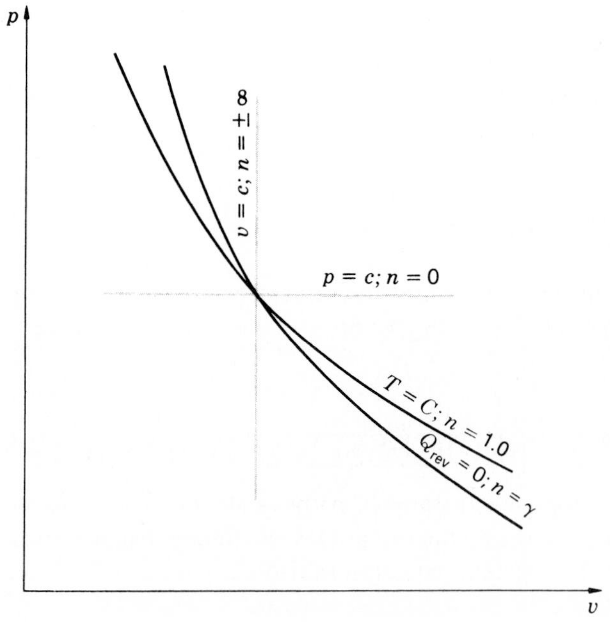

FIG. 2-10 Polytropic process, pv^n = const.

and we may compare Eqs. (2-34) and (2-35) with (2-36) and show that our four special processes mentioned above can be fitted to the polytropic relation if we use the following values of n:

$$T = \text{const.} \qquad n = 1.0$$

$$p = \text{const.} \qquad n = 0$$

$$v = \text{const.} \qquad n = \pm\infty$$

$$Q = 0 \qquad n = \gamma = \frac{c_p}{c_v}$$

These processes are plotted on a p-v diagram in Fig. 2-10, and the logarithmic relation of Eq. (2-35) is shown in Fig. 2-11. Note that $-n$ is the slope of the straight lines on this figure. We have not yet examined the adiabatic case to show that $n = \gamma$ but will do so in the following section.

We may write Eq. (2-34) between two end states 1 and 2 as

$$p_1 v_1^n = p_2 v_2^n = C \tag{2-37}$$

By manipulating this relation along with the ideal-gas law [Eq. (2-36)] we can obtain several alternate expressions for the end states in a polytropic process:

$$\frac{v_2}{v_1} = \left(\frac{p_1}{p_2}\right)^{1/n} \tag{2-38}$$

$$\frac{T_2}{T_1} = \left(\frac{p_2}{p_1}\right)^{(n-1)/n} \tag{2-39}$$

$$\frac{T_2}{T_1} = \left(\frac{v_1}{v_2}\right)^{n-1} \tag{2-40}$$

Please note that Eqs. (2-39) and (2-40) apply *only for ideal gases.*

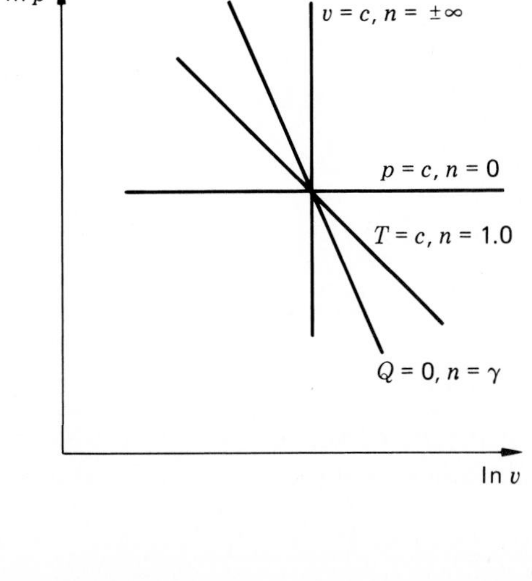

FIG. 2-11 Logarithmic representation of a polytropic process, $pv^n = \text{const.}$

We may now consider the general case of ideal-gas compression work in a piston-cylinder arrangement, which is given by

$$W = -\int p\, dv \tag{2-41}$$

Inserting the polytropic relation for p,

$$W = -\int_{v_1}^{v_2} Cv^{-n}\, dv = C\left(\frac{v^{1-n}}{n-1}\right)_{v_1}^{v_2} \tag{2-42}$$

But $C = p_1v_1^n = p_2v_2^n$, so that

$$W = \frac{p_2v_2 - p_1v_1}{n-1} \tag{2-43}$$

With Eq. (2-36) this may be written in an alternative form:

$$W = \frac{R(T_2 - T_1)}{n-1} \tag{2-44}$$

For the special case of a constant-temperature process, $n = 1.0$ and Eqs. (2-43) and (2-44) are indeterminate; however, we may return to Eq. (2-42) with $n = 1$ and write

$$W = -\int_{v_1}^{v_2} Cv^{-1}\, dv = C \ln \frac{v_1}{v_2}$$

For $T = \text{const.}$,

$$pv = C = RT$$

so that

$$W = p_1V_1 \ln \frac{v_1}{v_2} = RT \ln \frac{v_1}{v_2} \tag{2-45}$$

Again, because the temperature is constant, $v_2/v_1 = p_1/p_2$, so that

$$W = RT \ln \frac{p_2}{p_1} \tag{2-46}$$

2-11 THE QUASISTATIC-ADIABATIC PROCESS FOR AN IDEAL GAS

As promised in the preceding section, we shall now examine the interesting case of a quasistatic-adiabatic process for an ideal gas. Recall that the term *quasistatic* means a series of equilibrium states, and, in a practical sense, generally refers to slow-moving frictionless processes. Later on, we shall also use the term *reversible* to describe such processes. For an adiabatic process we have, with compression work $d'W = -p\,dv$ for a unit mass,

$$d'Q = 0 = du + p\,dv \tag{2-47}$$

and for the ideal gas

$$du = c_v\,dT$$

so that

$$0 = c_v\,dT + p\,dv$$

and

$$c_v\,dT = -p\,dv \tag{2-48}$$

Also, from the definition of enthalpy, and using Eq. (2-47),

$$dh = c_p\,dT = du + p\,dv + v\,dp$$

But $du + p\,dv = 0$, so that

$$c_p\,dT = v\,dp \tag{2-49}$$

Dividing Eq. (2-49) by Eq. (2-48) gives

$$\frac{c_p}{c_v} = -\frac{v}{p}\frac{dp}{dv}$$

or

$$\frac{dp}{p} + \frac{c_p}{c_v}\frac{dv}{v} = 0 \tag{2-50}$$

The solution to Eq. (2-50) is

$$pv^{\gamma} = \text{const.} \tag{2-51}$$

where $\gamma = c_p/c_v$ is the ratio of specific heats and is called the *isentropic exponent*. The meaning of the word *isentropic* will become clear in later chapters. For now we may regard it simply as the value of the polytropic exponent when the process is quasistatic and adiabatic. The adiabatic process is illustrated in Figs. 2-10 and 2-11.

As a matter of interest, the gas constant may be expressed in terms of the adiabatic exponent in the following ways:

$$R = c_p - c_v = (\gamma - 1)c_v \tag{2-52}$$

$$R = \frac{\gamma - 1}{\gamma} c_p \tag{2-53}$$

The work to compress an ideal gas quasistatically and adiabatically in a cylinder may be calculated by applying Eq. (2-43) or (2-44), with $n = \gamma$.

The compression work in a cylinder, which is discussed in the foregoing paragraphs, is the work for a closed system. The mass of gas is fixed within the system, and the calculation is performed with the indicated equations. In an open, or flow, system one may still encounter a compression or expansion process. High-pressure air or steam may be used to drive a turbine and produce a useful work output. Conversely, a reciprocating piston compressor or turbine arrangement may be employed to compress air flow from a low to high pressure. We can have a quasistatic adiabatic process in flow systems as well as in the closed piston-cylinder arrangement. The derivation of Eq. (2-51) did not specify the type of system—open or closed; therefore, it may be used to describe the quasistatic adiabatic process for both types of systems.

Compression and Expansion

Although the terms may be intuitively apparent at this time we should clarify the meaning of the words "compression" and "expansion." In a *compression* process the pressure of the system is raised, by whatever mechanism, while in an *expansion* process the pressure is lowered. Thus we shall apply these terms to both flow and nonflow processes, and in various work and heat interaction situations.

EXAMPLE 2-10 Work in a Cycle

A 1-g quantity of nitrogen undergoes the following sequence of quasistatic processes in a piston-cylinder arrangement:

1 An adiabatic expansion in which the volume doubles
2 A constant-pressure process in which the volume is reduced to its initial value
3 A constant-volume compression back to the initial state

The nitrogen is initally at 150°C and 5 atm pressure. Calculate the net work done on the gas in this sequence of processes.

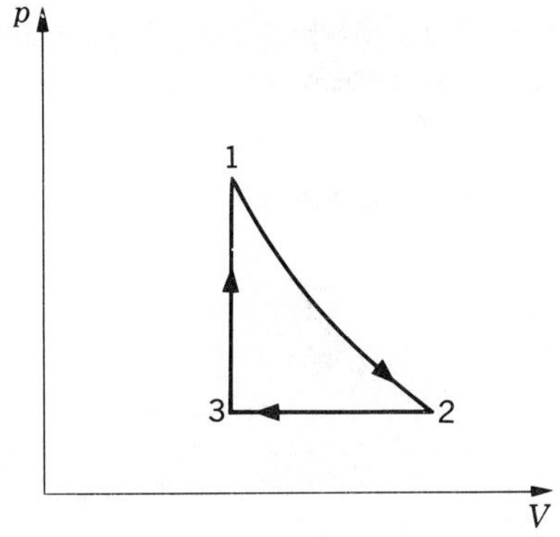

FIG. EXAMPLE 2-10

SOLUTION We take the nitrogen as the thermodynamic system. The processes under consideration are illustrated in the accompanying figure. We have

$$p_1 = 5 \text{ atm} = 5.066 \times 10^5 \text{ Pa}$$

$$T_1 = 150°\text{C} = 423 \text{ K}$$

$$m = 1 \text{ g} = 10^{-3} \text{ kg}$$

For nitrogen the gas constant can be calculated as

$$R = \frac{\mathcal{R}}{M} = \frac{8314}{28} = 297 \text{ J/kg·K}$$

or obtained directly from Table 2-2, and the initial volume is calculated as

$$V_1 = \frac{mRT_1}{p_1} = \frac{(0.001)(297)(423)}{5.066 \times 10^5} = 2.48 \times 10^{-4} \text{ m}^3$$

We also have $V_2 = 2V_1 = 4.96 \times 10^{-4}$ m^3. From Eq. (2-38),

$$p_2 = p_1 \left(\frac{V_1}{V_2}\right)^\gamma = (5.066 \times 10^5)\left(\frac{1}{2}\right)^{1.4} = 1.92 \times 10^5 \text{ Pa}$$

For process 1-2 the work done on the gas is calculated from Eq. (2-43):

$$W_{1-2} = \frac{p_2V_2 - p_1V_1}{\gamma - 1} = \frac{[(1.92)(4.96) - (5.066)(2.48)](10)}{1.4 - 1}$$

$$= -76.05 \text{ J}$$

where $\gamma = 1.4$ from Table 2-2. In process 2-3 the pressure is constant and the work is obtained from

$$W_{2\text{-}3} = -\int_{v_2}^{v_3} p\,dV = -p_2(V_3 - V_2)$$

$$= -(1.92 \times 10^5)(2.48 - 4.96)(10^{-4})$$

$$= 47.62 \text{ J}$$

There is no work done in process 3-1 because $V_3 = V_1$. The total work in the sequence of processes is therefore

$$W = W_{1\text{-}2} + W_{2\text{-}3} + W_{3\text{-}1}$$
$$= -76.05 + 47.62 + 0 = -28.43 \text{ J}$$

The negative sign indicates that there is a net work *output* from the three-process cycle.

We can make some additional calculations which illustrate an important point. From Eq. (2-19) the sum of the heat flows in the cycle should equal the negative of the net work added in the cycle. Let us calculate these heat flows. The temperatures of interest are

$$T_2 = T_1\left(\frac{V_1}{V_2}\right)^{\gamma-1} = (423)\left(\frac{1}{2}\right)^{0.4} = 320.6 \text{ K}$$

Because $p_2 = p_3$,

$$T_3 = T_2\frac{V_3}{V_2} = (320.6)\left(\frac{1}{2}\right) = 160.3 \text{ K}$$

From Table 2-2 the specific heats for nitrogen are

$$c_v = 0.177 \text{ Btu/lbm·°F} = 741.1 \text{ J/kg·°C}$$

$$c_p = 0.248 \text{ Btu/lbm·°F} = 1038.3 \text{ J/kg·°C}$$

Process 1-2 is adiabatic so $Q_{1-2} = 0$. The heat added in the constant-pressure process 2-3 is

$$Q_{2\text{-}3} = mc_p(T_3 - T_2)$$
$$= (0.001 \text{ kg})(1038.3 \text{ J/kg·°C})(160.3 - 320.6)$$
$$= -166.4 \text{ J}$$

Process 3-1 is constant volume so the heat added is

$$\begin{aligned} Q_{3\text{-}1} &= mc_v(T_1 - T_3) \\ &= (0.001)(741.1)(423 - 160.3) \\ &= 194.8 \text{ J} \end{aligned}$$

The net heat added in the cycle is therefore

$$\begin{aligned} Q &= Q_{1\text{-}2} + Q_{2\text{-}3} + Q_{3\text{-}1} \\ &= 0 - 166.4 + 194.8 = 28.4 \text{ J} \end{aligned}$$

and this, of course, is equal to the negative of the work added in the cycle.

EXAMPLE 2-11 Discharge from a Tank
A 0.5-m^3 tank contains air at $p = 7.0$ MPa and 250°C and is perfectly insulated from the surroundings. A valve on the tank is opened and the air discharged until the pressure drops to 400 kPa. Calculate the mass of air discharged from the tank. The system is indicated in the figure.

SOLUTION This problem requires some special assumptions. First let us observe that the mass of air discharged from the tank is clearly the difference between the initial mass m_1 and the final mass at the end of the discharge, m_2. Thus

$$\Delta m = m_1 - m_2 = \frac{p_1 V_1}{RT_1} - \frac{p_2 V_2}{RT_2} \qquad (a)$$

The initial conditions p_1, V_1, and T_1 are known so m_1 is easily calculated. Assuming the tank is rigid, $V_2 = V_1$, and p_2 is given, so we need only find T_2 to solve the problem.

FIG. EXAMPLE 2-11

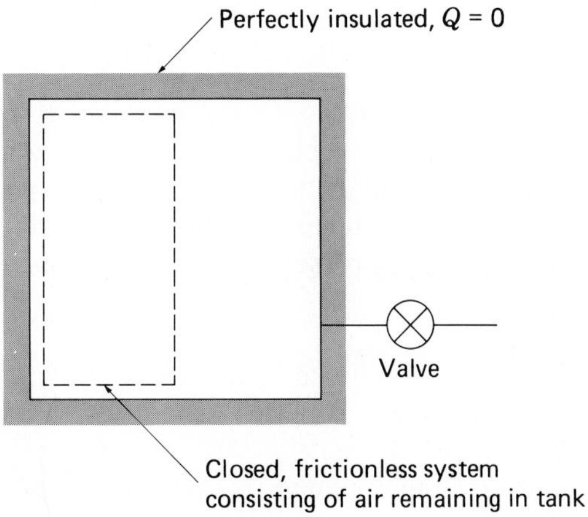

If one takes the system boundary as the inside surface of the tank, the system is *not* closed; i.e., mass crosses the boundary. We need a description of the process in order to find T_2. Consider the mass of air remaining in the tank (m_2) which will just expand to fill the tank. Using $T_1 = 250 + 273 = 523$ K, the initial mass is calculated as

$$m_1 = \frac{p_1 V}{RT_1} = \frac{(7 \times 10^6)(0.5)}{(287.1)(523)} = 23.31 \text{ kg} \qquad (b)$$

To calculate the final mass we must determine the final temperature in the tank. This mass undergoes an adiabatic process (because the tank is insulated) and there is no evidence of frictional effects, so we may assume that the process is quasistatic as well. Therefore we may employ Eq. (2-38) with $n = \gamma$ to calculate the final temperature:

$$\frac{T_2}{T_1} = \left(\frac{p_2}{p_1}\right)^{(\gamma-1)/\gamma}$$

$\gamma = 1.4$ for air, and so

$$T_2 = (523)\left(\frac{0.4}{7.0}\right)^{(1.4-1)/1.4} = 230.7 \text{ K} \qquad (c)$$

The final mass is now calculated as

$$m_2 = \frac{p_2 V}{RT_2} = \frac{(4 \times 10^5)(0.5)}{(287.1)(230.7)} = 3.02 \qquad (d)$$

and the mass lost is

$$\Delta m = m_1 - m_2 = 23.31 - 3.02 = 20.29 \text{ kg}$$

EXAMPLE 2-12

Helium at 20 atm and 40°C is contained in a small steel cylinder having a volume of 15 cm^3. The cylinder is placed in a large container having a volume of 1500 cm^3. The large container is perfectly evacuated and perfectly insulated. By an appropriate means the helium is allowed to discharge and fill the large container. Calculate the final pressure after the entire assembly reaches equilibrium.

SOLUTION The system is shown in the accompanying figure. We choose the system boundary as the inside surface of the large container and have

$Q = 0$ because the container is insulated

$W = 0$ because the boundary does not move

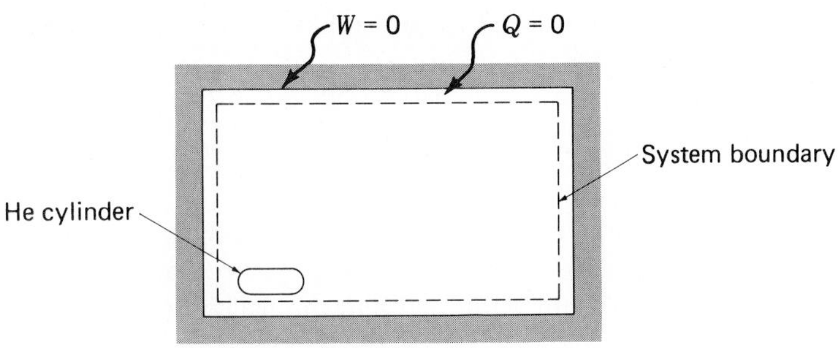

FIG. EXAMPLE 2-12

The first law then becomes

$$Q + W = 0 = \Delta U = m(u_2 - u_1) \qquad (a)$$

But helium is an ideal gas, so *internal energy is a function of temperature alone,* and this requires that

$$T_2 = T_1$$

We also have

$$mR = \frac{p_1 V_1}{T_1} = \frac{p_2 V_2}{T_2}$$

so that

$$p_2 = p_1 \frac{V_1}{V_2} = (20)\left(\frac{15}{1500}\right) = 0.2 \text{ atm}$$

COMMENT Had the cylinder contained a substance that was not an ideal gas (like high-pressure water), we would not have $T_2 = T_1$ because the internal energy would also depend on volume; that is, $u = f(T, v)$.

2-12 SUMMARY

This chapter has served to generalize the first law of thermodynamics as the conservation-of-energy principle. Work has been defined as energy expended by a force acting through a displacement. In thermodynamics this is a quantity which is delivered to the boundary of the system. The existence of a general internal-energy function has been demonstrated, which is dependent only on the state of the system; thus it is given a designation as a thermodynamic property. Heat has been defined as energy which may cross the boundary of a system but is not expressible as a force acting through a distance. Finally, all three types of energy have been incorporated into the conservation idea, and an analytical statement of the first law has been presented.

We have entered into a rather full discussion of ideal gases and specific heats and have stated that

1 Internal energy and enthalpy for ideal gases are functions of temperature alone.
2 The change in internal energy and enthalpy for ideal gases can be calculated over a rather wide range of pressures and temperatures, with c_v and c_p obtained from Table 2-2 and the following

$$\Delta u = c_v \, \Delta T$$

$$\Delta h = c_p \, \Delta T$$

REVIEW QUESTIONS

1 How is work defined?
2 What is meant by expansion work?
3 What is meant by the term *potential energy*?
4 What is the first law of thermodynamics?
5 What is the general meaning attached to the term *internal energy*?
6 How does heat differ from work?
7 Why cannot heat be contained in a system?
8 Why is the term *specific heat* a misnomer?
9 What is a cyclic process?
10 How is enthalpy defined?
11 How would you explain the first law of thermodynamics to a layperson?
12 How does work differ from power?
13 Upon what property is the internal energy and enthalpy of an ideal gas dependent?
14 Explain the sign convention for heat and work.
15 Why is work not a property of a system?
16 What is meant by a *closed* system?
17 What is an ideal gas?
18 Define the terms *isothermal* and *adiabatic*.
19 How are properties related in a quasistatic adiabatic process for an ideal gas?
20 How are specific heats and the gas constant related?
21 What is a polytropic process?
22 In Table 2-2 the units for specific heat are given as kJ/kg·°C. Explain why it would be equally suitable to use kJ/kg·K.
23 Why is kJ/kg·°C not a suitable set of units for the gas constant?
24 A perfectly insulated room contains a refrigerator. If the door to the refrigerator is left open, will the temperature of the room (*a*) rise, (*b*) fall, or (*c*) remain the same?

25 Why can the relation $\Delta U = mc_v \Delta T$ be used to calculate change in internal energy for an ideal gas in a *constant-pressure* process?

PROBLEMS (ENGLISH UNITS)

2-1 A pressure of 342 psia acts on a piston having a diameter of 7.5 cm. Calculate the work done when this piston moves through a displacement of 3 in.

2-2 A large rocket can produce a thrust of 600 000 lbf at a velocity of 600 mph. Calculate the horsepower of the rocket and the energy dissipated in 1 s expressed in Btu.

2-3 A small heat-power system receives a heat input of 100 000 Btu/h along with a work input of 1.5 hp. The system produces an electric energy output at the rate of 20 kW. Calculate the change in internal energy of the system during a 2-min time interval.

2-4 A certain battery is charged by applying a current of 40 A at 12 V for a 30-min time period. During the charging process the battery loses 200 Btu of heat to the surroundings. How much does the internal energy of the battery increase during the 30-min period?

2-5 Air expands behind a piston in a quasistatic adiabatic process from 155 psia and 200°F to a pressure of 25 psia. The initial volume of the air is 0.5 ft^3. Calculate the work and change in enthalpy for the process.

2-6 A piston-cylinder arrangement contains air at 20 psia and 50°F. A funny-shaped spring is attached to the piston such that the pressure varies as the square root of the volume. The initial volume is 0.2 ft^3. Heat is added to the cylinder until the temperature rises to 250°F. Calculate the input of heat.

2-7 A 1-ft^3 rigid enclosure contains air at 20 psia and 70°F. Heat is added to the container until the pressure reaches 50 psia. Calculate the heat added, assuming that air behaves an an ideal gas with constant specific heat.

2-8 Calculate the temperature rise for the tank of water in Example 2-3 if the tank loses heat at the rate of 800 Btu/h while the 1-hp stirring action is applied.

2-9 Nitrogen at 2000 psia and 100°F is contained in a rigid tank having a volume of 3.0 ft^3. The tank is insulated. A valve on the side of the tank is opened and nitrogen allowed to escape to the atmosphere until the tank pressure drops to 1200 psia. Calculate the mass of nitrogen which escapes from the tank.

2-10 Suppose the tank of Example 2-3 is uninsulated and loses heat to the surroundings according to $q = 18\ \Delta T$ Btu/h, where ΔT is the temperature difference between the tank and the surroundings and q is the heat transfer *rate*. Assuming that the stirring action starts when the tank and surroundings are at 70°F, calculate the equilibrium temperature of the water. Also calculate the temperature the water will attain after 15 min.

2-11 Air expands in a cylinder from 200 psia, 800°F, to 30 psia, 200°F. Calculate the work and heat transfer assuming that the process varies as $pv^n =$ const. and c_v for air is 0.1715 Btu/lbm·°F.

2-12 Air in the amount of 0.1 lbm is contained by a piston-cylinder arrangement at 25 psia and 100°F. Heat is added to the air so that it is allowed to expand at constant pressure until the volume doubles. Calculate the change in internal energy of the air and the heat added.

2-13 A small steel cylinder, V = 0.1 ft³, contains nitrogen at 2000 psia and 70°F. The cylinder is placed inside an evacuated and insulated chamber having a volume of 10 ft³. By some suitable means the small cylinder is allowed to discharge its contents into the large chamber. Calculate the final nitrogen pressure in the chamber when equilibrium conditions are established.

2-14 A certain 12-V battery carries a rating of 100 A·h. What is the energy stored in the battery expressed in Btu, J, and cal?

2-15 Heat is added to a closed system which is so constructed that the pressure will remain constant. Calculate the change in enthalpy of the system when 800 Btu of heat are added and the system produces a work output of 200 Btu.

2-16 A spring is available which requires an applied force of 100 lbf to compress it 1 in. Calculate the change in internal energy of the spring when it is compressed 3 in. Neglect heat transfer.

2-17 Nitrogen expands in a cylinder from 100 psia and 500°F to 30 psia and 100°F. Assuming ideal-gas behavior ($pV = mRT$), calculate the work done on the face of the piston per pound of nitrogen. Assume pressure and volume are related by $pV^n =$ const., where n is a suitable exponent.

2-18 A window air conditioner is designed to remove 12 000 Btu/h from the room, and 1.2 hp of electrical work must be delivered to the unit to accomplish this cooling. How much heat is dissipated to the surroundings?

2-19 A certain automobile consumes fuel at the rate of 12 mi/gal at a speed of 60 mi/h, traveling on a level road. The internal energy of the fuel is 140 000 Btu/gal and the overall efficiency of the engine and drive mechanism is 13 percent; i.e., only 13 percent of the energy of the fuel is delivered to the road in the form of work and the remainder is dissipated as heat. Calculate the overall drag force on the automobile using energy considerations. Suppose the same automobile, weighing 4400 lbm, is driven at the same speed up a grade inclined 5° with the horizontal. What will be its fuel consumption under these new conditions?

2-20 A 250-lb man is told that he can lose weight by increasing his energy consumption through exercise. He proposes to do this by climbing a 20-ft staircase several times a day. His current food consumption provides an energy input of 4000 kcal/day. What would you advise him regarding exercise versus a reduced food intake to effect a weight loss?

2-21 A thermodynamic system undergoes a cycle composed of a series of three processes for which $Q_1 = +10$ Btu, $Q_2 = +30$ Btu, $Q_3 = -5$ Btu.

For the first process $\Delta E = +20$ Btu, and for the third process, $\Delta E = -20$ Btu. What is the work in the second process, and the net work *output* of the cycle?

2-22 4 lbm of oxygen are heated from 100 to 500°F in a constant-volume process. Calculate the change in enthalpy and heat transfer.

2-23 5 g of carbon dioxide are contained in a vessel at 10 psia and 100°F. Calculate the heat which must be added to raise the pressure to a value of 2 atm.

2-24 Calculate the heat necessary to raise 1 g mol of each of the following gases from 200 to 500°F at a constant pressure of 15 psia:
(*a*) Hydrogen
(*b*) Helium
(*c*) Nitrogen
(*d*) Oxygen
(*e*) Argon
(*f*) Air

2-25 A balloon having a volume of 25 ft^3 is filled with air at 70°F and 14.696 psia. After exposure to solar radiation, the volume of the balloon increases to 30 ft^3 and the pressure remains constant since the air is loosely contained. Calculate the heat added.

2-26 Consider a constant-pressure process in which hydrogen is heated from 300 to 4000°F in an electric-arc process. Calculate the heat transfer per pound-mass. Suppose the specific heat of the hydrogen is evaluated at some mean temperature (2100°F) and this value is used to calculate the heat transfer using a constant specific heat analysis. What error would result if this latter technique were employed?

2-27 Water vapor is heated at a constant pressure of 1 psia from 150 to 1000°F. Assume that the vapor obeys the ideal-gas equation of state in this region, and calculate the heat transfer by using the data of Table 2-3.

2-28 One exotic energy-utilization scheme would employ a 5-by-5-mi solar collector in space to absorb the sun's energy in special solar cells. These cells convert the radiant energy directly into electric energy which can then be transmitted to the earth with a complicated microwave system. The efficiency of the solar cells is about 12 percent; i.e., only 12 percent of the incoming solar energy is converted to electricity, the remaining 88 percent is radiated to space. Estimate the electric power output of the device using the data of Prob. 1-28.

2-29 A 1-ft^3 rigid container is open to the atmosphere and heat is added to the bottom of the container until the air temperature inside reaches 850°F. The container is then quickly sealed, removed from the heating source, and allowed to cool to room temperature of 70°F. How much heat is lost during the cooling process?

2-30 An energy-conservationist group claims that a substantial amount of electric power can be saved by using more efficient lighting systems. An incandescent light bulb is only about 5 percent efficient in converting electric energy into visible light; the remaining 95 percent is dissipated as heat in the room. Fluorescent lights are about 20 percent efficient. In modern buildings the heat dissipation must be removed by the air-conditioning system, which requires further power. About 1 kW of electric power is required to produce 10 000 Btu/h of cooling. What reduction in power-consumption rate could be accomplished in a room currently using 2000 W of incandescent lights if they were all converted to fluorescents?

2-31 The *mean* specific heat for a gas over a given temperature range is defined as

$$c_{p,\text{mean}} = \frac{\int_{T_1}^{T_2} c_p \, dT}{T_2 - T_1}$$

Calculate such a mean specific heat for the hydrogen in Prob. 2-26. How does this value compare with that given in Table 2-2?

2-32 In a large majority of applicable problems air is assumed to behave as an ideal gas with constant specific heats for the temperature range 0 to 1000°F. Assuming that air behaves like nitrogen (air is 78 percent nitrogen), estimate the validity of this assumption.

PROBLEMS (METRIC UNITS)

2-1M A pressure of 2.4 MPa acts on a piston having a diameter of 7.5 cm. Calculate the work done when this piston moves through a displacement of 8 cm.

2-2M An electric potential of 110 V is impressed on a certain resistor such that a current of 12 A is drawn. Calculate the energy dissipated in a time of 3 min.

2-3M A large rocket can produce a thrust of 2.7 MN at a velocity of 200 m/s. Calculate the power of the rocket and the energy dissipated in 1 s expressed in kJ.

2-4M A rigid tank has a volume of 2.3 m^3 and is perfectly insulated. It contains helium at 2.0 MPa and 200°C. A valve on the side of the tank is opened and helium allowed to escape until half the mass has been lost from the tank. The valve is then shut off. Calculate the final pressure in the tank.

2-5M Air is compressed in a piston-cylinder arrangement from 20°C, 200 kPa, to 120°C, 300 kPa. Assuming the process is polytropic, calculate the heat added per kilogram of air.

2-6M A spring is available which requires an applied force of 445 N to compress it 2.5 cm. Calculate the change in internal energy of the spring when it is compressed 7.5 cm. Neglect heat transfer.

2-7M Nitrogen expands in a cylinder from 690 kPa and 260°C to 210 kPa and 40°C. Assuming ideal-gas behavior ($pV = mRT$), calculate the work done on the face of the piston per pound of nitrogen. Assume pressure and volume are related by pV^n = const., where n is a suitable exponent.

2-8M A 28-liter rigid enclosure contains air at 140 kPa and 20°C. Heat is added to the container until the pressure reaches 345 kPa. Calculate the heat added.

2-9M A spherical tank having a volume of 20 m^3 contains helium at 20 atm and 100°C. It is connected to a second tank of the same size which is perfectly evacuated. A valve connecting the tanks is opened and then closed when the pressure in the first tank reaches 10 atm. Both tanks are perfectly insulated. Calculate the final temperature in the tank which is initially evacuated.

2-10M A small heat-power system receives a heat input of 30 kW along with a work input of 1.1 kW. The system produces an electric energy output at the rate of 20 kW. Calculate the change in internal energy of the system during a 2-min time interval.

2-11M A certain battery is charged by applying a current of 40 A at 12 V for a 30-min time period. During the charging process the battery loses 200 kJ of heat to the surroundings. How much does the internal energy of the battery increase during the 30-min period?

2-12M Heat is added to a closed system which is constructed so that the pressure will remain constant. Calculate the change in enthalpy of the system when 850 kJ of heat are added and the system produces a work output of 200 kJ.

2-13M Air is contained in a frictionless piston-cylinder arrangement at 200 kPa and 100°C. 20 kJ/kg of heat are added to the air and the temperature rises to 200°C. If the process is quasistatic and polytropic, calculate the final pressure of the air.

2-14M Steam at 150 kPa and 120°C is contained in a piston-cylinder arrangement. Heat is added at constant pressure until the volume doubles. Calculate the heat added per unit mass.

2-15M A spherical tank having a diameter of 5 m is charged with air to conditions of 30 atm and 40°C. The tank is perfectly insulated. Air is allowed to escape from the tank until the pressure drops to 20 atm. Calculate the mass of air which escapes from the tank.

2-16M A small steel cylinder, $V = 0.28$ liter, contains nitrogen at 14 MPa and 20°C. The cylinder is placed inside an evacuated and insulated chamber having a volume of 280 liters. By some suitable means the small cylinder is allowed to discharge its contents into the large chamber. Calculate the final nitrogen pressure in the chamber when equilibrium conditions are established.

2-17M A window air conditioner is designed to remove 13 MJ/h from the room and 900 W of electrical work must be delivered to the unit to accomplish this cooling. How much heat is dissipated to the surroundings?

2-18M A certain automobile consumes fuel at the rate of 5 km/liter at a speed of 100 km/h, traveling on a level road. The internal energy of the fuel is 40 000 kJ/liter, and the overall efficiency of the engine and drive mechanism is 13 percent; i.e., only 13 percent of the energy of the fuel is delivered to the road in the form of work, and the remainder is dissipated as heat. Calculate the overall drag force on the automobile using energy considerations. Suppose the same automobile, weighing 2000 kg, is driven at the same speed up a grade inclined 5° with the horizontal. What will be its fuel consumption under these new conditions?

2-19M A 115-kg man is told that he can lose weight by increasing his energy consumption through exercise. He proposes to do this by climbing a 6-m staircase several times a day. His current food consumption provides an energy input of 17 kJ/day. What would you advise him regarding exercise versus a reduced food intake to effect a weight loss?

2-20M A 28-liter rigid container is open to the atmosphere, and heat is added to the bottom of the container until the air temperature inside reaches 450°C. The container is then quickly sealed, removed from the heating source, and allowed to cool to room temperature of 20°C. How much heat is lost during the cooling process?

2-21M An energy-conservationist group claims that a substantial amount of electric power can be saved by using more efficient lighting systems. An incandescent light bulb is only about 5 percent efficient in converting electric energy into visible light; the remaining 95 percent is dissipated as heat in the room. Fluorescent lights are about 20 percent efficient. In modern buildings the heat dissipation must be removed by the air-conditioning system, which requires further power. About 1 kW of electric power is required to produce 10 MJ/h of cooling. What reduction in power-consumption rate could be accomplished in a building currently using 2000 W of incandescent lights if they were all converted to fluorescents?

2-22M A thermodynamic system undergoes a cycle composed of a series of three processes for which $Q_1 = +10$ kJ, $Q_2 = +30$ kJ, $Q_3 = -5$ kJ. For the first process, $\Delta E = +20$ kJ, and for the third process, $\Delta E = -20$ kJ. What is the work in the second process, and the net work *output* of the cycle?

2-23M One exotic energy-utilization scheme would employ an 8-by-8-km solar collector in space to absorb the sun's energy in special solar cells. These cells convert the radiant energy directly into electric energy, which can then be transmitted to the earth with a complicated microwave system. The efficiency of the solar cells is about 12 percent; i.e., only 12 percent of the incoming solar energy is converted to electricity, with the remaining 88 percent being reradiated to space. Estimate the electric power output of the device using the data of Prob. 1-28.

2-24M 0.1 kg of nitrogen is compressed in a piston-cylinder arrangement in a polytropic process, with $n = 1.26$. The initial conditions are 100 kPa and 100°C, while the final pressure is 950 kPa. Calculate the heat transfer for this process.

2-25M Air is compressed in a piston-cylinder arrangement from 30°C, 110 kPa, to 130°C, 350 kPa. Assuming the process is polytropic, calculate the heat added per kilogram of air.

2-26M For the p-V diagram shown in the figure below, a certain closed thermodynamic system may execute a cyclic process from point 1 through point 2 in a clockwise or counterclockwise manner. In the clockwise process the heat is $(Q_{1\text{-}2})_{\text{CW}} = 85$ kJ and the work is $(W_{1\text{-}2})_{\text{CW}} = -40$ kJ. When the system goes from state 2 to state 1, the heat is the same for either CW or CCW directions and is equal to $(Q_{2\text{-}1})_{\text{CW}} = (Q_{2\text{-}1})_{\text{CCW}} = -55$ kJ. Taking the counterclockwise path from 1 to 2, it is found that $(Q_{1\text{-}2})_{\text{CCW}} = 75$ kJ. Determine $(W_{1\text{-}2})_{\text{CCW}}$ and $(W_{2\text{-}1})$.

2-27M Calculate the temperature rise for the tank of water in Example 2-3 if the tank loses heat at the rate of 850 kJ/h while the 1-hp stirring action is applied.

2-28M Suppose the tank of Example 2-3 is uninsulated and loses heat to the surroundings according to the following relation $q = 18\ \Delta T$ kJ/h, where ΔT is the temperature difference between the tank and the surroundings and q is the heat-transfer *rate*. Assume that the stirring action starts when the tank and the surroundings are at 20°C, and calculate the equilibrium temperature of the water. Also calculate the temperature the water will attain after 15 min.

2-29M Air expands in a cylinder from 1.4 MPa, 700 K, to 200 kPa, 370 K. Calculate the work and heat transfer assuming that the process varies as $pv^n = \text{const.}$, and c_v for air is 0.718 kJ/kg·°C.

2-30M 100 g of air are contained by a piston-cylinder arrangement at 170 kPa and 40°C. Heat is added to the air so that it is allowed to expand at constant pressure until the volume doubles. Calculate the change in internal energy of the air and the heat added.

2-31M Helium is contained in an insulated tank at 7 MPa and 70°C. A valve on the side of the tank is opened and 40 percent of the helium is allowed to escape. The valve is then closed. The total volume of the tank is 4.5 m^3. Calculate the final pressure in the tank.

2-32M The nuclear fission of one uranium 235 atom releases about 200 million electron volts (MeV) of energy. Assuming that 30 percent of this

FIG. P2-26M

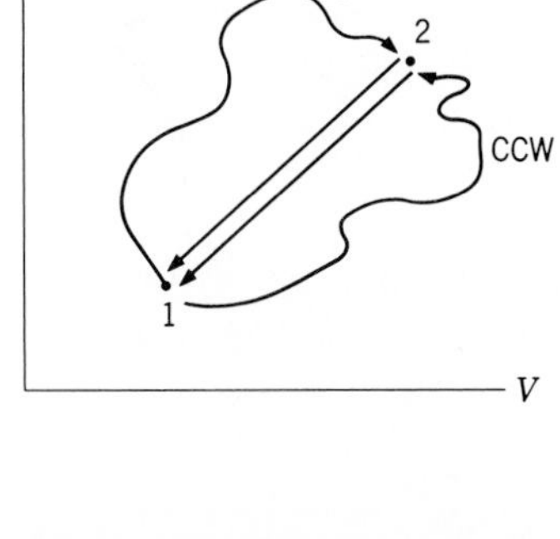

energy could be converted to electricity in a power plant, how much U^{235} would be consumed in a year to produce 1000 MW continuously? (See Appendix for conversion factors.)

2-33M A weather balloon is constructed of a nonstretchable plastic material to produce an inflated spherical shape 10 m in diameter. Initially the balloon is collapsed and is to be filled with helium at a location where the pressure is 1 atm. How much work is done on the surroundings during the inflation process?

2-34M A hydraulic lift for a service station uses compressed air for power. The air is stored at 7 atm and released into a cylinder which drives a 35-cm-diameter piston to raise the automobiles. Normally, the automobiles are lifted through a height of 2.3 m. The air supply is provided by a small electric-motor-driven compressor which may run continuously if needed. Assuming that an adequate air-storage tank is available, what horsepower motor would be required to perform 50 lifts per day on a 4000-lb vehicle?

2-35M Oxygen is compressed quasistatically and adiabatically in a piston-cylinder device. Calculate the work required for compression from 20°C, 1 atm to p = 500 kPa. Also calculate the final temperature.

2-36M A substance undergoes a certain quasistatic process behind a piston such that the pressure varies according to $p = a + bV$. Derive an expression for the work between two pressures p_1 and p_2.

2-37M A balloon contains air at 120 kPa and 20°C and has a diameter of 1.0 m. Heat is added to the balloon until the diameter increases to 1.4 m. During this process the pressure is directly proportional to the diameter. Calculate the heat added.

2-38M In a certain spring the force is related to displacement by $F = k_s x^2$. A force of 1000 N is required to compress the spring 1 cm. Calculate the work to compress the spring 3.0 cm. Express the answer in joules, ft·lbf, and Btu.

2-39M An experiment is conducted with a certain gas. It is found that when the gas is maintained in a constant-volume container 70 kJ/kg of energy are required to raise the temperature 83°C. It is known that γ = 1.50 for the gas. Calculate the values of c_p and R for the gas.

2-40M Hydroelectric power production is based on the conversion of gravitational potential energy of water into mechanical energy by means of large turbines. Suppose a dam facility is available whereby the water may fall through an elevation of 100 m. What water flow would be necessary to produce a power of 10 000 kW?

2-41M A large spring is to be designed to store energy which can be recovered at the rate of 300 W for a period of 10 min. The deformation of the spring is to be 30 cm. What must the spring constant be to accomplish this energy storage?

2-42M A 3.0-m³ tank contains helium at 2 MPa and 100°C and is perfectly insulated. A valve on the side of the tank is opened and the helium allowed to discharge to the atmosphere until the tank pressure reaches 500 kPa. Calculate the mass of helium removed from the tank.

2-43M An ideal gas undergoes a quasistatic adiabatic expansion from conditions of 800 kPa, 0.7 liter, and 100°C to a final pressure of 200 kPa. Calculate the work done if the ratio of specific heats is 1.39. Also calculate values of the specific heats based on total mass.

2-44M Proposals have been made to employ flywheels as energy-storage devices in the future. Consult whatever references you feel appropriate and comment on such proposals.

2-45M A tank having a volume of 0.1 m^3 contains air at 14 MPa and 50°C. It is connected through a valve to a larger tank having a volume of 15 m^3 which is completely evacuated. The entire assembly is completely insulated. The valve is opened and the gas allowed to come to equilibrium in both tanks. Calculate the final pressure.

2-46M Suppose the valve in Prob. 2-45M is only opened long enough for the pressure to drop to 3.0 MPa in the small tank and that the process occurs rapidly. Calculate the final temperature in each tank.

2-47M A certain thermodynamic cycle is represented as a 5-by-5-cm square on a p-v diagram in which 1 cm = 1 m^3 and 1 cm = 1 atm are the coordinate scales. If there is to be a net work input to the system during the cycle, calculate the overall heat transfer during the cycle.

2-48M An electric resistance heater is placed inside a 1-liter rigid container filled with nitrogen at 50°C and 10 atm pressure. What electric power rate must be supplied to the heater to make the pressure rise to 15 atm in a period of 10 min, if the container is perfectly insulated?

2-49M In an exotic power device, hydrogen is burned with oxygen to produce a very high combustion temperature. Prior to the burning, the oxygen is heated from 20 to 1200°C at constant pressure of 130 kPa. How much heat is required for an oxygen consumption of 1 kg?

2-50M Oxygen is compressed in a piston-cylinder arrangement from 160 kPa, 40°C, to 600 kPa, 90°C. Calculate the work done on the gas and the heat transfer if the process is quasistatic.

2-51M An ideal gas undergoes a quasistatic process at constant pressure of 3 atm in a piston-cylinder arrangement. During the process 25 kJ of heat are added and the volume is reduced from 0.15 to 0.06 m^3. Calculate the change in internal energy for this process.

2-52M 5 kg of nitrogen are compressed in such a manner that the internal energy increases by 160 kJ while 20 kJ of heat are added. Calculate the work that is supplied to the gas. Assume that the compression occurs in a closed system.

2-53M A 1-kW-resistance heater is placed in a 3000-liter container filled with air at 100 kPa and 20°C. The heater is allowed to operate for 10 min and the container is insulated. Calculate the temperature and pressure of the air at the end of this time period.

2-54M A balloon having a volume of 700 liters is filled with air at 20°C and 100 kPa. After exposure to solar radiation, the volume of the balloon increases to 850 liters and the pressure remains constant since the air is loosely contained. Calculate the heat added.

2-55M Consider a constant-pressure process in which hydrogen is heated from 150 to 2200°C in an electric-arc process. Calculate the heat transfer per kilogram-mass. Suppose the specific heat of the hydrogen is evaluated at some mean temperature (1200°C) and this value is used to calculate the heat transfer using a constant-specific-heat analysis. What error would result if this latter technique were employed?

2-56M A closed thermodynamic system is designed to raise a 4.0-kg mass in a standard gravitational field while exchanging heat with the surroundings and receiving an electrical work input. In one cycle operation, the system receives 350 kJ of heat during one part of the cycle and gives up 150 kJ during another part of the cycle. For the overall cycle 50 J of electric energy is delivered to the system. Calculate the displacement of the 4.0-kg mass.

2-57M 2 kg of oxygen are heated from 40 to 200°C in a constant-volume process. Calculate the change in enthalpy and heat transfer.

2-58M 5 g of carbon dioxide are contained in a vessel at 70 kPa and 40°C. Calculate the heat which must be added to raise the pressure to a value of ≈200 kPa.

2-59M Calculate the heat necessary to raise 1 kg mol of each of the following gases from 90 to 260°C at a constant pressure of 100 kPa:
(*a*) Hydrogen (*b*) Helium
(*c*) Nitrogen (*d*) Oxygen
(*e*) Argon (*f*) Air

2-60M In a large majority of applicable problems air is assumed to behave as an ideal gas with constant specific heats for the temperature range 255 to 810 K. Assuming that air behaves like nitrogen (air is 78 percent nitrogen), estimate the validity of this assumption.

2-61M 3 kg of argon are heated at constant pressure from 20 to 120°C. Calculate the heat required and the change in internal energy.

2-62M A certain gas undergoes a constant-pressure process in a piston-cylinder arrangement at 120 kPa. During this process the internal energy increases by 190 kJ/kg for a temperature rise of 75°C. At the same time the work delivered to the face of the piston is 100 kJ/kg. Calculate the values of c_v, c_p, and R for the gas.

2-63M 10.5 kg of air are heated from 15 to 100°C in a constant-pressure process. Calculate the change in internal energy.

2-64M An experiment is conducted with a closed system containing an unknown substance. As the system is allowed to change from state 1 to state 2, 15 kJ of heat are removed while the work output of the system is observed to be 38 kJ. The system is then restored back to state 1 in a different process in which 28 kJ of heat are added along with a work quantity. Calculate the value of the work in this second process.

2-65M A closed system undergoes a polytropic process in which the heat added is 17 kJ and the internal energy increases by 48 kJ. The initial state is 130 kPa and 0.15 m^3, and the final pressure is 800 kPa. Calculate the final volume in the process. If the system were air, what mass would be required?

2-66M 11.6 kg of air are compressed isothermally with a piston cylinder from 110 kPa, 30°C, to 450 kPa. Calculate the work required and the heat added.

REFERENCES

1 Callen, H. B.: "Thermodynamics," John Wiley & Sons, Inc., New York, 1960.
2 Hatsopoulos, G. N., and J. H. Keenan: "Principles of General Thermodynamics," John Wiley & Sons, Inc., New York, 1965.
3 Reynolds, W. C.: "Thermodynamics," 2d ed., McGraw-Hill Book Company, New York, 1968.
4 Sears, F. W.: "Thermodynamics," 2d ed., Addison-Wesley Publishing Company, Inc., Reading, Mass., 1953.
5 Van Wylen, G. J., and R. E. Sonntag: "Fundamentals of Classical Thermodynamics," John Wiley & Sons, Inc., New York, 1965.
6 Sweigert, R. L., and M. W. Beardsley: *Bull. No. 2,* Georgia School of Technology, 1938.

3

MACROSCOPIC PROPERTIES OF PURE SUBSTANCES

3-1 INTRODUCTION

The properties of a substance depend on its constitution. The number of phases, the homogeneous or heterogeneous nature of the substance, and the types of pertinent energy quantities all have a bearing on the number and types of properties which may be required for analysis purposes. If, for example, we want to calculate the energy release in the fission of a U^{235} atom, we need to know the proton-neutron makeup and the binding energy of the nucleus. To determine the energy necessary to boil water, we need to know the energy of vaporization for water. To analyze a substance where electrical or magnetic work is involved, we need information concerning the electrical or magnetic properties of the substance.

The number of properties required to specify the state of a substance obviously depends on the nature of the energy interactions to be studied. There can be, of course, only heat and work interactions with a substance. But the work interactions can take several different forms. Consider an ionized gas, for example. One way of doing work on the gas is to compress it. Another way is to subject it to an electric potential so that electrical work is done to move the charged particles. Still another possible way to perform work is to move the gas through a magnetic field or to alter the orientation of magnetic dipoles in the gas. To calculate the compression work, we would need information about pressure and volume at various temperatures; to calculate the electrical work, we would need information about the electric conductivity as a function of pressure and temperature (and possibly as a function of electric-field strength); and to calculate the magnetic work, we would need information on the conductivity and magnetic dipole moment of the gas. The total number of independent properties needed to specify the state of the gas under these circumstances may be obtained with just a little reasoning.

Intensive and Extensive Properties

We are basically interested in determining the internal thermal energy of the substance u. For each possible work mode (volume change, electrical, magnetic, etc.) there is associated a characteristic *intensive* property (pressure, electric potential, magnetic potential) which describes the ability of a unit mass or unit volume of the system to accomplish the particular work mode. For each work mode there is also some characteristic *extensive* property (volume, conductivity, magnetic dipole moment) which describes how much total work will be performed in each mode. The intensive property represents the *force* and the extensive property represents the *displacement* in each work mode. The internal energy is a function of the possible work interactions and the heat interaction. The intensive property that describes the heat interaction is the temperature.†

† We have omitted kinetic and gravitational potential energies in this consideration. They are important, of course, but are normally treated as separate terms in the general internal-energy function E.

In accordance with this brief discussion, we might conclude that in order to specify the state of a substance we need an intensive property for each work mode plus an intensive property to describe heat interactions. This principle is restricted to systems in equilibrium, which, of course, are our main concern. It is required that these be *independent* properties.

The State Principle

This brief discussion specifies the state principle of Chap. 1 in explicit form, which may be expressed as follows:

The number of independent properties required to specify the thermodynamic state of a system is equal to the number of possible work modes plus one.

There are other considerations involved in the state principle which are discussed by Kline and Koenig [1] and Hatsopoulos and Keenan [2].

A *simple substance* involves only one possible work mode. A nonionized gas with no magnetic dipole moment is such a substance. In this case we have only one possible work mode, a volume change, so that only two independent properties are required to specify the state of the gas. Any two of pressure, temperature, or specific volume will do.

The term *pure substance* designates a substance which is homogeneous and has the same chemical composition in all phases. Water is a pure substance since its chemical composition is the same in all phases, even for a mixture of liquid and vapor. A mixture of gases, like air, behaves as a pure substance, but if the mixture is cooled such that some of the components condense out into a liquid phase, the mixture may no longer be considered as a pure substance because it does not have the same chemical composition in all phases. The pure substance is of significant interest for many practical applications and, thus, deserves much of our attention.

The purpose of this chapter is to describe the thermodynamic properties of some characteristic simple substances. Specific numerical data are presented in the Appendix so that they may be applied to problem solutions in the subsequent chapters. At this point the properties are presented in a purely empirical sense; i.e., they are based on experimental measurements. The calculation of some types of thermodynamic properties from fundamental microscopic considerations will be discussed in Chap. 13.

Review Thoughts

What do the following terms mean: *work modes, intensive and extensive properties, state principle, simple substance, pure substance?*

3-2 PROPERTIES OF PURE SUBSTANCES

Let us now examine the behavior of a pure substance which has work interactions involving only the compression (volume change) work mode. For such a substance only two independent properties are required to define the state of the system. To illustrate the physical behavior of such substances, let us consider the behavior of one of the most common substances of all—ordinary water. The substance may exist in several *phases:*

1 A pure solid phase commonly known as ice
2 A pure liquid phase
3 A pure vapor phase commonly known as steam
4 An equilibrium mixture of liquid and vapor phases
5 An equilibrium mixture of liquid and solid phases
6 An equilibrium mixture of solid and vapor phases

Figure 3-1 is a pressure-temperature diagram illustrating these various phase regions for water.

Fusion, Vaporization, and Sublimation

When a solid changes to a liquid, we say that a *fusion* process has taken place; when a liquid changes to a vapor, there is *vaporization;* and, finally, when a solid changes directly to a vapor, a *sublimation* process is said to have been executed. In each of these processes energy must be added to the substance to effect the change in phase. The temperature at which these changes will occur is dependent on the pressure exerted on the substance. Carbon dioxide at normal atmospheric pressure will sublime; i.e., the solid will change directly to vapor with no intermediate change to liquid. The sublimation temperature for CO_2 at atmospheric pressure is approximately $-78°C$.

Water at atmospheric pressure may be made to exist in all three phases, as depicted by the constant-pressure line *AB* in Fig. 3-1. If ice is heated at con-

FIG. 3-1 Pressure-temperature diagram for a substance that expands on freezing.

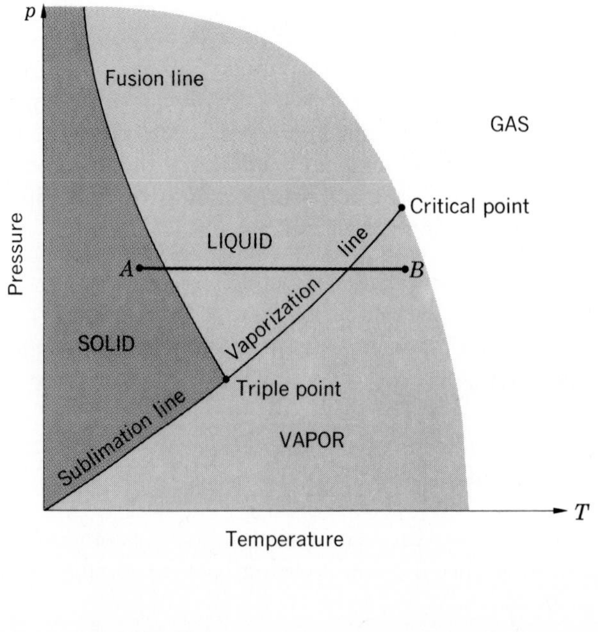

stant pressure, the temperature increases until the fusion line is encountered and melting starts. The melting process proceeds at constant temperature until all of the solid ice is transformed to liquid. Further heating then increases the liquid temperature until the vaporization line is reached. Further addition of heat progressively transforms the liquid into vapor, and the temperature remains constant in the vaporization process. Once all the liquid is converted to vapor, additional heating causes the temperature to rise once again. At atmospheric pressure the vaporization temperature for water is 100°C (212°F) and the fusion temperature is approximately 0°C (32°F). We also see from Fig. 3-1 that a change in pressure will alter the freezing or vaporization point of the liquid; for water, an increase in pressure raises the vaporization temperature and lowers the freezing temperature. It should be noted that Fig. 3-1 is not drawn to scale. In the actual case the fusion line would be almost vertical, and very large increases in pressure would be required to lower the melting temperature substantially.

The fusion line represents the solid-liquid mixture, the vaporization line the liquid-vapor mixture, and the sublimation line the solid-vapor mixture.

The *triple point* is the state where it is possible to maintain an equilibrium mixture of all three phases.

The *critical point* is the state where the pure vapor phase has identical properties with a pure liquid phase at the same pressure and temperature. It is not possible to observe a distinction between liquid and vapor phases at supercritical pressures and temperatures. In other words, the surface meniscus separating liquid and vapor phases vanishes at the critical point.

3-3 SATURATION REGIONS

The three equilibrium lines in Fig. 3-1 (fusion, vaporization, and sublimation) are said to designate *saturation* regions. Thus we may say that the vaporization line represents the saturation region between liquid and vapor. The vapor which is present in such a mixture is called a *saturated vapor,* and the liquid present in the mixture is called a *saturated liquid.*

To illustrate the properties further, consider the constant-pressure process AB of Fig. 3-1 as plotted on the p-v diagram of Fig. 3-2 and the surface of Fig. 3-3. Here we see the nature of the saturation regions more clearly. Each region represents a circumstance where pressure and temperature remain constant while there is a significant change in specific volume. In the liquid-vapor saturation region the increase in specific volume (volume per unit mass) results from an increase in the fraction of the total mass which is present in the vapor form.

Quality and Moisture

The *quality* x represents the fraction of mass present in the vapor phase, and the *moisture* represents the fraction of mass present in the liquid phase.

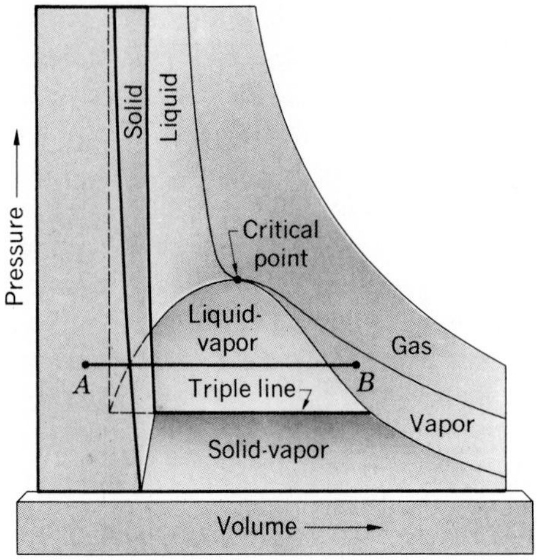

FIG. 3-2 Pressure-volume diagram for a substance that expands on freezing.

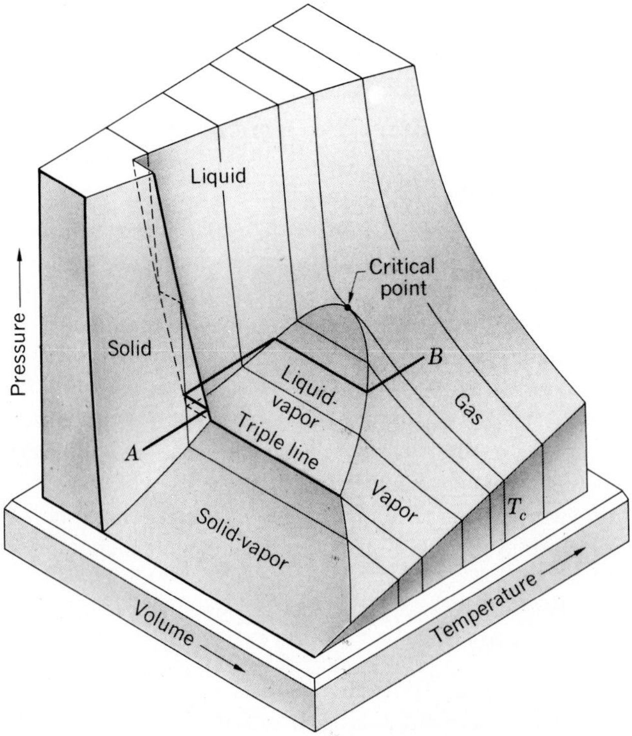

FIG. 3-3 *p-v-T* surface for a substance that expands on freezing.

Clearly,

$$\text{Moisture} = 1 - x \tag{3-1}$$

The term *wet mixture* is widely used to designate a liquid-vapor mixture with a quality of less than 100 percent. In designating the saturation properties of substances, the subscripts *i, f,* and *g* are used for solid, liquid, and vapor phases. Thus the specific volume of a saturated liquid would be v_f, and the specific volume of a saturated vapor would be v_g. The specific volume of an equilibrium mixture of liquid and vapor would be given by

$$v = \frac{V}{m} = \frac{1}{m}(m_f v_f + m_g v_g)$$

$$= \frac{m_f}{m} v_f + x v_g$$

$$= (1 - x)v_f + x v_g$$

Introducing the definition

$$v_{fg} = v_g - v_f \tag{3-2}$$

we obtain

$$v = v_f + x v_{fg} \tag{3-3}$$

Several other important terms are defined as follows:

A vapor is said to be *superheated* if it exists at a temperature greater than the saturation temperature corresponding to its pressure. A *subcooled liquid* or *compressed liquid* is one which exists at a temperature lower than the saturation temperature corresponding to its pressure *or* at a pressure greater than the saturation pressure corresponding to its temperature. The two terms are synonymous.

A three-dimensional *p-v-T* surface for a substance like water (one that expands upon freezing) is shown in Fig. 3-3. A similar figure for a substance which contracts upon freezing is shown in Fig. 3-4.

We have already said that only two independent properties are necessary to define the state of the simple, pure substance previously described. What two properties shall we use? Temperature is one logical choice. It should be obvious that the other choice cannot be pressure when the substance is in one of the saturation regions, because pressure and temperature are not independent in these regions; i.e., a line of constant temperature in a saturation region is also a line of constant pressure; however, specific volume is independent of pressure and temperature in this region, so it may be selected for the other independent property. In a single-phase region, temperature and pressure may be selected as the independent properties.

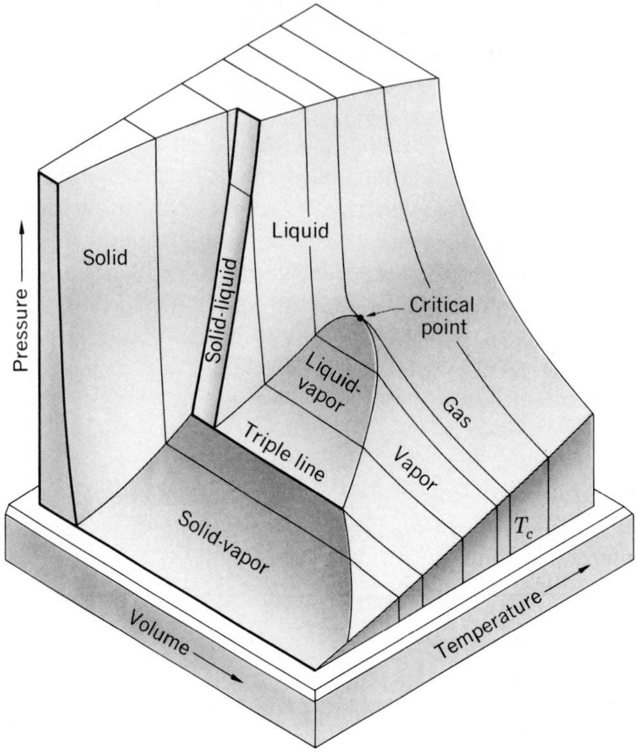

FIG. 3-4 *p-v-T* surface for a substance that contracts on freezing.

Energy Properties

We shall be interested in the energy properties of substances. The specific internal energy u and specific enthalpy h are properties commonly tabulated for pure substances. From the foregoing discussion we may anticipate that these energy properties may be expressed as functions of temperature and specific volume:

$$u = u(T, v)$$

$$h = h(T, v)$$

In the wet-mixture region these properties are determined in a fashion analogous to that used for specific volume, i.e.,

$$u = u_f + xu_{fg}$$

$$h = h_f + xh_{fg}$$

where $u_{fg} = u_g - u_f$ and $h_{fg} = h_g - h_f$ are the internal energy and enthalpy of vaporization, respectively.

Another property called the *entropy* s is useful in solution of practical problems, as we shall see in Chap. 5, and is commonly tabulated along with internal energy and enthalpy. In the two-phase evaporation region it, too, takes the characteristic form

$$s = s_f + xs_{fg}$$

where $s_{fg} = s_g - s_f$ is the entropy of vaporization.

Properties for several substances are tabulated in the Appendix, the most extensive tabulation being that for water and its vapor. In the compressed-liquid region the properties for water are tabulated in Table A-11 as corrections to the saturated-liquid properties at the liquid temperature. In general, the corrections are not large as we can see by looking at an example of compressed-liquid water at 100°F and 1000 psia. At 100°F we have (see Table A-7 or A-11)

$$v_f = 0.016\ 132 \text{ ft}^3/\text{lbm}$$

$$h_f = 67.97 \text{ Btu/lbm}$$

$$p_{\text{sat}} = 0.9492 \text{ psia}$$

Entering Table A-11 we find the corrections to these values as

$$(v - v_f) \times 10^5 = -5.2$$

$$h - h_f = +2.64$$

so that the properties for the compressed liquid at 100°F and 1000 psia are

$$v = 0.016\ 132 - 5.2 \times 10^{-5} = 0.016\ 08 \text{ ft}^3/\text{lbm}$$

$$h = 67.97 + 2.64 = 70.61 \text{ Btu/lbm}$$

In most cases it will be inconvenient to use Table A-11 for the compressed liquid, and Table A-10 furnishes a direct tabulation of compressed-liquid properties. It should be noted that the lowest pressure entry in Table A-11 is 200 psia. Below this value the corrections are so small that they can usually be neglected and the properties of the compressed liquid can be taken as those of the saturated liquid at the liquid temperature.

At this point we note that the properties of liquids are *primarily* a function of the temperature, with pressure exerting a minor effect, as was illustrated in the foregoing calculation.

EXAMPLE 3-1

Determine the specific volume v, enthalpy h, and entropy s of water under the following conditions:

(*a*) 7.5 MPa, 100°C
(*b*) 4 MPa, 360°C
(*c*) 150 kPa, 10 percent quality
(*d*) 300 kPa, 10 percent moisture
(*e*) Saturated liquid at 100 kPa
(*f*) Saturated vapor at 100 kPa

SOLUTION

(*a*) This is a compressed-liquid condition because 7.5 MPa is greater than the saturation pressure corresponding to 100°C (101.4 kPa). Therefore, Table A-10M is consulted to find

$$v = 1.0397 \times 10^{-3}\ \text{m}^3/\text{kg}$$

$$h = 424.62\ \text{kJ/kg}$$

$$s = 1.3011\ \text{kJ/kg·K}$$

(*b*) 360°C is greater than the saturation temperature for 4 MPa (250.4°C) so this is superheated and Table A-9M is consulted to find

$$v = 0.067\,88\ \text{m}^3/\text{kg}$$

$$h = 3117.2\ \text{kJ/kg}$$

$$s = 6.6215\ \text{kJ/kg·K}$$

(*c*) The 10 percent quality tells us that a saturation condition is encountered at 150 kPa so Table A-8M (saturation-*pressure* table) is consulted to obtain

$$v_f = 1.0528 \times 10^{-3} \qquad v_g = 1.159\ \text{m}^3/\text{kg}$$

$$h_f = 467.11 \qquad h_{fg} = 2226.5\ \text{kJ/kg}$$

$$s_f = 1.4336 \qquad s_g = 7.2233\ \text{kJ/kg·K}$$

We are given the quality as $x = 0.10$ so that

$$v = v_f + xv_{fg} = 1.0528 \times 10^{-3} + (0.10)(1.159 - 1.0528 \times 10^{-3})$$
$$= 0.116\,85\ \text{m}^3/\text{kg}$$

$$h = h_f + xh_{fg} = 467.11 + (0.1)(2226.5) = 689.76\ \text{kJ/kg}$$

$$s = s_f + xs_{fg} = 1.4336 + (0.1)(7.2233 - 1.4336)$$
$$= 2.0126\ \text{kJ/kg·K}$$

(*d*) This problem is like part (*c*) except we note that

$$\text{Moisture} = 1 - \text{quality} = 1 - x$$

so that

$$x = 1 - 0.1 = 0.9$$

We then enter Table A-8M at 300 kPa to determine the values as

$$v = 0.545\ 33\ \text{m}^3/\text{kg}$$

$$h = 2508.9\ \text{kJ/kg}$$

$$s = 6.4599\ \text{kJ/kg·K}$$

(e, f) These properties are read directly from Table A-8M at 100 kPa:

Saturated Liquid	Saturated Vapor
$v_f = 1.0432 \times 10^{-3}\ \text{m}^3/\text{kg}$	$v_g = 1.694\ \text{m}^3/\text{kg}$
$h_f = 417.46\ \text{kJ/kg}$	$h_g = 2675.5\ \text{kJ/kg}$
$s_f = 1.3026\ \text{kJ/kg·K}$	$s_g = 7.3594\ \text{kJ/kg·K}$

3-4 EQUATIONS OF STATE

An equation of state is an analytical expression which interrelates properties of substances. The ideal-gas expression derived in Sec. 1-14

$$pv = RT \tag{3-4}$$

is a very simple equation of state. Unfortunately, all equations of state are not so simple. In practice, the equation of state for a solid, liquid, or vapor is obtained by an empirical curve fit of experimental data. A different relationship is usually employed for each phase or saturation region, and each equation is quite long and complicated.

Tabular and graphical presentations of thermodynamic properties such as the ones given in the Appendix serve as substitutes for an analytical equation of state. We shall make use of this information for the present and defer further development of equations of state until Chap. 6. More extensive tabulations as well as equations of state suitable for computer use are given by Reynolds [5].

EXAMPLE 3-2

A rigid 1-ft^3 enclosure contains a mixture of liquid and vapor water at a pressure of 12 psia. The total mass of water in the container is 30 lbm. Calculate the heat added and quality of the mixture after sufficient heat is added to raise the pressure to 1000 psia.

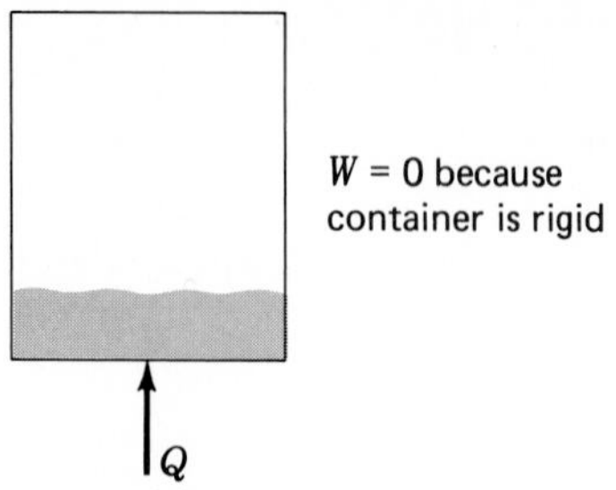

FIG. EXAMPLE 3-2

SOLUTION This problem requires an exercise of the first law of thermodynamics as well as information pertaining to properties of pure substances. Since the container is rigid, there can be no change in the system volume and, hence, no work done. The first-law statement then becomes

$$Q = \Delta U = m\,\Delta u \qquad (a)$$

To calculate the internal energy, we need to establish the thermodynamic state explicitly. Since we have a mixture of liquid and vapor, pressure and temperature are not independent parameters. The specific volume of the mixture is

$$v = \frac{V}{m} = \frac{1}{30} = 0.0333 \text{ ft}^3/\text{lbm}$$

This volume is then expressed as

$$v = v_f + xv_{fg} \qquad (b)$$

in accordance with Eq. (3-3). For the initial state, designated with subscript 1,

$$v_{f_1} = 0.016\,65 \qquad v_{fg_1} = 32.38 \text{ ft}^3/\text{lbm} \qquad (12 \text{ psia})$$

where the properties are obtained from the saturation-pressure table for steam given in Table A-8.

Inserting these values in Eq. (*b*) gives for the initial quality

$$x_1 = 0.000\,515$$

The initial internal energy is thus

$$\begin{aligned} u_1 &= u_{f_1} + x_1\,u_{fg_1} \\ &= 169.92 + (0.000\,515)(904.8) \\ &= 170.38 \text{ Btu/lbm} \end{aligned}$$

Since the container is rigid, the specific volume remains the same in the initial and final states. Designating the final state by subscript 2,

$$v_{f_2} = 0.0216 \qquad v_{fg_2} = 0.4240 \text{ ft}^3/\text{lbm} \qquad (1000 \text{ psia})$$

Again, the quality is calculated from Eq. (*b*) with

$$v_2 = v_1 = 0.0333 \text{ ft}^3/\text{lbm}$$

The result is

$$x_2 = 0.0292$$

The final internal energy is then calculated as

$$\begin{aligned} u_2 &= u_{f_2} + x_2\, u_{fg_2} \\ &= 538.4 + (0.0292)(571) \\ &= 555.0 \text{ Btu/lbm} \qquad (1.29 \times 10^6 \text{ J/kg}) \end{aligned}$$

The heat transfer is now calculated from Eq. (*a*) as

$$\begin{aligned} Q &= m(u_2 - u_1) \\ &= (30)(555.0 - 170.38) = 11\ 500 \text{ Btu} \qquad (1.213 \times 10^7 \text{ J}) \end{aligned}$$

EXAMPLE 3-3

Steam is contained at 100 psia and 50 percent quality in a 0.5-ft³ cylinder. The cylinder is fitted with a piston arrangement which maintains a constant pressure. Calculate the amount of heat which must be added to the steam to increase the volume to 2.0 ft³.

SOLUTION We first write the first law for this system:

$$Q + W = \Delta U \qquad (a)$$

For the constant-pressure process the work done *on* the gas is

$$\begin{aligned} W &= -\int_{V_1}^{V_2} p\, dV = -p(V_2 - V_1) \\ &= -(100)(144)(2.0 - 0.5) = -21\ 600 \text{ ft}\cdot\text{lbf} \qquad (-29\ 285 \text{ J}) \qquad (b) \end{aligned}$$

To apply Eq. (*a*), we need to calculate the change in internal energy for the steam. This requires that the mass be determined from

$$m = \frac{V_1}{v_1}$$

$$v_1 = v_{f_1} + x_1 v_{fg_1}$$

$$= 0.01774 + (0.5)(4.434 - 0.017\,74) = 2.226 \text{ ft}^3/\text{lbm}$$

$$m = \frac{0.5}{2.226} = 0.225 \text{ lbm} \qquad (0.1021 \text{ kg})$$

where the properties are obtained from the saturation-pressure table for steam at 100 psia. The final specific volume is

$$v_2 = \frac{V_2}{m} = \frac{2.0}{0.225} = 8.889 \text{ ft}^3/\text{lbm}$$

The final pressure is the same as the initial pressure, i.e., 100 psia. Consulting the saturation tables we find that v_g at 100 psia is only 4.434 ft^3/lbm. Because v_2 is greater than v_g, the final state must be superheated. We therefore consult the superheat tables to find the final temperature at 100 psia and $v = 8.889$,

$$T_2 = 1038°\text{F}$$

and the final internal energy is

$$u_2 = 1387.7 \text{ Btu/lbm}$$

The initial internal energy is calculated from

$$u_1 = u_{f_1} + x_1 u_{fg_1}$$

$$= 298.3 + (0.5)(1105.8 - 298.3) = 702.05 \text{ Btu/lbm}$$

The internal energies and work are now inserted into Eq. (*a*) to determine the heat added:

$$Q - \frac{21\,600}{778} = m(u_2 - u_1) = (0.225)(1387.7 - 702.05)$$

$$Q = 182 \text{ Btu} \qquad (192 \text{ kJ})$$

EXAMPLE 3-4

A rigid enclosure, 50 cm on each side, contains a wet mixture of water vapor at 90°C and 20 percent quality. Heat is added until the pressure is raised to 500 kPa. Determine the final state and the quantity of heat added.

SOLUTION For a rigid container there is no change in volume and therefore no work done. The first-law statement then becomes

$$Q = \Delta U = m\,\Delta u = m(u_2 - u_1) \qquad (a)$$

Since both the total volume V and the mass remain constant, the specific volume must also remain constant.

$$v = \frac{V}{m} = v_1 = v_2 \qquad (b)$$

At the initial state

$$v = v_f + xv_{fg} \qquad (c)$$

Consulting the saturation tables at 90°C, we have

$$v_{f_1} = 1.0360 \times 10^{-3}\ \text{m}^3/\text{kg} \qquad v_{g_1} = 2.361\ \text{m}^3/\text{kg}$$

$$u_{f_1} = 376.85\ \text{kJ/kg} \qquad u_{g_1} = 2494.5\ \text{kJ/kg}$$

Then, from Eq. (*c*),

$$v_1 = 0.001\,036 + (0.20)(2.361 - 0.001\,036) = 0.473\ \text{m}^3/\text{kg} \qquad (d)$$

We may now calculate the mass from Eq. (*b*) and the given total volume

$$m = \frac{V}{v} = \frac{(0.50)^3}{0.473} = 0.2643\ \text{kg} \qquad (e)$$

The final state is determined by consulting the steam tables at $v_2 = 473\ \text{m}^3/\text{kg}$ and $p = 500$ kPa = 5.0 bar. At 500 kPa we find $v_g = 374.9\ \text{m}^3/\text{kg}$. We have a specific volume greater than this value so the final state must be superheated. Consulting the superheat table at $p = 500$ kPa and interpolating yields

$$T_2 = 244.3°\text{C} \qquad u_2 = 2721.4\ \text{kJ/kg}$$

The initial internal energy is calculated from

$$\begin{aligned} u_1 &= u_{f_1} + x_1 u_{fg_1} \\ &= 376.85 + (0.2)(2494.5 - 376.85) \\ &= 800.38\ \text{kJ/kg} \end{aligned}$$

We now insert the numerical values in Eq. (*a*) to calculate the heat transfer:

$$\begin{aligned} Q &= (0.2643)(2721.4 - 800.38) \\ &= 507.7\ \text{kJ} \qquad (481.3\ \text{Btu}) \end{aligned}$$

EXAMPLE 3-5

A rigid container is filled with steam at 700 kPa and 200°C. At what temperature will the steam start to condense when the container is cooled? To what temperature must the container be cooled to condense 50 percent of the steam mass?

SOLUTION Because the container is rigid, the specific volume must remain constant during the cooling process. The initial state is superheated so, from Table A-9M,

$$v_1 = 0.2999 \text{ m}^3/\text{kg} \qquad (4.804 \text{ ft}^3/\text{lbm}) \tag{a}$$

When the steam just starts to condense, this will be the specific volume of saturated vapor

$$v_1 = v_2 = v_{g_2} = 0.2999 \tag{b}$$

Interpolating in Table A-7M, we find

$$T_2 = 161.1°\text{C}$$

$$p_2 = 636.9 \text{ kPa} \qquad (92.4 \text{ psia})$$

Cooling still further to some point 3 where $x_3 = 0.50$ means

$$v_3 = v_1 = 0.2999 = v_{f_3} + (0.50)v_{fg_3} \tag{c}$$

Most of the volume comes from the vapor so, very nearly,

$$v_3 \sim 0.5v_{g_3} \tag{d}$$

or

$$v_{g_3} \sim 0.5998$$

Again, consulting Table A-7M, we find by interpolation that

$$T_3 = 134.3°\text{C}$$

$$p_3 = 309.3 \text{ kPa} \qquad (44.2 \text{ psia})$$

EXAMPLE 3-6

A small capsule has a volume of 0.3 in^3 and contains liquid water at 1000 psia and 100°F. The capsule is placed inside a large container having a volume of 1 ft^3. The large container is subsequently evacuated, and the capsule is broken by some appropriate mechanism so that the water is allowed to evaporate and fill the large container. Calculate the final quality of the water-vapor mixture, assuming that the large container is in heat communication with its surroundings and reaches a final equilibrium temperature of 100°F. Also determine the heat exchange with the surroundings.

SOLUTION For this problem we take the boundary of the system as the inside surface of the large container. This surface is stationary, and thus there is no work done on the boundary of the system. From the first law,

$$Q + W = \Delta U \tag{a}$$

Since $W = 0$, we have

$$Q = \Delta U = m(u_2 - u_1) \tag{b}$$

We have $V_1 = 0.3\ \text{in}^3 = 1.735 \times 10^{-4}\ \text{ft}^3$. Using the compressed-liquid tables for water, we have

$$v_1 = 0.016\ 082\ \text{ft}^3/\text{lbm}$$

$$u_1 = 67.70\ \text{Btu/lbm}$$

The mass of water is

$$m = \frac{V_1}{v_1} = \frac{1.735 \times 10^{-4}}{1.608 \times 10^{-2}} = 0.0108\ \text{lbm}$$

The final specific volume is thus

$$v_2 = \frac{V_2}{m} = \frac{1}{0.0108} = 92.5\ \text{ft}^3/\text{lbm} \qquad (5.774\ \text{m}^3/\text{kg})$$

The final state is determined by entering the steam tables with the known values of T_2 (100°F) and v_2. At 100°F, $v_g = 350.0\ \text{ft}^3/\text{lbm}$, so that a mixture of vapor and liquid is obtained since $v_2 < v_g$. Thus

$$v_2 = v_{f_2} + x_2 v_{fg_2}$$

Using the properties from the saturation-temperature table (Table A-7),

$$92.5 = 0.016\ 13 + x_2(350.0 - 0.016\ 13)$$

and

$$x_2 = 0.264$$

The final internal energy is now obtained from this same table:

$$\begin{aligned} u_2 &= u_{f_2} + x_2 u_{fg_2} \\ &= 68.04 + (0.264)(1043.5 - 68.04) \\ &= 326\ \text{Btu/lbm} \end{aligned}$$

The heat transfer is now calculated with Eq. (*b*):

$$Q = (0.0108)(326 - 67.7)$$
$$= 2.79 \text{ Btu} \qquad (2943 \text{ J})$$

The positive sign on the heat transfer indicates that 2.79 Btu of heat must be supplied to the large container to maintain the vapor at 100°F during the expansion-evaporation process.

3-5 PROPERTIES FOR SOLID-VAPOR SATURATION REGION

Properties for the solid-vapor saturation region for water may also be expressed in tabular form, but in this region the vapor is at such a low pressure that it behaves very nearly as an ideal gas, and the properties may be expressed in fairly simple analytical form. Accordingly, the appropriate equations have

TABLE 3-1
Solid-Vapor Saturation Properties for Water

Property	Equation	Temperature Range	Equation Number
Saturation pressure	$p_g = 35.18 \exp\left(18.42 - \frac{6144}{T}\right)$ kPa; T in K	−40 to 0°C	3-5*a*
	$p_g = 5.103 \exp\left(18.42 - \frac{11\,059}{T}\right)$ psia; T in °R	−40 to 32°F	3-5*b*
Enthalpy and internal energy of saturated solid	$h_i = u_i = -334.6 + 1.96T$ kJ/kg; T in °C	−40 to 0°C	3-6*a*
	$h_i = u_i = -158.9 + 0.467T$ Btu/lbm; T in °F	−40 to 32°F	3-6*b*
Internal energy of saturated vapor	$u_g = 2374.9 + 1.403T$ kJ/kg; T in °C	−40 to 0°C	3-7*a*
	$u_g = 1010.3 + 0.335T$ Btu/lbm; T in °F	−40 to 32°F	3-7*b*
Enthalpy of saturated vapor	$h_g = 2501 + 1.863T$ kJ/kg; T in °C	−40 to 0°C	3-8*a*
	$h_g = 1061.0 + 0.445T$ Btu/lmb; T in °F	−40 to 32°F	3-8*b*
Specific volume of saturated solid	$v_i = 1.091 \times 10^{-3} + 2.5 \times 10^{-6}T$ m³/kg; T in °C	−40 to 0°C	3-9*a*
	$v_i = 0.017\,43 + 1.39 \times 10^{-6}T$ ft³/lbm; T in °F	−40 to 32°F	3-9*b*
Entropy of saturated vapor	$s_g = 1.863 \ln T + \frac{2833.6}{T} - 11.66$ kJ/kg·K; T in K	−40 to 0°C	3-10*a*
	$s_g = 0.445 \ln T + \frac{1218.2}{T} - 3.046$ Btu/lbm·°R; T in °R	−40 to 32°F	3-10*b*
Entropy of saturated solid	$s_i = -1.223 + 7.747 \times 10^{-3}T$ kJ/kg·K; T in °C	−40 to 0°C	3-11*a*
	$s_i = -0.325 + 0.001\,028T$ Btu/lbm·°R; T in °F	−40 to 32°F	3-11*b*

been assembled in Table 3-1 in a convenient format. These equations apply for the saturation region from −40 to 0°C (−40 to 32°F) with an accuracy of approximately 1 percent. In Table 3-1 we have also given equations for a property called the entropy s. The property will not be defined until Chap. 5, but we list it here for completeness and later reference.

Although these equations pertain specifically to the solid-vapor saturation region, they are applicable to other circumstances. Because the solid is only modestly compressible, the internal energies and enthalpies given by Eq. (3-6) can be used for pressures up to several atmospheres. Likewise, the specific volume of the solid is not too pressure-sensitive and Eq. (3-9) may be also used for higher pressures.

EXAMPLE 3-7

Using the equations given for the solid-vapor saturation region and the steam tables, estimate the energy required to melt 1 kg of ice at 0°C in a constant-pressure process.

SOLUTION At 0°C = 32°F water can exist in all three phases: solid, liquid, and vapor. From the steam tables at 32°F

$$h_f = 0.00 \text{ Btu/lbm}$$

and from Eq. (3-6)

$$h_i = -158.9 + (0.467)(32) = -143.96 \text{ Btu/lbm} \tag{a}$$

For a constant-pressure process we have

$$\begin{aligned} dQ_p &= dh_p \\ Q_p &= \Delta h_p = h_f - h_i = 0 - (-143.96) \qquad (b) \\ &= 143.96 \text{ Btu/lbm} \\ &= 317.09 \text{ Btu/kg} \\ &= 334.5 \text{ kJ/kg} \end{aligned}$$

3-6 SPECIFIC HEATS

In Chap. 2 the constant-volume and constant-pressure specific heats were defined as

$$c_v = \left(\frac{\partial u}{\partial T}\right)_v \tag{3-12}$$

$$c_p = \left(\frac{\partial h}{\partial T}\right)_p \tag{3-13}$$

Thus, once tabular information on enthalpy and internal energy is available, we can evaluate the specific heats rather easily by performing the indicated differentiations. The following example illustrates such a calculation.

EXAMPLE 3-8
Using data from the superheat region of the steam tables, calculate the specific heat at constant pressure for water vapor at 400 psia and 600°F.

SOLUTION We shall employ a numerical approximation to the partial derivative relation for c_p. From the superheat tables at 400 psia

$$h = 1245.2 \text{ Btu/lbm} \qquad \text{at } 500°\text{F}$$

$$h = 1362.5 \text{ Btu/lbm} \qquad \text{at } 700°\text{F}$$

Then we write, approximately,

$$c_p = \left(\frac{\partial h}{\partial T}\right)_p \approx \left(\frac{\Delta h}{\Delta T}\right)_p = \frac{h_{700} - h_{500}}{700 - 500}$$

$$= \frac{1362.5 - 1245.2}{700 - 500} = 0.587 \text{ Btu/lbm·°F}$$

In this calculation we assume the specific-heat value to apply midway between 500 and 700°F, or at the desired value of 600°F.

3-7 COEFFICIENT OF EXPANSION

Two additional properties of simple compressible substances are important. The *volume coefficient of expansion* β is defined as the rate of change of volume with temperature at constant pressure and per unit volume:

$$\beta = \frac{1}{v}\left(\frac{\partial v}{\partial T}\right)_p \tag{3-14}$$

The *isothermal compressibility* κ is defined as the rate of change of volume with pressure at constant temperature and per unit volume:

$$\kappa = -\frac{1}{v}\left(\frac{\partial v}{\partial p}\right)_T \tag{3-15}$$

These two properties are important because they may be determined fairly easily by experiment. For an ideal gas, very simple relations can be derived as follows. Since

$$v = \frac{RT}{p}$$

we have

$$\beta = \frac{1}{v}\left(\frac{\partial v}{\partial T}\right)_p = \frac{1}{v}\frac{R}{p} = \frac{1}{T} \qquad (3\text{-}16)$$

$$\kappa = -\frac{1}{v}\left(\frac{\partial v}{\partial p}\right)_T = -\frac{1}{v}\left(-\frac{RT}{p^2}\right) = \frac{1}{p} \qquad (3\text{-}17)$$

The isothermal compressibility is quite useful when employed in the calculation of work involved in the compression of liquids and solids. The work per unit mass is expressed as

$$w = -\int p\, dv \qquad (3\text{-}18)$$

For an isothermal process the volume change is expressed in terms of κ by

$$dv_T = -\kappa v\, dp_T$$

Substituting this relation in Eq. (3-18) gives

$$w_T = \int_{p_1}^{p_2} p\kappa v\, dp_T \qquad (3\text{-}19)$$

TABLE 3-2
Volume Coefficient of Expansion for Various Substances at 25°C

Substance	β	
	°F^{-1} × 10^6	°C × 10^6
Solids		
Aluminum	39.9	71.8
Copper	27.6	49.7
Gold	23.7	42.7
Iron	19.5	35.1
Magnesium	42.0	75.6
Nickel	21.6	38.9
Silver	33.0	59.4
Steel, AISI 304 sheet	29.7	53.5
Tungsten	7.5	13.5
Pyrex glass	5.4	9.7
Ice	84.9	152.8
Wood, oak	8.1	14.6
Concrete	24.0	43.2
Liquids		
Acetic acid	600	1080
Ethyl alcohol	610	1098
Benzene	770	1386
Glycerin	280	504
Hydrochloric acid	270	486
Petroleum crude, approximate	450	810
Water	115	207

For many liquids and solids κ is very nearly constant over a fairly wide range of pressures. If the specific volume is considered constant for purposes of evaluating the integral in Eq. (3-19), we have

$$w_T = \frac{\kappa v}{2}(p_2^2 - p_1^2) \tag{3-20}$$

Equation (3-20) may thus be used to obtain an estimate of the work involved in compressing liquids and solids to elevated pressures while maintaining a constant-temperature condition.

Values of some typical coefficients of expansion appear in Table 3-2.

EXAMPLE 3-9

75 kg of water are compressed in a closed container from 140 kPa to 7.5 MPa while the temperature is maintained constant at 40°C. Calculate the isothermal work in this process.

SOLUTION Equation (3-20) applies for this problem. To estimate the value of κ we make use of the compressed-liquid table and the approximation

$$\kappa = -\frac{1}{v}\left(\frac{\partial v}{\partial p}\right)_T \approx -\frac{1}{v}\left(\frac{\Delta v}{\Delta p}\right)_T \tag{a}$$

From the saturation-temperature table (Table A-7M, 40°C)

$$v_f = 1.0078 \text{ cm}^3/\text{g} = 1.0078 \times 10^{-3} \text{ m}^3/\text{kg}$$

From the compressed-liquid table we can evaluate the specific volume at 40°C and 7.5 MPa to obtain

$$v = 1.0045 \text{ cm}^3/\text{g}$$

Then applying the approximation of Eq. (*a*),

$$\kappa \approx -\frac{1}{1.0078}\left(\frac{1.0045 - 1.0078}{7500 \text{ kPa} - 140 \text{ kPa}}\right)_{40^\circ\text{C}}\left(\frac{1}{1000}\right)$$

$$= 4.45 \times 10^{-10} \text{ m}^2/\text{N} \qquad (\text{Pa}^{-1})$$

We now calculate the isothermal work per kilogram from Eq. (3-20):

$$w_T = \frac{\kappa v}{2}(p_2^2 - p_1^2)$$

$$= \frac{(4.45 \times 10^{-10} \text{ m}^2/\text{N})(1.0078 \times 10^{-3} \text{ m}^3/\text{kg})}{2}$$

$$\times [(7.5 \times 10^6)^2 - 140 \times 10^3)^2] \text{ N}^2/\text{m}^4$$

$$= 12.61 \text{ J/kg}$$

For the total mass of 75 kg the work is

$$W_T = mw_T = (75)(12.61) = 945.7 \text{ J}$$

EXAMPLE 3-10
At 300 K copper has the properties $\bar{v} = 70.062$ cm³/g mol and $\kappa = 0.776 \times 10^{-12}$ cm²/dyn, with $M = 63.54$. Calculate the work to compress a block 2.0 cm on a side from 1 to 1000 atm pressure while maintaining the temperature constant at 300 K.

SOLUTION Again, we employ Eq. (3-20). On a molal basis

$$W_T = \frac{\kappa \bar{v}}{2}(p_2^2 - p_1^2)$$

with

$$\bar{v} = 7.062 \text{ cm}^3/\text{g mol} = 7.062 \times 10^{-3} \text{ m}^3/\text{kg mol}$$

$$\kappa = 0.776 \times 10^{-2} \text{ cm}^2/\text{dyn} = 0.776 \times 10^{-11} \text{ m}^2/\text{N}$$

$$W_T = \frac{(0.776 \times 10^{-11} \text{ m}^2/\text{N})(7.062 \times 10^{-3} \text{ m}^3/\text{kg mol})}{2}$$

$$\times (1000^2 - 1^2)(1.0132 \times 10^5 \text{ N/m}^2)^2$$

$$= 281.3 \text{ J/kg mol}$$

The number of moles is

$$\eta = \frac{V}{v} = \frac{2.0^3 \times 10^{-6}}{7.062 \times 10^{-3}} = 1.133 \times 10^{-3} \text{ kg mol}$$

so the total work is

$$W_T = (1.133 \times 10^{-3} \text{ kg mol})(281.3 \text{ J/kg mol})$$

$$= 0.319 \text{ J}$$

REVIEW QUESTIONS

1 What is a *pure substance*?
2 Distinguish between *intensive* and *extensive* properties.
3 Distinguish between the terms *pure substance* and *simple substance*.
4 Describe the physical phenomenon of vaporization at constant pressure.
5 How does the phenomenon of fusion help an ice skater perform maneuvers?

6 How do you account for the fact that liquid CO_2 is not observed when dry ice (solid CO_2) is placed on a table at room temperature and pressure?
7 Why do you suppose foods cook faster in a pressure cooker (a device that allows the pressure to build up when water is boiled inside)?
8 What is meant by the term *saturation region*?
9 Distinguish between the terms *moisture* and *quality*. How do these terms relate to the terms *saturated liquid* and *saturated vapor*?
10 What is the quality of a superheated vapor?
11 Why are the terms *subcooled liquid* and *compressed liquid* synonymous?
12 What is an ideal gas?
13 Does an ideal gas have constant specific heats?
14 Is a wet mixture of water vapor an ideal gas? Why?
15 What is meant by the term *superheated vapor*?
16 What happens at the critical point?
17 What is the *state principle*?
18 What is a simple compressible substance?
19 How many properties are necessary to describe the state of a system?
20 What would be the specific heat of a wet mixture of steam?
21 Upon what variable are the properties of liquids primarily dependent?

PROBLEMS (ENGLISH UNITS)

3-1 Saturated steam vapor at 100 psia is contained in a 2.0-ft³ rigid tank. Calculate the amount of steam which will be condensed if the tank is cooled until the pressure drops to 20 psia.

3-2 A tank having a volume of 10 ft³ contains ammonia at 220 psia and 200 °F. The ammonia is discharged slowly from the tank until conditions inside the tank reach 80 psia and 100°F. Calculate the mass discharged from the tank.

3-3 A container having a volume of 3 ft³ is filled with steam at 30 psia and 90 percent quality. Calculate the quantity of heat which must be removed from the container in order to reduce the pressure to 20 psia.

3-4 10 lbm of water vapor are contained at 25 psia and 90 percent quality in a suitable enclosure. Calculate the heat which must be added in order to just produce a saturated vapor. What will the pressure be at the end of the heating process?

3-5 Calculate the percent error which would result from assuming that water at 100°F and 1500 psia has a density of 62.4 lbm/ft³.

3-6 Calculate the specific volume for Freon 12 at 10 psia and 50 percent quality.

3-7 Suppose heat is added to 0.5 lbm of the Freon in Prob. 3-6 during a constant-pressure process. The heating proceeds until a temperature of 140°F is attained. Calculate the heat required and the work produced, assuming that the Freon is contained by a suitable cylinder-piston arrangement.

3-8 Calculate the percent increase in volume when a block of copper is heated from 100 to 1000°F.

3-9 In Table 3-2, β for water is given as $115 \times 10^{-6}\ °F^{-1}$. Verify this value by direct calculation from data in the steam tables for water at approximately 100°F.

3-10 Steam is compressed in a constant-pressure process in a piston-cylinder arrangement from conditions of 60 psia and 400°F to a final state where half the vapor is condensed. Calculate the heat transfer per lbm of steam for this process.

3-11 A frost layer is formed in the freezing compartment of a refrigerator over a period of time so that it has a thickness of 1 cm. Because of its porous nature, its overall density is only 30 percent of that for solid ice. The layer covers an area of 0.5 m^2. How much hot water at 140°F would have to be added to the frost to just produce water at 32°F? Assume that the melting process occurs in an insulated container.

3-12 1 lbm of steam is heated at a constant pressure of 14.696 psia from a temperature of 220 to 350°F. Calculate the heat required.

3-13 Calculate the energy required to vaporize 1 lbm of water at normal atmospheric pressure (14.696 psia). Compare this quantity with the amount of energy to vaporize ammonia or Freon 12 at this same pressure.

3-14 Calculate the mass of H_2O contained in a 2-ft^3 vessel under the following conditions:
(*a*) 14.696 psia, 100°F
(*b*) 100 psia, 100°F
(*c*) 1 psia, 500°F
(*d*) 20 psia, 30 percent quality
(*e*) 1000 psia, 1000°F
(*f*) At the critical state

3-15 How much energy is required to vaporize 1 lbm of Freon 12 at a temperature of 70°F in a constant-pressure process? How does this compare with that required to vaporize water or ammonia under the same conditions?

3-16 Steam is contained in a piston-cylinder arrangement at 20 psia and 300°F. Heat is removed from the cylinder until the volume is reduced by half, while the pressure is maintained constant. Calculate the heat removed per pound of steam.

3-17 Saturated water vapor at 1 atm is contained in a piston-cylinder arrangement and the piston is restrained by a spring such that the pressure in the cylinder is proportional to the cylinder volume. How much heat must be added to the steam to raise the pressure to 20 psia?

3-18 With the use of the superheat steam tables, calculate the isothermal compressibility for water vapor under the following conditions:
(*a*) 1 psia, 300°F
(*b*) 500 psia, 500°F

Compare these values with that obtained by assuming that the steam obeys the ideal-gas equation of state. Comment on the comparison.

3-19 Repeat Prob. 3-18 for the volume coefficient of expansion.

3-20 Calculate the work required to compress 1 g of water from 50 psia, 100°F, to 1500 psia in an isothermal process.

3-21 A man who normally consumes energy at the rate of 85 kcal/h increases his consumption rate to 150 kcal/h during more vigorous activity. In a hot surrounding environment the body will perspire to cool the skin surface by evaporation. Assuming that the perspiration is evaporated at constant pressure and a constant skin temperature of 80°F, calculate the amount which must be evaporated per hour to make up for the difference in energy consumption rates stated above. (*Ans.:* 112 g)

3-22 A tank having a volume of 2.5 ft^3 contains liquid water and vapor in equilibrium at 100°F. The tank is sealed. How much heat must be added to just produce saturated vapor at 1 atm? Assume that the initial proportions of vapor and liquid are such that saturated vapor at 1 atm can be attained by the heating process.

3-23 Water vapor at 180 psia and 800°F is contained in a rigid box having a volume of 20 ft^3. How much heat must be removed from the container to condense 10 percent of the vapor?

3-24 Verify that it takes approximately 80 cal to melt 1 g of ice at 32°F.

3-25 Water vapor at 14.7 psia and 300°F is compressed isothermally until half the vapor has condensed. How much work must be performed on the steam in this compression process per lbm?

3-26 Water vapor at 14.7 psia and 300°F is contained in a rigid vessel. At what temperature will the vapor start to condense when the container is cooled slowly? How much heat will have been removed per lbm in this cooling process?

3-27 On a hot summer day in the south the solar irradiation on the earth is about 300 Btu/h·ft^2, and about 75 percent of this energy will be absorbed by green grass. Assuming that all the energy absorbed goes to evaporate water at constant pressure and an average temperature of 100°F, calculate the amount of watering required each week for a 100-by-160-ft plot of land to just replenish the water lost by evaporation. For this calculation assume that the sun's radiation occurs over a 9-h period. Express the answer in gallons per week.

PROBLEMS (METRIC UNITS)

3-1M 1 kg of steam is heated at a constant pressure of 100.0 kPa from a temperature of 100 to 200°C. Calculate the heat required.

3-2M Calculate the energy required to vaporize 1 kg of water at 100.0 kPa. Compare this quantity with the amount of energy needed to vaporize ammonia or Freon 12 at this same pressure.

3-3M Calculate the mass of H_2O contained in a 60-liter vessel under the following conditions:

(*a*) 100 kPa, 40°C	(*d*) 150 kPa, 30 percent quality
(*b*) 700 kPa, 40°C	(*e*) 8000 kPa, 520°C
(*c*) 6 kPa, 240°C	(*f*) At the critical state

3-4M Steam is contained in a rigid vessel at 100 kPa and 200°C. The volume is 2.3 m^3. Calculate the cooling required to condense half the steam.

3-5M A piston-cylinder arrangement contains steam at 200 kPa and 75 percent quality. The mass of steam is 500 g. Heat is added to the cylinder while maintaining the steam temperature constant. How much heat will be added when the steam becomes a saturated vapor?

3-6M A container having a volume of 85 liters is filled with steam at 200 kPa and 90 percent quality. Calculate the quantity of heat which must be removed from the container in order to reduce the pressure to 150 kPa.

3-7M 5 kg of water vapor are contained at 150 kPa and 90 percent quality in a suitable enclosure. Calculate the heat which must be added in order to just produce a saturated vapor. What will the pressure be at the end of the heating process?

3-8M Calculate the percent error which would result from assuming that water at 40°C and 10 MPa has a density of 999.6 kg/m^3.

3-9M Suppose heat is added to 0.5 kg of the Freon in Prob. 3-6 during a constant-pressure process. The heating proceeds until a temperature of 60°C is attained. Calculate the heat required and the work produced, assuming that the Freon is contained by a suitable piston-cylinder arrangement.

3-10M Steam at 20 kPa and 80 percent quality is contained in a frictionless piston-cylinder arrangement. It is subsequently compressed at constant pressure until the volume is reduced by half. Calculate the heat transfer per unit mass.

3-11M Steam is contained in a rigid box at 100 kPa and 200°C. The volume of the box is 1.3 m^3. How much heat must be removed from the box to condense half the steam?

3-12M Steam at 300 kPa and 160°C is contained in a piston-cylinder arrangement. A constant restraining force is maintained on the piston and the cylinder is cooled until half the vapor condenses. How much cooling must be supplied per unit mass of water?

3-13M A rigid container having a volume of 1.2 m^3 contains 10 percent by volume of liquid water at 100 kPa. The remaining volume is vapor at 100 kPa. How much heat must be added to the container to just vaporize all the liquid water?

3-14M Water vapor is heated at a constant pressure of 7 kPa from 65 to 550°C. Assume that the vapor obeys the ideal-gas equation of state in this region and calculate the heat transfer by using the data of Table 2-3. Check this calculation by using the tabulated properties in the steam tables. Comment on the agreement or disagreement of these two calculations.

3-15M Calculate the percent increase in volume when a block of copper is heated from 40 to 540°C.

3-16M With the use of the superheat steam tables, calculate the isothermal compressibility for water vapor under the following conditions:
(*a*) 7 kPa, 160°C
(*b*) 3 MPa, 240°C

3-17M Calculate the work required to compress 1 kg of water from 350 kPa, 40°C, to 10 MPa in an isothermal process.

3-18M Using SI units, plot a curve of saturation pressure versus v_g for water for temperatures between 0 and 250°C.

3-19M Using the appropriate tables for water and water vapor, fill in the blanks in the following table:

Temperature, °C	280		480		
Pressure, Pa		700 k		8.0 M	350 k
Mass, kg	1.0	2.0	1.3	3.5	3
Specific volume, m³/kg		0.3852			
Volume, m³					
Quality, %					60
Moisture, %				20	
Enthalpy, kJ/kg	2779.6				
Internal energy, kJ/kg			3005.4		

3-20M Calculate the value of c_p for steam at p = 500 kPa and 240°C using the steam tables.

3-21M A man who normally consumes energy at the rate of 350 kJ/h increases his consumption rate to 630 kJ/h during more vigorous activity. In a hot surrounding environment the body will perspire to cool the skin surface by evaporation. Assuming that the perspiration is evaporated at constant pressure and a constant skin temperature of 25°C, calculate the amount which must be evaporated per hour to make up for the difference in energy consumption rates stated above. (*Ans.*: 112 g)

3-22M Water is contained in a piston-cylinder arrangement at 200 kPa and 40°C. The cylinder has a diameter of 15 cm and a height of 10 cm. How much work would be required to compress the water to a pressure of 15 MPa in an isothermal process?

3-23M How much energy is required to vaporize 1 kg of Freon 12 at room temperature of 20°C in a constant-pressure process? How does this compare with that required to vaporize water or ammonia under the same conditions?

3-24M Saturated water vapor at 100 kPa is contained in a piston-cylinder arrangement, and the piston is restrained by a spring such that the pressure in the cylinder is proportional to the cylinder volume. How much heat must be added to the steam to raise the pressure to 150 kPa?

3-25M Liquid water is contained in a cylinder at 7.5 MPa and 40°C. The cylinder has a diameter of 3 cm and contains 15 g of water. A piston has a constant restraining force imposed on it to keep the pressure constant at 7.5 MPa. How much heat must be added to the cylinder to move the piston through a distance of 3.5 cm?

3-26M Verify that it takes approximately 335 J to melt 1 g of ice at 0°C.

3-27M Water vapor at 100 kPa and 150°C is compressed isothermally until half the vapor has condensed. How much work must be performed on the steam in this compression process per kilogram?

3-28M Water vapor at 100 kPa and 150°C is contained in a rigid vessel. At what temperature will the vapor start to condense when the container is cooled slowly? How much heat will have to be removed per kilogram in this cooling process?

3-29M At a certain state, steam has a temperature of 240°F and an entropy of 1.419 Btu/lbm·°F. Determine the enthalpy and specific volume at this point.

3-30M A rigid tank having a total volume of 0.1 m^3 is perfectly insulated and divided into two equal chambers separated by an adiabatic partition. One chamber contains steam at 100 kPa and 200°C and the other chamber contains steam at 1.5 MPa and 320°C. The partition is removed and the system allowed to reach equilibrium. Determine the pressure and temperature at the new equilibrium state.

3-31M On a hot summer day in the south, the solar irradiation on the earth is about 1 kW/m^2, and about 75 percent of this energy will be absorbed by green grass. Assuming that all the energy absorbed goes to evaporate water at constant pressure and an average temperature of 40°C, calculate the amount of watering required each week for a 30-by-50-m plot of land to just replenish the water lost by evaporation. For this calculation assume that the sun's radiation occurs over a 9-h period. Express the answer in liters per week.

3-32M Calculate the enthalpy of steam at 10.0 MPa and 600°C using tabulated values of u and v and compare with the tabulated value for h.

3-33M Water is contained in a piston-cylinder arrangement at 2 atm and 100°F. The cylinder has a diameter of 15 cm and a height of 10 cm. How much work would be required to compress the water to a pressure of 150 atm in an isothermal process?

3-34M A thermometer is constructed of Pyrex glass and ethyl alcohol. The "bulb" portion has a volume of 0.4 cm^3, and the column length is 20 cm. Assuming that only the bulb is used as the temperature sensor, what must be the cross-sectional area of the column in order for the thermometer to have a range of 10 to 40°C over the 20-cm length?

REFERENCES

1 Kline, S. J., and F. O. Koenig: The State Principle, *J. Appl. Mech.,* vol. 24, p. 29, 1957.
2 Hatsopoulos, G. N., and J. H. Keenan: "Principles of General Thermodynamics," John Wiley & Sons, Inc., New York, 1965.
3 Keenan, J. H., and F. G. Keyes: "Thermodynamic Properties of Steam," John Wiley & Sons, Inc., New York, 1936.
4 Van Wylen, G. J.: "Thermodynamics," John Wiley & Sons, Inc., New York, 1959.
5 Reynolds, W. C.: "Thermodynamic Properties in SI," Mech. Engr. Dept., Stanford University, Stanford, Calif. 1978.

4

ENERGY ANALYSIS IN OPEN SYSTEMS

4-1 INTRODUCTION

The two previous chapters have expressed the first law of thermodynamics in analytical form and presented methods for calculating thermodynamic properties of simple thermodynamic substances. In this chapter we wish to apply this information to the analysis of a wider class of problems, in particular, to open systems where mass may cross the boundaries. In this way, we will be able to establish a rather full exploitation of the methodology of energy analysis.

It is to our advantage to adopt a plan of attack for performing an energy analysis of thermodynamic systems. The general procedure is basically a method of answering the following questions:

1 What is the thermodynamic system to be analyzed—how is it defined?
2 What is the thermodynamic substance to be considered in the analysis? Are the necessary properties of this substance readily available?
3 What *type* of system shall be used for analysis—open or closed?
4 How shall we describe the thermodynamic process under consideration?
5 Are we considering a steady-state situation or is the system changing with time?

Once a clear answer is obtained for each of these questions, the solution of the problem is usually obtained through mathematical analysis; however, some experience is required to be efficient in performing this analysis. The questions are not necessarily posed in the order presented, and the answer to one question may be dependent on the answer to some of the other questions.

As has been our experience in the previous chapters, we shall find that the plan of attack described in the above questions can be developed into a more specific algorithm:

1 Sketch the system, and label all inlet and exit mass and energy quantities. Define the system boundary.
2 Write out the information which is available for analysis.
3 Write the appropriate energy balance in accordance with the labeling procedure followed in the system schematic.
4 Write the process description relations which apply.
5 Determine properties for use in the energy balance(s).
6 Combine the information to obtain the desired result.

4-2 ENERGY BALANCES AND SIGN CONVENTIONS

In Chap. 2 we developed general energy relationships for closed systems, i.e., systems which do not admit the transfer of mass across their boundaries. We adopted the sign convention that energy added to the system (in the form of

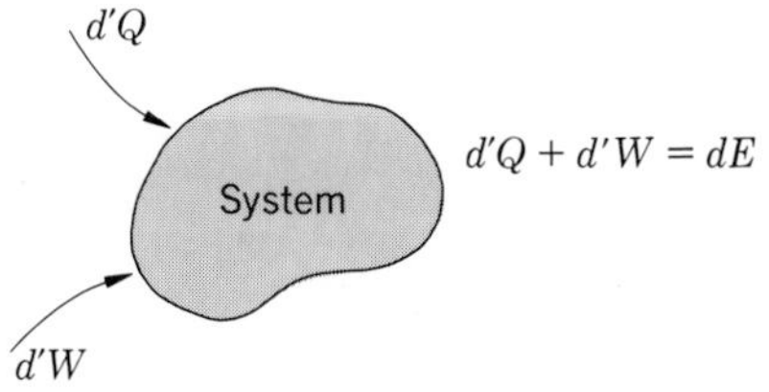

FIG. 4-1 Energy balance for closed system.

work or heat) is positive and energy given up by the system to other things is negative. The main consideration is always the fact that *energy is conserved* or

Energy which goes into system = energy which leaves system
+ energy which accumulates
in the system

Energy analysis, then, is basically an accounting procedure, and the equations of Chap. 2 express the analytical relations necessary to balance the books.

4-3 THE CLOSED SYSTEM

The general discussion of the first law of thermodynamics as presented in Chap. 2 pertained to closed thermodynamic systems. These systems do not admit the transfer of mass across their boundaries. The only essential point is that all types of work and internal energy be considered in the energy balance

$$d'Q + d'W = dE \tag{4-1}$$

where the energy quantities have been written as differential values in order to apply the equation to any type of process. The energy balance for the closed system is illustrated in Fig. 4-1. Closed and open systems are contrasted in Fig. 4-2.

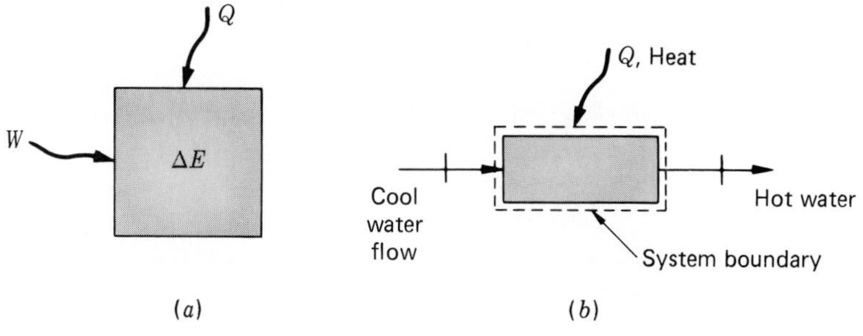

FIG. 4-2 (*a*) Closed container is a closed system with no mass transfer across the boundary. (*b*) Water flow through a heater is treated as an open system.

4-4 THE OPEN SYSTEM

We have seen that the closed system does not admit the transfer of mass across its boundaries; only energy may flow in and out of the system as it changes from one state to another. We now consider the open system, where mass may enter and leave some volume in space. This may at first seem inconsistent with the meaning that has been attached to a thermodynamic system so far, because we were considering a quantity of matter broken away from its surroundings. This apparent anomaly will be resolved, however, as we relate the analysis of open systems directly to that of closed systems. In any event, the reader must realize that we are discussing analysis techniques which may be used to apply the conservation-of-energy principle to a variety of situations.

First, let us show how to calculate the mass flow rate through a channel, as illustrated in Fig. 4-3. The rate at which the volume of fluid is swept through the channel is $A\ ds/d\tau$, where A is the flow cross-sectional area and ds is the displacement. The term $A\ ds$ is the volume displacement. The volume flow *rate* is designated by $\dot{V}$. The volume per unit mass is v, and so the rate at which the mass flows in the channel $\dot{m}$ is

$$\dot{m} = \frac{\dot{V}}{v} = \frac{1}{v}\frac{A\ ds}{d\tau}$$

But, $ds/d\tau = \bar{V}$ = channel flow velocity, so that

$$\dot{m} = \frac{A\bar{V}}{v} \tag{4-2}$$

We recall that the fluid *density* is defined as the inverse of the specific volume, $\rho = 1/v$, so that

$$\dot{m} = \rho A \bar{V} \tag{4-3}$$

The $\bar{V}$ symbol is used here to distinguish the velocity from the symbol for total volume.

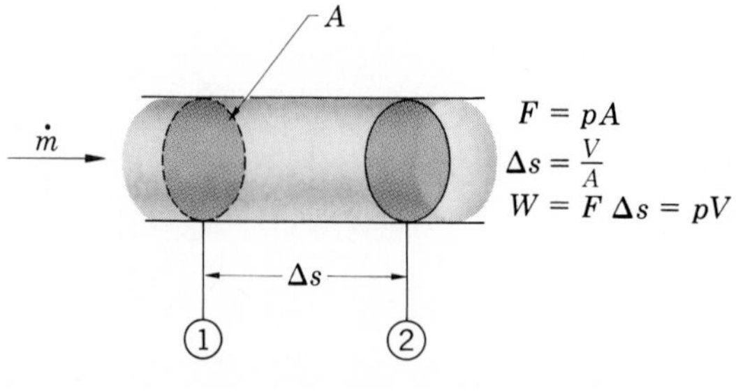

FIG. 4-3 Mass flow and flow work in a channel.

4-5 THE CONTROL VOLUME

To study open systems, we introduce the concept of *a control volume,* as shown in Fig. 4-4. This volume is region in space to be observed in respect to the matter and energy which cross its boundaries. Consider first the conservation-of-mass principle, which may be written as

Mass flow into control volume = mass flow out of control volume + increase in mass inside the control volume (4-4)

FIG. 4-4 Open system balances: (*a*) mass balance; (*b*) energy balance.

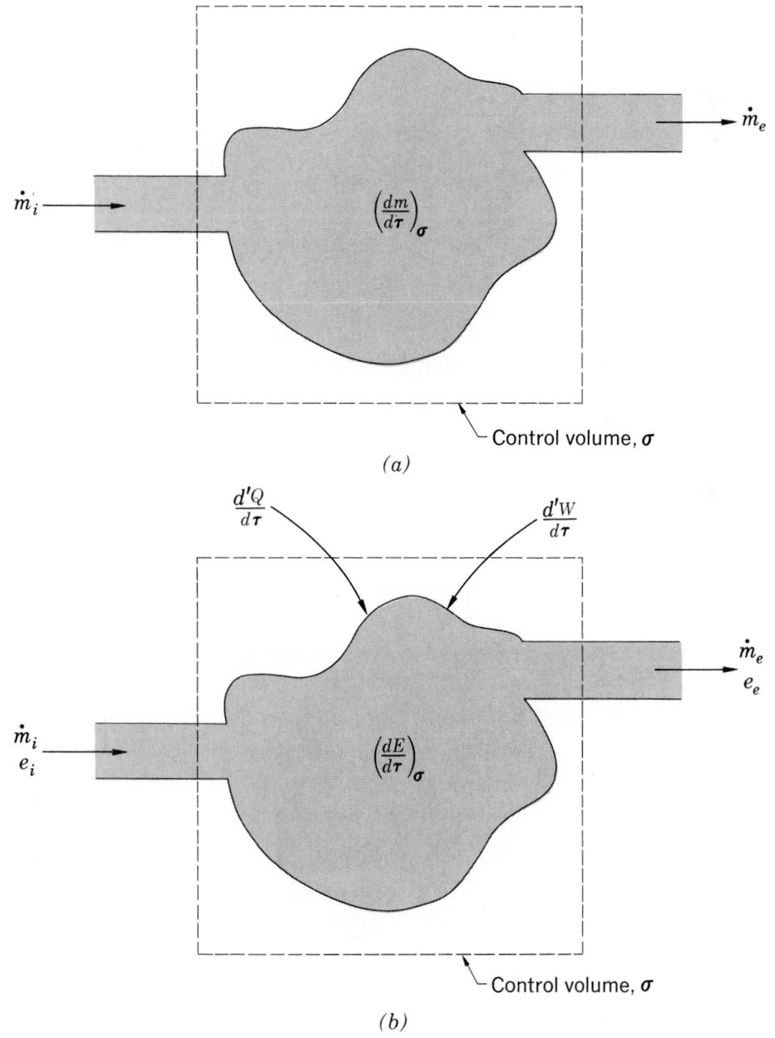

or that which goes in either stays in or goes out. In accordance with Fig. 4-4a, we may write the mass-conservation principle as

$$\dot{m}_i = \left(\frac{dm}{d\tau}\right)_\sigma + \dot{m}_e \tag{4-5}$$

where $\dot{m}_i$ is the mass flow per unit time into the control volume, $\dot{m}_e$ is the mass flow per unit time out of the control volume, and the subscript σ indicates the mass accumulation per unit time within the control volume.

If multiple inlet and exist streams are involved, then we must perform a summation over all such flows to determine the mass balance:

$$\sum_i \dot{m}_i = \left(\frac{dm}{d\tau}\right)_\sigma + \sum_e \dot{m}_e \tag{4-6}$$

As a preliminary to an energy analysis of a control volume, let us qualitatively consider what happens to a fixed quantity of mass as it moves through the control volume, i.e., the behavior of a *closed* thermodynamic system as it undergoes a process which causes it to move through the control volume. The closed thermodynamic system may undergo effects of pressure from the surroundings, the transfer of heat across its boundaries, and the action of various work-producing forces. The internal energy of the closed system could change as a result of its physical movement from one position to another and, perhaps, from a change in its velocity. Regardless of the effects observed, we could certainly analyze them with the conservation-of-energy principle. Furthermore, the total mass flow into or out of the control volume may be thought of as a group of mass elements dm, in a sense, a group of small, closed thermodynamic systems. We may therefore consider that the mass flow in and out of the control volume *transports* internal energy across the boundaries of the control volume. Thus the energy-conservation principle for this type system is

Transport of internal energy into the control volume + heat added to control volume + work done on all elements as they pass through the control volume = increase of internal energy in control volume + transport of internal energy out of the control volume (4-7)

In analytical form according to Fig. 4-4b,

$$\dot{E}_i + \frac{d'Q}{d\tau} + \frac{d'W}{d\tau} = \left(\frac{dE}{d\tau}\right)_\sigma + \dot{E}_e \tag{4-8}$$

where $\dot{E}_i$ and $\dot{E}_e$ represent the energy transport per unit time at entrance and outlet, respectively, and the subscript σ again designates the change within the boundaries of the control volume. The internal-energy transports may be written as

$$\dot{E}_i = \dot{m}_i e_i \tag{4-9}$$

$$\dot{E}_e = \dot{m}_e e_e \tag{4-10}$$

Flow Work

Equation (4-8) can be used to analyze open systems, but the work term is usually expressed in a more useful form for analysis purposes. As an intermediate step in developing such an expression, consider the very simple control volume, as shown in Fig. 4-3. Notice that, in order for the mass to move through the control volume, there must be some forcing function to push it through. This force is provided by the pressure of the system. We imagine a mass having an area of A and a length of Δs. To push this quantity of mass either in or out of the control volume, we must exert a force of pA through the distance of Δs. Regardless of the quantity of mass, Δs will be given by

$$\Delta s = \frac{V}{A}$$

so that the work to push the mass in or out is

$$W = \int F\, ds = F\, \Delta s = pA\frac{V}{A} = pV \tag{4-11}$$

The net work done on the system as it moves from station 1 to 2 in Fig. 4-3, exclusive of external work, is

$$W_{\text{net}} = p_1 V_1 - p_2 V_2$$

where the $p_1 V_1$ term is the work done on the fluid to force it into the control volume and the $p_2 V_2$ is the work to force the fluid out of the control volume. The difference in the two terms is the net work added.

The work term designated by pV is called *flow work*, and it is customary to consider it separately from the work delivered to things outside the control volume. The energy equation may then be written in the form

$$\dot{m}_i(e_i + p_i v_i) + \frac{d'Q}{d\tau} + \frac{d'W_{\text{ext}}}{d\tau} = \left(\frac{dE}{d\tau}\right)_\sigma + \dot{m}_e(e_e + p_e v_e) \tag{4-12}$$

where W_{ext} is the *work delivered to the control volume by external forces*. This quantity is sometimes called *shaft work* because in a practical sense it is frequently delivered by a turbine-shaft arrangement.

Steady State and Steady Flow

Equation (4-12) represents the general energy balance for an open system. If the open system is operating in *steady-state* conditions, then there are no changes inside the control volume with time; thus, $(dE/d\tau)_\sigma = 0$ and $(dm/d\tau)_\sigma = 0$. A *steady-flow* situation is that circumstance in which the flow rates $\dot{m}_i$ and $\dot{m}_e$ do not change with time. For the special case of steady flow *and* steady state there is no accumulation of mass or energy inside the control volume, and $\dot{m}_i = \dot{m}_e$ so that Eq. (4-12) reduces to

$$\frac{d'Q}{d\tau} + \frac{d'W_{\text{ext}}}{d\tau} = \dot{m}[(e_e + p_e v_e) - (e_i + p_i v_i)] \tag{4-13}$$

We note once again that the internal energy e is composed of the internal thermal energy u and the gravitational potential energy, kinetic energy, etc. For convenience, it is customary to introduce the property enthalpy h, defined in Sec. 2-8 as

$$h = u + pv \tag{4-14}$$

Enthalpy is a property because it is composed of other thermodynamic properties. Notice that the enthalpy has a physical meaning and significance when applied to open systems, but that this meaning is no longer valid when applied to a closed system because the pv term does not represent flow work in a closed system.

With the introduction of the enthalpy our general energy equation for an open system may be written as

$$\dot{m}_i(h_i + \text{KE}_i + \text{ChE}_i + \cdots) + \frac{d'Q}{d\tau} + \frac{d'W_{\text{ext}}}{d\tau} = \left(\frac{dE}{d\tau}\right)_\sigma + \dot{m}_e(h_e + \text{KE}_e + \text{ChE}_e + \cdots) \tag{4-15}$$

If multiple entrance and exit streams are involved, a simple summation is performed over all streams supplying or removing energy from the control volume, and we have

$$\sum_i \dot{m}_i(h_i + \text{KE}_i + \text{ChE}_i + \cdots) + \frac{d'Q}{d\tau} + \frac{d'W_{\text{ext}}}{d\tau} = \left(\frac{dE}{d\tau}\right)_\sigma + \sum_e \dot{m}_e(h_e + \text{KE}_e + \text{ChE}_e + \cdots) \tag{4-16}$$

The above equations for the general open system will appear rather cumbersome and foreboding to the reader becoming involved with thermodynamics for the first time. However, in most practical problems these equations will simplify considerably, and we will find that the energy-analysis technique will be quite easy. For example, in our discussions of ideal gases and water-steam systems there was no need to be concerned with chemical energy, and so it will be possible to leave that term out in many cases.

Significance of Kinetic Energy

At first glance, kinetic energy may seem to be a very important quantity in *flow* (open) systems. The fluid is moving so its kinetic energy must be important. For a large number of practical problems this is *not* the case because the kinetic energies are very small. To show that this is true let us calculate the velocity necessary to produce just 1.0 Btu/lbm or 1.0 kJ/kg of kinetic energy:

$$1.0 \text{ Btu/lbm} = 778 \text{ ft·lbf/lbm} = \frac{1}{2g_c} \bar{V}^2$$

With $g_c = 32.2$ lbm·ft/lbf·s^2 we obtain

$$\bar{V} = 224 \text{ ft/s}$$

In SI units, for KE = 1.0 kJ/kg,

$$1000 \text{ J/kg} = \frac{1}{2} \bar{V}^2$$

and

$$\bar{V} = 44.7 \text{ m/s}$$

These are much higher velocities than will be encountered in many practical problems. A notable exception is the flow in a jet engine or gas turbine; but, for flow of water, air, or steam in pipes, Freon in an air-conditioner coil, and many other examples, the flow velocity is low enough that we may neglect kinetic energies in comparison to the respective enthalpies or internal energies of the flow streams.

4-6 SUMMARY OF OPEN-SYSTEM SPECIAL CASES

To aid the reader in applying open-system analysis we now give the general equations in reduced form for a number of special cases in Table 4-1. The purpose of this table is not to invoke a cookbook philosophy in the reader, but rather to show the simplicity which the relations can assume for a number of practical problems.

TABLE 4-1

Special Cases for Open Systems

Steady Flow	Steady State	Multiple Inlet Streams	KE	ChE	$\frac{d'Q}{d\tau}$	$\frac{d'W_{ext}}{d\tau}$	Energy Equation
Yes	Yes	No	0	0	0	—	$\frac{d'W_{ext}}{d\tau} = \dot{m}(h_e - h_i)$
Yes	Yes	No	0	0	—	0	$\frac{d'Q}{d\tau} = \dot{m}(h_e - h_i)$
Yes	No	No	0	0	—	—	$\dot{m}_i h_i + \frac{d'Q}{d\tau} + \frac{d'W_{ext}}{d\tau} = \dot{m}_e h_e + \left(\frac{dU}{d\tau}\right)_\sigma$
Yes	Yes	Yes	0	0	0	0	$\Sigma \dot{m}_i h_i = \Sigma \dot{m}_e h_e$
Yes	Yes	Yes	0	0	—	0	$\frac{d'Q}{d\tau} = \Sigma \dot{m}_e h_e - \Sigma \dot{m}_i h_i$
No	No	No	0	0	0	0	$\dot{m}_i h_i = \dot{m}_e h_e + \left(\frac{dU}{d\tau}\right)_\sigma$

In summary, we may remark that the analysis of open systems deals with an energy balance, similar in many respects to the analysis for closed systems. The primary difference in the analysis for closed or open systems is that the work term is split into two parts in the open-system analysis; one part represents the work done external to the control volume, and the other part represents the work necessary to force the mass in and out of the control volume. The property of enthalpy is introduced as a matter of convenience for summing the internal thermal energy and flow work at a particular position. The following examples illustrate the use of the energy-conservation principle for analysis of open systems.

EXAMPLE 4-1 Steady-Flow Steam Turbine

1 lbm/s of steam enters a steady-flow turbine at 500 psia and 700°F with a velocity of 800 ft/s. The steam leaves the turbine at 5 psia and 200°F with a velocity of 100 ft/s. The turbine is well insulated so that the process may be assumed to be adiabatic. Calculate the horsepower output of the turbine. Assume the process to be steady flow, steady state.

FIG. EXAMPLE 4-1

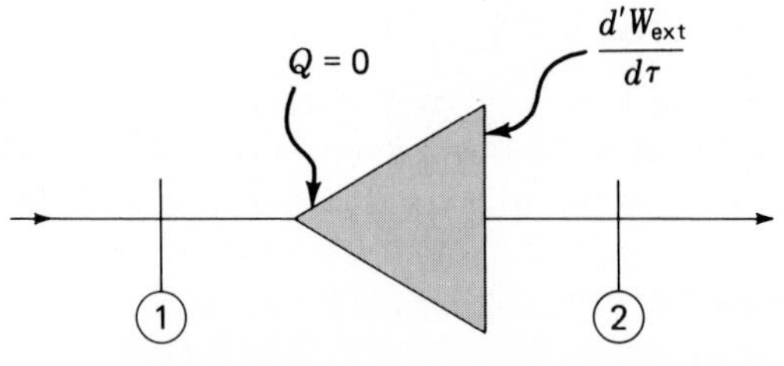

SOLUTION This is a steady-flow steady-state problem so $\dot{m}_i = \dot{m}_e$ and $(dE/d\tau)_\sigma = 0$. Omitting the chemical-energy terms from Eq. (4-15) and observing that $d'Q/d\tau = 0$, we have

$$\frac{d'W_{ext}}{d\tau} = \dot{m}(h_e - h_i) + \frac{\dot{m}}{2g_c}(V_e^2 - V_i^2)$$

From the superheat steam tables, we have

$h_i = 1357.0$ Btu/lbm 500 psia and 700°F

$h_e = 1148.8$ Btu/lbm 5 psia and 200°F

Then

$$\frac{d'W_{ext}}{d\tau} = (1)(1148.8 - 1357.0) + \frac{1}{(2)(32.2)(778)}(100^2 - 800^2)$$

$$= -221 \text{ Btu/s} = -7.96 \times 10^5 \text{ Btu/h}$$

Since 1 hp = 2545 Btu/h, the horsepower *output* is

$$\text{hp} = \frac{7.96 \times 10^5}{2545} = 313 \qquad (233.4 \text{ kW})$$

The negative sign on the work term above indicates that work is delivered to things external to the control volume.

EXAMPLE 4-2 Adiabatic Pump Work
Water at the rate of 80 lbm/min is compressed adiabatically from 100°F and 100 psia to 1000 psia in a steady-flow process. Calculate the horsepower required, assuming that the water is very nearly incompressible.

SOLUTION A schematic for this process is indicated in the accompanying figure. We neglect the kinetic energy of the flow stream since the flow velocities are probably small, and we write Eq. (4-15) for steady flow and steady state,

$$\dot{m}h_1 + W = \dot{m}h_2$$

FIG. EXAMPLE 4-2

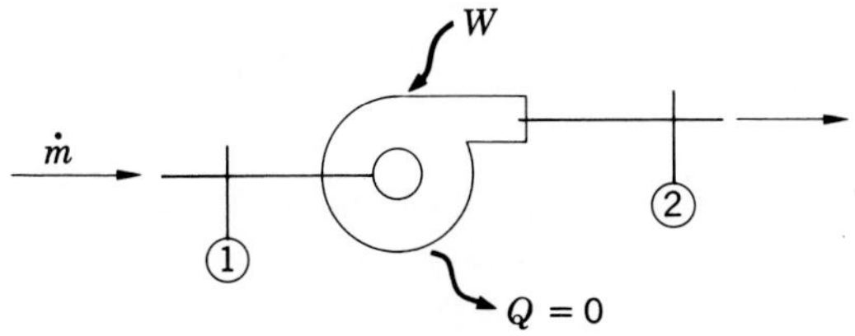

or

$$W = \dot{m}(h_2 - h_1) \qquad (a)$$

where W is the rate of doing work on the pump. From Chap. 2 we have

$$d'Q = du + p\,dv \qquad (b)$$

for a simple compressible substance. Also,

$$dh = du + p\,dv + v\,dp \qquad (c)$$

For an adiabatic process $d'Q = 0$, so that Eq. (*c*) becomes

$$dh_{\text{ad}} = v\,dp \qquad (d)$$

Integrating Eq. (*d*), we have

$$(h_2 - h_1)_{\text{ad}} = \int_{p_1}^{p_2} v\,dp$$

If the fluid is incompressible, $v \approx$ const. and

$$(h_2 - h_1)_{\text{ad}} = v(p_2 - p_1) \qquad (e)$$

The work is then calculated with Eq. (*a*) as

$$W = \dot{m}v(p_2 - p_1) \qquad (f)$$

The specific volume is taken as the value for saturated liquid at 100°F:

$$v = 0.01613 \text{ ft}^3/\text{lbm}$$

and the work becomes

$$\begin{aligned} W &= (80)(0.016\,13)(1000 - 100)(144) \\ &= 167\,000 \text{ ft·lbf/min} \\ &= 5.06 \text{ hp} \qquad (3775 \text{ W}) \end{aligned}$$

The incompressibility assumption may be checked by consulting the compressed-liquid tables wherein it will be found that the change in specific volume due to the elevated pressure is only about 0.3 percent. Note the sign convention employed in writing Eq. (*a*). Because we seek the work delivered to the pump, we draw the figure and take the work added as positive. The right side of the equation represents the difference between the energy leaving and entering the pump and thus must be equal to the work energy added.

EXAMPLE 4-3 Filling a Tank

An evacuated tank is connected through an appropriate valve to an infinite source of steam at 100 psia and 500°F, as shown in the accompanying sketch. The valve is opened and the steam is allowed to flow into the tank until the pressure reaches 100 psia. Calculate the temperature of the steam in the tank when the pressure just reaches 100 psia. Assume that the tank is insulated and that there is no shaft work delivered to the tank.

SOLUTION This is a non-steady-flow and non-steady-state process, since the flow rate into the tank and the mass of steam in the tank both vary with time. We choose the control volume as shown by the dashed line in the figure. Neglecting kinetic energies, we may write Eq. (4-15) in integral form as

$$\int \dot{m}_i h_i d\tau = \int \left(\frac{dE}{d\tau}\right)_\sigma d\tau = m_2 u_2 - m_1 u_1 \qquad (a)$$

where m_1 and m_2 are the initial and final masses in the tank. From the problem statement, $m_1 = 0$, but the final mass in the tank is just

$$m_2 = \int \dot{m}_i d\tau$$

and so Eq. (a) becomes

$$h_i = u_2$$

From the superheat steam tables we have

$$h_i = 1279.1 \text{ Btu/lbm} \qquad (2975 \text{ kJ/kg})$$

FIG. EXAMPLE 4-3

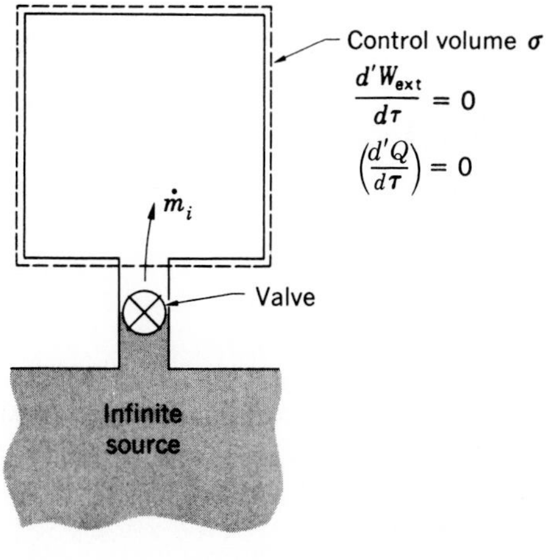

The final temperature corresponding to u_2 = 1279.1 Btu/lbm is found from Table A-9. The result is

$$T_2 = 767°\text{F} \qquad (408.6°\text{C})$$

EXAMPLE 4-4

The tank of Example 4-3 is reconstructed so that a piston is held in place by a charge of air at 20 psia. The piston is constructed of a non-heat-conducting material and the tank is insulated. A small cooling coil is placed in the cylinder to maintain the air temperature constant at 100°F. The total volume of the tank is 3 ft^3. The tank is again connected to an infinite source of steam at 100 psia and 500°F, and the valve is opened until the pressure in the tank reaches 100 psia. Calculate the final temperature of the steam under these conditions.

SOLUTION The control volume for this problem is shown as the dashed line in the accompanying sketch. Note that the thermodynamic system is chosen as the air, steam, and piston inside the cylinder. Even though the tank is insulated, heat is being removed by the action of the cooling coil. The amount of heat removed is calculated by determining the heat transfer when the given quantity of air is compressed isothermally from 20 to 100 psia. The mass of air is

$$m_a = \frac{pV}{RT} = \frac{(20)(144)(3)}{(53.35)(560)} = 0.289 \text{ lbm} \qquad (131 \text{ g})$$

The work done on the air is

$$W_a = \int -p\, dV = -\int_{V_1}^{V_2} mRT\frac{dV}{V} = -mRT \ln\frac{V_2}{V_1}$$

FIG. EXAMPLE 4-4

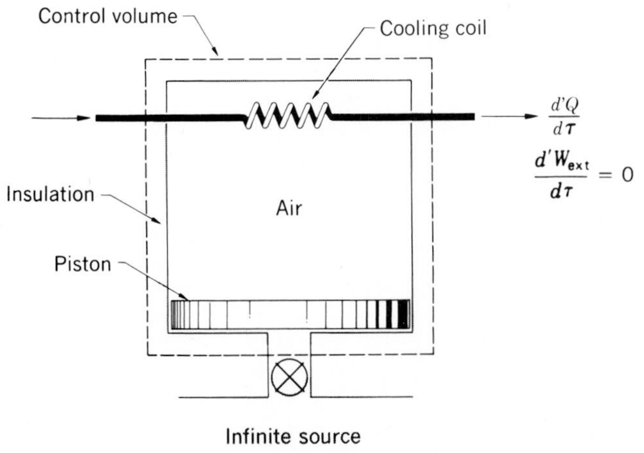

Since the temperature remains constant, $V_2/V_1 = p_1/p_2$, so that the work is

$$W_a = -mRT \ln \frac{p_1}{p_2}$$

$$= -(0.289)(53.35)(560) \ln \frac{20}{100}$$

$$= 13\ 850 \text{ ft·lbf} = 17.8 \text{ Btu} \qquad (18.89 \text{ kJ})$$

For the air as a closed system we may write the first law as

$$Q_a + W_a = \Delta U_a$$

But since T_a = const., $\Delta U_a = 0$ and

$$Q_a = -W_a = -17.8 \text{ Btu} \qquad (-18.89 \text{ kJ})$$

The term ''non-heat-conducting piston'' means that the piston is a heat-insulating barrier between the hot steam and the cooler air, and the steam, by itself, undergoes an adiabatic process. Thus, the heat transfer from the air is the total heat loss from the control volume. We may now write Eq. (4-15) in integral form as

$$\int \dot{m}_i h_i \, d\tau + Q = \int \left(\frac{dE}{d\tau}\right)_\sigma d\tau = m_{2s}u_{2s} - m_{1s}u_{1s} \qquad (a)$$

since there is no change in internal energy of the air in the control volume. As in Example 4-3 we find that

$$\int \dot{m}_i \, d\tau = m_{2s}$$

so that, finally, since $m_{1s} = 0$,

$$m_{2s}(h_i - u_{2s}) = -Q \qquad (b)$$

The final volume of the steam is

$$V_{2s} = 3 - V_{a2} = 3.0 - \frac{3}{5} = 2.4 \text{ ft}^3 \qquad (67.96 \text{ liters})$$

and the mass of steam is

$$m_{2s} = \frac{V_{2s}}{v_{2s}}$$

Equation (*b*) thus becomes, with h_i = 1279.1 Btu/lbm and $Q = -17.8$ Btu,

$$\frac{2.4}{v_{2s}}(1279.1 - u_{2s}) = 17.8 \qquad (c)$$

Equation (c) must be solved by iteration with the use of the superheat steam tables (Table A-9). The result is

$$T_{2s} = 536°\text{F} \qquad (280°\text{C})$$

The temperature of the steam is lower in this case than in Example 4-3 because part of the incoming energy is used to perform the compression work on the air. This compression work is subsequently lost from the control volume because of the cooling process.

EXAMPLE 4-5 Transient Heating of a Water Tank
Water at 80°F enters a tank at the rate of 10 lbm/min. The tank contains 300 lbm of water and is provided with an exit flow channel which discharges at the rate of 10 lbm/min. Water in the tank is well mixed. If the tank is insulated and an electric heater is suddenly placed in the tank generating heat at the constant rate of 20 000 Btu/h, calculate the time required for the exit temperature to reach 100°F. What will the exit temperature be when steady-state conditions are reached? Neglect the mass of the tank.

SOLUTION This is an example of a problem which is steady *flow* but *not* steady *state*.
According to the problem statement and the schematic,

$$T_1 = \text{const.} = 80°\text{F} \qquad (26.67°\text{C})$$

$$T_2 = 80°\text{F} \qquad \text{at } \tau = 0$$

$$T_\sigma = 80°\text{F} \qquad \text{at } \tau = 0 \qquad (\sigma \text{ refers to water inside tank})$$

$$\dot{m}_1 = \dot{m}_2 = 10 \text{ lbm/min} = 600 \text{ lbm/h} = \text{const.}$$

$$m_\sigma = 300 \text{ lbm} = \text{const.}$$

$$Q = 0 \qquad \text{(insulated tank)}$$

$$\frac{d'W_{ext}}{d\tau} = +20\,000 \text{ Btu/h starting at } \tau = 0$$

(electric work crossing boundary of control volume)

FIG. EXAMPLE 4-5

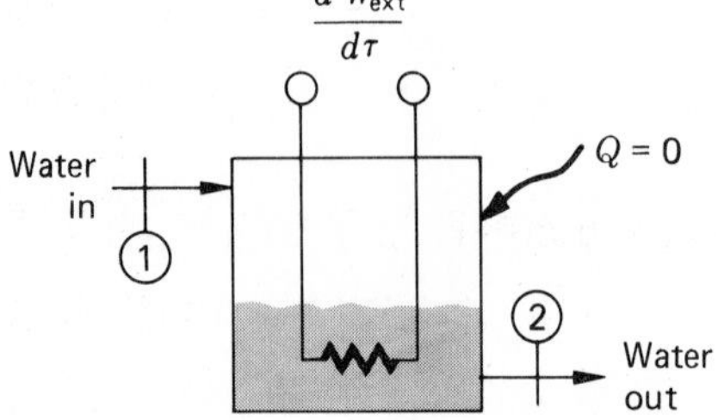

The energy inside the tank is

$$E_\sigma = m_\sigma e_\sigma = m_\sigma u_\sigma \qquad \text{(negligible KE)} \tag{a}$$

Because of the *well-mixed* condition in the tank, $T_\sigma = T_2$ at all times. For liquid water in this temperature range Δh and Δu are very closely approximated by

$$\Delta h \approx (1.0)\Delta T \qquad \text{Btu/lbm} \tag{b}$$

$$\Delta u \approx (1.0)\Delta T \qquad \text{Btu/lbm} \tag{c}$$

Checking Table A-7 at 80 and 100°F,

$$h_f \text{ at } 80°\text{F} = 48.09$$

$$h_f \text{ at } 100°\text{F} = 68.05$$

$$\Delta h = 68.05 - 48.09 = 19.96 \approx (1.0)(100 - 80) \tag{d}$$

A similar calculation applies for Δu. Because this is an unsteady-state problem, we will need to calculate $(du/d\tau)_\sigma$, and we can now write

$$\left(\frac{dE}{d\tau}\right)_\sigma = \frac{d(m_\sigma u_\sigma)}{d\tau} = m_\sigma(1.0)\frac{dT_\sigma}{d\tau} \tag{e}$$

The general energy balance is now written as

$$\dot{m}_1(1.0)T_1 + 20\,000 = \dot{m}_2(1.0)T_2 + m_\sigma(1.0)\frac{dT_\sigma}{d\tau} \qquad \text{Btu/h} \tag{f}$$

where τ is in hours because the energy input is per hour. Noting that $T_2 = T_\sigma$ and inserting the various numerical values gives the first-order differential equation

$$300\frac{dT_2}{d\tau} + 600T_2 = 68\,000 \tag{g}$$

The solution to this equation is

$$T_2 - 113.33 = C_1 e^{-2\tau} \tag{h}$$

and the constant C_1 is evaluated from the initial condition

$$T_2 = 80°\text{F} \qquad \text{at } \tau = 0 \tag{i}$$

$$C_1 = -33.33$$

so that

$$T_2 - 113.33 = -33.33e^{-2\tau} \tag{j}$$

with τ in hours. Equation (j) may then be solved for the time when T_2 attains a value of 100°F as

$$\tau = 0.458 \text{ h} \qquad \text{when } T_2 = 100°\text{F}$$

When steady-state conditions are reached, T_2 is no longer changing with time. Therefore, taking $dT_2/d\tau = 0$ in Eq. (g) yields

$$600T_2 = 68\,000 \qquad \text{(steady state)}$$
$$T_2 \text{ at steady state} = 113.33°\text{F} \qquad (k)$$

The same result can be obtained by evaluating Eq. (j) at a very long time, $\tau \rightarrow \infty$.

EXAMPLE 4-6 Closed Feedwater Heater

A closed feedwater heater for a steam power plant operates as shown in the accompanying figure. High-pressure liquid water at 10 MPa and 30°C enters a series of heating tubes. Steam at 1.5 MPa and 95 percent quality is sprayed over the tubes and allowed to condense. The saturated liquid is subsequently allowed to leave the device at station 4. We assume that the high-pressure water is heated up to the saturation temperature of the condensing steam. We wish to calculate the mass of steam required per unit mass of incoming liquid. The heater is assumed to be well insulated (adiabatic).

SOLUTION From the problem statement

$$m_1 = m_2 = 1.0$$
$$m_3 = m_4 = m_s$$
$$p_1 = p_2 = 10 \text{ MPa} \qquad (a)$$
$$p_3 = p_4 = 1.5 \text{ MPa} \qquad x_3 = 0.95$$
$$T_3 = T_4 = T_2 = T_{\text{sat}} \quad \text{at } 1.5 \text{ MPa} = 198.3°\text{C}$$

FIG. EXAMPLE 4-6

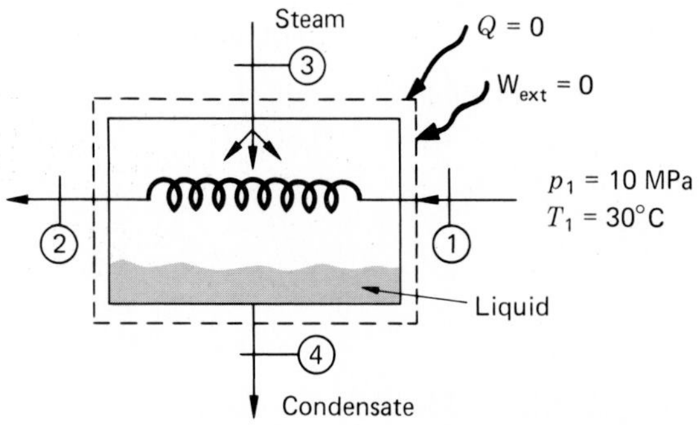

This is a steady-flow steady-state process and no external work is delivered to the control volume. Assuming negligible kinetic energies, the energy balance on the feedwater heater becomes

$$m_1 h_1 + m_3 h_3 = m_2 h_2 + m_4 h_4 \tag{b}$$

or

$$h_2 - h_1 = m_s(h_3 - h_4) \tag{c}$$

This equation simply states that the energy given up by the condensing steam is absorbed by the high-pressure water. We now determine the enthalpies from the steam table:

$$h_1 = 134.86 \text{ kJ/kg} \qquad \text{(compressed liquid, 10 MPa, 30°C)}$$

$$h_2 = 848.54 \text{ kJ/kg} \qquad \text{(compressed liquid, 10 MPa, 198.3°C)}$$

$$\begin{aligned} h_3 &= h_{f_3} + x_3 h_{fg_3} \\ &= 844.89 + (0.95)(1947.3) \\ &= 2694.8 \text{ kJ/kg} \qquad \text{(saturation pressure table at 1.5 MPa)} \end{aligned}$$

$$h_4 = h_f \text{ at 1.5 MPa} = 844.89 \text{ kJ/kg}$$

Inserting these values in Eq. (*c*) gives

$$m_s = 0.3858 \text{ kg steam/kg water}$$

EXAMPLE 4-7 Adiabatic Air Compressor

An air compressor takes in air at 1 atm and 20°C and discharges into a line having an inside diameter of 1 cm. The average air velocity in the line at a point close to the discharge is 7 m/s, and the discharge pressure is 3.5 atm. Assuming that the compression occurs quasistatically and adiabatically, calculate the work input to the compressor. Assume that the inlet air velocity is very small.

SOLUTION The schematic for the process is shown in the accompanying figure. We shall employ SI units for solution of the problem. The given proper-

FIG. EXAMPLE 4-7

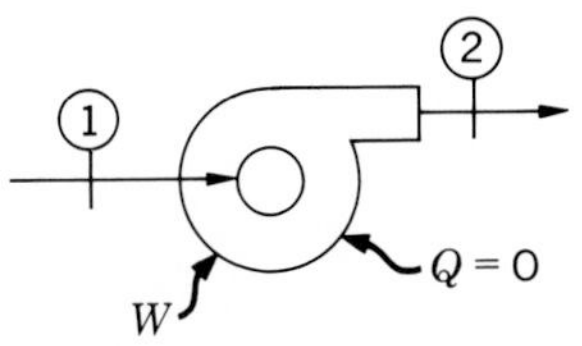

ties are

$$p_1 = 1 \text{ atm} = 1.0132 \times 10^5 \text{ Pa}$$
$$p_2 = 3.5 \text{ atm} = 3.546 \times 10^5 \text{ Pa}$$
$$T_1 = 20°\text{C} = 293 \text{ K}$$
$$V_2 = 7 \text{ m/s}$$
$$V_1 \approx 0$$
$$\gamma = 1.4 \text{ for air}$$
$$d_2 = 1.0 \text{ cm} = 10^{-2} \text{ m}$$

Because of the quasistatic-adiabatic assumption, with air as an ideal gas,

$$T_2 = T_1 \left(\frac{p_2}{p_1}\right)^{(\gamma-1)/\gamma}$$
$$= (293)(3.5)^{0.286} = 419 \text{ K}$$

At station 2 we may calculate the mass flow rate using Eq. (4-3), but first we need the air density at that point:

$$\rho_2 = \frac{1}{v_2} = \frac{p_2}{RT_2}$$

The gas constant for air is

$$R = \frac{\mathcal{R}}{M} = \frac{8315}{28.97} = 287 \text{ J/kg·K}$$

so that

$$\rho_2 = \frac{3.546 \times 10^5}{(287)(419)} = 2.949 \text{ kg/m}^3$$

The mass flow is then

$$\dot{m} = \rho_2 A_2 V_2$$
$$= \frac{(2.949)\pi(0.01)^2(7)}{4} \qquad (a)$$
$$= 1.621 \times 10^{-3} \text{ kg/s}$$

For steady-flow, steady-state, and zero heat flow the energy equation is

$$\dot{m}\left(h_1 + \frac{1}{2g_c}V_1^2\right) + \frac{d'W_{ext}}{d\tau} = \dot{m}\left(h_2 + \frac{1}{2g_c}V_2^2\right) \tag{b}$$

For air the value of c_p is

$$c_p = 0.24\ \text{Btu/lbm}\cdot{}^\circ\text{F} = 1005\ \text{J/kg}\cdot{}^\circ\text{C}$$

The KE at station 1 is zero by the problem statement; that at station 2 is

$$\text{KE}_2 = \frac{V_2^2}{2g_c} = \frac{(7\ \text{m/s})^2}{(2)(1.0\ \text{kg}\cdot\text{m/N}\cdot\text{s}^2)} = 24.5\ \text{J/kg}$$

Taking air as an ideal gas the enthalpy difference is

$$\begin{aligned} h_2 - h_1 &= c_p(T_2 - T_1) \\ &= (1005\ \text{J/kg}\cdot{}^\circ\text{C})(419 - 293) \qquad (c) \\ &= 1.266 \times 10^5\ \text{J/kg} \end{aligned}$$

From these calculations we see that the kinetic energy at station 2 is negligible compared to the enthalpy difference. Now, inserting the numerical values into Eq. (*b*) we obtain

$$\begin{aligned} \frac{d'W_{ext}}{d\tau} &= (1.621 \times 10^{-3}\ \text{kg/s})(1.266 \times 10^5 + 24.50)\ \text{J/kg} \\ &= 205.1\ \text{J/s} = 205.1\ \text{W} \end{aligned}$$

EXAMPLE 4-8 Helium Turbine

Helium is expanded quasistatically and adiabatically in a turbine from 400 kPa and 260°C to 100 kPa. The turbine is insulated and the inlet velocity is small. The exit velocity is 200 m/s. Calculate the work output of the turbine per unit mass of helium flow.

SOLUTION The schematic diagram indicates the nomenclature for the problem. For a steady-flow adiabatic process the energy equation becomes, assum-

FIG. EXAMPLE 4-8

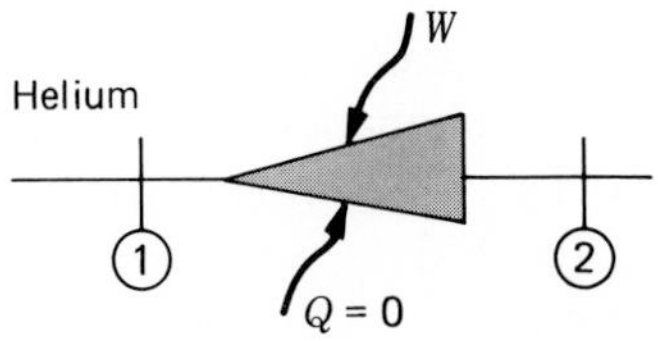

ing unit mass flow,

$$h_1 + \frac{1}{2g_c} V_1^2 + W = h_2 + \frac{1}{2g_c} V_2^2 \qquad (a)$$

Helium may be considered as an ideal gas, so that

$$h_2 - h_1 = c_p(T_2 - T_1) \qquad (b)$$

The final temperature may be found from Eq. (2-39) with $n = \gamma$ for the quasi-static-adiabatic process:

$$\frac{T_2}{T_1} = \left(\frac{p_2}{p_1}\right)^{(\gamma-1)/\gamma} \qquad (c)$$

with

$$T_1 = 260 + 273 = 533 \text{ K}$$

From Table 2-2, for helium,

$$c_p = 5.234 \text{ kJ/kg·°C} \qquad \gamma = 1.66$$

$$T_2 = (260 + 273)\left(\frac{100}{400}\right)^{(1.66-1)/1.66}$$

$$= 307 \text{ K}$$

Inserting the numerical values in Eq. (a), with $V_1 = 0$, gives

$$W = (5234)(307 - 533) + \frac{1}{(2)(1)}(200)^2$$

$$= -1.163 \text{ MJ/kg}$$

In writing Eq. (a) we have taken the work term as a positive quantity entering the control volume because it appears on the left side of the equation along with the enthalpy and kinetic energy entering the turbine. Thus the negative sign in the answer indicates that there is a net work *output* from the turbine.

EXAMPLE 4-9 Adiabatic Nozzle Expansion
Air is expanded quasistatically and adiabatically in a nozzle from 200 psia and 300°F to a pressure of 100 psia. The inlet velocity to the nozzle is very small, and the process occurs under steady-flow, steady-state conditions. Calculate the exit velocity from the nozzle.

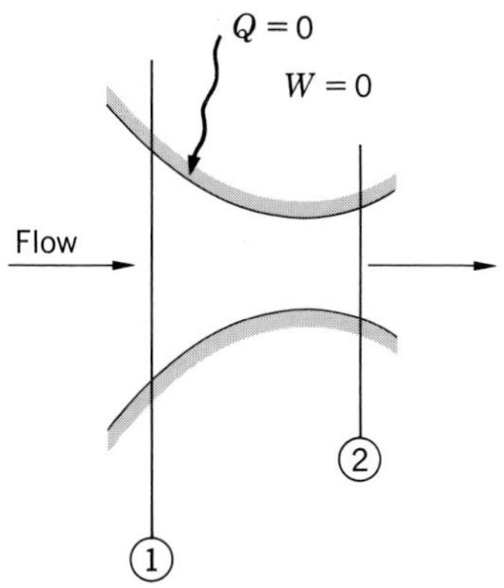

FIG. EXAMPLE 4-9

SOLUTION We again make use of the steady-flow energy equation. For no heat transfer and zero external work we have, for unit mass flow,

$$h_1 + \frac{1}{2g_c} V_1^2 = h_2 + \frac{1}{2g_c} V_2^2 \tag{a}$$

where the subscripts 1 and 2 refer to entrance and exit conditions, as shown in the accompanying sketch. Since $V_1 \approx 0$ and enthalpy for an ideal gas is a function of temperature alone, Eq. (*a*) may be solved for the exit velocity as

$$V_2 = [2g_c(h_1 - h_2)]^{1/2} = [2g_c c_p(T_1 - T_2)]^{1/2} \tag{b}$$

The final temperature is found by applying Eq. (2-39) with $n = \gamma$ to obtain

$$\frac{T_2}{T_1} = \left(\frac{p_2}{p_1}\right)^{(\gamma-1)/\gamma} \tag{c}$$

for the quasistatic adiabatic process. The final temperature is therefore

$$T_2 = (300 + 460)\left(\frac{100}{200}\right)^{(1.4-1)/1.4}$$

$$= 625°\text{R} = 165°\text{F} \qquad (73.9°\text{C})$$

Then the final velocity is calculated from Eq. (*b*) as

$$V_2 = [(2)(32.2)(778)(0.24)(300 - 165)]^{1/2}$$

$$= 1272 \text{ ft/s} \qquad (387.7 \text{ m/s})$$

EXAMPLE 4-10 Freon Evaporator

The "cooling coil" for an air-conditioning system is really a Freon evaporator which operates as shown in the accompanying sketch. Saturated liquid Freon

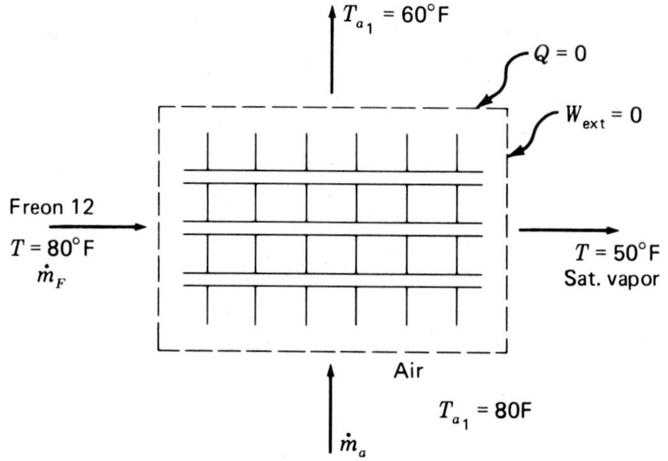

FIG. EXAMPLE 4-10

at 80°F enters a series of tubes and expands to a lower pressure such that a lower temperature of 50°F is attained. Fins are attached to the tubes and air at 80°F is blown across them. As a result, energy is given up by the air, causing the Freon to evaporate. We assume that it leaves the tubes as a saturated vapor. If the exit temperature of the air is 60°F and 60 000 Btu/h are removed from the air, calculate the mass flow rate of Freon required and the mass flow of air which is cooled. The outside of the cooling coil is perfectly insulated.

SOLUTION The process is assumed to be steady-flow, steady-state, with $Q = 0$ and $W_{ext} = 0$. We have from the problem statement,

$$T_{a_1} = 80°\text{F}$$

$$T_{a_2} = 60°\text{F}$$

$$T_{F_1} = 80°\text{F (saturated liquid)}$$

$$T_{F_2} = 50°\text{F (saturated vapor)}$$

The control volume is indicated by the dashed line and the energy balance is

$$\dot{m}_F h_{F_1} + \dot{m}_a h_{a_1} = \dot{m}_F h_{F_2} + \dot{m}_a h_{a_2} \tag{a}$$

or

$$\dot{m}_F(h_{F_2} - h_{f_1}) = \dot{m}_a(h_{a_1} - h_{a_2}) \tag{b}$$

This equation could also be stated as

Energy gained by Freon = energy lost by air = 60 000 Btu/h

or

$$\dot{m}_F(h_{F_2} - h_{F_1}) = 60\ 000 \text{ Btu/h}$$

From the Freon saturation table

$$h_{F_2} = h_g \text{ at } 50°\text{F} = 82.433 \text{ Btu/lbm}$$

$$h_{F_1} = h_f \text{ at } 80°\text{F} = 26.365 \text{ Btu/lbm}$$

so that

$$\dot{m}_F = \frac{60\ 000 \text{ Btu/h}}{82.433 - 26.365} = 1070 \text{ lbm/h} \qquad (0.135 \text{ kg/s})$$

Similarly,

$$60\ 000 \text{ Btu/h} = \dot{m}_a c_{p_a}(T_{a_1} - T_{a_2})$$

$$\dot{m}_a = \frac{60\ 000}{(0.24)(80 - 60)} = 12\ 500 \text{ lbm/h} \qquad (1.58 \text{ kg/s})$$

EXAMPLE 4-11 Steam Discharge from a Tank
A rigid tank having a volume of 0.2 m^3 contains water at 120°C in the saturation state, with 20 percent of the volume occupied by liquid. The tank is perfectly insulated. A valve on the top of the tank is opened, and vapor is allowed to escape until the pressure drops to 1 atm and the temperature to 100°C, at which time the valve is closed. Calculate the mass which leaves the tank.

FIG. EXAMPLE 4-11

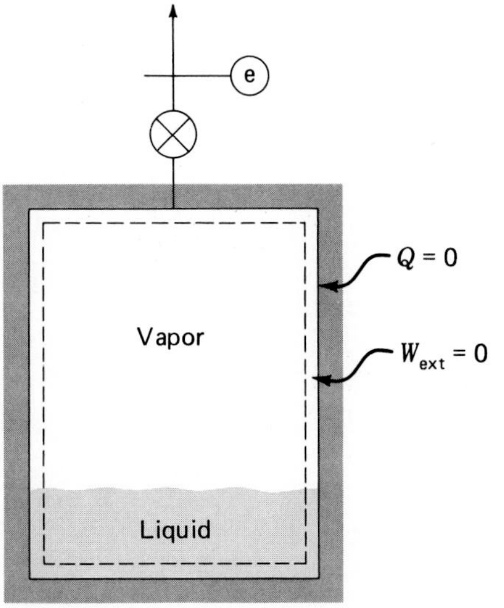

SOLUTION The control volume is shown as the dashed line in the accompanying sketch. We have

$T_1 = 120°C \qquad V = 0.2 m^3$

$T_2 = 100°C \qquad V_{f_1} = (0.2)(0.2) = 0.04\ m^3$

$V_{g_1} = (0.8)(0.2) = 0.16\ m^3$

$Q = 0$ because the tank is insulated

$W_{ext} = 0$ because there is no external shaft

Only vapor is allowed to escape, so the assumption is that a *saturated* vapor escapes but that the temperature varies. The first law is written with no inlet stream as

$$0 = \dot{m}_e h_e + \left(\frac{dU}{d\tau}\right)_\sigma \qquad (a)$$

At $T = 120°C$,

$$h_e = h_g = 2706.3\ kJ/kg$$

At $T = 100°C$,

$$h_e = h_g = 2676.1\ kJ/kg$$

Over this narrow range we may assume to a good approximation that the steam leaves at an average enthalpy of

$$\bar{h}_e = \frac{2706.3 + 2676.1}{2} = 2691.2\ kJ/kg$$

The mass which leaves the tank is clearly

$$m = m_1 - m_2 \qquad (b)$$

Inserting $\bar{h}_e$ in Eq. (*a*) and integrating gives

$$\int \dot{m}_e \bar{h}_e\, d\tau + \int dU = 0$$

or

$$\bar{h}_e(m_1 - m_2) + m_2 u_2 - m_1 u_1 = 0 \qquad (c)$$

Now let us examine these relations and see what is known or easily determined. The initial mass is

$$m_1 = m_{f_1} + m_{g_1} = \frac{V_{f_1}}{v_{f_1}} + \frac{V_{g_1}}{v_{g_1}} \tag{d}$$

The initial quality is

$$x_1 = \frac{m_{g_1}}{m_1} \tag{e}$$

and the initial internal energy is

$$u_1 = u_{f_1} + x_1 u_{fg_1} \tag{f}$$

At the final state

$$m_2 = \frac{V}{v_2} = \frac{V}{v_{f_2} + x_2 v_{fg_2}} \tag{g}$$

and

$$u_2 = u_{f_2} + x_2 u_{fg_2} \tag{h}$$

If all relations are entered back in Eq. (*c*) with properties from the steam tables we will find only one unknown, x_2, the final quality. First let us obtain the steam properties:
At 120°C,

$$v_{f_1} = 0.001\ 060\ 3\ \text{m}^3/\text{kg}$$

$$v_{g_1} = 0.8919\ \text{m}^3/\text{kg}$$

$$u_{f_1} = 503.50\ \text{kJ/kg}$$

$$u_{g_1} = 2529.3\ \text{kJ/kg}$$

At 100°C,

$$v_{f_2} = 0.001\ 043\ 5\ \text{m}^3/\text{kg}$$

$$v_{g_2} = 1.673\ \text{m}^3/\text{kg}$$

$$u_{f_2} = 418.94\ \text{kJ/kg}$$

$$u_{g_2} = 2506.5\ \text{kJ/kg}$$

Inserting in Eq. (*d*),

$$m_1 = \frac{0.04}{0.001\ 060\ 3} + \frac{0.16}{0.8919}$$

$$= 37.725 + 0.179 = 37.905 \text{ kg}$$

and

$$x_1 = \frac{0.179}{37.905} = 0.004\ 72$$

Then, from Eq. (*f*)

$$u_1 = 503.50 + (0.004\ 72)(2529.3 - 503.50)$$

$$= 513.07 \text{ kJ/kg}$$

Inserting all known quantities in Eq. (*c*), we have

$$m_1(\bar{h}_e - u_1) + m_2(u_2 - \bar{h}_e) = 0$$

or

$$37.905(2691.2 - 503.02) + \left[\frac{0.2}{0.001\ 043\ 5 + x_2(1.673 - 0.001\ 043\ 5)}\right]$$
$$\times [418.94 + x_2(2506.5 - 418.94) - 2691.2] = 0$$

which may be solved for x_2 to give

$$x_2 = 0.002\ 66$$

and the final mass is

$$m_2 = \frac{0.2}{0.001\ 043\ 5 + (0.002\ 66)(1.672)}$$

$$= 36.425 \text{ kg}$$

The mass which has escaped is thus

$$m = m_1 - m_2 = 37.905 - 36.425 = 1.48 \text{ kg}$$

EXAMPLE 4-12 Ammonia Heater

Hot combustion gases at 400°F are used to produce superheated ammonia at 260 psia and 140°F from liquid ammonia at the same pressure and 80°F. The heat-exchange device is to handle 50 lbm/min of the ammonia, and the exit

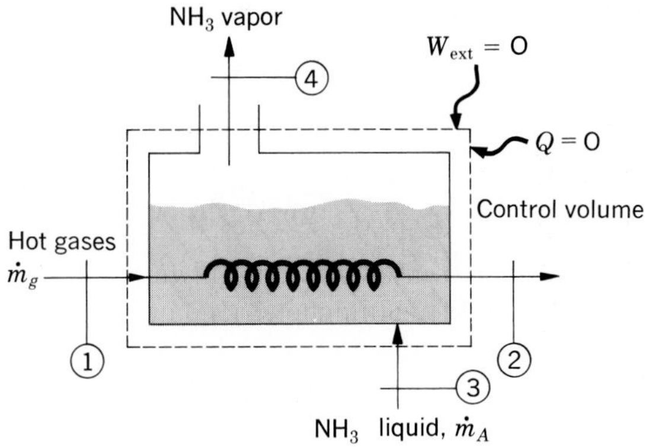

FIG. EXAMPLE 4-12

temperature of the combustion gases is to be 120°F. The velocity of the ammonia vapor leaving the device is not to exceed 20 ft/s, and the specific heat of the gases may be taken as 0.26 Btu/lbm·°F. Calculate the mass flow of combustion gases required and the diameter of the exit pipe for the ammonia vapor.

SOLUTION The schematic for this heat-exchange device is shown in the accompanying figure. If we assume that the overall heat-transfer device is insulated, then $Q = 0$. No mention is made of work, so $W_{ext} = 0$. Neglecting kinetic energies for the low velocities involved, we then write the energy balance as

Energy transported into heat exchanger = energy transported out of heat exchanger

or

$$\dot{m}_g h_1 + \dot{m}_A h_3 = \dot{m}_g h_2 + \dot{m}_A h_4 \qquad (a)$$

The ammonia properties are obtained as

$$h_3 = 132.0 \text{ Btu/lbm} \qquad \text{(Table A-12, sat. liq. at 80°F)}$$

$$h_4 = 655.6 \text{ Btu/lbm} \qquad \text{(Table A-14, 260 psia, 140°F)}$$

If we assume ideal-gas behavior for the combustion gases,

$$h_1 - h_2 = c_p(T_1 - T_2)$$

$$= (0.26 \text{ Btu/lbm·°F})(400 - 120)$$

$$= 72.8 \text{ Btu/lbm} \qquad (169.3 \text{ kJ/kg})$$

$\dot{m}_A$ is given as 50 lbm/min so we may insert the numerical values in Eq. (*a*) to obtain

$$\dot{m}_g(h_1 - h_2) = \dot{m}_A(h_4 - h_3) \tag{b}$$

$$\dot{m}_g = \frac{(50)(655.6 - 132.0)}{72.8} = 359.6 \text{ lbm/min} \qquad (163.1 \text{ kg/min})$$

The specific volume of the superheated ammonia at exit from the exchanger is

$$v_4 = 1.257 \text{ ft}^3/\text{lbm} \qquad (\text{Table A-14, 260 psia, 140°F})$$

We may now use Eq. (4-2) to calculate the flow area required for this vapor. We have $V_4 = 20$ ft/s so that

$$\dot{m}_A = \frac{A_4 V_4}{v_4} \tag{c}$$

and

$$A_4 = \frac{(50 \text{ lbm/min})(1.257 \text{ ft}^3/\text{lbm})}{(20 \text{ ft/s})(60 \text{ s/min})}$$

$$= 0.052\,38 \text{ ft}^2 \qquad (48.66 \text{ cm}^2)$$

Assuming a circular pipe at exit,

$$A_4 = \frac{\pi d_4^2}{4}$$

and

$$d_4 = 0.258 \text{ ft} = 3.10 \text{ in} \qquad (7.87 \text{ cm})$$

EXAMPLE 4-13 Solar Water Heater

A simple solar water heater is designed to heat water from 30 to 90°C by collecting the sun's energy in the glass-covered device shown. The short wavelength radiation from the sun is transmitted through the glass and trapped inside. The bottom portion of the device is insulated but some energy may be radiated back out by the glass top, which for these temperatures would be at about 200 W/m². On a clear day the irradiation from the sun is about 1000 W/m². Calculate the surface area required to heat the water, and express as m²·s/liter. How many square feet would be required to produce a hot-water flow of 10 gal/min?

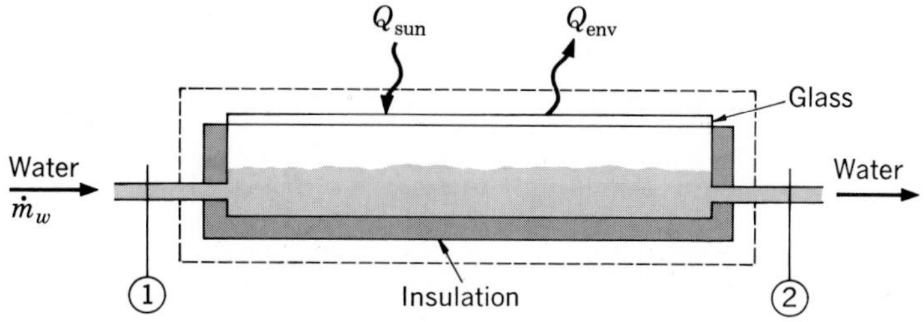

FIG. EXAMPLE 4-13

SOLUTION The control volume for this problem is shown as the dashed line in the figure and the energy balance becomes

$$\dot{m}_2 h_1 + Q_{\text{sun}} = \dot{m}_w h_2 + Q_{\text{env}} \tag{a}$$

where $Q_{\text{sun}} = 1000\ \text{W/m}^2$ and $Q_{\text{env}} = 200\ \text{W/m}^2$. The bottom of the device is insulated so there are no other energy transfers. We could obtain the water enthalpies from the saturated-liquid tables at 30 and 90°C, but it is a bit more convenient to observe that in this range the value of c_p for the liquid is very nearly constant at 1.0 kcal/kg·°C so that, to a good approximation,

$$\begin{aligned} h_2 - h_1 &= c_p(T_2 - T_1) \\ &= (1.0\ \text{kcal/kg·°C})(90 - 30) \\ &= 60\ \text{kcal/kg} \\ &= 2.51 \times 10^5\ \text{J/kg} \end{aligned} \tag{b}$$

At a mean temperature of 60°C = 140°F the specific volume of saturated liquid is obtained from Table A-7M as 1.0172 cm³/g or

$$v = 1.0172\ \text{liters/kg} \tag{c}$$

We may now calculate the water flow rate from Eq. (*a*): (*d*)

$$\dot{m}_w(h_2 - h_1) = Q_{\text{sun}} - Q_{\text{env}}$$

$$\dot{m}_w = \frac{(1000 - 200)\ \text{W/m}^2}{2.51 \times 10^5\ \text{J/kg}}$$

$$= 3.187 \times 10^{-3}\ \text{kg/s·m}^2$$

With the conversion to liters from Eq. (*c*), this becomes

$$\dot{m}_w = 3.235 \times 10^{-3}\ \text{liter/s·m}^2$$

The inverse of this is the quantity we seek, or,

$$\text{Area required} = 309.1 \text{ m}^2\cdot\text{s/liter} \qquad (e)$$

To produce a hot-water flow of 10 gal/min (1 gal = 231 in³) we would have

$$\dot{m}_w = 2310 \text{ in}^3/\text{min} = 37.85 \text{ liters/min} = 0.6309 \text{ liter/s}$$

Using the relation in Eq. (*e*), the area is

$$\begin{aligned} A &= (309.1 \text{ m}^2\cdot\text{s/liter})(0.6309 \text{ liter/s}) \\ &= 195 \text{ m}^2 \\ &= 2099 \text{ ft}^2 \end{aligned}$$

EXAMPLE 4-14 Free Expansion

Air at 750 kPa and 200°C is contained in a piston-cylinder arrangement, as shown in the accompanying figure. The cylinder is perfectly insulated. Stops on the piston hold it in place. The system is surrounded by atmospheric air at 20°C, and the piston is non-heat-conducting. The stops are removed and the piston allowed to suddenly expand to a new stop position such that $v_2 = 1.5v_1$. The piston-cylinder is frictionless. Calculate the final temperature of the air. Also calculate the final pressure of the air.

SOLUTION We take the air as the thermodynamic system and write the first law as

$$Q + W = \Delta U$$

Because the cylinder is insulated, $Q = 0$ and we have

$$W = \Delta U$$

FIG. EXAMPLE 4-14

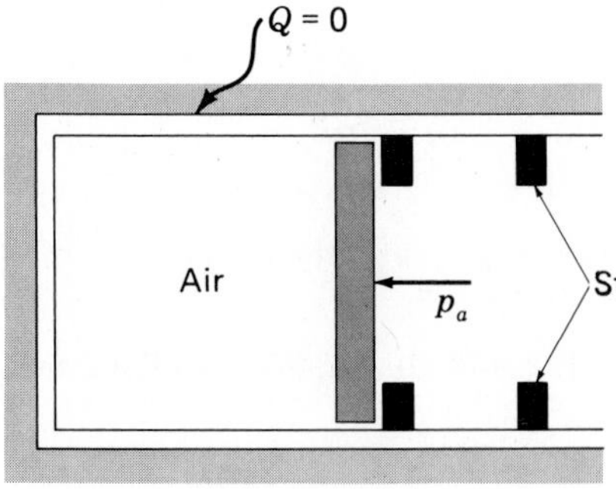

or, for unit mass,

$$W = \Delta U = c_v(T_2 - T_1) \qquad (a)$$

If we can calculate the work, we can then calculate T_2. We need to determine the process and observe that:

The process is *not* constant volume.
The process is *not* constant pressure.
The process is *not* constant temperature (isothermal).
The process *is adiabatic* ($Q = 0$) but *not necessarily* quasistatic, because the system may not remain in equilibrium for a sudden expansion.

We can observe that the only work at the boundary is that to displace the environment, which does remain at 1 atm. For unit mass the work done *on* the air in the cylinder is

$$W = -\int p(1\text{ atm})\,dv = p(1\text{ atm})(v_1 - v_2) \qquad (b)$$

The initial specific volume is calculated from the ideal-gas equation:

$$v_1 = \frac{RT_1}{p_1} = \frac{(287\text{ J/kg·K})(200 + 273)}{1.013 \times 10^5}$$
$$= 1.34\text{ m}^3/\text{kg}$$

and the final specific volume is

$$v_2 = 1.5v_1 = (1.5)(1.34) = 2.01\text{ m}^3/\text{kg}$$

From Eq. (*b*) the work per unit mass is

$$W = (1.0132 \times 10^5)(1.34 - 2.01) = -67880\text{ J/kg}$$

Inserting this value in Eq. (*a*) with $c_v = 718$ J/kg °C gives

$$-67\,880 = (718)(T_2 - 200)$$
$$T_2 = 105.4°\text{C} = 378\text{ K}$$

It is interesting to compare this with the value we would obtain if the process were quasistatic and adiabatic. From Eq. (2-40), with $n = \gamma = 1.4$, we have

$$\frac{T_2}{T_1} = \left(\frac{v_1}{v_2}\right)^{\gamma-1} = \left(\frac{1}{1.5}\right)^{0.4} = 0.874$$

and

$$T_2 = (0.874)(473) = 413 \text{ K}$$

which is substantially different from the 378 K obtained above.
The final pressure is calculated from

$$R = \frac{p_1 v_1}{T_1} = \frac{p_2 v_2}{T_2}$$

$$p_2 = (750)\left(\frac{1}{1.5}\right)\left(\frac{378}{473}\right) = 400 \text{ kPa} \qquad (c)$$

Had the process been quasistatic we could use

$$p_1 v_1^\gamma = p_2 v_2^\gamma$$

and

$$p_2 = (750)\left(\frac{1}{1.5}\right)^{1.4} = 425 \text{ kPa}$$

COMMENT In this problem we have made use of the fundamental definition of work as a force exerted through a displacement *at the boundary of the system.*

4-7 THROTTLING AND THE JOULE-THOMSON COEFFICIENT

Consider the flow situation shown in Fig. 4-5. A *real* gas flows through the constant-area duct. Between sections 1 and 2 a porous restriction is placed in the flow which causes an appropriate pressure drop. The process is called a *throttling process*, and if changes in kinetic and potential energies are negligi-

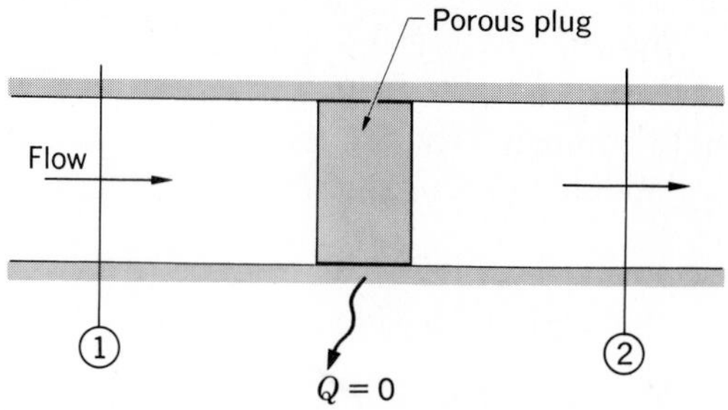

FIG. 4-5 Throttling process through a porous plug.

ble, the steady-flow energy equation will reduce to

$$h_1 = h_2$$

for an adiabatic process. Although the enthalpies at sections 1 and 2 are the same for the given conditions, the enthalpy of the flow may vary considerably as the fluid moves through the porous plug. This variation will depend on the size of the openings in the plug and the frictional resistance imparted to the fluid. The throttling process need not occur in a porous plug. It can also be observed as an adiabatic expansion of a fluid in a valve when the inlet and exit kinetic energies are negligible. If the kinetic energies are large enough to be considered in the analysis, then the complete energy equation must be used and the term *throttling* is no longer applied.

If experimental data for real gases are plotted on a T-p diagram, a family of curves like those in Fig. 4-6 will result. The locus of the maximum points in the constant-enthalpy curves is called the *inversion curve,* and the maximum point in each curve is called an *inversion point*. The slope of an isenthalpic (constant enthalpy) curve is called the *Joule-Thomson coefficient* μ_J:

$$\mu_J = \left(\frac{\partial T}{\partial p}\right)_h \tag{4-17}$$

As we shall show in Chap. 6 and as has been assumed in Chap. 2, the enthalpy for an ideal gas is a function of temperature alone, so that, for an ideal gas, a line of constant enthalpy would also be a line of constant temperature. Thus

$$\mu_J = 0 \qquad \text{for an ideal gas} \tag{4-18}$$

The importance of the Joule-Thomson coefficient is that it may be measured experimentally and subsequently used to calculate other thermodynamic

FIG. 4-6 Inversion points for porous-plug experiment of Fig. 4-5.

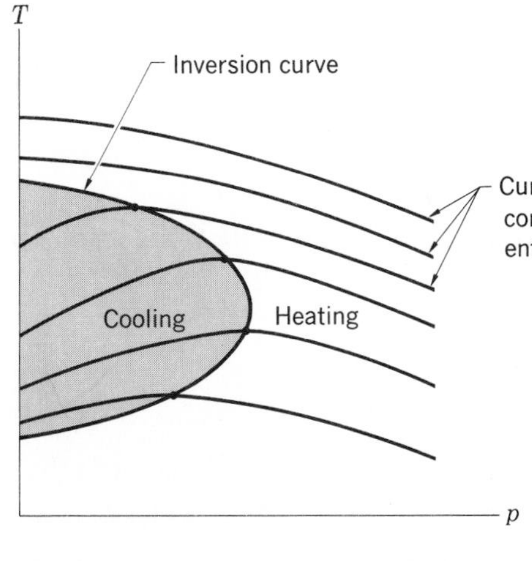

properties of interest. We shall explore the matter of calculation of properties more thoroughly in Chap. 6, indicating the utility of a Joule-Thomson type of experiment at this point. If we assume that the enthalpy may be expressed as a function of the two independent variables p and T, then from the partial differential calculus we have

$$dh = \left(\frac{\partial h}{\partial p}\right)_T dp + \left(\frac{\partial h}{\partial T}\right)_p dT \tag{4-19}$$

Introducing the definition of c_p, we have

$$dh = \left(\frac{\partial h}{\partial p}\right)_T dp + c_p\, dT \tag{4-20}$$

For the adiabatic throttling process $dh = 0$, and Eq. (4-20) may be rearranged to give

$$0 = \left(\frac{\partial h}{\partial p}\right)_T \left(\frac{\partial p}{\partial T}\right)_h + c_p$$

or

$$c_p = -\frac{1}{\mu_J}\left(\frac{\partial h}{\partial p}\right)_T \tag{4-21}$$

The constant-temperature coefficient μ_T is defined as

$$\mu_T = \left(\frac{\partial h}{\partial p}\right)_T \tag{4-22}$$

It is readily measured by performing the porous-plug experiment at constant temperature for various pressures. The heat supplied or removed from the constant-area section (e.g., by an electric heater) represents the change in enthalpy of the gas since

$$d'Q = dh$$

for the process (no change in KE, $W = 0$). Once μ_J and μ_T are determined, the constant-pressure specific heats may be evaluated with

$$c_p = -\frac{\mu_T}{\mu_J} \tag{4-23}$$

The knowledge of specific heats is a crucial matter in the calculation of thermodynamic properties, as we shall see in Chap. 6.

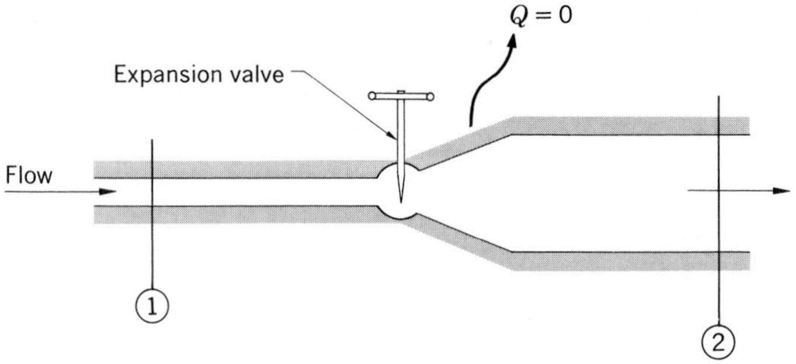

FIG. EXAMPLE 4-15

EXAMPLE 4-15 Throttling of Freon 12
Freon 12 is throttled in an expansion valve from saturated-liquid conditions at 150°F to a temperature of 40°F. Calculate the quality of the Freon vapor after the throttling process.

SOLUTION The schematic for this process is shown in the accompanying figure. For an adiabatic throttling process, we have

$$h_1 = h_2$$

From the Freon table (Table A-15),

$$h_1 = 43.850 \text{ Btu/lbm} \qquad \text{saturated liquid at } 150°\text{F}$$

This value is then used to calculate the quality at the exit (40°F):

$$h_2 = h_1 = h_{f_2} + x_2 h_{fg_2}$$

or

$$43.850 = 17.273 + x_2(64.163)$$

and

$$x_2 = 0.414$$

This process is widely used in air-conditioning and refrigeration systems, as we shall see in Chap. 9.

REVIEW QUESTIONS

1 Distinguish between open and closed systems.
2 Distinguish between the terms *steady flow* and *steady state*.
3 What is meant by the term *flow work*?

4 Why is the enthalpy function valuable in the analysis of open systems?
5 What is an ideal gas?
6 What equation may be used to describe a quasistatic adiabatic process for an ideal gas?
7 What is a throttling process? Under what conditions are the enthalpies at entrance and exit equal?
8 Why is it reasonable to assume that an expansion process in a nozzle occurs in an adiabatic manner?
9 In Example 4-3, the steam enters the tank from a source at 500°F. The tank is insulated so no heat is added. In a physical sense, how do you account for the rise in temperature to 767°F without heat addition or external work?

PROBLEMS (ENGLISH UNITS)

4-1 10 lbm of nitrogen are compressed in such a manner that the internal energy increases by 150 Btu while 20 Btu of heat are added. Calculate the work that is supplied to the gas. Assume that the compression occurs in a closed system.

4-2 Rework Prob. 4-1, assuming that the compression occurs in a steady-flow open system.

4-3 In a certain steady-flow compressor oxygen is to be compressed from 16 psia, 120°F, to a final pressure of 150 psia in a quasistatic adiabatic process. Calculate the power input to the compressor for a flow rate of 1.3 lbm/s.

4-4 A rigid tank contains air at 100°F and 75 psia. A valve on the side of the tank is opened and the air is discharged slowly until the pressure drops to 25 psia. If the temperature of the air in the tank remains constant during this slow process, calculate the heat transfer between the tank and the surroundings. The volume of the tank is 1.5 ft^3.

4-5 A 5.2-ft^3 tank contains air at 5 psia and 70°F and is exposed to the atmosphere at 14.7 psia, 70°F. A valve on the side of the tank is opened and air from the surroundings is allowed to enter the tank slowly. The process is slow enough that the tank temperature stays at 70°F. Calculate the heat exchange between the tank and the surroundings when the tank pressure reaches 14.7 psia.

4-6 A 1-kW-resistance heater is placed in a 100-ft^3 container filled with air at 14.7 psia and 70°F. The heater is allowed to operate for 10 min and the container is insulated. Calculate the temperature and pressure of the air at the end of this time period.

4-7 Air flows through a turbine in a quasistatic adiabatic process from 50 psia and 1000°F to 15 psia. The inlet velocity is 100 ft/s and the outlet velocity is 1000 ft/s. Calculate the work output of the turbine per pound of air.

4-8 Calculate the work required to pump 10 gal/min of water from 25 psia and 100°F to 1200 psia in an adiabatic process.

4-9 1 lbm of steam is heated in a constant-pressure process from saturated vapor at 100 psia to 500°F. Calculate the heat added in (*a*) a nonflow process in which the steam is contained behind a piston in a cylinder and (*b*) a steady-flow process for which there is no external work.

4-10 Rework Prob. 4-9 for a constant-volume process. Calculate the final pressure of the steam.

4-11 In a small power plant in west Texas there is a steam turbine which is adiabatic and produces a power output of 20 000 kW. Steam enters the turbine at 500 psia and 700°F and discharges at 1 atm and 97 percent quality. Calculate the steam flow rate required.

4-12 A large turbocompressor takes in air at 1 atm and 100°F and compresses it quasistatically and adiabatically to 75 psia in a steady-flow process. The power input to the compressor is 5000 hp. Calculate the volume flow rate of air at the inlet expressed in ft^3/s.

4-13 Steam at 300 psia is throttled to 14.7 psia and 260°F. Calculate the quality of the steam at the 300-psia condition.

4-14 Freon 12 vapor enters a steady-flow compressor as saturated vapor at 40°F. The outlet conditions are 250 psia and 200°F, and the process is assumed to be adiabatic. Calculate the horsepower required if the Freon flow rate is 10 lbm/min. What diameter inlet tubing to the compressor is required if the inlet velocity shall not exceed 15 ft/s?

4-15 A heat exchanger is designed as shown in the accompanying sketch. Steam at 20 psia and 300°F is used to heat air from 40 to 85°F. Calculate the flow rate of steam required to heat 100 lbm/min of air with the exit conditions of the steam at 20 psia and 75 percent quality.

4-16 A 15-ft^3 rigid tank is perfectly evacuated and perfectly insulated. A valve on the side of the tank is opened and local atmospheric air at 14.7 psia and 70°F is allowed to enter the tank. When the pressure in the tank reaches 14.7 psia, the valve is closed. Calculate the amount of air which enters the tank.

4-17 High-pressure steam is delivered to an adiabatic turbine at 800 psia and 1000°F with a velocity of 500 ft/s. Discharge from the turbine is at 40 psia and 500°F with a velocity of 1200 ft/s. Calculate the flow rate required to produce a power output of 20 MW.

4-18 Saturated vapor (steam) is contained in a tank at 80 psia. Calculate the

FIG. P4-15

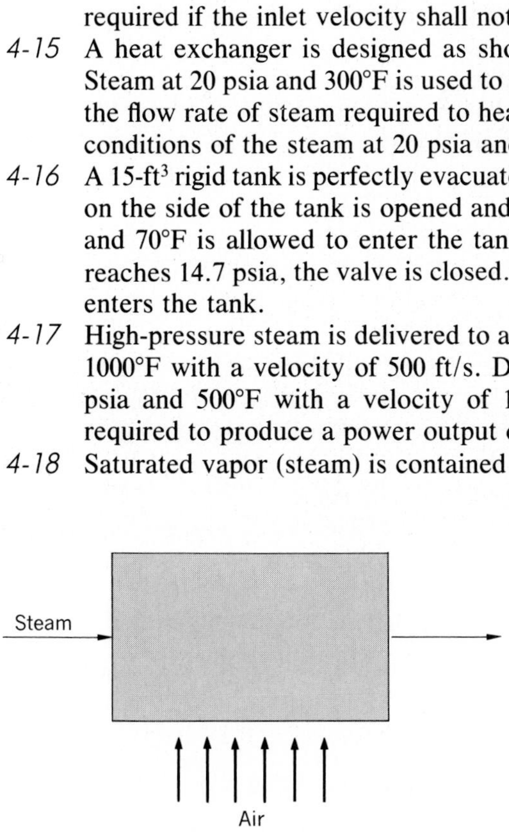

heat which must be removed to reduce the quality to 50 percent. Also calculate the final pressure for this process.

4-19 A 10-ft^3 tank contains air at 14.7 psia and 70°F. A high-pressure line is connected to the tank until the pressure reaches 300 psia. Assuming that the tank and the connecting valve are adiabatic, calculate the final temperature in the tank if the conditions in the high-pressure line remain constant at 300 psia and 100°F. How much cooling has to be supplied to the tank to maintain the air temperature constant at 70°F?

4-20 In an open feedwater heater for a steam power plant, saturated vapor at 100 psia is mixed with subcooled liquid at 80°F and 100 psia. Just enough steam is supplied to ensure that the mixed stream leaving the heater will be saturated liquid at 100 psia. Assuming that the heater is a box which is perfectly insulated from its surroundings and that the process is steady flow, calculate the required flow rate of vapor per pound of entering subcooled liquid.

4-21 Steam at the rate of 5 lbm/min at 100 psia and 500°F is mixed with 10 lbm/min of steam at 100 psia and 50 percent quality in an adiabatic steady-flow device. Calculate the temperature or quality of the outlet mixture.

4-22 Steam enters a small insulated tube as saturated vapor at 60 psia. Because of the frictional resistance in the tube, the pressure of the steam will decrease to 45 psia over a sufficiently long length of the tube. Calculate the temperature of the steam when the 45-psia condition is reached.

4-23 Steam enters a turbine at 1000 psia and 1000°F and leaves at 4 psia and 97 percent quality. The turbine is adiabatic and the power output is 25 000 hp. Calculate the diameter of the inlet line if the inlet velocity is 80 ft/s.

4-24 A steam turbine is adiabatic and receives steam at 700 psia and 900°F with a velocity of 100 ft/s. Discharge is at 14.7 psia and 300°F with a velocity of 500 ft/s. What flow rate is necessary to produce a power output of 15 MW?

4-25 A 15-ft^3 insulated tank contains helium at 500 psia and 400°F. A valve on the side of the tank is opened and helium is allowed to escape until the pressure drops to 100 psia at which time the valve is closed. Calculate the change in total energy inside the tank. State any assumption(s) you may think are necessary.

4-26 The horsepower of a jet engine may be roughly calculated in terms of the energy imparted to the air passing through it. Assuming that the exhaust gases have the same properties as air and that ideal-gas behavior is experienced, calculate the flow rate of air necessary to produce 15 000 hp under the following conditions:

$$\text{Inlet air temperature} = -50°\text{F}$$

$$\text{Inlet air velocity} = 700 \text{ ft/s}$$

Outlet air temperature = 2000°F

Outlet air velocity = 1000 ft/s

Also calculate the inlet flow area required for an inlet pressure of 3 psia.

4-27 A 0.5-hp fan is mounted inside a 100-ft^3 box filled with air at 14.7 psia and 70°F. The box is perfectly insulated, and the fan is allowed to run for 10 min. Calculate the final temperature and pressure of the air.

4-28 Calculate the heat which must be removed from the box in Prob. 4-27 to have a final temperature of 110°F.

4-29 1 lbm of a certain gaseous fuel is burned with 50 lbm of air in a constant-pressure steady-flow process. The chemical energy of the fuel is 24 000 Btu/lbm, and both the fuel and air enter the burner at 70°F and negligible velocity. The products of combustion may be assumed to have the same properties as air. Assuming ideal-gas behavior and an insulated burner, calculate the temperature of the exit gases. For this calculation take the enthalpy of incoming air and fuel as zero at 70°F.

4-30 A heat exchanger is designed to heat liquid water in a constant-pressure steady-flow process from 100°F, 50 psia, to superheated conditions of 400°F. The heating is accomplished by cooling hot air from 800 to 600°F in a steady-flow process. Calculate the mass flow of air required for a mass flow of water of 2.2 lbm/s.

4-31 An air heater is to be designed which will heat 5 lbm/s of air at 25 psia from 40 to 100°F by condensing steam from 20 psia and 300°F to saturated liquid conditions at 20 psia. What size inlet steam line is required if the inlet steam velocity is to be 16 ft/s?

4-32 The cooling coil in a certain air-conditioning system is designed to remove 60 000 Btu/h from 1500 ft^3/min of air at 85°F and 14.7 psia. The cooling is accomplished in a heat exchanger like that indicated in Prob. 4-15, except that Freon 12 is evaporated at constant pressure from saturated-liquid conditions at 40°F to superheat at 43°F. Calculate the mass flow rate of Freon required and the exit air temperature.

4-33 Air at 3 psia and −30°F enters a diffuser-nozzle section at 900 ft/s. The air goes through a quasistatic-adiabatic process until its velocity is reduced to 100 ft/s. Calculate the final temperature and pressure.

4-34 Oxygen is compressed in a piston-cylinder arrangement from 23 psia, 100°F, to 85 psia, 200°F. Calculate the work done on the gas and the heat transfer if the process is quasistatic.

4-35 Saturated-liquid water at 500°F is throttled to atmospheric pressure. Calculate the final quality if saturated or the final temperature if superheated. Calculate the velocity of the steam in the final state for a flow rate of 1 lbm/min through a 1-in-diameter tube.

4-36 Calculate the inlet and exit areas required in Prob. 4-33 to accommodate a flow rate of 1 lbm/s.

4-37 What horsepower is required to pump 100 gal/min of liquid water at 70°F through a rise in elevation of 100 ft? Assume isothermal flow and negligible change in kinetic energy.

4-38 Show that the mass rate of flow for a nozzle like that in Example 4-9 is given by

$$\dot{m}^2 = 2g_c A_2^2 \frac{\gamma}{\gamma - 1} \frac{p_1^2}{RT_1} \left[\left(\frac{p_2}{p_1}\right)^{2/\gamma} - \left(\frac{p_2}{p_1}\right)^{(\gamma+1)/\gamma} \right]$$

where A_2 is the exit area of the nozzle and $V_1 \approx 0$.

4-39 In the cooling section of an ammonia refrigerator the ammonia enters at 0°F saturated liquid and leaves as a superheated vapor at 20°F. During the steady-flow process, 20 kW of heat (the cooling) are added to the ammonia. Calculate the flow rate of ammonia. State any necessary assumption(s).

4-40 A steady-flow turbine uses argon as the working fluid and is to produce a power output of 100 kW. Inlet conditions are 100 psia and 230°F, and the exit pressure is 20 psia. The turbine may be assumed to be adiabatic. Calculate the diameter of the exit line if the exit velocity is to be 100 ft/s. State any assumption(s) necessary to solve the problem.

4-41 Two 1-ft^3 chambers are separated by a non-heat-conducting wall as shown. The left chamber is filled with air at 100 psia and 150°F, and the right chamber is evacuated. A small valve is opened which allows the air to flow from the left to right chamber. The valve is closed when the pressures in the two chambers are equal. Calculate the final pressure and temperature for the air in each chamber. State any assumptions necessary to solve the problem.

4-42 Repeat Prob. 4-41 for the case of a separating wall which is perfectly heat-conducting; i.e., the temperature is always the same on both sides of the wall. Again, state any assumptions carefully.

4-43 Oxygen is contained at 30 atm and 50°C in a tank having a volume of 10 m^3. A large valve is opened from the tank and the gas is suddenly discharged until the pressure drops to 15 atm. What is the quantity discharged in kilograms?

FIG. P4-41

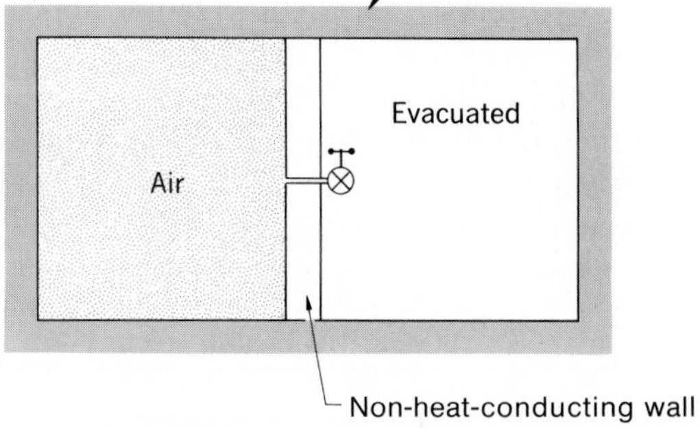

4-44 Steam enters a certain nozzle at 300 psia and 500°F with a velocity of 500 ft/s. During the expansion process the steam loses 15 Btu/lbm and the enthalpy drops to 1200 Btu/lbm. What is the exit velocity from the nozzle? What flow area at inlet would be required for a flow rate of 2 lbm/s?

4-45 A small tank having a volume of 500 cm^3 contains Freon 12 in a saturation state at 70°F. Only 20 percent of the tank is filled with liquid. A small nozzle on the top of the can is used to spray the Freon in a room at atmospheric pressure. After an extended period of spraying the temperature in the can has dropped to 50°F, and the liquid occupies only 10 percent of the volume. Assuming that the nozzle is an adiabatic throttling device, how much energy is added to the can during the discharge process?

4-46 Consider the system shown in the accompanying figure. The two chambers initially have equal volumes of 1 ft^3 and contain air and hydrogen, respectively. The chambers are separated by a frictionless piston which is non-heat-conducting. Both gases are initially at 20 psia and 100°F. Heat is added to the air side until the pressure of both gases reaches 40 psia. All outside walls of the chamber are insulated except for the surface where heat is added to the air. Calculate the final temperature of the air.

4-47 A small nozzle is constructed for use with liquid water at 100°F. The inlet pressure to the nozzle is 100 psia, and the discharge pressure is atmospheric. Assuming an adiabatic process, calculate the exit velocity. For an exit area of 0.05 in^2, calculate the flow rate in gal/min.

4-48 What horsepower pump would be required to produce the pressure differential and flow rate in Prob. 4-47?

4-49 An industrial hot water heater uses a gaseous fuel with an internal chemical energy content of about 23 000 Btu/lbm. The heater is designed so that 18 lb of air at 70°F and 1 atm are supplied for each pound of fuel, and it is designed to produce 10 gal/min of hot water at 180°F from cold water at 60°F. The hot discharge gases are at 230°F. Calculate the quantity of fuel required.

FIG. P4-46

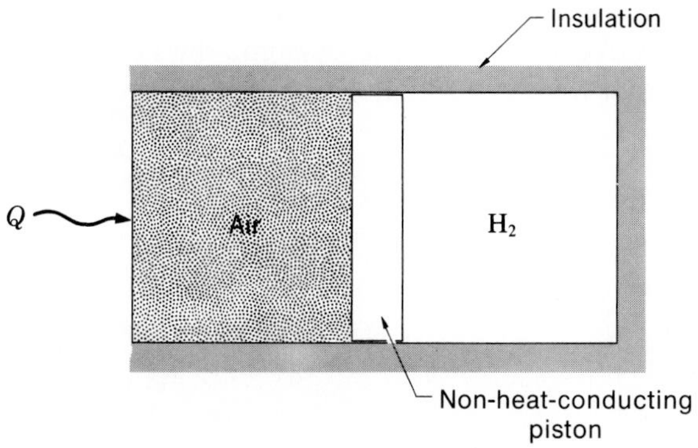

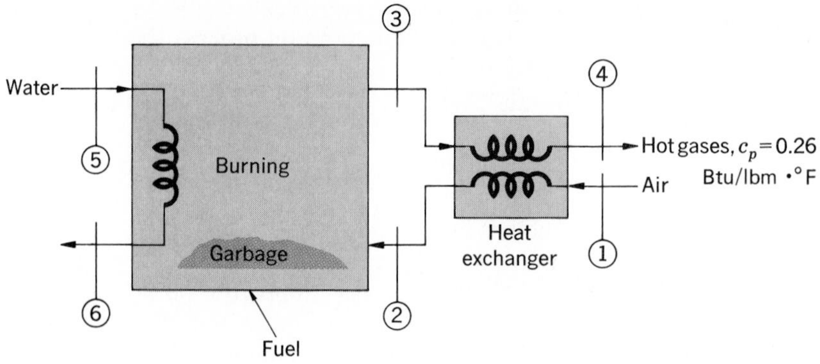

FIG. P4-50

4-50 A municipal incinerator is to burn garbage which has an average internal chemical energy of 3000 Btu/lbm. The hot gases from the burning process are then used to preheat the air as it enters the incinerator. The internal construction of the incinerator also allows it to be used as a high-pressure hot water heater for a nearby power plant, as shown in the sketch. The respective temperatures are T_1 = 100°F, T_3 = 1500°F, T_4 = 400°F, T_5 = 95°F, T_6 = 300°F. Assume that the hot gases behave as an ideal gas and that 15 lbm of air are required to burn each pound of garbage or each pound of an auxiliary fuel. The auxiliary fuel has an internal chemical energy of 20 000 Btu/lbm and must be used to get the burning temperature high enough for complete combustion. How much total energy in the form of garbage and fuel must be supplied to heat 3000 gal of water? If 1 ton of garbage is available for heating each 3000 gal of water, how much auxiliary fuel would be required?

4-51 A small auxiliary power device is designed as shown in the schematic. Nitrogen is contained at 2000 psia and 70°F in the tank which is connected through a valve to an adiabatic turbine which drives an electric generator. The power output of the device is to be 75 W, and the pressure at point 1 is maintained constant at 100 psia by a pressure regulator which operates as a throttling device. The nitrogen discharges from the turbine at atmospheric pressure, and the turbine process may be assumed adiabatic. What size tank is necessary to produce the power output for a period of 1 h? Assume the tank remains isothermal at 70°F and that power production ceases when the tank pressure reaches 100 psia.

FIG. P4-51

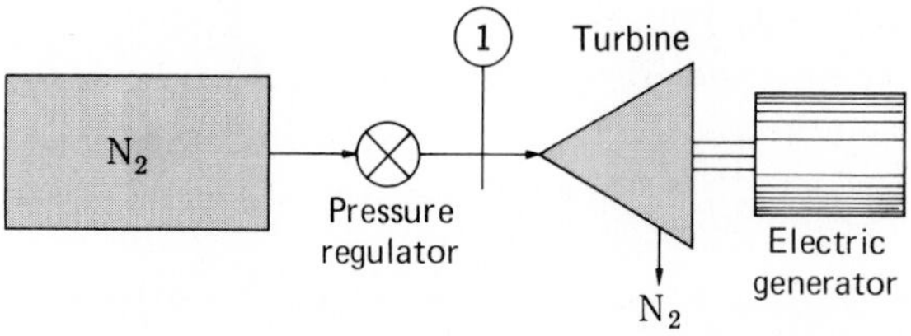

PROBLEMS (METRIC UNITS)

4-1M Air flows through a turbine in a quasistatic-adiabatic process from 350 kPa and 550°C to 100 kPa. The inlet velocity is 30 m/s and the outlet velocity is 300 m/s. Calculate the work output of the turbine per kilogram of air.

4-2M Calculate the work required to pump 40 liters/min of water from 170 kPa and 40°C to 7.5 MPa in an adiabatic process.

4-3M 1 kg of steam is heated in a constant-pressure process from saturated vapor at 700 kPa to 260°C. Calculate the heat added in (*a*) a nonflow process in which the steam is contained behind a piston in a cylinder and (*b*) a steady-flow process for which there is no external work.

4-4M Saturated vapor (steam) is contained in a tank at 600 kPa. Calculate the heat which must be removed to reduce the quality to 50 percent. Also calculate the final pressure for this process.

4-5M A large steam line carries steam at 700 kPa and 200°C. It is connected through a throttling valve to a tank having a volume of 3 m^3 which is perfectly insulated and initially evacuated. The valve is opened and steam allowed to enter the tank until the pressure in the tank reaches 700 kPa, at which time the valve is closed. Calculate the mass of steam which enters the tank. Assume that the steam line is an infinite source of steam.

4-6M A steady-flow adiabatic compressor is used to compress argon from 100 kPa and 27°C to a pressure of 1.2 MPa. The work required is 400 kW. If the inlet flow velocity is 10 m/s calculate the diameter of the inlet flow line.

4-7M Water at the rate of 5 kg/s, at 250 kPa and 40°C, is mixed with 3 kg/s of steam at 250 kPa and 80 percent quality in a steady-flow process at constant pressure. The container where the mixing occurs is insulated. Determine the state of the outlet flow stream.

4-8M Argon enters a frictionless adiabatic turbine at conditions of 1.2 MPa and 100°C and discharges at a pressure of 100 kPa. The work output of the turbine is 400 kW. If the inlet flow velocity is 15 m/s, calculate the diameter of the inlet flow channel (assuming it to be circular).

4-9M The cooling coil in a certain air-conditioning system is designed to remove 18 kW from 40 000 liters/min of air at 30°C and 1 atm. The cooling is accomplished in a heat exchanger like that indicated in Prob. 4-15M, except that Freon 12 is evaporated at constant pressure from saturated-liquid conditions at 40°F to superheat at 43°F. Calculate the mass flow rate of Freon required and the exit air temperature.

4-10M Air at 20 kPa and −35°C enters a diffuser-nozzle section at 280 m/s. The air goes through a quasistatic-adiabatic process until its velocity is reduced to 30 m/s. Calculate the final temperature and pressure.

4-11M Calculate the inlet and exit areas required in Prob. 4-10M to accommodate a flow rate of 1 kg/min.

4-12M Steam enters a small insulated tube as saturated vapor at 400 kPa.

Because of the frictional resistance in the tube, the pressure of the steam will decrease to 300 kPa over a sufficiently long length of the tube. Calculate the temperature of the steam when the 300-kPa condition is reached.

4-13M Steam at 2 MPa is throttled to 100 kPa and 120°C. Calculate the quality of the steam at the 2-MPa condition.

4-14M Freon 12 vapor enters a steady-flow compressor as saturated vapor at 40°F. The outlet conditions are 250 psia and 200°F, and the process is assumed to be adiabatic. Calculate the horsepower required if the Freon flow rate is 5 kg/min. What diameter inlet tubing to the compressor is required if the inlet velocity shall not exceed 5 m/s?

4-15M A heat exchanger is designed as shown in the accompanying sketch. Steam at 150 kPa and 160°C is used to heat air from 50 to 30°C. Calculate the flow rate of steam required to heat 50 kg/min of air with the exit conditions of the steam at 150 kPa and 75 percent quality.

4-16M A 350-W fan is mounted inside a 3000-liter box filled with air at 20 kPa and 20°C. The box is perfectly insulated, and the fan is allowed to run for 10 min. Calculate the final temperature and pressure of the air.

4-17M Calculate the heat which must be removed from the box in Prob. 4-16M to have a final temperature of 45°C.

4-18M What power is required to pump 380 liters/min of liquid water at 20°C through a rise in elevation of 30 m? Assume isothermal flow and negligible change in kinetic energy.

4-19M The power of a jet engine may be roughly calculated in terms of the energy imparted to the air passing through it. Assuming that the exhaust gases have the same properties as air and ideal-gas behavior is experienced, calculate the flow rate of air necessary to produce 10 MW under the following conditions:

$$\begin{aligned} \text{Inlet air temperature} &= -45°\text{C} \\ \text{Inlet air velocity} &= 200 \text{ m/s} \\ \text{Outlet air temperature} &= 1100°\text{C} \\ \text{Outlet air velocity} &= 300 \text{ m/s} \end{aligned}$$

FIG. P4-15M

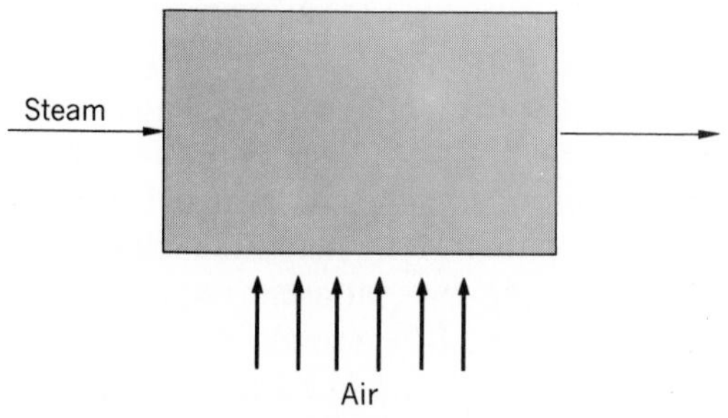

Also calculate the inlet flow area required for an inlet pressure of 20 kPa.

4-20M A high-speed flow of helium is to be heated in a duct. The gas enters the duct at 3 atm, 20°C, and a velocity of 150 m/s. While traversing the duct, the pressure drops 0.15 atm and the velocity is reduced to 130 m/s. The exit temperature of the gas is 140°C. How much heat must be added for a mass flow of 100 g/s? What are the inlet and exit flow areas for the given conditions?

4-21M High-pressure air at 5 atm and 30°C is to be used to drive a small turbine for powering a high-speed drill. The drill requires a power input of 746 W (1 hp), and the air may be assumed to discharge from the turbine at 1.8 atm. If the turbine process is quasistatic and adiabatic, calculate the mass flow rate of air required. If the inlet air line is 1 cm in diameter, calculate the entering air velocity.

4-22M A 280-liter tank contains air at 100 kPa and 20°C. A high-pressure line is connected to the tank until the pressure reaches 2 MPa. Assuming that the tank and the connecting valve are adiabatic, calculate the final temperature in the tank if the conditions in the high-pressure line remain constant at 2 MPa and 40°C. How much cooling has to be supplied to the tank to maintain the air temperature constant at 20°C?

4-23M In an open feedwater heater for a steam power plant, saturated water vapor at 700 kPa is mixed with subcooled liquid at 25°C and 700 kPa. Just enough steam is supplied to ensure that the mixed stream leaving the heater will be saturated liquid at 700 kPa. Assuming that the heater is a box which is perfectly insulated from its surroundings and that the process is steady flow, calculate the required flow rate of vapor per kilogram of entering subcooled liquid.

4-24M Steam at the rate of 2 kg/min at 700 kPa and 260°C is mixed with 4 kg/min of steam at 700 kPa and 50 percent quality in an adiabatic steady-flow device. Calculate the temperature or quality of the outlet mixture.

4-25M Steam at the rate of 3 kg/s at 200 kPa and 90 percent quality is mixed with 2 kg/s of water at 100°C and 200 kPa in a steady-flow process at constant pressure. The container where the mixing occurs is insulated. Determine the state of the outlet flow stream.

4-26M Liquid water at the rate of 5 kg/s at 300 kPa and 50°C is mixed with steam at 300 kPa and 240°F in a constant-pressure adiabatic steady-flow process to produce an outlet flow stream at 100°C. Calculate the amount of steam required.

4-27M Helium enters an adiabatic nozzle at 1 MPa and 150°C with a velocity of 250 m/s. It is to be expanded to a velocity of 900 m/s. Calculate the exit flow area for a mass flow of 2.3 kg/s.

4-28M A steady-flow adiabatic compressor is used to compress 15 g/s of argon from 150 kPa and 35°C to a final pressure of 1.5 MPa. If the process is quasistatic, calculate the work required.

4-29M Helium enters a nozzle at a low velocity and conditions of 500 kPa and

200°C. The nozzle is insulated, and the helium expands to a pressure at exit of 100 kPa. Calculate the exit diameter of the nozzle if the flow rate is 0.56 kg/s.

4-30M Steam is throttled from conditions of 700 kPa and 5 percent moisture content to a pressure of 100 kPa. Determine the final state of the steam.

4-31M 1 kg of a certain gaseous fuel is burned with 50 kg of air in a constant-pressure steady-flow process. The chemical energy of the fuel is 60 MJ/kg, and both the fuel and air enter the burner at 22°C and negligible velocity. The products of combustion may be assumed to have the same properties as air. Assuming ideal-gas behavior and an insulated burner, calculate the temperature of the exit gases. For this calculation take the enthalpy of the incoming air and fuel as zero at 22°C.

4-32M A small nozzle is constructed for use with liquid water at 40°C. The inlet pressure to the nozzle is 700 kPa, and the discharge pressure is 100 kPa. Assuming an adiabatic process, calculate the exit velocity. For an exit area of 0.3 cm^2, calculate the flow rate in liters/min.

4-33M Two 28-liter chambers are separated by a non-heat-conducting wall as shown in the figure. The left chamber is filled with air at 700 kPa and 65°C, and the right chamber is evacuated. A small valve is opened which allows the air to flow from the left to the right chamber. The valve is closed when the pressures in the two chambers are equal. Calculate the final pressure and temperature for the air in each chamber. State assumptions necessary to solve the problem.

4-34M Oxygen is contained at 3 MPa and 50°C in a tank having a volume of 10 m^3. A large valve is opened from the tank and the gas is suddenly discharged until the pressure drops to 1.5 MPa. What is the quantity discharged in kilograms?

4-35M Steam enters a certain nozzle at 2 MPa and 280°C with a velocity of 150 m/s. During the expansion process the steam loses 35 kJ/kg and the enthalpy drops to 2.8 MJ/kg. What is the exit velocity from the nozzle? What flow area at the inlet would be required for a flow rate of 1 kg/s?

FIG. P4-33M

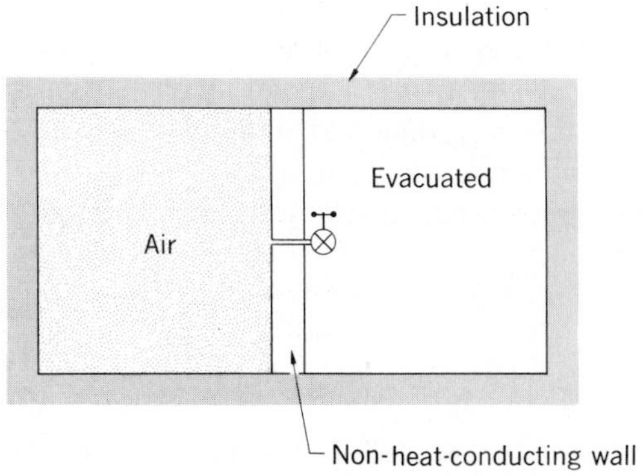

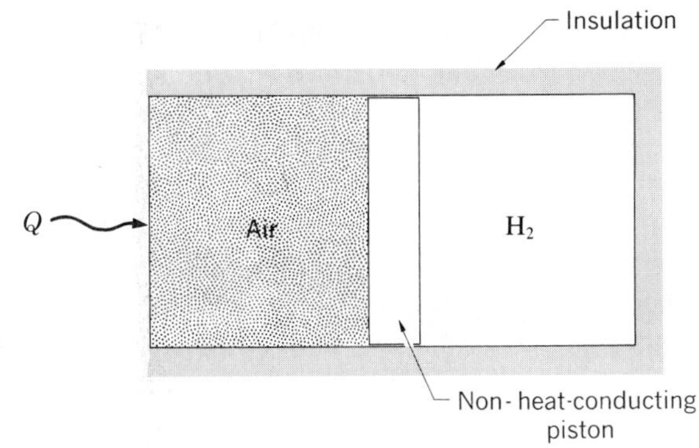

FIG. P4-36M

4-36M Consider the system shown in the accompanying figure. The two chambers initially have equal volumes of 28 liters and contain air and hydrogen, respectively. The chambers are separated by a frictionless piston which is non-heat-conducting. Both gases are initially at 140 kPa and 40°C. Heat is added to the air side until the pressure of both gases reaches 280 kPa. All outside walls of the chamber are insulated except for the surface where heat is added to the air. Calculate the final temperature of the air.

4-37M A tank having a volume of 200 liters contains air at 1.0 MPa and 20°C. A valve on the side of the tank is opened and the air is allowed to escape slowly such that the temperature of the tank remains constant at 20°C. The valve is closed when the tank pressure reaches 125 kPa. Calculate the heat absorbed by the tank.

4-38M Liquid water at 1.0 MPa and 20°C is mixed with steam at 1.0 MPa and 100 percent quality to produce hot water at 120°C. The mixing occurs in a steady-flow process inside an insulated box. If the inlet water flow at 20°C is 7.2 kg/s, calculate the quantity of steam required.

4-39M A high-speed flow of helium is to be heated in a duct. The gas enters the duct at 300 kPa, 20°C, and a velocity of 150 m/s. While traversing the duct, the pressure drops 15 kPa and the velocity is reduced to 130 m/s. The exit temperature of the gas is 140°C. How much heat must be added for a mass flow of 100 g/s? What are the inlet- and exit-flow areas for the given conditions?

4-40M High-pressure air at 500 kPa and 30°C is to be used to drive a small turbine for powering a high-speed drill. The drill requires a power input of 746 W (1 hp), and the air may be assumed to discharge from the turbine at 180 kPa. If the turbine process is quasistatic and adiabatic, calculate the mass flow rate of air required. If the inlet air line is 1 cm in diameter, calculate the entering air velocity.

4-41M A small turbine is designed for operation with helium at inlet conditions of 1.0 MPa and 45°C. Discharge is at atmospheric pressure, and the turbine process may be assumed as quasistatic and adiabatic.

Steady-flow steady-state may also be assumed. The inlet flow velocity is 10 m/s. Calculate the diameter of the inlet line if the power output of the turbine is 3 kW.

4-42M Helium enters an adiabatic nozzle at 1.4 MPa and 140°C, with a velocity of 20 m/s. It is expanded to a pressure of 110 kPa at the nozzle's exit. Calculate the exit velocity.

4-43M An industrial hot water heater uses a gaseous fuel with an internal chemical energy content of about 54 MJ/kg. The heater is designed so that 16 kg of air at 20°C and 1 atm are supplied for each kilogram of fuel, and it is designed to produce 38 liters/min of hot water at 80°C from cold water at 15°C. The hot discharge gases are at 110°C. Calculate the quantity of fuel required.

4-44M A municipal incinerator is to burn garbage which has an average internal chemical energy of 7 MJ/kg. The hot gases from the burning process are then used to preheat the air as it enters the incinerator. The internal construction of the incinerator also allows it to be used as a high-pressure hot water heater for a nearby power plant, as shown in the sketch. The respective temperatures are $T_1 = 40°C$, $T_3 = 815°C$, $T_4 = 205°C$, $T_5 = 35°C$, $T_6 = 150°C$. Assume that the hot gases behave as an ideal gas and that 14 kg of air are required to burn each kilogram of garbage or each kilogram of an auxiliary fuel. The auxiliary fuel has an internal chemical energy of 46 MJ/kg and must be used to get the burning temperature high enough for complete combustion. How much total energy in the form of garbage and fuel must be supplied to heat 11 000 liters of water? If 900 kg of garbage are available for heating each 11 000 liters of water, how much auxiliary fuel would be required?

4-45M The velocity of a stream of air is to be slowed by passing it through a flow channel which allows it to go from 50 kPa, −10°C, and 700 m/s to 150 m/s. The process may be assumed to be quasistatic and adiabatic. Calculate the entrance and exit flow areas for a mass flow rate of 1.2 kg/s.

FIG. P4-44M

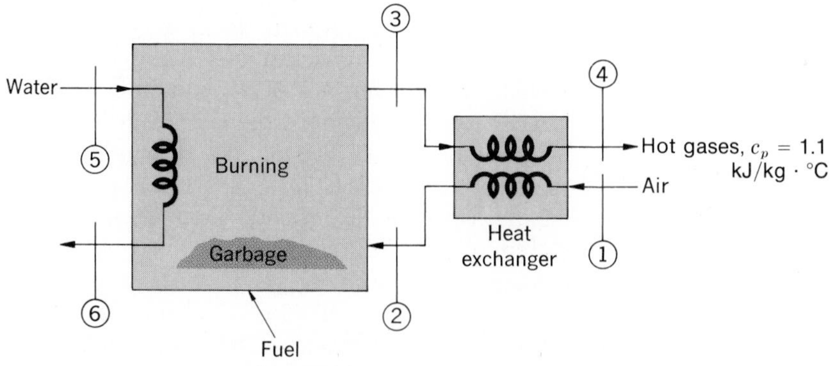

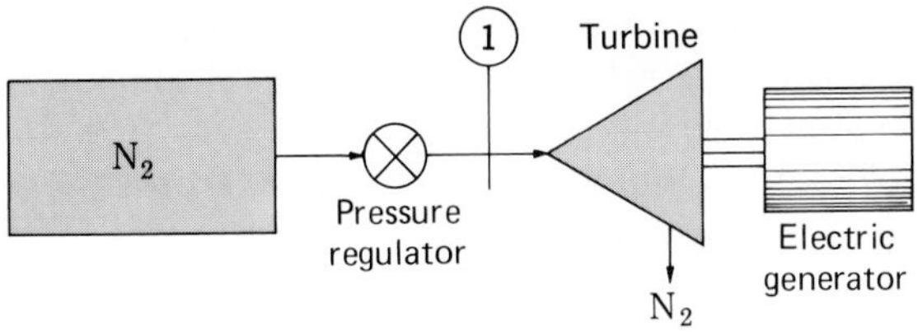

FIG. P4-46M

4-46M A small auxiliary power device is designed as shown in the schematic. Nitrogen is contained at 14 MPa and 20°C in the tank which is connected through a valve to an adiabatic turbine which drives an electric generator. The power output of the device is to be 75 W, and the pressure at point 1 is maintained constant at 700 kPa by a pressure regulator which operates as a throttling device. The nitrogen discharges from the turbine at 100 kPa, and the turbine process may be assumed adiabatic. What size tank is necessary to produce the power output for a period of 1 h? Assume the tank remains isothermal at 20°C and that power production ceases when the tank pressure reaches 700 kPa.

4-47M A spherical tank is perfectly insulated and evacuated and exposed to the surrounding air at 1 atm and 35°C. A valve on the tank is opened on the side of the tank and air is allowed to enter slowly. What is the temperature of the air in the tank when its pressure just reaches 1 atm?

4-48M A 1.0-m^3 rigid vessel contains steam at 2.0 MPa and 320°C. A valve on the vessel is opened and steam is allowed to escape until the pressure drops to 100 kPa. During the process heat is added to maintain the temperature constant. Calculate the quantity of heat required.

4-49M Steam flows in a line at 1.5 MPa and 320°C. Connected to the line is an adiabatic cylinder with a frictionless piston as shown in the figure. One side of the cylinder contains 0.05 kg of air at 1 atm and 105°C, while the other side contains 0.03 kg of steam at 1 atm and 105°C. The piston is adiabatic. The valve is opened and steam is allowed to enter the cylinder until the pressure reaches 1.5 MPa, at which time the valve is closed. Calculate the mass of steam added, the final state of the steam in the cylinder, and the final air temperature.

4-50M A fluid undergoes a steady-flow process in which its enthalpy increases by 4.5 kJ/kg. The initial state of the fluid is 2 atm and 100°C,

FIG. P4-49M

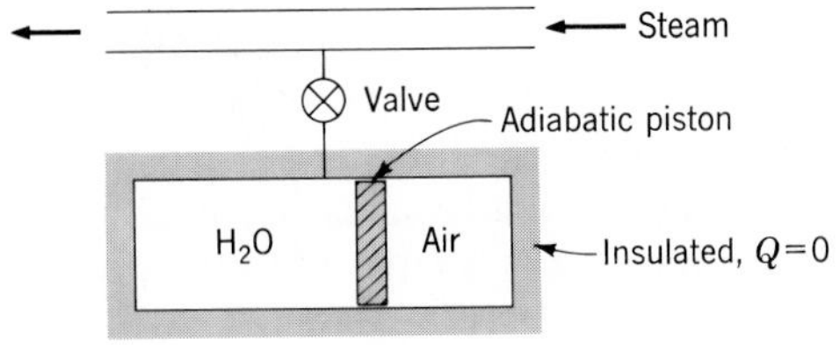

and the final state is 10 atm and 300°C. The initial and final velocities are 36 and 12 m/s, respectively, and the final state is at an elevation 25 m above the initial state. 25 MJ of heat are supplied for a flow rate of 0.75 kg/s. Calculate the work for the process and express in units of W, Btu, and hp.

4-51M Steam undergoes an adiabatic and steady-flow process in a turbine from 6.0 MPa, 500°C, to a pressure of 10 kPa and a quality of 90 percent. Calculate the work output for a flow of 1.0 kg/s. What flow rate would be required to produce a work output of 400 kW?

4-52M Steam undergoes a process at constant enthalpy from 1.5 MPa, 360°C, to a pressure of 10 MPa. Find the final temperature in this process and the change in volume.

4-53M Saturated-liquid water at 260°C is throttled to 100 kPa. Calculate the final quality if saturated or the final temperature if superheated. Calculate the velocity of the steam in the final state for a flow rate of 1 kg/min through a 5-cm-diameter tube.

4-54M A tank having a volume of 0.1 m^3 contains nitrogen at a pressure of 15 atm and 100°C and is perfectly insulated from the surroundings. A valve on the tank is opened and N_2 is discharged until the pressure drops to 2 atm. Calculate the mass of N_2 discharged from the tank.

4-55M A centrifugal blower receives air at 1 atm and 20°C in a volume flow rate of 0.7 m^3/s. The air enters at 1.0 m/s and discharges at 10.0 m/s, and temperature is essentially constant. Calculate the input power requirements. State any assumptions.

4-56M An alternative to an adiabatic throttling device for measuring steam quality is a setup in which the sample of steam is allowed to flow through a chamber where a measured quantity of electric heating (known) is applied. The flow rate of exit steam from the calorimeter is also measured. In one application, steam at 2.0 MPa is sampled at a flow rate of 3.8 g/s. Electric power in the amount of 380 W is applied to the device, and the outlet conditions are measured as $p = 150$ kPa and $T = 180$°C. Calculate the quality of the steam at the 2.0-MPa condition.

4-57M A closed system undergoes a quasistatic process between two end states such that the enthalpy decreases by 110 kJ and the system gives up 10 kJ of heat to the surroundings. Calculate the work for this process. Then suppose the same system undergoes a steady-flow process between the same end states and also experiences an increase in kinetic energy of 5 kJ while experiencing the same heat loss. Determine the external work for this process.

4-58M Oxygen flows in a line at 70 m/s with conditions of 120 kPa and 15°C. Connected to the line is a 6.0-liter tank containing oxygen at 5°C and 1 atm. A valve is opened allowing oxygen to flow into the tank until it contains five times the original mass, at which time the valve is closed. The tank is adiabatic. Calculte the final pressure and temperature in the tank.

REFERENCES

1 Hatsopoulos, G. N., and J. H. Keenan: "Principles of General Thermodynamics," John Wiley & Sons, Inc., New York, 1965.
2 Reynolds, W. C.: "Thermodynamics," 2d ed., McGraw-Hill Book Company, New York, 1968.
3 Sears, F. W.: "An Introduction to Thermodynamics, The Kinetic Theory of Gases, and Statistical Mechanics," Addison-Wesley Publishing Company, Inc., Reading, Mass., 1950.
4 Van Wylen, G. J., and R. E. Sonntag: "Fundamentals of Classical Thermodynamics," John Wiley & Sons, Inc., New York, 1965.

5

THE SECOND LAW OF THERMODYNAMICS

5-1 INTRODUCTION

We have already noted in Chap. 1 that the second law of thermodynamics relates to the direction of energy-exchange processes. Some energy transformations are allowed, whereas others are not. In Chap. 2 a more precise explanation of the conservation of energy principle has been given and related to previous studies in mechanics. Insofar as the first law of thermodynamics is concerned, all forms of energy are equally useful; no preference is given for the different types of energy, and the only concern is that in an isolated system energy must be conserved. Even though no preference is given, the different types of energy are distinguished from one another: Work is a force acting through a distance and represents an interaction between systems; internal energy is a function of state, the change of which is defined in terms of work; and heat is subsequently defined in terms of work and internal energy. Nothing in the conservation of energy principle states that heat and work are not equivalent forms of energy; in fact, the cyclic equation

$$\oint d'Q + \oint d'W = 0 \tag{5-1}$$

carries a casual implication that they *are* equivalent. The second law of thermodynamics recognizes the fact that heat and work are not equivalent and eventually establishes a set of formal relationships which may be used to supplement the first law in the study of thermodynamic systems.

5-2 PHYSICAL DESCRIPTION OF THE SECOND LAW

A brief summary of a set of phenomena falling into the realm of the second principle has been given in Chap. 1. These phenomena are restated here for convenience.

1. Heat flows from a high temperature to a low temperature in the absence of other effects. This means that a hot body will cool down when brought into contact with a body at a lower temperature, and not the opposite.
2. Two gases, when placed in an isolated chamber, will mix uniformly throughout the chamber but will not separate spontaneously once mixed.
3. A battery will discharge through a resistor releasing a certain amount of energy, but it is not possible to make the reverse of this happen, i.e., to add energy to the resistor by heating and, thus, cause the battery to charge itself.
4. It is not possible to construct a machine or device which will operate continuously while receiving heat from a single reservoir and producing an equivalent amount of work.

Let us examine statements 1 and 4 in detail. The circumstance in statement 1 is in accordance with our everyday experience, and at first thought it would seem to be a direct consequence of the conservation of energy. It is true that the energy given up by the hot block goes to raise the energy level of the cooler block such that the total energy of the set of blocks remains constant. But the total energy would also remain constant in the case where energy taken from the cooler block was used to raise the energy level of the hot block by an equivalent amount. This second process does not occur, even though it does not violate the conservation of energy principle. In another example, we do not expect a pan of water to boil when placed in a refrigerator as a result of heat transferred to it *from* the cold surroundings. This simply does not happen.

In statement 4 the considerations are a bit more abstract but still in the realm of normal experience. We would not expect to be able to build an engine that would operate continuously and in cyclic processes, extracting energy from some readily available source such as the ocean or earth and producing an equivalent amount of mechanical work. If it were possible to build such an engine, it would be a very fine thing indeed, for we would only have to connect it to the ocean and produce all the power needed for some time to come. Such an engine is called a *perpetual-motion machine of the second kind* because it succeeds in converting heat from a single reservoir into work on a continuous basis. A perpetual-motion machine of the first kind is one which *creates* energy in violation of the first law of thermodynamics.

In accordance with this brief discussion, we may tentatively conclude that the second law of thermodynamics has its primary emphasis in an acknowledgment of the unidirectional nature of heat transfer and certain types of energy conversion.

Our overall objective in this chapter is to develop satisfactory analytical descriptions of the second law of thermodynamics based strictly on macroscopic arguments which accept statement 4 as an empirical axiom. This macroscopic analysis ignores the microscopic structure of matter. For those readers who choose to study the microscopic thermodynamic presentations of Chaps. 12 and 13, it will be shown that the two types of analysis can be tied together neatly but that such material is unnecessary for an understanding of this chapter.

5-3 CLAUSIUS AND KELVIN-PLANCK STATEMENTS

Two commonly accepted statements of the second law of thermodynamics relate to the foregoing statements 1 and 4.

Clausius Statement It is impossible to construct a device which operates in a cycle and whose sole effect is to transfer heat from a cooler body to a hotter body.

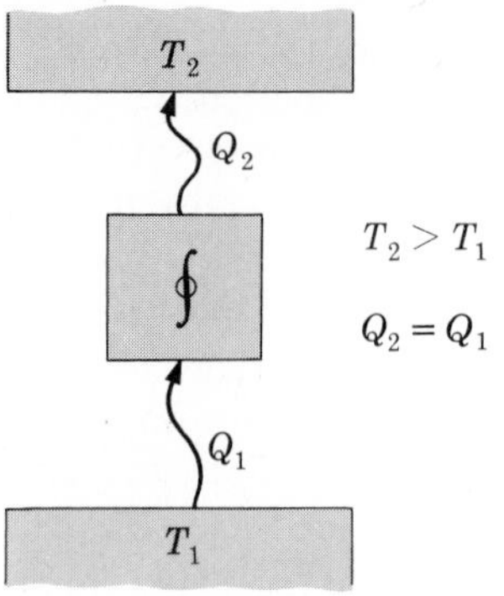

FIG. 5-1 It is impossible to construct a device which operates in a cycle and whose sole effect is to transfer heat from a cooler body to a hotter body.

Kelvin-Planck Statement It is impossible to construct a device which operates in a cycle and produces no other effect than the production of work and exchange of heat with a single reservoir.

These statements are depicted in Figs. 5-1 and 5-2.

5-4 REVERSIBLE PROCESSES AND CYCLES

Let us now consider an idealized state path called the *reversible* process. A process is reversible if the initial state of the system can be restored with no observable effects in the system *and* its surroundings. Consider the expansion of a gas behind a piston in a cylinder. If the expansion occurs slowly and frictionlessly, the system (the gas) will remain essentially in equilibrium, and we might call the process *quasistatic*, or a succession of equilibrium states, as discussed in Chap. 1. To find out if this process is reversible, we ask the question: Is it possible to run the process backward with no extra effects observed in the system or surroundings? In other words, could we compress the gas back to its original state by applying *exactly* the same amount of work that was removed in the expansion process and adding *exactly* the same amount of heat that was removed in the expansion process? It turns out that the quasistatic process is, indeed, a reversible process. A general criterion for reversibility may be stated as follows:

If a hypothetical reversal of a thermodynamic process can be performed which does not violate the second law of thermodynamics, then the process is said to have been reversible.

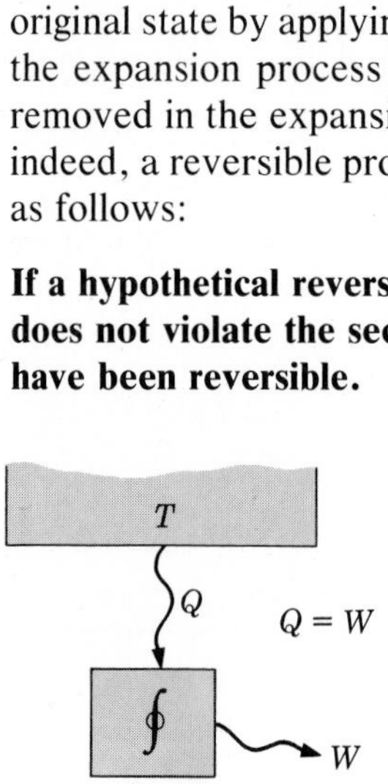

FIG. 5-2 It is impossible to construct a device which operates in a cycle and produces no other effect than the production of work and exchange of heat with a single reservoir.

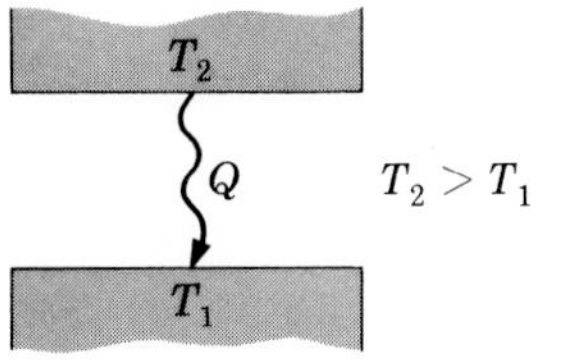

FIG. 5-3 Heat transfer across a finite temperature difference is irreversible.

Two obvious applications of this criterion are shown in Figs. 5-3 and 5-4. If we try to reverse the heat-transfer process in Fig. 5-3, a violation of the Clausius statement of the second law results. If we try to reverse the situation in Fig. 5-4, the Kelvin-Planck statement is violated. We may conclude that transfer of heat across a finite temperature difference and the conversion of work into heat during a cyclic process are both irreversible processes. To make this conclusion, we define an *irreversible process as one which is not reversible*. This definition is, of course, rather obvious.

A *reversible cycle* is a succession of reversible processes such that the system periodically returns to its initial state. As previously mentioned, the cyclic engine, or device, depicted in Fig. 5-2 is called a perpetual-motion machine of the second kind. The term *second kind* means that it violates the second law of thermodynamics.

Cyclic Heat Engines

At this point we may digress for a moment to discuss the meaning of a *cyclic heat engine*. Work is a very useful form of energy in that it can be adapted to practical applications in a variety of ways. Electrical work can power numerous devices, and mechanical work can drive automobiles, machines, etc. Heat is not necessarily so useful, and the purpose of a heat engine is to convert heat into work. The notion of the cycle and cyclic process is introduced because energy-conversion processes of interest are those which can operate continuously. The internal-combustion engine is a cyclic engine which sequentially inhales fuel and air, compresses and combusts the mixture, and produces work output while exhausting the products of combustion to the surroundings. A *power cycle* is one which produces a net work output (with some net heat input), and a *refrigeration cycle* involves a net work *input* and net heat output. A *reversible* power cycle can be changed to a reversible refrigeration cycle by just reversing all the heat flow and work quantities. To fix the proper directions

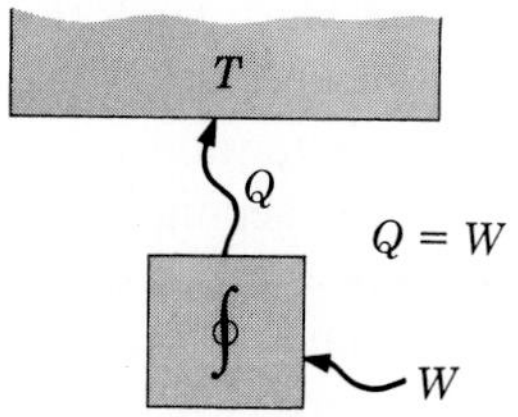

FIG. 5-4 Conversion of work into heat is irreversible.

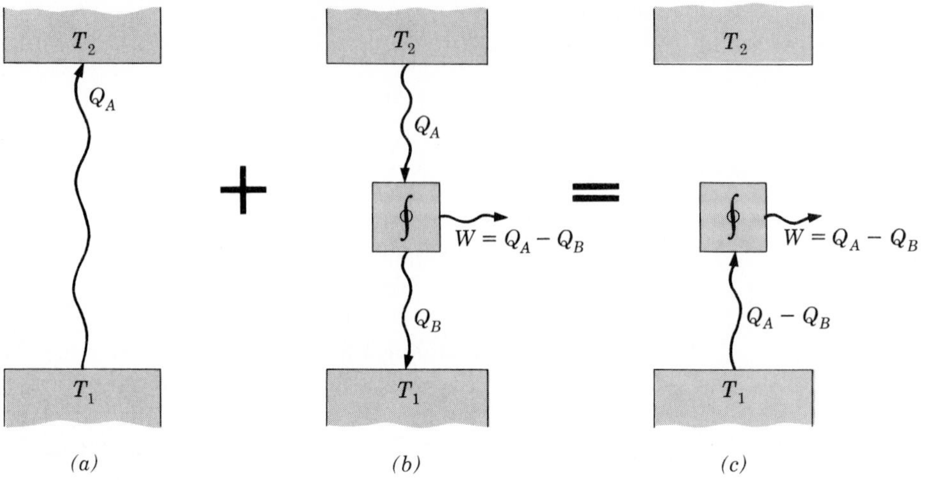

FIG. 5-5 Equivalence of the Clausius and Kelvin-Planck statements of the second law.

and signs for the work and heat quantities we shall rely on proper schematic diagrams rather than on a rigid sign convention.

Equivalence of Second-Law Statements

The equivalence of the Clausius and Kelvin-Planck statements may be demonstrated rather quickly, as shown in Fig. 5-5. Let us suppose that it would be possible to transfer heat as in Fig. 5-5*a* in violation of the Clausius statement. To the system seen in Fig. 5-5*a* we could add the reversible engine as in Fig. 5-5*b*. From energy conservation, the work output of this engine is $W = Q_A - Q_B$. The second law, of course, does not prohibit this type of engine. However, the net result of adding the heat engine in Fig. 5-5*b* is that no net heat is exchanged with the reservoir at T_2, and the arrangement is equivalent to the one in Fig. 5-5*c*. The engine in Fig. 5-5*c* is just the type of device excluded by the Kelvin-Planck statement. Thus we have shown that a violation of the Clausius statement results in a violation of the Kelvin-Planck statement as well.

5-5 SIGN CONVENTION FOR HEAT AND WORK

In our discussion of the first law of thermodynamics and energy balances we employed a sign convention for heat and work such that energy added to the boundary of a system (heat or work) is positive. The equations that represent the energy balances were based on this convention. Now, as we start to examine thermal power cycles in the light of the second law, the reader will notice that schematic diagrams are drawn with heat and work "arrows" so that energies are both added to and removed from the cycles. The purpose of such diagrams is to recognize, in advance, the directions of such energy transfers so that a better appreciation may be gleaned of the processes involved. From an energy-balance standpoint there is no change in fundamental concepts. We still say that

Energy added to system = energy accumulated in system + energy removed from system

By drawing the arrows in certain directions we are simply focusing our attention on energy quantities of practical interest. In a *power cycle* the object is to produce a work *output* as a result of a heat input at a high temperature; thus, we draw a schematic diagram such that the work is coming out of the cycle and heat is added. In a *refrigeration cycle* the object is to remove heat from some source at low temperature by adding work; so we draw the schematic diagram with a heat addition at low temperature and a work *input*.

The important point of this chapter is that the second law places very definite restrictions on allowable conversions of heat into work, but the restrictions do not abrogate the basic notion of energy conservation in any way. As we have stated many times, energy-conversion analysis is basically an accounting procedure, and as long as it is performed correctly, the sign convention does not matter. As our exposition of the second law develops, the reader should concentrate on understanding the physical phenomena involved; with this concentration, sign convention will take care of itself.

5-6 THERMODYNAMIC TEMPERATURE

Let us suppose that we have at our disposal two reversible heat engines which operate in cyclic processes between temperature reservoirs, as shown in Fig. 5-6*a*. The thermal efficiency η_t of a heat engine is defined as

$$\eta_t = \frac{\text{useful energy effect}}{\text{energy that costs}} = \frac{\text{work output}}{\text{heat added}} \tag{5-2}$$

According to this definition the "energy that costs" is the heat added at the high temperature because this heat must be supplied by some fuel. Thus

$$\eta_t = \frac{W}{Q_2} \tag{5-3}$$

So far, we have said nothing about such an efficiency. The reversible engines A and B could be constructed differently and, at first glance, we might expect them to have different efficiencies. We are about to prove, however, that they have the *same* efficiency. To show this, let us tentatively assume that engine A has a greater efficiency than B and set

$$Q_{2A} = Q_{2B}$$

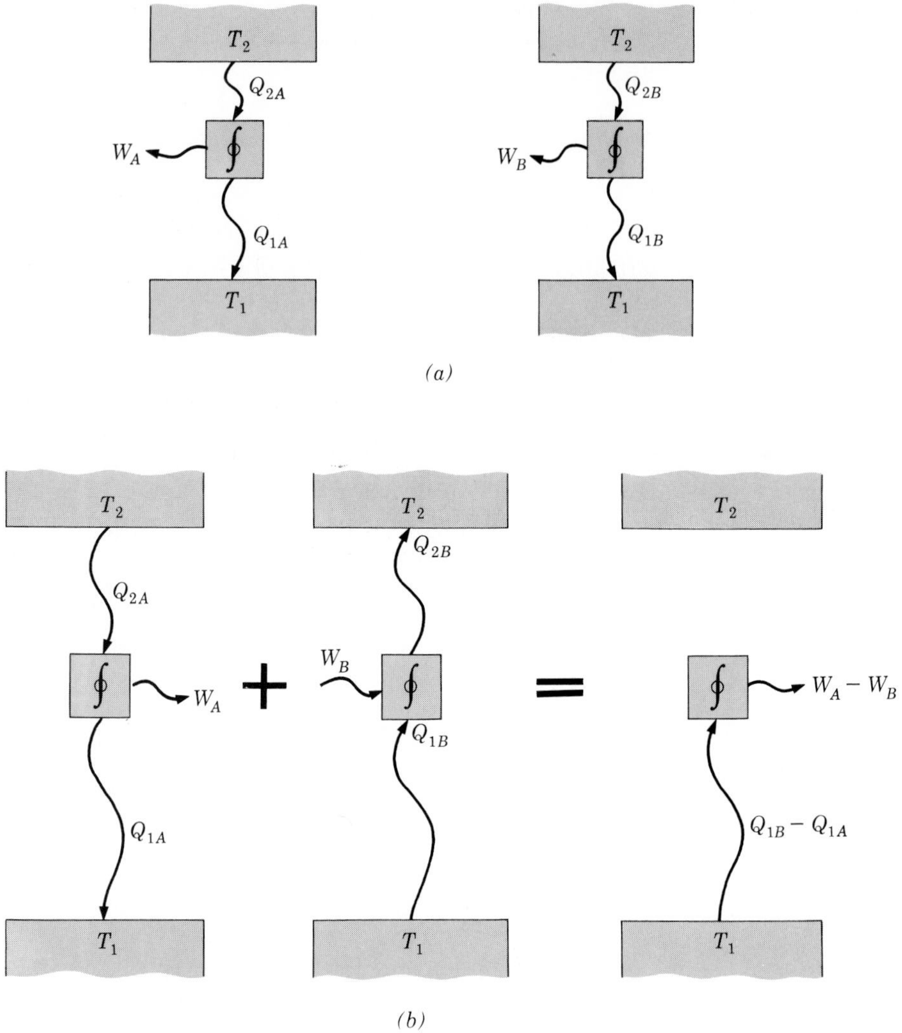

FIG. 5-6 All reversible engines operating between the same temperature limits have the same efficiency.

Then

$$W_A > W_B \qquad \text{and} \qquad Q_{1A} < Q_{1B}$$

Because the engines are reversible, engine *B* could be reversed and combined with engine *A*, as shown in Fig. 6-6*b*. But because $Q_{2A} = Q_{2B}$,

$$W_A - W_B = Q_{1B} - Q_{1A}$$

The arrangement in Fig. 5-6*b* is a violation of the Kelvin-Planck statement of the second law, so the assumption that *A* has a greater efficiency than *B* must be false. We are, therefore, justified in concluding that *all reversible* *engines operating between the same temperature limits have the same efficiency.* This efficiency may be written as

$$\eta_t = \frac{W}{Q_2} = \frac{Q_2 - Q_1}{Q_2} = 1 - \frac{Q_1}{Q_2} \tag{5-4}$$

Because the efficiency is the same for all reversible engines operating between the same temperature limits, it follows that the heat ratio Q_1/Q_2 must be a function of these two temperatures:

$$\frac{Q_1}{Q_2} = f(T_1, T_2) \tag{5-5}$$

Suppose that some intermediate temperature T_3 is inserted between T_1 and T_2 with two new reversible heat engines as shown in Fig. 5-7. Because the two reversible engines A and B comprise a larger reversible engine operating between T_2 and T_1, the combination must have the same efficiency as engine C. From Eq. (5-5) we know that the heat ratios are functions of the temperature limits of the cycles:

$$\frac{Q_1}{Q_2} = f(T_1, T_2)$$

$$\frac{Q_3}{Q_2} = f(T_3, T_2)$$

$$\frac{Q_1}{Q_3} = f(T_1, T_3)$$

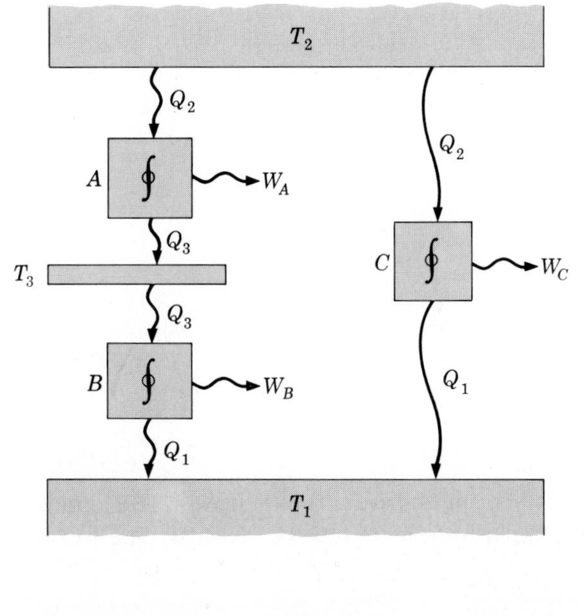

FIG. 5-7 Thermodynamic temperature.

Combining these relations,

$$\frac{Q_1}{Q_2} = f(T_1, T_2) = \frac{Q_1}{Q_3}\frac{Q_3}{Q_2} = f(T_1, T_3) \times f(T_3, T_2) \tag{5-6}$$

or

$$f(T_1, T_2) = f(T_1, T_3) \times f(T_3, T_2) \tag{5-7}$$

Because the left side of Eq. (5-7) does not involve T_3, we conclude that the f function must be of such form that the T_3 will cancel out of the product on the right side of Eq. (5-7). This is possible if

$$f(T_1, T_3) = \frac{\phi(T_1)}{\phi(T_3)}$$

$$f(T_3, T_2) = \frac{\phi(T_3)}{\phi(T_2)}$$

where ϕ is a new function. The net result is that we select for the heat ratios the functional form

$$\frac{Q_1}{Q_2} = \frac{\phi(T_1)}{\phi(T_2)} \tag{5-8}$$

It is possible, of course, to satisfy this equation with a number of temperature functions. The proposal was made by Kelvin that the temperature function be taken as

$$\frac{|Q_1|}{|Q_2|} = \frac{T_1}{T_2} \tag{5-9}$$

and that Eq. (5-9) serve as a *definition* of an *absolute thermodynamic temperature scale*. The efficiency of a reversible heat engine which operates between two reservoirs and exchanges heat only with these two reservoirs is thus

$$\eta_t = 1 - \frac{T_L}{T_H} \tag{5-10}$$

where T_L designates the low-temperature reservoir and T_H designates the high-temperature reservoir. This temperature scale also corresponds to the temperature measured by an ideal-gas thermometer as discussed in Chap. 1.

5-7 THE INEQUALITY OF CLAUSIUS

Consider the arrangement shown in Fig. 5-8. An amount of heat $d'Q_R$ from the constant-temperature reservoir T_R is transferred to the reversible engine and

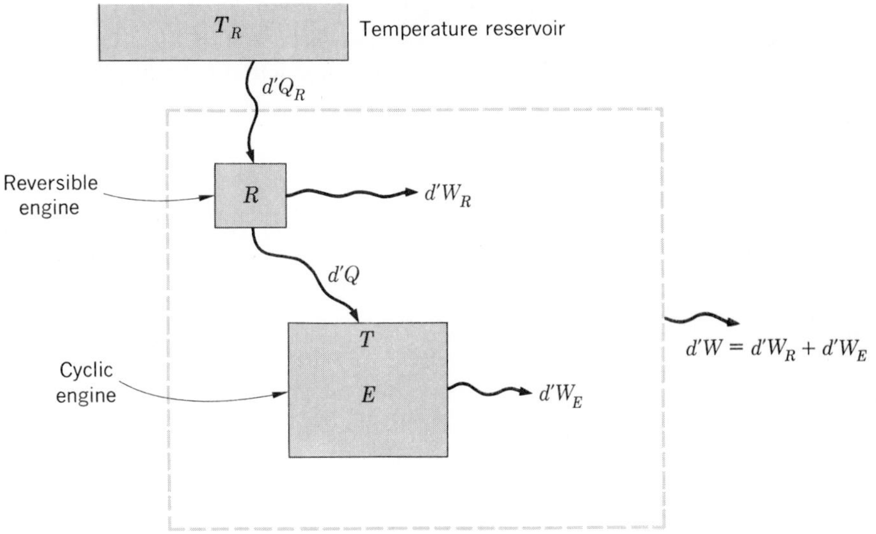

FIG. 5-8 Diagram for proof of inequality of Clausius.

produces the amount of work $d'W_R$. The heat rejected from this engine is used to supply the other cyclic engine producing the work output $d'W_E$. Taking *both* engines as a system, the total work *ouput* $d'W$ is

$$d'W = d'W_R + d'W_E$$

For a process with the engine we have

$$d'Q = dU + d'W_E$$

The total work output is thus

$$d'W = d'W_R + d'Q - dU$$

If we now allow the whole assembly to undergo a cycle

$$\oint d'W = \oint d'W_R + \oint d'Q - 0 \tag{5-11}$$

But

$$d'W_R = d'Q_R\left(1 - \frac{T}{T_R}\right) = d'Q\left(\frac{T_R}{T} - 1\right)$$

so that Eq. (5-11) becomes, for the complete cycle,

$$\oint d'W = \oint d'Q\left(\frac{T_R}{T} - 1 + 1\right) = T_R \oint \frac{d'Q}{T} \tag{5-12}$$

The arrangement in Fig. 5-8 cannot produce a net work *output* because this would violate the second law as given by the Kelvin-Planck statement. Accordingly, the only way the arrangement can operate is with a cyclic work *input* and a cyclic heat flow into the reservoir. Mathematically, this means that

$$\oint d'W \leqslant 0 \tag{5-13}$$

where $d'W$ is the work *output*. Using Eq. (5-12), this condition becomes

$$\oint \frac{d'Q}{T} \leqslant 0 \tag{5-14}$$

The relation given in Eq. (5-14) is called the *inequality of Clausius*.

In the foregoing derivation no mention was made of the reversibility of engine E. Suppose this engine is reversible. Suppose further that

$$\oint d'W < 0 \tag{5-15}$$

for this circumstance. Since the engine is assumed reversible, we could reverse it and obtain

$$\oint d'W > 0$$

But this is impossible according to the second law because a perpetual-motion machine of the second kind would be created. The net result is that the inequality of Eq. (5-15) cannot hold for the case of a reversible engine E. It follows that the equality sign of Eq. (5-14) must hold in the reversible case, and we may write

$$\oint \left(\frac{d'Q}{T}\right)_{\text{rev}} = 0 \tag{5-16}$$

5-8 MACROSCOPIC DEFINITION OF ENTROPY

Mathematically, any quantity which may be represented as an exact differential is a property or point function; i.e., it is a function of the *state* of the system. The mathematical condition for an exact differential is that the cyclic integral of the differential be zero; i.e., if

$$\oint dx = 0$$

then x is a property for the region over which the cyclic integral is taken. In accordance with this definition, Eq. (5-16) yields

$$dS = \left(\frac{d'Q}{T}\right)_{\text{rev}} \tag{5-17}$$

where S is a property which we choose to call the *entropy*. *Equation (5-17) is the macroscopic definition of entropy.* Note that entropy is defined only for reversible processes and a change in entropy may be calculated with

$$\Delta S = S_2 - S_1 = \int_1^2 \left(\frac{d'Q}{T}\right)_{\text{rev}} \tag{5-18}$$

Only *changes* in entropy are defined by these relations. We have not yet established an absolute scale for entropy.

5-9 PRINCIPLE OF INCREASE OF ENTROPY

Consider the two state points shown in Fig. 5-9 and the two processes which comprise a cycle operating between the two points. R is a reversible process and I is an irreversible process, so that the overall cycle is irreversible. From the inequality of Clausius, we have

$$\oint \frac{d'Q}{T} = \int_1^2 \frac{d'Q_I}{T} + \int_2^1 \frac{d'Q_R}{T} < 0 \tag{5-19}$$

We have used the inequality sign because the cycle is irreversible. Observing that

$$\int_2^1 \frac{d'Q_R}{T} = S_1 - S_2$$

Eq. (5-19) may be written as

$$\int_1^2 \frac{d'Q_I}{T} + S_1 - S_2 < 0$$

or

$$S_2 - S_1 > \int_1^2 \frac{d'Q_I}{T}$$

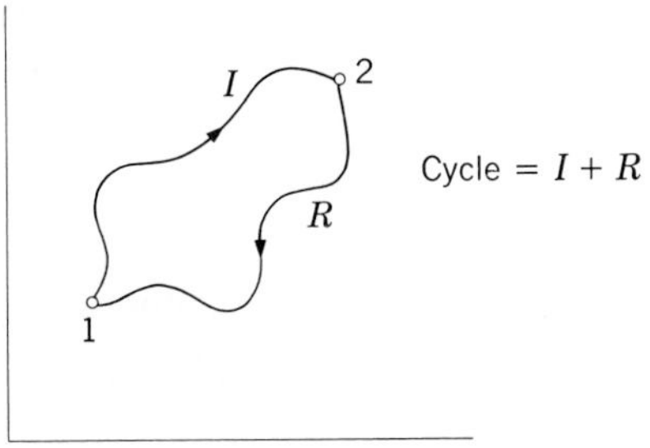

FIG. 5-9 Irreversible cycle.

In general then, we may write

$$S_2 - S_1 \geqslant \int_1^2 \frac{d'Q}{T} \tag{5-20}$$

where the equality holds for a reversible process and the inequality for an irreversible process. Equation (5-20) is called the *principle of increase of entropy*.

Suppose Eq. (5-20) is written for adiabatic processes, $d'Q = 0$:

$$S_2 - S_1 \geqslant 0 \qquad \text{adiabatic process} \tag{5-21}$$

If the process is a reversible adiabatic one, the entropy change will be zero, and we call it a constant-entropy or *isentropic* process. If the process is irreversible and adiabatic, then the entropy must increase. Suppose, furthermore, that we consider an isolated system. Then

$$\Delta S_{\text{isolated system}} \geqslant 0 \tag{5-22}$$

or the entropy of an isolated system must increase if irreversible changes are taking place inside it. If only reversible changes are taking place within the isolated system, the entropy will remain constant.

No *real* process is reversible. All heat transfers occur across a finite temperature difference. Frictional effects are always present to render processes involving work irreversible. Because of these facts we are forced to conclude that the entropy of the universe (presumably an isolated system) is always increasing. The postulate that the universe is "filling up with entropy" is a metaphysical matter of serious concern to philosophers and theologians. We shall not worry about the consequences of this, however, since it probably will not seriously affect engineering problems for some time.

Entropy is a property defined by Eq. (5-17). Its change may be calculated with Eq. (5-18), and that is all we need to know for most macroscopic considerations.

Entropy, a Thermodynamic Property

The foregoing discussion has demonstrated that entropy is a property of a system. Thus, it may be expressed in terms of other thermodynamic properties and tabulated just like enthalpy or internal energy. It is indeed tabulated as a function of pressure and temperature for a wide variety of substances. The properties given in the Appendix furnish examples of such tabulations.

The Reversible Adiabatic and Isothermal Processes

It is clear from Eq. (5-17) that for a reversible adiabatic process $dQ_{\text{rev}} = 0$ and hence $dS = 0$ or the entropy remains constant. This type of process is thus given the name *isentropic,* designating constant entropy. Similarly, the *revers-*

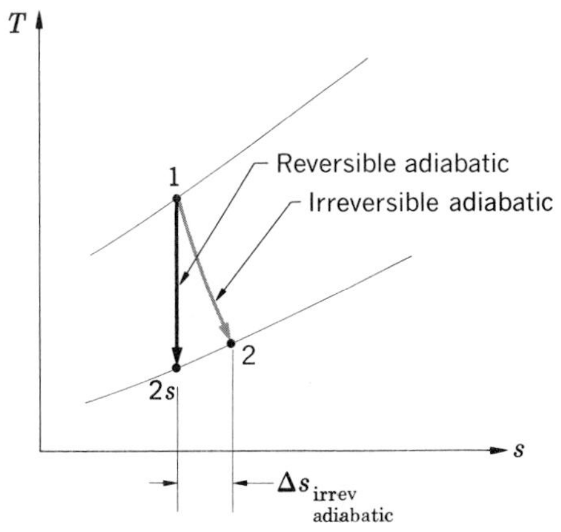

FIG. 5-10 Reversible and irreversible adiabatic processes on a temperature-entropy diagram. The greater the increase in entropy, the larger the irreversibility.

ible isothermal process can be represented by

$$Q_{\text{rev isothermal}} = T\,\Delta S \tag{5-23}$$

by an integration of Eq. (5-17).

The *T-S* Diagram and Irreversible Processes

It is clear from Eq. (5-21) that the entropy for an *irreversible* adiabatic process must increase, but the amount of increase is indefinite. Physically, the reversible process is quasistatic; i.e., it is a succession of equilibrium states, and we are able to specify the state of the system each step along the way. The irreversible process is one in which the system departs substantially from equilibrium because of frictional effects, substantial temperature gradients within the system, or other effects. In the irreversible process the entropy change must be calculated by devising some reversible process between the given end states. The degree to which friction or temperature gradients affect the change of state of a system must be determined apart from conventional thermodynamic analysis, and is usually treated as a subject in transport phenomena. For the adiabatic process the reversible and irreversible processes will appear as shown in Fig. 5-10. The greater the increase in entropy for the irreversible process, the greater we say is its "irreversibility." Note that the irreversible process is drawn as a shaded line because it is a series of nonequilibrium states and thus is not strictly representable on a diagram of thermodynamic properties of equilibrium states.

5-10 PRACTICAL CAUSES OF IRREVERSIBILITIES

We have already shown two examples of irreversible processes in Figs. 5-3 and 5-4. Thus, if a hot block of metal is brought into thermal communication with a

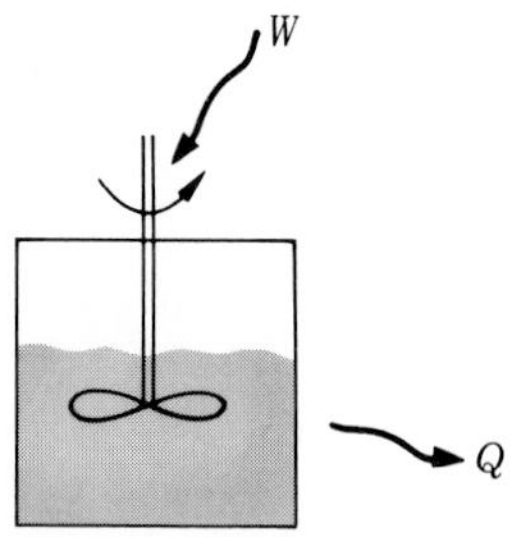

FIG. 5-11 Paddle-wheel work is irreversible.

cooler block, and the two allowed to come to temperature equilibrium, we will experience an irreversible process. If the two blocks were taken as an overall adiabatic system, an increase in entropy would be observed for the combination. If one adds work in a paddle-wheel process as shown in Fig. 5-11, this too is an irreversible process because we cannot reverse the process by adding heat to the container. The block sliding against a frictional resistance shown in Fig. 5-12 also represents the same type of irreversible process because the work W done by the force F is eventually dissipated as heat Q into either the plane or the surrounding atmosphere. Again, we cannot reverse the situation and expect the block to move when heat is added to it. Fluid friction in a pipe is another similar example of an irreversible process. Work is required to move the fluid through the pipe against a wall frictional resistance, and we cannot accomplish the same effect by adding heat to the pipe.

Thus, whenever heat is transferred across a finite temperature difference or friction is encountered, the process will be rendered irreversible to some extent. All *real* processes are irreversible, some more than others. Later on, we shall find that certain reversible processes represent the best-hoped-for situation in energy-conversion and power-production devices and thus merit our study and attention. The degree to which we are able to approach such ideal processes is a matter of the skill of engineering design in various items of equipment and the cost factors which can be tolerated in a particular economic climate.

5-11 ENTROPY OF A PURE SUBSTANCE

We have shown that entropy is a property of a system. It is an extensive property like the total internal energy or total enthalpy which may be calcu-

FIG. 5-12 Sliding block is irreversible.

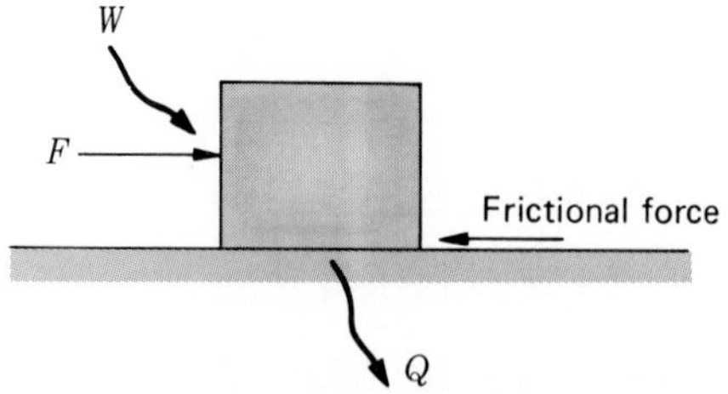

lated from *specific* entropies based on either a unit mass or unit molal quantity of the system so that

$$S = ms = n\bar{s}$$

For pure substances values of the specific entropy may be tabulated along with enthalpy, specific volume, and other thermodynamic properties of interest, as illustrated in the various tables of the Appendix. In the liquid-vapor saturation region the specific entropy is obtained from the saturation properties and quality in the same manner as other properties so that

$$s = s_f + xs_{fg} \qquad s_{fg} = s_g - s_f$$

$$s = (1 - x)s_f + xs_g$$

We have seen that the temperature-entropy diagram is a useful vehicle for studying thermodynamic processes, and the T-s diagram for a pure substance will be particularly helpful when we study practical power cycles. Figure 5-13 shows such a diagram for water and is illustrative of the general appearance of such diagrams for other substances. The figure shows the saturation line as a dome-shaped curve with the critical point at the peak, the saturated-liquid line to the left, and the saturated-vapor line to the right. All of the area under the dome is the *wet-mixture* region with appropriate lines of constant quality indicated. The region to the left of the saturated-liquid line is compressed liquid, while that to the right of the saturated-vapor line is the superheated vapor region. At temperatures above the cricial point there is no distinction between liquid and vapor. Some typical lines of constant pressure, constant volume, and constant enthalpy are also indicated to show the general shape of such curves.

FIG. 5-13 Temperature-entropy diagram for water.

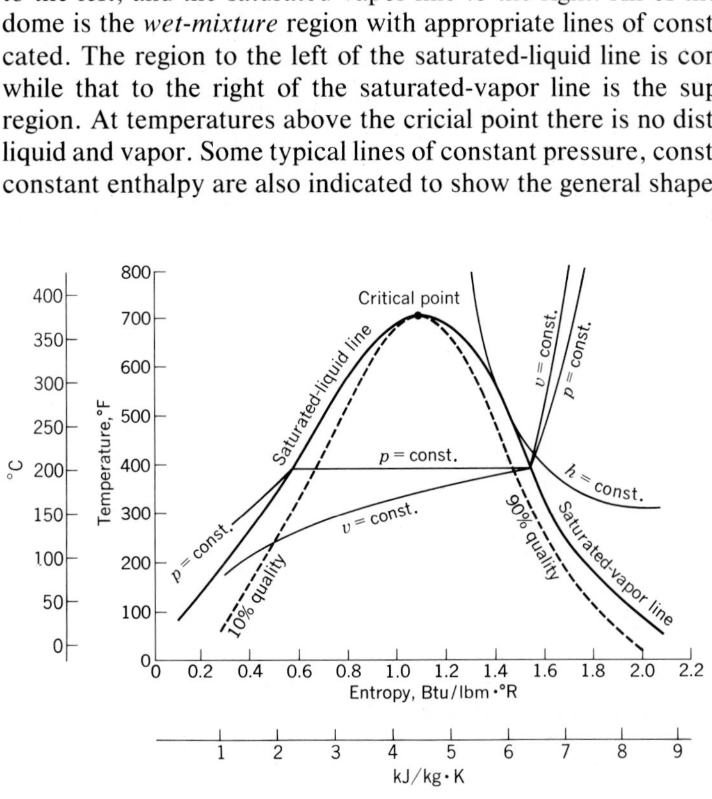

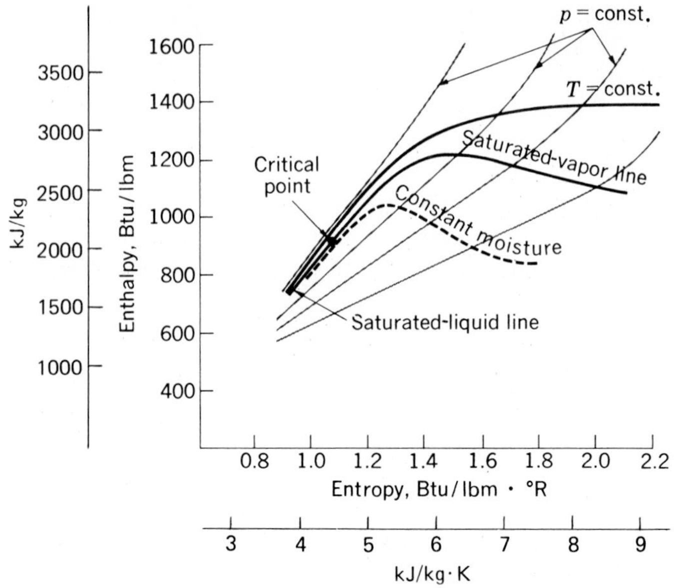

FIG. 5-14 Enthalpy-entropy diagram for water.

It should be noted that the line of constant pressure is also a line of constant temperature in the saturation region.

The Mollier Diagram

The *Mollier diagram* is a plot of enthalpy versus entropy, and a plot for steam is illustrated in Fig. 5-14. A large Mollier diagram for calculation purposes is presented as Fig. A-4 of the Appendix. The h-s diagram has an entirely different shape from the T-s diagram. The saturation line has a dome shape, but the critical point is located on the left side of the dome, with the saturated-vapor line extending over the top and to the right of the dome. Lines of constant quality also have a dome shape, and lines of constant temperature in the superheat region behave as indicated. The general shape of constant pressure lines is also shown. The h-s diagram in the Appendix covers mainly the saturation and superheat regions since they are of most interest in practical calculations.

Formulas for calculation of the entropy for solid and vapor water below 32°F were given in Chap. 3 as Eqs. (3-10) and (3-11).

5-12 CHANGE IN ENTROPY FOR AN IDEAL GAS

Equation (5-18) furnishes a basis for calculating changes in entropy between any two end states, provided a suitable reversible process can be devised between these states. An ideal gas offers a good example of the types of calculation which may be performed. The reader will recall that, for an ideal gas, enthalpy and internal energy are functions of temperature alone (we shall prove this in Chap. 6) and may be written as

$$du = c_v\, dT \tag{5-24}$$

$$dh = c_p\, dT \tag{5-25}$$

For a reversible process we may write the heat flow from Eq. (5-17) as

$$d'Q = T\, ds \tag{5-26}$$

An ideal gas is one class of simple compressible substance and thus we may write

$$d'Q = du + p\, dv = T\, ds$$

Thus

$$ds = \frac{du}{T} + \frac{p}{T}\, dv \tag{5-27}$$

From the ideal-gas equation of state we have

$$\frac{p}{T} = \frac{R}{v} \tag{5-28}$$

so that, making use of Eqs. (5-24) and (5-28), Eq. (5-27) becomes

$$ds = c_v \frac{dT}{T} + R \frac{dv}{v} \tag{5-29}$$

For constant specific heats, Eq. (5-29) may be integrated between two end states to give

$$s_2 - s_1 = c_v \ln \frac{T_2}{T_1} + R \ln \frac{v_2}{v_1} \tag{5-30}$$

We may develop an alternative expression by noting that

$$d'Q = dh - v\, dp = T\, ds$$

Again using the ideal-gas equation of state along with Eq. (5-25),

$$ds = c_p \frac{dT}{T} - R \frac{dp}{p} \tag{5-31}$$

For constant specific heat this relation may be integrated to give

$$s_2 - s_1 = c_p \ln \frac{T_2}{T_1} - R \ln \frac{p_2}{p_1} \tag{5-32}$$

It should be evident that the change in entropy given by Eqs. (5-30) and (5-32) is calculated for a reversible process between the given end states (p_1, v_1, T_1) and (p_2, v_2, T_2). However, since entropy is a property, its change between any two end states must be independent of the process; thus these relations apply to *any* ideal-gas process between these end states.

We have already shown in Chap. 2 that the *isentropic* (reversible adiabatic) process for an ideal gas with constant specific heats is given as

$$pv^\gamma = \text{const.}$$

where $\gamma = c_p/c_v$.

5-13 THE CARNOT CYCLE

We have already shown that the efficiency of all reversible cycles operating between the same temperature limits is the same and is given by the relation in Eq. (5-10). A particular cycle for which this relation applies is the *Carnot cycle,* composed of the following four reversible processes indicated in Fig. 5-15:

1 A reversible isothermal heat addition, *ab*
2 A reversible adiabatic process in which work is done by the system, *bc*
3 A reversible isothermal heat rejection, *cd*
4 A reversible adiabatic process in which work is done on the system, *da*

The cycle may involve any working substance whatsoever, just as long as these four processes are involved in sequence such that the system always returns to its original state. On a temperature-entropy diagram the Carnot cycle has a rectangular shape, as illustrated in Fig. 5-15*a*. In this diagram *da* and *bc* are the isentropic processes, while *ab* and *cd* are the reversible heat-addition and -rejection processes, respectively. From our isothermal relation derived above, the heat flows may be written

$$Q_H = T_H \, \Delta S$$

$$Q_L = T_L \, \Delta S$$

The net work output of the cycle is the difference between the heat added Q_H and the heat rejected Q_L:

$$W_{\text{net}} = Q_H - Q_L = (T_H - T_L) \, \Delta S \tag{5-33}$$

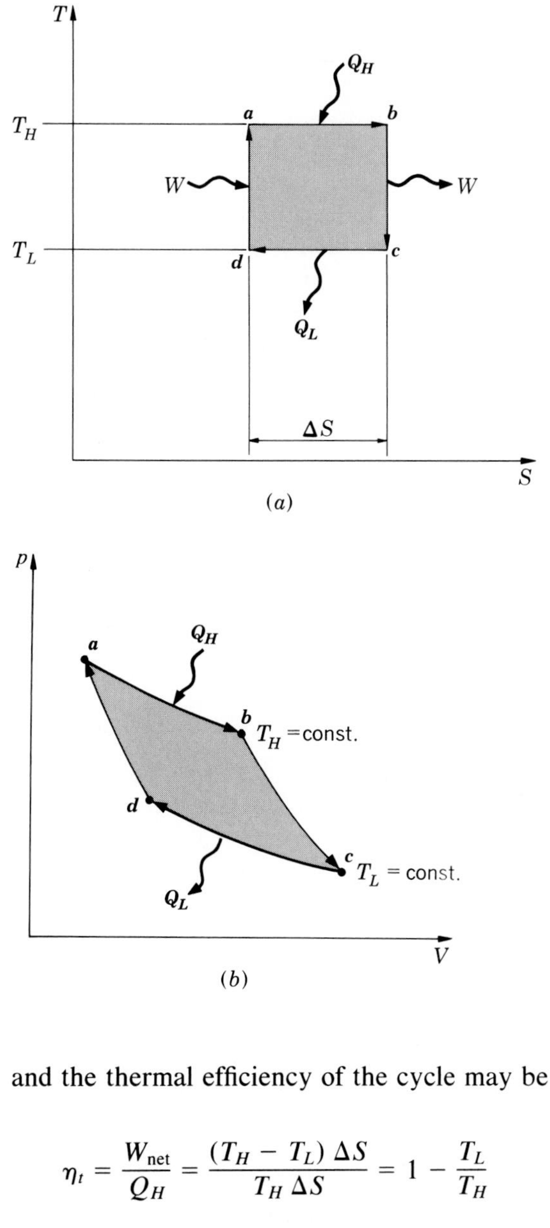

FIG. 5-15 Graphical representation of a Carnot cycle (*a*) *T-S* diagram; (*b*) *p-V* diagram for ideal gas.

and the thermal efficiency of the cycle may be written as

$$\eta_t = \frac{W_{\text{net}}}{Q_H} = \frac{(T_H - T_L)\,\Delta S}{T_H\,\Delta S} = 1 - \frac{T_L}{T_H}$$

This result agrees, of course, with Eq. (5-10), because our definition of entropy follows from the definition of the absolute temperature used to arrive at Eq. (5-10).

According to Eq. (5-33) the net work output of the Carnot cycle is given graphically as the enclosed area of the rectangle on a *T-S* diagram. It is easy to extend this statement to say that the net work output of *any reversible* cycle is the area enclosed on its corresponding *T-S* diagram.

If one were to execute the Carnot cycle with air or some other ideal gas as the working fluid, the pressure-volume diagram for the cycle would appear as in Fig. 5-15*b*. Heat addition would occur in the isothermal process from *a* to *b*, followed by the isentropic expansion from *b* to *c*. For a piston-cylinder arrangement there is work done on the face of the piston in both of these processes, and therefore a work *output*, because the force on the piston is moving in the same direction as the displacement. Carrying through the cycle, process *c-d* is the isothermal heat rejection and *d-a* is the isentropic compression. In these latter two processes there is a work *input* to the gas. For a piston-cylinder arrangement the net work output of the cycle will be the enclosed area on the *p-V* diagram and must correspond to the enclosed area on the *T-S* diagram.

EXAMPLE 5-1 Entropy Change in a Heating Process
5 lbm of steam are heated from saturated liquid at 50 psia to a temperature of 500°F. The heating process occurs at constant pressure. Calculate the change in entropy.

SOLUTION The solution to this problem involves the simple procedure of selecting values of entropy from the steam tables. The total change in entropy is then

$$\Delta S = m(s_2 - s_1)$$

At 50 psia, saturated liquid, $s_1 = 0.4113$ Btu/lbm·°R, and at 500°F the entropy is $s_2 = 1.7909$ Btu/lbm·°R. The change in entropy is

$$\Delta S = (5)(1.7909 - 0.4113) = 6.898 \text{ Btu/°R} \qquad (13.10 \text{ kJ/K})$$

EXAMPLE 5-2 Reversible Isothermal Heating
Water is evaporated in a reversible isothermal process at 200 kPa. Calculate the heat transfer per pound-mass in this process.

SOLUTION For the reversible isothermal process we have

$$Q = T\,\Delta S$$

and, for evaporation, this expression becomes

$$Q = T(s_g - s_f) = Ts_{fg}$$

For saturation conditions at 200 kPa, we obtain from the steam tables:

$$T = 120.2\text{°C} = 393.35 \text{ K}$$

$$s_{fg} = s_g - s_f = 7.1271 - 1.5301 = 5.597 \text{ kJ/kg·K}$$

The heat transfer is thus

$$Q = (393.35)(5.597) = 2201.6 \text{ kJ/kg}$$

We may note that the heat added in a constant-pressure process is also equal to the change in enthalpy which for this problem would be h_{fg}. Consulting the steam tables, we find $h_{fg} = 2201.9$ kJ/kg. The difference in the two numbers results from round-off in the tables.

EXAMPLE 5-3 Polytropic Compression of Air
Air is compressed in a reversible polytropic process from an initial state of 15 psia and 100°F to a final pressure of 50 psia. The polytropic exponent is $n = 1.3$. Calculate the change in entropy per pound-mass of air.

SOLUTION We may calculate the change in entropy from Eq. (5-32) once the final temperature is known. For the polytropic process we have,

$$pv^n = \text{const.}$$

Using the ideal-gas relation $v = RT/p$, this equation may be written as

$$\frac{T_2}{T_1} = \left(\frac{p_2}{p_1}\right)^{(n-1)/n}$$

For the conditions of this problem

$$T_2 = (100 + 460)\left(\frac{50}{15}\right)^{(1.3-1)/1.3}$$

$$= 739°\text{R} = 279°\text{F} \qquad (137.2°\text{C})$$

Inserting this temperature in Eq. (5-32) gives

$$s_2 - s_1 = 0.24 \ln \frac{739}{560} - \frac{53.35}{778} \ln \frac{50}{15}$$

$$= -0.0159 \text{ Btu/lbm}\cdot°\text{R} \qquad (-0.0666 \text{ kJ/kg}\cdot\text{K})$$

EXAMPLE 5-4 Throttling Process
Air is throttled through a porous plug like that shown in Fig. 4-5. The inlet conditions are 30 psia and 85°F and the outlet pressure is 15 psia. The process is adiabatic and kinetic energies are negligible. Calculate the change in entropy per pound of air.

SOLUTION This is an irreversible process, but the change in entropy may be calculated as if there were any reversible process between the given end states.

For an ideal gas, enthalpy is a function of temperature alone, so that the relation

$$h_1 = h_2$$

reduces to

$$T_1 = T_2$$

We may then apply Eq. (5-32) as

$$s_2 - s_1 = c_p \ln 1.0 - R \ln \frac{p_2}{p_1}$$
$$= -R \ln \frac{p_s}{p_1}$$

Using the given data, this equation becomes

$$s_2 - s_1 = -\frac{53.35}{778} \ln \frac{15}{30}$$
$$= +0.0475 \text{ Btu/lbm}\cdot{}^\circ\text{R} \qquad (0.1989 \text{ kJ/kg}\cdot\text{K})$$

Note that the entropy change is positive, as it must be for an irreversible adiabatic process.

EXAMPLE 5-5 Heat Transfer Across a Finite Temperature Difference
A heat reservoir at 1000 K is brought into thermal communication with a heat reservoir at 500 K. 1000 kJ of heat are transferred from the high-temperature reservoir to the low-temperature reservoir. Calculate the change in entropy of the universe resulting from this heat-exchange process.

FIG. EXAMPLE 5-5

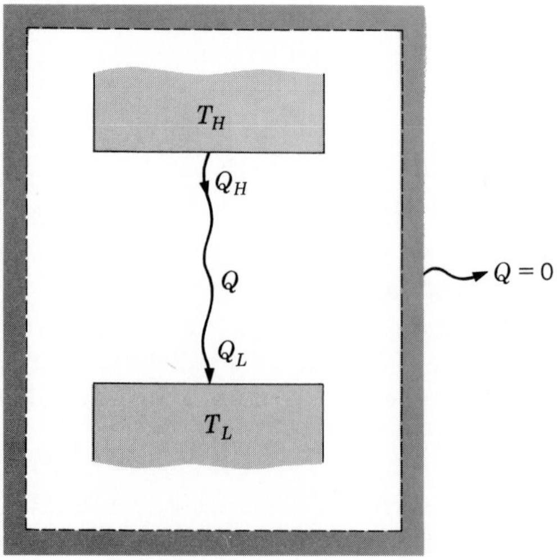

SOLUTION We may treat each of the two reservoirs as a subsystem and calculate their changes in entropy separately. The change in entropy of the universe is then the algebraic sum of these quantities. We assume an isothermal process for both reservoirs. Thus

$$\Delta S_H = \frac{Q_H}{T_H} \qquad \text{for } T_H = 1000 \text{ K}$$

$$\Delta S_L = \frac{Q_L}{\mathrm{T}_H} \qquad \text{for } T_L = 500 \text{ K}$$

$$\Delta S_H = \frac{-1000}{1000} = -1.0 \text{ kJ/k}$$

$$\Delta S_L = +\frac{1000}{500} = +2.0 \text{ kJ/K}$$

The change in entropy of the universe is

$$\Delta S_{\text{universe}} = \Delta S_H + \Delta S_L = -1.0 + 2.0$$
$$= +1.0 \text{ kJ/K}$$

The schematic diagram for this process shows that the two reservoirs taken together compose an adiabatic system. The heat-transfer process taking place inside this system is irreversible because it involves heat transfer across a finite temperature difference. We therefore have an irreversible adiabatic process insofar as the universe is concerned and experience an increase in entropy.

EXAMPLE 5-6 Isentropic Steam Turbine
Steam is expanded reversibly and adiabatically in a turbine from 400 psia and 800°F to 50 psia. Calculate the work done per pound of steam in a steady-flow process. Neglect changes in kinetic energy.

SOLUTION For a reversible adiabatic process we have

$$s_1 = s_2$$

The steady-flow energy equation for this process becomes

$$W = h_2 - h_1 \tag{a}$$

The values of h_1 and s_1 may be obtained directly from the superheat steam tables.

$$h_1 = 1416.6 \text{ Btu/lbm} \qquad (3295.0 \text{ kJ/kg})$$

$$s_1 = 1.6844 \text{ Btu/lbm·°R} \qquad (7.0514 \text{ kJ/kg·K})$$

The final state of the flow process is now determined by the two properties

$$s_2 = s_1 = 1.6844$$
$$p_2 = 50 \text{ psia}$$

We thus consult the steam tables again (superheat tables) and find that

$$T_2 = 315.2°\text{F} \qquad (157.3°\text{C})$$
$$h_2 = 1192.1 \text{ Btu/lbm} \qquad (2773 \text{ kJ/kg})$$

The work is now calculated from Eq. (*a*) as

$$\begin{aligned} W &= h_2 - h_1 \\ &= 1192.1 - 1416.6 \\ &= -224.5 \text{ Btu/lbm} \qquad (-522 \text{ kJ/kg}) \end{aligned} \tag{b}$$

The negative sign indicates that the work is delivered to things external to the turbine.

5-14 AVAILABLE AND UNAVAILABLE ENERGY

We have seen that there is some maximum efficiency for a heat engine operating between given temperature limits. For a given heat added at the upper temperature limit there is some maximum amount of work which we can obtain as an output from the device. Of course, we can increase the work output by either raising the upper temperature limit of the cycle or by reducing the lower temperature limit. In a practical sense, the lower temperature limit is not a quantity which can be reduced indefinitely; to reduce it below the surrounding atmospheric temperature we would have to provide a refrigeration device with a work input which would penalize the performance of our overall system. For this reason it is customary to speak of a lowest available temperature T_0 which is accessible to us for operation of a heat engine. This temperature is usually the local atmospheric temperature or that of some large heat sink like a lake or river.

Given some lowest available temperature T_0 at which heat can be rejected, a small reversible engine receiving heat at temperature T will produce at most a work *output* of

$$d'W = \left(1 - \frac{T_0}{T}\right) d'Q \tag{5-34}$$

where $d'Q$ is the heat added at the upper temperature. The *available energy* of a system is the portion of the heat added to the system which could be converted to work in a series of reversible engines operating between the temperature of the system and the lowest available temperature T_0. This quantity is obtained by integration of Eq. (5-34) and is designated by W_{max}:

$$W_{\text{max output}} = \int \left(1 - \frac{T_0}{T}\right) d'Q \tag{5-35}$$

The *unavailable energy* is the difference between the total heat added and the available energy.

Now let us suppose that we wish to determine the change in available energy of a system as it changes from some state 1 to state 2. We would then integrate Eq. (5-35) between these states to find the maximum increase in work output. The maximum work output occurs when heat is added reversibly so we choose a reversible process between the states to evaluate the integral

$$W_{\text{max output}} = \int_1^2 \left(1 - \frac{T_0}{T}\right) d'Q_{\text{rev}}$$

We may then recall the definition of entropy and write

$$W_{\text{max output}} = Q - T_0(S_2 - S_1) \tag{5-36}$$

where S_1 and S_2 are the initial and final entropies, respectively, and $W_{\text{max output}}$ represents the increase in available energy resulting from a reversible process between states 1 and 2.

Of course, the overriding notion in this discussion is the fact that a larger portion of heat energy is available for producing work, the higher the temperature of the system. From Eq. (5-36) we see that the *increase* in the *unavailable* energy is represented by the term $T_0(S_2 - S_1)$.

EXAMPLE 5-7 Available Energy "Loss"
How much available energy is "lost" in the process of Example 5-5, assuming $T_0 = 20°\text{C} = 293$ K?

SOLUTION If the 1000 kJ leaving the high-temperature reservoir (1000 K) were added to a reversible engine operating between that temperature and 293 K, the work produced would be

$$W = \left(1 - \frac{T_0}{T_H}\right) Q_H = \left(1 - \frac{293}{1000}\right)(1000) = 707 \text{ kJ}$$

The 707 kJ is the "available" portion of the 1000 kJ at the high-temperature reservoir. However, if this same 1000 kJ were added to a reversible engine operating between 500 and 293 K, the work produced would be only

$$W = \left(1 - \frac{T_0}{T_L}\right) Q_L = \left(1 - \frac{293}{500}\right)(1000) = 414 \text{ kJ}$$

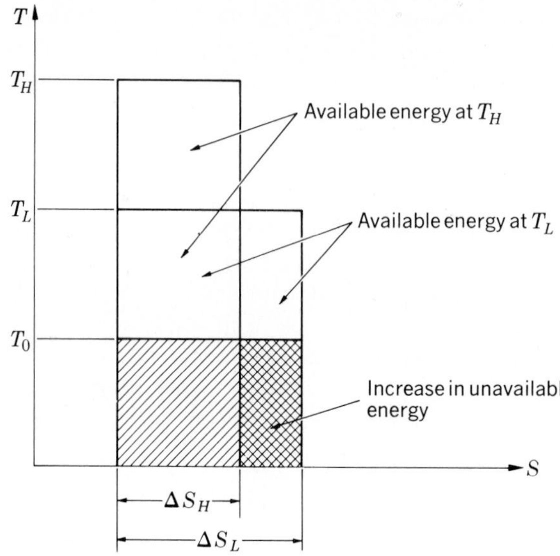

FIG. EXAMPLE 5-7

Thus, as a result of the transfer of 1000 kJ of heat across the 500 K temperature difference there has been a ''loss'' of available energy in the amount of 707 − 414, or 293 kJ. The available and unavailable energies at these two temperatures are displayed graphically in the accompanying figure. Of course, if the 1000 kJ were transferred directly to T_0 the total available energy would be lost.

EXAMPLE 5-8 Heating Increases Available Energy
How much is the available energy of 1 kg of air increased by heating it reversibly at a constant pressure of 1.5 atm from 40 to 250°C with a lowest available temperature of 20°C?

SOLUTION For this problem we find the increase in available energy by performing an integration of Eq. (5-35). For the constant-pressure process and unit mass,

$$d'Q = c_p\, dT$$

Then

$$W_{\text{max}} = \int_{T_1}^{T_2} \left(1 - \frac{T_0}{T}\right) c_p\, dT$$

$$= c_p(T_2 - T_1) - c_p T_0 \ln \frac{T_2}{T_1}$$

where we recognize that T_0 is a constant. We have $T_0 = 20°\text{C} = 293$ K; $T_1 = 40°\text{C} = 313$ K; $T_2 = 250°\text{C} = 523$ K; and c_p for air is 0.24 Btu/lbm·°F =

1005 J/kg·°C. Thus

$$W_{max} = (1005)\left[(523 - 313) - (293)\ln\frac{523}{313}\right]$$
$$= 5.988 \times 10^4 \text{ J/kg}$$

The total heat added is $c_p(T_2 - T_1) = 2.11 \times 10^5$ J/kg, so the increase in unavailable energy is $(21.1 - 5.988) \times 10^4 = 15.11 \times 10^4$ J/kg. Note that the heat-addition process causes an increase in both the available and unavailable energy of the air.

EXAMPLE 5-9 Work Addition Increases Available Energy
A 10-ft³ insulated rigid box contains air at 14.7 psia and 100°F. Paddle-wheel work is added to the box until the temperature rises to 400°F. Calculate the change in available energy of the air for $T_0 = 70$°F.

SOLUTION This example illustrates the important point that the available energy of a system may be changed by either heat *or* work additions. In other words, any change in the state of a system produces a change in the maximum amount of work which could be obtained when operating between the temperature of the system and T_0, the lowest available temperature. Let us first calculate some needed numerical quantities. The mass of the air is

$$m = \frac{p_1 V_1}{RT_1} = \frac{(14.7)(144)(10)}{(53.35)(560)} = 0.709 \text{ lbm} \qquad 0.322 \text{ kg})$$

From the first law

$$Q + W = \Delta U = mc_v(T_2 - T_1)$$

where W is the work *added*. The box is insulated so $Q = 0$ and

$$W = (0.709)(0.1715)(400 - 100) = 36.48 \text{ Btu} \qquad (38.49 \text{ kJ})$$

For constant-volume ideal-gas behavior

$$\frac{p_2}{p_1} = \frac{T_2}{T_1}$$

and

$$p_2 = (14.7)\left(\frac{860}{560}\right) = 22.58 \text{ psia} \qquad (155.7 \text{ kPa})$$

The entropy change for the air can be calculated with Eq. (5-30):

$$s_2 - s_1 = c_v \ln \frac{T_2}{T_1} + R \ln \frac{v_2}{v_1}$$

$$= 0.1715 \ln \frac{860}{560} = 0.0736 \text{ Btu/lbm·°R}$$

$$S_2 - S_1 = m(s_2 - s_1) = (0.709)(0.0736) = 0.0522 \text{ Btu/°R}$$

Now, to evaluate the change in available energy we imagine a *reversible* process between the initial and final states, and then evaluate Eq. (5-36) accordingly. Clearly, the reversible process of interest is a constant-volume heating process in which

$$Q = mc_v(T_2 - T_1) = 36.48 \text{ Btu} \qquad (38.49 \text{ kJ})$$

Then, evaluating Eq. (5-36),

$$\begin{aligned} W_{\text{max output}} &= Q - T_0(s_2 - s_1) \\ &= 36.48 - (530)(0.0522) \\ &= 8.814 \text{ Btu} \qquad (9.299 \text{ kJ}) \end{aligned}$$

The important point of this example is the fact that the available energy of a system may be changed by any action which changes the state of the system, whether the action occurs reversibly or not. In this case the change in state results from the irreversible paddle-wheel work addition, but the increase in available energy is the same as would have been experienced by a like amount of heat addition.

In summary, the 8.814 Btu represents the maximum amount of work which could be produced if the air were cooled reversibly and at constant volume from 400 to 100°F, in conjunction with a lowest available temperature of 70°F.

5-15 SECOND-LAW ANALYSIS FOR A CONTROL VOLUME

We have already discussed first-law energy analyses for a control volume (open system) in Chap. 4 and illustrated calculation techniques which may be applied to such systems. Let us now examine the control volume from the viewpoint of the second law as illustrated in Fig. 5-16. In this diagram the fluid moves through the control volume from section i to section e while work is delivered external to the control volume. We assume that the boundary of the control volume is at the environment temperature T_0 and that all of the heat transfer Q occurs at this boundary. We have already noted that entropy is a property, so it may be transported just like enthalpy or internal energy. The entropy flow into the control volume resulting from mass transport is therefore $\dot{m}_i s_i$, and the

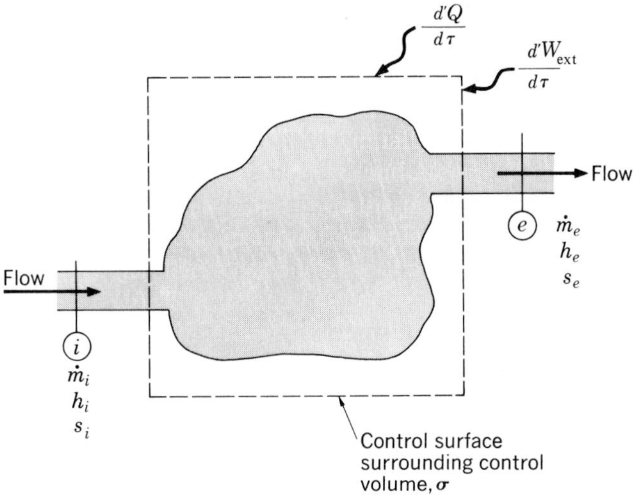

FIG. 5-16 Control volume for second-law analysis.

entropy flow out of the control volume is $\dot{m}_e s_e$, assuming that the properties are uniform at sections i and e. Entropy may also be added to the control volume because of heat transfer at the boundary of the control volume. Realizing that heat may be added at a number of locations, and that the temperature may vary, we write

$$\begin{array}{l}\text{Entropy inflow}\\ \text{from heat transfer}\\ \text{at surface of}\\ \text{control volume}\end{array} = \int_{\text{surface}} \frac{1}{T_i}\frac{d'Q_1}{d\tau} \tag{5-37}$$

where the integral is taken over all the surface of the control volume where heat transfer is present and T_i is the temperature of the surface corresponding to each $d'Q_i$. In addition, there can be an increase in entropy in the control volume resulting from mass accumulation or internal irreversibilities, fluid friction, etc. We shall designate this storage or accumulation term as $(dS/d\tau)_\sigma$ using the σ subscript as before to designate a quantity internal to the control volume. Finally, we can recognize that there may be several streams entering or leaving the control volume, and we perform the following sum over all such streams to obtain the total *entropy production* for the control volume:

$$\left(\frac{dS}{d\tau}\right)_{\text{prod}} = \underbrace{\sum_{\text{out}} \dot{m}_e s_e}_{\text{entropy outflow}} - \underbrace{\sum_{\text{in}} \dot{m}_i s_i - \int_{\text{surface}} \frac{1}{T_i}\frac{d'Q_i}{d\tau}}_{-\ \text{entropy inflow}} + \underbrace{\left(\frac{dS}{d\tau}\right)_\sigma}_{+\ \text{entropy accumulation}} \tag{5-38}$$

If one imagines this flow process as happening over a small length of time $d\tau$, Eq. (5-38) is concerned with the mass elements involved at that instant, and the right side of Eq. (5-38) is equivalent to writing for a closed system consisting of all the mass elements

$$(dS)_{\text{prod}} = \Delta S - \int \frac{d'Q}{T} \tag{5-39}$$

From the second law of thermodynamics this quantity must be greater than or equal to zero. Therefore, for Eq. (5-38) we write

$$\left(\frac{dS}{d\tau}\right)_{\text{prod}} \geqslant 0 \tag{5-40}$$

The equal sign applies to reversible processes while the inequality is applicable to irreversible processes. For steady flow and steady state there is no change with time inside the control volume, so $(dS/d\tau)_\sigma = 0$; $\dot{m}_i = \dot{m}_e$, and

$$\sum_{\text{out}} \dot{m}_e s_e - \sum_{\text{in}} \dot{m}_i s_i \geqslant \int_{\text{surface}} \frac{1}{T_i} \frac{d'Q_i}{d\tau} \tag{5-41}$$

For an adiabatic, steady-flow, steady-state process, $d'Q_i/d\tau = 0$ and Eq. (5-41) becomes

$$s_e \geqslant s_i \tag{5-42}$$

5-16 AVAILABILITY AND IRREVERSIBILITY IN STEADY FLOW

We have already seen the applicability of steady-flow energy analysis to practical power-production systems. The foregoing sections have shown how the principle of increase of entropy can place restrictions on the allowed directions of energy transfer in both flow and nonflow processes. Let us now use these principles to predict the maximum amount of work output we can obtain from a steady-flow device, as illustrated in Fig. 5-17. We have drawn the diagram with the work assumed positive leaving the control volume because we are interested in maximizing the power output. The steady-flow energy equation is therefore

$$\dot{m}(h_1 + \text{KE}_1 + \text{PE}_1 + \cdot \cdot \cdot) + \dot{Q}_0 = \dot{m}(h_2 + \text{KE}_2 + \text{PE}_2 + \cdot \cdot \cdot) + \dot{W} \tag{5-43}$$

and the entropy production rate is obtained from Eq. (5-38) as

$$\frac{dS}{d\tau} = \dot{m}s_2 - \left(\dot{m}s_1 + \frac{\dot{Q}_0}{T_0}\right) \geqslant 0 \tag{5-44}$$

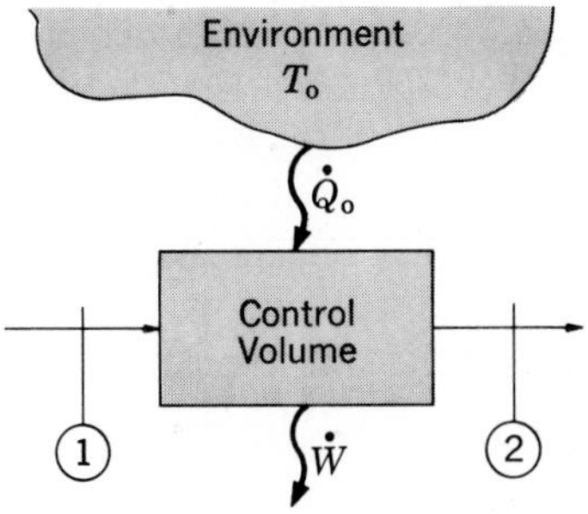

FIG. 5-17 Availability in steady flow.

where we assume that all the heat Q_0 is obtained from the environment at T_0 and this is the temperature at the boundary of the control volume. Combining Eq. (5-43) and (5-44) gives

$$\dot{m}(s_2 - s_1)T_0 - \dot{m}[(h_2 - h_1) + \text{KE}_2 - \text{KE}_1 + \text{PE}_2 - \text{PE}_1 + \dot{W}] \geq 0$$

or

$$\dot{W} \leq \dot{m}[(h + \text{KE} + \text{PE} - T_0 s)_1 - (h + \text{KE} + \text{PE} - T_0 s)_2] \tag{5-45}$$

In many practical problems the kinetic and potential energy changes are negligible and we obtain

$$\dot{W} \leq \dot{m}[(h - T_0 s)_1 - (h - T_0 s)_2] \tag{5-46}$$

The quantity $h - T_0 s$ is called the *steady-flow availability function b*

$$b = h - T_0 s \tag{5-47}$$

and the maximum work output per unit mass will be equal to the decrease in this function,

$$\dot{W}_{\text{max output}} = \dot{m}(b_1 - b_2) \tag{5-48}$$

The actual work output will be less than this value. The *irreversibility* is defined as the difference between the maximum possible work output and the actual work output, or

$$\dot{I} = \dot{W}_{\text{max}} - \dot{W}_{\text{act}} = \dot{m}(b_1 - b_2) - \dot{W}_{\text{act}} \tag{5-49}$$

If the process is adiabatic and kinetic energies are negligible, the actual work output must be equal to the decrease in enthalpy, or

$$\dot{W}_{\text{act}} = \dot{m}(h_1 - h_2) \tag{5-50}$$

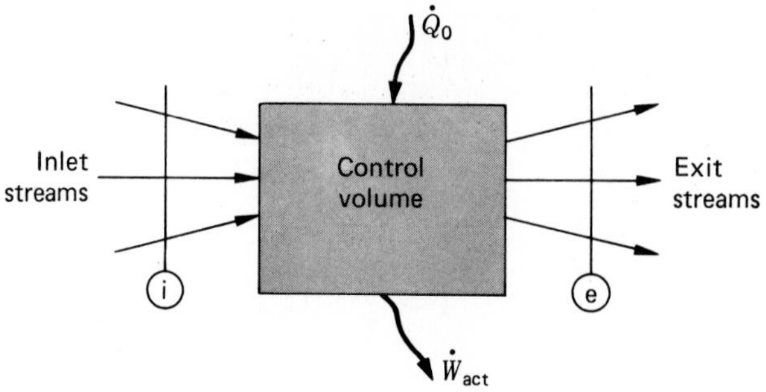

FIG. 5-18 Irreversibilities for multiple inlet-exit streams.

Combining the above relations for an *adiabatic* process yields

$$\begin{aligned}\dot{I} &= \dot{m}[(h - T_0 s)_1 - (h - T_0 s)_2] - \dot{m}(h_1 - h_2) \\ &= \dot{m}T_0(s_2 - s_1)\end{aligned} \tag{5-51}$$

When multiple streams enter and leave the control volume, as shown in Fig. 5-18, a summation must be performed over all the availability functions to yield

$$\dot{W}_{max} = \Sigma\, \dot{m}_i b_i - \Sigma\, \dot{m}_e b_e \tag{5-52}$$

and

$$\dot{I} = \dot{W}_{max} - \dot{W}_{act} = \Sigma\, \dot{m}_i b_i - \Sigma\, \dot{m}_e b_e - \dot{W}_{act} \tag{5-53}$$

For an adiabatic process this reduces to

$$\dot{I} = T_0(\Sigma\, \dot{m}_e s_e - \Sigma\, \dot{m}_i s_i) \tag{5-54}$$

If kinetic and potential energies are significant, they can be included in the availability function.

The availability function is useful for analyzing real processes to see how closely they approximate an ideal reversible process. We should note that this quantity is a function of both the fluid properties and the surroundings tempera-

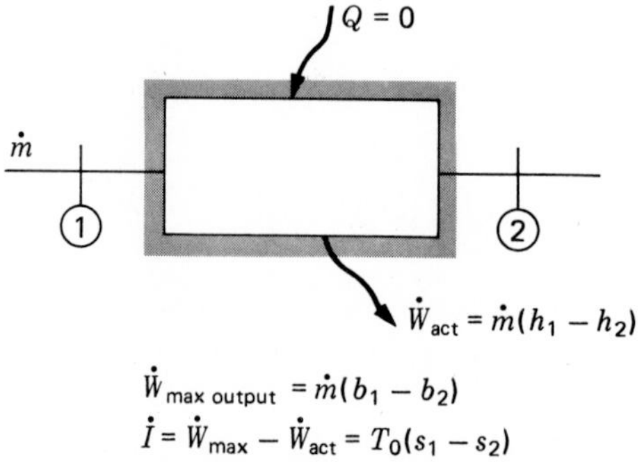

FIG. 5-19 Availability decrease and irreversibility for adiabatic steady flow process with KE ≈ 0.

ture, which comes into play because any heat transfer across a finite temperature difference is irreversible and generates an increase in entropy at the boundary of the control volume. In a reversible adiabatic process there is no heat transfer ($s_1 = s_2$), and the irreversibility is zero.

5-17 PRACTICAL CONSIDERATIONS WITH AVAILABILITY

A few remarks concerning the practicality of irreversibility analysis are in order at this point. In choosing between alternatives for power-production or energy-conversion devices several factors must be brought into play. Naturally, one must be able to design and build the device so that it operates in a reliable fashion. Beyond this general statement, there must be considerations of initial cost of the equipment as well as the annual operating cost, which is usually a strong function of energy costs. Capital costs depend on many factors such as material and fabrication expenses and, not the least, the care with which engineering design of equipment is effected. Costs of energy consumption are somewhat different. We can make a fairly general statement that energy-consumption rates in power-producing devices are generally lower, the lower the generation of irreversibility.

Thus we may anticipate that a process or device which produces the least irreversibility will probably involve the least energy cost. This is not an absolute principle by any means, but it does provide us with a technique for evaluation of technical alternatives.

First- and Second-Law Efficiencies

Consider the adiabatic turbine and compressor shown in Fig. 5-20. For negligible kinetic energies the actual work *output* of the turbine is

$$\dot{W}_{act} = \dot{m}(h_1 - h_2)$$

and the work *input* to the compressor is

$$\dot{W}_{act} = \dot{m}(h_2 - h_1)$$

FIG. 5-20 (*a*) Turbine; (*b*) compressor.

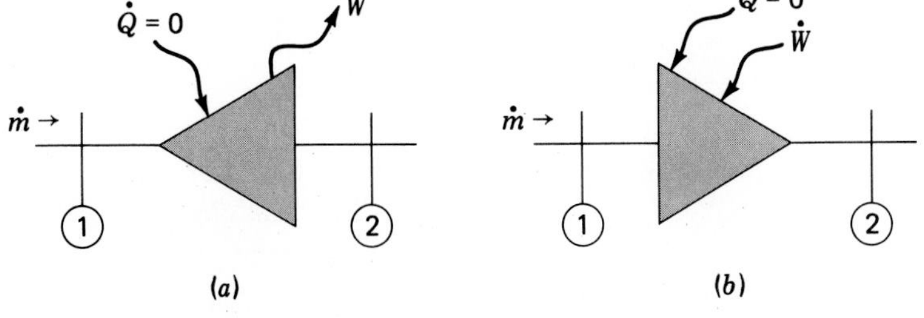

If the processes were also reversible, i.e., isentropic, the respective work terms would be

$$\dot{W}_{\text{isen out}} = \dot{m}(h_1 - h_{2s})$$

$$\dot{W}_{\text{isen in}} = \dot{m}(h_{2s} - h_1)$$

where the s subscript refers to an isentropic process, and we could define first-law efficiencies for the devices as

$$\eta_{\text{turb}} = \frac{\text{actual work output}}{\text{isentropic work output}} = \frac{h_1 - h_2}{h_1 - h_{2s}} \tag{5-55}$$

$$\eta_{\text{comp}} = \frac{\text{isentropic work input}}{\text{actual work input}} = \frac{h_{2s} - h_1}{h_2 - h_1} \tag{5-56}$$

These are so-called *first-law efficiencies,* which compare the actual process with an idealized process.

In the above discussions we have seen that given two end states there is a maximum work output which may be obtained which is calculated by the change in the availability function. A *second-law efficiency* may be defined which compares this maximum work with the actual work. Thus, for the turbine

$$\eta_{\text{2d law}} = \frac{\text{actual work output}}{\begin{array}{c}\text{maximum work output}\\ \text{for end states}\end{array}} = \frac{h_1 - h_2}{b_1 - b_2} \quad \text{for a turbine} \tag{5-57}$$

$$\eta_{\text{2d law}} = \frac{\text{reversible work input}}{\text{actual work input}} = \frac{b_2 - b_1}{h_2 - h_1} \quad \text{for a compressor} \tag{5-58}$$

Obviously, there are many other factors to consider in the design of practical equipment including ease of maintainance, labor skills required for operation, and the physical availability of a particular energy source. For example, it would make no sense to consider natural gas as an energy source in some region where natural gas is unavailable.

The examples that follow illustrate how the concepts of availability and irreversibility may be used for analysis purposes.

EXAMPLE 5-10

A steady-flow compressor is used to compress air from 1 atm, 25°C to 8 atm in an adiabatic process. The first-law efficiency for the process is 87 percent. Calculate the irreversibility for the process and the second-law efficiency. Take T_0 = 20°C = 293 K.

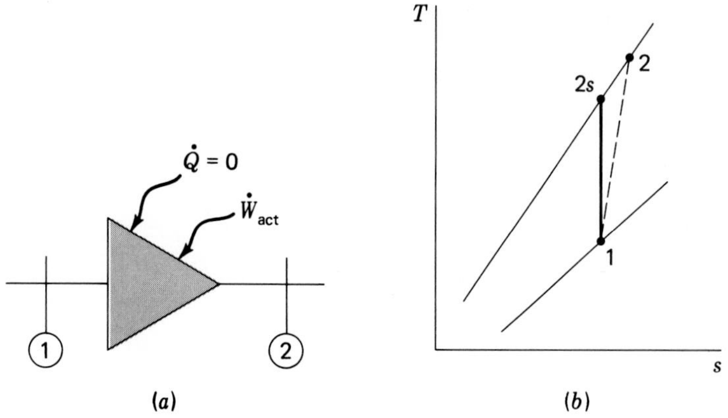

FIG. EXAMPLE 5-10

SOLUTION The compressor is shown in the accompanying figure along with a *T-S* diagram for the process.

We have $T_1 = 25°C = 298$ K, $p_1 = 1$ atm, $p_2 = 8$ atm. For unit mass the actual work input is

$$\dot{W}_{act} = h_2 - h_1 = c_p(T_2 - T_1) \qquad (a)$$

Were the process reversible, i.e., isentropic, the work input would be

$$\dot{W}_{isen} = h_{2s} - h_1 = c_p(T_{2s} - T_1) \qquad (b)$$

Taking air as an ideal gas T_{2s} is calculated from

$$T_{2s} = T_1 \left(\frac{p_2}{p_1}\right)^{(\gamma-1)/\gamma} = 298 \left(\frac{8}{1}\right)^{(1.4-1)/1.4} = 540 \text{ K}$$

and the isentropic work is

$$\dot{W}_{isen} = (1.005 \text{ kJ/kg°C})(540 - 298)$$
$$= 243.3 \text{ kJ/kg}$$

From the given first-law efficiency,

$$\eta_{comp} = \frac{\dot{W}_{isen}}{\dot{W}_{act}} = 0.87 \qquad (c)$$

so that

$$\dot{W}_{act} = \frac{243.3}{0.87} = 280 \text{ kJ/kg}$$

Then from Eq. (*a*) we find

$$T_2 = \frac{W_{\text{act}}}{c_p} + T_1$$
$$= \frac{280}{1.005} + 298 = 576 \text{ K}$$

The maximum work output is calculated as the change in the availability function:

$$\dot{W}_{\max} = b_1 - b_2 = h_1 - h_2 - T_0(s_1 - s_2) \qquad (d)$$

and the change in entropy is given by Eq. (5-32):

$$s_1 - s_2 = c_p \ln \frac{T_1}{T_2} - R \ln \frac{p_1}{p_2}$$
$$= (1005 \text{ J/kg}\cdot{}^\circ\text{C}) \ln \frac{298}{576} - (287 \text{ J/kg}\cdot\text{K}) \ln \frac{1}{8}$$
$$= -66.06 \text{ J/kg}\cdot\text{K}$$

so that from Eq. (*d*)

$$\dot{W}_{\max} = -280000 - (293)(-66.06)$$
$$= -260\,644 \text{ J/kg}$$

Of course, this is equivalent to a work *input* of +260.6 kJ/kg. The irreversibility from Eq. (5-51) for an adiabatic process is

$$\dot{I} = T_0(s_2 - s_1) = (293)(66.06) = +19\,356 \text{ J/kg}$$

Using Eq. (5-58), the second-law efficiency is calculated as

$$\eta_{\text{2d law}} = \frac{b_2 - b_1}{h_2 - h_1} = \frac{260.6 \text{ kJ/kg}}{280 \text{ kJ/kg}}$$
$$= 0.931$$

EXAMPLE 5-11
The closed feedwater heater of Example 4-6 is to be analyzed in respect to the irreversibility which is generated in the heat-exchange process. We shall make the calculation for a lowest available temperature of 20°C.

SOLUTION The schematic diagram of Example 4-6 is reproduced here for convenience. The actual work for this process is zero and the maximum work output can be expressed in terms of the availability functions as

$$\dot{W}_{\max} = \dot{m}_1(b_1 - b_2) + \dot{m}_s(b_3 - b_4) \qquad (a)$$

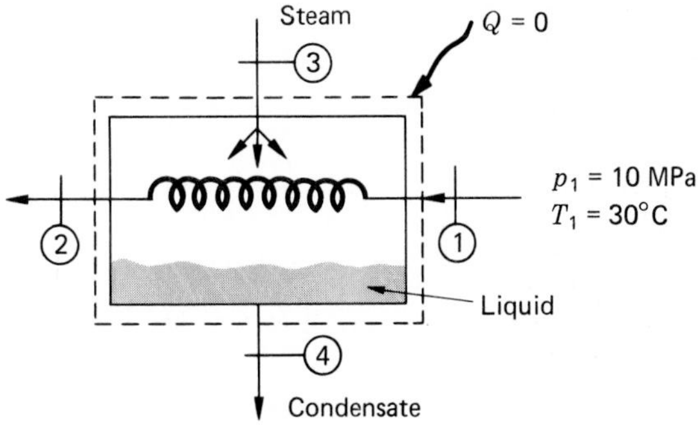

FIG. EXAMPLE 5-11

The entropies for these calculations are obtained from the steam tables as

$s_1 = 0.4316$ kJ/kg·K (compressed liquid, 10 MPa, 30°C)

$s_2 = 2.3016$ kJ/kg·K (compressed liquid, 10 MPa, 198.3°C)

$$\begin{aligned} s_3 &= s_{f_3} + x_3 s_{fg_3} \\ &= 2.3150 + (0.95)(6.4448 - 2.3150) \\ &= 6.2383 \text{ kJ/kg·°C} \end{aligned}$$

$$s_4 = s_f \text{ at } 1.5 \text{ MPa} = 2.3150 \text{ kJ/kg·°C}$$

Equation (*a*) may now be evaluated with $m_1 = 1.0$, $m_s = 0.3858$, and $T_0 = 20°C = 293$ K:

$$\begin{aligned} \dot{W}_{max} &= [(h_1 - h_2) - T_0(s_1 - s_2)] + \dot{m}_s[(h_3 - h_4) - T_0(s_3 - s_4)] \\ &= [(134.86 - 848.54) - (293)(0.4316 - 2.3016)] \\ &\quad + (0.3858)[(2694.8 - 844.89) - (293)(6.2383 - 2.3150)] \\ &= 104.44 \text{ kJ/kg water} \end{aligned} \qquad (b)$$

Because $\dot{W}_{act} = 0$, the irreversibility is

$$\dot{I} = \dot{W}_{max} - \dot{W}_{act} = 104.44 \text{ kJ/kg water}$$

EXAMPLE 5-12 Alternatives for An Air Heater

We wish to evaluate the heating of 1.0 lbm/s of air in a constant-pressure process from 50 to 120°F using four types of heaters:

1. Condensing steam at 1 atm with a change from saturated vapor to saturated liquid
2. Using hot water entering the heat exchanger at 200°F and leaving at 150°F
3. Using hot water entering the heat exchanger at 150°F and leaving at 100°F
4. Using an electric heater

For all the alternatives it is assumed that there are no heat losses from the heat exchanger. The schematic of the system is shown in the accompanying figure.

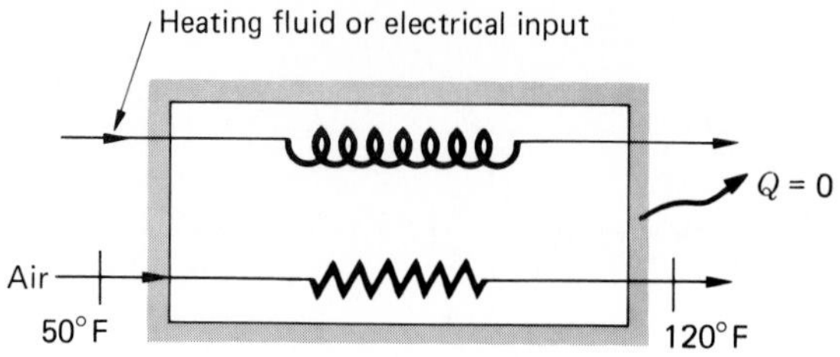

FIG. EXAMPLE 5-12

We wish to evaluate the overall irreversibility for each alternative and comment on the advisability of using each in a practical situation. The lowest available temperature is taken as $T_0 = 70°F = 530°R$.

SOLUTION The heat which must be transferred to the air in each system is

$$q = \dot{m}_a(h_2 - h_1)_a = (1)(0.24)(120 - 50) = 16.8 \text{ Btu/s} \qquad (17.7 \text{ kW})$$

In this process the decrease of the availability of the air is

$$\dot{m}_a(-\Delta b_a) = \dot{m}_a[(h_1 - T_0 s) - (h_2 - T_0 s)]_a \qquad (a)$$

The entropy change for the air in a constant-pressure process is

$$\begin{aligned} s_2 - s_1 &= c_p \ln \frac{T_2}{T_1} \\ &= 0.24 \ln \frac{580}{510} = 0.03087 \text{ Btu/lbm·°R} \\ &= 0.1292 \text{ kJ/kg·K} \end{aligned} \qquad (b)$$

Then, the decrease in the availability is

$$\begin{aligned} -\Delta b_a &= (1.0)[(0.24)(50 - 120) - (530)(-0.030\,87)] \\ &= -0.4389 \text{ Btu/s} \qquad (-0.463 \text{ kW}) \end{aligned} \qquad (c)$$

This result must now be combined with the decrease in availability for the heater or heating fluid to evaluate the overall performance.

Case 1: Condensing Steam The properties of steam at 1 atm of interest are

$$h_1 - h_2 = -h_{fg} = 970.4 \text{ Btu/lbm}$$

$$s_1 - s_2 = s_{fg} = 1.4446 \text{ Btu/lbm·°R}$$

The energy gained by the air is equal to the energy lost by the steam, and so

$$q = 16.8 = \dot{m}_s h_{fg} \qquad (d)$$

and

$$\dot{m}_s = 0.017\,31 \text{ lbm/s} \qquad (0.00785 \text{ kg/s})$$

The decrease in availability of the steam is then

$$\begin{aligned} \dot{m}_s(-\Delta b_s) &= (0.0173)[970.4 - (530)(1.4446)] \\ &= +3.5424 \text{ Btu/s} \qquad (3.736 \text{ kW}) \end{aligned}$$

The overall decrease in availability is

$$\begin{aligned} \dot{m}_a(-\Delta b_a) + \dot{m}_s(-\Delta b_s) &= -0.4389 + 3.5424 \\ &= +3.1035 \text{ Btu/s} \end{aligned}$$

This is the maximum work *output*. The *actual* work output is zero, so the irreversibility is

$$\begin{aligned} \dot{I} = \dot{W}_{\text{max}} - \dot{W}_{\text{act}} = 3.1035 - 0 &= +3.1035 \text{ Btu/s} \\ &= 3.274 \text{ kW} \end{aligned} \qquad (e)$$

Case 2: Hot Water Source of Heating For this case the appropriate water properties are

$$h_1 = h_f \text{ at } 200°\text{F} = 168.07 \text{ Btu/lbm}$$

$$h_2 = h_f \text{ at } 150°\text{F} = 117.96 \text{ Btu/lbm}$$

$$s_1 = s_f \text{ at } 200°\text{F} = 0.2940 \text{ Btu/lbm·°R}$$

$$s_2 = s_f \text{ at } 150°\text{F} = 0.2150 \text{ Btu/lbm·°R}$$

The energy balance enables a calculation of the water mass flow from

$$q = 16.8 \text{ Btu/s} = \dot{m}_w(h_1 - h_2)_w$$

$$\dot{m}_w = 0.3353 \text{ lbm/s} \qquad (0.1521 \text{ kg/s})$$

The corresponding decrease in the availability of the water is

$$\begin{aligned} \dot{m}_w = (-\Delta b_w) &= (0.3353)[(168.07 - 117.96) - (530)(0.2940 - 0.2150)] \\ &= +2.7629 \text{ Btu/s} \qquad (2.91 \text{ kW}) \end{aligned}$$

Now the overall decrease in the availability is the sum of that for the water and that for the air, or

$$-0.4389 + 2.7629 = 2.3240 \text{ Btu/s} \qquad (2.451 \text{ kW})$$

As in case 1, this is the irreversibility:

$$\dot{I} = +2.3240 \text{ Btu/s} \qquad (2.451 \text{ kW}) \qquad (f)$$

Case 3: Hot Water Heating at Lower Temperature For this case we should remark that a larger heat exchanger would be required because of the lower temperature difference between the hot water and air. The water properties of interest are

$$h_1 = h_f \text{ at } 150°\text{F} = 117.96 \text{ Btu/lbm}$$

$$h_2 = h_f \text{ at } 100°\text{F} = 68.05 \text{ Btu/lbm}$$

$$s_1 = s_f \text{ at } 150°\text{F} = 0.2150 \text{ Btu/lbm·°R}$$

$$s_2 = s_f \text{ at } 100°\text{F} = 0.1296 \text{ Btu/lbm·°R}$$

As in case 2, we can use an energy balance and obtain the water mass flow as

$$\dot{m}_w = 0.3366 \text{ lbm/s}$$

The decrease in the availability for the water is then

$$\begin{aligned} \dot{m}_w(-\Delta b_w) &= \dot{m}_w[(h_1 - T_0 s_1) - (h_2 - T_0 s_2)]_w \\ &= (0.3366)[(117.96 - 68.05) - (530)(0.2150 - 0.1296)] \qquad (g) \\ &= +1.5645 \text{ Btu/s} \qquad (1.650 \text{ kW}) \end{aligned}$$

The overall decrease in availability is then the sum of that for the air and that for the water, and this is equal to the irreversibility:

$$-0.4389 + 1.5645 = +1.1256 \text{ Btu/s} = \dot{I} \qquad (h)$$

Case 4: Electric Heating For this case there is no heating fluid and the overall decrease in availability is that for the air, or

$$\dot{m}_a(-\Delta b_a) = -0.4389 \text{ Btu/s} \qquad (-0.463 \text{ kW})$$

This represents the maximum work output for this heater system. Unlike the three previous cases, however, the actual work is no longer zero. Now there is an electrical work *input* of 16.8 Btu/s (or an actual work *output* of *minus* 16.8 Btu/s). The irreversibility is thus

$$\begin{aligned} \dot{I} &= \dot{W}_{max} - \dot{W}_{act} \\ &= -0.4389 - (-16.8) \qquad (i) \\ &= +16.3611 \text{ Btu/s} \qquad (+17.258 \text{ kW}) \end{aligned}$$

The four cases are summarized in the tabulation below. The first three cases show a drop in irreversibility as the temperature of the heating fluid is dropped. This indicates that lower irreversibility is experienced when there is a smaller temperature difference across which the heat is to be transferred. The last case shows the high irreversibility which results from a conversion of work (in this case electrical work) to heat.

Example 5-12 illustrates one approach to a second-law analysis of energy-transfer schemes. We must caution the reader, however, that the analysis is incomplete. One should also evaluate the irreversibility which results from producing the steam or hot water and the losses and irreversibility creation which can take place at the power plant where electricity is generated. The purpose here has been to employ availability and irreversibility analysis as a tool to indicate the relative second-law effectiveness of heating schemes.

Heater	Decrease in Availability, Btu/s	Actual Work Output, Btu/s	Irreversibility	
			Btu/s	kW
Steam, 1 atm, 212°F	3.1035	0	3.1035	3.274
Hot water, 200 to 150°F	2.3240	0	2.3240	2.451
Hot water, 150 to 100°F	1.1256	0	1.1256	1.187
Electric heater	−0.4389	−16.8	16.3611	17.258

EXAMPLE 5-13 Irreversibility in a Steam Turbine
The adiabatic turbine of Example 5-6 is operated so that the steam enters at 800°F and 400 psia and exits at 50 psia and 400°F. Calculate the maximum work output which could be obtained between these end states and the irreversibility. Neglect kinetic energies. The lowest available temperature is 70°F. Also calculate the second-law efficiency.

SOLUTION The properties of interest are

$$h_1 = 1416.6 \text{ Btu/lbm} \qquad \text{at } 800°\text{F and } 400 \text{ psia}$$

$$s_1 = 1.6844 \text{ Btu/lbm·°R}$$

$$h_2 = 1235.0 \text{ Btu/lbm} \qquad \text{at } 400°\text{F and } 50 \text{ psia}$$

$$s_2 = 1.7370 \text{ Btu/lbm·°R}$$

The actual work output is

$$\dot{W}_{act} = h_1 - h_2 = 181.6 \text{ Btu/lbm} \qquad (422.4 \text{ kJ/kg})$$

while the maximum possible work output is obtained as the decrease in the availability function:

$$\begin{aligned}\dot{W}_{\text{max output}} &= (h - T_0 s)_1 - (h - T_0 s)_2 \\ &= h_1 - h_2 - T_0(s_1 - s_2) \\ &= 181.6 - (530)(1.6844 - 1.7370) \\ &= 209.48 \text{ Btu/lbm} \qquad (487.25 \text{ kJ/kg})\end{aligned}$$

The irreversibility is therefore

$$\dot{I} = \dot{W}_{max} - \dot{W}_{act}$$
$$= 209.48 - 181.6 = 27.88 \text{ Btu/lbm} \qquad (64.85 \text{ kJ/kg})$$

The second-law efficiency is

$$\eta_{2d\ law} = \frac{\dot{W}_{act}}{\dot{W}_{max}} = \frac{181.6}{209.48} = 0.867$$

5-18 AVAILABILITY IN CLOSED SYSTEMS

Consider a closed system as shown in Fig. 5-21 which is exposed to local atmospheric conditions of p_0 and T_0. We wish to determine the maximum work output exclusive of work to displace the environment. Because the system is not, in general, in equilibrium with the environment we connect a reversible engine E between the two. The maximum useful work output, exclusive of work to displace the environment, is thus

$$d'W_{max\ useful} = d'W - (-p_0 dV) + d'W_E \qquad (5\text{-}59)$$

For a reversibie process of the closed system

$$d'W_{rev} = dE - d'Q_{E_{rev}}$$

For the reversible engine operating in a cycle,

$$d'W_E = -d'Q_E - d'Q_0$$

From the definition of the absolute thermodynamic temperature scale

$$\frac{d'Q_0}{T_0} = -\frac{d'Q_E}{T}$$

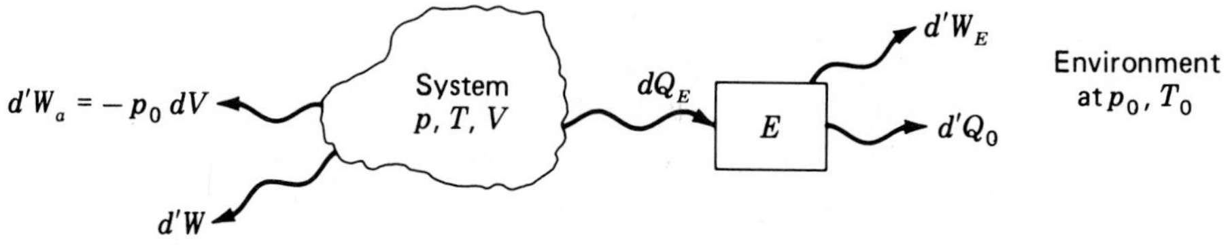

FIG. 5-21 Availability in closed systems.

For a reversible process $d'Q_{E,\mathrm{rev}} = T\,ds$ so that

$$d'W_E = d'Q_{E,\mathrm{rev}} - d'Q_{E,\mathrm{rev}}\left(\frac{T_0}{T}\right) = d'Q_{E,\mathrm{rev}} - T_0\,ds$$

Equation (5-59) then becomes

$$\begin{aligned} d'W_{\text{max useful}} &= dE + p_0\,dV + (d'Q_{E,\mathrm{rev}} - T_0\,ds) - d'Q_{E,\mathrm{rev}} \\ &= dE + p_0\,dV - T_0\,ds \end{aligned} \tag{5-60}$$

If we now allow the whole assembly to undergo a process between two states 1 and 2, we obtain

$$W_{\text{max useful}} = E_2 - E_1 + p_0(V_2 - V_1) - T_0(s_2 - s_1) \tag{5-61}$$

If the internal thermal energy U is the only important variable in E, then we may write the equation on a unit mass basis as

$$W = (u_2 - u_1) + p_0(v_2 - v_1) - T_0(s_2 - s_1) \tag{5-62}$$

Let us now allow the system to proceed from its initial state at p, v, T to a point where it is in equilibrium with the environment at p_0, v_0, T_0. This final state is called the *dead state*. The maximum useful work output is then

$$W_{\text{max useful}} = (u_0 - u) + p_0(v_0 - v) - T_0(s_0 - s)$$

Common practice is to define a closed system availability ϕ as

$$\phi = (u + p_0 v - T_0 s) - (u_0 + p_0 v_0 - T_0 s) \tag{5-63}$$

which we note, of course, is a function of both the system and its surroundings.

For a change between given states 1 and 2, the maximum work output is thus

$$W_{\text{max useful}} = \Delta\phi = \phi_2 - \phi_1 \tag{5-64}$$

As with our earlier concept of available energy, the availability of a closed system can be increased by adding work to the system or by adding heat.

EXAMPLE 5-14

2 kg of air are contained in a rigid vessel at 500 kPa and 700 K. The dead state is taken as 20°C and 100 kPa. Calculate the maximum useful work (*a*) if the system were to change to the dead state and (*b*) when the air is cooled to 400 K in the rigid container.

SOLUTION For part (*a*) the maximum useful work is equal to the availability function. We first calculate the needed properties:

$$v_1 = \frac{RT_1}{p_1} = \frac{(287)(700)}{500 \times 10^3} = 0.4018 \text{ m}^3/\text{kg}$$

$$v_0 = \frac{RT_0}{p_0} = \frac{(287)(293)}{100 \times 10^3} = 0.8409 \text{ m}^3/\text{kg}$$

$$\begin{aligned} s - s_0 &= c_p \ln \frac{T_1}{T_0} - R \ln \frac{p_1}{p_0} \\ &= 1005 \ln \frac{700}{298} - 287 \ln \frac{500}{100} \\ &= 396.3 \text{ J/kg·K} \end{aligned}$$

$$u_1 - u_0 = c_v(T_1 - T_0) = (718)(700 - 293) = 2.922 \times 10^5 \text{ J/kg}$$

and then

$$\begin{aligned} \Phi_1 &= m[(u_1 - u_0) + p_0(v_1 - v_0) - T_0(s_1 - s_0)] \\ &= 2[2.922 \times 10^5 + (100 \times 10^3)(0.4018 - 0.8409) - (293)(396.3)] \\ &= 264.3 \text{ kJ} \end{aligned}$$

For part (*b*) the maximum useful work is

$$W_{\text{max useful}} = m(\phi_2 - \phi_1)$$

Because the container is rigid, $v_2 = v_1$ and

$$p_2 = p_1 \frac{T_2}{T_1} = 500 \frac{400}{700} = 286 \text{ kPa}$$

The change in entropy is thus

$$\begin{aligned} s_2 - s_1 &= 1005 \ln \frac{400}{700} - 287 \ln \frac{286}{500} \\ &= -402.1 \text{ J/kg K} \end{aligned}$$

and because $v_2 = v_1$

$$\begin{aligned} W_{\text{max useful}} &= m[(u_2 - u_1) - T_0(s_2 - s_1)] \\ &= 2[(718)(400 - 700) - 293)(-402.1)] \\ &= -195.2 \text{ kJ} \end{aligned}$$

Thus the maximum useful work output is 195.2 kJ.

5-19 MICROSCOPIC INTERPRETATION OF ENTROPY

For those readers who choose to study the principles of microscopic thermodynamics it can be shown that there is a characteristic function of state which behaves like the macroscopic entropy we have described in this chapter. In an isolated system of particles (no mass or energy added to the system) it is found that the system tends to move toward a most probable distribution of the particles among the allowable energy states. It is also shown that deviations from the most probable state are very unlikely indeed. This is equivalent to saying that the entropy of an isolated system tends to increase to its maximum value, and it is highly improbable that it will decrease from this maximum. A microscopic definition of entropy thus furnishes the same information as Eq. (5-22), except that the principle of increase of entropy becomes one which represents a *most probable* behavior. A further implication of the microscopic definition is that the second law of thermodynamics becomes a statement of what will most probably happen. But we are no longer permitted to use the word "impossible."

A further study of the distribution of energy at the microscopic level will show that our simplified kinetic theory model of an ideal gas presented in Chap. 1 turns out to be correct. For gases with wide spacing between molecules we are able to identify the temperature of the gas with the mean kinetic energy of the molecules, even when one takes into account the allowable quantum energy states of the molecules.

5-20 MAXWELL'S DEMON AND INFORMATION THEORY

An interesting sidelight to the second law was given by J. C. Maxwell in 1871. Imagine the container shown in Fig. 5-22 with the two compartments *A* and *B* filled with identical gases at the same temperatures. Some of the molecules move fast and some move slowly, but the *average* kinetic energy of the molecules is the same in both chambers, in accordance with our simple discussion in Sec. 1-14. Maxwell imagined what might happen if there were a small door in the separating partition and a small "demon" intelligent and quick enough to open and close the door very rapidly so that the faster molecules in section *A*

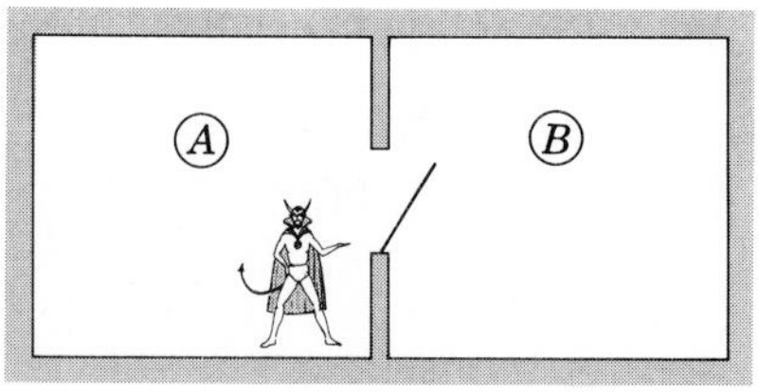

FIG. 5-22 Maxwell's demon "fast" molecules to one side, "slow" molecules to the other side.

would be allowed to pass through to section *B*, while the slower molecules from *B* would be allowed to pass through into *A*. It is easy to see that the demon would be a remarkable person indeed if he were able to perform this task because the separation of fast and slow molecules would result in a temperature difference between the chambers. Such an occurrence is a violation of the second law of thermodynamics because it is, in effect, a transfer of heat "uphill on the temperature scale."

The concept of the Maxwell demon has been debated for years. The general tenor of the arguments is that the demon must be considered as part of the overall thermodynamic system under analysis. By some means or other, the demon must be able to recognize whether a molecule is moving "fast" or "slow." This requires that some measurement be performed and information about the measurement be communicated to the demon. The conclusion of a considerable body of opinion is that either (1) the demon will be unable to distinguish between fast and slow molecules because he is subjected to a random molecular bombardment or (2) the entropy of the demon must change to account for the "information" he receives in the measurement process. This latter argument pertaining to information has led to the use of the term *entropy* in the science of information theory, and has even prompted some authors to develop the entire subject of thermodynamics on the basis of concepts from information theory.† In this development, entropy is taken to be a measure of our "increasing lack of information."

In our discussion we shall prefer to think of the principles of thermodynamics as experimental axioms which are based on observations of the physical behavior of nature. We have no means of determining *why* nature behaves the way it does, but it would not seem reasonable to assume this behavior is caused by our lack of information. Such reasoning is analogous to an individual saying, "I wrote some checks which I forgot to list in my checkbook, so I do not have enough information to determine my bank balance." This individual's lack of information does not alter or affect in any way the accounting procedure at the bank, and if he or she tries to violate that procedure or "principle" by writing too many checks there will be trouble, regardless of any personal "lack of information" about the balance. If the individual wants to get the true story, he or she should experimentally observe the accounting procedure at the bank and duplicate it very carefully in the checkbook. Such an observation-and-duplication process is exactly what we try to accomplish in thermodynamic analysis, although it is not so easy to perform as the checkbook accounting process.

REVIEW QUESTIONS

1 What is the first law of thermodynamics?
2 Why is heat a different form of energy from work?

† See Ref. [7].

3 What is a perpetual motion machine of the second kind?
4 How would you describe the second law of thermodynamics in simple terms to a layperson?
5 What is a reversible process? What is the meaning of the term in a physical sense?
6 Why is the concept of a heat engine important?
7 Why is the second law necessary for a definition of an absolute temperature scale?
8 Is it possible to attain zero temperature on the absolute thermodynamic temperature scale? Why?
9 What is the *macroscopic* definition of entropy?
10 If entropy change is defined for a reversible process, how is it possible to make a calculation for irreversible processes?
11 Give an analytical statement of the second law in terms of entropy.
12 Why is a reversible adiabatic process one of constant entropy?
13 What is a Carnot cycle? What other kinds of cycles can be more efficient than the Carnot when operating between the same temperature limits?
14 What is meant by the term *available energy*? How does this differ from the concept of availability in steady-flow systems?
15 Is entropy conserved in an isolated system?
16 How would you explain entropy to a layperson?
17 How is work defined?
18 What is meant by the thermal efficiency of a heat engine?
19 What is a coefficient of performance of a refrigeration process?
20 What is meant by "a series of equilibrium states"?
21 What is a cycle?
22 What is the inequality of Clausius and how is it related to entropy?
23 What is a property?
24 How is the state of a system defined?
25 Is entropy a property?
26 How is reversible heat flow represented on a T-S diagram?
27 What is irreversibility in a steady-flow system?

PROBLEMS (ENGLISH UNITS)

5-1 In a system which undergoes a reversible process, the work done is 5 Btu and the heat rejected is 7 Btu. The change in entropy is
(*a*) Positive
(*b*) Negative
(*c*) Zero
(*d*) Can't tell

5-2 In a system which undergoes an irreversible process, the work done is 5 Btu and the heat rejected is 7 Btu. The change in entropy is
(*a*) Positive
(*b*) Negative

(c) Zero
(d) Can't tell

5-3 In a system which undergoes an irreversible process, the work done is 5 Btu and the heat added is 7 Btu. The change in entropy is
(a) Positive
(b) Negative
(c) Zero
(d) Can't tell

5-4 Steam expands in an adiabatic turbine from 900 psia and 900°F to an exit pressure of 1 atm and a quality of 96 percent. The power output of the turbine is 9000 kW. Calculate the flow rate and the change in entropy.

5-5 Steam expands isentropically in a nozzle from 100 psia and 500°F to atmospheric pressure. If the inlet velocity is small, calculate the exit velocity from the nozzle.

5-6 Steam at 100 psia and 10 percent moisture content is throttled adiabatically to atmospheric pressure. Calculate the change in entropy for a flow rate of 10 lbm/min.

5-7 A Carnot cycle received 600 Btu of heat at 1000°F and has an efficiency of 50 percent. Calculate the work output and the temperature at which heat is rejected.

5-8 A Carnot cycle receives 1000 Btu of heat at 800°C and rejects heat at 300°F. Calculate the work output.

5-9 A reversed Carnot cycle is to be used as a refrigerator operating between the temperature limits of 0 and 70°F. Calculate the horsepower input required to remove 12 000 Btu/h from the low-temperature reservoir.

5-10 A certain heat pump for a house may be approximated by a reversed Carnot cycle operating between the outside air temperature of 10°F and the inside air temperature of 80°F. Calculate the horsepower input required to the cycle to deliver 100 000 Btu/h to the high-temperature reservoir (the house).

5-11 1 lbm of saturated liquid at 100 psia is mixed adiabatically with 1 lbm of saturated vapor at 100 psia. What is the change in entropy of the universe resulting from this process? Repeat for a liquid at 200°F, 100 psia, and a saturated vapor at 100 psia. Assume steady flow.

5-12 2 lbm of water at 150°F is mixed with 5 lbm of water at 100°F in an adiabatic process. Calculate the change in entropy of the universe.

5-13 Two blocks of steel ($c_p = 0.11$ Btu/lbm·°F), one at 800°F and the other at 300°F, are brought into contact inside an insulated container and allowed to reach thermal equilibrium. What is the change in entropy of the universe?

5-14 Freon 12 is compressed isentropically from 25 psia, 40°F, to 150 psia in a piston-cylinder arrangement. Calculate the work per lbm of fluid. Repeat the calculation for compression in a steady-flow compressor device.

5-15 Calculate the change in entropy resulting from the heat-exchange process of Prob. 4-15. Assume that the air is heated at a constant pressure of 15 psia.

5-16 A 5-lbm block slides along a horizontal surface with a velocity of 10 ft/s. The coefficient of friction is 0.2 and the temperature of the surroundings is 70°F. Calculate the rate of entropy generation in this process. State assumptions.

5-17 Helium is contained in a rigid container at 20 psia and 100°F. The volume of the container is 10 ft^3. Heat is added to the container until the pressure rises to 35 psia. Calculate the change in entropy of the helium.

5-18 Helium at the rate of 5.0 lbm/min is compressed reversibly and isothermally in a piston-cylinder arrangement from conditions of 100°F and 20 psia to 100 psia. Calculate the heat transfer and change in entropy.

5-19 Steam at the rate of 2 lbm/s at 200 psia and 450°F is mixed with 4 lbm/s of liquid water at 200 psia and 150°F in a steady-flow adiabatic process. Calculate the change in entropy resulting from this process.

5-20 Prospective inventors frequently devise machines which are in violation of the second law. Suppose an inventor asks you to analyze his "new" device which will receive heat from boiling water at atmospheric pressure (212°F) and reject heat at 70°F. He claims the device could achieve an efficiency of 75 percent. How do you rate this claim? What is the maximum efficiency you would think is realistic?

5-21 A Carnot cycle operates with air as the working fluid. The temperature limits of the cycle are 800 and 100°F, respectively, and the pressure at the start of heat addition is 50 psia. Calculate the mass of air required to produce a work output of 1 hp if the cycle is repeated 5000 times per minute. What is the lowest pressure occurring in the cycle if the volume doubles in the heat addition?

5-22 Steam is expanded reversibly and adiabatically in a nozzle from 800 psia, 700°F, to 20 psia. Calculate the exit velocity of the steam if the inlet velocity is very small.

5-23 Steam is expanded in a reversible isothermal process from 100 psia, 400°F, to 50 psia in a piston-cylinder arrangement. Calculate the work and heat transfer for this process.

5-24 Calculate the change in entropy resulting from the mixing process of Prob. 4-20.

5-25 Steam enters a turbine at 500 psia, 500°F, and a velocity of 600 ft/s. The process is isentropic and the exit velocity is 200 ft/s. Calculate the work output of the turbine per lbm of steam for discharge at 1 atm.

5-26 Helium is compressed in a polytropic process with $n = 1.24$ in a piston-cylinder arrangement. The initial and final pressures are 15 psia and 85 psia and the initial temperature is 100°F. Calculate the heat transfer and entropy change per lbm of helium.

5-27 A rigid vessel contains steam and water at 500 psia. Ten percent of the vessel is filled with liquid water and the vessel is perfectly insulated. A valve on the top of the vessel is opened and steam (pure vapor) is allowed to escape slowly until the pressure drops to 200 psia. If the total volume of the vessel is 1 ft^3, calculate the mass of steam removed in this process.

5-28 Freon 12 at the rate of 1.2 kg/s is throttled from saturated-liquid conditions at 120°F to a temperature of 30°F. Calculate the change in entropy.

5-29 A Carnot cycle receives heat from a constant-temperature reservoir at 800°F and rejects heat to 10 lbm of water which is initially saturated liquid at 15 psia. As the engine operates, the water is heated at constant pressure until its temperature reaches 300°F. Calculate the work output of the Carnot engine.

5-30 Repeat Prob. 5-29, assuming that the reservoir from which heat is supplied is 10 lbm of steam initially at 800°F and 800 psia, and that its pressure remains constant during the heat-removal process.

5-31 A 1-lbm piece of putty is dropped from a height of 10 ft onto a flat surface. It immediately sticks to the surface. Calculate the change in entropy of the universe. State assumptions.

5-32 Show that the irreversibility for an adiabatic steady-flow process is

$$I = \dot{m}T_0(s_2 - s_1)$$

5-33 A Carnot cycle is to be designed to operate with an efficiency of 30 percent with steam as the working fluid. Heat addition occurs at 500°F and is accomplished by changing the water from saturated liquid to saturated vapor. Calculate the temperature, entropy, and quality at each point in the cycle and sketch the cycle on a *T-S* diagram. What is the work output per lbm of steam?

5-34 Carbon dioxide in the amount of 0.5 lbm is compressed reversibly and isothermally in a piston-cylinder arrangement from 150°F and 15 psia to 85 psia. Calculate the change in entropy and heat transfer.

5-35 Air flows through an adiabatic throttling valve from 100 psia, 90°F, to atmospheric pressure. Calculate the change in entropy of the air for a flow rate of 3.0 lbm/s.

5-36 10 lbm of air at 50 psia and 300°F are mixed with 5 lbm of air at 30 psia and 500°F. The mixing takes place in an adiabatic chamber. Calculate the change in entropy if the final pressure is 25 psia.

5-37 Calculate the total change in entropy for the process in Prob. 4-46.

5-38 Calculate the total change in entropy for the process in Prob. 4-41.

5-39 Steam at 20 psia and 300°F enters an insulated capillary. The pressure drops to 16 psia as a result of fluid friction. Calculate the change in entropy per lbm of steam.

5-40 Air is expanded adiabatically in a turbine from 1000°F and 200 psia to 400°F. If the process had been reversible, the final temperature would have been 350°F for discharge at the same pressure. Calculate the change in entropy and the work produced per lbm of air.

5-41 A 1-kW electric heater is placed in a 300-ft^3 insulated room filled with air at 14.7 psia and 70°F. The heater is allowed to operate for 15 min. Calculate the change in entropy of the air.

5-42 An electric heater is used to heat 30 gal of water in an insulated container. The temperature of the heater is constant at 200°F and the water is heated from 60 to 120°F. Calculate the change in entropy resulting from this heating process.

5-43 A special air turbine is to be employed for powering drills. The power output is to be 2.0 hp and the turbine has an efficiency of 89 percent. Air enters the turbine at 100 psia and 70°F and leaves at atmospheric pressure. What diameter inlet line is required if the inlet velocity is 20 ft/s? State any assumptions.

5-44 Steam is expanded reversibly and adiabatically from 500 psia and 780°F to 1 atm pressure. What flow rate is required to produce a work output of 15 000 kW if the device is a steady-flow turbine?

5-45 In a steady-flow mixing device 1 lbm/s of steam at 20 psia and 20 percent quality is mixed with 2.6 lbm/s of steam at 20 psia and 80 percent quality, and a sufficient quantity of heat is added to produce an outlet flow at 1 atm pressure and 300°F. The heat is obtained from a reservoir maintained at a constant temperature of 350°F. Calculate the change in entropy resulting from this process.

5-46 Steam is contained in a perfectly insulated 10-ft^3 tank at 700°F and 50 psia. A valve on the side of the tank is opened and the steam allowed to discharge to the atmosphere. The valve is closed when the pressure in the tank just reaches 14.7 psia. Calculate the mass of steam which escapes in this process and the final temperature in the tank.

5-47 A small spray can having a volume of 20 in^3 contains Freon 12 at 70°F, and 30 percent of the volume is occupied by liquid. A small nozzle is attached to the top of the can which operates as a throttling device for discharge to the surroundings at 14.7 psia and 70°F. For a slow discharge process the temperature in the can remains constant at 70°F. Calculate the heat added to the can and change in entropy inside the can for a slow discharge process in which half the initial mass is lost. Also calculate the change in entropy of the vapor which is lost.

5-48 A 2000-ft^3 tank contains water in the saturated-vapor state at 300 psia. Suppose the tank is to serve as a heat source, giving up energy as the steam is condensed to the liquid state. How much "available energy" does it have relative to an environment temperature of 70°F?

5-49 Water vapor is cooled at constant pressure of 200 psia from the saturated-vapor state to the saturated-liquid state. How much has the available energy decreased for this system, assuming a lowest available temperature of 70°F?

5-50 A tank of high-pressure steam having a volume of 200 ft^3 is available to drive a small turbine which may be assumed to operate reversibly and adiabatically. Initially, the steam is at 1000 psia and 1000°F, and the tank is perfectly insulated. Discharge from the turbine is at 100 psia. A pressure-control valve on the discharge from the turbine throttles the steam to atmospheric pressure. The turbine is allowed to run until the tank pressure reaches 300 psia, at which time the discharge valve is shut off. Calculate the work output of the turbine for this process.

5-51 A heat exchanger is designed which will condense steam from the saturated-vapor to saturated-liquid state at 15 psia. Air serves as the cooling medium and enters and leaves the exchanger at 70 and 150°F, respectively. The device is insulated so that all of the heat lost by the steam is gained by the air. Steady-flow operation may be assumed. For each lbm of steam condensed calculate:
(*a*) The mass flow of air required
(*b*) The loss of availability of the steam
(*c*) The gain of availability of the air (assume $T_0 = 70°F$)
(*d*) The steady-flow irreversibility of the process
(*e*) The change in entropy resulting from the process

5-52 Steam at 20 psia undergoes a constant-pressure steady-flow heating process from a quality of 50 percent to superheated vapor at 500°F. Heat for the process is obtained from a constant-temperature reservoir at 600°F, and the flow rate of steam is 25 lbm/min. Calculate:
(*a*) The change in entropy of the steam
(*b*) The change in entropy of the reservoir
(*c*) The overall change in entropy
(*d*) The change in available energy of the reservoir for $T_0 = 70°F$
(*e*) The steady-flow irreversibility for the steam for $T_0 = 70°F$

5-53 Ammonia at the rate of 2 kg/s, at 140 psia and 120°F, is mixed with 3 kgs of ammonia at 140 psia and 60°F in an adiabatic steady-flow device. Calculate the irreversibility for an outlet pressure of 140 psia. Take $T_0 = 70°F$.

5-54 In a large power plant combustion gases ($c_p = 0.26$ Btu/lbm·°F) are cooled from 2300 to 1800°F at constant pressure while giving up heat to steam at 1200 psia operating between saturated liquid and 1400°F. For each pound of steam, calculate:
(*a*) The mass flow of gas required
(*b*) The change in entropy for both the gas and steam
(*c*) The change in the availability function for each flow stream, using $T_0 = 70°F$
(*d*) The irreversibility for the overall process

5-55 Calculate the change in available energy resulting from the process in Prob. 5-13.

5-56 An inventor proposes a novel work-energy storage device which will operate by providing a work input to a Carnot refrigerator which adds heat to liquid water at 14.7 psia at the high temperature and extracts heat from liquid water at 14.7 psia at the low temperature. When the device is first turned on both water reservoirs are at 70°F and they each have equal masses of 500 lbm of liquid water. Calculate:
(*a*) The work energy input to just produce solid ice at 32°F in the low-temperature reservoir
(*b*) The final state of the high-temperature reservoir for this work input
(*c*) The change in available energy of both reservoirs with $T_0 = 70°F$

5-57 For a certain refrigeration machine Freon 12 is compressed from saturated vapor at 10°F to 200 psia and 180°F in an adiabatic steady-flow process. Calculate the work input per unit mass of Freon, the change in entropy, and irreversibility for $T_0 = 70$°F.

5-58 Steam flows in a pipe at 100 psia and 500°F and is connected by a valve to a 30-ft³ tank initially containing steam at 20 psia and 300°F. The valve is opened and steam allowed to flow into the tank. The valve is closed when the pressure just reaches 100 psia. If the tank is perfectly insulated, calculate:
(*a*) The mass added to the tank
(*b*) The change in entropy of the contents of the tank
(*c*) The change in entropy of the mass which *enters* the tank
(*d*) The change in entropy resulting from the process

5-59 An inventor claims that she can store large quantities of energy for conversion to work by just storing high-pressure steam at 5000 psia and 1600°F. For a storage tank having a volume of 200 ft³ she claims that she could produce a work output of 8×10^8 J. Evaluate this claim by analyzing some ideal device which could be connected to the steam tank for producing work.

5-60 Ammonia vapor is compressed adiabatically from saturated vapor at 20°F to 240 psia in a steady-flow device. Because of frictional effects and other factors, the actual work input is 17 percent greater than the work which would have been required in an isentropic process with the same discharge pressure. For each lbm of ammonia compressed calculate:
(*a*) The actual work required
(*b*) The actual exit temperature from the compressor
(*c*) The irreversibility for $T_0 = 70$°F

5-61 Freon 12 undergoes an adiabatic throttling process from saturated-liquid conditions at 100°F to a state where the final temperature is 40°F. For each lbm of Freon calculate:
(*a*) The change in entropy of the liquid
(*b*) The irreversibility of the process with $T_0 = 70$°F

PROBLEMS (METRIC UNITS)

5-1M In a system which undergoes a reversible process, the work done is 5 kJ and the heat rejected is 7 kJ. The change in entropy is
(*a*) Positive
(*b*) Negative
(*c*) Zero
(*d*) Can't tell

5-2M In a system which undergoes an irreversible process, the work done is 5 kJ and the heat rejected in 7 kJ. The change in entropy is
(*a*) Positive
(*b*) Negative

(*c*) Zero
(*d*) Can't tell

5-3M In a system which undergoes an irreversible process, the work done is 5 kJ and the heat added is 7 kJ. The change in entropy is
(*a*) Positive
(*b*) Negative
(*c*) Zero
(*d*) Can't tell

5-4M 5 kg/sec of liquid water at 200 kPa and 50°C are mixed with steam at 150 kPa and 80 percent quality to produce an outlet flow of liquid water at 150 kPa and 100°C. The process is adiabatic. Calculate the steam flow required. Also calculate the change in entropy.

5-5M A Carnot cycle receives 630 kJ of heat at 540°C and has an efficiency of 50 percent. Calculate the work output and temperature at which heat is rejected.

5-6M A Carnot cycle receives 1 MJ of heat at 800°C and rejects heat at 150°C. Calculate the work output.

5-7M A reversed Carnot cycle is to be used as a refrigerator operating between the temperature limits of −18 and 20°C. Calculate the horsepower input required to remove 3.5 kW from the low-temperature reservoir.

5-8M A certain heat pump for a house may be approximated by a reversed Carnot cycle operating between the outside air temperature of −10°C and the inside air temperature of 30°C. Calculate the power input required to the cycle to deliver 30 kW to the high-temperature reservoir (the house).

5-9M Prospective inventors frequently devise machines which are in violation of the second law. Suppose an inventor asks you to analyze his "new" device which will receive heat from boiling water at 100 kPa (99.6°C) and reject heat at 20°C. He claims the device could achieve an efficiency of 75 percent. How do you rate this claim? What is the maximum efficiency you would think is realistic?

5-10M A 5.3-m^3 rigid container is perfectly insulated and filled with steam at 10 MPa and 600°C. A valve on the side of the tank is opened and steam allowed to escape until the pressure drops to 700 kPa. Calculate the quantity of steam which escapes from the tank.

5-11M Helium at the rate of 2 kg/s is heated in a steady-flow process from 200 kPa and 150°C to conditions of 100 kPa and 200°C. During this process the velocity increases from 150 to 300 m/s because of a change in the area of the flow channel. Calculate the change in entropy of the helium.

5-12M Air is compressed in a polytropic process in a piston-cylinder arrangement from 150 kPa and 20°C to 450 kPa and 70°C. Calculate the change in entropy per unit mass.

5-13M 2 kg of water at 65°C are mixed with 5 kg of water at 40°C in an adiabatic process. Calculate the change in entropy.

5-14M Two blocks of steel (c_p = 460 J/kg·°C), one at 700 K and the other at 420 K, are brought into contact inside an insulated container and allowed to reach thermal equilibrium. What is the change in entropy?

5-15M Calculate the change in entropy resulting from the heat-exchange process of Prob. 4-15M. Assume that the air is heated at a constant pressure of 100 kPa.

5-16M A 5-kg block slides along a horizontal surface with a velocity of 3 m/s. The coefficient of friction is 0.2 and the temperature of the surroundings is 20°C. Calculate the rate of entropy generation in this process. State assumptions.

5-17M Steam enters a perfectly insulated pipe at 10.0 MPa and 500°C. At an extended distance in the pipe, frictional effects cause the pressure to drop to 8.0 MPa. Calculate the change in entropy for the steam and irreversibility with T_0 = 20°C.

5-18M Water at 1 MPa and 30°C enters an open feedwater heater where steam at 1 MPa and 80 percent quality is added to produce an output flow at 1 MPa and 160°C. The heater is insulated. Calculate the change in entropy per unit mass of cold water entering the heater.

5-19M Steam is expanded adiabatically in a nozzle from conditions of 1.5 MPa and 400°C to an exit pressure of 1 atm. The inlet velocity is low, and the process may be assumed to be quasistatic. Calculate the exit velocity. Also calculate the exit diameter of the nozzle for a flow rate of 1.2 kg/s.

5-20M A Carnot cycle operates with air as the working fluid. The temperature limits of the cycle are 500 and 310 K, respectively, and the pressure at the start of heat addition is 350 kPa. Calculate the mass of air required to produce a work output of 1 kW if the cycle is repeated 5000 times per minute. What is the lowest pressure occurring in the cycle if the volume doubles in the heat addition?

5-21M Steam is expanded reversibly and adiabatically in a nozzle from 6 MPa, 360°C, to 150 kPa. Calculate the exit velocity of the steam if the inlet velocity is very small.

5-22M Steam is expanded in a reversible isothermal process from 700 kPa, 200°C, to 300 kPa in a piston-cylinder arrangement. Calculate the work and heat transfer for the process.

5-23M Steam enters a turbine at 3 MPa, 260°C, and a velocity of 180 m/s. The process is isentropic and the exit velocity is 60 m/s. Calculate the work output of the turbine per kilogram of steam for discharge at 100 kPa.

5-24M 1 kg of saturated liquid water at 700 kPa is mixed adiabatically with 1 kg of saturated vapor at 700 kPa. What is the change in entropy of the universe resulting from this process? Repeat, but for a liquid at 95°C, 700 kPa, and saturated vapor at 700 kPa. Assume steady flow.

5-25M Helium at the rate of 2.3 kg/s is cooled in a steady-flow process from 200 kPa and 200°C to a temperature of 132°C. Because of frictional effects in the pipe, the pressure drops to 180 kPa. Calculate the cooling required and the change in entropy of the helium. The local atmospheric temperature is 25°C.

5-26M An electric heater is used to heat 110 liters of water in an insulated container. The temperature of the heater is constant at 370 K and the water is heated from 15 to 50°C. Calculate the change in entropy resulting from this heating process.

5-27M A Carnot cycle receives heat from a constant-temperature reservoir at 700 K and rejects heat to 5 kg of water which is an initially saturated liquid at 100 kPa. As the engine operates, the water is heated at constant pressure until its temperature reaches 150°C. Calculate the work output of the Carnot engine.

5-28M 3 kg/s of water are heated from 30°C to 75°C in a steady-flow device. Two alternatives are to be considered: (*a*) electric heating and (*b*) heating with a contant-temperature source at 450°C. Calculate the overall change in entropy and irreversibility for both cases for $T_0 =$ 20°C.

5-29M A throttling calorimeter is connected to a steam line to sample steam at $p = 1.5$ MPa. After steam undergoes an adiabatic throttling process to $p = 100$ kPa, the temperature is measured as 105°C. Determine the quality of the steam at 1.5 MPa in the line.

5-30M Steam is contained in a rigid 4.5-m^3 box which is perfectly insulated. Conditions are 300 kPa and 600°C. A valve on the side of the tank is opened and steam allowed to escape until the steam in the tank just reaches saturation conditions. Calculate the mass of steam which leaves the tank.

5-31M Steam enters a turbine at 6.0 MPa and 600°C and leaves at 100 kPa and 120°C. The flow is adiabatic. The power output of the turbine is 50 MW. Calculate the change in entropy of the steam as it flows through the turbine.

5-32M Steam at the rate of 2.0 kg/s, at 20 MPa and 500°C, is expanded reversibly and adiabatically in a nozzle from a low velocity to an exit pressure of 200 kPa. What exit area is required for this flow rate?

5-33M Carbon dioxide at the rate of 1.1 kg/s is throttled from 700 kPa and 100°C to atmospheric pressure. Calculate the change in entropy.

5-34M Steam enters a turbine at 8.0 MPa and 560°C and leaves at atmospheric pressure. The expansion is reversible and adiabatic. Calculate the flow rate necessary for a power output of 500 MW.

5-35M A tank of nitrogen at 4.0 MPa and 60°C is used to force liquid out of another tank. The two tanks are connected by a pressure-regulating valve which maintains the nitrogen-discharge pressure into the liquid tank at 700 kPa. A discharge valve on the liquid tank is opened, and 0.25 m^3 of liquid is allowed to escape as a result of the nitrogen entry into the liquid tank. At the end of the process the pressure in the nitrogen tank has dropped to 700 kPa. Both tanks and the connecting valve are insulated. Calculate the volume of the nitrogen tank and the final nitrogen temperature in both tanks.

5-36M Nitrogen in employed as a working fluid in a Carnot cycle. In the heat-addition process the volume *increases* from 1.0 to 1.44 liters, while in the heat-rejection process the volume *decreases* from 22.43 to 15.88 liters. Calculate the thermal efficiency of the cycle.

5-37M Steam is throttled from 700 kPa, 75 percent quality, to 100 kPa. Calculate the increase in entropy.

5-38M 10 kg of air at 350 kPa and 420 K are mixed with 5 kg of air at 200 kPa and 530 K. The mixing takes place in an adiabatic chamber. Calculate the change in entropy if the final pressure is 170 kPa.

5-39M Steam at 150 kPa and 160°C enters an insulated capillary. The pressure drops to 100 kPa as a result of fluid friction. Calculate the change in entropy per kilogram of steam.

5-40M Air is expanded adiabatically in a turbine from 800 K and 1.4 MPa to 480 K. If the process had been reversible, the final temperature would have been 450 K for discharge at the same pressure. Calculate the change in entropy and the work produced per kilogram of air.

5-41M A 1-kW electric heater is placed in an 8500-liter insulated room filled with air at 1 atm and 22°C. The heater is allowed to operate for 15 min. Calculate the change in entropy of the air.

5-42M A spherical tank having a diameter of 2.0 m contains steam at 10 MPa and 600°C. The tank is perfectly insulated. A valve on the side of the tank is opened and the steam allowed to escape until the pressure drops to 2 MPa. The valve is then closed. Calculate the amount of steam which escapes from the tank.

5-43M Hydrogen at 10 atm and 400°C is expanded in a turbine to 2 atm and 200°C. The turbine is perfectly insulated and kinetic energies may be neglected. If the power output is 10 kW, calculate the total change in entropy of the hydrogen as it flows through the turbine.

5-44M A small turbine is designed to use helium at 1.0 MPa and 40°C to produce a power output of 2.0 kW. The turbine may be assumed adiabatic with an efficiency of 85 percent and discharge pressure of 1 atm. If the exit velocity is 30 m/s, calculate the diameter of the exit flow line.

5-45M An inventor claims that he has designed an engine which will produce a power output of 170 kW while consuming 0.5 kg/min of fuel having an energy content (heating value) of 42 000 kJ/kg. He states that the upper and lower temperatures of the engine are 670 and 330 K, respectively. Do you believe his claim?

5-46M Determine the change in entropy for the system in Problem 3-30M.

5-47M Repeat Prob. 5-27M, assuming that the reservoir from which heat is supplied is 5 kg of steam initially at 700°K and 6 MPa, and that its pressure remains constant during the heat-removal process.

5-48M A 1-kg piece of putty is dropped from a height of 3 m onto a flat surface. It immediately sticks to the surface. Calculate the change in entropy. State assumptions.

5-49M Emergency power is to be supplied to a computer center by employing an air turbine operating from a high-pressure tank. Inlet conditions to the turbine are 3.0 MPa and 45°C at a velocity of 20 m/s. The exit pressure from the turbine is 600 kPa and the turbine may be assumed to be adiabatic. Calculate the diameter of the inlet pipe if the power output from the turbine is 2.5 MW.

5-50M Air in the amount of 5.2 kg is compressed from 1 atm and 40°C to a pressure of 500 kPa in a reversible isothermal process. Calculate the change in entropy.

5-51M A 1.0-m^3 tank contains water at 20°C and 1 atm pressure. Work is added to the water by a large paddle wheel at the rate of 2.5 kW. At the same time heat escapes to the surroundings at a rate of 0.75 kW. The surroundings temperature is 10°C. Calculate the change in entropy of the water when its temperature reaches 35°C, with the pressure remaining constant at 1 atm.

5-52M A Carnot cycle is to be designed to operate with an efficiency of 30 percent with steam as the working fluid. Heat addition occurs at 260°C and is accomplished by changing the water from saturated liquid to saturated vapor. Calculate the temperature, entropy, and quality at each point in the cycle, and sketch the cycle on a *T-S* diagram. What is the work output per kilogram of steam?

5-53M A reversed Carnot cycle is to be used as a refrigeration device with Freon 12 as the working fluid. The device is to operate between 50 and 120°F, and heat rejection (at the *upper* temperature) occurs as the fluid is changed from saturated vapor to saturated liquid. Calculate the entropy and quality at each point in the cycle, the refrigeration effect, and the work input per kilogram of Freon.

5-54M Steam is contained in a perfectly insulated 280-liter tank at 360°C and 300 kPa. A valve on the side of the tank is opened and the steam allowed to discharge to the atmosphere. The valve is closed when the pressure in the tank just reaches 1 atm. Calculate the mass of steam which escapes in this process and the final temperature in the tank.

5-55M A constant-temperature reservoir at 2000 K delivers 2.0 MJ of heat to a Carnot engine which receives the heat at 500 K and rejects heat at 300 K. Calculate the change in entropy of the universe, and the work output of the Carnot engine.

5-56M A small spray can having a volume of 0.3 liter contains Freon 12 at 70°F, and 30 percent of the volume is occupied by liquid. A small nozzle is attached to the top of the can which operates as a throttling device for discharge to the surroundings at 1 atm and 20°C. For a slow-discharge process the temperature in the can remains constant at 70°F. Calculate the heat added to the can and change in entropy inside the can for a slow-discharge process in which half the initial mass is lost. Also calculate the change in entropy of the vapor which is lost.

5-57M A 55 000-liter tank contains water in the saturated-vapor state at 2 MPa. Suppose the tank is to serve as a heat source, giving up energy as the steam is condensed to the liquid state. How much "available energy" does it have relative to an environment temperature of 20°C?

5-58M A reversible process is executed with a closed system such that 40 kJ of work are delivered by the system while it receives 40 kJ of heat. The change in entropy of the system is

(*a*) Positive
(*b*) Negative
(*c*) Zero
(*d*) Can't tell

5-59M 1.0 MJ of heat is transferred from a high-temperature reservoir at 800 K to a low-temperature reservoir at 400 K. Calculate the overall change in entropy for this process and the loss in available energy for $T_0 = 20°C$.

5-60M In a certain steam condenser, steam enters at 6 kPa and 120°C and leaves as a saturated liquid at the same pressure. Cooling water flows through the device to remove energy from the steam and effect the condensation. The cooling water enters at 25°C and leaves at 30°C, and the overall device is well insulated. Calculate the cooling water required to condense 28 kg/s of steam and the overall irreversibility for the device for $T_0 = 20°C$.

5-61M Air at the rate of 50 m^3/min, at 1 atm and 10°C, is heated to 30°C by condensing steam at 200 kPa in a steady-flow device. The steam enters as a saturated vapor and leaves as a saturated liquid. Calculate the irreversibility for this process. Take $T_0 = 25°C$.

5-62M Calculate the irreversibility for the process in Prob. 5-40M. Take $T_0 = 25°C$.

5-63M 2 kg/s of water at 200 kPa are heated from 30 to 50°C in a steady-flow process inside a tube. Calculate the irreversibility for (*a*) heating with an electric-resistance element and (*b*) condensing steam at 100 kPa. Take $T_0 = 25°C$.

5-64M 3 kg of water at 100 kPa and 20°C are mixed in a closed adiabatic container with 5 kg of water at 80°C and 100 kPa. Calculate the change in entropy resulting from this process.

5-65M Water vapor is cooled at constant pressure of 1.5 MPa from the saturated-vapor state to the saturated-liquid state. How much has the available energy decreased for this sytem, assuming a lowest available temperature of 20°C?

5-66M A tank of high-pressure steam having a volume of 5500 liters is available to drive a small turbine which may be assumed to operate reversibly and adiabatically. Initially, the steam is at 7 MPa and 540°C, and the tank is perfectly insulated. Discharge from the turbine is at 700 kPa. A pressure-control valve on the discharge from the turbine throttles the steam to 1 atm. The turbine is allowed to run until the tank pressure reaches 2 MPa at which time the discharge valve is shut off. Calculate the work output of the turbine for this process.

5-67M A Carnot cycle operates between temperature limits of 900 and 300 K. The heat supplied at the high temperature is 4.0 MJ. Calculate the change in entropy during the heat-addition and -rejection processes, the heat rejection, and the work output.

5-68M An electric-resistance heater is employed to heat a flow of air from 10 to 50°C at a constant pressure of 1 atm. The power input to the heater is 25 kW. Calculate the change in entropy of the air and the irreversibility for the process for $T_0 = 20°C$.

5-69M Large power plants employ "economizers," devices which use the hot exhaust gases from the boiler to preheat the water before entering the boiler. In one application, 68 kg/s of gases ($c_p = 1.05$ kJ/kg·°C) is used to heat 50 kg/s of water from 120 to 200°C in a steady-flow device. Assuming that all the energy lost by the gas goes to heat the water, compute the change in availability of the gas, water, and overall system. What is the irreversibility for the overall system for $T_0 = 20°C$? T_{gas} at inlet = 400°C.

5-70M Calculate the change in entropy for the steam in Prob. 4-56M.

5-71M A heat exchanger is designed which will condense steam from the saturated-vapor to saturated-liquid state at 100 kPa. Air serves as the cooling medium and enters and leaves the exchanger at 290 and 340 K, respectively. The device is insulated so that all of the heat lost by the steam is gained by the air. Steady-flow operation may be assumed. For each kilogram of steam condensed, calculate:

(*a*) The mass flow of air required
(*b*) The loss of availability of the steam
(*c*) The gain of availability of the air (assume $T_0 = 290$ K)
(*d*) The steady-flow irreversibility of the process
(*e*) The change in entropy resulting from the process

5-72M Steam at 6.0 MPa and 360°C is contained in a rigid vessel having a volume of 1.2 m^3. Calculate the availability in respect to 1 atm and $T_0 = 25°C$.

5-73M Steam at 150 kPa undergoes a constant-pressure steady-flow heating process from a quality of 50 percent to superheated vapor at 260°C. Heat for the process is obtained from a constant-temperature reservoir at 300°C, and the flow rate of steam is 10 kg/min. Calculate:

(*a*) The change in entropy of the steam
(*b*) The change in entropy of the reservoir
(*c*) The change in entropy of the universe
(*d*) The change in available energy of the reservoir for $T_0 = 20°C$
(*e*) The steady-flow irreversibility for the steam for $T_0 = 20°C$

5-74M In a large power plant, combustion gases ($c_p = 1.1$ kJ/kg·°C) are cooled from 1600 to 1300 K at constant pressure while giving up heat to steam at 8 MPa operating between saturated liquid and 740°C. For each kilogram of steam calculate:

(*a*) The mass flow of gas required
(*b*) The change in entropy for both the gas and steam

(*c*) The change in the availability function for each flow system using $T_0 = 20°C$
(*d*) The irreversibility for the overall process

5-75M Steam flows in a pipe at 700 kPa and 260°C and is connected by a valve to an 850-liter tank initially containing steam at 150 kPa and 160°C. The valve is opened and steam allowed to flow into the tank. The valve is closed when the pressure just reaches 700 kPa. If the tank is perfectly insulated, calculate:
(*a*) The mass added to the tank
(*b*) The change in entropy of the contents of the tank
(*c*) The change in entropy of the mass which *enters* the tank
(*d*) The change in entropy resulting from the process

5-76M An inventor claims he can store large quantities of energy for conversion to work by just storing high-pressure steam at 32 MPa and 900°C. For a storage tank having a volume of 5600 liters he claims he could produce a work output of 8×10^8 J. Evaluate this claim by analyzing some ideal device which could be connected to the steam tank for producing work.

5-77M An inventor proposes a novel work-energy storage device which will operate by providing a work input to a Carnot refrigerator which adds heat to liquid water at 100 kPa at the high temperature and extracts heat from liquid water at 100 kPa at the low temperature. When the device is first turned on, both water reservoirs are at 20°C and they each have equal masses of 200 kg of liquid water. Calculate:
(*a*) The work energy input to just produce solid ice at 0°C in the low-temperature reservoir
(*b*) The final state of the high-temperature reservoir for this work input
(*c*) The change in available energy of both reservoirs with $T_0 = 20°C$
Assume that the processes for both reservoirs occur at constant pressure.

5-78M Helium at the rate of 1.8 kg/s is expanded in an adiabatic turbine from 1.0 MPa and 300°C to 100 kPa. The turbine efficiency is 92 percent. Calculate the work output of the turbine, the second-law efficiency (effectiveness), and the irreversibility. Take $T_0 = 20°C$.

5-79M 3 kg/s of water at 100 kPa and 20°C are mixed with 5 kg/s of water at 80°C and 100 kPa in an adiabatic steady-flow device. Calculate the irreversibility of the process if the outlet pressure is also 100 kPa. Take $T_0 = 20°C$.

5-80M Air is contained in a piston-cylinder arrangement at 100 kPa and 30°C. Calculate the minimum work input to compress the air to 150°C and 350 kPa. Take $T_0 = 20°C$ and $p_0 = 100$ kPa.

5-81M A rigid tank having a volume of 750 cm^3 contains air at 1.0 MPa and 250°C. Calculate the maximum useful work when it is cooled to 50°C. Take $T_0 = 20°C$ and $p_0 = 100$ kPa.

5-82M Saturated water vapor at 200 kPa is contained in a piston-cylinder arrangement. Calculate the maximum useful work when it is cooled at constant pressure until half the vapor is condensed. Take $T_0 = 20°C$ and $p_0 = 100$ kPa.

5-83M A tank having a volume of 25 liters contains air at 1.2 MPa and 200°C. Calculate the availability for $p_0 = 100$ kPa and $T_0 = 25°C$.

5-84M A tank having a volume of 2.5 m^3 contains steam at 300 kPa and 200°C. Calculate the availability of steam for $p_0 = 100$ kPa and $T_0 = 20°C$.

5-85M Air is contained in a tank at 1.5 MPa and 100°C. Calculate the tank volume which would be required to produce a maximum useful work of 3400 kJ in conjunction with $p_0 = 100$ kPa and $T_0 = 25°C$.

5-86M A rigid container having a volume of 0.1 m^3 is perfectly insulated and separated into two equal chambers by a partition. One chamber contains water at 100°C and 200 kPa while the other chamber contains water at the same pressure and 40°C. The partition is removed and the assembly allowed to come to equilibrium. Calculate the maximum useful work for this process with $p_0 = 100$ kPa and $T_0 = 20°C$.

5-87M Saturated-water vapor at a temperature of 260°C is expanded isothermally until the pressure reaches 100 kPa. Using the Mollier diagram determine the change in enthalpy. Use the steam tables to also determine the change in volume.

5-88M 2 kg of steam are expanded isentropically from 3.0 MPa and 360°C to 40 kPa. Using the Mollier diagram determine the total change in enthalpy for the system. Also specify the final state.

5-89M Calculate the change in entropy for the air which enters the tank in Prob. 4-47M if the sphere has a diameter of 1.0 m.

5-90M An insulated tank contains steam at 8.0 MPa and 400°C. A valve is opened on the side and steam allowed to escape until the steam inside just reaches the saturated-vapor state. Calculate the fractional part of the steam removed when this state is attained.

5-91M A small cylinder having a volume of 15 cm^3 contains nitrogen at 70 atm and 20°C. This cylinder is placed in an insulated chamber having a volume of 10 000 cm^3 and punctured so that the nitrogen is allowed to fill the entire chamber. Assuming that the nitrogen behaves as an ideal gas, calculate:

(*a*) The change in entropy for the nitrogen
(*b*) The change in available energy of the nitrogen for $T_0 = 20°C$
(*c*) The work necessary to reversibly compress the nitrogen back to the initial state in a piston-cylinder arrangement

5-92M A novel reversible heat engine plots on the T-S diagram as a circle. The maximum and minimum temperatures are 1100 and 200 K, respectively, and the maximum entropy change in the cycle is 2000 J/K. Calculate the heat added to the cycle, the heat rejected, the net work output, and the thermal efficiency of the cycle.

5-93M A steady-flow compressor operates in an adiabatic manner to compress air from 1 atm, 20°C, to 5 atm, 210°C. Calculate:

(*a*) The work required per kilogram of air

(*b*) The change in entropy for the process

(*c*) The steady-flow irreversibility for the process, assuming T_0 = 20°C

5-94M A large tank having a volume of 15 m^3 contains air at 30 atm and 30°C. A valve is opened which allows the air to escape slowly to the atmosphere at 1 atm and 25°C. The tank is located outdoors in the sun, and the process occurs so slowly that the temperature in the tank remains constant at 30°C. The valve is closed when the pressure in the tank reaches 6 atm. Calculate the change in entropy resulting from this process.

REFERENCES

1 Callen, H. B.: "Thermodynamics," John Wiley & Sons, Inc., New York, 1960.
2 Lewis, G. N., and M. Randall: "Thermodynamics," 2d ed. (revised by K. S. Pitzer and L. Brewer), McGraw-Hill Book Company, New York, 1961.
3 Reynolds, W. C.: "Thermodynamics," 2d ed., McGraw-Hill Book Company, New York, 1968.
4 Sears, F. W.: "An Introduction to Thermodynamics, The Kinetic Theory of Gases, and Statistical Mechanics," Addison-Wesley Publishing Company, Inc., Reading, Mass., 1950.
5 Hatsopoulos, G. N., and J. H. Keenan: "Principles of General Thermodynamics," John Wiley & Sons, Inc., New York, 1965.
6 Van Wylen, G. J., and R. E. Sonntag: "Fundamentals of Classical Thermodynamics," John Wiley & Sons, Inc., New York, 1965.
7 Tribus, M.: "Thermostatics and Thermodynamics," D. Van Nostrand Company, Inc., Princeton, N.J., 1961.

6

EQUATIONS OF STATE AND GENERAL THERMODYNAMIC RELATIONS

6-1 INTRODUCTION

We now want to establish some general thermodynamic relations which may be applied to the development of equations of state for different substances. The property entropy will be quite useful in this development. No macroscopic thermodynamic analysis can be used to develop a realistic equation of state without the use of experimental data. The analysis may, however, be effectively utilized so that a minimum number of measurements are necessary to establish all properties of interest. We shall see which experimental measurements are necessary for this development.

6-2 SOME MATHEMATICAL PRELIMINARIES

It is worthwhile to review a few principles of the partial differential calculus to make most efficient use of later developments. Suppose that z is given as a function of the two independent variables x and y:

$$z = z(x, y) \tag{6-1}$$

Then, according to the calculus, we may write

$$dz = \frac{\partial z}{\partial x}\,dx + \frac{\partial z}{\partial y}\,dy \tag{6-2}$$

where dz is called an *exact differential*. Equation (6-1) is quite correct in a mathematical sense; however, the partial derivatives are usually written in the following way in thermodynamic presentations:

$$\frac{\partial z}{\partial x} = \left(\frac{\partial z}{\partial x}\right)_y = M \qquad \frac{\partial z}{\partial y} = \left(\frac{\partial z}{\partial y}\right)_x = N \tag{6-3}$$

Here the parentheses and subscripts are employed to show very clearly which variable is being held constant in the differentiation. Also, each one of these notations shows that x and y are the independent variables. The second mixed partial derivatives are equal, so that

$$\frac{\partial^2 z}{\partial x\,\partial y} = \frac{\partial^2 z}{\partial y\,\partial x}$$

or

$$\frac{\partial M}{\partial y} = \frac{\partial N}{\partial x} \tag{6-4}$$

where it is understood that Eq. (6-2) is written as

$$dz = M\,dx + N\,dy \tag{6-5}$$

It is implied from Eq. (6-1) that, in principle, either x or y could be expressed explicitly in terms of the other two variables; i.e., there exist relations of the form

$$x = x(y, z) \qquad y = y(x, z) \tag{6-6}$$

From these two relations we obtain

$$dx = \left(\frac{\partial x}{\partial y}\right)_z dy + \left(\frac{\partial x}{\partial z}\right)_y dz \tag{6-7}$$

$$dy = \left(\frac{\partial y}{\partial x}\right)_z dx + \left(\frac{\partial y}{\partial z}\right)_x dz \tag{6-8}$$

After some algebraic manipulation we find that

$$\left(\frac{\partial x}{\partial y}\right)_z \left(\frac{\partial y}{\partial z}\right)_x \left(\frac{\partial z}{\partial x}\right)_y = -1 \tag{6-9}$$

We shall refer to Eq. (6-9) as a *cyclical* relation.

In thermodynamic relations we must be careful in handling products of partial derivatives. Suppose we have

$$r = r(x, y)$$

$$z = z(x, r)$$

We might form the derivatives

$$\frac{\partial r}{\partial x} = \left(\frac{\partial r}{\partial x}\right)_y$$

$$\frac{\partial z}{\partial x} = \left(\frac{\partial r}{\partial x}\right)_r$$

but it is important to note that

$$\frac{\partial r/\partial x}{\partial z/\partial x} \neq \frac{\partial r}{\partial z}$$

It is permissible, of course, to write

$$\left(\frac{\partial z}{\partial x}\right)_y = \frac{1}{(\partial x/\partial z)_y}$$

To avoid confusion, the parentheses and subscripts will *always* be employed in our discussions of partial derivatives. In this way, it will be possible to keep the independent variables identified throughout the differentiation processes.

6-3 HELMHOLTZ AND GIBBS FUNCTIONS

Let us define two new properties: the *Helmholtz function a*,

$$a = u - Ts \tag{6-10}$$

and the *Gibbs function g*,

$$g = h - Ts \tag{6-11}$$

At this point we shall not be concerned with the physical significance of these properties, except to say that they are of value in the analysis of system equilibrium.

6-4 THE ENERGY EQUATION

In Chap. 3 it was shown that the energy-conservation principle for a pure substance having only a compression work mode could be written in differential form as

$$d'Q = du + p\,dv \tag{6-12}$$

If we now assume that the system is in equilibrium during the differential variations described in Eq. (6-12), then the system is presumed to experience a quasistatic or reversible process. Accordingly, the heat transfer $d'Q$ may be written in terms of the entropy, so that

$$d'Q_{\text{rev}} = T\,ds = du + p\,dv \tag{6-13}$$

6-5 THE MAXWELL RELATIONS

The definition of the enthalpy may be recalled as

$$h = u + pv \tag{6-14}$$

Differentiating Eqs. (6-10), (6-11), and (6-14) and rearranging Eq. (6-13),

$$da = du - T\,ds - s\,dT$$

$$dg = dh - T\,ds - s\,dT$$

$$dh = du + p\,dv + v\,dp$$

$$du = T\,ds - p\,dv$$

or, making further use of Eq. (6-13),

$$da = -s\,dT - p\,dv \tag{6-15a}$$

$$dg = -s\,dT + v\,dp \tag{6-15b}$$

$$dh = T\,ds + v\,dp \tag{6-15c}$$

$$du = T\,ds - p\,dv \tag{6-15d}$$

Each of these equations is of the same form as Eq. (6-5). We may therefore equate mixed partial derivatives in accordance with Eq. (6-4) and define the coefficients of each differential in accordance with Eq. (6-3). The following results are obtained:

$$-s = \left(\frac{\partial a}{\partial T}\right)_v \qquad -p = \left(\frac{\partial a}{\partial v}\right)_T \qquad \left(\frac{\partial s}{\partial v}\right)_T = \left(\frac{\partial p}{\partial T}\right)_v \tag{6-16a}$$

$$-s = \left(\frac{\partial g}{\partial T}\right)_p \qquad v = \left(\frac{\partial g}{\partial p}\right)_T \qquad \left(\frac{\partial s}{\partial p}\right)_T = -\left(\frac{\partial v}{\partial T}\right)_p \tag{6-16b}$$

$$T = \left(\frac{\partial h}{\partial s}\right)_p \qquad v = \left(\frac{\partial h}{\partial p}\right)_s \qquad \left(\frac{\partial T}{\partial p}\right)_s = \left(\frac{\partial v}{\partial s}\right)_p \tag{6-16c}$$

$$T = \left(\frac{\partial u}{\partial s}\right)_v \qquad -p = \left(\frac{\partial u}{\partial v}\right)_s \qquad \left(\frac{\partial T}{\partial v}\right)_s = -\left(\frac{\partial p}{\partial s}\right)_v \tag{6-16d}$$

The four sets of relations between the partial derivatives as obtained from equating the second mixed partial derivatives are called the *Maxwell relations*.

Two other relations are of particular interest at this time:

$$T = \left(\frac{\partial u}{\partial s}\right)_v \tag{6-17a}$$

$$T = \left(\frac{\partial h}{\partial s}\right)_p \tag{6-17b}$$

These relations may be used to define temperature in terms of other thermodynamic properties, particularly the entropy. For example, if we can obtain expressions for u and s from microscopic considerations, then it should be possible to calculate the temperature of the system by using Eq. (6-17*a*). As it turns out, this relation serves as a crucial link between macroscopic and microscopic thermodynamic analysis and, in fact, provides the analytical facility for obtaining a definition of the temperature of a microscopic system. We shall examine the microscopic analysis later, but, for now, let us concentrate our attention on the macroscopic relations.

6-6 ENTHALPY, INTERNAL ENERGY, AND ENTROPY

Let us assume that the internal energy of a pure substance may be expressed as a function of temperature and volume (see Sec. 3-2). Then $u = u(T, v)$ and

$$du = \left(\frac{\partial u}{\partial T}\right)_v dT + \left(\frac{\partial u}{\partial v}\right)_T dv \tag{6-18}$$

Recalling the definition

$$c_v = \left(\frac{\partial u}{\partial T}\right)_v$$

we have

$$du = c_v\, dT + \left(\frac{\partial u}{\partial v}\right)_T dv$$

From Eq. (6-13),

$$\begin{aligned} ds &= \frac{du}{T} + \frac{p}{T} dv \\ &= c_v \frac{dT}{T} + \left[\frac{1}{T}\left(\frac{\partial u}{\partial v}\right)_T + \frac{p}{T}\right] dv \end{aligned} \tag{6-19}$$

But we may write the entropy as a function of the independent variables T and v so that $s = s(T, v)$ and

$$ds = \left(\frac{\partial s}{\partial T}\right)_v dT + \left(\frac{\partial s}{\partial v}\right)_T dv \tag{6-20}$$

Equating the coefficients of dT and dv in Eqs. (6-19) and (6-20) gives

$$\frac{c_v}{T} = \left(\frac{\partial s}{\partial T}\right)_v \tag{6-21}$$

$$\left(\frac{\partial s}{\partial v}\right)_T = \frac{1}{T}\left[\left(\frac{\partial u}{\partial v}\right)_T + p\right] \tag{6-22}$$

Now the Maxwell relation of Eq. (6-16a) allows us to write

$$\left(\frac{\partial s}{\partial v}\right)_T = \left(\frac{\partial p}{\partial T}\right)_v$$

Thus Eq. (6-22) may be rewritten as

$$\left(\frac{\partial u}{\partial v}\right)_T = T\left(\frac{\partial p}{\partial T}\right)_v - p \tag{6-23}$$

and the relation for the internal energy [Eq. (6-18)] becomes

$$du = c_v\,dT + \left[T\left(\frac{\partial p}{\partial T}\right)_v - p\right]dv \tag{6-24}$$

Similarly, we may write for the entropy

$$ds = c_v\frac{dT}{T} + \left(\frac{\partial p}{\partial T}\right)_v dv \tag{6-25}$$

A similar development may be made for the enthalpy. The result is

$$dh = c_p\,dT + \left[v - T\left(\frac{\partial v}{\partial T}\right)_p\right]dp \tag{6-26}$$

where the definition of the specific heat at constant pressure will be recalled as

$$c_p = \left(\frac{\partial h}{\partial T}\right)_p \tag{6-27}$$

Taking the entropy as a function of pressure and temperature, and considering these independent variables instead of volume and temperature, results in

$$ds = c_p\frac{dT}{T} - \left(\frac{\partial v}{\partial T}\right)_p dp \tag{6-28}$$

The final equations for internal energy, Eq. (6-24), enthalpy, Eq. (6-26), and entropy, Eq. (6-25), involve five quantities: p, v, T, c_v, and c_p. We may develop a relationship between the specific heats in the following way. The two expressions for entropy, Eqs. (6-25) and (6-28), may be equated to give

$$c_v\frac{dT}{T} + \left(\frac{\partial p}{\partial T}\right)_v dv = c_p\frac{dT}{T} - \left(\frac{\partial v}{\partial T}\right)_p dp$$

Rearranging,

$$c_p - c_v = T\left[\left(\frac{\partial p}{\partial T}\right)_v\frac{dv}{dT} + \left(\frac{\partial v}{\partial T}\right)_p\frac{dp}{dT}\right] \tag{6-29}$$

For a variation at constant volume, $dv = 0$ and

$$(c_p - c_v)_v = T \left(\frac{\partial v}{\partial T}\right)_p \left(\frac{\partial p}{\partial T}\right)_v$$

or for a variation at constant pressure, $dp = 0$ and

$$(c_p - c_v)_p = T \left(\frac{\partial p}{\partial T}\right)_v \left(\frac{\partial v}{\partial T}\right)_p$$

These relations are identical. Using the cyclical relation of Eq. (6-9),

$$\left(\frac{\partial p}{\partial T}\right)_v = -\left(\frac{\partial v}{\partial T}\right)_p \left(\frac{\partial p}{\partial v}\right)_T$$

we obtain the final relation

$$c_p - c_v = -T\frac{(\partial v/\partial T)_p^2}{(\partial v/\partial p)_T} \tag{6-30}$$

By deriving Eq. (6-30) we have thus reduced the determination of u, h, and s to the specification of only four properties: p, v, T, and either c_v or c_p. Since the Gibbs and Helmholtz functions are expressed in terms of u, h, and s, we thus have also specified the information necessary for their determination.

The foregoing information is most important because it establishes the experimental measurements which must be performed to describe fully all the thermodynamic properties of a pure substance.

If we are concerned only with a single phase of the substance (i.e., liquid or vapor phase), we may presume that there exists an equation of state relating pressure, volume, and temperature:

$$v = v(T, p) \tag{6-31}$$

The p, v, T measurements may be employed to obtain an empirical relationship of the form of Eq. (6-31). This relation may then be combined with specific heat data to calculate other thermodynamic properties. We should recall at this point the utility of the Joule-Thomson porous-plug experiments in determining values of c_p from isothermal and adiabatic flow measurements. These experiments were discussed in Sec. 4-7.

6-7 SPECIFIC-HEAT RELATIONS

Additional useful expressions relating specific heats may be developed in the following manner. Equation (6-28) is an exact differential equation of the form of Eq. (6-5). We may thus equate the second mixed partial derivatives to obtain

$$\left[\frac{\partial(c_p/T)}{\partial p}\right]_T = -\left[\frac{\partial}{\partial T}\left(\frac{\partial v}{\partial T}\right)_p\right]_p$$

or

$$\left(\frac{\partial c_p}{\partial p}\right)_T = -T\left(\frac{\partial^2 v}{\partial T^2}\right)_p \tag{6-32}$$

Using Eq. (6-25) as a starting point, we obtain a similar expression for the variation of c_v with specific volume. Equating second mixed partial derivatives, there results

$$\left[\frac{\partial(c_v/T)}{\partial v}\right]_T = \left[\frac{\partial}{\partial T}\left(\frac{\partial p}{\partial T}\right)_v\right]_v$$

or

$$\left(\frac{\partial c_v}{\partial v}\right)_T = T\left(\frac{\partial^2 p}{\partial T^2}\right)_v \tag{6-33}$$

Equations (6-32), (6-33), and (6-23) provide very useful results when applied to an ideal gas. We have

$$v = \frac{RT}{p} \qquad p = \frac{RT}{v}$$

Differentiating and inserting these expressions in Eqs. (6-32), (6-33), and (6-23),

$$\left(\frac{\partial c_p}{\partial p}\right)_T = -T(0) = 0$$

$$\left(\frac{\partial c_v}{\partial v}\right)_T = T(0) = 0$$

$$\left(\frac{\partial u}{\partial v}\right)_T = T\left(\frac{R}{v}\right) - p = 0$$

That is, the specific heats and internal energy of an ideal gas are functions of temperature alone. This result has already been used in Chap. 2 to perform calculations for processes involving ideal gases.

6-8 THE GAS TABLES

We have noted that the enthalpy and internal energy of a gas which obeys the equation of state $pv = RT$ are functions only of temperature and, thus, may be calculated with appropriate relations for specific heat as given in Table 2-3. Such calculations have been performed by Keenan and Kaye [3] and tabulated in the *gas tables*. For these tables the zero reference level for enthalpy and internal energy is taken as zero degrees absolute.

For an ideal gas we may write the change in entropy for a process as [see Eq. (5-31)]

$$s_2 - s_1 = \int_{T_1}^{T_2} c_p \frac{dT}{T} - R \ln \frac{p_2}{p_1} \tag{6-34}$$

Let us choose a reference state for entropy such that

$$s = 0 \qquad \text{at } T = 0,\ p = 1 \text{ atm}$$

The entropy at any particular temperature and pressure is then

$$s = \int_0^T c_p \frac{dT}{T} - R \ln p \tag{6-35}$$

where, now, the pressure is expressed in units of atmospheres. The integral in this equation is a function of temperature alone and is tabulated in the *gas tables* as ϕ (see Tables A-17 and A-18):

$$\phi = \int_0^T c_p \frac{dT}{T} \tag{6-36}$$

The change in entropy between two given states, Eq. (6-34), may thus be written as

$$s_2 - s_1 = \phi_2 - \phi_1 - R \ln \frac{p_2}{p_1} \tag{6-37}$$

Now let us consider a reversible adiabatic process involving an ideal gas with variable specific heats. From Eq. (6-15*c*) we have

$$T\,ds = dh - v\,dp = 0$$

or

$$c_p\,dT = \frac{RT}{p}\,dp$$

Separating variables,

$$\frac{dp}{p} = \frac{c_p\,dT}{RT} \tag{6-38}$$

If this equation is now integrated between a reference state at T_0 and p_0, there results

$$\ln \frac{p}{p_0} = \frac{1}{R} \int_{T_0}^{T} \frac{c_p \, dT}{T} \tag{6-39}$$

Using the same reference temperature state as in the gas tables gives

$$\ln P_r = \frac{\phi}{R} \tag{6-40}$$

where the *relative pressure* is defined as

$$P_r = \frac{p}{p_0}$$

For an isentropic process it is apparent that the relative pressure is a function of temperature alone since ϕ is dependent only on temperature. Furthermore, for the isentropic process,

$$\frac{p_1}{p_2} = \left(\frac{P_{r_1}}{P_{r_2}}\right)_{s=\text{const.}} \tag{6-41}$$

A *relative specific volume* for the isentropic process may be developed in a similar manner such that

$$\frac{v_1}{v_2} = \left(\frac{v_{r_1}}{v_{r_2}}\right)_{s=\text{const.}} \tag{6-42}$$

v_r is also temperature dependent and is expressed by

$$\ln v_r = \ln \frac{v}{v_0} = -\frac{1}{R} \int_{T_0}^{T} \frac{c_v \, dT}{T} \tag{6-43}$$

Values of the relative pressure and relative specific volume are tabulated for air in Table A-17.

EXAMPLE 6-1 Isentropic Expansion
Air expands from 700 kPa, 650 K, to 200 kPa in an isentropic process. Calculate the final temperature and change in enthalpy with the use of the gas tables.

SOLUTION Using the gas tables, we have for the initial state

$$T_1 = 650 \text{ K}$$

$$h_1 = 659.84 \text{ kJ/kg}$$

$$P_{r_1} = 21.86$$

Using Eq. (6-41),

$$P_{r_2} = P_{r_1} \frac{P_2}{P_1}$$

$$= (21.86) \left(\frac{200}{700}\right) = 6.246$$

Entering the gas tables again with $P_{r_2} = 6.246$, we find

$$T_2 = 460 \text{ K}$$

$$h_2 = 462.01 \text{ kJ/kg} \qquad (198.6 \text{ Btu/lbm})$$

and the change in enthalpy is

$$\Delta h = h_2 - h_1 = 462.01 - 659.84 = -197.83 \text{ kJ/kg} \qquad (-85 \text{ Btu/lbm})$$

EXAMPLE 6-2 Irreversible Expansion
The expansion process of Example 6-1 occurs between the same pressure limits, but because of irreversible effects the final temperature is 480 K. Calculate the change in entropy for the process.

SOLUTION The final temperature is given as $T_2 = 480$ K. We may therefore obtain the values of ϕ from the gas tables as

$$\phi_1 = 3.3069$$

$$\phi_2 = 2.9909$$

From Eq. (6-37) the change in entropy is

$$s_2 - s_1 = \phi_2 - \phi_1 - R \ln \frac{p_2}{p_1}$$

$$= 2.9909 - 3.3069 - \frac{287.1}{1000} \ln \frac{200}{700}$$

$$= 0.04367 \text{ kJ/kg·K} \qquad (0.01043 \text{ Btu/lbm·°R})$$

EXAMPLE 6-3
Using the air tables, calculate the change in entropy for air under the following conditions:

(*a*) Compression from 1 atm, 300 K, to 10 atm, 550 K
(*b*) Constant-pressure heat addition at 340 kPa from 300 to 550 K
(*c*) Adiabatic throttling process from 1.4 MPa, 420 K, to 0.7 MPa
(*d*) 10 kJ of heat addition to a fixed mass inside a 0.058-m^3 rigid container from 300 K., 1 atm, to final state

SOLUTION

(a) The entropy change may be calculated directly from Eq. (6-37):

$$s_2 - s_1 = \phi_2 - \phi_1 - R \ln \frac{p_2}{p_1} \tag{a}$$

Using the air tables (Table A-17M),

$$T_1 = 300 \text{ K} \qquad \phi_1 = 2.5153 \text{ kJ/kg·K}$$

$$T_2 = 550 \text{ K} \qquad \phi_2 = 3.1314 \text{ kJ/kg·K}$$

From Eq. (a),

$$s_2 - s_1 = 3.1314 - 2.5153 - 0.287 \ln \frac{10}{1}$$

$$= -0.04474 \text{ kJ/kg·K} \tag{b}$$

(b) Equation (a) may still be used, but since $p_1 = p_2 = 340$ kPa, the last term is zero, and

$$s_2 - s_1 = 3.1314 - 2.5153 = 0.6161 \text{ kJ/kg·K} \tag{c}$$

(c) For adiabatic throttling with negligible kinetic energies $h_1 = h_2$; therefore $T_1 = T_2$. This means that $\phi_1 = \phi_2$, and Eq. (a) becomes

$$s_2 - s_1 = 0 - 287 \ln \frac{0.7}{1.4}$$

$$= 198.9 \text{ J/kg·K} = 0.1989 \text{ kJ/kg·K} \tag{d}$$

(d) The mass of air is calculated from the initial state:

$$p_1 = 1 \text{ atm} = 101.32 \text{ kPa} \qquad T_1 = 300 \text{ K} \qquad V_1 = 0.058 \text{ m}^3$$

so that, from the ideal-gas law,

$$m = \frac{pV}{RT} = \frac{(1.0132 \times 10^5)(0.058)}{(287)(300)} = 0.068\ 25 \text{ kg} \tag{e}$$

Because the container is rigid there is no work done on the air and the first law becomes

$$Q = U_2 - U_1 = m(u_2 - u_1) \tag{f}$$

Using the air tables (Table A-17M) at 300 K yields

$$u_1 = 214.09 \text{ kJ/kg} \qquad \phi_1 = 2.5153$$

Inserting the numerical values in Eq. (*f*) gives

$$\begin{aligned} 10 \text{ kJ} &= 0.068\,25(u_2 - 214.09) \\ u_2 &= 3.0358 \text{ kJ/kg·K} \end{aligned} \tag{g}$$

From Table A-17M the value of T_2 corresponding to this value of u_2 is

$$T_2 = 511.5 \text{ K} \qquad \phi_2 = 3.0358 \text{ kJ/kg·K} \tag{h}$$

Because the volume remains constant, the final pressure is calculated from

$$\frac{p_2}{p_1} = \frac{T_2}{T_1} \tag{i}$$

$$p_2 = (1.0)\left(\frac{511.5}{300}\right) = 1.705 \text{ atm}$$

Equation (*a*) is again used to calculate the change in entropy per unit mass:

$$s_2 - s_1 = 3.0358 - 2.5153 - \left(0.287 \ln \frac{1.705}{1.0}\right) = 0.3674 \tag{j}$$

The total entropy change is then

$$\begin{aligned} S_2 - S_1 &= m(s_2 - s_1) = (0.068\,25)(0.3674) \\ &= 0.025\,08 \text{ kJ/K} \end{aligned} \tag{k}$$

EXAMPLE 6-4 Isentropic Compression
Air is compressed isentropically in a certain steady-flow device from 300 K and 100 kPa to a final temperature of 530 K. Calculate the work per kilogram of air using the gas tables and compare the result with the value which would be obtained if the air were assumed to have constant specific heats.

SOLUTION We have

$$T_1 = 300 \text{ K} \qquad T_2 = 530 \text{ K}$$

For a steady-flow process in which kinetic energy effects are negligible, the adiabatic work done on the air is

$$W = h_2 - h_1$$

From the gas tables,

$$h_2 = 533.98 \text{ kJ/kg}$$

$$h_1 = 300.19 \text{ kJ/kg}$$

and

$$W = 533.98 - 300.19 = 233.8 \text{ kJ/kg} \qquad (100.5 \text{ Btu/lbm})$$

For a constant specific heat of $c_p = 1.005$ kJ/kg·°C, the change in enthalpy would be

$$\begin{aligned} h_2 - h_1 &= c_p(T_2 - T_1) \\ &= (1.005)(530 - 300) \\ &= 231.2 \text{ kJ/kg} \qquad (99.4 \text{ Btu/lbm}) \end{aligned}$$

Thus, in this example, a constant specific heat assumption would be accurate within approximately 1 percent.

6-9 SPECIFIC HEATS AT ELEVATED PRESSURES

Equation (6-32) is particularly useful in the development of property data for various substances. We may integrate with respect to pressure to obtain

$$(c_p)_{T,p} - (c_p)_{T,p=0} = \int_{p=0}^{p} T \left(\frac{\partial^2 v}{\partial T^2}\right)_p dp_T \tag{6-44}$$

Thus we may obtain the specific heat at any pressure and temperature once we know the specific heat at zero (very low) pressure and the (p, v, T) equation of state. Once the specific heats are determined, the enthalpy, entropy, and internal energies may be calculated. In summary, the properties of a single-phase pure substance may be calculated once empirical data are obtained which yield:

1 Specific heat behavior at zero pressure
2 (p, v, T) behavior over the property range of interest

Low-pressure specific heats of several gases were tabulated in Tables 2-2 and 2-3.

6-10 THE CLAUSIUS-CLAPEYRON EQUATION

The relations given in the foregoing obviously do not provide a basis for calculations of properties in a two-phase region. A useful relation for such calculations is the Clausius-Clapeyron equation which we shall now derive. Equation (6-16*a*) is rewritten for convenience:

$$\left(\frac{\partial p}{\partial T}\right)_v = \left(\frac{\partial s}{\partial v}\right)_T \tag{6-16a}$$

As a pure substance changes from a saturated-liquid state to a saturated-vapor state, it undergoes a process at constant temperature. Furthermore, the pressure and temperature are independent of volume in the saturation region. Thus the derivatives in Eq. (6-16*a*) may be written as

$$\left(\frac{\partial p}{\partial T}\right)_v = \frac{dp}{dT}$$

$$\left(\frac{\partial s}{\partial v}\right)_T = \frac{s_g - s_f}{v_g - v_f} = \frac{s_{fg}}{v_{fg}}$$

We also have the following relation for the heat added in the constant-pressure vaporization process:

$$\begin{aligned} Q &= \Delta u - W \\ &= u_g - u_f + p(v_g - v_f) \\ &= h_g - h_f = h_{fg} \end{aligned}$$

But $Q = Ts_{fg}$ in the constant-temperature process, so that

$$s_{fg} = \frac{h_{fg}}{T} \tag{6-45}$$

With the foregoing relations, Eq. (6-16*a*) becomes

$$\frac{dp}{dT} = \frac{h_{fg}}{Tv_{fg}} \tag{6-46}$$

Equation (6-46) is called the *Clausius-Clapeyron equation*. The derivative dp/dT represents the slope of the vapor-pressure curve, and the utility of the relation is clear. Presumably, saturated-liquid properties can be determined from some appropriate equation of state and an arbitrarily assigned reference state. For water, saturated liquid at 0°C is assigned a zero enthalpy and entropy. Experimental data may be used to establish an empirical equation for the vapor-pressure curve, so that the slope of this curve may be obtained from Eq. (6-46). Then v_{fg} is determined from the separate values of v_g and v_f, so that Eq. (6-46) may be used to calculate the value of h_{fg} and then the value of h_g. Once the saturated-vapor properties are established, the various equations of Sec. 6-6 may be employed to determine properties of the super-heated vapor.

EXAMPLE 6-5 Solid-Vapor Equilibrium
Calculate the saturation pressure for an equilibrium mixture of solid-vapor water at −80°F if $p_{sat} = 0.0019$ psia at −40°F and h_{ig} is essentially constant at 1221.2 Btu/lbm.

SOLUTION We shall employ the Clausius-Clapeyron equation for solution of this problem. At these low pressures the water vapor behaves very nearly as an ideal gas and

$$v_g = \frac{RT}{p} \tag{a}$$

Also $v_g \gg v_i$, so that $v_{ig} \cong v_g$ and Eq. (6-46) becomes, for the solid-vapor equilibrium line,

$$\frac{dp}{dT} = \frac{ph_{ig}}{RT^2} \tag{b}$$

where Eq. (*a*) has been substituted for v_g. Equation (*b*) may be integrated between two temperatures to give

$$\int_{p_1}^{p_2} \frac{dp}{p} = \int_{T_1}^{T_2} \frac{h_{ig}\,dT}{RT^2}$$

or

$$\ln \frac{p_2}{p_1} = \frac{h_{ig}}{R}\left(\frac{1}{T_1} - \frac{1}{T_2}\right) \tag{c}$$

Using the given data, $T_1 = 420°\text{R}$, $T_2 = 380°\text{R}$, and

$$\ln \frac{p_2}{p_1} = \frac{(1221.2)(778)}{85.6}\left(\frac{1}{420} - \frac{1}{380}\right)$$
$$= -2.78$$

or $p_2/p_1 = 0.062$ and $p_2 = (0.062)(0.0019) = 1.178 \times 10^{-4}$ psia (0.812 Pa).

6-11 DEVELOPMENT OF TABLES OF THERMODYNAMIC PROPERTIES

The overall procedure for developing tables of thermodynamic properties may now be summarized in outline form. The outline comments refer to the points and regions shown in Fig. 6-1.

A Experimental data required
- **1** Low-pressure specific heat data for liquid and vapor
- **2** *p-v-T* data for liquid, vapor, and supercritical-gas region
- **3** Experimental data for *p-T* variation of vaporization curve

B Direct empirical relations to be obtained from data
- **1** Low-pressure specific heat as a function of temperature
- **2** (p, v, T) equations of state for liquid, vapor, and supercritical regions
- **3** Empirical relation for vaporization curve
- **4** Arbitrarily assigned reference state for enthalpy and entropy

C Property determination procedures

1 Liquid properties, including saturated-liquid properties, are determined from (p, v, T) equation of state for liquid, liquid specific heats, and assigned reference state. See points 0, 3, and 4 in Fig. 6-1.

2 p-v-T properties of vapor, including v_g, are obtained from (p, v, T) equation of state for vapor. See points 1, 2, and 5 in Fig. 6-1.

3 Saturated-vapor enthalpies and entropies (h_g and s_g) are determined from saturated-liquid properties, v_g, and Eqs. (6-45) and (6-46) in conjunction with vaporization-curve data. See points 1 and 6 in Fig. 6-1.

4 Vapor enthalpies and entropies are determined from h_g, s_g, (p, v, T) equation of state for vapor, and specific heat equation for vapor. See points 2 and 5 in Fig. 6-1.

5 A separate equation of state may be required for the supercritical region.

It is well to note at this point that the development of tables of thermodynamic properties is a tedious task and requires much more physical insight than may be implied from the foregoing discussion. Experimental measurements of properties of substances can involve considerable uncertainties and the equation of state which is to be developed from these measurements must take the uncertainties into proper account. The purpose of this comment is not to alarm the reader, but rather to stress the fact that the relatively simple calculation procedure described must, in practice, be coupled with a vast amount of experimental data, physical insight, patience, and computer time to produce an accurate table of thermodynamic properties for a particular substance.

FIG. 6-1 Schematic of regions for thermodynamic property determinations.

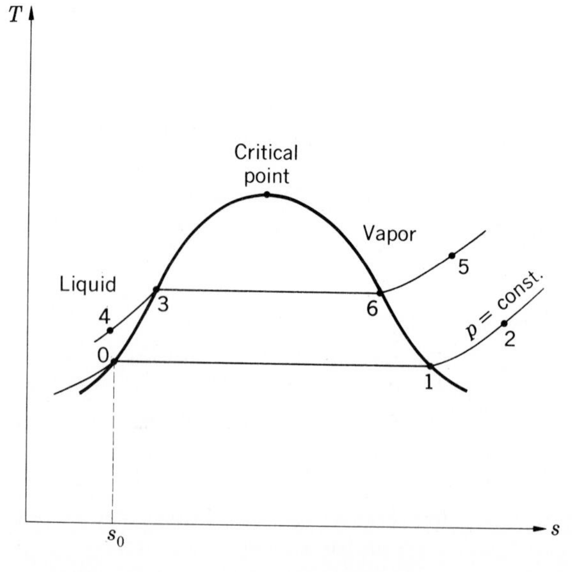

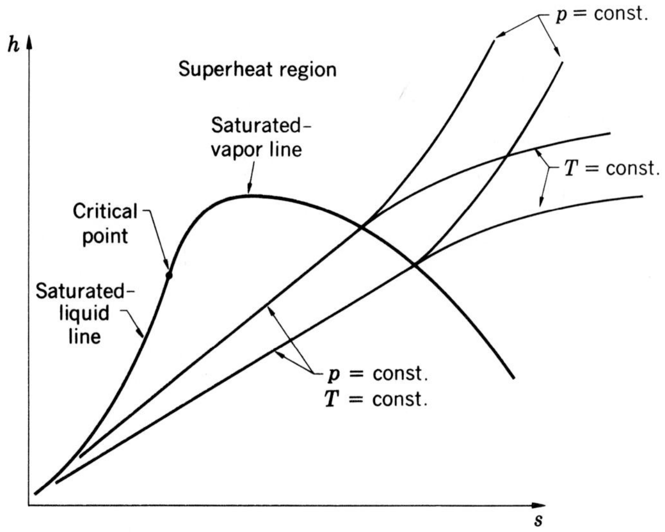

FIG. 6-2 Schematic of Mollier diagram.

The Mollier Diagram

A diagram of practical utility is the Mollier chart or plot of enthalpy versus entropy, shown in schematic form in Fig. 6-2. An interesting feature of this diagram is that isotherms (lines of constant temperature) plot as straight lines in the two-phase region as a result of the relation given in Eq. (6-17*b*). The term $(\partial h/\partial s)_p$ is clearly the slope of a constant-pressure line on the Mollier diagram, but in the wet-mixture region a line of constant pressure is also a line of constant temperature so that Eq. (6-17*b*) predicts a constant slope in this region. As mentioned before, a Mollier diagram for steam is given in Fig. A-4 of the Appendix and will be very useful in conjunction with the steam-power-cycle analysis of Chap. 9.

6-12 EQUATIONS OF STATE

With the foregoing theoretical background in mind, it is now worthwhile to examine some particular forms of equation of state which are employed in practice.

The van der Waals Equation of State

In accordance with the simple discussion of Sec. 1-14, we may consider an ideal gas as one in which:

1 The molecules are very widely spaced and behave as point masses.
2 The molecular force fields exert very little influence on collision processes because of the wide molecular spacing.

3 The volume occupied by the molecules is very small compared with the total volume occupied by the gas, again because of the wide molecular spacing.

When a gas is compressed, the molecules are forced into closer proximity to one another and nonideal-gas behavior will eventually develop. Finally, of course, there can be such strong force-field interaction that the gas is converted to a liquid. In 1879 van der Waals proposed an equation of state to account for nonideal-gas behavior. It has the form of

$$\left(p + \frac{a}{\bar{v}^2}\right)(\bar{v} - b) = \mathcal{R}T \tag{6-47}$$

where a is a constant which takes account of the molecular force-field interactions, and b is a constant which compensates for the volume occupied by the molecules. The reasoning used to obtain the $a/\bar{v}^2$ term is as follows. When a group of molecules strikes a surface, the pressure force exerted is proportional to the molecular density (molecules per unit volume, N/V), because it is this quantity that governs the number of molecular impacts. For molecule-molecule impacts the interaction force is reasoned to be proportional to $(N/V)^2$ since the net effect results from two molecules striking each other. The molecular density is clearly proportional to the macroscopic gas density which, in turn, is proportional to $1/\bar{v}$. Thus the net interaction force is proportional to $1/\bar{v}^2$ and a is the proportionality constant. The constant b simply takes account of the effective volume occupied by the molecule.

To use the van der Waals equation, values of the constants a and b must be determined experimentally and a summary of values for various gases is given in Table 6-1.

A conventional technique used to estimate the van der Waals constants is based on the behavior of pure substances at the critical point. The isotherm

TABLE 6-1
Van der Waals Constants

	a		b		
Gas	**kN·m⁴/(kg mol)²**	**atm·ft⁶/(lbm mol)²**	**m³/kg mol**	**ft³/lbm mol**	$Z_c = p_c v_c / RT_c$
Air	135.8	343.8	0.0365	0.585	0.284
O_2	138.0	349.5	0.0318	0.510	0.29
N_2	136.7	346	0.0386	0.618	0.291
H_2O	551.7	1397.1	0.0304	0.487	0.23
CH_4	228.6	578.9	0.0427	0.684	0.29
CO	147.9	374.5	0.0393	0.630	0.293
CO_2	365.6	926	0.0428	0.686	0.276
NH_3	424.9	1076	0.0373	0.598	0.242
H_2	24.8	62.9	0.0266	0.426	0.304
He	3.42	8.66	0.0235	0.376	0.30

passing through the critical point on a p-v diagram encounters a point of inflection at (T_c, p_c, v_c). Thus

$$\left(\frac{\partial p}{\partial v}\right)_{T_c} = 0 \qquad \text{at } T = T_c \tag{6-48}$$

$$\left(\frac{\partial^2 p}{\partial v^2}\right)_{T_c} = 0 \qquad \text{at } T = T_c \tag{6-49}$$

Solving Eq. (6-47) for p, we obtain

$$p = \frac{\mathcal{R}T}{\bar{v} - b} - \frac{a}{\bar{v}^2} \tag{6-50}$$

Applying Eqs. (6-48) and (6-49),

$$\frac{-\mathcal{R}T_c}{(\bar{v}_c - b)^2} + \frac{2a}{\bar{v}^3} = 0 \tag{6-51}$$

$$\frac{2\mathcal{R}T_c}{(\bar{v}_c - b)^3} - \frac{6a}{\bar{v}_c^4} = 0 \tag{6-52}$$

Equations (6-51) and (6-52) may now be solved for the constants a and b to yield

$$a = 3p_c\bar{v}_c^2 = \frac{9}{8}\mathcal{R}T_c\bar{v}_c = \frac{27}{64}\frac{\mathcal{R}^2T_c^2}{p_c} \tag{6-53}$$

$$b = \frac{\bar{v}_c}{3} = \frac{\mathcal{R}T_c}{8p_c} \tag{6-54}$$

Additionally, we obtain for a van der Waals gas

$$\frac{p_c\bar{v}_c}{\mathcal{R}T_c} = \frac{3}{8} \tag{6-55}$$

The critical properties for several gases are given in Table 6-2 as determined from experimental measurements. The accuracy of the van der Waals equation at the critical point can be checked very quickly by using these data and Eq. (6-55). For water,

$$T_c = 647.27\text{K} \qquad p_c = 22.105 \text{ MPa} \qquad \bar{v}_c = 0.0568 \text{ m}^3/\text{kg mol}$$

$$\frac{p_c\bar{v}_c}{\mathcal{R}T_c} = \frac{(22.105 \times 10^6)(0.0568)}{(8314)(647.27)} = 0.233$$

This value is in substantial disagreement with Eq. (6-55), and we would find, in general, that a similar calculation for many other substances would also pro-

TABLE 6-2
Some Selected Critical Constants

Gas	T_c K	T_c °R	p_c MPa	p_c atm	$\bar{v}_c$ (m³/kg mol) × 10²	$\bar{v}_c$ ft³/lbm mol
Air	132.41	238.34	3.774	37.25	8.30	1.33
O_2	154.78	278.6	5.080	50.14	7.43	1.19
N_2	126.2	227.16	3.398	33.54	8.99	1.44
H_2O	647.27	1165.3	22.105	218.167	5.68	0.91
CH_4	190.7	343.26	4.640	45.8	9.93	1.59
CO	132.91	239.24	3.496	34.53	9.30	1.49
CO_2	304.20	547.56	7.386	72.90	9.55	1.53
NH_3	405.4	729.72	11.277	111.3	7.24	1.16
H_2	33.24	59.83	1.297	12.8	6.49	1.04
He	5.19	9.34	0.229	2.26	5.81	0.93

duce disagreement. The reader should not be alarmed by this discrepancy, however. The simple proportionality relation for the force-field interaction $(a/\bar{v}^2)$ may only be expected to apply for modest deviations from ideal-gas behavior, i.e., for moderate- to long-range molecular spacings. For relatively close molecular spacings, as at the critical point, the interaction forces cannot be expressed by this simple relation, and we would not expect the van der Waals relation to give satisfactory results in this region.

A simpler equation of state is the *Clausius equation,* which is a modified form of the van der Waals equation that neglects interaction forces and is written as

$$p(\bar{v} - b) = \mathcal{R}T \tag{6-56}$$

The Beattie-Bridgeman Equation of State

A widely used equation of state with good accuracy is the Beattie-Bridgeman equation:

$$p = \frac{\mathcal{R}T}{\bar{v}^2}(1 - e)(\bar{v} + B) - \frac{A}{\bar{v}^2} \tag{6-57}$$

where

$$A = A_0\left(1 - \frac{a}{\bar{v}}\right)$$

$$B = B_0\left(1 - \frac{b}{\bar{v}}\right)$$

$$e = \frac{c}{\bar{v}T^3}$$

TABLE 6-3
Constants for Beattie-Bridgeman Equation of State

	a		A_0		b	b	B_0		$c \times 10^{-6}$	
Gas	(m^3/kg mol) × 10^2	ft^3/lbm mol	kN·m^4/(kg mol)^2	atm·ft^6/(lbm mol)^2	(m^3/kg mol) × 10^2	ft^3/lbm mol	(m^3/kg mol) × 10^2	ft^3/lbm mol	m^3·K^3/kg mol	ft^3·°R^3/lbm mol
Air	1.93	0.309	131.9	334.1	−4.43	−0.716	4.61	0.739	434	4.05
O_2	2.56	0.410	151.0	382.5	0.42	0.067 4	4.63	0.741	480	4.48
N_2	2.62	0.419	136.2	344.9	−0.692	−0.111	5.04	0.808	419.3	3.917
CH_4	1.85	0.297	230.8	584.6	−1.58	−0.254	5.59	0.895	1 282	11.98
C_3H_8	7.32	1.173	120.7	305.8	4.29	0.688	18.1	2.90	12 002	112.12
C_4H_{10}	12.16	1.948	180.2	456.5	9.43	1.51	24.62	3.944	35 005	327.02
CO	2.62	0.419	136.2	344.9	−0.069	−0.011 05	5.04	0.808	419	3.92
CO_2	7.14	1.143	507.4	1 284.9	7.24	1.159	10.48	1.678	6 599	61.65
NH_3	17.04	2.729	242.4	613.9	19.12	3.062	3.41	0.547	47 698	445.6
H_2	−0.506	−0.0811	19.97	50.57	−4.35	−0.698	2.10	0.336	5.04	0.047 1
He	5.98	0.958	2.19	5.55	0	0	1.40	0.224	0.399	0.003 73

There are five constants—a, A_0, b, B_0, and c—to be determined empirically. Table 6-3 lists values of these constants for several substances.

The Beattie-Bridgeman equation of state predicts properties accurately (within 2 percent) in regions where the density is less than 0.8 times the critical density.

Two other equations of state in use are the *Bertholet equation:*

$$p = \frac{\Re T}{\bar{v} - b} - \frac{a}{T\bar{v}^2} \tag{6-58}$$

and the *Dieterici equation:*

$$p = \frac{\Re T}{\bar{v} - b}\, e^{-a/\Re T\bar{v}} \tag{6-59}$$

The Dieterici equation predicts properties very well in the neighborhood of the critical point and on the critical isotherm. It is much less satisfactory in other regions and may produce large errors when used away from the critical region. Following the same procedure as with van der Waals' equation, the values of the constants are obtained as

$$a = \frac{4\Re^2 T_c^2}{p_c e^2} \qquad b = \frac{\Re T_c}{p_c e^2}$$

In addition, we obtain

$$\frac{p_c \bar{v}_c}{\Re T_c} = 0.271 = \frac{2}{e^2}$$

a value which is much closer to the experimental values for substances than the 3/8 predicted by the van der Waals relation.

The Berthelot equation is similar to van der Waals equation except for the inclusion of the temperature in the denominator of the second term. For low pressures it is accurate within 1 percent using the value of a given by Eq. (6-53) and $b = 9\mathcal{R}T_c/28p_c$.

For high-density regions up to approximately two times the critical density a complicated empirical relation involving eight experimentally determined constants is the equation of Benedict-Webb-Rubin [10] (Table 6-4):

$$p = \frac{\mathcal{R}T}{\bar{v}} + \frac{\mathcal{R}TB_0 - A_0 - C_0/T^2}{\bar{v}^2} + \frac{\mathcal{R}Tb - a}{\bar{v}^3} + \frac{a\alpha}{\bar{v}^6} + \frac{c}{\bar{v}^3T^2}\left(1 + \frac{\gamma}{\bar{v}^2}\right) e \qquad (6\text{-}60)$$

where A_0, B_0, C_0, a, b, c, α, and γ are the empirical constants.

Still another equation of state is that of Redlich and Kwong:

$$p = \frac{\mathcal{R}T}{\bar{v} - b} - \frac{a}{\bar{v}(\bar{v} + b)T^{1/2}} \qquad (6\text{-}61)$$

with the constants given in terms of critical state properties:

$$a = 0.427\,48\,\frac{\mathcal{R}^2T_c^{5/2}}{p_c} \qquad (6\text{-}62)$$

$$b = 0.086\,64\,\frac{\mathcal{R}T_c}{p_c} \qquad (6\text{-}63)$$

Virial Equation of State

The viral equation of state represents an expansion of the $p\bar{v}$ product in infinite-series form:

$$p\bar{v} = \mathcal{R}T\left(1 + \frac{B}{\bar{v}} + \frac{C}{\bar{v}^2} + \frac{D}{\bar{v}^3} + \cdots\right) \qquad (6\text{-}64)$$

The constants B, C, D, . . . are called *virial coefficients* and are temperature dependent. The word *virial* arises from the Latin word for force, and each term in the series provides a successive correction to the ideal-gas behavior, or no-interaction-force condition. In some cases it is possible to calculate values of the virial coefficients from statistical thermodynamic considerations together with empirical relations for molecular force fields. Although the calculation is quite complicated, a virial formulation does accurately predict p-v-T behavior over a wide range of densities.

TABLE 6-4

Constants for Benedict-Webb-Rubin Equation of State (SI units)

Constant	N_2	CO	CO_2	CH_4	C_4H_{10} (*n*-butane)
a, $N\cdot m^7/(kg\ mol)^3$	2540	3710	13 860	5000	19 070
A_0, $N\cdot m^4/(kg\ mol)^2$	1.0673×10^5	13 587	27 730	18 790	10 220
b, $m^6/(kg\ mol)^3$	232.8	263.2	721	338	4000
B_0, $m^3/kg\ mol$	4074	5454	4991	4260	12 440
c, $N\cdot m^7\cdot K^2/(kg\ mol)^3$	7.379×10^7	1.054×10^8	1.511×10^9	2.578×10^8	3.20×10^{10}
C_0, $N\cdot m^4\cdot K^2/(kg\ mol)^2$	8.164×10^8	8.673×10^8	1.404×10^{10}	2.286×10^9	1.006×10^{11}
α, $m^9/(kg\ mol)^3$	1.272×10^{-4}	8.47×10^{-5}	1.35×10^{-4}	1.24×10^{-4}	1.101×10^{-3}
γ, $m^6/(kg\ mol)^2$	0.0053	0.0060	0.005 39	0.0060	0.0340

It is sometimes more convenient to express the virial equation with pressure as the independent variable; i.e.,

$$p\bar{v} = \mathcal{R}T(1 + B'p + C'p^2 + D'p^3 + \cdots) \tag{6-65}$$

There are many other equations of state which have been developed and tailored to match various types of substances. No single equation will be satisfactory for all ranges of temperatures and pressures, as the foregoing discussion has illustrated. The interested reader may wish to consult Refs. [2, 4, 6] for more detailed information on the applicability of the different equations of state.

6-13 THE GENERALIZED COMPRESSIBILITY FACTOR

Because the ideal-gas equation of state is so simple, it is not unnatural to seek a means of modifying it in order to match with nonideal-gas behavior. The technique employed is to define a factor Z, called the *compressibility factor,* such that

$$Z = \frac{pv}{RT} \tag{6-66}$$

Clearly, $Z = 1.0$ for an ideal gas, and the various virial coefficients in Eq. (6-64) simply provide the series corrections to the ideal-gas behavior. It is possible to give some generalized information for Z, as indicated in the following paragraphs.

Let us first consider a van der Waals gas and define new variables called the *reduced pressure, reduced volume,* and *reduced temperature:*

$$p_r = \frac{p}{p_c} = \text{reduced pressure}$$

$$v_r = \frac{v}{v_c} = \text{reduced specific volume}$$

$$T_r = \frac{T}{T_c} = \text{reduced temperature}$$

Some algebraic manipulation of the van der Waals equation, Eq. (6-47), along with the relations for the constants, Eqs. (6-53) and (6-54), gives

$$Z^3 - \left(\frac{p_r}{8T_r} + 1\right) Z^2 + \left(\frac{27p_r}{64T_r^2}\right) Z - \frac{27p_r^2}{512T_r^3} = 0 \tag{6-67}$$

where Z is the compressibility factor defined in Eq. (6-66).

Clearly, Eq. (6-67) can be solved for Z explicitly in terms of p_r and T_r and consequently we postulate the existence of a relation of the form

$$Z = f(p_r, T_r) \tag{6-68}$$

Equation (6-68) is called the *law of corresponding states*. Our previous discussions have shown that the van der Waals equation is inaccurate and, hence, at first glance, we would be tempted to question the validity of Eq. (6-68). However, experimental measurements show that a relation like Eq. (6-68) does indeed exist for a wide variety of substances, although the functional variation is quite different from that given by a solution to Eq. (6-67). Figure 6-3 presents a display of experimental data indicating the general existence of a relation like Eq. (6-68). Figures 6-4 to 6-6 present the Nelson-Obert calculation charts that have been derived from the various experimental data [2]. It turns out that it is convenient to plot a pseudo-reduced volume v_r', instead of v_r, which is defined by

$$v_r' = \text{pseudo-reduced volume} = \frac{v}{RT_c/p_c} \tag{6-69}$$

which may be expressed in the alternate forms

$$\begin{aligned} v_c' &= \frac{v}{v_c} \frac{v_c}{RT_c/p_c} = v_r Z_c \\ &= \frac{p}{RT} \frac{T/T_c}{p/p_c} = Z \frac{T_r}{p_r} \end{aligned} \tag{6-70}$$

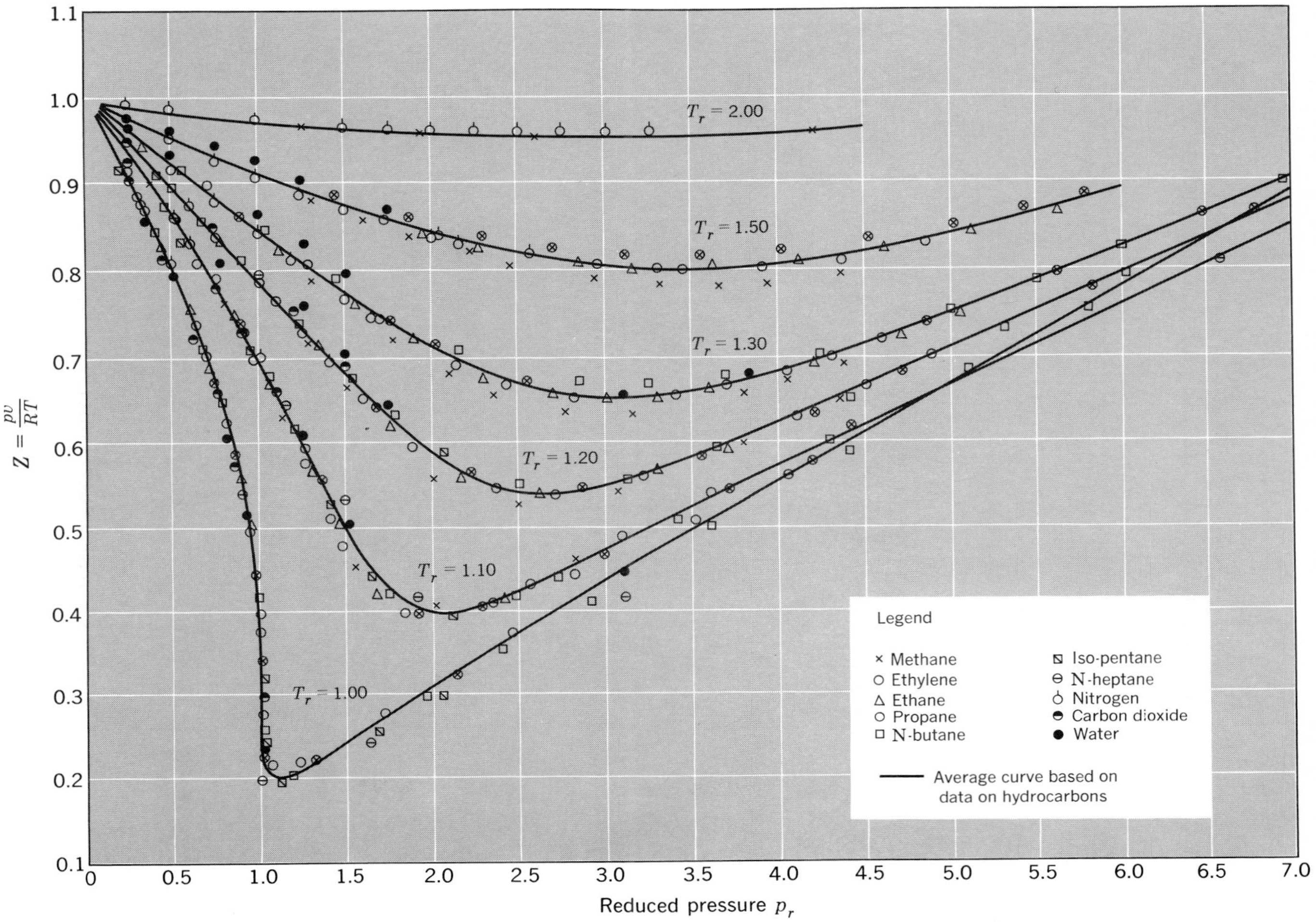

FIG. 6-3 Experimental data showing generalized compressibility relation from Ref. [1].

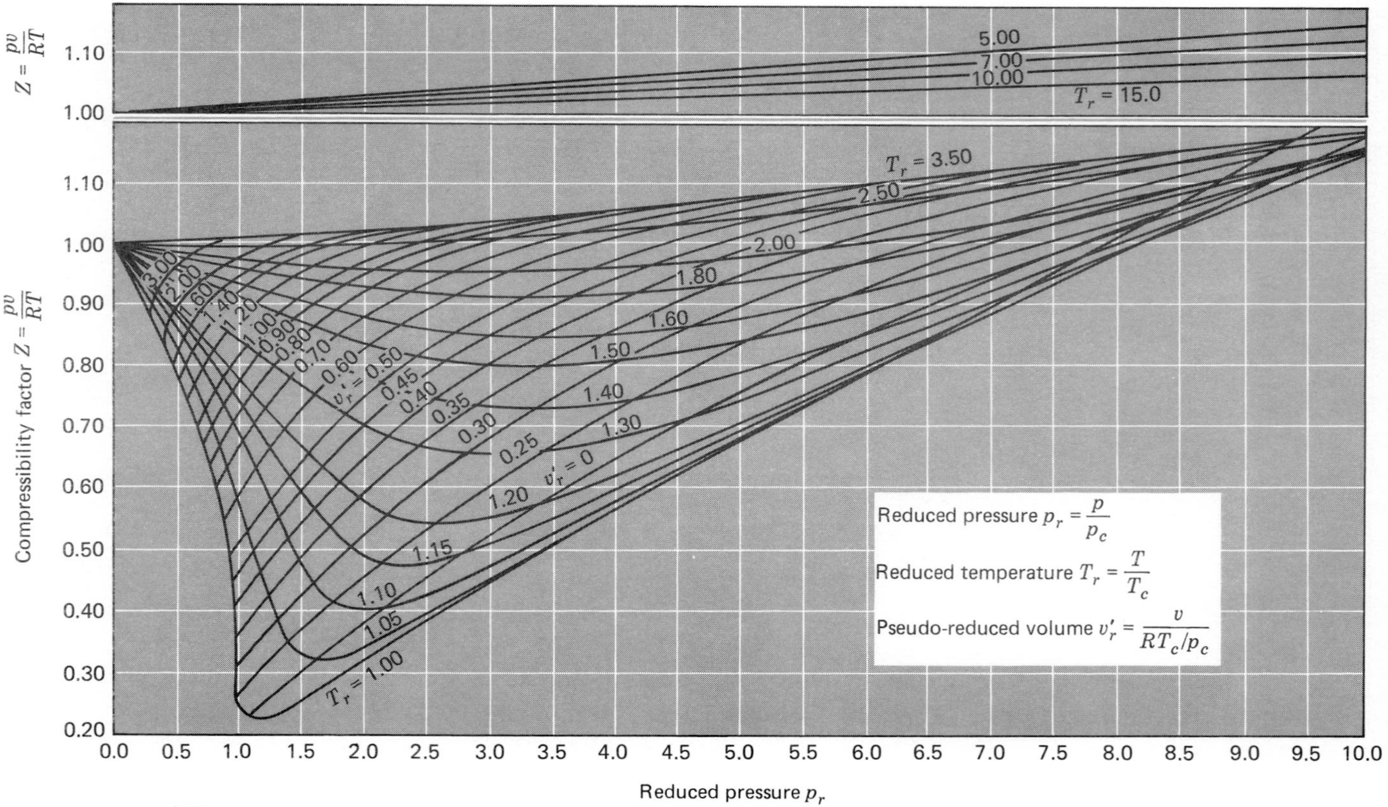

FIG. 6-4 Nelson-Obert generalized compressibility chart according to Ref. [2].

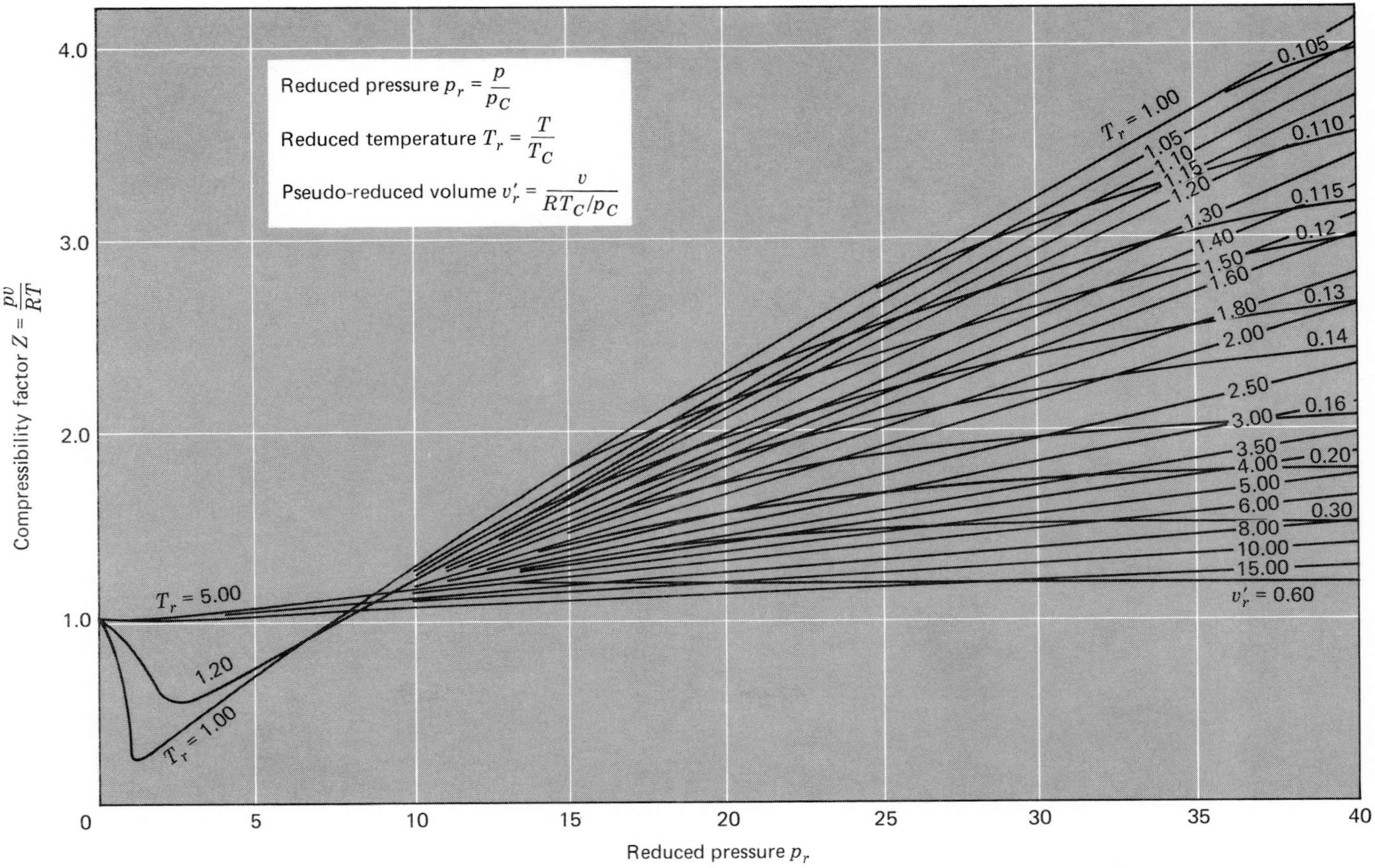

FIG. 6-5 Nelson-Obert generalized compressibility chart, corrected according to Ref. [2].

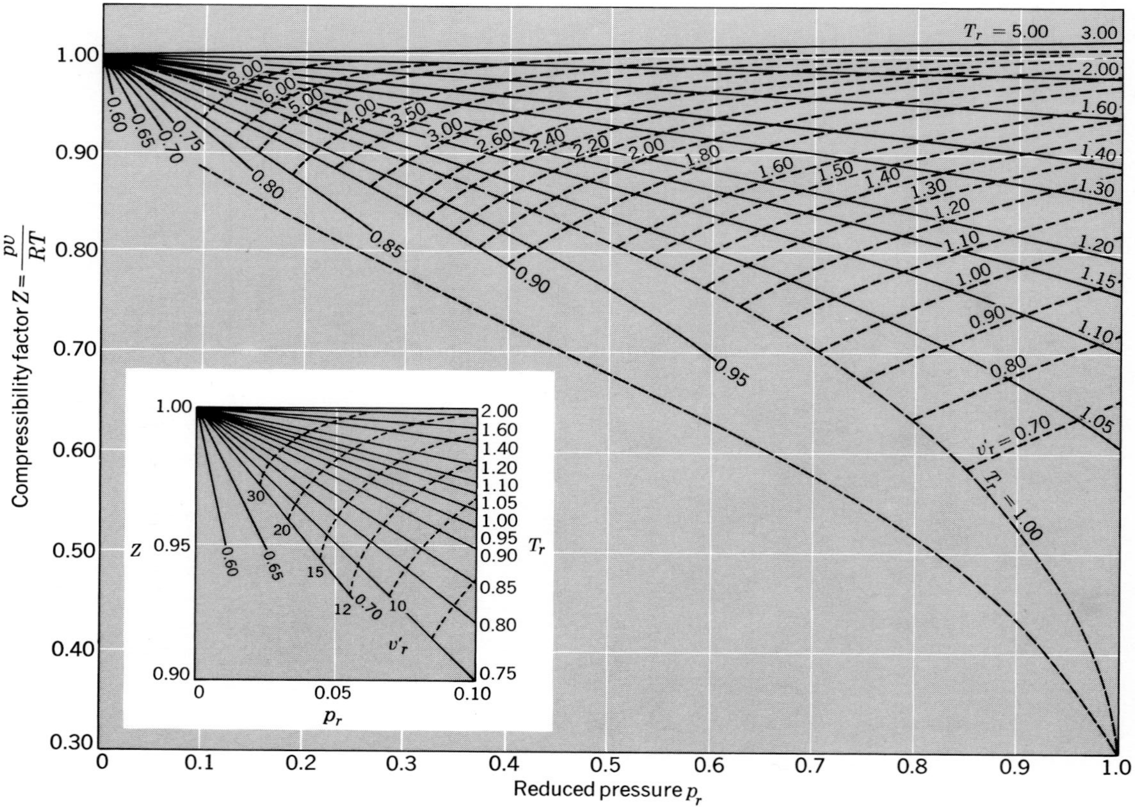

FIG. 6-6 Nelson-Obert generalized compressibility chart, low pressure region, according to Ref. [2].

The generalized compressibility charts provide us with a more explicit basis for interpreting ideal-gas behavior than has been previously at our disposal. The ideal gas is one for which $Z = 1.0$. Consequently, we need only examine Fig. 6-5 to predict the ranges of pressures and temperatures in which a substance will behave like an ideal gas. The two general notions we obtain are:

Ideal-gas behavior exists when

1 p_r becomes small compared to 1.0.
2 T_r becomes large compared to 1.0.

As an example, we might try to anticipate the behavior of nitrogen at 300 K and 100 kPa. From the critical constants in Table 6-2, we obtain

$$p_r = \frac{100}{3398} = 0.0294$$

$$T_r = \frac{300}{126.2} = 2.38$$

The reduced pressure is very small, so we would anticipate ideal-gas behavior, as we have assumed several times in previous chapters.

EXAMPLE 6-6 Comparison of Equations of State
A mass of CO_2 having a molal specific volume of 2.8 ft³/lbm mol is contained in a rigid vessel at 200°F. Estimate the pressure exerted by the gas using (a) the ideal-gas equation, (b) the van der Waals equation, and (c) the Beattie-Bridgeman equation.

SOLUTION We have $T = 200°\text{F} = 600°\text{R}$,

$$\bar{v} = 2.8 \text{ ft}^3/\text{lbm mol} \qquad (0.1748 \text{ m}^3/\text{kg mol})$$

(a) For the ideal-gas equation of state,

$$p = \frac{\mathcal{R}T}{\bar{v}} = \frac{(1545)(660)}{(2.8)(144)} = 2529 \text{ psia} = 172.1 \text{ atm}$$

(b) From Table 6-1 the van der Waals constants for CO_2 are

$$a = 926 \text{ atm}\cdot\text{ft}^6/(\text{lbm mol})^2$$

$$b = 0.686 \text{ ft}^3/\text{lbm mol}$$

Using Eq. (6-50), we then compute the pressure as

$$p = \frac{\mathcal{R}T}{\bar{v} - b} - \frac{a}{\bar{v}^2}$$

$$= \frac{(1545)(660)}{(2.8 - 0.686)(144)} - \frac{(926)(14.696)}{(2.8)^2}$$

$$= 1614 \text{ psia} = 109.8 \text{ atm}$$

(c) Finally, we use the Beattie-Bridgeman equation [Eq. (6-57)] with the constants obtained from Table 6-3:

$a = 1.143$ ft³/lbm mol $\quad A_0 = 1284.9$ atm·(ft³/lbm mol)²

$b = 1.159$ ft³/lbm mol $\quad B_0 = 1.678$ ft³/lbm mol

$c = 61.65 \times 10^6$ ft³·°R³/lbm mol

Then,

$$A = A_0\left(1 - \frac{a}{\bar{v}}\right) = 1284.9\left(1 - \frac{1.143}{2.8}\right) = 760.4$$

$$B = B_0\left(1 - \frac{b}{\bar{v}}\right) = 1.678\left(1 - \frac{1.159}{2.8}\right) = 0.9384$$

$$e = \frac{c}{\bar{v}T^3} = \frac{61.56 \times 10^6}{(2.8)(660)^3} = 0.0766$$

Then, using Eq. (6-57),

$$p = \frac{\mathcal{R}T}{\bar{v}^2}(1 - e)(\bar{v} + B) - \frac{A}{\bar{v}^2}$$

$$= \frac{(1545)(660)(1 - 0.0766)(2.8 + 0.9384)}{(2.8)^2(144)} - \frac{(760.4)(14.696)}{(2.8)^2}$$

$$= 1692 \text{ psia} = 115.2 \text{ atm}$$

From these calculations we see that there are substantial differences in the answers obtained from the three equations of state. Based on our discussions we would anticipate that the Beattie-Bridgeman calculation is the most accurate and we will verify this fact shortly.

EXAMPLE 6-7 Corresponding States

Calculate the pressure exerted by the CO_2 in Example 6-6 using the principle of corresponding states and the generalized compressibility charts.

SOLUTION From Table 6-2 the critical constants for CO_2 are

$$T_c = 548°\text{R}$$

$$p_c = 72.9 \text{ atm} = 1071 \text{ psia}$$

$$\bar{v}_c = 1.53 \text{ ft}^3/\text{lbm mol}$$

We may then compute $\bar{v}_r$ and T_r for use with Fig. 6-4:

$$T_r = \frac{T}{T_c} = \frac{600}{548} = 1.204$$

$$\bar{v}_r = \frac{\bar{v}}{\mathcal{R}T_c/p_c} = \frac{(2.8)(1071)(144)}{(1545)(548)} = 0.510$$

Using these values we read from Fig. 6-4,

$$p_r = 1.6 \qquad Z = 0.67$$

The pressure is then

$$p = p_r p_c = (1.6)(72.9) = 116.6 \text{ atm}$$

With this calculation we see that the Beattie-Bridgeman determination of Example 6-6 was indeed quite accurate, that is, 116.6 versus 115.2 atm. These numbers certainly agree within the accuracy with which one may read the charts. Checking back, we find the other two calculation methods in significant disagreement:

$$\text{Ideal-gas error} = \frac{172.1 - 116.6}{116.6} = +47.6\%$$

$$\text{van der Waals error} = \frac{109.8 - 116.6}{116.6} = -5.8\%$$

EXAMPLE 6-8 Properties from Corresponding States
Estimate the density of nitrogen at $-150°$F and 600 psia.

SOLUTION From Table 6-2 the critical constants are

$$p_c = 33.5 \text{ atm} = 482 \text{ psia}$$

$$T_c = 227°\text{R}$$

$$\bar{v}_c = 1.44 \text{ ft}^3/\text{lbm mol}$$

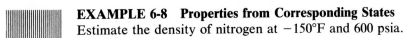

We compute the reduced pressure and temperatures as

$$p_r = \frac{600}{482} = 1.245$$

$$T_r = \frac{310}{227} = 1.365$$

Consulting Fig. 6-3 or 6-4, we find

$$Z = 0.83$$

The specific volume is now calculated from Eq. (6-66) as

$$v = \frac{ZRT}{p} = \frac{(0.83)(55.12)(310)}{(600)(144)} = 0.164 \text{ ft}^3/\text{lbm}$$

The density is the reciprocal of the specific volume, so

$$\rho = \frac{1}{0.164} = 6.1 \text{ lbm/ft}^3 \qquad (97.7 \text{ kg/m}^3)$$

EXAMPLE 6-8M Properties from Corresponding States
Estimate the density of nitrogen at 170 K and 4 MPa.

SOLUTION From Table 6-2 the critical constants are

$$p_c = 3.398 \text{ MPa}$$

$$T_c = 126.2 \text{ K}$$

$$\bar{v}_c = 0.0899 \text{ m}^3/\text{kg mol}$$

We compute the reduced pressure and temperature as

$$p_r = \frac{4}{3.398} = 1.177$$

$$T_r = \frac{170}{126.2} = 1.347$$

Consulting Fig. 6-3 or 6-4, we find

$$Z = 0.85$$

The specific volume is now calculated from Eq. (6-66) as

$$v = \frac{ZRT}{p} = \frac{(0.85)(296.6)(170)}{4 \times 10^6} = 0.0107 \text{ m}^3/\text{kg}$$

The density is the reciprocal of the specific volume, so

$$\rho = \frac{1}{0.0107} = 93.3 \text{ kg/m}^3 \qquad (5.8 \text{ lbm/ft}^3)$$

EXAMPLE 6-9

Calculate the pressure at which $v = 0.004$ m^3/kg and $T = 170$ K for nitrogen.

SOLUTION The critical constants are given in Example 6-8M for nitrogen. With the foregoing data we have for the reduced temperature and volume

$$T_r = \frac{170}{126.2} = 1.347$$

$$v_r' = \frac{v}{RT_c/p_c} = \frac{0.004}{(296.6)(126.2)/(3.398 \times 10^6)} = 0.363$$

Consulting Fig. 6-4, we find

$$p_r = 2.6$$

The pressure is therefore

$$p = p_r p_c = (2.6)(3.398) = 8.83 \text{ MPa}$$

6-14 ENERGY PROPERTIES OF REAL GASES

The law of corresponding states offers a means of calculating energy properties of substances in addition to the obvious information it provides on p-v-T properties. Consider the schematic chart for enthalpy behavior indicated in Fig. 6-7. At very low reduced pressures the enthalpy is a function of temperature alone since the gas obeys the $pv = RT$ equation of state. At higher pressures the enthalpy can be substantially dependent on pressure. The variation of enthalpy at low values of p_r can be obtained with the calculation methods outlined in previous sections. We shall define the enthalpy of this low-pressure state as $\bar{h}^*$. Then, for some higher pressure, we may integrate Eq. (6-26) to obtain

$$(\bar{h} - \bar{h}^*)_T = \int_{p=0}^{p} \left[\bar{v} - T\left(\frac{\partial \bar{v}}{\partial T}\right)_p \right] dp_T \tag{6-71}$$

But

$$\bar{v} = \frac{Z\mathcal{R}T}{p} \tag{6-72}$$

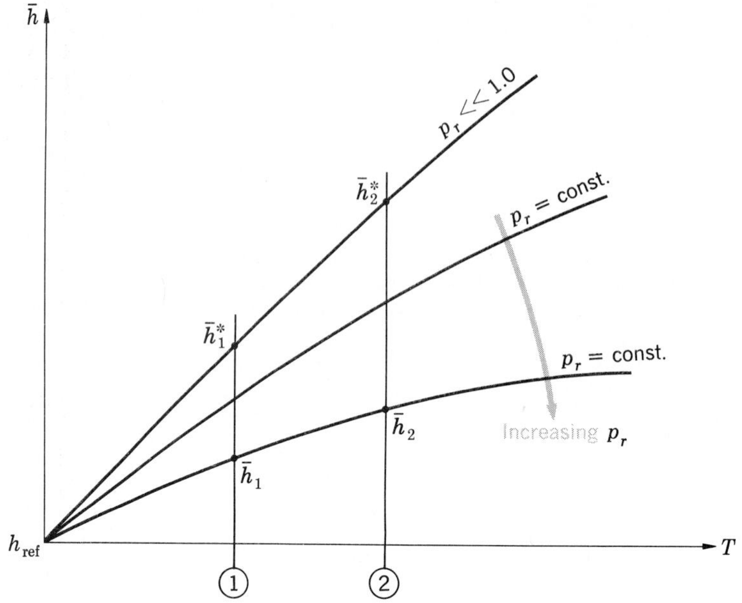

FIG. 6-7 Schematic representation of enthalpy variation for pure substance.

and

$$\left(\frac{\partial \bar{v}}{\partial T}\right)_p = \frac{Z\mathcal{R}}{p} + \frac{\mathcal{R}T}{p}\left(\frac{\partial Z}{\partial T}\right)_p \tag{6-73}$$

Introducing Eqs. (6-72) and (6-73) into Eq. (6-71), and expressing the variables in terms of reduced properties,

$$(\bar{h} - \bar{h}^*)_T = -\mathcal{R}T_c \int_0^{p_r} T_r^2 \left(\frac{\partial Z}{\partial T_r}\right)_p \frac{dp_r}{p_r} \tag{6-74}$$

In a constant-temperature process T_r = const., so that finally,

$$\frac{(\bar{h}^* - \bar{h})}{T_c} = \mathcal{R}T_r^2 \int_0^{p_r} \left(\frac{\partial Z}{\partial T_r}\right)_{p_r} \frac{dp_r}{p_r} \tag{6-75}$$

The right side of Eq. (6-75) is a function of T_r and p_r and may be obtained by a numerical integration of the generalized compressibility charts. The result of such a calculation is presented in Fig. 6-8.

A development similar to the previous one can be used to obtain generalized charts for entropy. The resulting equation is

$$\bar{s}_p^* - \bar{s}_p = \mathcal{R} \int_0^{p_r} \left[Z - 1 + T_r \left(\frac{\partial Z}{\partial T_r}\right)_{p_r}\right]_{T_r} \frac{dp_r}{p_r} \tag{6-76}$$

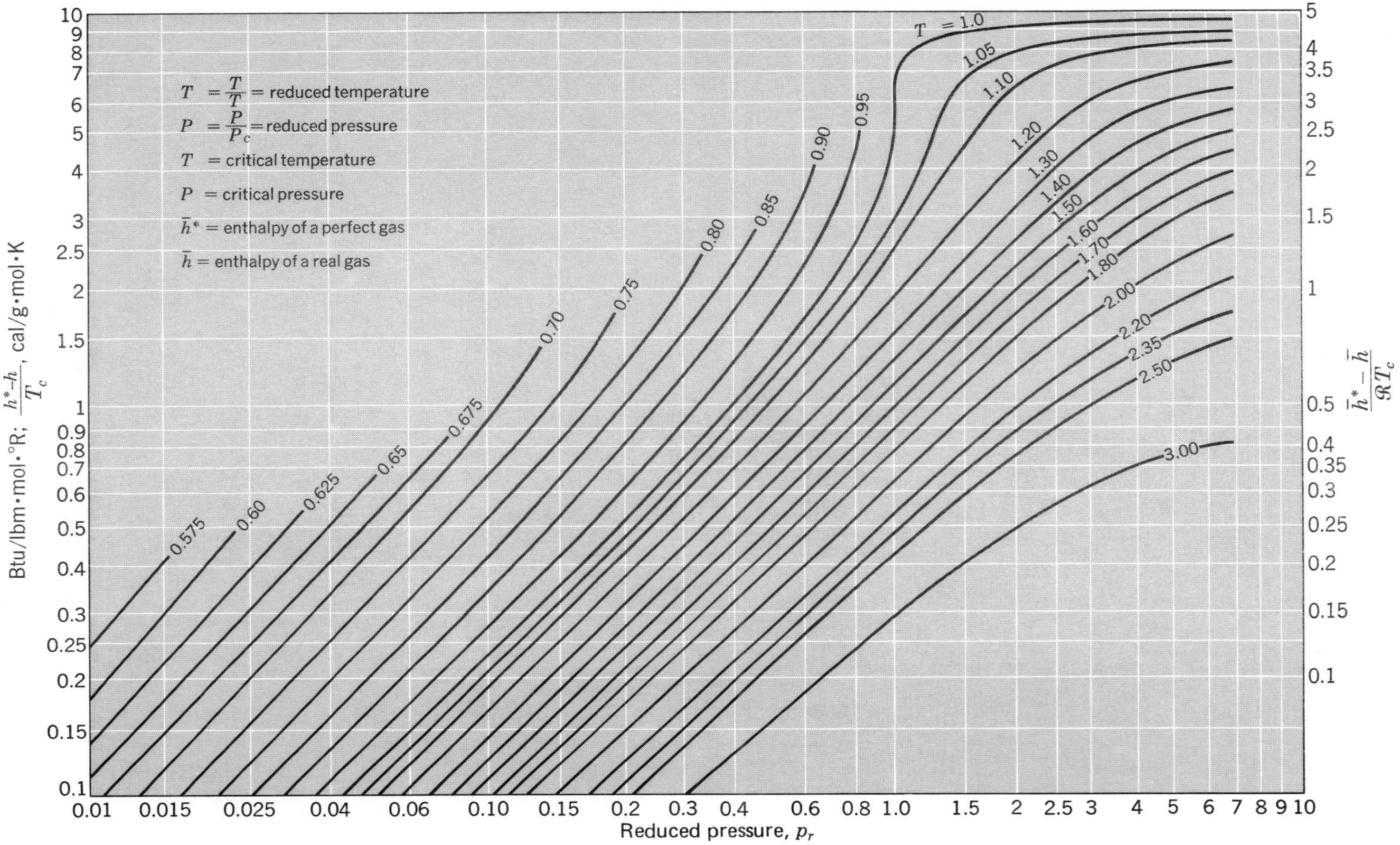

FIG. 6-8 Generalized enthalpy chart according to Ref. [4].

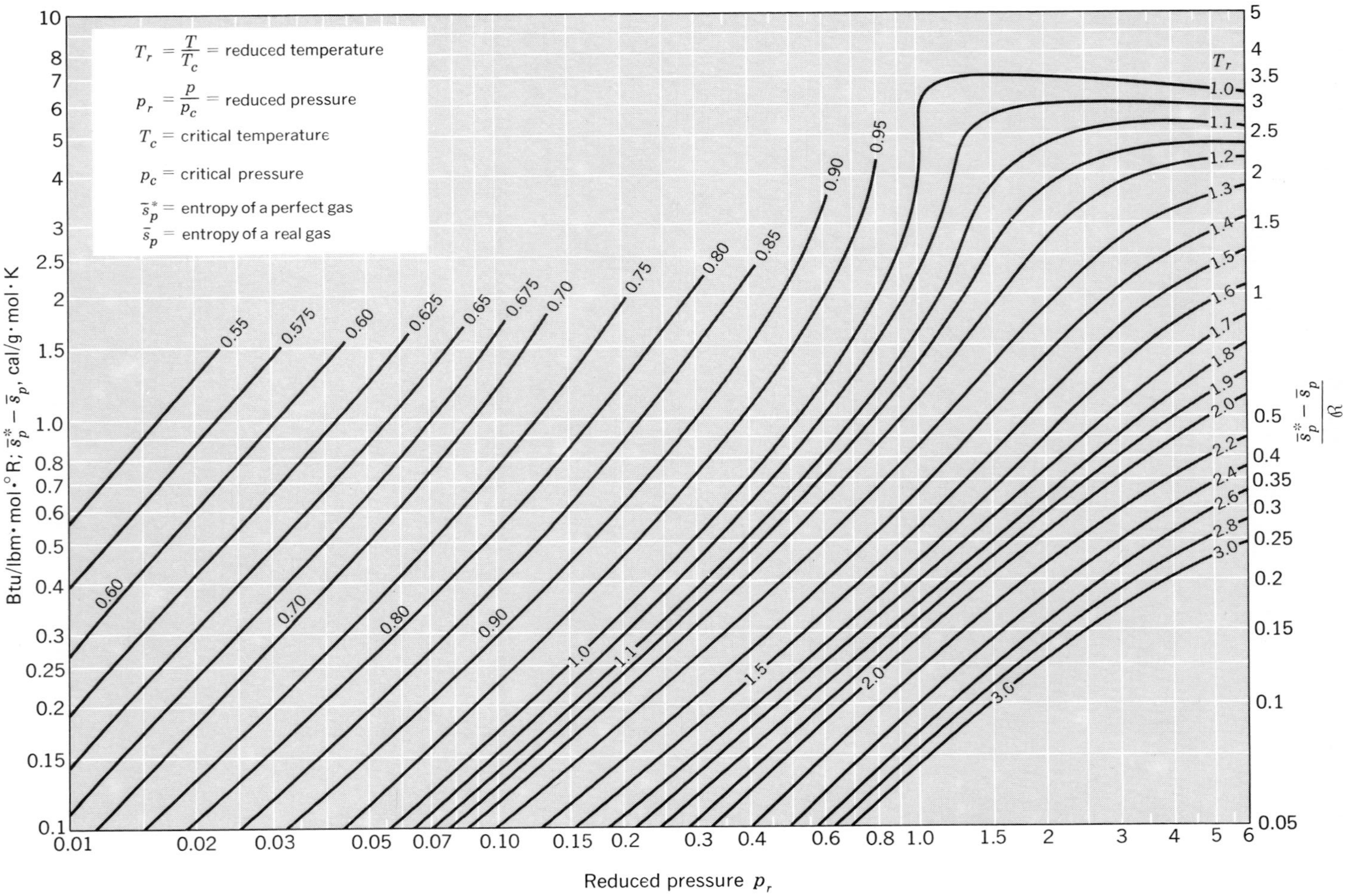

FIG. 6-9 Generalized chart for entropy according to Ref. [4].

where

$\bar{s}_p^*$ = entropy of ideal gas ($Z = 1.0$) at p and T in respect to some reference state

$\bar{s}_p$ = entropy of real gas at p and T in respect to same reference state

The generalized chart for entropy is presented in Fig. 6-9.

EXAMPLE 6-10 Isothermal Compression
Methane (CH_4) is compressed in a reversible isothermal process from 20 psia, 100°F, to 500 psia. The process occurs under steady-flow conditions. Calculate the work required per pound mass of methane.

SOLUTION If kinetic energies are negligible, the steady-flow energy equation for this process becomes

$$Q + W + h_1 = h_2 \tag{a}$$

where Q and W are the heat and work added, respectively. Since this is a reversible isothermal process, the heat transfer may be calculated from

$$Q = T(s_2 - s_1) \tag{b}$$

The work is therefore

$$W = h_2 - h_1 - T(s_2 - s_1) \tag{c}$$

We therefore need to obtain the values of the enthalpy and entropy from the generalized charts. The critical properties of methane are

$$T_c = 344°\text{R} \qquad (191.1\text{ K})$$

$$p_c = 45.8\text{ atm} = 673\text{ psia}$$

The reduced properties of interest are

$$p_{r_1} = \frac{20}{673} = 0.0297$$

$$p_{r_2} = \frac{500}{673} = 0.743$$

$$T_r = \frac{560}{344} = 1.63$$

From Fig. 6-8 we have

$$\frac{\bar{h}_1^* - \bar{h}_1}{T_c} \approx 0.03 \qquad \text{by extrapolation}$$

$$\frac{\bar{h}_2^* - \bar{h}_2}{T_c} = 0.68 \text{ Btu/lbm·mol·°R}$$

Since $T_1 = T_2$, $\bar{h}_1^* = \bar{h}_2^*$ and we have

$$\begin{aligned}\bar{h}_2 - \bar{h}_1 &= (0.03 - 0.68)T_c \\ &= -(0.65)(344) = -233 \text{ Btu/lbm mol} \qquad (-519 \text{ kJ/kg mol})\end{aligned}$$

From Fig. 6-9 we have

$$\bar{s}_{p_1}^* - \bar{s}_{p_1} \approx 0.018 \qquad \text{by extrapolation}$$

$$\bar{s}_{p_2}^* - \bar{s}_{p_2} = 0.35 \text{ Btu/lbm·mol·°R} \qquad (1.465 \text{ kJ/kg·mol·K})$$

In general, for the change in ideal-gas entropies we would write

$$\bar{s}_{p_2}^* - \bar{s}_{p_1}^* = \bar{\phi}_2 - \bar{\phi}_1 - \mathcal{R} \ln \frac{p_2}{p_1} \tag{d}$$

In this process $T_1 = T_2$ so that $\bar{\phi}_2 = \bar{\phi}_1$, and only the pressure ratio is retained. We have

$$\begin{aligned}\bar{s}_{p_2} - \bar{s}_{p_1} &= (\bar{s}_{p_1}^* - \bar{s}_{p_1}) - (\bar{s}_{p_2}^* - \bar{s}_{p_2}) + (\bar{s}_{p_2}^* - \bar{s}_{p_1}^*) \\ &= 0.018 - 0.35 - \mathcal{R} \ln \frac{p_2}{p_1} \\ &= -0.332 - \frac{1545}{778} \ln \frac{500}{20} \\ &= -6.71 \text{ Btu/lbm·mol·°R} \qquad (-28.09 \text{ kJ/kg·mol·K})\end{aligned}$$

We may now insert these properties back into Eq. (*c*) to obtain the work:

$$\begin{aligned}W &= -223 - (560)(-6.71) \\ &= 3537 \text{ Btu/lbm mol} \qquad (8227 \text{ kJ/kg mol})\end{aligned}$$

Since $M = 16$ for methane, the work per pound-mass is

$$W = \frac{357}{16} = 221 \text{ Btu/lbm} \qquad (514 \text{ kJ/kg})$$

EXAMPLE 6-10M Isothermal Compression
Methane (CH_4) is compressed in a reversible isothermal process from 140 kPa, 300 K, to 3.5 MPa. The process occurs under steady-flow conditions. Calculate the work required per kilogram of methane.

SOLUTION If kinetic energies are negligible, the steady-flow energy equation for this process becomes

$$Q + W + h_1 = h_2 \qquad (a)$$

where Q and W are the heat and work added, respectively. Because this is a reversible isothermal process, the heat transfer may be calculated from

$$Q = T(s_2 - s_1) \qquad (b)$$

The work is therefore

$$W = h_2 - h_1 - T(s_2 - s_1) \qquad (c)$$

We therefore need to obtain the values of the enthalpy and entropy from the generalized charts. The critical properties of methane are

$$T_c = 190.7 \text{ K} \qquad (343°\text{R})$$

$$P_c = 4.640 \text{ MPa} \qquad (673 \text{ psia})$$

The reduced properties of interest are

$$p_{r_1} = \frac{140}{4640} = 0.0302$$

$$p_{r_2} = \frac{3.5}{4.640} = 0.754$$

$$T_r = \frac{300}{190.7} = 1.57$$

From Fig. 6-8 we have

$$\frac{\bar{h}_1^* - \bar{h}_1}{\Re T_c} \approx 0.016 \qquad \text{by extrapolation}$$

$$\frac{\bar{h}_2^* - \bar{h}_2}{\Re T_c} = 0.39$$

Since $T_1 = T_2$, $\bar{h}_1^* = \bar{h}_2^*$ and we have

$$\begin{aligned}\bar{h}_2 - \bar{h}_1 &= (0.016 - 0.39)\mathcal{R}T_c \\ &= -(0.374)(8.314)(190.7) \\ &= -593 \text{ kJ/kg mol} \qquad (-255 \text{ Btu/lbm mol})\end{aligned}$$

From Fig 6-9 we have

$$\frac{\bar{s}_{p_1}^* - \bar{s}_{p_1}}{\mathcal{R}} \approx 0.011 \qquad \text{by extrapolation}$$

$$\frac{\bar{s}_{p_1}^* - \bar{s}_{p_2}}{\mathcal{R}} = 0.13$$

In general, for the change in ideal-gas entropies we would write

$$\bar{s}_{p_2}^* - \bar{s}_{p_1}^* = \bar{\phi}_2 - \bar{\phi}_1 - \mathcal{R} \ln \frac{p_2}{p_1} \tag{d}$$

In this process $T_1 = T_2$, so that $\phi_2 = \phi_1$, and only the pressure ratio is retained. We have

$$\begin{aligned}\bar{s}_{p_2} - \bar{s}_{p_1} &= (\bar{s}_{p_1}^* - \bar{s}_{p_1}) - (\bar{s}_{p_2}^* - \bar{s}_{p_2}) + (\bar{s}_{p_2}^* - \bar{s}_{p_1}^*) \\ &= \mathcal{R}\left(0.011 - 0.13 - \ln \frac{p_2}{p_1}\right)\end{aligned}$$

$$\begin{aligned}\bar{s}_{p_2} - \bar{s}_{p_1} &= (8.314)\left(-0.119 - \ln \frac{3500}{140}\right) \\ &= -27.75 \text{ kJ/kg·mol·K} \qquad (-6.63 \text{ Btu/lbm·mol·°R})\end{aligned}$$

We may now insert these properties back into Eq. (*c*) to obtain the work:

$$\begin{aligned}W &= -593 - (300)(-27.75) \\ &= 7732 \text{ kJ/kg mol} \qquad (3324 \text{ Btu/lbm mol})\end{aligned}$$

Since $M = 16$ for methane, the work per kilogram is

$$W = \frac{7732}{16} = 483.3 \text{ kJ/kg} \qquad (208 \text{ Btu/lbm})$$

EXAMPLE 6-11 Constant-Pressure Heating
Carbon dioxide (CO_2) is heated at a constant pressure of 1000 psia from 100 to 800°F in a steady-flow process. Calculate the heating required per pound-mass of CO_2.

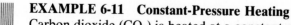

SOLUTION If kinetic energies are neglected, the heat transfer for this process is given by

$$Q = h_2 - h_1 \qquad (a)$$

and the problem reduces to one of determining the change in enthalpy from the generalized charts. From Table 6-2 the critical constants for CO_2 are

$$p_c = 72.9 \text{ atm} = 1070 \text{ psia}$$

$$T_c = 548°\text{R} \qquad (304 \text{ K})$$

The reduced properties of interest are therefore

$$p_{r_1} = p_{r_2} = \frac{1000}{1070} = 0.935$$

$$T_{r_1} = \frac{560}{548} = 1.02$$

$$T_{r_2} = \frac{1260}{548} = 2.30$$

From Fig. 6-8 we have

$$\frac{\bar{h}_1^* - \bar{h}_1}{T_c} = 3.5 \text{ Btu/lbm·mol·°R}$$

$$\frac{\bar{h}_2^* - \bar{h}_2}{T_c} = 0.48 \text{ Btu/lbm·°R}$$

We evaluate $\bar{h}_2^* - \bar{h}_1^*$ either by integrating the specific heat relation of Table 2-3 or by consulting the gas tables. We choose to consult the gas tables and obtain

$$\bar{h}_1^* = 4235.8 \text{ Btu/lbm mol}$$

$$\bar{h}_2^* = 11661.0 \text{ Btu/lbm mol}$$

We then have

$$\begin{aligned}\bar{h}_2 - \bar{h}_1 &= T_c(3.5 - 0.48) + \bar{h}_2^* - \bar{h}_1^* \\ &= (548)(3.5 - 0.48) + 11\,661.0 - 4235.8 \\ &= 9081 \text{ Btu/lbm mol}\end{aligned}$$

Since the molecular weight of CO_2 is 44, the heat transfer per pound-mass is

$$Q = h_2 - h_1 = \frac{9081}{44} = 207 \text{ Btu/lbm} \qquad (481.5 \text{ kJ/kg})$$

It is interesting to compare this value with that which would be obtained by assuming ideal-gas behavior with constant specific heats. From Table 2-2,

$$c_p \text{ for } CO_2 = 0.202 \text{ Btu/lbm·°F} \qquad (0.846 \text{ kJ/kg·°C})$$

Assuming constant specific heats, we would calculate the heat transfer as

$$Q = c_p(T_2 - T_1) = (0.202)(800 - 100) = 141 \text{ Btu/lbm} \qquad (328 \text{ kJ/kg})$$

This value is in error by approximately 30 percent.

EXAMPLE 6-11M Constant-Pressure Heating
Carbon dioxide (CO_2) is heated at a constant pressure of 7 MPa from 300 to 700 K in a steady-flow process. Calculate the heating required per kilogram of CO_2.

SOLUTION If kinetic energies are neglected, the heat transfer for this process is given by

$$Q = h_2 - h_1 \tag{a}$$

and the problem reduces to one of determining the change in enthalpy from the generalized charts. From Table 6-2 the critical constants for CO_2 are

$$p_c = 7.386 \text{ MPa}$$

$$T_c = 304.20 \text{ K} \qquad (548°\text{R})$$

The reduced properties of interest are therefore

$$p_{r_1} = p_{r_2} = \frac{7}{7.386} = 0.948$$

$$T_{r_1} = \frac{300}{304.2} = 0.986$$

$$T_{r_2} = \frac{700}{304.2} = 2.30$$

From Fig. 6-8 we have

$$\frac{\bar{h}_1^* - \bar{h}_1}{\mathcal{R}T_c} = 2.35$$

$$\frac{\bar{h}_2^* - \bar{h}_2}{\mathcal{R}T_c} = 0.23$$

We evaluate $\bar{h}_2^* - \bar{h}_1^*$ either by integrating the specific heat relation of Table 6-3 or by consulting the gas tables. We choose to consult the gas tables and obtain

$$\begin{aligned}\bar{h}_2^* - \bar{h}_1^* &= (\bar{h}_{700} - \bar{h}_{298}) - (\bar{h}_{300} - \bar{h}_{298}) \\ &= 17\,761 - 67 = 17\,694 \text{ kJ/kg mol}\end{aligned}$$

We then have

$$\begin{aligned}\bar{h}_2 - \bar{h}_1 &= \mathcal{R}T_c(2.35 - 0.23) + \bar{h}_2^* - \bar{h}_1^* \\ &= (8.314)(304.2)(2.12) + 17\,694 \\ &= 23\,056 \text{ kJ/kg mol}\end{aligned}$$

Since the molecular weight of CO_2 is 44, the heat transfer per kilogram is

$$Q = h_2 - h_1 = \frac{23\,056}{44} = 524 \text{ kJ/kg} \qquad (225 \text{ Btu/lbm})$$

It is interesting to compare this value with that which would be obtained by assuming ideal-gas behavior with constant specific heats. From Table 2-2,

$$c_p \text{ for } CO_2 = 0.846 \text{ kJ/kg}\cdot{}^\circ\text{C} \qquad (0.202 \text{ Btu/lbm}\cdot{}^\circ\text{F})$$

Assuming constant specific heats, we would calculate the heat transfer as

$$Q = c_p(T_2 - T_1) = (0.846)(700 - 300) = 338.4 \text{ kJ/kg} \qquad (145 \text{ Btu/lbm})$$

This value is in error by approximately 35 percent.

EXAMPLE 6-12 Adiabatic Throttling
Nitrogen at 500 psia and -100°F is throttled adiabatically to 50 psia in a Joule-Thomson porous-plug apparatus. Calculate the final temperature. Also estimate the Joule-Thomson coefficient in this region.

SOLUTION For this process we assume there is no appreciable kinetic energy or external work, as described in Sec. 4-7. Therefore

$$h_1 = h_2$$

If the nitrogen behaved as an ideal gas, this would mean that the initial and final temperatures would be the same because enthalpy is a function of temperature alone for an ideal gas. From Table 6-2 the critical constants are

$$p_c = 33.5 \text{ atm} = 492.5 \text{ psia}$$

$$T_c = 227\text{°F}$$

so that

$$p_{r_1} = \frac{500}{492.5} = 1.015$$

$$T_{r_1} = \frac{360}{227} = 1.586$$

$$p_{r_2} = \frac{50}{492.5} = 0.1015$$

In this temperature range we shall assume that the low pressure or ideal-gas enthalpies $\bar{h}^*$ may be calculated with a constant specific heat from Table 2-2. So, for nitrogen,

$$c_p = 0.248 \text{ Btu/lbm·°F} \qquad (1.038 \text{ kJ/kg·°C})$$

$$\bar{c}_p = 6.944 \text{ Btu/lbm·mol°F}$$

With the enthalpy reference taken as zero at 0°R,

$$\bar{h}^* = 6.944T$$

and

$$\bar{h}_1^* = (6.944)(360) = 2500 \text{ Btu/lbm mol}$$

Using the values of p_{r_1} and T_{r_1} we obtain from Fig. 6-8

$$\frac{\bar{h}_1^* - \bar{h}_1}{T_c} = 1.05$$

so that

$$\bar{h}_1 = 2500 - (1.05)(227) = 2262 \text{ Btu/lbm mol}$$

Unfortunately, there are no lines of constant enthalpy on Fig. 6-8 so we must adopt an iterative procedure which is directed toward obtaining T_2 so that $\bar{h}_1 = \bar{h}_2 = 2262$. The trials are as follows, using $p_{r_2} = 0.1015$ and Fig. 6-8:

T_2	$\bar{h}_2^*$	T_{r_2}	$\dfrac{\bar{h}_2^* = \bar{h}_2}{T_c}$	$\bar{h}_2$
360	2500	1.586	0.11	2475
340.5	2364	1.5	0.12	2336
317.8	2206	1.4	0.13	2177
329.8	2290	1.453	0.13	2261

It is worthwhile to point out, of course, that at the low pressure of 50 psia the nitrogen is beginning to behave more nearly like an ideal gas. From this calculation we see that about a 30° temperature drop is experienced (360 − 329.8) in the throttling process and the final temperature is

$$T_2 = 329.8°\text{R} = -130.2°\text{F} \qquad (-90.1°\text{C})$$

The Joule-Thomson coefficient is estimated from

$$\mu_J = \left(\frac{\partial T}{\partial p}\right)_h \approx \left(\frac{\Delta T}{\Delta p}\right)_h = \frac{329.8 - 360}{50 - 500}\left(\frac{14.7 \text{ psia}}{\text{atm}}\right)$$
$$= 0.987°\text{F/atm} \qquad (0.548°\text{C/atm})$$

From this example we can see that graphical plots of the Joule-Thomson coefficient would be very useful for throttling calculations.

6-15 THERMODYNAMICS OF MAGNETISM

Our discussion of thermodynamic properties in Chap. 3 and in this chapter has been concerned primarily with the so-called *pure substance* or, more explicitly, the simple compressible substance. We now want to examine the behavior of a simple magnetic substance, or one in which only two properties are required to fix the thermodynamic state. We shall take the internal energy and the magnetic dipole moment M as the two independent variables. For this simple substance we assume that the $p\,dv$ work mode is negligible (constant volume), so that only two properties are necessary to establish the thermodynamic state of the system. The magnetic flux density B is given by

$$B = \mu_0(H + M) \tag{6-77}$$

where H is the magnetic intensity of the external magnetic field, M is the magnetic dipole moment or magnetization vector in the direction of the applied field, and μ_0 is the permeability of free space. In accordance with Eq. (2-11), we write for the magnetic work per unit mass

$$\begin{aligned} d'W &= vH\,dB \\ &= \mu_0 vH\,dH + \mu_0 vH\,dM \end{aligned} \tag{6-78}$$

The first term represents the work done on free space and the second term expresses the work done by the magnetic substance. We may write the energy equation as

$$du = d'Q + d'W = T\,ds + \mu_0 vH\,dM \tag{6-79}$$

where we leave out of consideration the work done on free space, since we are analyzing the magnetic substance as the thermodynamic system. In Eq. (6-79) the internal energy is expressed in terms of the independent variables ds and dM, so we may write

$$du = \left(\frac{\partial u}{\partial s}\right)_M ds + \left(\frac{\partial u}{\partial M}\right)_s dM \tag{6-80}$$

Comparing Eq. (6-79) with (6-80) we obtain

$$\left(\frac{\partial u}{\partial s}\right)_M = T \tag{6-81}$$

$$\left(\frac{\partial u}{\partial M}\right)_s = \mu_0 v H \tag{6-82}$$

Equation (6-81) may be regarded as the definition of temperature for simple magnetic substances.

It should be noted at this point that the term $\mu_0 \, vH \, dM$ plays the same role as $p \, dv$ for a simple compressible substance; magnetic intensity H is analogous to pressure and $\mu_0 v \, dM$ is analogous to a volume change.

Specific heats for the simple magnetic substance may be appropriately defined by rewriting Eq. (6-79) as

$$d'Q = du - \mu_0 v H \, dM \tag{6-83}$$

Let us consider the constant H and constant M process:

$$d'Q_H = c_H \, dT_H = du_H - \mu_0 v H \, dM_H \tag{6-84}$$

If we define a magnetic enthalpy h_m as

$$h_m = u - \mu_0 v H M \tag{6-85}$$

then Eq. (6-84) becomes

$$c_H \, dT_H = (dh_m)_H$$

or

$$c_H = \left(\frac{\partial h_m}{\partial T}\right)_H \tag{6-86}$$

For the process at constant M

$$c_M \, dT_M = du_M$$

and

$$c_M = \left(\frac{\partial u}{\partial T}\right)_M \tag{6-87}$$

c_H and c_M are thus analogous to the specific heats at constant pressure and constant volume for a simple compressible substance.

Simple magnetic substances, like simple compressible substances, have equations of state and are expressed in the functional form

$$f(H,M,T) = 0$$

A *Curie substance* is one that has the very simple equation of state

$$M = C\frac{H}{T} \tag{6-88}$$

where C is called the *Curie constant*. Rewriting Eq. (6-79) and substituting $H = MT/C$,

$$ds = \frac{1}{T}\,du - \frac{\mu_0 v M}{C}\,dM \tag{6-89}$$

Because this is an exact differential equation, we may equate second mixed partial derivatives to obtain

$$\left[\frac{\partial(1/T)}{\partial M}\right]_u = \frac{-\mu_0 v}{C}\left(\frac{\partial M}{\partial u}\right)_M = 0 \tag{6-90}$$

We therefore have the result that the temperature is a function only of u, or $u = f(T)$. This result is analogous to the fact that the internal energy for an ideal gas is a function of temperature alone. The behavior of this particular type of magnetic substance is therefore quite simple, just as the ideal gas is a very elementary type of compressible substance. Finally, we may write the following equations for a Curie substance in view of the result in Eq. (6-90):

$$du = c_M\,dT \tag{6-91}$$

$$ds = c_M\frac{dT}{T} - \frac{\mu_0 v}{C}M\,dM \tag{6-92}$$

6-16 MAGNETOCALORIC EFFECT

An expansion of the foregoing theory shows that we may express the entropy of a simple magnetic substance in terms of T and H with the following relation:

$$ds = \frac{c_H}{T}\,dT \quad \mu_0 v\left(\frac{\partial M}{\partial T}\right)_H dH \tag{6-93}$$

For paramagnetic materials, experiments indicate that $(\partial M/\partial T)_H$ is *always negative*. As a result, we see from Eq. (6-93) that

1 A reversible isothermal increase in the magnetic field causes a rejection of heat; that is, $T\,ds$ is negative when dH is positive.
2 A reversible isothermal decrease in the magnetic field causes an absorption of heat; that is, $T\,ds$ is positive when dH is negative.

For a reversible adiabatic process it is clear from Eq. (6-93) that a decrease in the magnetic field causes a drop in temperature. This phenomenon is called the *magnetocaloric effect* and is widely used for achieving very low temperatures, of the order of 1 K. These low-temperature processes are fairly simple: A paramagnetic salt is cooled to a few degrees Kelvin with liquid helium and a strong magnetic field is applied. The energy from the field is subsequently dissipated in the liquid helium. After an equilibrium state is reached, the magnetic substance is isolated and the magnetic field is reduced to zero. As a result, the temperature of the substance drops to a very low value.

REVIEW QUESTIONS

1 Why is a thermodynamic property a point function?
2 Define the Helmholtz and Gibbs functions.
3 What minimum information is necessary to formulate a table of thermodynamic properties for a pure substance?
4 How is the Joule-Thomson experiment useful in determining thermodynamic properties?
5 What is known about the specific heat and internal-energy behavior for an ideal gas?
6 What basic assumptions are involved in formulating the gas tables?
7 What is the meaning of P_r, v_r, and ϕ as used in the gas tables?
8 What practical value is the Clausius-Clapeyron relation?
9 Why is a constant-pressure line a straight line in the wet-mixture region on a Mollier diagram?
10 Under what conditions, in terms of critical temperature and pressure, does one normally experience near-ideal-gas behavior?
11 What is the law of corresponding states?
12 Describe a simple magnetic substance. How does it differ from a simple compressible substance?
13 Describe the magnetocaloric effect.
14 What thermodynamic property relation is used to link microscopic and macroscopic thermodynamic analyses?
15 What experimental measurements are required to determine thermodynamic properties?
16 What is a van der Waals gas? What physical model is involved?
17 What is the generalized compressibility factor?

PROBLEMS (ENGLISH UNITS)

6-1 Air expands isentropically from a low velocity and 50 psia to atmospheric pressure. Calculate the final velocity for initial temperatures of (*a*) 1000°F, (*b*) 100°F. Compare with the final velocity which would be obtained if the air had constant specific heats.

6-2 Suppose the change in entropy for the two expansion processes in Prob. 6-1 is 0.008 Btu/lbm·°R. Calculate the exit velocity under these conditions, assuming that the process is still adiabatic, though irreversible.

6-3 Calculate the exit area of the nozzle for the conditions of Probs. 6-1 and 6-2 assuming a mass flow of 0.1 lbm/s.

6-4 Air is compressed from 10°F, 15 psia, to 500°F, 60 psia. Calculate the change in entropy.

6-5 Air expands in a turbine from 2000°F, 55 psia, to 20 psia. The initial-flow velocity is 200 ft/s and the exit velocity is 800 ft/s. Calculate the work output of the turbine if the process occurs isentropically.

6-6 100 lbm of carbon dioxide are contained at 100°F and 1500 psia in a rigid container which may exchange heat with the surroundings. A valve on the side of the container is opened and the CO_2 allowed to escape very slowly until the pressure drops to 700 psia, at which time the valve is closed. At completion of the process the temperature in the tank is measured as 70°F. Calculate the mass of CO_2 while escapes from the tank and the heat transfer to the tank from the surroundings using (*a*) ideal-gas relations and (*b*) generalized compressibility charts.

6-7 How much mass would have escaped and what would have been the final temperature in the tank of Prob. 6-6 if it were perfectly insulated?

6-8 Ethane (C_2H_6) is throttled adiabatically from 850 psia, 750°F, to 100 psia. Calculate the final temperature and the change in entropy per unit mass.

6-9 Carbon dioxide at 1000 psia and 800°F is to be expanded isentropically in a nozzle to a pressure of 500 psia. The inlet velocity to the nozzle is low and steady flow may be assumed. Calculate the exit temperature and exit velocity using (*a*) ideal gas with constant specific heats and (*b*) the generalized compressibility charts.

6-10 Suppose the carbon dioxide with the same initial conditions as in Prob. 6-9 is to be expanded to a velocity of 1200 ft/s in an isentropic nozzle. What exit flow area would be required for a mass flow of 0.4 lbm/s?

6-11 Calculate the pressure of water vapor at 1000°F and $v = 0.1143$ ft³/lbm using (*a*) the ideal-gas law, (*b*) the van der Waals equation, (*c*) the Beattie-Bridgeman equation, and (*d*) the steam tables.

6-12 Compare the pressure predicted by the Dieterici equation with the steam tables for (*a*) $T = 700$°F, $v = 0.0287$ ft³/lbm and (*b*) $T = 1400$°F, $v = 1.0893$ ft³/lbm. How do the results of this comparison jibe with the comments in the text regarding the applicability of the Dieterici equation?

6-13 Compare the ideal-gas equation, the van der Waals equation, and the Berthelot equation with tubular values for ammonia at (*a*) 200°F, $v = 13.73$ ft³/lbm and (*b*) 360°F, $v = 2.265$ ft³/lbm.

6-14 Compare the ideal-gas equation and the Dieterici equation with tabular values for Freon 12 at 240°F and v = 0.273 23 ft^3/lbm.

6-15 Verify that carbon dioxide and nitrogen may be considered to behave as ideal gases at room temperature and pressure (1 atm, 20°C).

6-16 Calculate the mass of hydrogen which would be contained in a 1000-cm^3 vessel at 3700 psia and 60°C.

6-17 Using the compressibility charts calculate the specific volume of each of the following substances at the given conditions:
(*a*) Air at 70°F and 500 psia
(*b*) Butane (C_4H_{10}) at 100 psia and 900°F
(*c*) Carbon dioxide at 500 K and 50 atm
(*d*) Nitric oxide (NO) at 100 atm and 200 K
(*e*) Water at 800 K and 50 atm

6-18 Using tabulated values of vapor pressure for water, calculate the enthalpy of vaporization for water at 100 psia. Compare this calculation with the value given in the steam tables.

6-19 Estimate the vapor pressure of Freon 12 at −150°F.

6-20 Estimate the vapor pressure of ammonia at −100 and 140°F.

6-21 Steam expands isentropically from 500 psia, 500°F, to 10 psia. With the use of the Mollier diagram, obtain the change in enthalpy and the final temperature, if superheated, or the final quality, if saturated.

6-22 Repeat Prob. 6-21 for expansion from 100 psia, 1000°F, to 2 psia.

6-23 Steam is heated at constant pressure of 20 psia from saturated vapor to 500°F. With the use of the Mollier diagram, calculate the heat required in a steady-flow process.

6-24 How much energy would be required to heat 1 lbm of air from 200 to 1500°F at a constant pressure of 20 psia? What percent error would result if the calculation were performed assuming constant specific heats?

6-25 A rigid container having a volume of 10 ft^3 is filled with air at 100°F and 15 psia. Heat is added until the temperature reaches 1800°F. What is the final pressure and the change in entropy? How much energy is added?

6-26 Calculate the energy which must be added to the air in Prob. 6-25 to raise the pressure to 100 psia. What is the final temperature?

6-27 Calculate the work per pound-mass when air expands isentropically from 100 psia, 1500°F, to 30 psia in a cylinder behind a piston. What percent error would result from assuming constant specific heat behavior?

6-28 Using saturated-water-vapor properties at 40°F, calculate the vapor pressure at 35°F.

6-29 The enthalpy of fusion for water is 143.3 Btu/lbm at 32°F. Estimate the melting temperature of ice at a pressure of 3000 psia.

6-30 Obtain an expression for $(u_2 - u_1)_T$ for a van der Waals gas. How does this answer differ from that obtained for an ideal gas? Can you offer an explanation for the difference on a physical basis?

6-31 Estimate the specific volume of oxygen and nitrogen at 500 atm and 1500°F.

6-32 Derive an expression for the Joule-Kelvin coefficient for a van der Waals gas.

6-33 Using the critical-state data given in the Appendix for ethane (C_2H_6), calculate the density at 300°F and 500 psia.

6-34 Calculate the work required to compress methane in a reversible isothermal process from 50 psia, 200°F, to 700 psia.

6-35 Nitrogen undergoes an adiabatic throttling process from 2000 psia and −150°F to 20 psia. Calculate the final temperature in the process.

6-36 Carbon dioxide is compressed in an isentropic steady-flow process from 100 to 1500 psia. The initial temperature is 70°F. Calculate the work necessary to accomplish the compression and the final temperature. Compare the results with that obtained by assuming that CO_2 behaves like an ideal gas with constant specific heats.

6-37 A tank contains oxygen at 1000 psia and −150°F. Calculate the mass contained if the volume of the tank equals 1 gal (231 in^3).

6-38 The tank in Prob. 6-37 is heated until its temperature reaches −100°F. Calculate the heat transfer.

6-39 How much heat must be added to the tank in Prob. 6-37 to increase the pressure to 1500 psia?

6-40 Steam at 1500 psia and 650°F is compressed isentropically in a steady-flow process to a pressure of 5500 psia. Calculate the work per lbm, using the generalized charts and compare with an answer obtained from the steam tables.

6-41 Propane (C_3H_8, Table A-5) is compressed adiabatically from 150°F, 25 psia, to 750°F, 500 psia. Assuming a steady-flow process, calculate the work input and change in entropy per pound mass using (*a*) ideal gas, constant specific heats and (*b*) the generalized compressibility charts.

6-42 Ammonia is contained in a 15-ft^3 rigid vessel at 280 psia and 360°F. Calculate the heat which must be removed to lower the temperature to 180°F using (*a*) ideal-gas relations, (*b*) the generalized compressibility charts, and (*c*) the ammonia tables. What is the error in each determination?

6-43 Steam is expanded isentropically in a turbine from 5000 psia, 1600°F, to 500 psia in a steady-flow process. Calculate the work output and final temperature of the steam using (*a*) ideal-gas relations with constant specific heats, (*b*) the generalized compressibility charts, and (*c*) the steam tables.

PROBLEMS (METRIC UNITS)

6-1M Determine the density and compressibility factor for the following fluids at $p = 10.0$ MPa and $T = 650$ K:

(*a*) Ammonia
(*b*) Nitrogen
(*c*) Oxygen
(*d*) Heptane

6-2M Using the generalized compressibility charts, calculate the pressure at which density calculated with the ideal-gas equation will be in error by 5 percent for nitrogen at 45°C.

6-3M A volume flow of 4.5 m^3/s of methane flows in a pipe at $p = 4.0$ MPa and $T = 30$°C. Determine the mass flow rate using both the ideal-gas equation and the compressibility factor.

6-4M Nitrogen in the amount of 0.5 m^3 is contained in a tank at $p = 60.0$ MPa and $T = 450$ K. Calculate the mass using (*a*) the ideal-gas equation, (*b*) the van der Waals equation, and (*c*) compressibility factors.

6-5M Calculate the value of $h^* - h$ for steam at 3.0 MPa and 360°C using both the generalized charts and the steam tables.

6-6M Repeat Problem 6-5M for a pressure of 10.0 MPa.

6-7M Calculate the enthalpy of nitrogen at 200 K and 7.0 MPa assuming the ideal-gas enthalpy is zero at 0 K.

6-8M Calculate the absolute entropy of nitrogen at 200 K and 7.0 MPa using the entropy deviation charts.

6-9M Nitrogen is throttled adiabatically from conditions of 20 MPa and 200 K to 1.5 MPa. Calculate the final temperature, change in entropy, and irreversibility for $T_0 = 20$°C.

6-10M Using the generalized compressibility charts, calculate the energy required to heat steam at a constant pressure of 20.0 MPa from the saturated-vapor state to a temperature of 640°C. Compare with appropriate values obtained from the steam tables.

6-11M 1 kg of methane is heated in a constant-pressure process from 20 to 100°C at a pressure of 6.0 MPa. Calculate the heating required.

6-12M Show that lines of constant pressure in the wet-mixture region of a Mollier diagram are straight lines and nonparallel.

6-13M Ethane is expanded adiabatically in a gas turbine from 200°C and 30.0 MPa to 70°C and 6.0 MPa. The volume flow at inlet is 0.1 m^3/s. Using the low-pressure specific heat of Table 2-2, calculate the power output of the turbine.

6-14M 1 kg/s of methane is compressed in a steady-flow process from 1.5 MPa and 60°C to 20.0 MPa and 300°C. Using the generalized charts calculate the change in enthalpy and change in entropy.

6-15M Estimate the enthalpy of vaporization of water at 100, 120, and 180°C using the Clausius-Clapyron equation and compare with tabulated values in the steam tables.

6-16M Nitrogen is compressed isothermally from 300 K and low pressure to 6.0 MPa. Calculate $h^* - h$ using the generalized charts.

6-17M Steam is expanded reversibly and adiabatically in a turbine from 24.0 MPa and 480°C to 10.0 MPa. Calculate the work output using the generalized charts for a flow of 1.5 kg/s. Check the calculation with the steam tables.

6-18M 2 kg of CO_2 are compressed isentropically in a steady-flow device from 500 kPa and 300 K to 3.5 MPa. Calculate the work input.

6-19M Show that the following relations hold for a van der Waals gas:

$$(\bar{h}_2 - \bar{h}_1)_T = p_2\bar{v}_2 - p_1\bar{v}_1 + a\left(\frac{1}{\bar{v}_1} - \frac{1}{\bar{v}_2}\right)$$

$$(\bar{s}_2 - \bar{s}_1)_T = \mathcal{R} \ln \frac{\bar{v}_2 - b}{\bar{v}_1 - b}$$

6-20M Derive the relation

$$\frac{c_p}{c_v} = \frac{(\partial p/\partial v)_s}{(\partial p/\partial v)_T}$$

6-21M Obtain an expression for $(\partial u/\partial p)_T$ that involves only p, v, and T.

6-22M Obtain an expression for $(\partial h/\partial v)_T$ that involves only p, v, and T.

6-23M Show that, for a van der Waals gas,

$$\beta = \frac{\mathcal{R}\bar{v}^2(\bar{v} - b)}{\mathcal{R}T\bar{v}^3 - 2a(\bar{v} - b)^2}$$

$$\kappa = \frac{\bar{v}^2(\bar{v} - b)^2}{\mathcal{R}T\bar{v}^3 - 2a(\bar{v} - b)^2}$$

6-24M Show that for a van der Waals gas undergoing a reversible adiabatic process, the following relations apply:

$$T(\bar{v} - b)^{\mathcal{R}/c_v} = \text{const.}$$

$$\left(p + \frac{a}{\bar{v}^2}\right)(\bar{v} - b)^{1+(\mathcal{R}/c)_v} = \text{const.}$$

6-25M Show that for a van der Waals gas

$$\bar{c}_p - \bar{c}_v = \frac{\mathcal{R}}{1 - 2a(v - b)^2/\mathcal{R}T\bar{v}^3}$$

6-26M Show that the Joule-Thomson coefficient for a van der Waals gas is

$$\mu_J = \frac{\bar{v}}{\bar{c}_p} \frac{2a(\bar{v} - b)^2 - \mathcal{R}Tb\bar{v}^2}{\mathcal{R}T\bar{v}^3 - 2a(\bar{v} - b)^2}$$

Subsequently show that the inversion curve on a p-v diagram could be expressed as

$$p = \frac{a}{b\bar{v}}\left(2 - \frac{3b}{\bar{v}}\right)$$

6-27M Show that the van der Waals equation can be expressed in the following generalized form:

$$p_r = \frac{T_r}{\bar{v}_r - (1/8)} - \frac{27/64}{\bar{v}_r^2}$$

From this show that the compressibility factor can be obtained as

$$Z = \frac{\bar{v}}{\bar{v}_r - (1/8)} - \frac{27/64}{T_r \bar{v}_r}$$

Subsequently, show that the $[\bar{v}_r - (1/8)]^{-1}$ term can be expanded in a power series so that the van der Waals equation can be expressed in the virial form

$$Z = 1 + \left(\frac{1}{8} - \frac{27/64}{T_r}\right)\frac{1}{\bar{v}_r} + \frac{1}{64\bar{v}_r^2} + \frac{1}{512\bar{v}_r^3} + \cdots$$

6-28M Show that

$$c_p = T\left(\frac{\partial p}{\partial T}\right)_s \left(\frac{\partial v}{\partial T}\right)_p$$

6-29M Show that

$$c_v = -T\left(\frac{\partial p}{\partial T}\right)_v \left(\frac{\partial v}{\partial T}\right)_s$$

6-30M Using Eq. (6-16*a*) and the cyclical relation of Eq. (6-9), derive the other three Maxwell relations.

6-31M Show that the heat transfer per unit mass in an isothermal process of a Curie substance is

$$Q_T = -\frac{\mu_0 C v}{2T}(H_2^2 - H_1^2)$$

where H_1 and H_2 are the initial and final magnetic intensities, respectively.

6-32M Show that the following relation applies to an isentropic change of magnetic field for a Curie substance:

$$T_2^2 = T_1^2 - \frac{2Q_{T_1}T_1}{C_H}$$

where T_1 and T_2 are the initial and final temperatures, respectively, and Q_{T_1} is the heat transfer for an isothermal process with the same change of field at temperature T_1. The value of Q_{T_1} is given in Prob. 6-31M.

6-33M Show that the Joule-Thomson coefficient can be expressed as

$$\mu = \left(\frac{\partial T}{\partial p}\right)_h = (T\beta - 1)\frac{v}{c_p}$$

6-34M Using the relation given in Prob. 6-33M, determine the Joule-Thomson coefficient for (*a*) an ideal gas and (*b*) a van der Waals gas.

6-35M Using appropriate general relations show that the following results for a van der Waals gas:

$$d\bar{s} = \bar{c}_v \frac{dT}{T} + \frac{\mathcal{R}\, d\bar{v}}{\bar{v} - b}$$

$$d\bar{u} = \bar{c}_v\, dT + \frac{a\, d\bar{v}}{\bar{v}^2}$$

6-36M Oxygen is cooled from 10.0 MPa, 250 K to 160 K, in a steady-flow constant-pressure process. Using the generalized charts, calculate the cooling required, change in entropy, and final specific volume of the oxygen.

6-37M Methane is contained in a rigid tank at 20 MPa and 0°C. Cooling is applied to the tank until the pressure drops to 10 MPa. Calculate the cooling required per unit mass of methane.

6-38M Oxygen undergoes a steady-flow adiabatic compression process from 5.08 MPa, 200K, to 25.4 MPa, 360 K. Calculate the work required and irreversibility per unit mass for $T_0 = 20$°C.

6-39M Calculate the saturation pressure for an ice-water mixture in equilibrium at −2°C.

6-40M Ethylene (C_2H_4, Table A-5) is to be discharged into a 10-m³ tank which is initially evacuated. The discharge comes from a high-pressure line containing the gas at 125 atm and 325 K. The filling process occurs slowly and the tank is cooled so that the temperature of the gas inside remains constant at 300 K. Calculate the mass of gas in the tank when the pressure just reaches 100 atm and the heat which has been removed up to that point.

6-41M Air at 1 atm is heated from 300 to 800 K in a constant-pressure process. Calculate the heating required for a mass of 4.0 kg. Also determine the change in internal energy.

REFERENCES

1 Gouq-Jen Su: Modified Law of Corresponding States for Real Gases, *Ind. Engr. Chem.*, vol. 38, p. 803, 1946.

2 Obert, E. F.: "Concepts of Thermodynamics," McGraw-Hill Book Company, New York, 1960.

3 Keenan, J. H., and J. Kaye: "Gas Tables," John Wiley & Sons, Inc., New York, 1948.

4 Hougen, O. A., K. M. Watson, and R. A. Ragatz: "Chemical Process Principles," pt. II, John Wiley & Sons, Inc., New York, 1947.

5 Keenan, J. H., and F. G. Keyes: "Thermodynamic Properties of Steam," John Wiley & Sons, Inc., New York, 1936.

6 Lewis, G. N., and M. Randall: "Thermodynamics," 2d ed. (revised by K. S. Pitzer and L. Brewer), McGraw-Hill Book Company, New York, 1961.

7 Van Wylen, G. J., and R. E. Sonntag: "Fundamentals of Classical Thermodynamics," John Wiley & Sons, Inc., New York, 1965.

8 Hilsenrath, J., et al.: Tables of Thermal Properties of Gases, *Natl. Bur. Std. Circ. 564*, U.S. Government Printing Office, Washington, D.C., 1955.

9 Beattie, J. A., and O. C. Bridgeman: A New Equation of State for Fluids, *Proc. Am. Acad. Arts Sci.*, vol. 63, p. 229, 1928.

10 Benedict, M., G. Webb, and L. Rubin: An Empirical Equation for the Thermodynamic Properties of Light Hydrocarbons and Their Mixtures, *J. Chem. Physics*, vol. 8, p. 334, 1940.

11 Nelson, L. C., and E. F. Obert: Generalized *pvT* Properties of Gases, *Trans. ASME*, p. 1057, October 1954.

7

GASEOUS MIXTURES

7-1 INTRODUCTION

Previous chapters have been concerned with the calculation and use of thermodynamic properties for substances which require only two independent properties to define the thermodynamic state. In this chapter we shall examine the behavior of mixtures of ideal gases and ideal-gas–vapor mixtures which require additional information pertaining to composition in order to establish the thermodynamic state. This presentation will then form a basis for studies of chemical equilibrium and combustion reactions to be developed in Chap. 8. Calculations involving gas-vapor mixtures are particularly applicable to air-conditioning processes and thus will merit examination in this chapter. Following the presentation of ideal-gas mixtures we shall discuss some methods for calculating properties of real-gas mixtures and relate the procedures to the equation-of-state material given in Chap. 6.

7-2 THE GIBBS-DALTON LAW

Let us begin our discussion by considering some basic definitions applicable to a mixture of ideal gases. The total mass of the mixture is clearly the sum of the masses of each component so that

$$m = m_1 + m_2 + \cdots + m_i = \sum_{i=1} m_i \tag{7-1}$$

The *mole fraction* of a component x_i is defined by

$$x_i = \frac{n_i}{n} \tag{7-2}$$

where n_i is the number of moles of the ith component and n is the total number of moles of the mixture. The sum of the fractions is equal to the whole, or

$$\Sigma x_i = 1.0 \tag{7-3}$$

The mass of each component can be calculated from

$$m_i = n_i M_i \tag{7-4}$$

where M_i is the molecular weight of the ith component. Now, if Eq. (7-4) is inserted into Eq. (7-1), we obtain

$$m = \sum_i n_i M_i \tag{7-5}$$

An equivalent molecular weight for the particular mixture may be defined as

$$M = \frac{m}{n} = \sum \frac{n_i}{n} M_i$$

or

$$M = \Sigma x_i M_i \tag{7-6}$$

An equivalent gas constant for the particular mixture may then be calculated by

$$R = \frac{\mathcal{R}}{M} \tag{7-7}$$

As long as ideal-gas behavior is studied, the assumption is made that each gaseous component of the mixture behaves as if it existed at the temperature of the mixture and filled the entire volume occupied by the mixture. This assumption is a statement of the *Gibbs-Dalton law*. In a physical sense, an ideal gas is one in which there is negligible influence of molecular force fields; hence, in a mixture of ideal gases, we would not expect appreciable interference between the different molecular components.

The *partial pressure* of the ith component is defined as

$$p_i = \frac{n_i \mathcal{R} T}{V} \tag{7-8}$$

In other words, this is the pressure the component would exert if it occupied the entire volume of the mixture V at the temperature of the mixture T. The *total pressure* of the mixture is defined as

$$p = \frac{n \mathcal{R} T}{V} \tag{7-9}$$

Dividing Eq. (7-8) by Eq. (7-9),

$$\frac{p_i}{p} = \frac{n_i}{n} = x_i \tag{7-10}$$

Applying Eq. (7-3),

$$\Sigma p_i = \Sigma x_i p = p \tag{7-11}$$

or the sum of the partial pressures equals the total pressure.

Now suppose that we separate a mixture of ideal gases into its individual components and place each component in a container such that it is maintained at the pressure and temperature of the total mixture. This separation process is

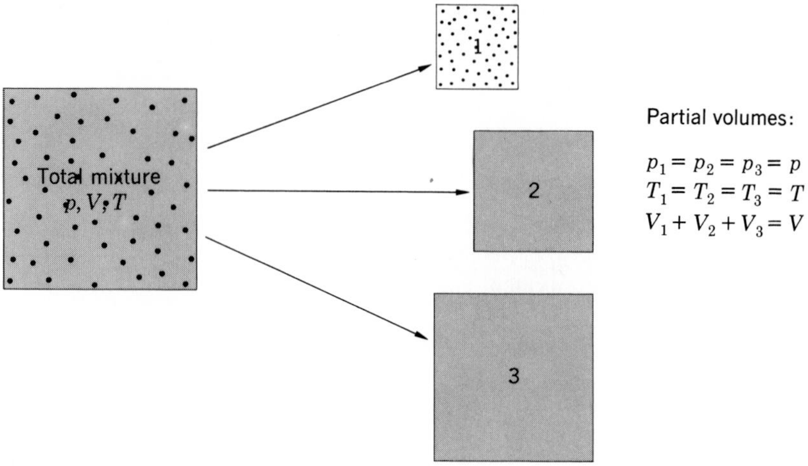

FIG. 7-1 Illustration of Amagat-Leduc law of additive volumes.

indicated in Fig. 7-1. The volume occupied by each component in this case would be

$$V_i = \frac{n_i \mathcal{R} T}{p} \tag{7-12}$$

The total volume of the mixture at conditions p and T is

$$V = \frac{n \mathcal{R} T}{p} \tag{7-13}$$

Dividing Eq. (7-12) by Eq. (7-13),

$$\frac{V_i}{V} = \frac{n_i}{n} = x_i \tag{7-14}$$

Applying Eq. (7-3),

$$\Sigma V_i = \Sigma x_i V = V \tag{7-15}$$

Equation (7-15) is called the *Amagat-Leduc law of additive volumes,* or the total volume of a mixture of ideal gases is the sum of the volumes each component gas would occupy if contained at the pressure and temperature of the mixture. V_i is usually called the *partial volume* of the ith component.

The concentrations of the constituents in a mixture may be expressed on either a mass or molal basis. Using a molal basis, the partial pressures or partial volumes are obtained very easily from either Eq. (7-10) or (7-14). Sometimes the concentrations in a mixture are expressed on a "volume" basis. The meaning of this term is that the concentrations are expressed in terms of the partial volumes. So, the term *volume basis,* or *volume fraction,* means the same thing as molal basis, or mole fraction, in accordance with Eq. (7-14).

EXAMPLE 7-1

A rigid tank contains 2 lbm of nitrogen at 300 psia and 100°F. A sufficient quantity of oxygen is added to increase the pressure to 400 psia while the temperature remains constant at 100°F. Calculate the mass of oxygen added.

SOLUTION In this process the number of moles of nitrogen remains constant at

$$n_{N_2} = \frac{m}{M} = \frac{2}{28} = 0.0714$$

After the oxygen is added, the partial pressure of the nitrogen will be 300 psia since the volume does not change. We thus have

$$\begin{aligned} p_{O_2} &= p_{\text{total}} - p_{N_2} \\ &= 400 - 300 \\ &= 100 \text{ psia} \qquad (6.89 \times 10^5 \text{ Pa}) \end{aligned}$$

From Eq. (7-8) we can write, using a ratio,

$$\frac{n_{O_2}}{n_{N_2}} = \frac{p_{O_2}}{p_{N_2}}$$

Using the numerical values given in the foregoing,

$$n_{O_2} = \frac{(0.0714)(100)}{300} = 0.0238$$

Finally, the mass of oxygen is calculated as

$$\begin{aligned} m_{O_2} &= M_{O_2} n_{O_2} = (32)(0.0238) \\ &= 0.761 \text{ lbm} \qquad (0.345 \text{ kg}) \end{aligned}$$

EXAMPLE 7-2

Calculate the volumes which would be occupied by the oxygen and nitrogen in Example 7-1 if each gas were contained separately at the temperature and pressure of the mixture, i.e., 100°F and 400 psia.

SOLUTION The volume of nitrogen is calculated from the ideal-gas law (Eq. 7-8):

$$\begin{aligned} V_{N_2} &= \frac{n\mathfrak{R}T}{p} = \frac{(0.0714)(1545)(560)}{(400)(144)} \\ &= 1.071 \text{ ft}^3 \qquad (0.0303 \text{ m}^3) \end{aligned}$$

For the oxygen,

$$V_{O_2} = \frac{(0.0238)(1545)(560)}{(400)(144)}$$
$$= 0.357 \text{ ft}^3 \qquad (0.0101 \text{ m}^3)$$

The total volume of the mixture would be

$$V = 1.071 + 0.357$$
$$= 1.428 \text{ ft}^3 \qquad (0.0404 \text{ m}^3)$$

It should be observed that the volume of each constituent is in direct proportion to the number of moles of that constituent.

7-3 ENERGY PROPERTIES OF MIXTURES

The principles and definitions of the preceding section may be easily extended to facilitate the calculation of energy properties of mixtures. Consider the internal energy. We may write

$$U = U_1 + U_2 + U_3 + \cdots = \Sigma U_i = \Sigma m_i u_i \tag{7-16}$$

or on a molal basis

$$U = \Sigma n_i \bar{u}_i \tag{7-17}$$

where $\bar{u}_i$ designates the internal energy per mole for the ith component. The specific internal energy for the mixture is then calculated with

$$u = \frac{1}{m} \sum m_i u_i \tag{7-18}$$

or the molal specific internal energy is

$$\bar{u} = \frac{1}{n} \sum n_i \bar{u}_i = \sum x_i \bar{u}_i \tag{7-19}$$

Relations for the enthalpy and entropy of the mixture are developed in a similar manner with the following results:

$$H = \Sigma H_i = \Sigma m_i h_i \tag{7-20a}$$

$$H = \Sigma n_i \bar{h}_i \tag{7-20b}$$

$$\bar{h} = \Sigma x_i \bar{h}_i \tag{7-21}$$

$$S = \Sigma S_i = \Sigma m_i s_i \tag{7-22a}$$

$$S = \Sigma n_i \bar{s}_i \tag{7-22b}$$

$$\bar{s} = \Sigma x_i \bar{s}_i \tag{7-23}$$

Specific heats for the mixture are obtained by applying the definitions

$$c_v = \left(\frac{\partial u}{\partial T}\right)_v \qquad \text{and} \qquad c_p = \left(\frac{\partial h}{\partial T}\right)_p$$

to the foregoing relations for internal energy and enthalpy. For the specific heat at constant volume, we have

$$\begin{aligned} c_v &= \left[\frac{\partial}{\partial T}\left(\frac{1}{m}\sum m_i u_i\right)\right]_v \\ &= \frac{1}{m}\sum m_i \left(\frac{\partial u_i}{\partial T}\right)_v \\ c_v &= \frac{1}{m}\sum m_i c_{v_i} \end{aligned} \tag{7-24}$$

On a molal basis,

$$\begin{aligned} \bar{c}_v &= \left(\frac{\partial \bar{u}}{\partial T}\right)_v = \sum x_i \left(\frac{\partial \bar{u}_i}{\partial T}\right)_{v_i} \\ \bar{c}_v &= \Sigma x_i \bar{c}_{v_i} \end{aligned} \tag{7-25}$$

The specific heat at constant pressure is developed in a similar fashion with the results

$$c_p = \frac{1}{m}\sum m_i c_{pi} \tag{7-26}$$

$$\bar{c}_p = \Sigma x_i \bar{c}_{p_i} \tag{7-27}$$

In the study of thermodynamic *processes* involving ideal-gas mixtures, it is of interest to calculate *changes* in the energy properties of the mixture. Such a calculation proceeds on the basic notion incorporated in the Gibbs-Dalton law that each component of the mixture occupies the entire volume of the mixture at the temperature of the mixture. The change of internal energy of the mixture would be calculated by

$$\Delta U = \Sigma\, \Delta U_i = \Sigma m_i\, \Delta u_i = \Sigma m_i c_{v_i}\, \Delta T \tag{7-28}$$

or, making use of Eq. (7-24),

$$\Delta U = m c_v\, \Delta T \tag{7-29}$$

where now c_v is the mass specific heat of the mixture. ΔU could also be calculated from

$$\Delta U = n\bar{c}_v \, \Delta T \tag{7-30}$$

Similarly, the change in enthalpy for the mixture could be written as

$$\Delta H = \Sigma \, \Delta H_i = \Sigma m_i \, \Delta h_i = \Sigma m_i c_{p_i} \, \Delta T \tag{7-31}$$

$$\Delta H = mc_p \, \Delta T \tag{7-32}$$

$$\Delta H = n\bar{c}_p \, \Delta T \tag{7-33}$$

The computation of the change in entropy of a mixture is a bit more complicated since the entropy is not a function of temperature alone. From Eq. (5-32) the change in entropy for an ideal gas with constant specific heats is

$$s_2 - s_1 = c_p \ln \frac{T_2}{T_1} - R \ln \frac{p_2}{p_1} \tag{7-34}$$

where the subscripts 1 and 2 pertain to the initial and final states, respectively. The change in entropy for a mixture of ideal gases is calculated by applying Eq. (7-34) to each component and summing the results. Thus

$$\Delta S = S_2 - S_1 = \sum m_i c_{pi} \ln \frac{T_2}{T_1} - \sum m_i R_i \ln \frac{p_{i_2}}{p_{i_1}} \tag{7-35}$$

where p_{i_1} and p_{i_2} are the initial and final *partial pressures* of the ith component, respectively. An alternate expression for the entropy change may be written on a molal basis. Thus

$$\Delta S = S_2 - S_1 = \sum n_i \bar{c}_{pi} \ln \frac{T_2}{T_1} - \sum n_i \mathcal{R} \ln \frac{p_{i_2}}{p_{i_2}} \tag{7-36}$$

The following example illustrates the use of these relations.

EXAMPLE 7-3

2 lbm of CO_2 at 100°F and 20 psia are mixed with 5 lbm of N_2 at 300°F and 15 psia to form a mixture at a final pressure of 10 psia. The process occurs adiabatically in a steady-flow apparatus. Calculate the final temperature of the mixture and the change in entropy.

SOLUTION The process is shown schematically in the accompanying figure. The steady-flow energy balance is

$$(mh_1)_{N_2} + (mh_1)_{CO_2} = (mh_2)_{\text{mixture}}$$
$$= (mh_2)_{N_2} + (mh_2)_{CO_2}$$

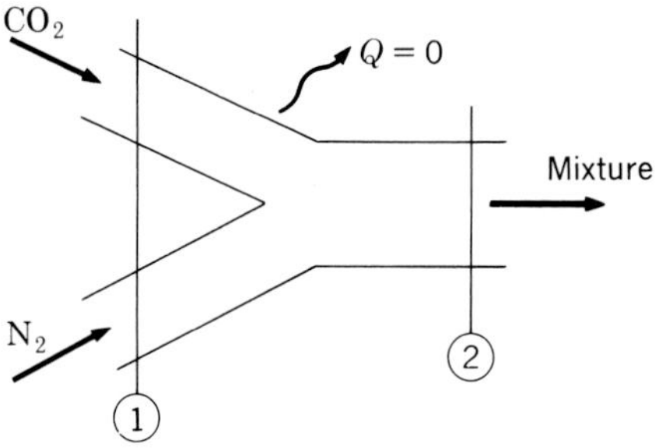

FIG. EXAMPLE 7-3

Assuming ideal-gas behavior with constant specific heats, this equation becomes

$$[mc_p(T_1 - T_2)]_{N_2} + [mc_p(T_1 - T_2)]_{CO_2} = 0$$

Inserting the appropriate numerical values,

$$(5)(0.248)(300 - T_2) + (2)(0.203)(100 - T_2) = 0$$

since the final temperature for both constituents is the same, i.e., the temperature of the mixture. This equation yields

$$T_2 = 250°\text{F} = 710°\text{R} \qquad (394\ \text{K})$$

The change in entropy may be calculated with either Eq. (7-35) or (7-36). For either equation the final partial pressures of the constituents must be computed. The molal quantities are

$$n_{CO_2} = \frac{2}{44} = 0.0455$$

$$n_{N_2} = \frac{5}{28} = 0.1785$$

$$n_{\text{total}} = 0.1785 + 0.0455$$
$$= 0.224$$

The final partial pressures may now be calculated from Eq. (7-10):

$$(p_2)_{CO_2} = \frac{(10)(0.0455)}{0.224} = 2.03\ \text{psia} \qquad (14\ \text{kPa})$$

$$(p_2)_{N_2} = \frac{(10)(0.1785)}{0.224} = 7.97\ \text{psia} \qquad (55\ \text{kPa})$$

The change in entropy is now calculated with Eq. (7-35):

$$S_2 - S_1 = \left(mc_p \ln \frac{T_2}{T_1} - mR \ln \frac{p_2}{p_1}\right)_{CO_2} + \left(mc_p \ln \frac{T_2}{T_1} - mR \ln \frac{p_2}{p_1}\right)_{N_2}$$

Inserting the appropriate numerical values,

$$S_2 - S_1 = (2)\left(0.203 \ln \frac{710}{560} - \frac{35.1}{778} \ln \frac{2.03}{20}\right)$$
$$+ (5)\left(0.248 \ln \frac{710}{760} - \frac{55.2}{778} \ln \frac{7.97}{15}\right)$$
$$= 0.443 \text{ Btu/°R} \qquad (841 \text{ J/K})$$

7-4 MIXTURES OF AN IDEAL GAS AND A VAPOR

Let us now consider an important type of gaseous mixture, that of an ideal gas and a condensable vapor. This class of mixture is encountered daily in the air we breathe and in the heating and air-conditioning processes associated therewith. The formation of dew on the grass on a still night, the drying and humidification of gases for various applications, and all air-conditioning calculations depend on an understanding of the behavior of gas-vapor mixtures.

We shall be concerned only with mixtures involving a single condensable vapor and shall focus our attention on air–water-vapor mixtures because of the wide range of practical circumstances in which they are encountered.

Dew-Point, Dry-Bulb, and Wet-Bulb Temperatures

The *dew point* of a mixture is the temperature at which the vapor starts to condense when the mixture is cooled at constant pressure. The *dry-bulb temperature* of a mixture is the temperature indicated by an ordinary thermometer placed in the mixture. The *wet-bulb temperature* is the temperature indicated by a thermometer covered with a wicklike material saturated with liquid, after the arrangement has been allowed to reach evaporation equilibrium with the mixture. These temperatures are indicated schematically in Fig. 7-2.

A gas-vapor mixture is said to be *saturated* when a reduction in temperature would cause part of the vapor to condense. Therefore, a saturated mixture does exist at the dew point, and the partial pressure of the vapor in such a mixture is the saturation pressure corresponding to the temperature of the mixture.

The *relative humidity* ϕ is defined as the ratio of the actual mass of vapor to the mass of vapor required to produce a saturated mixture at the same tempera-

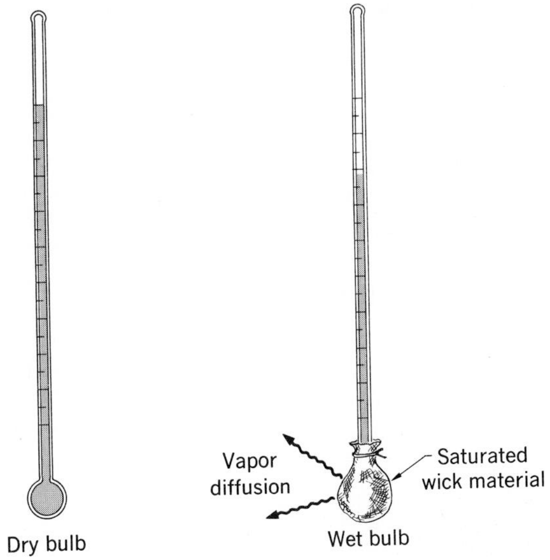

FIG. 7-2 Measurement of dry-bulb and wet-bulb temperature.

ture. If the vapor behaves like an ideal gas, we can write

$$\phi = \frac{m_v}{m_{\text{sat}}} = \frac{p_v V / R_v T}{p_g V / R_v T} = \frac{p_v}{p_g} \tag{7-37}$$

where p_v is the actual partial pressure of the vapor and p_g is the saturation pressure of the vapor at the temperature of the mixture (dry-bulb temperature).

Specific Humidity or Absolute Humidity

The *specific humidity* or humidity ratio ω is defined as the ratio of the mass of vapor to the mass of noncondensable gas. Our primary concern in this chapter will be with air–water-vapor mixtures, so that for these cases

$$\omega = \frac{m_v}{m_a} \tag{7-38}$$

where m_a designates the mass of dry air. If the assumption is made that the vapor behaves like an ideal gas, we can write

$$\omega = \frac{p_v V M_v / \mathcal{R} T}{p_a V M_a / \mathcal{R} T} = \frac{M_v}{M_a} \frac{p_v}{p_a} \tag{7-39}$$

For air–water-vapor mixtures this expression reduces to

$$\omega = 0.622 \frac{p_v}{p_a} = 0.622 \frac{p_v}{p - p_v} \tag{7-40}$$

Equation (7-37) may be introduced to obtain

$$\phi = \frac{\omega p_a}{0.622 p_g} \tag{7-41}$$

Another quantity of interest is the *degree of saturation* μ, defined as the ratio of the actual specific humidity to the specific humidity of saturated air at the dry-bulb temperature:

$$\mu = \left.\frac{\omega}{\omega_{\text{sat}}}\right|_T \tag{7-42}$$

For ideal-gas behavior this may be expressed as

$$\mu = \frac{0.622 p_v/(p - p_v)}{0.622 p_g/(p - p_g)} = \frac{p_v(p - p_g)}{p_g(p - p_v)} = \phi \frac{p - p_g}{p - p_v} \tag{7-43}$$

Because both p_v and p_g are small compared to the total pressure p, the degree of saturation is very nearly equal to the relative humidity for air–water-vapor mixtures at normal room temperatures and pressures.

In accordance with our discussion so far we see that a process which does not add or remove moisture from the mixture occurs at constant specific humidity. If the process occurs at constant total pressure also, then from Eq. (7-40) the partial pressure of the vapor must also remain constant. From these facts we can interpret the dew point of the mixture in a specific way. Recalling that the dew point is the temperature at which vapor will just start to condense when cooled at constant pressure, the vapor partial pressure at the dew point must be the saturation pressure at that temperature. But this is equal to the actual vapor pressure before the mixture is cooled, so

$$p_v = p_g \qquad \text{(evaluated at dew-point temperature)}$$

When the mixture is fully saturated (ϕ = 100 percent) the dry-bulb, wet-bulb, and dew-point temperatures are all the same.

EXAMPLE 7-4 Air–Water-Vapor Mixture
An air–water-vapor mixture exists at 25 psia, 140°F, and 50 percent relative humidity. Calculate the mass fraction of water vapor in the mixture.

SOLUTION The specific humidity is the quantity of interest in this problem. The saturation pressure at 140°F is

$$p_g = 2.892 \text{ psia} \qquad (1.994 \times 10^4 \text{ Pa})$$

From Eq. (7-37) the actual vapor pressure is

$$p_v = \phi p_g$$
$$= (0.5)(2.892) = 1.446 \text{ psia} \qquad (9970 \text{ Pa})$$

The partial pressure of the air is therefore

$$p_a = 25 - 1.446 = 23.554 \text{ psia} \qquad (1.624 \times 10^5 \text{ Pa})$$

The specific humidity may now be calculated with Eq. (7-40):

$$\omega = \frac{(0.622)(1.446)}{23.554}$$
$$= 0.0382 \text{ lbm vapor/lbm dry air}$$

The mass fraction of water vapor is

$$\frac{m_v}{m_{\text{total}}} = \frac{m_v}{m_a + m_v} = \frac{\omega}{1 + \omega}$$
$$= 0.0368 \text{ lbm vapor/lbm mixture}$$

7-5 ADIABATIC SATURATION

Because so many air–water-vapor processes involve the concepts of humidity and saturation, it is important to examine the energy balances encountered in a simple adiabatic saturation process. The system to be analyzed is shown schematically in Fig. 7-3.

An air–water-vapor mixture is blown over a pool of water until it leaves the container in a saturated state. Fresh liquid is continuously supplied at the exit temperature to compensate for water picked up by the air in the saturation

FIG. 7-3 Adiabatic saturation process.

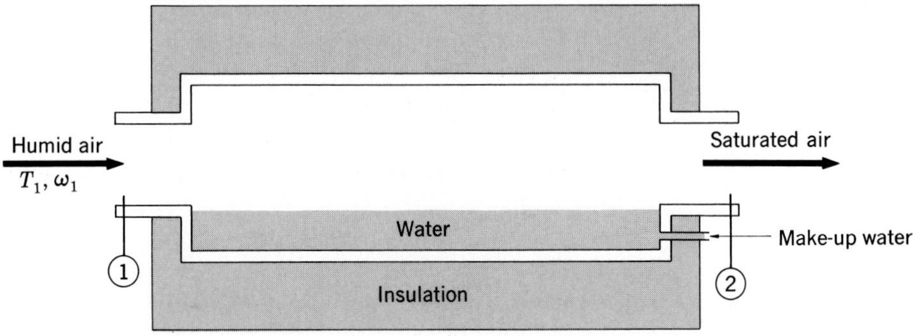

process. The container is insulated so that no heat exchange with the surroundings occurs during the process.

A steady-flow energy balance may now be applied to the process. We have

Enthalpy of mixture in + enthalpy of liquid supplied
= enthalpy of mixture out

or,

$$m_a h_{a_1} + m_{v_1} h_{v_1} + (m_{v_2} - m_{v_1}) h_{f_2} = m_a h_{a_2} + m_{v_2} h_{v_2}$$

Dividing by the air flow rate m_a,

$$h_{a_1} + \omega_1 h_{v_1} + (\omega_2 - \omega_1) h_{f_2} = h_{a_2} + \omega_2 h_{v_2} \tag{7-44}$$

If ideal-gas behavior is assumed,

$$h_{a_1} - h_{a_2} = c_{pa}(T_1 - T_2)$$

Furthermore,

$$h_{v_2} - h_{f_2} = h_{fg_2}$$

so that Eq. (7-44) may be solved for ω_1 to yield

$$\omega_1 = \frac{c_{pa}(T_2 - T_1) + \omega_2 h_{fg_2}}{h_{v_1} - h_{f_2}} \tag{7-45}$$

Equation (7-45) is useful in that the specific humidity of a mixture may be determined through a measurement of T_1 and T_2 and a calculation with Eq. (7-45). The wet-bulb temperature of a mixture is very nearly equal to the adiabatic saturation temperature. It differs slightly because of the mass diffusion processes which occur at the surface of the "wick" and the temperature gradients associated therewith.

The foregoing discussion leads to the conclusion that a specification of the wet-bulb and dry-bulb temperatures of a mixture serves to define the humidity state of the mixture. This information may then be used to calculate the energy properties of the mixture.

For most psychrometric mixtures the partial pressure of the water vapor is low and simplified relations may be used to calculate thermodynamic properties. For water vapor at low pressures almost ideal-gas behavior is experienced and the enthalpy and internal energy are essentially functions of temperature alone. Using the same reference levels as for the steam tables (0°C, saturated liquid), it is possible to give a summary of low-pressure water properties as shown in Table 7-1. Some of these relations have already been given in Chap. 3 for the solid-vapor saturation region.

TABLE 7-1
Some Approximate Formulas for Properties of Water at Low Pressures

Property	Units	Temp. Range	Equation Number
$h_f = u_f = T - 32$ Btu/lbm	T in °F	32 to 100°F	7-46*a*
$h_f = u_f = 4.19T$ kJ/kg	T in °C	0 to 40°C	7-46*b*
$h_g = 1061 + 0.445T$ Btu/lbm	T in °F	−40 to 100°F	7-47*a*
$h_g = 2501 + 1.863T$ kJ/kg	T in °C	−40 to 40°C	7-47*b*
$u_g = 1010.3 + 0.335T$ Btu/lbm	T in °F	−40 to 100°F	7-48*a*
$u_g = 2374.9 + 1.403T$ kJ/kg	T in °C	−40 to 40°C	7-48*b*
$u_i = h_i = -158.9 + 0.467T$ Btu/lbm	T in °F	−40 to 32°F	7-49*a*
$u_i = h_i = -334.6 + 1.96T$ kJ/kg	T in °C	−40 to 0°C	7-49*b*
$p_g = 5.103 \exp\left(18.42 - \frac{11\,059}{T}\right)$ psia	T in °R	−40 to 32°R	7-50*a*
$\ln \frac{p_g}{0.3390} = 12\,221\left(\frac{1}{527.67} - \frac{1}{T}\right)$	T in °R	32 to 100°F	7-51*a*
$- 5.031 \ln \frac{T}{527.67}$	p_g in psia		
$p_g = 35.18 \exp\left(18.42 - \frac{6144}{T}\right)$ kPa	T in K	−40 to 0°C	7-50*b*
$\ln \frac{p_g}{2337} = 6789\left(\frac{1}{293.15} - \frac{1}{T}\right)$	T in K	0 to 40°C	7-51*b*
$-5.031 \ln \frac{T}{293.15}$	p_g in Pa		

Below freezing temperatures it is possible, of course, to have frost or ice formation on cooling coils, and an energy balance on the system must take proper account of the energy required to effect the change of phase from water vapor to solid ice. It is interesting to note that in the sublimation region between −40 and 0°C the enthalpy of sublimation is very nearly constant at

$$h_g - h_i = h_{ig} = 2838 \text{ kJ/kg} = 1220 \text{ Btu/lbm} \tag{7-52}$$

Using a relation like Eq. (7-47), Eq. (7-45) may be modified to enable a calculation of actual vapor pressure directly from dry-bulb and wet-bulb temperatures. The resulting relation is called *Carrier's equation* [6] and is given as

$$p_v = p_{g_w} - \frac{(p - p_{g_w})(T_{DB} - T_{WB})}{K_w - T_{WB}} \tag{7-53}$$

where p_v = actual vapor pressure

p_{g_w} = saturation pressure corresponding to wet-bulb temperature

p = total pressure of mixture

T_{DB} = dry-bulb temperature, °F or °C

T_{WB} = wet-bulb temperature, °F or °C

K_w = 2800°F or 1537.8°C

For wet-bulb temperatures below 32°F, Doolittle [1] suggests that the following relation be employed:

$$p_v = p_{g_w} - \frac{p(T_{DB} - T_{WB})}{K_L}$$
$$K_L = 3160°\text{F or } 1756°\text{C} \tag{7-54}$$

EXAMPLE 7-5 Closed-System Cooling of an Air–Water-Vapor Mixture
An air–water-vapor mixture is contained in a 4000-ft³ tank at 30 psia and 100°F. The relative humidity is 80 percent. How much cooling is required to lower the temperature to 40°F, and what is the final pressure of the air–water-vapor mixture?

SOLUTION At 100°F, p_g = 0.9503 psia, so the initial vapor pressure is obtained from

$$\phi_1 = \frac{p_{v_1}}{p_{g_1}} \qquad p_{v_1} = (0.9503)(0.8) = 0.7602 \text{ psia} \qquad (5241 \text{ Pa})$$

At 40°F, p_g = 0.1217 psia, so the final state must be saturated. The initial partial pressure of the air is

$$p_{a_1} = p_1 - p_{v_1}$$
$$= 20 - 0.7602 = 19.2398 \text{ psia} \qquad (1.327 \times 10^5 \text{ Pa})$$

During the cooling process the volume remains constant so the final partial pressure of the air may be calculated from

$$\frac{p_{a_2}}{p_{a_1}} = \frac{T_2}{T_1} \qquad p_{a_2} = \frac{(19.2398)(460 + 40)}{460 + 100}$$
$$= 17.178 \text{ psia} \qquad (1.184 \times 10^5 \text{ Pa})$$

The final total pressure is therefore

$$p_2 = p_{a_2} + p_{v_2}$$
$$= 17.178 + 0.1217 = 17.3 \text{ psia} \qquad (1.193 \times 10^5 \text{ Pa})$$

The mass of air is calculated from

$$m_a = \frac{p_{a_1}V}{RT_1}$$
$$= \frac{(19.2398)(144)(4000)}{(53.35)(560)} = 371 \text{ lbm} \qquad (168 \text{ kg}) \qquad (a)$$

and the specific humidities are

$$\omega = 0.622\,\frac{p_v}{p_a}$$
$$\omega_1 = \frac{(0.622)(0.7602)}{19.2398} = 0.0246 \qquad (b)$$
$$\omega_2 = \frac{(0.622)(0.1217)}{17.178} = 0.004\ 41$$

Because the tank is a closed system the cooling is calculated from the following:

$$Q = U_1 - U_2$$
$$= m_a(u_{a_1} + \omega_1 u_{v_1}) - m_a(u_{a_2} + \omega_2 u_{v_2}) - m_a(\omega_1 - \omega_2)u_{f_2} \qquad (c)$$

The value of u_{f_2} is obtained from Table A-7:

$$u_{f_2} = 8.02 \text{ Btu/lbm} \qquad (\text{saturated liquid at } 40°\text{F})$$

and the vapor internal energies from Eq. (7-48):

$$u_{v_1} = 1010.3 + (0.335)(100) = 1043.8 \text{ Btu/lbm}$$
$$u_{v_2} = 1010.3 + (0.335)(40) = 1023.7 \text{ Btu/lbm}$$

We now insert the numerical values in Eq. (c) and obtain

$$Q = (371)[(0.1715)(100 - 40) + (0.0246)(1043.8) - (0.004\ 41)(1023.7) - (0.0246 - 0.004\ 41)(8.02)]$$
$$= (371)(10.29 + 25.68 - 4.51 - 0.16)$$
$$= 11\ 612 \text{ Btu} \qquad (12\ 251 \text{ kJ})$$

EXAMPLE 7-6

Calculate the relative humidity of an air stream having $T_{DB} = 100°\text{F}$, $T_{WB} = 80°\text{F}$, and a total pressure of 14.696 psia.

SOLUTION We shall employ Eq. (7-53) to calculate the vapor partial pressure and then determine ϕ. The quantities of interest are

$$p_{g_w} = 0.5073 \text{ psia} \qquad (3498 \text{ Pa})$$

$$p = 14.696 \text{ psia} \qquad (1.0132 \times 10^5 \text{ Pa})$$

Then, from Eq. (7-53),

$$p_v = 0.5073 - \frac{(14.696 - 0.5073)(100 - 80)}{2800 - 80}$$

$$= 0.4030 \text{ psia} \qquad (2778 \text{ Pa})$$

The saturation pressure at 100°F is

$$p_g = 0.9503 \text{ psia} \qquad (6552 \text{ Pa})$$

so that, from Eq. (7-37),

$$\phi = \frac{p_v}{p_g} = \frac{0.4030}{0.9503}$$

$$= 42.4 \text{ percent}$$

EXAMPLE 7-7 Helium–Water-Vapor Mixture
A mixture of helium and water vapor at a total pressure of 1 atm and a dry-bulb temperature of 35°C has a dew point of 20°C. Calculate the relative humidity and specific humidity of the mixture.

SOLUTION Assuming ideal-gas behavior, the actual vapor pressure will be the saturation pressure evaluated at the dew point or

$$p_v = p_g \text{ at } 20°\text{C} = 2.339 \text{ kPa} \qquad (a)$$

The saturation pressure at the dry-bulb temperature is

$$p_g = 5.628 \text{ kPa} \qquad \text{at } 35°\text{C}$$

so the relative humidity is

$$\phi = \frac{p_v}{p_g} = \frac{2.339}{5.628} = 41.6 \text{ percent} \qquad (b)$$

The specific humidity may be calculated from Eq. (7-39) by replacing the molecular weight and partial pressure for air with those of helium:

$$\omega = \frac{M_v}{M_{\text{He}}} \frac{p_v}{p_{\text{He}}} \qquad (c)$$

The helium partial pressure is

$$p_{He} = p - p_v = 101.32 - 2.339 = 98.981 \text{ kPa}$$

Using $M_{He} = 4.0$ and inserting the values in (*c*) gives

$$\omega = \frac{(18)(2.339)}{(4)(98.981)} = 0.106 \text{ kg vapor/kg dry helium}$$

EXAMPLE 7-8 Iterative Calculation
An air–water-vapor mixture at a total pressure of 150 kPa has a dry-bulb temperature of 35°C and a relative humidity of 60 percent. Calculate the wet-bulb temperature and dew point.

SOLUTION At 35°C the saturation pressure is

$$p_g = 5.628 \text{ kPa}$$

so, from Eq. (7-37) we may calculate the actual vapor pressure as

$$p_v = \phi p_g = (0.6)(5.628) = 3.3768 \text{ kPa}$$

The dew point is the saturation temperature corresponding to this pressure or

$$T_{DP} = 25.96°\text{C}$$

To determine the wet-bulb temperature we must employ the Carrier relation of Eq. (7-53) and an iterative procedure. Rewriting Eq. (7-53) in the form

$$p_v - p_{g_w} + \frac{(p - p_{g_w})(T_{DB} - T_{WB})}{1537.8 - T_{WB}} = 0 = f(T_{WB})$$

we see that with p_v and T_{DB} known, the equation involves only one unknown, T_{WB}, because p_{g_w} is a function of T_{WB}. This is a nonlinear function so we must use tabular values and iterate until we obtain $f(T_{WB}) = 0$. The trials are shown below.

T_{WB}, °C	p_{g_w}, kPa	$f(T_{WB})$
20	2.339	2497.1
25	3.169	1178.4
30	4.246	−385.9
28.77		0

Thus we find $T_{WB} = 28.77°\text{C}$.

7-6 THE PSYCHROMETRIC CHART

Although the equations of the foregoing sections provide an adequate basis for calculations involving air–water-vapor mixtures, it is desirable to have a calculation chart which may be used in those cases where only modest accuracy is required but with the obvious utility of a speedier computation procedure. The display of interest is called a *psychrometric chart* and is shown schematically in Fig. 7-4.

Three large charts drawn to scale are given in the Appendix.

1 English units for dry-bulb temperatures of 20 to 110°F
2 English units for dry-bulb temperatures of −20 to +50°F
3 SI units for dry-bulb temperatures of −10 to +55°C

All three of these charts are for a mixture total pressure of 1 atm. On the English unit charts the specific humidity is given in both pound-mass of vapor per pound-mass of dry air and *grains* of vapor per pound-mass of dry air. The conversion factor is

1 lbm = 7000 grains

The volume lines on the chart below give the volume of air–water-vapor *mixture* per unit mass of dry air in the mixture. We shall designate this quantity with the symbol v_a:

$$v_a = \frac{\text{volume mixture}}{\text{mass dry air in mixture}} = \frac{V}{m_a} \tag{7-55}$$

FIG. 7-4 Schematic of psychrometric chart.

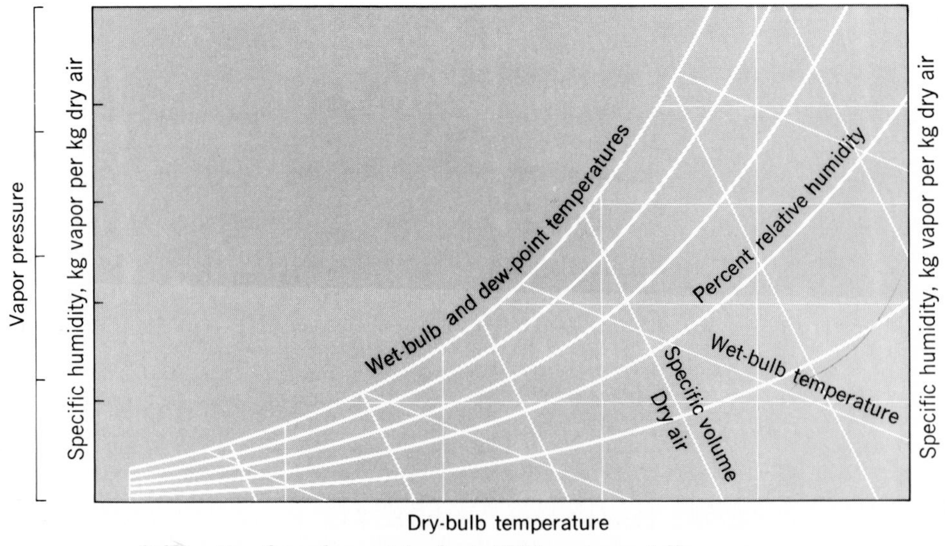

Schematic of psychrometric chart. 7000 grains = 1 lbm.

The true specific volume of the mixture may be obtained by manipulating Eq. (7-55) along with

$$m = m_a + m_v = m_a(1 + \omega) \tag{7-56}$$

The true specific volume is $v = V/m$ so that

$$v = \frac{m_a v_a}{m_a(1 + \omega)} = \frac{v_a}{1 + \omega} \tag{7-57}$$

We should note that in most air–water-vapor processes m_a is the invariant quantity, and thus v_a is of greatest practical utility in solving problems. Therefore, it is not surprising that it is the volumetric quantity displayed on the psychrometric chart.

Reference Levels for Psychrometric Charts

Values of the mixture enthalpies per unit mass of dry air in the mixture are also shown in the chart with the following zero base levels assumed:

English units
Enthalpy of dry air = 0 at 0°F
Enthalpy of saturated-liquid water = 0 at 32°F

SI units
Enthalpy of dry air = 0 at 0°C
Enthalpy of saturated-liquid water = 0 at 0°C

The mixture enthalpy per unit mass of dry air in the mixture is defined as

$$h_m = h_{\text{mix}} = h_a + \omega h_v \tag{7-58}$$

It turns out that, to a good approximation, lines of constant wet-bulb temperature are very nearly lines of constant mixture enthalpy and are so indicated on the chart. The error in this approximation is usually less than 1 percent for mixtures at 1 atm.

For those readers wishing to avoid the approximation, the charts in the Appendix have a scale for mixture enthalpies at saturation, i.e., at 100 percent relative humidity, and dashed lines indicating deviations from these enthalpies at lower humidities and the same wet-bulb temperature. As an example, consider a mixture at 90°F dry bulb and 30 percent relative humidity. The mixture enthalpy at saturation is 31.8 Btu/lbm dry air, while the deviation at 30 percent humidity is −0.19 Btu/lbm dry air, or a deviation of only 0.6 percent.

For a given total pressure of the mixture the partial pressure of the water vapor is a direct function, through Eq. (7-40), of the specific humidity, and so a line of constant ω is also a line of constant p_v and constant dew point. The dew point is found by first locating the point in question and then following a line of constant ω (horizontal line) until 100 percent relative humidity is encountered. The temperature when saturation is reached is the dew point.

It is very important to note that a given psychrometric chart is applicable only for the total-pressure conditions indicated; for the charts in the Appendix a total pressure of 1 atm applies.

7-7 AIR-CONDITIONING PROCESSES

To illustrate the application of the principles of the foregoing sections, we now shall consider some typical air-conditioning processes and show how they may be represented on the psychrometric diagram. Although we will examine a number of different applications the analysis procedure will usually take the following form:

1 Draw a sketch of the system and label with known information.
2 Sketch the process(es) on the psychrometric chart.
3 Write a mass balance on the water (liquid, vapor, or ice).
4 Write a mass balance on the "dry air" stream(s).
5 Write an energy balance for the system.
6 Determine properties for use in the mass and energy balances either by calculation or, if at 1 atm, by using the psychrometric chart(s).
7 Solve for the required quantities.

Dehumidification by Cooling

Consider first the simple cooling process illustrated in Fig. 7-5. The humid air stream is passed across cooling coils which lower the mixture temperature below the dew point. As a result of the cooling process, a portion of the vapor is condensed on the coils and removed as liquid. The process is shown schemati-

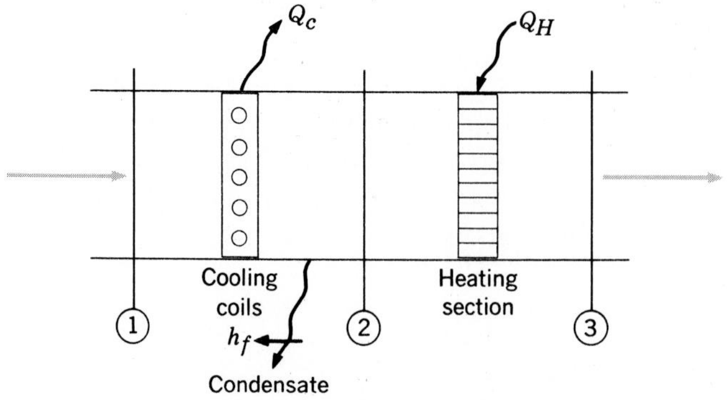

FIG. 7-5 Simple cooling and reheat process for air–water-vapor mixture.

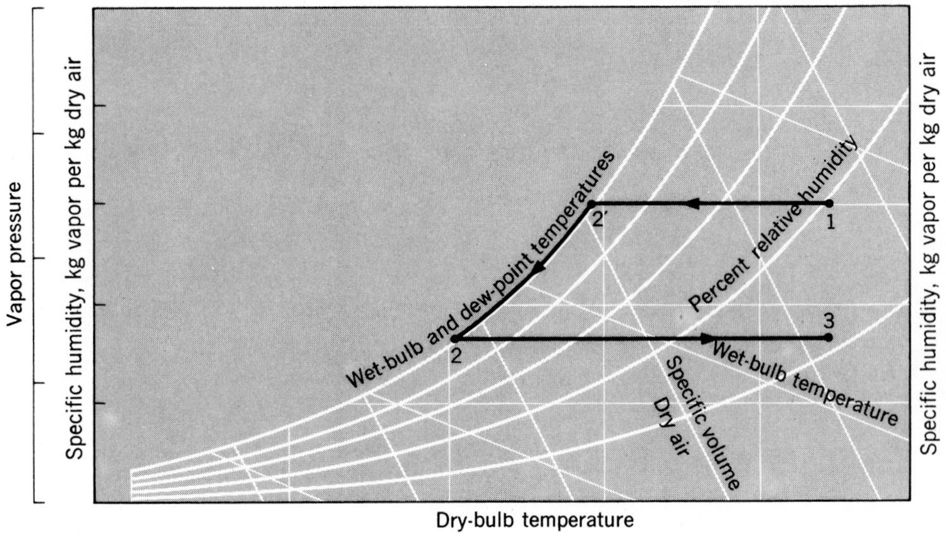

FIG. 7-6 Psychrometric diagram for the process of Fig. 7-5.

cally on the psychrometric chart of Fig. 7-6. For constant-pressure cooling the stream is first cooled at constant specific humidity from point 1 to 2′, at which point the water vapor starts to condense. Further cooling reduces the specific humidity to point 2. An energy balance on the system yields

$$Q_c = m_a(h_{a_1} - h_{a_2} + \omega_1 h_{v_1} - \omega_2 h_{v_2}) - m_a(\omega_1 - \omega_2)h_f \tag{7-59}$$

or, in terms of mixture enthalpies,

$$Q_c = \dot{m}_a(h_{m_1} - h_{m_2}) - \dot{m}_a(\omega_1 - \omega_2)h_f \tag{7-59a}$$

The coefficient of h_f results from a mass balance on the vapor and liquid water as

$$\dot{m}_{v_1} = \dot{m}_{v_2} + \dot{m}_f$$

$$\dot{m}_f = \dot{m}_{v_1} - \dot{m}_{v_2} = \dot{m}_a(\omega_1 - \omega_2)$$

Q_c is the heat removed by the cooling coils. The dry air may be reheated back to the original dry-bulb temperature if desired. Such a reheat process would proceed to point 3 in Figs. 7-5 and 7-6 and the heating required would be

$$Q_H = \dot{m}_a(h_{a_3} + \omega_3 h_{v_3}) - \dot{m}_a(h_{a_2} + \omega_2 h_{v_2})$$

or

$$Q_H = \dot{m}_a(h_{m_3} - h_{m_2})$$

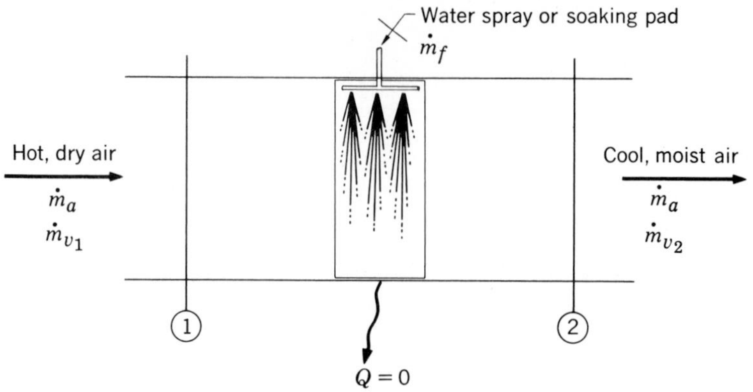

FIG. 7-7 Adiabatic humidification or evaporative cooling process.

Evaporative Cooling

A cooling technique which finds considerable application in desert climates is that of evaporative cooling, as shown in Fig. 7-7. Relatively dry air is delivered to the chamber at point 1 and is either sprayed with water or allowed to pass across a soaked pad which is continuously replenished with water. The process is illustrated on the psychrometric chart of Fig. 7-8, and is assumed to occur adiabatically so that a drop in temperature results from the evaporation process.

The mass balance on the vapor and liquid is

$$\dot{m}_{v_1} + \dot{m}_f = \dot{m}_{v_2}$$

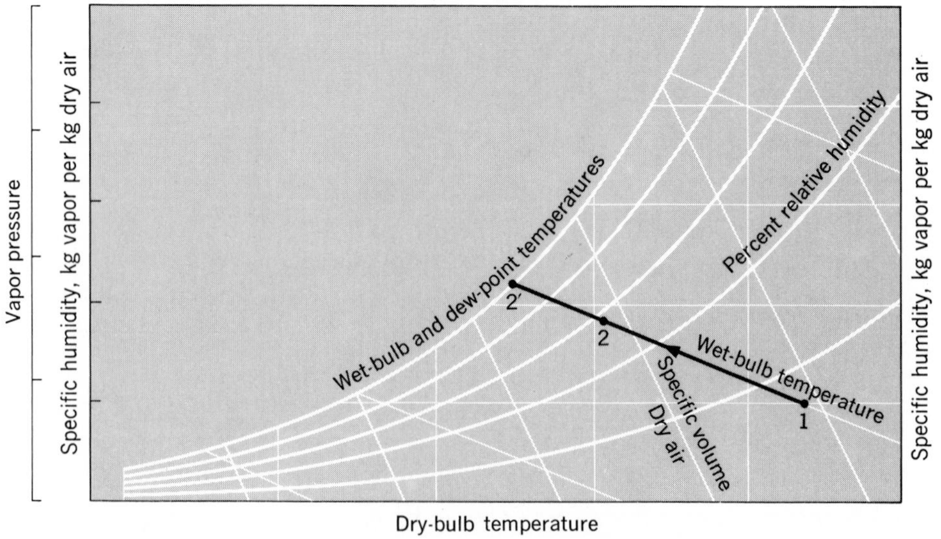

FIG. 7-8 Psychrometric diagram for the process of Fig. 7-7.

or

$$\dot{m}_f = \dot{m}_{v_2} - \dot{m}_{v_1} = \dot{m}_a(\omega_2 - \omega_1)$$

and the energy balance is

$$m_a(h_{a_1} + \omega_1 h_{v_1}) + m_a(\omega_2 - \omega_1)h_f = m_a(h_{a_2} + \omega_2 h_{v_2}) \tag{7-60}$$

or, in terms of mixture enthalpies

$$\dot{m}_a h_{m_1} + \dot{m}_a(\omega_2 - \omega_1)h_f = \dot{m}_a h_{m_2} \tag{7-60a}$$

This means that the inlet mixture enthalpy is very nearly equal to the exit mixture enthalpy.

There is, of course, some minimum temperature which may be achieved by this process, designated by point 2′ in Fig. 7-8. It is fairly easy to see that the evaporative cooling process is an adiabatic saturation process which is carried more or less to completion and, therefore, follows a line of constant wet-bulb temperature. The amount of make-up water which must be supplied is very small and its temperature does not exert a significant influence on the process.

Humidification with Heating

When cool air is heated at constant specific humidity, the relative humidity is reduced and the effect is to produce rather "dry" air if the heating process occurs over a substantial temperature range. This circumstance is observed in heating applications in extremely cold weather. Very dry air can cause discomfort such as chapping of lips, etc., to building occupants so that many heating systems include provisions for humidification. The heating system for a hospital operating room or maternity ward is a particular case in point. A typical humidification process is shown in Fig. 7-9. The additional water vapor is supplied either by a water or steam spray—the steam having the advantage that

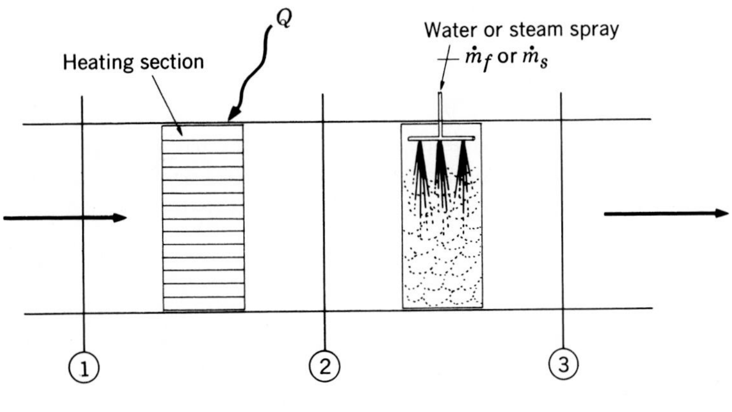

FIG. 7-9 Humidification-heating process.

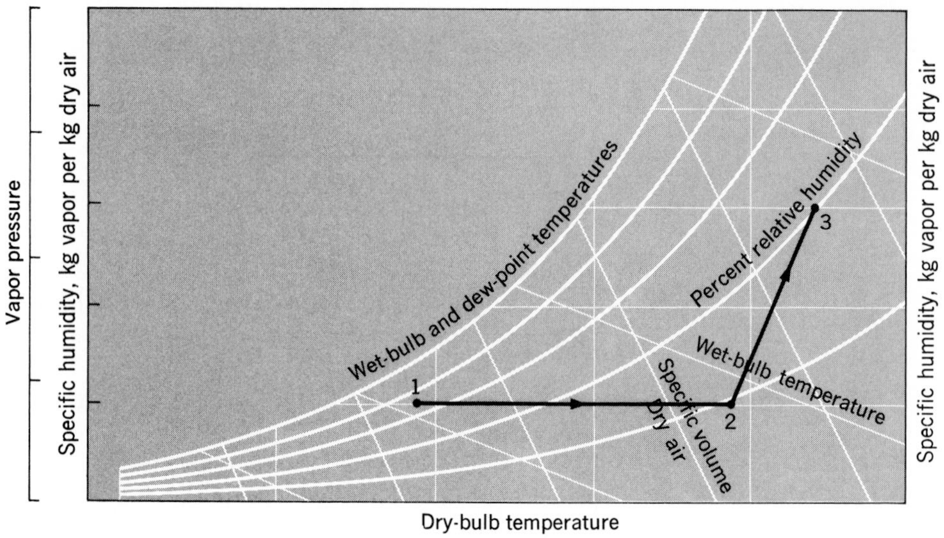

FIG. 7-10 Psychrometric diagram for the process of Fig. 7-9.

it provides heating as well as an increased vapor content and thereby reduces the quantity of external heat which must be supplied to the process. A disadvantage of steam humidification is that an objectionable odor is experienced unless the purity of the steam is carefully controlled. The heating-humidification process is illustrated on the psychrometric diagram of Fig. 7-10.

Between points 1 and 3 the mass balance on vapor is

$$\dot{m}_{v_1} + \dot{m}_f = \dot{m}_{v_3}$$

or

$$\dot{m}_f = \dot{m}_{v_3} - \dot{m}_{v_1} = \dot{m}_a(\omega_3 - \omega_1)$$

and the energy balance, in terms of mixture enthalpies, is

$$\dot{m}_a h_{m_1} + Q + \dot{m}_f h_f = \dot{m}_a h_{m_2}$$

Adiabatic Mixing of Two Streams

Consider the arrangement shown in Fig. 7-11. Two humid airstreams 1 and 2 are mixed adiabatically in the chamber to produce the outlet stream 3. The air and water-vapor mass balances for this system yield the following relations:

$$m_{a_1} + m_{a_2} = m_{a_3} \tag{7-61}$$

$$\dot{m}_{v_1} + \dot{m}_{v_2} = \dot{m}_{v_3}$$

$$m_{a_1}\omega_1 + m_{a_2}\omega_2 = m_{a_3}\omega_3 \tag{7-62}$$

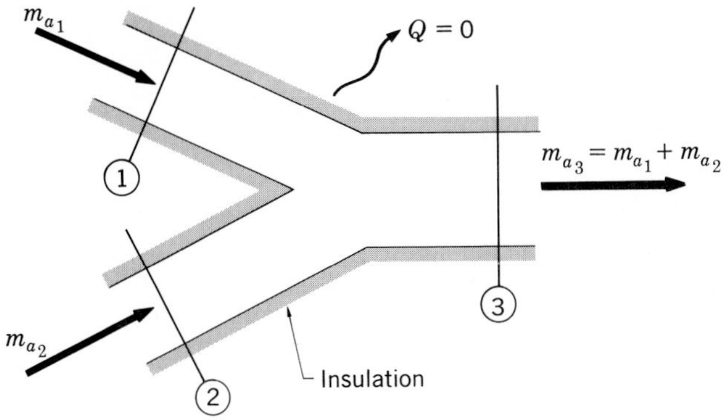

FIG. 7-11 Adiabatic mixing of two air–water-vapor systems.

For adiabatic mixing the energy balance for steady flow is

$$m_{a_1}(h_{a_1} + \omega_1 h_{v_1}) + m_{a_2}(h_{a_2} + \omega_2 h_{v_2}) = m_{a_3}(h_{a_3} + \omega_3 h_{v_3}) \tag{7-63}$$

or, in terms of mixture enthalpies

$$\dot{m}_{a_1} h_{m_1} + \dot{m}_{a_2} h_{m_2} = \dot{m}_{a_3} h_{m_3}$$

Introducing Eq. (7-61) into Eq. (7-62) gives

$$\frac{m_{a_1}}{m_{a_2}} = \frac{\omega_3 - \omega_2}{\omega_1 - \omega_3} \tag{7-64}$$

If Eq. (7-61) is applied to Eq. (7-63), the result is

$$\frac{m_{a_1}}{m_{a_2}} = \frac{(h_{a_3} + \omega_3 h_{v_3}) - (h_{a_2} + \omega_2 h_{v_2})}{(h_{a_1} + \omega_1 h_{v_1}) - (h_{a_3} + \omega_3 h_{v_3})} = \frac{h_{m_3} - h_{m_2}}{h_{m_1} - h_{m_3}} \tag{7-65}$$

Equations (7-64) and (7-65) may be combined to yield

$$\frac{m_{a_1}}{m_{a_2}} = \frac{h_{m_3} - h_{m_2}}{h_{m_1} - h_{m_3}} = \frac{\omega_3 - \omega_2}{\omega_1 - \omega_3} \tag{7-66}$$

We may now examine the different physical conditions which can persist when two streams are mixed. In deriving the foregoing relations the tacit assumption has been made that no vapor will be condensed. If vapor is condensed, we would need to add another exit stream for the liquid condensate. The possibility exists that the relative humidity of the exit stream may exceed 100 percent and that the excess water will be carried along as small droplets in a foglike flow. If we assume that the total enthalpy is a function primarily of wet-bulb temperature, then the mixing process may be represented on a psychrome-

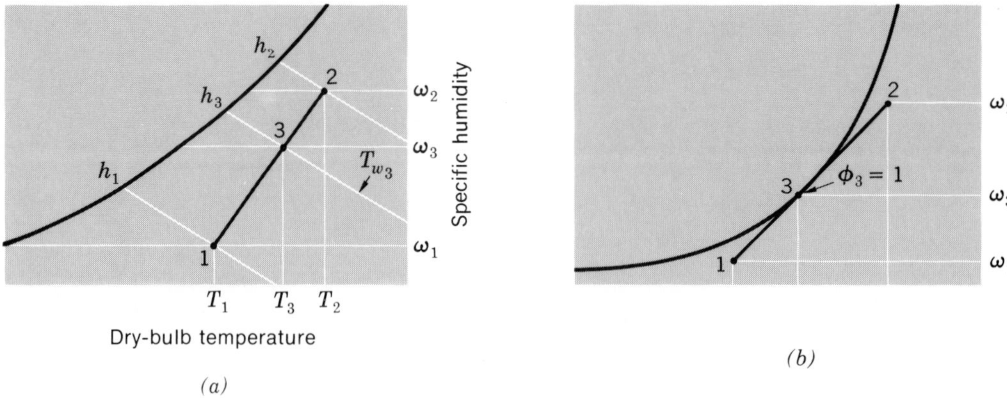

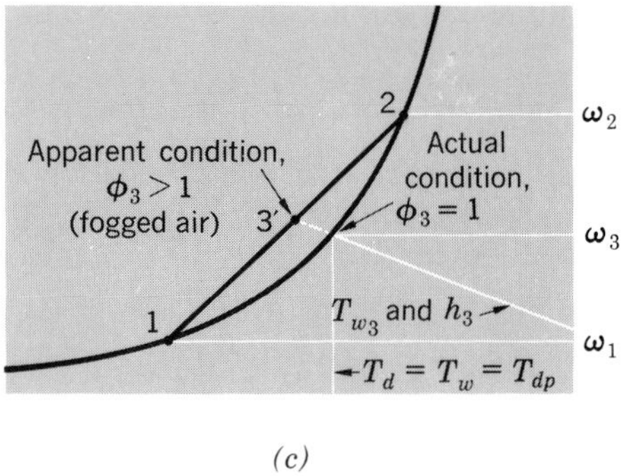

FIG. 7-12 Psychrometric diagrams for the process of Fig. 7-11: (*a*) $\phi_3 < 1$, no vapor condensed; (*b*) $\phi_3 = 1$, outlet airstream is saturated; (*c*) $\phi_3 > 1$, vapor condenses producing a foglike flow.

TABLE 7-2

Process Relations for Air–Water-Vapor Mixtures

	Mixture Properties				
Process or Property	**Dry-Bulb Temp., T_{DB}**	**Wet-Bulb Temp., T_{WB}**	**Dew-Point Temp., T_{DP}**	**Specific Humidity, ω**	**Mixture Enthalpy, $h_m = h_a + \omega h_v$**
Simple heating, no moisture addition	Increases	Increases	Constant	Constant	Increases
Cooling, no dehumidification	Decreases	Decreases	Constant	Constant	Decreases
Cooling, with dehumidification	Decreases	Decreases	Constant to 100% R.H., then decreases	Constant to 100% R.H., then decreases	Decreases
Evaporative cooling	Decreases	Approx. constant	Increases	Increases	Approx. constant
Heating, with humidification	Increases	Increases	Increases	Increases	Increases
Cooling, dehumidification, reheat	Variable, detailed mass and energy balances required				
Mixture enthalpy per mass of dry air	Primarily function of wet-bulb temperature				
Constant dew point	Same as constant p_v and ω				
Wet-bulb temperature	Approximately same as adiabatic saturation temperature				
Dry-bulb temperature	*Actual* temperature of mixture				

tric chart as shown in Fig. 7-12, where three possibilities are indicated:

a $\phi_3 < 1$ no vapor condensed
b $\phi_3 = 1$ outlet air stream is saturated
c $\phi_3 > 1$ vapor condenses, producing foglike flow

The foregoing theory for adiabatic mixing is useful in the design of air-conditioning systems which receive and mix air from multiple sources.

Table 7-2 gives a convenient summary of the different types of air-conditioning and psychrometric mixture problems discussed in the foregoing sections.

EXAMPLE 7-9 Cooling-Reheat
The air of Example 7-6 enters a cooler-reheat section like that shown in Fig. 7-5. The outlet airstream is to be at 75°F and 50 percent relative humidity. Calculate the heat which must be removed in the cooling coils and the heat which must be added in the heating section.

SOLUTION We shall employ the psychrometric chart for solution of this problem. The processes are shown in schematic form in Fig. 7-6. From the chart we have

$$\omega_1 = \omega_2' = 123 \text{ grains} = 0.0176 \text{ lbm} \qquad (7.98 \text{ g})$$

$$\omega_2 = \omega_3 = 65 \text{ grains} = 0.0093 \text{ lbm} \qquad (4.22 \text{ g})$$

$$T_2' = 73°\text{F} \qquad (22.8°\text{C})$$

$$T_2 = 55°\text{F} \qquad (12.8°\text{C})$$

The energy balance for the cooling section is given in Eq. (7-59), and the vapor enthalpy can be taken from Eq. (7-47) or, as a good approximation, saturated vapor enthalpies from the steam tables at the given temperatures may be used. We shall use Eq. (7-47):

$$h_{v_1} = 1061 + (0.445)(100) = 1106 \text{ Btu/lbm} \qquad (2572 \text{ J/g})$$

$$h_{v_2} = 1061 + (0.445)(55) = 1085 \text{ Btu/lbm} \qquad (2523 \text{ J/g})$$

$$h_f \text{ at } 55°\text{F} = 23.06 \text{ Btu/lbm} \qquad (53.63 \text{ kJ/kg})$$

Now inserting these values in Eq. (7-59) gives

$$Q_c = m_a[(0.24)(100 - 55) + (0.0176)(1106) - (0.0093)(1085)] - m_a(0.0176 - 0.0093)(23.06)$$

or

$$Q_c = 20 \text{ Btu/lbm dry air} \qquad (46.5 \text{ kJ/kg})$$

In the heating section there is no water lost or gained, and the energy balance yields

$$Q_H = m_a[h_{a_3} - h_{a_2} + \omega_3(h_{v_3} - h_{v_2})]$$

where Q_H is the heat added. With the proper numerical values added, this relation becomes

$$Q_H = m_a[(0.24)(75 - 55) + (0.0093)(0.445)(75 - 55)]$$

or

$$Q_H = 4.88 \text{ Btu/lbm dry air} \qquad (11.35 \text{ kJ/kg})$$

A somewhat easier approach is to read the mixture enthalpies directly from the psychrometric chart. We have

$$h_{m_1} = 43.7 \text{ Btu/lbm dry air} \qquad (T_{\text{DB}} = 100°\text{F}, T_{\text{WB}} = 80°\text{F})$$

$$h_{m_2} = 23.1 \text{ Btu/lbm dry air} \qquad (\phi = 100\% \text{ at } 55°\text{F})$$

$$h_{m_3} = 28.4 \text{ Btu/lbm dry air} \qquad (T_{\text{DB}} = 75°\text{F}, \phi = 50\%)$$

We also have

$$m_f = \omega_1 - \omega_3 = 0.0083 \text{ lbm/lbm dry air}$$

so the energy balance is

$$\begin{aligned} Q_c &= \dot{m}_a(h_{m_1} - h_{m_2}) - \dot{m}_a(\omega_1 - \omega_2)h_f \\ &= 43.7 - 23.1 - (0.0083)(23.06) \\ &= 20.41 \text{ Btu/lbm dry air} \end{aligned}$$

The difference between this and the 20 Btu/lbm dry air obtained above results from inaccuracies in reading the psychrometric chart.

For the heating process we obtain

$$Q_H = \dot{m}_a(h_{m_3} - h_{m_2}) = 28.4 - 23.1 = 5.3 \text{ Btu/lbm dry air}$$

Again, the numbers differ because of inaccuracies in reading the psychrometric chart.

EXAMPLE 7-10 Evaporative Cooling

Determine the minimum temperature which may be achieved by using an evaporative cooling process with desert air at 115°F and 5 percent relative humidity.

SOLUTION We could obtain this minimum temperature either by using a psychrometric chart as illustrated in Fig. 7-8 or by calculating the adiabatic saturation temperature. From the psychrometric chart we read, immediately, for $T_{DB} = 115°F$, $\phi = 5$ percent,

$$T_{WB} = 67°F$$

The usefulness of evaporative coolers in desert climates is quite obvious.

EXAMPLE 7-11 Adiabatic Mixing
2 lb of air at 100°F and 70 percent humidity are mixed adiabatically with 3 lbm of air at 70°F and 50 percent humidity. Calculate the temperature of the mixture if the process occurs under steady-flow conditions at a constant pressure of 14.696 psia. Also determine the relative humidity of the mixture.

SOLUTION From the psychrometric chart we have

$$T_1 = 100°F \qquad \omega_1 = 206 \text{ grains} = 0.029 \text{ lbm/lbm dry air}$$

$$T_2 = 70°F \qquad \omega_2 = 55 \text{ grains} = 0.0079 \text{ lbm/lbm dry air}$$

$$m_{a_1} = 2$$

$$m_{a_2} = 3$$

We may now calculate the exit specific humidity from Eq. (7-64):

$$\frac{m_{a_1}}{m_{a_2}} = \frac{\omega_3 - \omega_2}{\omega_1 - \omega_3}$$

$$\frac{2}{3} = \frac{\omega_3 - 55}{206 - \omega_3}$$

$$\omega_3 = 115 \text{ grains} = 0.0021 \text{ lbm/lbm dry air}$$

A line is now drawn on the psychrometric chart shown schematically in Fig. 7-12*a*. The final state is found at the intersection of this line and $\omega_3 = 115$ grains. We obtain

$$T_{DB_3} = 82°F$$

$$T_{WB_3} = 74.5°F$$

$$\phi_3 = 70 \text{ percent}$$

EXAMPLE 7-12 Low-Temperature Cooling
A special insulated room is designed to store meat at −20°F, and it is estimated that the leakage of outside air through the entry doors amounts to 500 ft^3/min at 95°F and 40 percent relative humidity. How much cooling is required to compensate for this leakage assuming that the air in the room is saturated at −20°F?

SOLUTION The total energy removed from the air is

$$Q = \dot{m}_a(h_1 - h_2) \tag{a}$$

where h_1 is the enthalpy of the outside air–water-vapor mixture and h_2 is the enthalpy of the air–water-vapor–frost mixture inside the room. For the outside air we can use the psychrometric chart to obtain

$$\omega_1 = 101 \text{ grains} = 0.014\,42 \text{ lbm/lbm dry air}$$

$$T_{DB_1} = 95°\text{F}$$

$$v_{a_1} = 14.3 \text{ ft}^3\text{/lbm dry air} \qquad (0.893 \text{ m}^3\text{/kg})$$

The mass flow of dry air is therefore

$$\dot{m}_a = \frac{500 \text{ ft}^3\text{/min}}{14.3 \text{ ft}^3\text{/lbm}} = 34.96 \text{ lbm/min} \qquad (15.86 \text{ kg/min}) \tag{b}$$

The vapor pressure at $-20°$F (440°R) is calculated from Eq. (7-50*a*):

$$p_{g_2} = 5.103 \exp\left(18.42 - \frac{11\,059}{440}\right) = 0.006\,19 \text{ psia} \qquad (42.67 \text{ Pa}) \tag{c}$$

so the specific humidity in the room is

$$\omega_2 = 0.622\frac{p_{g_2}}{p_{a_2}} = \frac{(0.622)(0.006\,19)}{14.696 - 0.006\,19} = 2.62 \times 10^{-4} \tag{d}$$

The mass of inlet vapor converted to frost in the room is $\dot{m}_a(\omega_1 - \omega_2)$ so the enthalpy of the room mixture is

$$h_2 = h_{a_2} + \omega_2 h_{v_2} + (\omega_1 - \omega_2)h_{i_2} \tag{e}$$

while the inlet enthalpy is

$$h_1 = h_{a_1} + \omega_1 h_{v_1} \tag{f}$$

The vapor and frost enthalpies are now determined as

$$h_{v_1} = 1061 + (0.445)(95) = 1103.3 \text{ Btu/lbm}$$

$$h_{v_2} = 1061 + (0.445)(-20) = 1052.1 \text{ Btu/lbm}$$

$$h_{i_2} = -158.9 + (0.467)(-20) = -168.24 \text{ Btu/lbm}$$

We may now insert the numerical values in Eq. (*a*) to determine the heat removal:

$$\begin{aligned} Q &= 34.96[(0.24)(95 + 20) + (0.014\,43)(1103.3) \\ &\quad -(2.62 \times 10^{-4})(1052.1) \\ &\quad -(0.014\,43 - 2.62 \times 10^{-4})(-168.24)] \\ &= 34.96(27.6 + 15.92 - 0.276 + 2.384) \\ &= 1595 \text{ Btu/min} = 95\,709 \text{ Btu/h} \qquad (28.04 \text{ kW}) \end{aligned}$$

This represents a cooling requirement of 7.98 tons of refrigeration, where a "ton" is defined as 12 000 Btu/h. We could also work the problem by obtaining mixture enthalpies directly from the psychrometric chart. We may read

$$h_{m_1} = 38.8 \text{ Btu/lbm dry air} \qquad (T_{DB} = 95°\text{F}, \phi = 40\%)$$

$$h_{m_2} = -4.5 \text{ Btu/lbm dry air} \qquad (T_{DB} = -20°\text{F}, \phi = 100\%)$$

The energy balance would then be

$$Q = \dot{m}_a[(h_{m_1} - h_{m_2}) - (\omega_1 - \omega_2)h_{i_2}]$$

or

$$\begin{aligned} Q &= 34.96[38.8 - (-4.5) - (0.0142)(-168.24)] \\ &= 1597 \text{ Btu/min} \end{aligned}$$

EXAMPLE 7-13 Air Conditioning of an Office Space

In a small office space it is determined that a heat removal of 10 kW will suffice for air conditioning. The room is to be maintained at 22°C, 50 percent relative humidity, and 25 percent of the moisture is picked up in the room. The air is completely recirculated and the cooling system must remove the moisture added in the room as indicated in the schematic diagram. Calculate the volume of air flow required under the condition that the air is cooled just enough to remove the moisture.

FIG. EXAMPLE 7-13

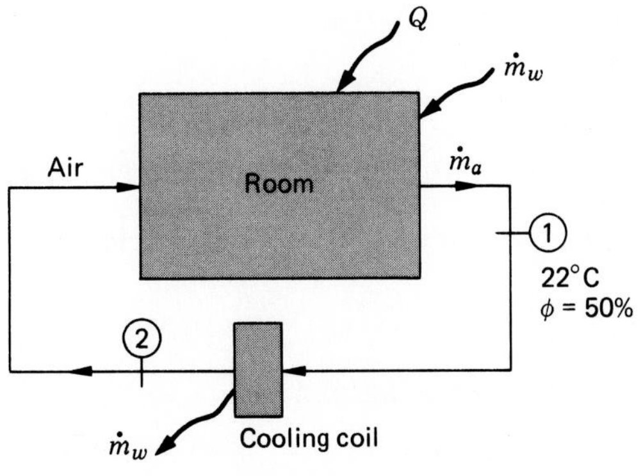

SOLUTION From the problem statement

$$T_1 = 22°\text{C} \qquad \phi_1 = 50\% \qquad Q = 10 \text{ kW}$$
$$\dot{m}_w = 0.25\dot{m}_{v_1} \qquad \phi_2 = 100\%$$

The water mass balance thus requires that

$$\dot{m}_{v_2} = \dot{m}_{v_1} - \dot{m}_w = 0.75\dot{m}_{v_1} \tag{a}$$

From the psychrometric chart

$$\omega_1 = 0.008\ 20 \text{ kg/kg dry air} \tag{b}$$

State 2 is saturated according to the problem statement and has

$$\omega_2 = 0.75\omega_1 = 0.006\ 15 \tag{c}$$

The temperature at point 2 is then obtained from the psychrometric chart as

$$T_2 = T_{\text{DB}_2} = T_{\text{WB}_2} = 7°\text{C} \tag{d}$$

The energy balance on the room is now written

$$\dot{m}_a(h_2 + \omega_2 h_{v_2}) + \dot{m}_w h_w + Q = \dot{m}_a(h_1 + \omega_1 h_{v_1}) \tag{e}$$

The vapor enthalpies are calculated from Eq. (7-47*b*):

$$h_v = 2501 + 1.863T$$
$$h_{v_1} = 2501 + (1.863)(22) = 2542 \text{ kJ/kg}$$
$$h_{v_2} = 2501 + (1.863)(7) = 2514 \text{ kJ/kg}$$

A reasonable estimate for h_w is

$$h_w = h_f \text{ at } 22°\text{C} = 92.33 \text{ kJ/kg} \tag{f}$$

Noting that $\dot{m}_w = \dot{m}_a(\omega_1 - \omega_2)$, Eq. (*e*) becomes

$$\dot{m}_a[(1.005)(7 - 22) + (0.006\ 15)(2514) - (0.008\ 20)(2542) + (0.008\ 20 - 0.006\ 15)(92.33)] + 10 = 0 \tag{g}$$

which may be solved to give

$$\dot{m}_a = 0.4934 \text{ kg/s} = 29.6 \text{ kg/min} \tag{h}$$

From the psychrometric chart the specific volume leaving the cooling coils is

$$v_{a_2} = 0.802 \text{ m}^3/\text{kg dry air} \tag{i}$$

so the volume flow rate at that point would be

$$\dot{V}_2 = \dot{m}_a v_{a_2} = (29.6)(0.802) = 23.7 \text{ m}^3/\text{min} \tag{j}$$

At the inlet conditions to the cooling coil the specific volume is

$$v_{a_1} = 0.847 \text{ m}^3/\text{kg dry air}$$

so the volume flow at that point would be

$$\dot{V}_1 = \dot{m}_a v_{a_1} = (29.6)(0.847) = 25.1 \text{ m}^3/\text{min} \tag{k}$$

EXAMPLE 7-14
It is common practice in many air-conditioning systems to mix a small amount of fresh outside air with the cool air coming off the refrigerant coils. In a Texas application the air from the cooling coils is at 10°C and 100 percent humidity, and it is to be mixed with 20 percent outside air at 38°C and 30 percent humidity. Calculate the final temperature and relative humidity of the mixture.

SOLUTION For this problem we consult the psychrometric chart and read, using Fig. 7-12 as a guide

$$T_1 = 10°\text{C} \qquad \phi_1 = 100\% \qquad \omega_1 = 0.0077 \text{ kg/kg dry air}$$

$$T_2 = 38°\text{C} \qquad \phi_2 = 30\% \qquad \omega_2 = 0.0123 \text{ kg/kg dry air}$$

From Eq. (7-64)

$$\frac{\dot{m}_{a_1}}{\dot{m}_{a_2}} = \frac{\omega_3 - \omega_2}{\omega_1 - \omega_3} \tag{a}$$

The term *20 percent outside air* means $\dot{m}_{a_1}/\dot{m}_{a_2} = 1/0.2 = 5$, so that

$$5 = \frac{\omega_3 - 0.0123}{0.0077 - \omega_3} \tag{b}$$

and

$$\omega_3 = 0.0085$$

We now draw a line between points 1 and 2 on the psychrometric chart, and determine T_3 at the intersection of $\omega = 0.0085$. Thus,

$$T_3 = 14.5°\text{C}$$

$$\phi_3 = 83\%$$

EXAMPLE 7-15 Heating with Humidification
Air at the rate of 1600 ft³/min, at 40°F and 50 percent relative humidity, is mixed with 1500 ft³/min of air at 70°F and 40 percent relative humidity in a steady-flow process with enough heat also added to bring the exit temperature to 80°F. Sufficient liquid water at 60°F is sprayed into the stream to produce 50 percent relative humidity at the 80°F condition. Calculate the heating required.

SOLUTION The schematic for the process is shown in the accompanying sketch. The data corresponding to the figure are

$$T_{DB_1} = 40°\text{F} \qquad \phi_1 = 50\% \qquad \dot{V}_1 = 600\ \text{ft}^3/\text{min}$$

$$T_{DB_2} = 70°\text{F} \qquad \phi_2 = 40\% \qquad \dot{V}_2 = 1500\ \text{ft}^3/\text{min}$$

$$T_3 = 60°\text{F} \qquad \text{Sat. liquid water}$$

$$T_{DB_4} = 80°\text{F} \qquad \phi_4 = 50\%$$

We now make the following balances:

Mass balance on dry air:

$$\dot{m}_{a_1} + \dot{m}_{a_2} = \dot{m}_{a_4} \tag{a}$$

Mass balance on water and vapor:

$$\dot{m}_{v_1} + \dot{m}_{v_2} + \dot{m}_{f_3} = \dot{m}_{v_4}$$

FIG. EXAMPLE 7-15

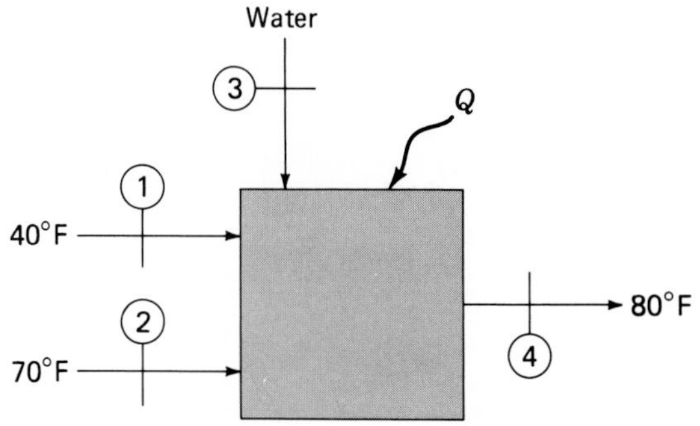

or

$$\dot{m}_{f_3} = \dot{m}_{a_4}\omega_4 - \dot{m}_{a_1}\omega_1 - \dot{m}_{a_2}\omega_2 \tag{b}$$

Energy balance on entire system:

$$\dot{m}_{a_1}h_{m_1} + \dot{m}_{a_2}h_{m_2} + \dot{m}_{f_3}h_{f_3} + Q = \dot{m}_{a_4}h_{m_4} \tag{c}$$

The mass flow of dry air is obtained from

$$\dot{m}_a = \frac{\dot{V}}{v_a} \tag{d}$$

We now consult the psychrometric chart and obtain values of the various parameters:

$$v_{a_1} = 12.6 \text{ ft}^3\text{/lbm dry air}$$
$$v_{a_2} = 13.5 \text{ ft}^3\text{/lbm dry air}$$
$$h_{m_1} = 12.3 \text{ Btu/lbm dry air}$$
$$h_{m_2} = 23.7 \text{ Btu/lbm dry air}$$
$$h_{m_4} = 31.5 \text{ Btu/lbm dry air}$$
$$\omega_1 = 0.0028$$
$$\omega_2 = 0.0063$$
$$\omega_4 = 0.0111$$
$$h_{f_3} = 60 - 32 = 23 \text{ Btu/lbm}$$

The mass flows of air are then calculated from Eq. (*d*):

$$\dot{m}_{a_1} = \frac{600}{12.6} = 47.62 \text{ lbm/min}$$

$$\dot{m}_{a_2} = \frac{1500}{13.5} = 111.1 \text{ lbm/min}$$

Then, from Eq. (*a*)

$$m_{a_4} = 47.62 + 111.1 = 158.73 \text{ lbm/min}$$

and from Eq. (*b*)

$$\dot{m}_{f_3} = (158.73)(0.0111) - (47.62)(0.0028) - (111.1)(0.0063)$$
$$= 0.93 \text{ lbm/min}$$

Finally, all the quantities are inserted in Eq. (c) to obtain the heating:

$$(47.62)(12.3) + (111.1)(23.7) + (0.93)(28) + Q = (158.73)(31.5)$$

and

$$Q = 1755 \text{ Btu/min} \qquad (30.9 \text{ kW})$$

7-8 COOLING TOWERS

As we shall see in Chap. 9 large power and refrigeration cycles require the dissipation of considerable quantities of heat to their atmospheric surroundings. Typically, water is the cooling medium for the power or refrigeration cycle, and a *wet cooling tower* is employed to transfer the energy to the atmosphere, as shown schematically in Fig. 7-13.

We assume that there is little heat transferred from the surroundings. The inlet warm water is sprayed into the upward flow of relatively dry air. As the air is saturated the temperature drops, similar to the evaporative cooler, and the cooler liquid is collected in the bottom of the cooling tower. Because of the water removal with the air, make-up water must be supplied in order for the cool-water return to equal the warm-water inlet. Thus, we assume

$$\dot{m}_{w_1} = \dot{m}_{w_2} \tag{7-67}$$

We may now make the appropriate mass and energy balances on the system. Clearly, the dry air at inlet is equal to that at exit. The mass balance on vapor

FIG. 7-13 Schematic of cooling tower.

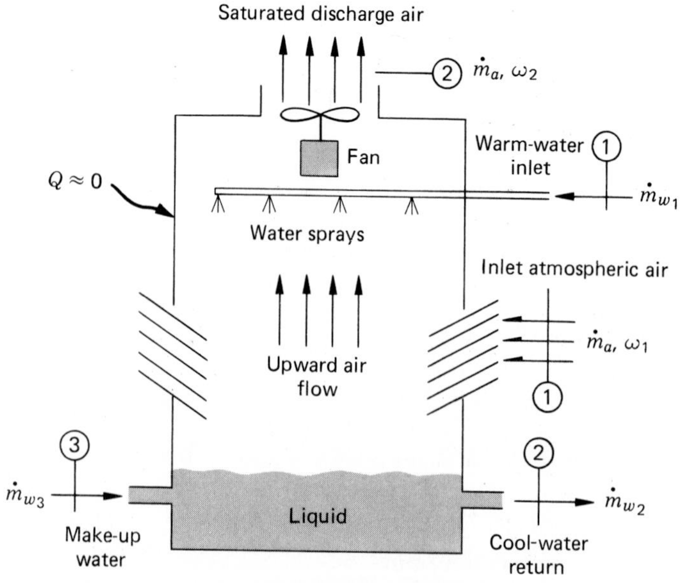

and liquid water is

$$\dot{m}_{w_1} + \dot{m}_{v_1} + \dot{m}_{w_3} = \dot{m}_{w_2} + \dot{m}_{v_2} \tag{7-68}$$

Using Eq. (7-67), this reduces to

$$\dot{m}_{w_3} = \dot{m}_{v_2} - \dot{m}_{v_1} \tag{7-69}$$

and, using specific humidities, we obtain

$$\dot{m}_{w_3} = \dot{m}_a(\omega_2 - \omega_1) \tag{7-70}$$

The energy balance is then

$$\dot{m}_a(h_{a_1} + \omega_1 h_{v_1}) + \dot{m}_{w_1}h_{w_1} + \dot{m}_{w_3}h_{w_3} = \dot{m}_a(h_{a_2} + \omega_2 h_{v_2}) + \dot{m}_{w_2}h_{w_2} \tag{7-71}$$

or, using mixture enthalpies and taking the water enthalpies as those of saturated liquid,

$$\dot{m}_a[h_{m_1} - h_{m_2} + h_{f_3}(\omega_2 - \omega_1)] = \dot{m}_{w_1}(h_{f_2} - h_{f_1}) \tag{7-72}$$

An example illustrates the use of these concepts.

EXAMPLE 7-16 Cooling Tower for a Power Plant
A power plant requires that 12 000 kg/s of water be cooled from 40 to 30°C. Inlet air for a cooling tower is available at 25°C and 35 percent relative humidity and may be assumed to leave the tower at 35°C and 90 percent relative humidity. Make-up water is available at 20°C. Calculate the mass flow rates of air and make-up water required.

SOLUTION We first compute the vapor pressures from Eq. (7-37) and the saturation pressures:

$$\begin{aligned} p_{g_1} &= 3.169 \text{ kPa} \qquad \text{at } T_{a_1} = 25°\text{C} \\ p_{g_2} &= 6.268 \text{ kPa} \qquad \text{at } T_{a_2} = 35°\text{C} \end{aligned} \tag{a}$$

$$\begin{aligned} p_{v_1} &= \phi_1 p_{g_1} = (3.169)(0.35) = 1.109 \text{ kPa} \\ p_{v_2} &= \phi_2 p_{g_2} = (5.268)(0.90) = 4.741 \text{ kPa} \end{aligned} \tag{b}$$

Then, the specific humidities are

$$\begin{aligned} \omega_1 &= 0.622\frac{p_{v_1}}{p_{a_1}} = \frac{(0.622)(1.109)}{101.32 - 1.109} = 6.88 \times 10^{-3} \\ \omega_2 &= \frac{(0.622)(4.741)}{101.32 - 4.741} = 0.0305 \text{ kg/kg dry air} \end{aligned} \tag{c}$$

The liquid enthalpies are obtained from the steam tables as

$$h_{f_1} = 167.57 \text{ kJ/kg} \qquad \text{at } 40°\text{C}$$
$$h_{f_2} = 125.79 \text{ kJ/kg} \qquad \text{at } 30°\text{C} \tag{d}$$
$$h_{f_3} = 83.96 \text{ kJ/kg} \qquad \text{at } 20°\text{C}$$

The vapor enthalpies are calculated with Eq. (7-47*b*):

$$\begin{aligned} h_{v_1} &= 2501 + (1.863)(25) = 2548 \text{ kJ/kg} \\ h_{v_2} &= 2501 + (1.863)(35) = 2566 \text{ kJ/kg} \end{aligned} \tag{e}$$

The difference in air enthalpies is calculated as $c_{pa}(T_1 - T_2)_a$ with $c_{pa} = 1.005$ kJ/kg·°C.

All of the properties may now be inserted in Eq. (7-72) to give

$$\begin{aligned} \dot{m}_a[(1.005)(25 - 35) &+ (6.88 \times 10^{-3})(2548) - (0.0305)(2566) \\ &+ (0.0305 - 6.88 \times 10^{-3})(83.96)] \\ &= (12\,00 \text{ kg/s})(125.79 - 167.57) \end{aligned} \tag{f}$$

The dry air flow required is thus obtained as

$$\dot{m}_a = 7287 \text{ kg/s} \tag{g}$$

The make-up water required is thus

$$\begin{aligned} \dot{m}_{w_3} &= \dot{m}_a(\omega_2 - \omega_1) \\ &= (7287)(0.0305 - 6.88 \times 10^{-3}) \\ &= 172.1 \text{ kg/s} \end{aligned} \tag{h}$$

We should note that the total air and water-vapor flow at inlet is

$$\begin{aligned} \dot{m}_{\text{tot}} &= \dot{m}_a + \dot{m}_{v_1} = \dot{m}_a(1 + \omega_1) \\ &= (7287)(1 + 6.88 \times 10^{-3}) \\ &= 7337 \text{ kg/s} \end{aligned} \tag{i}$$

7-9 REAL-GAS MIXTURES

All of the previous discussions in this chapter have been concerned with the behavior of gaseous mixtures in which each constituent behaves as an ideal gas, independent of its composition in the mixture. Physically, this behavior will be observed when the constituent gases exist at low enough pressures that their molecules are spaced far apart and molecular force fields do not influence the pressure. In terms of the generalized compressibility factors of Chap. 6 this means that the gaseous constituents will exist at $T_r \gg 1.0$ and $p_r \ll 1.0$. Our concern in this section is to develop a technique for calculating the behavior of gaseous mixtures in temperature and pressure regions where the ideal-gas law no longer applies. Regretfully, such calculations can be rather complicated.

Let us first consider two simplified approaches to the problem. We might assume that the total pressure of the mixture could be calculated as a linear sum of the pressures exerted by each constituent at the temperature and total volume of the mixture. Thus

$$p_m = \Sigma p_i \tag{7-73}$$

where p_m is the total pressure of the mixture. This is called *Dalton's rule of additive pressures*, and differs from the *law* of partial pressures in that we now must calculate the p_i from some real-gas equation of state. A similar relation is the *Amagat rule of additive volumes*, which assumes that the total volume of the mixture can be calculated as a sum of constituent volumes, each evaluated at the temperature and pressure of the mixture:

$$V_m = \Sigma V_i \tag{7-74}$$

The V_i must be evaluated with a real-gas equation of state. The reader must note carefully that the p_i and V_i are *not* the partial pressure and volume discussed in Sec. 7-2 because they are not evaluated with the ideal-gas equation of state.

If we use the generalized compressibility equation of state for the constituents,

$$p_i = \frac{Z_i n_i \mathcal{R} T}{V_m} \tag{7-75}$$

and

$$V_i = \frac{Z_i n_i \mathcal{R} T}{p_m} \tag{7-76}$$

where p_m and V_m are the total pressure and volume of the mixture, respectively, which we assume are related to a Z value for the mixture through

$$p_m V_m = n_m Z_m \mathcal{R} T \tag{7-77}$$

Inserting these compressibility relations back into either Eq. (7-75) or (7-76) gives

$$n_m Z_m = \Sigma n_i Z_i$$

or

$$Z_m = \Sigma x_i Z_i$$

However, we must go back and note that the conditions for evaluating the Z_i are different, depending on whether the additive-pressure or additive-volume rule is employed. For the additive-pressure rule

$$Z_m = \Sigma\,(x_i Z_i)_{T,v} \tag{7-78}$$

while for the additive-volume rule

$$Z_m = \Sigma(x_i Z_i)_{T,p} \tag{7-79}$$

For strict ideal-gas behavior $Z_i = 1.0$ and the two rules would be equivalent, as we already know.

We can anticipate some errors in the use of the above relations by noting that the Dalton rule neglects interaction effects between the constituent molecules because it forces a calculation at the total volume of the mixture; consequently, we anticipate that it would be most useful for low gas densities. The Amagat rule creates an error in just the opposite direction because it does allow for a molecular force interaction effect by forcing a calculation at the total pressure of the mixture. Consequently, we expect that the Amagat rule might overcompensate for molecular interactions when the mixture exists at low densities.

There are many relations which have been developed for the calculation of real-gas mixtures in addition to the simple additive rules given above. A simple procedure was developed by Kay [5] which is accurate within 10 percent over a wide range of temperatures and pressures. The method assumes that the mixture may be treated as a pseudo-pure substance with some empirical artifice used to determine the pseudo-critical constants for the mixture. Kay uses a simple linear combination of the critical properties so that

$$p_{cm} = \Sigma x_i p_{ci} \tag{7-80}$$

$$T_{cm} = \Sigma x_i T_{ci} \tag{7-81}$$

where p_{cm} and T_{cm} are the pseudo-critical pressure and temperature of the mixture. Once these pseudo-critical constants are determined the calculation can then proceed based on the generalized compressibility charts.

EXAMPLE 7-17 High-Pressure Air-Water-Vapor Mixture
Air in the amount of 0.5 kg is saturated with water vapor and contained in a rigid tank at 160°C and 1 MPa. Calculate the cooling required to lower the temperature to 80°C. How much moisture is condensed in this process?

SOLUTION In this process we may anticipate that the air will behave as an ideal gas, while the water vapor will exist at a pressure high enough to deviate from the ideal-gas equation of state. However, the pressure is low enough compared to the critical pressure for water to anticipate that an additive-pressure rule might be employed. For the saturated mixture the vapor pressure must be

$$p_v = p_g \quad \text{at } 160°\text{C}$$
$$= 617.8 \text{ kPa (Table A-7M)}$$
$$v_g = 0.3071 \text{ m}^3/\text{kg}$$

The pressure of the air is then

$$p_a = p_m - p_v = 1000 - 617.8 = 382.2 \text{ kPa} \qquad (55.4 \text{ psia})$$

and the mixture volume is calculated from

$$V_m = \frac{m_a R_a T}{p_a} = \frac{(0.5)(287.1)(433)}{382.2 \times 10^3} = 0.1626 \text{ m}^3 \qquad (5.743 \text{ ft}^3)$$

Carrying through with the assumption of no interaction between the gases, the mass of vapor at the total mixture volume is calculated as

$$m_v = \frac{V_m}{v_g} = \frac{0.1626}{0.3071} = 0.529 \text{ kg} \qquad (1.167 \text{ lbm})$$

At the 80°C condition,

$$p_g = 47.39 \text{ kPa} \qquad v_g = 3.407 \text{ m}^3/\text{kg}$$

For the same container volume, the mass of vapor at this condition is

$$m_v = \frac{0.1626}{3.407} = 0.0477 \text{ kg} \qquad (0.105 \text{ lbm})$$

For the constant-volume process the final air pressure can be calculated from

$$p_{a_2} = p_{a_1} \frac{T_2}{T_1} = \frac{(382.2)(353)}{433} = 311.6 \text{ kPa} \qquad (45.19 \text{ psia})$$

and the final mixture pressure is

$$p_{m_2} = p_{a_2} + p_{v_2} = 311.6 + 47.39 = 359.0 \text{ kPa} \qquad (52.07 \text{ psia})$$

For the process at constant volume here there is no work and the heat transfer is calculated from

$$Q = U_2 - U_1 = (U_2 - U_1)_{\text{air}} + (U_2 - U_1)_{\text{vapor}} \qquad (a)$$

We may evaluate the change in internal energy separately for the air and vapor. For the air,

$$U_2 - U_1 = mc_v(T_2 - T_1) = (0.5)(0.718)(80 - 160)$$
$$= -28.72 \text{ kJ} \qquad (-27.22 \text{ Btu})$$

The internal energy for the water vapor is obtained from the steam tables:

$$u_{v_1} = 2568.4 \text{ kJ/kg} \qquad (1104.2 \text{ Btu/lbm}) \qquad \text{sat. vapor at } 160°\text{C}$$

The mass of water condensed is

$$m_w = m_{v_1} - m_{v_2} = 0.529 - 0.0477 = 0.4813 \text{ kg} \tag{b}$$

The final vapor and liquid internal energies are

$$u_{v_2} = 2482.2 \text{ kJ/kg} \qquad (1067 \text{ Btu/lbm}) \qquad \text{sat. vapor at } 80°\text{C}$$
$$u_{w_2} = 334.86 \text{ kJ/kg} \qquad (144 \text{ Btu/lbm}) \qquad \text{sat. liquid at } 80°\text{C}$$

The total change in internal energy of the water is

$$U_2 - U_1 = m_{v_2}u_{v_2} + m_{w_2}u_{w_2} - m_{v_1}u_{v_1}$$
$$= (0.0477)(2482.2) + (0.4813)(334.86) - (0.529)(2568.4) \tag{c}$$
$$= -1079.1 \text{ kJ} \qquad (-1023 \text{ Btu})$$

Adding together the changes in internal energy for air and vapor in Eq. (*a*) gives the heat transfer as

$$Q = -28.72 - 1079.1 = -1107.8 \text{ kJ} \qquad (-1050 \text{ Btu})$$

The negative sign means that heat is removed from the system during the cooling process.

It is interesting to see what would happen in the calculation if we applied an additive-volume rule instead of the additive-pressure rule. The calculation of the volume of air at the temperature and pressure of the mixture is an easy matter:

$$V_a = \frac{m_a R_a T}{p} = \frac{(0.5)(287.1)(433)}{1 \times 10^6} = 0.06216 \text{ m}^3 \qquad (2.195 \text{ ft}^3)$$

But if one evaluates the volume of the water at 1 MPa and 160°C, we find that a compressed-liquid state is encountered, which of course makes no sense in this problem. So, as discussed earlier, a use of the additive-volume concept at low pressures may result in substantial errors.

EXAMPLE 7-18 Comparison of Calculation Techniques
A mixture of 1.0 lbm mol and equal molal parts of CO_2 and N_2 is contained at 100°F in a vessel having a volume of 1.5 ft^3. Calculate the total pressure of the mixture using (*a*) the ideal-gas law, (*b*) the rule of additive pressures and the compressibility charts, (*c*) the rule of additive volumes and the generalized compressibility charts, and (*d*) Kay's pseudo-critical method.

SOLUTION We have 1 mol of total mixture so

$$x_{N_2} = 0.5 \qquad \text{and} \qquad x_{CO_2} = 0.5$$

Also, $T_m = 100°\text{F} = 560°\text{R}$, $V_m = 1.5 \text{ ft}^3$, and $n_m = 1.0$ lbm mol.

(*a*) For ideal-gas behavior the total pressure is calculated immediately from

$$p_m = \frac{n_m \mathcal{R} T_m}{V_m} = \frac{(1.0)(1545)(560)}{(1.5)(144)} = 4005 \text{ psia}$$

(*b*) The critical properties for CO_2 and N_2 are obtained from Table 6-2 as

$$T_{c,N_2} = 277°\text{R} \qquad T_{c,CO_2} = 548°\text{R}$$

$$p_{c,N_2} = 33.5 \text{ atm} \qquad p_{c,CO_2} = 72.9 \text{ atm}$$

For the rule of additive pressures we must evaluate the pressure of each constituent at the temperature and volume of the total mixture. Because we have 0.5 mol of each constituent the molal specific volume of each is

$$\bar{v} = \frac{V_m}{n} = \frac{1.5}{0.5} = 3.0 \text{ ft}^3/\text{lbm mol}$$

The pseudo-reduced volume for each constituent is thus

$$\bar{v}_r' = \frac{\bar{v}}{\mathcal{R}T_c/p_c}$$

$$\bar{v}_{r,CO_2}' = \frac{3.0}{(1545)(548)/(72.9)(14.7)(144)} = 0.547$$

$$\bar{v}_{r,N_2}' = \frac{3.0}{(1545)(227)/(33.5)(14.7)(144)} = 0.607$$

and the reduced temperatures are

$$T_{r,CO_2} = \frac{560}{548} = 1.022$$

$$T_{r,N_2} = \frac{560}{227} = 2.467$$

We may consult the generalized compressibility charts of Chap. 6 to obtain the reduced pressures as

$$p_{r,CO_2} = 1.0 \qquad p_{r,N_2} = 4.25$$

The pressures to be used in the additive-pressure rule are therefore

$$p_{CO_2} = (1.0)(72.9) = 72.9 \text{ atm} = 1071 \text{ psia}$$

$$p_{N_2} = (4.25)(33.5) = 142.4 \text{ atm} = 2092 \text{ psia}$$

The total pressure of the mixture is then

$$p_m = p_{N_2} + p_{CO_2} = 2092 + 1071 = 3163 \text{ psia}$$

This value is substantially lower than that predicted by the ideal-gas law.

(*c*) To employ the additive-volume rule an iterative procedure is required because we do not know the total mixture pressure for evaluation of the volumes. The easiest way to proceed is to assume a value for the total pressure, calculate the Z_i's for insertion in Eq. (7-79), compute Z_m, and then calculate the total pressure from this. The procedure continues until agreement is reached between the assumed and calculated values. For these calculations the values of T_r are the same as in part (*b*). As a first trial we might take

$$p_m = 3500 \text{ psia} = 238.2 \text{ atm}$$

Then

$$p_{r,CO_2} = \frac{238.2}{72.9} = 3.27$$

$$p_{r,N_2} = \frac{238.2}{33.5} = 7.11$$

Consulting the generalized charts,

$$Z_{CO_2} = 0.48 \qquad Z_{N_2} = 1.09$$

Then, from Eq. (7-78)

$$Z_m = \Sigma x_i Z_i = (0.5)(0.48) + (0.5)(1.09) = 0.785$$

and the mixture pressure is calculated from

$$p_m = \frac{Z_m n_m \mathcal{R} T_m}{V_m} = \frac{(0.785)(1.0)(1545)(560)}{(1.5)(144)} = 3144 \text{ psia}$$

This differs substantially from our 3500 psia assumption so we try

$$p_m = 3100 \text{ psia} = 210.9 \text{ atm}$$

$$p_{r,CO_2} = \frac{210.9}{72.9} = 2.89$$

$$p_{r,N_2} = \frac{210.9}{33.5} = 6.30$$

Consulting the generalized charts again,

$$Z_{CO_2} = 0.43 \qquad Z_{N_2} = 1.075$$

and

$$Z_m = \Sigma x_i Z_i = (0.5)(0.43) + (0.5)(1.075) = 0.753$$

Then

$$p_m = \frac{Z_m n_m \mathcal{R} T_m}{V_m} = \frac{(0.753)(1.0)(1545)(560)}{(1.5)(144)} = 3014 \text{ psia}$$

Further iteration would give agreement between the assumed and calculated values at

$$p_m = 2970 \text{ psia} = 202.1 \text{ atm}$$

This value is also in substantial disagreement with the ideal-gas calculation.

(*d*) Kay's rule requires that we calculate pseudo-critical properties for the mixture using Eqs. (7-80) and (7-81):

$$p_{cm} = \Sigma x_i p_{ci} = (0.5)(33.5) + (0.5)(72.9) = 53.2 \text{ atm}$$

$$T_{cm} = \Sigma x_i T_{ci} = (0.5)(227) + (0.5)(548) = 387.5°\text{R}$$

The mixture-reduced properties for use with the compressibility charts are then calculated as

$$T_{rm} = \frac{T_m}{T_{cm}} = \frac{560}{387.5} = 1.445$$

$$\bar{v}'_{rm} = \frac{\bar{v}}{\mathcal{R} T_{cm}/p_{cm}} = \frac{1.5}{(1545)(387.5)/(53.2)(14.7)(144)}$$

$$= 0.282$$

Consulting Fig. 6-4, we obtain

$$Z_m = 0.775$$

so that

$$p_m = \frac{Z_m n_m \mathcal{R} T_m}{V_m} = \frac{(0.775)(1.0)(1545)(560)}{(1.5)(144)}$$
$$= 3104 \text{ psia} = 211.12 \text{ atm}$$

The true value may only be obtained from experimental data, but we can compare the answers obtained by the different methods, taking the result obtained from Kay's rule as a reference point. The deviations from this value are then

Ideal-gas law: 29 percent high

Additive-pressure rule: 1.9 percent high

Additive-volume rule: 4.3 percent low

In this particular problem the additive rules and Kay's rule are in fairly close agreement.

EXAMPLE 7-18M Comparison of Calculation Techniques
A mixture of 1.0 kg mol and equal molal parts of CO_2 and N_2 is contained at 40°C in a vessel having a volume of 85 liters. Calculate the total pressure of the mixture using (*a*) the ideal-gas law, (*b*) the rule of additive pressures and the compressibility charts, (*c*) the rule of additive volumes and the generalized compressibility charts, and (*d*) Kay's pseudo-critical method.

SOLUTION We have 1 mol of total mixture so

$$x_{N_2} = 0.5 \qquad \text{and} \qquad x_{CO_2} = 0.5$$

Also, $T_m = 40°\text{C} = 313$ K, $V_m = 0.085$ m^3, and $n_m = 1.0$ kg mol.

(*a*) For ideal-gas behavior the total pressure is calculated immediately from

$$p_m = \frac{n_m \mathcal{R} T_m}{V_m} = \frac{(1.0)(8314.41)(313)}{0.085} = 30.62 \text{ MPa}$$

(*b*) The critical properties for CO_2 and N_2 are obtained from Table 6-2 as

$$T_{c,N_2} = 126.2 \text{ K} \qquad T_{c,CO_2} = 304.20 \text{ K}$$

$$p_{c,N_2} = 3.398 \text{ MPa} \qquad p_{c,CO_2} = 7.386 \text{ MPa}$$

For the rule of additive pressures we must evaluate the pressure of each constituent at the temperature and volume of the total mixture. Because we have 0.5 mol of each constituent, the molal specific volume of each is

$$v = \frac{V_m}{n} = \frac{0.085}{0.5} = 0.170 \text{ m}^3/\text{kg mol}$$

The pseudo-reduced volume for each constituent is thus

$$\bar{v}'_r = \frac{\bar{v}}{\mathcal{R}T_c/P_c}$$

$$\bar{v}'_{r,CO_2} = \frac{0.170}{(8314.41)(304.20)/(7.386 \times 10^6)} = 0.496$$

$$\bar{v}'_{r,N_2} = \frac{0.170}{(8314.41)(126.2)/(3.398 \times 10^6)} = 0.551$$

and the reduced temperatures are

$$T_{r,CO_2} = \frac{313}{304.2} = 1.029$$

$$T_{r,N_2} = \frac{313}{126.2} = 2.480$$

We may consult the generalized compressibility charts of Chap. 6 to obtain the reduced pressures as

$$p_{r,CO_2} = 1.05 \qquad p_{r,N_2} = 4.78$$

The pressures to be used in the additive-pressure rule are therefore

$$p_{CO_2} = (1.05)(7.386) = 7.755 \text{ MPa}$$

$$p_{N_2} = (4.78)(3.398) = 16.24 \text{ MPa}$$

The total pressure of the mixture is then

$$p_m = p_{N_2} + p_{CO_2} = 7.755 + 16.24 = 24.00 \text{ MPa}$$

This value is substantially lower than that predicted by the ideal-gas law.

(*c*) To employ the additive-volume rule an iterative procedure is required because we do not know the total mixture pressure for evaluation of the volumes. The easiest way to proceed is to assume a value for the total pressure, calculate the Z_i's for insertion in Eq. (7-79), compute Z_m, and then calculate the total pressure from this. The procedure continues until agreement is reached between the assumed and calculated values. For these calculations the values of T_r are the same as in part (*b*). As a first trial we might take

$$p_m = 30.0 \text{ MPa}$$

Then

$$p_{r,CO_2} = \frac{30}{7.386} = 4.06$$

$$p_{r,N_2} = \frac{30}{3.398} = 8.83$$

Consulting the generalized charts,

$$Z_{CO_2} = 0.57 \qquad Z_{N_2} = 1.15$$

Then, from Eq. (7-78),

$$Z_m = \Sigma x_i Z_i = (0.5)(0.57) + (0.5)(1.15) = 0.860$$

and the mixture pressure is calculated from

$$p_m = \frac{Z_m n_m \mathcal{R} T_m}{V_m} = \frac{(0.860)(1.0)(8314.41)(313)}{0.085} = 26.33 \text{ MPa}$$

This differs substantially from our 30.0 MPa assumption, and so we try

$$p_m = 25.0 \text{ MPa}$$

$$p_{r,CO_2} = \frac{25}{7.386} = 3.38$$

$$p_{r,N_2} = \frac{25}{3.398} = 7.36$$

Consulting the generalized charts again,

$$Z_{CO_2} = 0.49 \qquad Z_{N_2} = 1.10$$

and

$$Z_m = \Sigma x_i Z_i = (0.5)(0.49) + (0.5)(1.10) = 0.795$$

Then

$$p_m = \frac{Z_m n_m \mathcal{R} T_m}{V_m} = \frac{(0.795)(1.0)(8314.41)(313)}{0.085} = 24.34 \text{ MPa}$$

Further iteration would give agreement between the assumed and calculated values at

$$p_m = 24.22 \text{ MPa}$$

This value is also in substantial disagreement with the ideal-gas calculation.

(*d*) Kay's rule requires that we calculate pseudo-critical properties for the mixture using Eqs. (7-80) and (7-81):

$$p_{cm} = \Sigma x_i p_{ci} = (0.5)(3.398) + (0.5)(7.386) = 5.392 \text{ MPa}$$

$$T_{cm} = \Sigma x_i T_{ci} = (0.5)(126.2) + (0.5)(304.2) = 215.2 \text{ K}$$

The mixture-reduced properties for use with the compressibility charts are then calculated as

$$T_{rm} = \frac{T_m}{T_{cm}} = \frac{313}{215.2} = 1.454$$

$$\bar{v}'_{rm} = \frac{\bar{v}}{\mathcal{R}T_{cm}/p_{cm}} = \frac{0.085}{(8314.41)(215.2)/(5.392 \times 10^6)} = 0.256$$

Consulting Fig. 6-4, we obtain

$$Z_m = 0.78$$

so that

$$p_m = \frac{Z_m n_m \mathcal{R} T_m}{V_m} = \frac{(0.78)(1.0)(8314.41)(313)}{0.085} = 23.88 \text{ MPa}$$

The true value can only be obtained from experimental data, but we can compare the answers obtained by the different methods, taking the result obtained from Kay's rule as a reference point. The deviations from this value are then

Ideal-gas law: 28 percent high

Additive-pressure rule: 0.5 percent high

Additive-volume rule: 1.4 percent high

In this particular problem the additive rules and Kay's rule are in fairly close agreement.

REVIEW QUESTIONS

1 State the Gibbs-Dalton law. Why is it applicable to ideal gases?

2 What is the Amagat-Leduc law?

3 It is sometimes stated that the composition of a gaseous mixture on a volume basis is the same as a mole basis. Explain.
4 How is the term *partial pressure* defined?
5 Show that the entropy change of a mixture in a process at constant total pressure is a function of only the initial and final temperatures and the mixture specific heat.
6 Define the terms *dew point, dry- and wet-bulb temperature, relative humidity, specific humidity,* and *degree of saturation.*
7 Why is it possible to assume that water vapor behaves as an ideal gas for purposes of making psychrometric mixture calculations?
8 Why cannot the psychrometric charts in the Appendix be used for air–water-vapor mixtures at 25 psia?
9 What is meant by the term *evaporative cooling*?
10 What is a *grain*?
11 How is wet-bulb temperature measured?
12 What is a cooling tower and how does it work?
13 What is the rule of additive pressures? When does it apply?
14 What is the rule of additive volumes? When does it apply?
15 What is Kay's rule for real-gas mixtures? When does it apply?
16 How is specific humidity related to dew point?
17 How is mixture enthalpy related to wet-bulb temperature for an air–water-vapor mixture?
18 How do you know when to expect that moisture will be condensed from an air–water-vapor stream in an air-conditioning process?
19 What is meant by the term *mole fraction*?
20 What is an ideal gas?
21 What is a *saturated mixture* in air–water-vapor problems?

PROBLEMS (ENGLISH UNITS)

7-1 A certain mixture has the following composition on a mass basis:

N_2 = 50 percent
He = 10 percent
CO = 12 percent
O_2 = 15 percent
CO_2 = 13 percent

Calculate the specific heats for the mixture and the gas constant. Also calculate the mole fraction of each constituent.

7-2 The mixture of Prob. 7-1 is compressed from 15 psia, 70°F, to 50 psia, 200°F, in a reversible nonflow process. Calculate the work per pound-mass of mixture and the heat transfer.

7-3 A mixture of 55 percent CO_2, 15 percent O_2, and 30 percent methane (CH_4) by volume exists at a total pressure of 3 atm and 110°F. Calculate the mass fraction of each constituent.

7-4 The following gaseous mixture is on a volume basis:

CO_2 = 25 percent
He = 15 percent
CH_4 (methane) = 35 percent
A (argon) = 25 percent

Calculate the value of c_p for the mixture in units of Btu/lbm · °F. Also calculate the gas constant for the mixture.

7-5 A mixture of carbon dioxide at 25 percent, helium at 30 percent, and oxygen at 45 percent, with a total flow rate of 2.3 lbm/s is cooled from 100°F and 30 psia to 50°F in a steady-flow process. Calculate the cooling required and the gas constant of the mixture. Proportions are on mol basis.

7-6 The mixture of Prob. 7-1 is compressed isentropically from 15 psia, 70°F, to 50 psia. Calculate the work done and the change in entropy for each constituent.

7-7 A certain mixture has the following composition on a *volumetric* basis:

CO = 10 percent
CO_2 = 25 percent
O_2 = 10 percent
N_2 = 55 percent

Calculate the mass fractions of each constituent, the gas constant, apparent molecular weight, and specific heats.

7-8 1 lbm of N_2 at 100°F and 4 psia is mixed with 2 lbm of H_2 at 200°F and 5 psia in a steady-flow process. The mixture leaves at 3 psia and the process is adiabatic. Calculate the change in entropy.

7-9 Complete the following table for an air–water-vapor mixture at atmospheric pressure. Use both analytical calculations and the psychrometric chart.

T_{DB}, °F	T_{WB}, °F	ϕ, percent	Dew Point, °F	p_v
100		30		
80	60			
		50	55	
	80		60	
90			60	
	70	40		

7-10 Repeat Prob. 7-9 for an air–water-vapor mixture at a total pressure of 30 psia.

7-11 The partial pressure of water vapor in a certain psychrometric mixture is 0.3 psia and the total pressure of the mixture is 14.696 psia. Calculate the relative humidity when the dry-bulb temperature is 80, 90, and 100°F.

7-12 A rigid tank contains 1 lbm of H_2 at 200 psia and 50°F. A tap is connected to a high-pressure nitrogen line at 150°F, and a sufficient quantity is drawn off until the total pressure in the tank reaches 400 psia. The process is adiabatic. Calculate the final temperature in the tank, the mass of nitrogen added, and the change in entropy.

7-13 Repeat Prob. 7-12 for a *slow* filling process, where the temperature of the gas in the tank remains constant at 50°F.

7-14 Two 1.0-ft³ tanks are connected by a pipe and valve. One tank is filled with N_2 at 15 psia and 100°F, and the other tank is filled with argon at 100°F. The valve is opened and the two gases are allowed to mix adiabatically. Calculate the change in entropy when the initial argon pressure is (*a*) 15 psia, (*b*) 30 psia.

7-15 Air at the rate of 15 000 ft³/min, at 1 atm, 90°F, and 70 percent relative humidity, is cooled to 25°F in a steady-flow process. Calculate the cooling required.

7-16 Air at the rate of 1300 ft³/min, at 1 atm, 40°F, and 30 percent relative humidity, is heated in a steady-flow process to a temperature of 90°F. During the heating process a sufficient quantity of water at 60°F is added such that the relative humidity at the 90°F exit condition will be 40 percent. Calculate the heat added.

7-17 Atmospheric air at the rate of 1000 ft³/min, at 40°F and 30 percent relative humidity, flows into a heater-humidifier where water at 70°F is sprayed into the mixture and sufficient heat is added to produce exit conditions of 90°F and 60 percent relative humidity. Calculate the change in entropy of the flow, defined as exit entropy minus inlet entropy. If you need to make assumptions (realistic ones, please), be sure to state them clearly.

7-18 A certain home has a heat load of 60 000 Btu/h in the summer. Room air conditions are at 75°F and 60 percent relative humidity and the air is cooled to 50°F as it passes over the cooling coils. Assume that the air picks up the heat load while in the house. Calculate the volume flow of air which must be delivered to the room. Also calculate the amount of water removed.

7-19 In infant intensive-care wards a *supersaturated* air condition is frequently maintained. Suppose air is available from the regular heating system at 80°F and 40 percent relative humidity. The air flow rate for the ward is 1200 ft³/min. How much water must be added to the air per hour, in fine droplet form, to produce an effective humidity of 105 percent at 75°F? How much additional heat must be supplied?

7-20 A certain precision-gage room must be maintained at low-humidity conditions to ensure proper accuracy of the instruments. A cooling-reheat air-conditioning process is to be employed. The incoming air is at 90°F and 50 percent relative humidity and is to be delivered to the room at 65°F and 20 percent humidity. Air at the rate of 1000 ft³/min is required. Calculate the total cooling and heating requirements.

7-21 Air at the rate of 1400 ft^3/min, at atmospheric pressure and 30°F and with 60 percent relative humidity, is to be heated to 90°F. Just enough steam at 20 psia and 15 percent moisture is added to make the final relative humidity of the air 80 percent. Calculate the heating required.

7-22 Air at the rate of 1350 ft^3/min, at 90°F and 45 percent relative humidity, is mixed with 400 ft^3/min of air at 40°F and 70 percent relative humidity in a steady-flow adiabatic process at atmospheric pressure. Determine the outlet temperature and humidity.

7-23 In a certain northern city a heating system is designed to heat outside air at −10°F and 100 percent relative humidity to 80°F. Calculate the heat which must be supplied under these circumstances per pound-mass of air. What is the final relative humidity of the mixture?

7-24 The heating system in Prob. 7-23 is to be modified so that the final humidity of the mixture will be 50 percent. For this purpose a quantity of steam at 20 psia, 80 percent quality, is mixed with the air during the heating process. How much steam must be supplied per pound-mass of air and what are the heating requirements under these new conditions?

7-25 The process in Prob. 7-24 is to be accomplished by spraying the heated air with liquid water at 180°F. What quantity of water will be required per pound mass of air, and what are the total heating requirements?

7-26 A small cooling system is designed solely for dehumidification purposes. The system is designed to remove 1 gal of moisture per hour from air at 80°F and 80 percent relative humidity. What air flow must the system handle, and what must its cooling capacity be? (*Note:* There are several correct answers; specify two suitable choices.)

7-27 An evaporative cooler is used to deliver 5000 ft^3/min of air, operating in a desert climate where T_{DB} = 110°F, ϕ = 10 percent. Calculate the quantity of water which must be supplied to the unit if the air leaves the unit at 90 percent relative humidity.

7-28 An air–water-vapor mixture at a total pressure of 25 psia is at 30°F and 90 percent relative humidity. For a volume flow rate of 1300 ft^3/min, calculate the heating required to raise the temperature to 100°F.

7-29 In Denver, Colorado, the standard atmospheric pressure is only 85 percent of that at sea level. For an air–water-vapor mixture at 95°F and 40 percent relative humidity, calculate the percent change in the mixture enthalpy from the value it would have at sea level. Also calculate the dew point of the mixture under Denver conditions.

7-30 Air at the rate of 2000 ft^3/min, at 1 atm, 30°F, and 80 percent relative humidity, is sprayed with hot water at 200°F, and heat is supplied so that the final conditions are 100°F and 60 percent relative humidity. Calculate the heating required.

7-31 An office complex requires an air-conditioning system with a capacity of 100 tons (1 ton = 12 000 Btu/h). The interior space is maintained at 78°F and 50 percent relative humidity. This air is cooled to 52°F as it passes over the cooling coils. After leaving the cooling coils, the air picks up the heat load in the office. Calculate the water flow rate removed from the air and the volume flow in ft^3/min which must be delivered to the room.

7-32 An air–water-vapor mixture is compressed isentropically from 100°F, 14.7 psia, and 50 percent relative humidity to 100 psia. Calculate the temperature at the end of compression. If the mixture is then cooled at constant pressure, calculate the amount of moisture removed when the high-pressure gas is cooled to (*a*) 80°F, (*b*) 100°F.

7-33 A cold-water pipe at 55°F passes through a room with atmospheric air at 80°F. What is the maximum relative humidity the air may have such that condensation does not occur on the pipe?

7-34 Meats and other food products will dry out when exposed directly to the very-low-temperature air in cold-storage vaults. Explain this phenomenon on the basis of information contained in this chapter. For calculation purposes, assume the air temperature is at 0°F.

7-35 Air at the rate of 1000 ft^3/min, at 110°F and 15 percent relative humidity, is mixed adiabatically with 500 ft^3/min of air at 65°F and 90 percent humidity. Calculate the final temperature and humidity.

7-36 The output air from the evaporative cooler of Prob. 7-27 is mixed with 1000 ft^3/min of the outside air. What is the resultant temperature and humidity?

7-37 Air at the rate of 1000 ft^3/min, at 100°F and 80 percent relative humidity, is mixed with an air flow at 55°F. What flow rate of the cooler air and humidity is necessary just to produce a mixture at 100 percent relative humidity? What is the final mixture temperature?

7-38 4 lbm of air at 25 psia, 40°F, and 80 percent relative humidity are mixed with 8 lbm of air at 25 psia, 100°F, and 20 percent relative humidity in a steady-flow adiabatic process. Calculate the final mixture temperature and humidity for a final pressure of (*a*) 25 psia, (*b*) 20 psia.

7-39 An inexpensive humidifier for home heating systems operates by atomizing liquid water and spraying it directly into the air duct system. One particular device is to be designed to handle the following air flows:

Input to heater-humidifier section: 300 ft^3/min at 20°F, $\phi = 80\%$
plus 1200 ft^3/min at 60°F, $\phi = 80\%$
plus liquid water at 60°F

Output from heater-humidifier section: $T_{DB} = 85°F$, $\phi = 35$ percent

Calculate the amount of heating required and the amount of liquid water which must be supplied to the humidifier.

7-40 A small building in a desert climate requires comfort cooling of 120 000 Btu/h, and an evaporative cooler is to be designed to perform the task. Outside air is available at 110°F and $\phi = 5$ percent. What inlet air flow rate is required, and how much water must be supplied to the cooler?

7-41 In Long Island, New York, it is quite muggy in the summer. On one summer day the outside temperature is 92°F and the relative humidity is

50 percent. Dryer air at 80°F and 20 percent relative humidity is to be produced with a cooling-and-reheat operation. If 5500 ft^3/min is needed at the dry conditions, calculate the cooling and heating requirements.

7-42 A humidifier is designed to take in atmospheric air at 20°F and 50 percent relative humidity and spray it with water at 55°F. Enough heat is also added so that the outlet air stream from the humidifier is at 80°F and 40 percent humidity. The outlet flow rate is 1700 ft^3/min. Calculate the heating required and the quantity of water required per hour.

7-43 A high degree of moisture removal is to be accomplished by cooling 2500 ft^3/min of air at 80°F and 80 percent relative humidity to a low temperature of −10°F. Calculate the cooling required and the amount of moisture removed per hour.

7-44 One way to produce very dry air is to cool it to very low temperatures and then reheat to the desired operating temperature. Suppose a mixture is available at 1 atm, 75°F, and 80 percent relative humidity. What final relative humidity would result if the mixture is cooled to −30°F and then reheated to 75°F? How much cooling and heating would be required for a flow rate of initial mixture of 1700 ft^3/min?

7-45 The "chiller" section in Prob. 7-44 must be defrosted when 50 lbm of frost or ice has accumulated. Assuming that the water removed from the air at temperatures above 35°F is able to run off freely, calculate the time required to accumulate 50 lbm of ice. Repeat the calculation for no water runoff.

7-46 A special vaporizer device for use with small children in cold, dry weather is designed to mix one part of a supersaturated mixture at 60°F and 110 percent effective relative humidity with two parts of room air at 75°F and 15 percent relative humidity. What are the humidity and temperature of the outlet mixed stream if the extra 10 percent humidity in the supersaturated stream is in the form of small liquid droplets? Assume the mixture parts are on a dry-air basis.

7-47 An air–water-vapor mixture at a total pressure of 23.5 psia has a relative humidity of 80 percent and a dry-bulb temperature of 100°F. Calculate the amount of cooling required to cool 1100 ft^3/min of this mixture to 50°F.

7-48 A certain home has 3500 ft^2 of floor space with an 8-ft ceiling height. How much total water vapor is contained in the house at 75°F and 50 percent relative humidity?

7-49 Calculate the mass specific heats for a mixture of 1 kg of helium and 1 kg of nitrogen. If such a mixture is heated and compressed from 1 atm, 20°C, to 5 atm, 150°C, in a steady-flow device, calculate the heat added, the shaft work input, and change in entropy of the mixture. Repeat for heating and compressing in a closed system.

7-50 Repeat Prob. 7-49 for the same quantities of helium and nitrogen but with enough water vapor added so that the initial relative humidity is 20 percent.

7-51 A large cooling tower for a power plant is designed on the evaporation cooling concept. In one installation the heat which must be absorbed in the cooling tower is 1000 MW, and this is to be accomplished by spraying air at 90°F, $\phi = 30$ percent, with water at 80°F. The cooling-tower design assumes that the air leaves at 70°F saturated. What inlet air and water flows are needed to accomplish the required cooling rate?

7-52 Calculate the mixture enthalpies for air–water-vapor mixtures at 1 atm and (*a*) $T_{DB} = 100°F$, $T_{WB} = 60°F$, (*b*) $T_{DB} = 70°F$, $T_{WB} = 60°F$. What percent error would result from assuming that the mixture enthalpy is a function only of wet-bulb temperature?

7-53 How much cooling is required to cool air at 90°F and 60 percent relative humidity to −30°F in a constant-pressure process at (*a*) 1 atm and (*b*) 2 atm? How much moisture will be removed in these processes?

7-54 Air at the rate of 1500 ft^3/min, at 1 atm, 40°F, and 80 percent relative humidity, is heated to a temperature of 100°F while enough water at 60°F is added to cause the final relative humidity to be 50 percent. Calculate the quantities of heat and water which must be added.

7-55 In a munitions plant very dry air is produced by cooling 3500 ft^3/min of outside air at 85°F and 70 percent relative humidity to a temperature of −5°F and then reheating. Calculate the amount of cooling required in tons. Do not neglect any terms.

7-56 An evaporative cooler is to be used on hot days in Dallas, Texas. If the outside air is at 110°F and 20 percent relative humidity, what is the lowest temperature that can be attained? If the outlet volume flow from such a cooler is 10 000 ft^3/min, what quantity of water would be required for minimum temperature at outlet?

7-57 Outside air at the rate of 2500 ft^3/min and at 40°F and 100 percent relative humidity is mixed with 5000 ft^3/min of inside air at 60°F and 50 percent relative humidity. During the mixing process sufficient heat is added to produce an outlet air temperature of 80°F. Calculate the heat added and the outlet relative humidity.

7-58 A mixture of one part nitrogen and two parts carbon dioxide by mass contains enough water vapor to produce 50 percent relative humidity at 100°F and 1 atm total pressure. How much cooling is required to cool 1 lbm of mixture to 40°F at constant pressure? How much water is removed in the process?

7-59 A mixture containing 1 lbm oxygen and 1 lbm helium has enough water vapor to produce 10 percent relative humidity at 105°F. The mixture is to be cooled in an adiabatic saturation process. What is the lowest temperature which can be achieved in this manner, and how much water must be supplied per lbm of initial mixture? What is the mass specific heat of the mixture before and after the saturation process?

7-60 A mixture of methane (CH_4) and air is to be produced which has mass proportions of 1 part methane to 15 parts dry air. The mixing is accomplished in a steady-flow process with both methane and air entering at 80°F. The inlet air has $\phi = 70$ percent and the gases are heated during the mixing process to produce an outlet temperature of 200°F. The process occurs at constant pressure of 1 atm. How much heating must be supplied for an inlet volume flow of methane of 100 ft^3/h, and what is the relative humidity of the outlet mixture?

7-61 A tank having a volume of 20 ft^3 contains an air–water-vapor mixture at 14.7 psia, 80°F, and $\phi = 50$ percent. A high-pressure line with dry air at 120°F and 70 psia is used to fill the tank with additional air. The filling process occurs slowly so that the temperature inside the tank remains constant at 80°F. How much air must be added to raise the tank pressure to 50 psia, what will be the relative humidity under the new conditions, and how much heat will have been lost by the tank in the process?

7-62 An air–water-vapor mixture has 25 percent water vapor on a molal basis and is contained at 500 psia and 800°F. Calculate the specific volume of the mixture using (*a*) the ideal-gas law, (*b*) the rule of additive pressures, (*c*) the rule of additive volumes, and (*d*) Kay's pseudo-critical rule.

7-63 How much heat per pound-mass of mixture would be required to heat the mixture of Prob. 7-62 to 1000°F in a constant-pressure, steady-flow process?

7-64 A mixture containing 30 percent CO_2, 10 percent CO, 40 percent N_2, and 20 percent CH_4 on a molal basis is contained at 4000 psia and 200°F in a 5-ft^3 rigid vessel. Calculate the heat which must be removed to lower the temperature to 100°F using (*a*) the ideal-gas law and (*b*) Kay's rule.

7-65 Oxygen is contained in a rigid 1-ft^3 vessel at 50 atm and 100°F. Nitrogen from a high-pressure line at 4000 psia and 100°F is bled slowly into the tank until the mixture proportions in the tank approximate those for air, i.e., 78 percent N_2 and 22 percent O_2. During the filling process cooling is supplied to the tank to maintain the temperature at 100°F. Calculate the heat removed from the tank and the mass of nitrogen added. Use Kay's rule.

7-66 A mixture of equal molal parts of methane (CH_4) and ethane (C_2H_6) is compressed reversibly and isothermally from 150°F, 100 psia, to 700 psia in a steady-flow process. Calculate the work required per lbm of mixture.

7-67 A mixture of 40 percent argon and 60 percent nitrogen on a molal basis is to be heated from −100°F and 700 psia to 200°F in a steady-flow heat exchanger device. Because of fluid friction the pressure at exit of the device drops to 650 psia. Flow velocities are low enough to neglect kinetic energies. Calculate the heating which must be supplied per lbm of mixture.

7-68 A mixture of equal molal parts of air and water vapor at 14.7 psia and 300°F is heated to 500°F in a constant pressure process. How much heat is required per lbm of mixture?

7-69 Estimate the temperature at which the vapor in the mixture of Prob. 7-68 will start to condense if the mixture is cooled at constant pressure.

7-70 A mixture of air and water vapor is saturated at 200 psia and 300°F.

Calculate the cooling required to cool the mixture to 150°F in a steady-flow process at constant pressure.

7-71 Assuming that air may be considered as a mixture of 78 percent nitrogen and 22 percent oxygen on a molal basis, calculate the critical constants using Kay's rule and compare with tabulated values.

7-72 1 lbm · mol of a mixture of 50 percent CO_2 and 50 percent CH_4 on a molal basis is contained in a rigid vessel at 300°F and 500 psia. Calculate the volume of the mixture using (*a*) the ideal-gas law, (*b*) the rule of additive volumes, and (*c*) the Kay pseudo-critical method.

7-73 Calculate the energy required to heat the mixture of Prob. 7-72 to 500°F in a constant-pressure steady-flow process using (*a*) the ideal-gas law and (*b*) the pseudo-critical method.

7-74 A mixture of 60 percent methane (CH_4) and 40 percent ethane (C_2H_6) on a molal basis is to be compressed from 150°F, 100 psia, to 350 psia in a steady-flow device. The mass flow rate is 10 lbm/min. Assuming that the process is reversible and adiabatic, calculate the work input using (*a*) ideal-gas relations and (*b*) the pseudo-critical method.

PROBLEMS (METRIC UNITS)

7-1M The mixture of Prob. 7-1 is compressed from 100 kPa, 295 K, to 350 kPa, 370 K, in a reversible nonflow process. Calculate the work per kilogram of mixture and the heat transfer.

7-2M The mixture of Prob. 7-1 is compressed isentropically from 100 kPa, 295 K, to 350 kPa. Calculate the work done and the change in entropy for each constituent.

7-3M 1 kg of N_2 at 310 K and 28 kPa is mixed with 2 kg of H_2 at 370 K and 35 kPa in a steady-flow process. The mixture leaves at 20 kPa and the process is adiabatic. Calculate the change in entropy.

7-4M A rigid tank contains 1 kg of H_2 at 1.4 MPa and 280 K. A tap is connected to a high-pressure nitrogen line at 340 K, and a sufficient quantity is drawn off until the total pressure in the tank reaches 2.8 MPa. The process is adiabatic. Calculate the final temperature in the tank, the mass of nitrogen added, and the change in entropy.

7-5M Repeat Prob. 7-4M for a *slow* filling process, where the temperature of the gas in the tank remains constant at 280 K.

7-6M Two 28-liter tanks are connected by a pipe and valve. One tank is filled with N_2 at 100 kPa and 310 K, and the other tank is filled with argon at 310 K. The valve is opened and the two gases are allowed to mix adiabatically. Calculate the change in entropy when the initial argon pressure is (*a*) 100 kPa and (*b*) 200 kPa.

7-7M Complete the following table for an air–water-vapor mixture at atmospheric pressure. Use both analytical calculations and the psychrometric chart.

T_{DB}, °C	T_{WB}, °C	ϕ, percent	Dew Point, °C	p_v
38	30			
27	15			
		50	13	
	27		15	
32			15	
	20	40		

7-8M Repeat Prob. 7-7M for an air–water-vapor mixture at a total pressure of 200 kPa.

7-9M A mixture has the following analysis on a volume basis: argon, 20 percent; nitrogen, 25 percent, hydrogen, 40 percent; and helium, 15 percent. The total pressure is 1 atm. Calculate the value of c_p for the mixture.

7-10M Argon at the rate of 1.5 kg/sec, at 120 kPa and 10°C, is mixed with 1.0 kg/s of air at 120 kPa and 60°C in an adiabatic steady-flow process. The outlet pressure is 90 kPa. Calculate the change in entropy.

7-11M 1 kg/s of air at 20°C and 1 atm is mixed with 1.5 kg/s of helium at 1 atm and 50°C in an adiabatic process. The outlet pressure of the mixture is 90 kPa. Calculate the change in entropy.

7-12M A mixture contains 10 percent helium, 30 percent CO_2, 15 percent argon, and 45 percent nitrogen on a mass basis at conditions of 85 kPa and 30°C. Calculate the gas constant of the mixture.

7-13M Air at the rate of 1.2 kg/s, at 1 atm and 40°C, is mixed adiabatically with 0.85 kg/s of helium at 1 atm and 10°C in a steady-flow device such that the discharge pressure of the mixed gases is 0.9 atm. Calculate the overall change in entropy for this process.

7-14M In a certain northern city a heating system is designed to heat outside air at −25°C and 100 percent relative humidity to 30°C. Calculate the heat which must be supplied under these circumstances per kilogram of air. What is the final relative humidity of the mixture?

7-15M The heating system in Prob. 7-14M is to be modified so that the final humidity of the mixture will be 50 percent. For this purpose a quantity of steam at 140 kPa, 80 percent quality, is mixed with the air during the heating process. How much steam must be supplied per kilogram of air, and what are the heating requirements under these new conditions?

7-16M The process in Prob. 7-15M is to be accomplished by spraying the heated air with liquid water at 80°C. What quantity of water will be required per kilogram of air, and what are the total heating requirements?

7-17M A small cooling system is designed solely for dehumidification purposes. The system is designed to remove 4 liters of moisture per hour from air at 27°C and 80 percent relative humidity. What air flow must the system handle, and what must its cooling capacity be? (*Note:* There are several correct answers; specify two suitable choices.)

7-18M An insulated box has two equal-size chambers having volumes of 0.2 m^3, separated by an insulated wall. One chamber contains nitrogen at 100°C and 200 kPa, and the other chamber contains helium at 100 kPa and 100°C. The separating wall is removed and the two gases are allowed to mix together until they reach equilibrium. Calculate the change in entropy resulting from the mixing process.

7-19M 1 kg/s of N_2 at 1 atm and 400 K is mixed with 3 kg/s of argon at 1 atm and 200 K in an adiabatic process. The final mixture pressure is also 1 atm. Calculate the change in entropy.

7-20M A mixture of 32 percent helium, 40 percent nitrogen, and 38 percent hydrogen, on a mol basis, is contained in a rigid box having a volume of 5.6 m^3 at conditions of 500 kPa and 100°C. Calculate the change in entropy if the mixture is cooled to 20°C.

7-21M A mixture of 20 percent CO_2, 30 percent argon, 25 percent helium, and 25 percent hydrogen by volume is heated at constant pressure from 100°C to 150°C. Calculate the heat added per unit mass. If this mixture is contained in a tank at 300 kPa and 140°C, calculate the mass in the tank if the tank volume is 5.6 m^3.

7-22M A mixture of 25 percent CO_2, 25 percent argon, 30 percent helium, and 20 percent hydrogen by volume is heated at constant pressure from 75°C to 150°C. Calculate the heat added per unit mass. If this mixture is contained in a tank at 200 kPa and 130°C, calculate the mass in the tank for a tank volume of 6.5 m.

7-23M A mixture of 25 percent nitrogen, 40 percent carbon dioxide, and 35 percent hydrogen on a mol basis is contained in a rigid box having a volume of 10 m^3 at conditions of 400 kPa and 120°C. Calculate the amount of cooling required to lower the temperature to 40°C. Also calculate the change in entropy.

7-24M The partial pressure of water vapor in a certain psychrometric mixture is 2 kPa, and the total pressure of the mixture is 100 kPa. Calculate the relative humidity when the dry-bulb temperature is 30, 35, and 40°C.

7-25M An air–water-vapor mixture is compressed isentropically from 40°C, 1 atm, and 50 percent relative humidity to 700 kPa. Calculate the temperature at the end of the compression. If the mixture is then cooled at constant pressure, calculate the amount of moisture removed when the high-pressure gas is cooled to (*a*) 30°C and (*b*) 40°C.

7-26M A cold-water pipe at 13°C passes through a room with atmospheric air at 27°C. What is the maximum relative humidity the air may have such that condensation does not occur on the pipe?

7-27M Meats and other food products will dry out when exposed directly to the very-low-temperature air in cold-storage vaults. Explain this phenomenon on the basis of information contained in this chapter. For calculation purposes, assume the air temperature is at −18°C.

7-28M An evaporative cooler is used to deliver 140 m³/min of air, operating in a desert climate where T_{DB} = 43°C, ϕ = 10 percent. Calculate the quantity of water which must be supplied to the unit if the air leaves the unit at 90 percent relative humidity.

7-29M In infant intensive-care wards a *supersaturated* air condition is frequently maintained. Suppose air is available from the regular heating system at 27°C and 40 percent relative humidity. The air flow rate for the ward is 34 m³/min. How much water must be added to the air per hour, in fine droplet form, to produce an effective humidity of 105 percent at 24°C? How much additional heat must be supplied?

7-30M A certain precision-gage room must be maintained at low-humidity conditions to ensure proper accuracy of the instruments. A cooling-reheat air-conditioning process is to be employed. The incoming air is at 32°C and 50 percent relative humidity and is to be delivered to the room at 18°C and 20 percent humidity. Air at the rate of 28 m³/min is required. Calculate the total cooling and heating requirements.

7-31M Air at the rate of 28 m³/min, at 43°C and 15 percent relative humidity, is mixed adiabatically with 14 m³/min of air at 18°C and 90 percent humidity. Calculate the final temperature and humidity.

7-32M The output air from the evaporative cooler of Prob. 7-28M is mixed with 28 m³/min of the outside air. What is the resultant temperature and humidity?

7-33M 30 m³/min of air at 33°C and 90 percent relative humidity are mixed with an air flow at 15°C. What flow rate of the cooler air and humidity is necessary just to produce a mixture at 100 percent relative humidity? What is the final mixture temperature?

7-34M An inexpensive humidifier for home heating systems operates by atomizing liquid water and spraying it directly into the air duct system. One particular device is to be designed to handle the following air flows:

Input to heater-humidifier section: 8.5 m³/min at −7°C, ϕ = 80%
plus 34 m³/min at 15°C, ϕ = 80%
plus liquid water at 15°C

Output from heater-humidifier section: T_{DB} = 30°C, ϕ = 35%

Calculate the amount of heating required and the amount of liquid water which must be supplied to the humidifier.

7-35M A small building in a desert climate requires comfort cooling of 35 kW and an evaporative cooler is to be designed to perform the task. Outside air is available at 45°C and ϕ = 5 percent. What inlet air flow rate is required, and how much water must be supplied to the cooler?

7-36M A certain home has 325 m² of floor space with a 2.5-m ceiling height.

How much total water vapor is contained in the house at 24°C and 50 percent relative humidity?

7-37M Calculate the mass specific heats for a mixture of 1 kg of helium and 1 kg of nitrogen. If such a mixture is heated and compressed from 1 atm, 20°C, to 5 atm, 150°C, in a steady-flow device, calculate the heat added, the shaft work input, and change in entropy of the mixture. Repeat for heating and compressing in a closed system.

7-38M 2 kg of air at 170 kPa, 5°C, and 80 percent relative humidity are mixed with 4 kg of air at 170 kPa, 40°C, and 20 percent relative humidity in a steady-flow adiabatic process. Calculate the final mixture temperature and humidity for a final pressure of (*a*) 170 kPa and (*b*) 140 kPa.

7-39M One way to produce very dry air is to cool it to very low temperatures and then reheat to the desired operating temperature. Suppose a mixture is available at 1 atm, 25°C, and 80 percent relative humidity. What final relative humidity would result if the mixture is cooled to −35°C and then reheated to 25°C? How much cooling and heating would be required for a flow rate of initial mixture of 48 m^3/min?

7-40M The "chiller" section in Prob. 7-39M must be defrosted when 25 kg of frost or ice has accumulated. Assuming that the water removed from the air at temperatures above 2°C is able to run off freely, calculate the time required to accumulate 25 kg of ice. Repeat the calculation for no water runoff.

7-41M A special vaporizer device for use with small children in cold, dry weather is designed to mix one part of a supersaturated mixture at 15°C and 110 percent effective relative humidity with two parts of room air at 24°C and 15 percent relative humidity. What are the humidity and temperature of the outlet mixed stream if the extra 10 percent humidity in the supersaturated stream is in the form of small liquid droplets? Assume the mixture parts are on a dry-air basis.

7-42M A mixture of methane (CH_4) and air is to be produced which has mass proportions of 1 part methane to 15 parts dry air. The mixing is accomplished in a steady-flow process with both methane and air entering at 27°C. The inlet air has $\phi = 70$ percent, and the gases are heated during the mixing process to produce an outlet temperature of 93°C. The process occurs at constant pressure of 1 atm. How much heating must be supplied for an inlet volume flow of methane of 28 m^3/min, and what is the relative humidity of the outlet mixture?

7-43M A tank having a volume of 570 liters contains an air–water-vapor mixture at 1 atm, 27°C, and $\phi = 50$ percent. A high-pressure line with dry air at 50°C and 480 kPa is used to fill the tank with additional air. The filling process occurs slowly so that the temperature inside the tank remains constant at 27°C. How much air must be added to raise the tank pressure to 350 kPa, what will the relative humidity be under the new conditions, and how much heat will have been lost by the tank in the process?

7-44M A mixture of equal molal parts of air and water vapor at 1 atm and

150°C is heated to 260°C in a constant-pressure process. How much heat is required per kilogram of mixture?

7-45M Estimate the temperature at which the vapor in the mixture of Prob. 7-44M will start to condense if the mixture is cooled at constant pressure.

7-46M A mixture of air and water vapor is saturated at 1.4 MPa and 150°C. Calculate the cooling required to cool the mixture to 65°C in a steady-flow process at constant pressure.

7-47M A mixture of equal molal parts of methane (CH_4) and ethane (C_2H_6) is compressed reversibly and isothermally from 65°C, 700 kPa, to 5 MPa in a steady-flow process. Calculate the work required per kilogram of mixture.

7-48M A large cooling tower for a power plant is designed on the evaporation cooling concept. In one installation the heat which must be absorbed in the cooling tower is 1000 MW, and this is to be accomplished by spraying air at 32°C, $\phi = 30$ percent, with water at 27°C. The cooling-tower design assumes that the air leaves at 20°C saturated. What inlet air and water flows are needed to accomplish the required cooling rate?

7-49M Calculate the mixture enthalpies for air–water-vapor mixtures at 1 atm and (*a*) $T_{DB} = 40$°C, $T_{WB} = 16$°C and (*b*) $T_{DB} = 20$°C, $T_{WB} = 16$°C. What percent error would result from assuming that the mixture enthalpy is a function only of wet-bulb temperature?

7-50M How much cooling is required to cool air at 32°C and 60 percent relative humidity to −35°C in a constant-pressure process at (*a*) 1 atm and (*b*) 2 atm? How much moisture will be removed in these processes?

7-51M A mixture of one part nitrogen and two parts carbon dioxide by mass contains enough water vapor to produce 50 percent relative humidity at 40°C and 1 atm total pressure. How much cooling is required to cool 1 kg of mixture to 5°C at constant pressure? How much water is removed in the process?

7-52M A mixture containing 1 kg of oxygen and 1 kg of helium has enough water vapor to produce 10 percent relative humidity at 40°C. The mixture is to be cooled in an adiabatic saturation process. What is the lowest temperature which can be achieved in this manner, and how much water must be supplied per kilogram of initial mixture? What is the mass specific heat of the mixture before and after the saturation process?

7-53M 1 kg of a mixture of 50 percent CO_2 and 50 percent CH_4 on a molal basis is contained in a rigid vessel at 150°C and 3.5 MPa. Calculate the volume of the mixture using (*a*) the ideal-gas law, (*b*) the rule of additive volumes, and (*c*) the Kay pseudo-critical method.

7-54M Calculate the energy required to heat the mixture of Prob. 7-53M to 260°C in a constant-pressure, steady-flow process using (*a*) the ideal-gas law and (*b*) the pseudo-critical method.

7-55M A mixture of 60 percent methane (CH_4) and 40 percent ethane (C_2H_6) on a molal basis is to be compressed from 65°C, 700 kPa, to 2.4 MPa in a steady-flow device. The mass flow rate is 5 kg/min. Assuming that the process is reversible and adiabatic, calculate the work input using (*a*) ideal-gas relations and (*b*) the pseudo-critical method.

7-56M A mixture containing 30 percent CO_2, 10 percent CO, 40 percent N_2, and 20 percent CH_4 on a molal basis is contained at 28 MPa and 90°C in a 140-liter rigid vessel. Calculate the heat which must be removed to lower the temperature to 40°C using (*a*) the ideal-gas law and (*b*) Kay's rule.

7-57M Oxygen is contained in a rigid 30-liter vessel at 50 atm and 40°C. Nitrogen from a high-pressure line at 28 MPa and 40°C is bled slowly into the tank until the mixture proportions in the tank approximate those for air, i.e., 78 percent N_2 and 22 percent O_2. During the filling process cooling is supplied to the tank to maintain the temperature at 40°C. Calculate the heat removed from the tank and the mass of nitrogen added. Use Kay's rule.

7-58M A mixture of 40 percent argon and 60 percent nitrogen on a molal basis is to be heated from −70°C, 4.8 MPa, to 90°C in a steady-flow heat-exchanger device. Because of fluid friction, the pressure at exit of the device drops to 4.5 MPa. Flow velocities are low enough to neglect kinetic energies. Calculate the heating which must be supplied per kilogram of mixture.

7-59M A mixture of equal molal parts of argon and CO_2 is contained in a rigid vessel at 20 atm and 25°C. How much heat must be supplied per kilogram to raise the temperature to 1500°C?

7-60M An air–water-vapor mixture has 25 percent water vapor on a molal basis and is contained at 3.5 MPa and 430°C. Calculate the specific volume of the mixture using (*a*) the ideal-gas law, (*b*) the rule of additive pressures, (*c*) the rule of additive volumes, and (*d*) Kay's pseudo-critical rule.

7-61M How much heat per kilogram of mixture would be required to heat the mixture of Prob. 7-60M to 540°C in a constant-pressure, steady-flow process?

7-62M 4000 liters/s of air at 10°C, 100 percent relative humidity, are mixed in a steady-flow adiabatic process with 3000 liters/s of air at 33°C, 60 percent relative humidity, at a pressure of 1 atm. Calculate the properties for the final mixture.

7-63M 800 liters/s of air at 1 atm, 5°C, and 20 percent relative humidity are delivered to a heating chamber where the air is also sprayed with water at 25°C. The air discharges from the chamber at 25°C, 50 percent relative humidity. Calculate the amount of heat which must be supplied and the mass of water spray needed.

7-64M Hot air at 1 atm, 35°C, and 60 percent relative humidity is cooled by spraying with liquid water at 3°C. The air leaves the spray chamber saturated with a specific humidity of 0.009 kg/kg dry air, and the water leaves the chamber at 9°C. The inlet air flow is 2800 liters/s. Calculate the flow rate of spray water required assuming the chamber is adiabatic.

7-65M A mixture of CO_2 and N_2 has equal proportions by volume. Determine the mass fraction of each constituent and the gas constant for the mixture.

7-66M 1 kg of O_2 is mixed with 4 kg of another ideal gas, and the total mixture occupies a volume of 0.6 m^3 at a total pressure of 320 kPa and 50°C. Determine the gas constant of the other gas and the mole fraction of each constituent.

7-67M 500 liters/s of air at 1 atm, 20°C, and 50 percent relative humidity are cooled to 5°C in a steady-flow process at constant pressure. Calculate the cooling required and amount of moisture removed.

7-68M 5 kg of an air-steam mixture at 5.0 atm and 35°C are expanded isothermally until the partial pressure of the water vapor is 1.228 kPa. The mixture is saturated at the initial condition. Find the total pressure for the final state and the heat transfer.

7-69M 500 liters of an air–water-vapor mixture at 1 atm, 35°C, and 90 percent relative humidity are cooled at constant volume to 5°C. Calculate the cooling required, amount of water vapor condensed, and temperature at which condensation starts.

7-70M 2300 liters/s of air at 1 atm, 30°C dry bulb, and 35 percent relative humidity enter an adiabatic chamber where water is added to the flow in the amount of 8 g for each kilogram of dry air in the mixture. During the process the wet-bulb temperature remains constant. Calculate the dry-bulb temperature and relative humidity of the exit flow. How much total water is added per hour?

7-71M 700 liters/s of air at 20°C and 100 percent relative humidity are heated to 35°C in a steady-flow process. Determine the final relative humidity and the heating required.

7-72M 700 liters/s of atmospheric air at 32°C and 60 percent relative humidity are to undergo a cooling process to remove moisture and then a heat-addition process to bring the mixture to 24°C, 40 percent relative humidity. Calculate the amount of cooling and heating required and the moisture removed.

7-73M For a sample of air at 100 kPa the dry-bulb temperature is 34°C and the wet-bulb temperature is 20°C. Determine the specific and relative humidities, dew point, mole fractions of water vapor and dry air, and volume of mixture per kilogram of dry air.

7-74M Air and water vapor are contained in a rigid tank at 30°C. The water vapor is in the saturated-vapor state, and there is only 0.1 kg of dry air per kilogram of vapor. Calculate the specific humidity and total pressure of the mixture. Also calculate the mole fractions of air and water vapor.

7-75M 8 kg/s of dry air at 105°C and 1 atm are mixed with 0.5 kg/s of steam at 1 atm and 90 percent quality in an adiabatic steady-flow device at constant pressure. Calculate the absolute humidity of the final mixture and the irreversibility for the process for $T_0 = 20°C$.

7-76M Equal mass fractions of CH_4 and CO_2 are contained at 700 kPa and 30°C in a tank having a volume of 0.2 m^3. 1 kg of O_2 is added to the mixture while maintaining the total volume and temperature constant. Calculate the molal analysis of the new mixture and the final pressure.

7-77M A certain mixture has the following molal analysis: 40 percent CH_4, 35 percent CO_2, and 25 percent N_2. Determine the mass fractions of each constituent and the gas constant for the mixture.

7-78M A 3.5-kg mixture of CO_2 and N_2 is contained in a rigid vessel at 300 kPa and 70°C. The mole fraction of each gas is 50 percent. 1 kg of N_2 is added to the vessel while maintaining the temperature constant at 70°C. Determine the mass analysis of the new mixture, its molecular weight, and total pressure.

7-79M In an air-conditioning application for a building, 4.7 MW of energy must be dissipated to the surroundings using a wet-cooling tower. Inlet air is at 35°C and 30 percent relative humidity, and the air leaves at 38°C and 90 percent relative humidity. Make-up water enters the cooling tower at 23°C. Cooling water from the air-conditioning unit enters at 40°C and is returned to the unit at 30°C. Calculate the mass flow of make-up water and inlet volume flow of air required.

7-80M A power plant operates in the winter with outside air at 0°C, 80 percent relative humidity. 1 GW of energy is to be dissipated by a wet-cooling tower which cools liquid water from 30 to 20°C. Make-up water is available at 10°C, and the exit air from the tower is at 20°C, 80 percent relative humidity. Calculate the mass flow of make-up water and inlet volume flow of air required.

REFERENCES

1 Doolittle, J. S.: "Thermodynamics for Engineers," 2d ed., International Textbook Company, Scranton, Pa., 1964.
2 Obert, E. F.: "Concepts of Thermodynamics," McGraw-Hill Book Company, New York, 1960.
3 Van Wylen, G. J., and R. E. Sonntag: "Fundamentals of Classical Thermodynamics," John Wiley & Sons, Inc., New York, 1965.
4 Hall, N. A., and W. E. Ibele: "Engineering Thermodynamics," Prentice-Hall, Inc., Englewood Cliffs, N.J., 1960.
5 Kay, W. B.: Density of Hydrocarbon Gases and Vapors, *Ind. Eng. Chem.*, vol. 28, p. 1014, 1936.
6 Carrier, W. H.: Rational Psychrometric Formulae, *Trans. ASME*, vol. 33, p. 1005, 1911.

8

CHEMICAL THERMO-DYNAMICS AND EQUILIBRIUM

8-1 INTRODUCTION

The thermodynamics of chemical processes is an important part of engineering practice in a number of industries. This chapter presents an introduction to such processes. The exposition begins with a discussion of simple combustion reactions from the standpoint of mass balances. Next, techniques for describing chemical energy properties are examined, followed by application to combustion reactions. A chemical reaction is always somewhat incomplete in that reactants are never fully converted to products. The degree to which a reaction is completed depends on temperature, pressure, and other factors, and falls under the general subject of thermodynamic equilibrium. In the latter sections of this chapter we shall develop some general relations for thermodynamic equilibrium and apply them to chemical processes. Finally, we shall state the third law of thermodynamics and note its relation to the concepts of chemical equilibrium and statistical thermodynamics.

8-2 COMBUSTION REACTIONS

A combustion process involves the burning of a fuel with oxygen or a substance containing oxygen such as air. The fuel and oxidizer are called the *reactants,* and the constituents resulting from the combustion process are called the *products*. The combustion of carbon, for example, involves the simple reaction

$$C + O_2 \rightarrow CO_2 \tag{8-1}$$

When the combustion process involves the burning of a hydrocarbon fuel, the products will normally contain carbon dioxide and water. Thus, for methane, CH_4, we have the reaction

$$CH_4 + 2O_2 \rightarrow CO_2 + 2H_2O \tag{8-2}$$

The numbers that precede the chemical symbols in these reactions are called the *stoichiometric coefficients*. When a fuel is burned with air, all constituents in normal atmospheric air, other than oxygen, are assumed to be inert insofar as the reaction is concerned, unless very high temperatures are encountered such that chemical dissociation takes place. For normal air the composition is approximately 78 percent nitrogen, 21 percent oxygen, and 1 percent argon. For calculation purposes the argon is usually ignored and the air is assumed to be 79 percent nitrogen and 21 percent oxygen. Consequently, there are 79/21 = 3.76 mol of nitrogen for each mole of oxygen in the air. For the burning of methane with air we would have

$$CH_4 + 2O_2 + (2)(3.76)N_2 \rightarrow CO_2 + 2H_2O + (2)(3.76)N_2 \tag{8-3}$$

The quantity of air necessary to balance this equation is called the *stoichiometric air*. It is not uncommon, however, to have a reaction where too much air is supplied for the quantity of fuel involved. In such cases we say that *excess air* is used, and the excess is usually expressed as a percentage of the stoichiometric air. The amount of excess air simply carries through the reaction. Using the combustion of methane as an example again, 10 percent excess air would mean that (2.0)(1.1) mol of oxygen were involved in the reactants. In the products (0.1)(2.0) mol of oxygen would be left unused, and the chemical equation would be written as

$$CH_4 + (1.1)(2)O_2 + (1.1)(2)(3.76)N_2 \rightarrow CO_2 + 2H_2O + (0.1)(2)O_2 + (1.1)(2)(3.76)N_2 \tag{8-4}$$

The term *theoretical air* is sometimes used instead of stoichiometric air.

A combustion reaction for a general hydrocarbon C_nH_y burning with the theoretical proportion of air may be obtained as follows:

$$C_nH_y + xO_2 + 3.76xN_2 \rightarrow aCO_2 + bH_2O + 3.76xN_2 \tag{8-5}$$

From the carbon balance

$$n = a$$

and the hydrogen balance yields

$$y = 2b$$

Finally, the oxygen balance is

$$2x = 2n + \frac{y}{2}$$

$$x = n + \frac{y}{4}$$

The final balance is then

$$C_nH_y + \left(n + \frac{y}{4}\right) O_2 + 3.76 \left(n + \frac{y}{4}\right) N_2 \rightarrow nCO_2 + \left(\frac{y}{2}\right) H_2O + 3.76 \left(n + \frac{y}{4}\right) N_2 \tag{8-6}$$

If less than the stoichiometric air is supplied, the reaction does not go to completion; i.e., all of the carbon does not burn and appear in the form of CO_2 in the products. Instead a portion of the carbon will usually appear in the form of carbon monoxide, CO. For methane an incomplete reaction might appear as

$$CH_4 + xO_2 + 3.76xN_2 \rightarrow aCO_2 + bCO + 2H_2O + 3.76xN_2 \tag{8-7}$$

In this case we would expect that $x < 2.0$. The oxygen balance on the equation would yield

$$2x = 2a + b + 2 \tag{8-8}$$

Further information concerning the temperature of the reaction and chemical equilibrium considerations would be necessary to predict the exact split between CO_2 and CO, i.e., the exact values of the coefficients a and b. If enough O_2 were added to Eq. (8-7) to bring it up to stoichiometric proportions, the CO would then react with the oxygen to form additional CO_2 in accordance with the reaction equation:

$$CO + \tfrac{1}{2}O_2 \rightarrow CO_2 \tag{8-9}$$

Even when excess air is supplied, real reactions do not usually go to completion because of various factors involved in the combustion process. In a practical sense, therefore, it is usually necessary to make experimental measurements of the composition of the products of combustion to balance the equation.

The Orsat Apparatus

Figure 8-1 illustrates a simple *Orsat apparatus* which is used to analyze products of combustion. It consists of a measuring burette and three reagent pipettes which may be used to successively absorb carbon dioxide, oxygen, and carbon monoxide in the mixture. First, a sample of the flue gas is drawn into the measuring burette. Next, the sampling manifold is shut off from the flue gas and the sample forced into the first reagent pipette where the carbon dioxide is absorbed. The sample is then brought back into the measuring burette and the reduction in volume noted. The procedure is then repeated for the other two pipettes which absorb the O_2 and CO successively. In the measuring process the combustion products are saturated with water vapor, and the procedure is

FIG. 8-1 Orsat apparatus.

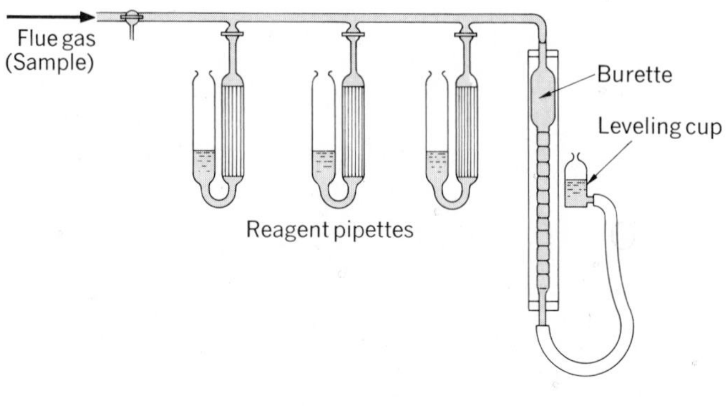

conducted so that the volumetric proportions of the products are obtained on a so-called dry basis, i.e., exclusive of the water vapor which is present.

Gas chromatograph and infrared absorption techniques for measuring products of combustion are described in Ref. [13].

It turns out that even when excess air is supplied for the combustion process both CO and CO_2 may be obtained in the products because of chemical reactions occurring at the high temperatures of combustion. In actual practice then, the excess air reaction of Eq. (8-4) could contain both CO and O_2 in the products. This matter will be clarified when we discuss chemical equilibrium in later sections of this chapter.

A parameter of interest in practical combustion problems is the *air/fuel ratio,* which is the ratio of air to fuel in the reactants expressed on either a mass or molal basis.

EXAMPLE 8-1 Burning of Octane

Octane (C_8H_{18}) is burned with 150 percent theoretical air. Calculate the air/fuel ratio and the molal analysis of the products of combustion. What is the dew point of the products for a total pressure of 14.696 psia?

SOLUTION For theoretical air the combustion equation is

$$C_8H_{18} + 12.5O_2 + (3.76)(12.5)N_2 \rightarrow 8CO_2 + 9H_2O + (3.76)(12.5)N_2$$

For 150 percent theoretical air the corresponding reaction would be

$$C_8H_{18} + (1.5)(12.5)O_2 + (1.5)(3.76)(12.5)N_2 \rightarrow$$
$$8CO_2 + 9H_2O + (0.5)(12.5)O_2 + (1.5)(3.76)(12.5)N_2$$

We therefore have

$$m_f = (1)(114) = 114 \text{ lbm/mol fuel}$$

$$m_a = (1.5)(12.5)(1 + 3.76)(28.96) = 2580 \text{ lbm/mol fuel}$$

and the air/fuel ratio is

$$\text{AF} = \frac{m_a}{m_f} = \frac{2580}{114} = 22.6 \text{ lbm air/lbm fuel}$$

The total number of moles of products is

$$n_p = 8 + 9 + (0.5)(12.5) + (1.5)(3.76)(12.5)$$
$$= 93.75 \text{ mol/mol fuel}$$

so that the mole fractions of the constituents are

$$x_{CO_2} = \frac{8}{93.75} = 8.53 \text{ percent}$$

$$x_{H_2O} = \frac{9}{93.75} = 9.60 \text{ percent}$$

$$x_{O_2} = \frac{6.25}{93.75} = 6.66 \text{ percent}$$

$$x_{N_2} = \frac{70.5}{93.75} = 75.21 \text{ percent}$$

The dew point of the products is the saturation temperature corresponding to the partial pressure of the water vapor. We have

$$\begin{aligned} p_{H_2O} &= x_{H_2O}\, p_{\text{total}} \\ &= (0.096)(14.696) \\ &= 1.41 \text{ psia} \qquad (9.72 \text{ kPa}) \end{aligned}$$

The saturation temperature corresponding to 1.41 psia is

$$T_{\text{dew point}} = 113.5°\text{F} \qquad (45.3°\text{C})$$

Thus moisture would condense if the products were cooled below 113°F. This explains, of course, why water drips from the tailpipe of an automobile on a cold morning before the exhaust system has warmed up.

8-3 ENTHALPY OF FORMATION

In the following sections we shall be interested in performing energy balances on chemical reactions in order to determine the quantity of heat which may be liberated in a particular combustion process, the highest temperature attainable for a given fuel when burning with air, etc. For these energy balances we require a knowledge of the enthalpies of the substances involved in the reaction. Tables which give the properties as a function of temperature are available for various substances. To formulate such tables, however, it is necessary to establish a reference state for the properties which takes proper account of the internal chemical energy of the substance.

The base (zero level) energy level is chosen so that the energy of all *elements* is zero at 25°C (77°F) and 1-atm pressure. The energy of a compound is not zero at this reference state, but is defined in accordance with the following reasoning as applied to CO_2 at 25°C and 1 atm. To form CO_2 in a steady-flow process, we would have

$$C + O_2 \rightarrow CO_2 \tag{8-10}$$

and the energy balance

$$H_R + Q = H_P \tag{8-11}$$

where H_R is the total enthalpy of the reactants and H_P is the total enthalpy of the products. Q is the heat transfer necessary to make the reaction go. Using the above reference state, $H_R = 0$ since all the reactants are elements and $H_P = Q$. The value of Q may be determined experimentally and has the following value for CO_2:

$$Q = H_p = -169\ 297 \text{ Btu/lbm mol} \qquad (-393\ 766 \text{ kJ/kg mol})$$

In other words, the reaction in Eq. (8-10) *liberates* 169 297 Btu of energy for each mole of carbon burned. The enthalpy of CO_2 at 25°C and 1 atm is therefore less than the reference state by this amount. This quantity is called the *enthalpy of formation* and has the symbol $\bar{h}_f^\circ$, where the bar designates *per mole of substance*. Values for typical substances are given in Table A-6 in the Appendix. A reaction in which heat is liberated is called *exothermic*, with the term *endothermic* being employed to describe reactions which absorb heat.

It is now fairly easy to see how we may calculate the enthalpies of substances with respect to the zero element state of 298 K and 1 atm. We employ the tables in the Appendix to compute the change from the reference state and add this to the enthalpy of formation. Thus the total molal enthalpy at temperature T is

$$\bar{h}(T)_{\text{total}} = \bar{h}_f^\circ + \bar{h}_T - \bar{h}_{298} \tag{8-12}$$

8-4 REFERENCE LEVELS FOR TABLES

As we have seen before, it is possible to choose different zero levels, or reference states, for tabulations of thermodynamic properties. For the steam tables we chose the saturated-liquid state at 0°C, while for the air tables we chose 0 K and a pressure of 1 atm. For use with Eq. (8-12) tables are employed using a reference temperature of 0 K as well as the standard state temperature of 298 K (77°F). The 0 K state is perhaps more common for English units (as in Table A-18) while the 298 K reference is employed for SI tabulations (as in Table A-18M). The modern approach is also to use a reference pressure of 0.1 MPa instead of 1 atm (0.101 32 MPa). This small difference has essentially no effect on the enthalpy or internal energies.

8-5 GIBBS FUNCTION OF FORMATION AND ABSOLUTE ENTROPY

As with enthalpy of formation it is possible to establish a Gibbs function of formation $\bar{g}_f^\circ$ with the zero level set at 298 K and 1 atm for all elements. The values for a number of compounds are given in Table A-6. Values of absolute entropies are set in accordance with the third law of thermodynamics discussed in Sec. 8-16 which states that

The entropy of any pure substance in thermodynamic equilibrium approaches zero as the absolute temperature approaches zero.

Using this zero level, one may integrate between zero Kelvin and the standard state of 25°C and 1 atm to obtain the values of absolute entropy given in Table A-6. For those substances which follow ideal-gas behavior, the entropy variation with pressure is given through an application of Eq. (6-37), written in terms of molal properties

$$\bar{s}_2 - \bar{s}_1 = \bar{\phi}_2 - \bar{\phi}_1 - \mathscr{R} \ln\left(\frac{p_2}{p_1}\right) \tag{8-13}$$

For ideal gases the absolute entropy at any pressure and temperature can thus be written as

$$\bar{s}^\circ(p, T) = \bar{s}^\circ + \bar{\phi}_T - \bar{\phi}_{298} - R \ln\left(\frac{p}{p_0}\right) \tag{8-14}$$

where $\bar{s}^\circ$ is taken from Table A-6 and p_0 is the reference pressure of 1 atm. Table A-18M gives the values of $\bar{s}^\circ$ for gases at various temperatures and a pressure of 0.1 MPa. Absolute values for entropy are required for certain second-law analyses of combustion reactions as we shall see in later sections.

The important point to remember is that reference states must be carefully specified when dealing with combustion reactions. The tables in the Appendix specify these states and indicate some appropriate conversion equations.

To show how enthalpy of formation, absolute entropy, and Gibbs function of formation are related we may calculate $\bar{g}_f^\circ$ for CO_2 from the other quantities in Table A-6. We have

$$C + \tfrac{1}{2}O_2 \rightarrow CO_2$$

and from Table A–6 at 1 atm and 25°C

$$\bar{h}_f^\circ(CO_2) = -393\ 520\ \text{kJ/kg} \cdot \text{mol} \qquad \bar{s}^\circ(CO_2) = 213.64\ \text{kJ/kg} \cdot \text{mol}$$
$$\bar{s}^\circ(O_2) = 205.03\ \text{kJ/kg} \cdot \text{mol} \qquad \bar{h}_f^\circ = 0 \text{ for C and } O_2$$
$$\bar{s}^\circ(C) = 5.74\ \text{kJ/kg} \cdot \text{mol}$$

The temperature is 25 + 233.15 = 298.15 K. The value of $\bar{g}_f^\circ$ for CO_2 is thus

$$\bar{g}_f^\circ = G_p - G_R = \sum_p (\bar{h}_f^\circ - Ts^\circ) - \sum_R (\bar{h}_f^\circ - Ts^\circ)$$

$$= 393\ 520 - 0 - 0 - (298.15)(213.64 - 5.74 - 205.03)$$
$$= -394\ 376 \text{ Btu/lbm} \cdot \text{mol}$$

in good agreement with the tabular value of $\bar{g}_f^\circ(CO_2) = -394\ 360$. The difference is due to round-off in tabular values.

EXAMPLE 8-2 Calculation of Excess Air
An analysis is made of the products of combustion when methane (CH_4) is burned with atmospheric air. The mole fractions of the *dry* products (those other than water) are as follows:

$CO_2 = 9.00$ percent $\qquad O_2 = 1.50$ percent

$CO = 1.20$ percent $\qquad N_2 = 88.30$ percent

Calculate the percent theoretical air which is used in the combustion process.

SOLUTION We shall assume 100 mol of dry products and balance the combustion equation on this basis. We have

$$aCH_4 + xO_2 + 3.76xN_2 \rightarrow 9.00CO_2 + 1.2CO + 1.5O_2 + bH_2O + 88.3N_2$$

Carbon balance: $\quad a = 9.00 + 1.2 = 10.2$

Hydrogen balance: $\quad 4a = 2b$

$$b = 2a = 20.4$$

Oxygen balance: $\quad 2x = (2)(9.00) + 1.2 + (2)(1.5) + 20.4$

$$x = 21.3$$

Nitrogen balance: $\quad 3.76x = 88.3$

$$x = 23.5$$

Note that the two determinations of x do *not* agree. In actual practice, the nitrogen content of the products is usually determined by subtracting the other constituents from the total; i.e., it is not measured directly. For this reason, the value of x determined from the nitrogen balance is not believed to be as reliable as the other calculation. Using $x = 21.3$ gives

$$\frac{\text{Mol } O_2}{\text{Mol fuel}} = \frac{21.3}{10.2} = 2.09$$

According to Eq. (8-3) there are 2.00 mol of O_2 required for each mole of fuel when the theoretical air is burned. The desired result is therefore

$$\text{Percent theoretical air} = \frac{2.09}{2.00}(100) = 104.5 \text{ percent}$$

EXAMPLE 8-3
Calculate the molal enthalpy of CO_2 at 440°F.

SOLUTION We employ Eq. (8-12) for this calculation along with Table A-6 and the gas tables (Table A-18):

$$\bar{h}_{CO_2} = -169\ 290 \text{ Btu/lbm mol}$$

$$\bar{h}_{440} = 7597.6 \text{ Btu/lbm mol} \qquad (17\ 671 \text{ kJ/kg mol})$$

$$\bar{h}_{77} = 4030.2 \text{ Btu/lbm mol} \qquad (9373.8 \text{ kJ/kg mol})$$

Thus

$$\begin{aligned}\bar{h}_{\text{total}} &= -169\ 290 + 7597.6 - 4030.2\\ &= -165\ 722 \text{ Btu/lbm mol}\\ &= -385\ 451 \text{ kJ/kg mol}\end{aligned}$$

EXAMPLE 8-4
Ethane (C_2H_6) is burned with the theoretical amount of oxygen. Determine the heat transfer for reactants and products, both at 298 K and 1 atm.

SOLUTION The combustion equation is

$$C_2H_6 + 3.5O_2 \rightarrow 2CO_2 + 3H_2O(l)$$

From Eq. (8-11) the heat transfer is

$$Q = H_P - H_R$$

Also,

$$\begin{aligned}H_R &= \bar{h}^{\circ}_{C_2H_6} = -36\ 420 \text{ Btu}\\ H_P &= 2\bar{h}^{\circ}_{CO_2} + 3\bar{h}^{\circ}_{H_2O}\\ &= (2)(-169\ 290) + (3)(-122\ 970)\\ &= -707\ 490 \text{ Btu} \qquad (-7.464 \times 10^5 \text{ kJ})\end{aligned}$$

The heat transfer is therefore

$$\begin{aligned}Q &= -707\ 490 - (-36\ 420)\\ &= -671\ 070 \text{ Btu/mol fuel} \qquad (-1.56 \times 10^6 \text{ kJ/kg mol})\end{aligned}$$

The negative sign indicates that heat is liberated during the process.

8-6 HEAT OF REACTION AND HEATING VALUE

The energy liberated in a chemical reaction is given by the following relations. For a constant-volume reaction,

$$Q_V = (\Delta U)_v = (U_P - U_R)_v = \bar{u}_{RP} \tag{8-15}$$

and, at constant pressure,

$$Q_P = (\Delta H)_P = (H_P - H_R)_P = \bar{h}_{RP} \tag{8-16}$$

The quantities given in Eqs. (8-15) and (8-16) are called the *heats of reaction* or the *internal energy and enthalpy of combustion*. If the heats are expressed per mole of fuel, then the symbols $\bar{u}_{RP}$ and $\bar{h}_{RP}$ are appropriately used to designate the energies indicated in the equation.

The term *heating value* for a combustion process is in wide use and is a synonym for heat of reaction. When a hydrocarbon fuel is burned in a combustion process, water will appear in the products. The maximum energy release will be obtained when all the water in the products *due to combustion* is in the liquid state; in such cases the heat of reaction is called a *higher heating value* (HHV). Similarly, a lower heat of reaction will be experienced when all the water in the products due to combustion is in the vapor state; in this case we say that a *lower heating value* (LHV) is obtained. In reality, both liquid and vapor will be present in many combustion processes, so that a heating value between the HHV and LHV can be observed. A tabulation of the heating values of several common fuels is given in Table 8-1. The examples illustrate the use of these values.

Table 8-1 gives the heating values of fuels for combustion at the convenient reference temperature of 25°C (77°F). Most combustion processes certainly do not occur at this temperature, and it is of interest to show how these tabular values may be used to calculate heats of combustion at other temperatures. The combustion process at some arbitrary temperature T can be expressed in terms of $\bar{h}_{RP}$ for the standard state by means of the simple sketch shown in Fig. 8-2. The reactants are first cooled to 25°C, combustion is allowed to proceed at 25°C, and the products are then heated back to the temperature T. In terms of the diagram, the heat of reaction may be expressed as

$$\bar{h}_{RP_T} = H_4 - H_1 = (H_4 - H_3) + (H_3 - H_2) + (H_2 - H_1)$$

This relation may be rewritten as

$$\bar{h}_{RP_T} = (H_T - H_{25})_P + \bar{h}_{RP_{25}} + (H_{25} - H_T)_R \tag{8-17}$$

If property data for the various constituents of the reactants and products are available, the value of $\bar{h}_{RP_T}$ is easily calculated.

TABLE 8-1
Higher Heating Values for Fuels with Combustion at 25°C (77°F) and 1 atm. Liquid Water Is in the Products of Combustion.

Fuel	Formula	HHV, Btu/lbm mol	kJ/kg mol	Vaporization Enthalpy $\bar{h}_{fg}$ for Fuel, Btu/lbm mol	kJ/kg mol
Hydrogen	$H_2(g)$	−122 970	−285 840		
Carbon	$C(s)$	−169 290	−393 520		
Carbon monoxide	$CO(g)$	−121 750	−282 990		
Methane	$CH_4(g)$	−383 040	−890 360		
Acetylene	$C_2H_2(g)$	−559 120	−1 299 600		
Ethylene	$C_2H_4(g)$	−607 010	−1 410 970		
Ethane	$C_2H_6(g)$	−671 080	−1 559 900		
Propylene	$C_3H_6(g)$	−885 580	−2 058 500		
Propane	$C_3H_8(g)$	−955 070	−2 220 000	6 480	15 060
n-butane	$C_4H_{10}(g)$	−1 237 800	−2 877 100	9 060	21 060
n-pentane	$C_5H_{12}(g)$	−1 521 300	−3 536 100	11 360	26 410
n-hexane	$C_6H_{14}(g)$	−1 804 600	−4 194 800	13 563	31 530
n-heptane	$C_7H_{16}(g)$	−2 088 000	−4 853 500	15 713	36 520
n-octane	$C_8H_{18}(g)$	−2 371 400	−5 512 200	17 835	41 460
Benzene	$C_6H_6(g)$	−1 420 300	−3 301 500	14 552	33 830
Toluene	$C_7H_8(g)$	−1 698 400	−3 947 900	17 176	39 920
Methyl alcohol	$CH_3OH(g)$	−328 700	−764 540	16 092	37 900
Ethyl alcohol	$C_2H_5OH(g)$	−606 280	−1 409 300	18 216	42 340

Source: JANAF Thermochemical Tables [1], Selected Values of Chemical Thermodynamic Properties [2], and Selected Values of Physical and Thermodynamic Properties of Hydrocarbons and Related Compounds [3].

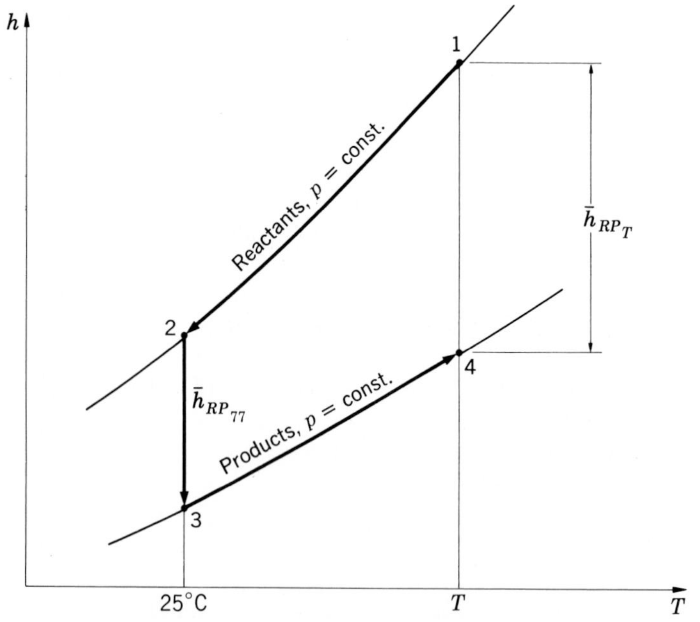

FIG. 8-2 Calculation of heating values.

Of course, it is not necessary to use the concept of heating values for energy analysis of combustion reactions. One might use the enthalpies of formation directly to carry out the desired calculations.

EXAMPLE 8-5
Calculate the enthalpy of combustion for gaseous methane at 440°F.

SOLUTION We assume that the water in the products is in the vapor phase. The higher heating value for methane at 25°C is given in Table 8-1 as

$$\bar{h}_{RP_{25}} = -383\ 040 \text{ Btu/lbm mol}$$

The combustion equation is

$$CH_4 + 2O_2 \rightarrow CO_2 + 2H_2O$$

For the modest temperature range involved we may assume the specific heat for CH_4 as a constant of 0.532 Btu/lbm·°F (from Table 2-2). The enthalpies of other constituents in the reactants and products are determined from the gas tables. We shall employ Eq. (8-17) for the calculation of enthalpy of combustion. The quantities of interest are

$$\begin{aligned}(H_{77} - H_T)_R &= n_{CH_4}(\bar{h}_{77} - \bar{h}_T)_{CH_4} + n_{O_2}(\bar{h}_{77} - \bar{h}_T)_{O_2} \\ &= (1)(16)(0.532)(77 - 440) + (2)(3725.1 - 6337.9) \\ &= -8315 \text{ Btu}\end{aligned}$$

$$\begin{aligned}(H_T - H_{77})_P &= n_{CO_2}(\bar{h}_T - \bar{h}_{77})_{CO_2} + n_{H_2O}[\bar{h}_T(g) - \bar{h}_{77}(l)]_{H_2O} \\ &= (1)(7597.6 - 4030.2) + (2)(18)(1260.3 - 45.02) \\ &= 47\ 317 \text{ Btu} \qquad (49\ 922 \text{ kJ})\end{aligned}$$

Note that the enthalpy of water at 77°F is taken as the value for saturated liquid, since we will be using the higher heating value for $\bar{h}_{RP_{77}}$.

The foregoing values may now be inserted in Eq. (8-17) to yield

$$\begin{aligned}\bar{h}_{RP_T} &= 47\ 317 + (-383\ 040) + (-8315) \\ &= -344\ 038 \text{ Btu/lbm mol fuel}\end{aligned}$$

EXAMPLE 8-5M
Calculate the enthalpy of combustion for gaseous methane at 500 K.

SOLUTION We assume that the water in the products is in the vapor phase. The higher heating value for methane at 298 K is given in Table 8-1 as

$$\bar{h}_{RP_{298}} = -890\ 360 \text{ kJ/kg mol}$$

The combustion equation is

$$CH_4 + 2O_2 \rightarrow CO_2 + 2H_2O$$

For the modest temperature range involved we may assume the specific heat for CH_4 as a constant of 2.227 kJ/kg·°C (from Table 2-2). The enthalpies of other constituents in the reactants and products are determined from the gas tables. We shall employ Eq. (8-17) for the calculation of enthalpy of combustion. The quantities of interest are

$$\begin{aligned}(H_{298} - H_T)_R &= n_{CH_4}(\bar{h}_{298} - \bar{h}_T)_{CH_4} + n_{O_2}(\bar{h}_{298} - \bar{h}_T)_{O_2} \\ &= (1)(16)(2.227)(298 - 500) + (2)(-6088) \\ &= -19\,374 \text{ kJ/kg mol fuel}\end{aligned}$$

$$\begin{aligned}(H_T - H_{298})_P &= n_{CO_2}(\bar{h}_T - \bar{h}_{298})_{CO_2} + n_{H_2O}[\bar{h}_T(g) - \bar{h}_{298}(g) + \bar{h}_{fg_{298}}]_{H_2O} \\ &= (1)(8314) + (2)[6920 + (18)(2442.3)] \\ &= 110\,077 \text{ kJ/kg mol} \qquad (47\,325 \text{ Btu/lbm mol})\end{aligned}$$

Note that the enthalpy of water at 298 K is taken as the value for saturated liquid, since we will be using the higher heating value for $\bar{h}_{RP_{298}}$. The foregoing values may now be inserted in Eq. (8-17) to yield

$$\begin{aligned}\bar{h}_{RPT} &= 110\,077 + (-890\,360) + (-19\,374) \\ &= -799\,657 \text{ kJ/kg mol fuel} \qquad (-343\,791 \text{ Btu/lbm mol})\end{aligned}$$

EXAMPLE 8-6

A certain gaseous fuel consists of 35 percent methane (CH_4), 25 percent propane (C_3H_8), and 40 percent pentane (C_5H_{12}), where the percentages are on a molal basis. Calculate the mass fraction of each constituent and the gas constant for the mixture. Also calculate the higher heating value for this fuel when burned with the theoretical air at 25°C.

SOLUTION For 1 mol of mixture the mass and mass fraction of each constituent are

		Mass fraction
$m(CH_4) = (0.35)(16) =$	5.6	0.123
$m(C_3H_8) = (0.25)(44) =$	11	0.242
$m(C_5H_{12}) = (0.40)(72) =$	28.8	0.635
Total	45.4	1.000

The molecular weight of the mixture is

$$M = \frac{45.4}{1.0} = 45.4 \tag{a}$$

so that the gas constant of the mixture is

$$R = \frac{\mathcal{R}}{M} = \frac{8314.41}{45.4} = 183.14 \text{ J/kg·K} \tag{b}$$

To calculate the heating value we first construct the combustion equation. Considering 1 mol of fuel, the equation would be

$$0.35CH_4 + 0.25C_3H_8 + 0.40C_5H_{12} + xO_2 + 3.76xN_2 \rightarrow aCO_2 + bH_2O + 3.76xN_2 \qquad (c)$$

The carbon balance then gives

$$0.35 + (0.25)(3) + (0.40)(5) = a = 3.10 \qquad (d)$$

From the hydrogen balance we obtain

$$(0.35)(4) + (0.25)(8) + (0.40)(12) = 2b \qquad b = 4.1 \qquad (e)$$

Finally, the oxygen balance yields

$$2x = 2a + b = (2)(3.10) + 4.1 \qquad x = 5.15 \qquad (f)$$

The heating value is now obtained from

$$Q_p = H_p - H_R \qquad (g)$$

where all of the enthalpies are evaluated at 25°C. This means that we can use the formation enthalpies obtained from Table A-6:

$$CH_4\text{: } \overline{h_f^\circ} = -74\ 850 \text{ kJ/kg mol}$$

$$C_3H_8\text{: } \overline{h_f^\circ} = -103\ 850 \text{ kJ/kg mol}$$

$$C_5H_{12}\text{: } \overline{h_f^\circ} = -146\ 440 \text{ kJ/kg mol}$$

$$O_2\text{: } \overline{h_f^\circ} = 0$$

$$N_2\text{: } \overline{h_f^\circ} = 0$$

$$CO_2\text{: } \overline{h_f^\circ} = -393\ 520 \text{ kJ/kg mol}$$

$$H_2O\ (l)\text{: } \overline{h_f^\circ} = -285\ 830 \text{ kJ/kg mol}$$

We then obtain for the enthalpies of products and reactants,

$$H_R = (0.35)(-74\ 850) + (0.25)(-103\ 850) + (0.40)(-146\ 440) = -110\ 736 \text{ kJ/mol fuel} \qquad (h)$$

$$H_P = (3.10)(-393\ 520) + (4.10)(-285\ 830) = -2\ 391\ 815 \text{ kJ/mol fuel} \qquad (i)$$

The heating value is obtaining by inserting the values into Eq. (g).

$$Q = -2\,391\,815 - (-110\,736) = -2\,281\,079 \text{ kJ/mol fuel} \qquad (j)$$

8-7 ADIABATIC FLAME TEMPERATURE

Consider the combustion process shown in Fig. 8-3. If the process is allowed to proceed adiabatically, the final temperature which the products will attain is called the *adiabatic flame temperature*. In this instance, all of the heat of reaction is converted to internal energy and is manifested in the form of a high temperature for the products of combustion. For the temperatures involved in most reactions, the combustion process usually will not go to completion, and equations of chemical equilibrium must be employed to determine the final temperature.

To compute the adiabatic flame temperature an iterative procedure must be employed because of the complex relation between the products temperature and their constituent enthalpies. The following example illustrates such a procedure.

EXAMPLE 8-7 Calculation of the Adiabatic Flame Temperature
Calculate the adiabatic flame temperature for liquid octane burning with 200 percent theoretical air at 77°F.

SOLUTION The combustion equation for this reaction is

$$C_8H_{18}(l) + (2)(12.50)O_2 + (2)(12.5)(3.76)N_2 \rightarrow 8CO_2 + 9H_2O + (2)(12.5)(3.76)N_2 + (12.5)O_2$$

Because the process is adiabatic, we must have

$$H_R = H_P \qquad (a)$$

The final temperature of the products is obtained by an iterative technique outlined in the following. The total enthalpy of the reactants must be that of liquid octane at 77°F. For octane in the gaseous state the enthalpy of formation

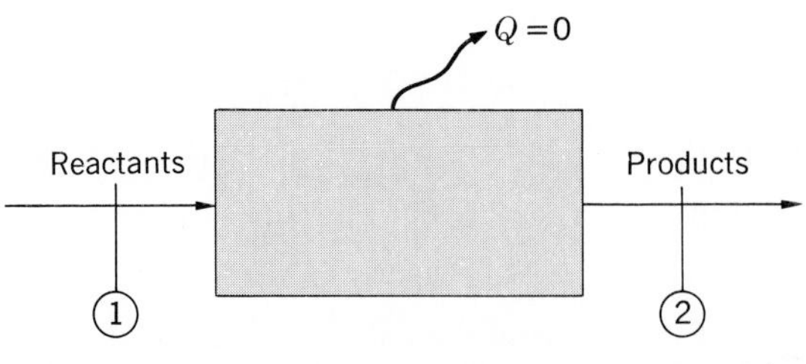

FIG. 8-3 Adiabatic combustion process.

is given in Table A-6 as −89 680 Btu/lbm mol. The enthalpy of vaporization is given in Table 8-1 as 17 835 Btu/lbm mol, so the enthalpy in the liquid state would be

$$H_R = \bar{h}^\circ_{C_8H_{18}} = -89\,680 - 17\,835 = -107\,515 \text{ Btu/mol fuel}$$

and the relation for the enthalpy of the products is

$$\begin{aligned} H_P &= 8(\bar{h}^\circ + \bar{h}_T - \bar{h}_{537})_{CO_2} \\ &\quad + 9(\bar{h}^\circ + \bar{h}_T - \bar{h}_{537})_{H_2O} \\ &\quad + 94(\bar{h}^\circ + \bar{h}_T - \bar{h}_{537})_{N_2} \\ &\quad + 12.5(\bar{h}_T - \bar{h}_{537})_{O_2} \\ &= 8(-169\,290 + \bar{h}_{TCO_2} - 4030) \\ &\quad + 9(-104\,040 + \bar{h}_{TH_2O} - 4268) \\ &\quad + 94(\bar{h}_T - 3730)_{N_2} + 12.5(\bar{h}_T - 3745)_{O_2} \end{aligned}$$

We now substitute H_R and H_P in Eq. (a) and assume values of T until the equation is satisfied. The iterations are

$$T = 3000°\text{R} \qquad \begin{aligned} H_P &= 8(-173\,320 + 34\,806) \\ &\quad + 9(-108\,308 + 28\,386) \\ &\quad + 94(22\,761 - 3730) \\ &\quad + 12.5(23\,817 - 3725) \\ &= +310\,000 \end{aligned}$$

This value is too high since $H_R = -107\,530$. We therefore try a lower temperature for the products:

$$T = 2000°\text{R} \qquad \begin{aligned} H_P &= 8(-173\,320 + 21\,018) \\ &\quad + 9(-108\,308 + 17\,439) \\ &\quad + 94(14\,534 - 3730) \\ &\quad + 12.5(15\,164 - 3725) \\ &= -877\,673 \end{aligned}$$

This value is too low, so we try an intermediate value determined by interpolation:

$$T = 2600°\text{R} \qquad \begin{aligned} H_P &= 8(-173\,320 + 29\,187) \\ &\quad + 9(-108\,308 + 23\,869) \\ &\quad + 94(19\,415 - 3730) \\ &\quad + 12.5(20\,311 - 3725) \\ &= -231\,000 \end{aligned}$$

This value is still too low, but a quick interpolation yields

$$T = 2700°\text{R} \qquad (1500 \text{ K})$$

An additional iteration would show this value to be in good agreement with $H_R = -107\ 515$ Btu ($-113\ 434$ kJ).

It is well to remark this temperature would not be achieved in practice because the products would dissociate. Section 8-10 discusses the chemical equilibrium conditions which must be applied in such circumstances.

EXAMPLE 8-7M Calculation of the Adiabatic Flame Temperature
Calculate the adiabatic flame temperature for liquid octane burning with 200 percent theoretical air at 298 K.

SOLUTION The combustion equation for this reaction is

$$C_8H_{18}(l) + (2)(12.50)O_2 + (2)(12.5)(3.76)N_2 \rightarrow 8CO_2 + 9H_2O + (2)(12.5)(3.76)N_2 + (12.5)O_2$$

Since the process is adiabatic, we must have

$$H_R = H_P \tag{a}$$

The final temperature of the products is obtained by an iterative technique outlined in the following. The total enthalpy of the reactants must be that of liquid octane at 298 K. For octane in the gaseous state the enthalpy of formation is given in Table A-6 as −280 450 kJ/kg mol. The enthalpy of vaporization is given in Table 8-1 as 41 460 kJ/kg mol, and so the enthalpy in the liquid state would be

$$H_R = \bar{h}^\circ_{C_8H_{18}} = -208\ 450 - 41\ 460 = -249\ 910 \text{ kJ/kg mol fuel}$$

and the relation for the enthalpy of the products is

$$\begin{aligned} H_p &= 8(\bar{h}^\circ + \bar{h}_T - \bar{h}_{298})_{CO_2} \\ &\quad + 9(\bar{h}^\circ + \bar{h}_T - \bar{h}_{298})_{H_2O} \\ &\quad + 94(\bar{h}^\circ + \bar{h}_T - \bar{h}_{298})_{N_2} \\ &\quad + 12.5(\bar{h}_T - \bar{h}_{298})_{O_2} \\ &= 8[-393\ 520 + (\bar{h}_T - \bar{h}_{298})_{CO_2}] \\ &\quad + 9[-241\ 820 + (\bar{h}_T - \bar{h}_{298})_{H_2O}] \\ &\quad + 94(\bar{h}_T - \bar{h}_{298})_{N_2} + 12.5(\bar{h}_T - \bar{h}_{298})_{O_2} \end{aligned}$$

We now substitute H_R and H_P in Eq. (a) and assume values of T until the equation is satisfied. The iterations are

$$T = 1700 \text{ K} \qquad \begin{aligned} H_P &= 8(-393\,520 + 73\,492) \\ &\quad + 9(-241\,820 + 57\,685) \\ &\quad + 94(45\,430) + 12.5(47\,970) \\ &= +652\,606 \end{aligned}$$

This value is too high since $H_R = -249\,910$. We therefore try a lower temperature for the products:

$$T = 1100 \text{ K} \qquad \begin{aligned} H_P &= 8(-393\,520 + 38\,894) \\ &\quad + 9(-241\,820 + 30\,167) \\ &\quad + 94(24\,757) + 12.5(26\,217) \\ &= -2\,087\,015 \end{aligned}$$

This value is too low, so we try an intermediate value determined by interpolation:

$$T = 1400 \text{ K} \qquad \begin{aligned} H_P &= 8(-393\,520 + 55\,907) \\ &\quad + 9(-241\,820 + 43\,447) \\ &\quad + 94(34\,936) + 12.5(36\,966) \\ &= -740\,202 \end{aligned}$$

This value is still too low, but a quick interpolation yields

$$T = 1506 \text{ K} \qquad (2712°\text{R})$$

An additional iteration would show this value to be in good agreement with

$$H_R = 249\,910 \text{ kJ/kg mol} \qquad (107\,442 \text{ Btu/lbm mol})$$

It is well to remark that this temperature would not be achieved in practice because the products would dissociate. Section 8-10 discusses the chemical equilibrium conditions which must be applied in such circumstances.

Values of some adiabatic flame temperatures which do take into account dissociation are given in Table 8-2.

TABLE 8-2
Sample Adiabatic Flame Temperatures, Reactants at 298 K, 1 atm, Stoichiometric Mixture

Fuel	Oxygen K	Oxygen °R	Air K	Air °R
H_2	3079	5542	2384	4291
CH_4	3054	5497	2227	4009
C_8H_{18}	3108	5594	2277	4098

8-8 GENERAL PRODUCT ENTHALPIES

Tabulations of product enthalpies for a general hydrocarbon $(CH_2)_n$ in combustion with 200 and 400 percent theoretical air have been assembled by Keenan and Kaye [12] and are given in Table A-18. Use of these tables can speed computations considerably. It should be noted that the molal enthalpies listed do not include enthalpies of formation for CO_2 or H_2O in the products and assume that the fuel is in the gaseous state. Although these tables have been developed for the $(CH_2)_n$ hydrocarbon, it turns out that they give a very satisfactory representation for the products of combustion with other hydrocarbons as well. Example 8-8 illustrates their application to an adiabatic flame temperature calculation for octane (C_8H_{18}).

EXAMPLE 8-8

Use the general combustion products table to calculate adiabatic flame temperature for the octane in Example 8-7.

SOLUTION We have already calculated the enthalpy of the reactants as

$$H_R = -249\ 910 \text{ kJ/kg mol} = -107\ 442 \text{ Btu/lbm mol} \tag{a}$$

The enthalpy of the products is the sum of the formation enthalpies for CO_2 and H_2O and the product enthalpies to be taken from Table A-18:

$$H_P = H_P^\circ + n_P[\bar{h}_P(T) - \bar{h}_P(537)] \tag{b}$$

Thus we calculate

$$\begin{aligned} H_P^\circ &= (n\bar{h}_f^\circ)_{CO_2} + (n\bar{h}_f^\circ)_{H_2O} \\ &= (8)(-169\ 300) + (9)(-104\ 040) \\ &= -2\ 290\ 760 \text{ Btu} \end{aligned} \tag{c}$$

The total number of moles of products is

$$n_P = 8 + 9 + 94 + 12.5 = 123.5 \text{ mol/mol fuel} \tag{d}$$

and, from Table A-18, for 200 percent excess air,

$$\bar{h}_P(537) = 3774.9 \text{ Btu/lbm mol}$$

Inserting the numerical values in Eq. (*b*) gives

$$\bar{h}_P(T) = 21\ 454 \text{ Btu/lbm mol}$$

and, from Table A-18, the corresponding temperature for 200 percent excess air is

$$T_P = 2718°\text{R} = 1510 \text{ K}$$

Thus we see that there is good agreement between this calculation and that of Example 8-7. We may anticipate that the combustion products tables can be used to advantage where a more speedy calculation is desired.

EXAMPLE 8-9

Using the general combustion products table, estimate the adiabatic flame temperature for propylene (gas) C_3H_6 burning with 200 percent excess air at 1 atm. Assume reactants at 77°F (25°C).

SOLUTION The general reaction equation for a hydrocarbon of the form $(CH_2)_n$ with 200 percent excess air is

$$(CH_2)_n + 3xO_2 + 3.76(3x)N_2 \rightarrow aCO_2 + bH_2O + 2xO_2 + 3.76(3x)N_2 \quad (a)$$

from the carbon balance $a = n$ and from the hydrogen balance $b = n$. Then, the oxygen balance yields $x = 3n/2$. The total number of moles of products is therefore

$$n_P = n + n + 2\left(\frac{3n}{2}\right) + (3.76)(3)\left(\frac{3n}{2}\right) = 21.92n \text{ mol product/mol fuel} \quad (b)$$

For propylene $n = 3$ and $n_P = 65.76$ mol/mol fuel. The enthalpy of the reactants is just the formation enthalpy for propylene or

$$H_R = \bar{h}_f^\circ(C_3H_6) = 8790 \text{ Btu/lbm mol}$$

For the products, we first calculate the component of enthalpy due to the formation enthalpies of CO_2 and $H_2O(g)$:

$$\begin{aligned} H_P^\circ &= (n\bar{h}_f^\circ)_{CO_2} + (n\bar{h}_f^\circ)_{H_2O} \\ &= (3)(-169\,290) + (3)(-104\,040) \\ &= -819\,990 \text{ Btu/mol fuel} \end{aligned} \quad (c)$$

The other component of the enthalpy can be obtained from Table A-18 for 200 percent excess air. Thus, the total product enthalpy is

$$\begin{aligned} H_P &= H_{P_0}^\circ + H_P(T) - H_P(537) \\ &= H_P^\circ + n_P[\bar{h}_P(T) - \bar{h}_P(537)] \end{aligned} \quad (d)$$

Because we seek the adiabatic flame temperature,

$$H_P = H_R = 8790 \text{ Btu/mol fuel}$$

From Table A-18,

$$\bar{h}_P(537) = 3744.9 \text{ Btu/lbm mol}$$

Inserting the numerical values in Eq. (*d*), we have

$$8790 = -819\,990 + (65.76)[\bar{h}_P(T) - 3774.9]$$

and

$$\bar{h}_P(T) = 16\,378 \text{ Btu/lbm mol}$$

Consulting Table A-18 again, we find the corresponding temperature for the products to be

$$T_P = 2140°\text{R}$$

EXAMPLE 8-9M Irreversibility in a Burner
Gaseous octane (C_8H_{18}) is burned with air at a pressure of 4 atm. Both the fuel and air enter at 4 atm. Fuel entrance temperature is 298 K and the air enters at 710 K. The products leave the combustion chamber at 1100 K, and the overall process is adiabatic. Calculate the fuel/air ratio and the irreversibility for the process with $T_0 = 25°\text{C}$.

SOLUTION This process is what is normally experienced in the burner for a gas turbine cycle (see Sec. 9-12). Because the process is adiabatic we have

$$H_R = H_P \tag{a}$$

The combustion reaction for 1 mol of fuel is written as

$$C_8H_{18} + xO_2 + 3.76xN_2 \rightarrow aCO_2 + 6H_2O + cO_2 + 3.76xN_2 \tag{b}$$

Now x is unknown because we do not know how much excess air is supplied. The molal balances yield

C: $\quad 8 = a$

H: $\quad 18 = 2b \qquad b = 9$

O: $\quad 2x = 2a + b + 2c \qquad x = 12.5 + c$

or $c = x - 12.5$

To determine the AF ratio we combine Eq. (*a*), the energy balance, with Eq. (*b*) to solve for x and thence the AF ratio. For the reactants

$$H_R = (1)\bar{h}^\circ_{f\,C_8H_{18}} + (4.76x)(28.97)(h_{710} - h_{298})_{air}$$

where we are referencing to 298 K. From Tables A-6 and A-17M this gives

$$\begin{aligned} H_R &= (1)(-208\,450) + (4.76x)(28.97)(713.27 - 298.19) \\ &= -208\,450 + 57\,239x \qquad \text{kJ/kg mol fuel} \end{aligned} \tag{c}$$

For the products referenced to 298 K

$$\begin{aligned} H_P = {} & (8)(\bar{h}^\circ_f + \bar{h}_{1100} - \bar{h}_{298})_{CO_2} + (9)(\bar{h}^\circ_f + \bar{h}_{1100} - \bar{h}_{298})_{H_2O} \\ & + (x - 12.5)(\bar{h}_{1100} - \bar{h}_{298})_{O_2} + (3.76x)(\bar{h}_{1100} - \bar{h}_{298})_{N_2} \end{aligned}$$

Consulting Tables A-6 and A-18M we obtain

$$\begin{aligned} H_P &= (8)(-393\,520 + 38\,894) + (9)(-241\,820 + 30\,167) \\ &\quad + (x - 12.5)(26\,217) + (3.76x)(24\,757) \\ &= -5\,069\,598 + 119\,303x \qquad \text{kJ} \end{aligned} \tag{d}$$

Equating (*c*) and (*d*) gives

$$-208\,450 + 57\,239x = -5\,069\,598 + 119\,303x$$

and

$$x = 78.32$$

$$c = x - 12.5 = 65.82$$

For stoichiometric combustion $x = 12.5$, so the percent theoretical air is 78.32/12.5 = 626 percent. The air/fuel ratio is thus

$$\text{AF} = \frac{(4.76x)(28.97)}{114} = 94.74 \text{ kg air/kg fuel}$$

The change in availability is

$$\Delta B = B_P - B_R = H_P - H_R - T_0(S_P - S_R) \tag{e}$$

But $H_P = H_R$ so that

$$\Delta B = -T_0(S_P - S_R) \tag{e}$$

The reaction occurs at 4 atm, so we must change the tabular values of entropies accordingly. We shall assume ideal-gas behavior. For the reactants

$$C_8H_{18}: \qquad \bar{s} = \bar{s}^\circ - \mathcal{R} \ln\left(\frac{4.0}{1.0}\right)$$

$$= 466.73 - (8.314) \ln 4 = 455.20 \text{ kJ/kg·mol·K}$$

$$\text{Air:} \qquad \bar{s} = (28.97)[\phi_{710} - (0.287) \ln (4) + 4.1869]$$

$$= (28.97)[3.4014 - (0.287) \ln 4 + 4.1869]$$

$$= (208.31 \text{ kJ/kg·mol·K}$$

$$S_R = (1)(455.20) + (4.76)(78.32)(208.31)$$

$$= 78\ 113 \text{ kJ/K} \qquad (f)$$

To determine the entropy for the products we must first determine the mole fractions and partial pressures of the constituents. The total number of moles of products is

$$n_p = a + b + c + 3.76x = 377.30 \qquad (g)$$

The mole fractions are then

$$x_{CO_2} = \frac{8}{377.30} = 0.0212 \qquad x_{O_2} = \frac{65.82}{377.30} = 0.174\ 45$$

$$x_{H_2O} = \frac{9}{377.30} = 0.023\ 85 \qquad x_{N_2} = \frac{294.48}{377.30} = 0.780\ 49$$

The partial pressures are then these values multiplied by the total pressure of 4 atm. In using Table A-18M we note that the reference pressure is 0.1 MPa and *not* 1 atm (0.101 32 MPa). The absolute entropies are already included in these tables, and we need only correct for the actual partial pressures. We therefore use the relation

$$\bar{s} = \bar{s}\text{ (table)} - \mathcal{R} \ln \frac{p}{p_{\text{ref}}}$$

So, for the products at 1100 K,

$$CO_2: \qquad \bar{s} = 274.55 - 8.314 \ln \left[\frac{(4)(0.101\ 32)(0.0212)}{0.1}\right]$$

$$= 294.96 \text{ kJ/kg·mol·K}$$

$$H_2O: \qquad \bar{s} = 236.694 - 8.314 \ln \left[\frac{(4)(0.101\ 32)(0.023\ 85)}{0.1}\right]$$

$$= 256.12 \text{ kJ/kg·mol·K}$$

$$O_2: \quad \bar{s} = 246.922 - 8.314 \ln \left[\frac{(4)(0.101\,32)(0.174\,45)}{0.1}\right]$$
$$= 249.80 \text{ kJ/kg·mol·K}$$

$$N_2: \quad \bar{s} = 231.302 - 8.314 \ln \left[\frac{(4)(0.101\,32)(0.780\,49)}{0.1}\right]$$
$$= 221.73 \text{ kJ/kg·mol·K}$$

The total entropy of the products is thus

$$\begin{aligned} S_P &= S_{CO_2} + S_{H_2O} + S_{O_2} + S_{N_2} \\ &= (8)(294.96) + (9)(256.12) + (65.82)(249.80) + (294.48)(221.73) \\ &= 86\,402 \text{ kJ/K} \end{aligned}$$

Then from Eq. (*e*),

$$\Delta B = -(298)(86\,402 - 78\,113) = -2\,470\,122 \text{ kJ}$$

The maximum work *output* is the negative of this quantity, or

$$\dot{W}_{max} = +2\,470\,122 \text{ kJ}$$

The *actual* work output is zero, so the irreversibility is

$$\dot{I} = \dot{W}_{max} - \dot{W}_{act} = +2\,470\,122 \text{ kJ/kg mol fuel}$$

For 1 mol of fuel the mass of reactants is

$$m_R = (1)(114) + (4.76)(78.32)(28.97) = 10\,914 \text{ kg reactants/kg mol fuel}$$

so an alternate answer for the irreversibility would be

$$\dot{I} = \frac{2\,470\,122}{10\,914} = 226.31 \text{ kJ/kg reactants} = 97.26 \text{ Btu/lbm reactants}$$

EXAMPLE 8-10 Irreversibility in Combustion Reaction
Gaseous methane at 77°F is burned with the theoretical air at 77°F and used to heat another fluid. The discharge temperature for the combustion products is 340°F. Calculate the irreversibility of this process. Assume the process occurs at 1 atm and the lowest available temperature is 77°F.

SOLUTION The theoretical combustion equation for methane is

$$CH_4 + 2O_2 + (2)(3.76)N_2 \rightarrow CO_2 + 2H_2O + (2)(3.76)N_2$$

The enthalpy of the reactants for 1 mol of CH_4 is

$$H_R = \bar{h}^\circ_{CH_4} = -32\,210 \text{ Btu} \qquad (-33\,981 \text{ kJ}) \tag{a}$$

For the products,

$$\begin{aligned} H_P &= H_{CO_2} + H_{H_2O} + H_{N_2} \\ &= (1)[-169\,290 + 6552.9 - 4030.2]_{CO_2} \\ &\quad + (2)[-104\,040 + 6396.9 - 4258.3]_{H_2O} \\ &\quad + (7.52)[0 + 5564.4 - 3729.5]_{N_2} \\ &= -166\,767 - 203\,803 + 13\,798 \\ &= -356\,771 \text{ Btu} \qquad (-376\,390 \text{ kJ}) \end{aligned} \tag{b}$$

The heat *released* is thus

$$\begin{aligned} Q = H_R - H_P &= 324\,562 \text{ Btu/lbm mol fuel} \\ &= 20\,285 \text{ Btu/lbm fuel} \end{aligned} \tag{c}$$

To calculate the irreversibility we must evaluate the availability function for the inlet and exit streams; i.e., we must evaluate

$$b = h - T_0 s \tag{d}$$

where the enthalpy includes the energy of formation. Entropies must be calculated from

$$\bar{s} = \bar{s}^\circ - \mathcal{R} \ln \frac{p}{p_{ref}} \tag{e}$$

The reference pressure for tabular values of $\bar{s}^\circ$ is 1 atm. We assume that both the methane and air enter at 1 atm. The entropy of methane is thus

$$\bar{s}_{CH_4} = \bar{s}^\circ \text{ at } 77^\circ \text{ F} = 44.50 \text{ Btu/lbm·mol·°R}$$

We may compute the entropy for the air from the mole fractions of oxygen and nitrogen and their respective reference entropies. For 1 mol of air we have

$$x_{O_2} = \frac{1}{4.76} = 0.21 \qquad x_{N_2} = \frac{3.76}{4.76} = 0.79$$

and the entropy at 77°F is calculated using Eq. (*e*):

$$\begin{aligned} \bar{s} &= 0.21\bar{s}_{O_2} + 0.79\bar{s}_{N_2} \\ &= 0.21(48.982 - 1.986 \ln 0.21) \\ &\quad + 0.79(45.743 - 1.986 \ln 0.79) \\ &= 47.444 \text{ Btu/lbm·mol·°R} \end{aligned}$$

We assume that the *total* products are at 1 atm and that each constituent therein exists at its individual partial pressure. The partial pressure is calculated in terms of mole fractions. The total number of moles of products is

$$n_{\text{Tot}} = 1 + 2 + 7.52 = 10.52$$

so that the mole fractions are

$$x_{CO_2} = \frac{1}{10.52} = 0.0951$$

$$x_{H_2O} = \frac{2}{10.52} = 0.1901$$

$$x_{N_2} = \frac{7.52}{10.52} = 0.7148$$

The molal entropies of the product constituents at 340°F (800°R) are now calculated with the use of Table A-18 and Eq. (*e*):

$$CO_2: \quad \bar{s} = 54.839 - 1.986 \ln 0.0951 = 59.512 \text{ Btu/lbm·mol·°R}$$

$$H_2O: \quad \bar{s} = 48.316 - 1.986 \ln 0.1901 = 51.613$$

$$N_2: \quad \bar{s} = 48.552 - 1.986 \ln 0.7148 = 49.219$$

We may now evaluate the availability functions for the various streams in reactants and products.

Reactants

$$CH_4: \quad 1(\bar{h} - T_0\bar{s}) = 1[-32\,210 - (537)(44.50)] = -56\,107$$

$$\text{Air:} \quad (2)(4.76)(\bar{h} - T_0\bar{s}) = 9.52[0 - (537)(47.444)] = -242\,545$$

$$\text{Total reactants } B_R = -298\,652 \text{ Btu/lbm mol fuel} = -694\,660 \text{ kJ/kg mol fuel}$$

Products

$$CO_2: \quad 1(\bar{h} - T_0\bar{s}) = 1[-166\,767 - (537)(59.512)] = -198\,725$$

$$H_2O: \quad 2(\bar{h} - T_0\bar{s}) = 2[-101\,901 - (537)(51.613)] = -259\,234$$

$$N_2: \quad 7.52(\bar{h} - T_0\bar{s}) = 7.52[1834.9 - (537)(49.219)] = -184\,960$$

$$\text{Total products } B_P = -642\,919 \text{ Btu/lbm mol fuel} = -1\,495\,542 \text{ kJ/kg mol fuel}$$

The overall *decrease* in availability is therefore

$$B_R - B_P = 344\ 267 \text{ Btu/lbm mol fuel}$$
$$= 800\ 765 \text{ kJ/kg mol fuel}$$

Because there is no actual work in the process this decrease is equal to the irreversibility

$$I = +344\ 267 \text{ Btu/lbm mol fuel}$$
$$= 21\ 516 \text{ Btu/lbm fuel}$$
$$= 50\ 046 \text{ kJ/kg fuel}$$

8-9 EQUILIBRIUM

The second law of thermodynamics and the principle of increase of entropy furnish the basic notions for analysis of problems in equilibrium. In an isolated system the entropy always tends toward its maximum value; for this reason we shall say that in an isolated system the most stable state is one which maximizes the entropy. For processes which are nonadiabatic the considerations are not so simple. Let us first consider a simple compressible substance undergoing a general process in which both heat and work may be exchanged with the surroundings. In this process the total entropy change of the universe is expressed as a sum of the entropy change of the system and its surroundings. Thus, from the principle of increase of entropy [Eq. (5-22)],

$$dS_0 + dS_{sys} > 0 \tag{8-18}$$

In this process we assume that the system is in heat communication with the surroundings at a constant temperature of T_0, as indicated in Fig. 8-4. The

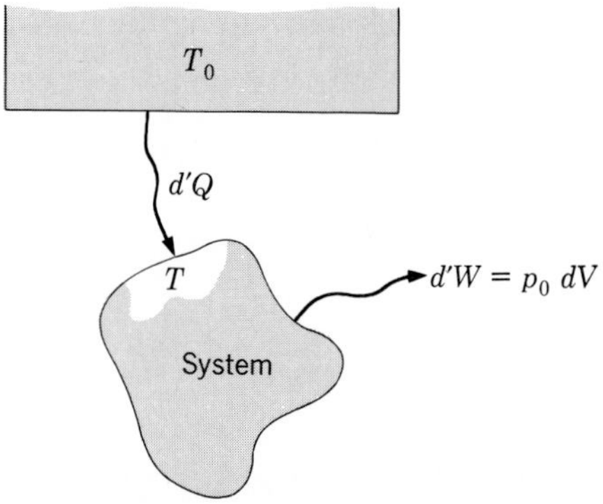

FIG. 8-4 Equilibrium of simple compressible substance.

entropy change of the surroundings is

$$dS_0 = -\frac{d'Q}{T_0} \tag{8-19}$$

where $d'Q$ is the heat transferred to the system from the surroundings. The work which the system does against the surroundings is $p_0\,dV$, where p_0 is the pressure of the surroundings and dV represents the change in volume of the system. The heat transfer to the system is then written as

$$d'Q = dU + p_0\,dV \tag{8-20}$$

Introducing Eqs. (8-19) and (8-20) into Eq. (8-18) gives for the inequality,

$$dU_{\text{sys}} + p_0\,dV_{\text{sys}} - T_0\,dS_{\text{sys}} < 0 \tag{8-21}$$

The inequality expressed by Eq. (8-21) presents a fairly general criterion for determining the direction that an irreversible process will take to seek the most probable equilibrium state. For an isolated system there is no work done ($dV_{\text{sys}} = 0$) and no internal energy change ($dU_{\text{sys}} = 0$), and the relation reduces to the simple increase of entropy principle.

Let us consider a somewhat simpler case in which the system remains in temperature and pressure equilibrium with its surroundings; that is, $T = T_0$, $p = p_0$. Then we have

$$dU + p\,dV - T\,dS < 0 \tag{8-22}$$

where, now, all properties refer to the system. In effect, we have removed the irreversibilities resulting from interactions with the surroundings by writing the inequality in this manner. Equation (8-22) is just a refined way of restating the simple relation

$$dS \geqslant \frac{d'Q}{T}$$

for an irreversible process. Equation (8-22) may be more conveniently written in terms of the Helmholtz and Gibbs functions. Recalling that

$$A = U - TS \tag{8-23}$$

$$G = H - TS = U + pV - TS \tag{8-24}$$

and

$$dA = dU - T\,dS - S\,dT \tag{8-25}$$

$$dG = dU + p\,dV + V\,dp - T\,dS - S\,dT \tag{8-26}$$

then

$$dU - T\,dS = dA + S\,dT$$
$$= dG - p\,dV - V\,dp + S\,dT$$

and Eq. (8-22) may be written as

$$dA + S\,dT + p\,dV < 0 \qquad (8\text{-}27)$$

$$dG + S\,dT - V\,dp < 0 \qquad (8\text{-}28)$$

For a process at constant temperature and constant volume it is clear that the Helmholtz function must approach its minimum value. Similarly, for a process at constant pressure and temperature, the Gibbs function must approach its minimum value. Thus our criterion for approach to equilibrium may be expressed as

$$(dA)_{T,V} < 0 \qquad (8\text{-}29)$$

$$(dG)_{T,P} < 0 \qquad (8\text{-}30)$$

At this point the reader is cautioned to remember that the only work modes we have considered are those involving $p\,dV$ type of work, i.e., work done by a simple compressible substance.

The equilibrium criterion expressed by Eq. (8-30) is an important one. It states, simply, that in a process at constant temperature and pressure the Gibbs function will decrease and seek its minimum value. Once this minimum is attained, the process will cease and equilibrium will be attained with no further change in the Gibbs function. Thus at equilibrium,

$$(dG)_{T,p} = 0 \qquad (8\text{-}31)$$

The net result of Eq. (8-31) is that, if a system is to change from one equilibrium state to another equilibrium state *at the same pressure and temperature,* the Gibbs function must remain constant. A classic example of this principle is the equilibrium of phases of a pure substance. If the system is to change from a saturated liquid to a saturated vapor, for example, we would obtain

$$g_f = g_g \qquad (8\text{-}32)$$

Expanding the Gibbs functions,

$$h_f - Ts_f = h_g - Ts_g$$

or

$$h_{fg} = Ts_{fg}$$

a relation we have utilized previously. We may also employ the equilibrium principle to rederive the Clausius-Clapeyron relation. As long as a system of vapor and liquid remains in equilibrium, we may write

$$dg_f = dg_g \tag{8-33}$$

when the system changes from one equilibrium saturation state to another. From Eq. (8-26),

$$dg = v\,dp - s\,dT$$

and

$$dg_f = v_f\,dp - s_f\,dT = dg_g = v_g\,dp - s_g\,dT \tag{8-34}$$

Collecting terms, we obtain

$$\frac{dp}{dT} = \frac{s_g - s_f}{v_g - v_f} = \frac{s_{fg}}{v_{fg}} = \frac{h_{fg}}{Tv_{fg}} \tag{8-35}$$

The application of Eq. (8-35) to single-component systems is fairly straightforward, as illustrated in previous chapters.

To obtain a criterion for equilibrium for a multicomponent mixture we simply rewrite Eq. (8-31) as

$$d\left(\sum_i n_i\bar{g}_i\right)_{T,p} = 0 \tag{8-36}$$

where now n_i is the number of moles of the ith constituent and $\bar{g}_i$ is the molal Gibbs function of that constituent. Equation (8-36) will serve as our basis for determining equilibrium concentrations in chemical reactions.

To calculate the Gibbs function of a substance we must choose a standard reference state for the entropy as well as the enthalpy. The usual practice is to take the same reference we used for enthalpy of formation, i.e., 1-atm pressure and 25°C (77°F). Then

$$\bar{g}_f^\circ = \bar{h}_f^\circ - T_0\bar{s}^\circ \tag{8-37}$$

where T_0 is taken as 25°C. Values of both $\bar{g}_f^\circ$ and $\bar{s}^\circ$ are given in Table A-6 of the Appendix for several substances. One also can use the symbol $\bar{g}^\circ$ to designate the Gibbs function at some reference state to be defined.

8-10 CHEMICAL EQUILIBRIUM OF IDEAL GASES

Consider a simple reaction involving four ideal gases:

$$\nu_1 A_1 + \nu_2 A_2 \leftrightharpoons \nu_3 A_3 + \nu_4 A_4 \tag{8-38}$$

where A_1 and A_2 represent the reactants and A_3 and A_4 represent the products. The values of the ν's are to be determined in accordance with the requirements for equilibrium at the temperature and pressure for the reaction. To evaluate these molal quantities we shall employ the equilibrium condition given by Eq. (8-36). Let us first show how the Gibbs function for an ideal gas is calculated. We have

$$dg = v\,dp - s\,dT \tag{8-39}$$

If we assume that pressure is expressed in atmospheres and that values of the Gibbs function at the standard state $\bar{g}^\circ$ are known, Eq. (8-38) may be integrated to give

$$\bar{g} - \bar{g}^\circ = \int_{p=1}^{p} \bar{v}\,dp - \int_{T_{\text{ref}}}^{T} \bar{s}\,dT \tag{8-40}$$

The Gibbs function *at the reference temperature* is then written as

$$\bar{g} - \bar{g}^\circ = \int_{p=1}^{p} \bar{v}\,dp = \int_{p=1}^{p} \mathcal{R}T\frac{dp}{p} = \mathcal{R}T\ln p \tag{8-41}$$

In both Eq. (8-40) and (8-41) we are choosing the reference state at 1-atm pressure.

Now let us consider the physical circumstance represented by Eq. (8-38). The reaction proceeds toward the right, and constituents 1 and 2 are steadily depleted while the products 3 and 4 build up. During this process the Gibbs function of the mixture is steadily decreasing. Eventually, the Gibbs function attains its minimum value and equilibrium is established. If the reaction proceeds at constant temperature and pressure, Eq. (8-36) describes the equilibrium condition, and this relation is equivalent to stating that, *at equilibrium*,

$$-dG_{\text{reactants}} = dG_{\text{products}}$$

or

$$dG_{\text{total}} = 0$$

Degree of Reaction or Dissociation

When a reaction involving stoichiometric proportions goes to completion, only A_3 and A_4 will appear and A_1 and A_2 are completely consumed. If stoichiometric proportions are not present, some of the initial reactants may remain even though the reaction goes to completion. The term *degree of reaction* ϵ indicates how far the reaction has gone to completion, and is usually defined in terms of one of the initial constituents. Based on constituent 1,

$$\epsilon = \frac{n_{1,\max} - n_1}{n_{1,\max} - n_{1,\min}} \tag{8-42}$$

where n_1 is the number of moles of constituent 1 present in the mixture, $n_{1,max}$ is usually the number of moles at the start of the reaction, and $n_{1,min}$ is the number at completion. If only one initial constituent is involved, ϵ is called the *degree of dissociation*. If a single initial constituent is ionized, ϵ is the *degree of ionization*.

Again, it is important to realize that at completion the reactants are not necessarily completely consumed. Consider the stoichiometric reaction for combustion of carbon:

$$C + O_2 \rightarrow CO_2 \tag{8-43}$$

It is clear that all the O_2 will not be consumed if more than 1 mol is supplied for each mole of carbon. We could have, for example,

$$C + 2O_2 \rightarrow CO_2 + O_2 \tag{8-44}$$

In this case the maximum and minimum number of moles of O_2 would be 2 and 1, respectively, and these values would be employed in conjunction with Eq. (8-42) to calculate the degree of reaction.

Returning now to the matter of equilibrium, we can envision the reaction proceeding, at constant temperature, as shown in Fig. 8-5, until

$$\left(\frac{\partial G_{total}}{\partial n_1}\right)_T = 0 \tag{8-45}$$

If Eq. (8-38) is a stoichiometric mixture, then the ν's are so arranged that $n_{1,max}$ is the initial number of moles and $n_{1,min}$ is zero, since the original constituents are completely consumed. Some algebraic manipulation will show that the change in the number of moles of each constituent is a linear function of the

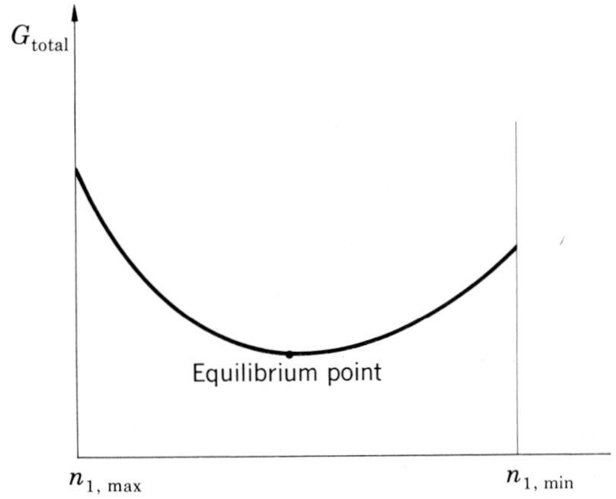

FIG. 8-5 Behavior of Gibbs function at equilibrium.

initial number of moles, the ν's, and the change in the degree of reaction. Thus

$$dn_1 = -\nu_1\, d\epsilon$$

$$dn_2 = -\nu_2\, d\epsilon$$

$$dn_3 = \nu_3\, d\epsilon$$

$$dn_4 = \nu_4\, d\epsilon$$

Accordingly, Eq. (8-36) may be written as

$$d\left(\sum_i n_i \bar{g}_i\right)_{T,p} = \left(\sum_i \bar{g}_i\, dn_i\right)_{T,p} = 0$$

or

$$(-\bar{g}_1\nu_1 - \bar{g}_2\nu_2 + \bar{g}_3\nu_3 + \bar{g}_4\nu_4)d\epsilon = 0 \tag{8-46}$$

Free Energy

Making use of Eq. (8-41) for the Gibbs function at the reference temperature, this relation becomes

$$-\mathcal{R}T(\nu_3 \ln p_3 + \nu_4 \ln p_4 - \nu_1 \ln p_1 - \nu_2 \ln p_2) = \nu_3\bar{g}_3^\circ + \nu_4\bar{g}_4^\circ - \nu_1\bar{g}_1^\circ - \nu_2\bar{g}_2^\circ \tag{8-47}$$

The grouping of terms on the right side of this equation is clearly the difference in standard Gibbs functions of the products and reactants, and is sometimes called the *free-energy change* for the reaction designated by the symbol ΔG°. We should note once again that $\bar{g}^\circ$ is defined for some specified reference state, which in this case we shall choose as 1-atm pressure and the temperature of the mixture. In reality then

$$\Delta G^\circ = f(T, p_0)$$

where p_0 is the reference pressure taken as 1 atm.

Equation (8-47) is now written more compactly as

$$-\mathcal{R}T \ln \frac{p_3^{\nu_3} p_4^{\nu_4}}{p_1^{\nu_1} p_2^{\nu_2}} = \Delta G^\circ \tag{8-48}$$

The Equilibrium Constant

The *equilibrium constant* K_p is defined as

$$K_p = \frac{p_3^{\nu_3} p_4^{\nu_4}}{p_1^{\nu_1} p_2^{\nu_2}} \tag{8-49}$$

so that

$$\Delta G^\circ = -\mathcal{R}T \ln K_p \tag{8-50}$$

TABLE 8-3

Logarithms to the Base 10 of the Equilibrium Constant K_p. $K_p = p_3^{\nu_3}p_4^{\nu_4}/p_1^{\nu_1}p_2^{\nu_2}$ for reaction $\nu_1A_1 + \nu_2A_2 \leftrightharpoons \nu_3A_3 + \nu_4A_4$

T, K	$H_2 \leftrightharpoons 2H$	$O_2 \leftrightharpoons 2O$	$N_2 \leftrightharpoons 2N$	$H_2O(g) \leftrightharpoons H_2 + \frac{1}{2}O_2$	$H_2O(g) \leftrightharpoons OH + \frac{1}{2}H_2$	$CO_2 \leftrightharpoons CO + \frac{1}{2}O_2$	$\frac{1}{2}O_2 + \frac{1}{2}N_2 \leftrightharpoons NO$	$CO_2 + H_2 \leftrightharpoons CO + H_2O(g)$
298	−71.228	−81.208	−159.600	−40.048	−46.054	−45.066	−15.171	−5.018
500	−40.318	−45.880	−92.672	−22.886	−26.130	−25.025	−8.783	−2.139
1000	−17.292	−19.614	−43.056	−10.062	−11.280	−10.221	−4.062	−0.159
1500	−9.514	−10.790	−26.434	−5.725	−6.284	−5.316	−2.487	+0.409
1800	−6.896	−7.836	−20.874	−4.270	−4.613	−3.693	−1.962	+0.577
2000	−5.582	−6.356	−18.092	−3.540	−3.776	−2.884	−1.699	+0.656
2200	−4.502	−5.142	−15.810	−2.942	−3.091	−2.226	−1.484	+0.716
2400	−3.600	−4.130	−13.908	−2.443	−2.520	−1.679	−1.305	+0.764
2500	−3.202	−3.684	−13.070	−2.224	−2.270	−1.440	−1.227	+0.784
2600	−2.836	−3.272	−12.298	−2.021	−2.038	−1.219	−1.154	+0.802
2800	−2.178	−2.536	−10.914	−1.658	−1.624	−0.825	−1.025	+0.833
3000	−1.606	−1.898	−9.716	−1.343	−1.265	−0.485	−0.913	+0.858
3200	−1.106	−1.340	−8.664	−1.067	−0.951	−0.189	−0.815	+0.878
3500	−0.462	−0.620	−7.312	−0.712	−0.547	+0.190	−0.690	+0.902
4000	+0.400	+0.340	−5.504	−0.238	−0.011	+0.692	−0.524	+0.930
4500	+1.074	+1.086	−4.094	+0.133	+0.408	+1.079	−0.397	+0.946
5000	+1.612	+1.686	−2.962	+0.430	+0.741	+1.386	−0.296	+0.956

Source: JANAF Thermochemical Tables [1] and Selected Values of Chemical Thermodynamic Properties [2].

Equation (8-50) is called the *law of mass action*. Tables 8-3 and 8-4 list values of the equilibrium constant as a function of temperature for several simple reactions, and Fig. 8-6 gives a graphical presentation of the equilibrium constant for some of these reactions.

For a constant pressure circumstance ΔG° is a function of only the properties of the constituents at temperature T, so it follows that the equilibrium

TABLE 8-4

Logarithms to the Base 10 of the Equilibrium Constants for Ionization of Na and Cs, as Calculated from Saha Equation

T, K	$Na \leftrightharpoons Na^+ + e^-$	$Cs \leftrightharpoons Cs^+ + e^-$
298	−32.3	−25.1
400	−24.3	−17.5
600	−14.6	−10.0
800	−9.58	−6.15
1000	−6.54	−3.79
1200	−4.47	−2.18
1400	−2.97	−1.010
1600	−1.819	−0.108
1800	−0.913	+0.609
2000	−0.175	+1.194
2200	+0.438	+1.682
2400	+0.956	+2.098
2600	+1.404	+2.46
2800	+1.792	+2.77
3000	+2.13	+3.05
3200	+2.44	+3.29
3500	+2.84	+3.62
4000	+3.38	+4.07
4500	+3.82	+4.43
5000	+4.18	+4.73

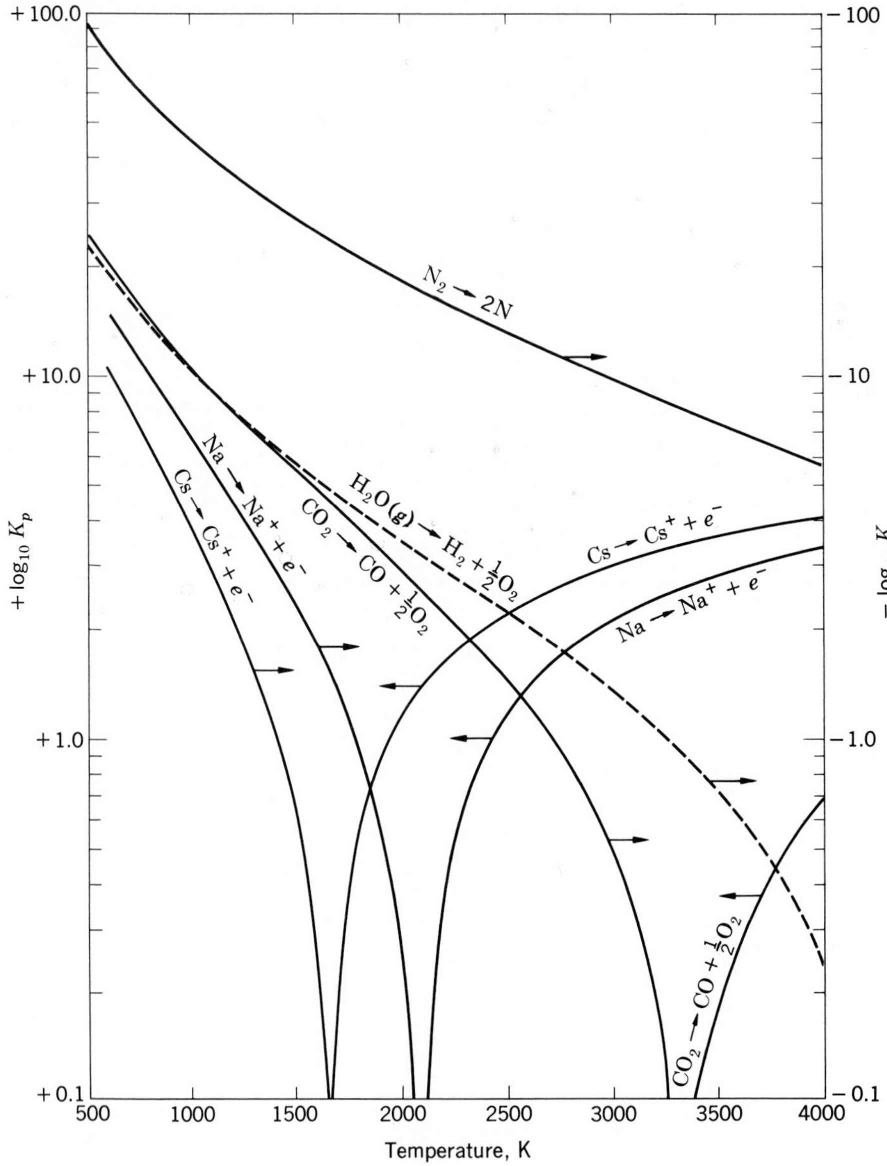

FIG. 8-6 Equilibrium constants for several reactions.

constant is also a function of temperature alone. We reemphasize that these equations are for perfect gases. In this development the reader should note that

1. Pressure is expressed in *atmospheres*.
2. The values of the ν's employed for the calculation of the equilibrium constant are the stoichiometric values for the particular reaction.

At equilibrium the partial pressures of the constituents may be expressed in terms of the mole fractions of these constituents and the total pressure. Thus

$$p_i = x_i p_{\text{total}}$$

and the relation for the equilibrium constant becomes

$$K_p = \frac{x_3^{\nu_3} x_4^{\nu_4}}{x_1^{\nu_1} x_2^{\nu_2}} p_{\text{total}}^{\nu_3+\nu_4-\nu_1-\nu_2} \tag{8-51}$$

A quick inspection of Table 8-3 shows that for the simple reactions indicated the equilibrium constant increases with temperature. In terms of Eq. (8-38) this means that these reactions proceed further toward completion as the temperature is elevated.

It is clear that the value of ΔG° depends on temperature and, therefore, the equilibrium constant pertains to equilibrium of the reaction for the temperature at which ΔG° is evaluated and on the pressure of the mixture.

EXAMPLE 8-11

Determine the equilibrium compositions for dissociation of CO_2 into CO and O_2 at 3200 K and 1 atm. Also determine the degree of reaction (or degree of dissociation).

SOLUTION The chemical equation for dissociation is

$$CO_2 \leftrightharpoons CO + \tfrac{1}{2}O_2 \tag{a}$$

It is to be noted that this expression indicates the stoichiometric coefficients employed for calculation of the equilibrium constant. Thus, if we rewrite Eq. (*a*) as

$$\nu_1 CO_2 \leftrightharpoons \nu_3 CO + \nu_4 O_2 \tag{b}$$

we may express the equilibrium constant as

$$K_p = \frac{x_{CO}^{\nu_3} \times x_{O_2}^{\nu_4}}{x_{CO_2}^{\nu_1}} p_{\text{total}}^{\nu_3+\nu_4-\nu_1}$$

where $\nu_1 = 1$, $\nu_3 = 1$, $\nu_4 = \frac{1}{2}$, and $p_{\text{total}} = 1$ atm. Then

$$K_p = \frac{x_{CO} x_{O_2}^{1/2}}{x_{CO_2}} (1)^{1/2} \tag{c}$$

From Table 8-2 we have $\log K_p = -0.189$ and $K_p = 0.647$. As the reaction proceeds we do *not* have stoichiometric proportions present in the mixture; that is, $\nu_1 \neq 1$, $\nu_3 \neq 1$, etc. Instead, we have the CO_2 dissociated to some degree

ϵ, so that

$$CO_2 \rightarrow (1 - \epsilon)CO_2 + \epsilon CO + \frac{\epsilon}{2} O_2$$

The total number of moles in the equilibrium mixture is therefore

$$n_{\text{total}} = (1 - \epsilon) + \epsilon + \frac{\epsilon}{2} = 1 + \frac{\epsilon}{2} \qquad (d)$$

so that the equilibrium mole fractions are

$$x_{CO_2} = \frac{1 - \epsilon}{1 + \epsilon/2}$$

$$x_{CO} = \frac{\epsilon}{1 + \epsilon/2}$$

$$x_{O_2} = \frac{\epsilon/2}{1 + \epsilon/2}$$

Using Eq. (c) we obtain

$$K_p = \frac{\dfrac{\epsilon}{1 + \epsilon/2}\left(\dfrac{\epsilon/2}{1 + \epsilon/2}\right)^{1/2}}{\dfrac{1 - \epsilon}{1 + \epsilon/2}}$$

or

$$K_p^2 = \frac{\epsilon^3}{(1 - \epsilon)^2(2 + \epsilon)} = (0.647)^2 \qquad (e)$$

Equation (e) may be solved by iterative methods to yield

$$\epsilon = 0.578$$

The equilibrium concentrations are

For CO_2,

$$\frac{1 - \epsilon}{1 + \epsilon/2} = 0.327$$

For CO,

$$\frac{\epsilon}{1 + \epsilon/2} = 0.448$$

For O_2,

$$\frac{\epsilon/2}{1 + \epsilon/2} = 0.225$$

EXAMPLE 8-12

CO_2 is heated at constant pressure of 1 atm from 77°F to 3200 K. Calculate the heat required per mole and compare with the heating energy which would be required if there were no dissociation.

SOLUTION We have already determined the degree of dissociation for CO_2 at this temperature in Example 8-11. The basic energy balance is

$$H_R + Q = H_P \tag{a}$$

where the reactant is 1 mol of CO_2 and the products are the concentrations of CO_2, CO, and O_2 which may be determined from Example 8-11. The total moles of products are $1 + \epsilon/2 = 1 + 0.578/2 = 1.289$ and the number of moles of each constituent of the products is

$$n_{CO_2} = 1 - \epsilon = 0.422$$

$$n_{CO} = \epsilon = 0.578$$

$$n_{O_2} = \frac{\epsilon}{2} = 0.289$$

Because the reactants are at 77°F,

$$H_R = \bar{h}^\circ_{CO_2} = -169\ 290 \text{ Btu/lbm mol}$$

The temperature of the products is 3200 K = 5760°R, and the highest temperature entry in the gas tables A-18 is 5300°R; however, the specific heats may be assumed very nearly constant between these two temperatures so that the tables may be extrapolated to give at 5760°R

$$\bar{h}_{CO_2} = 75\ 357 \text{ Btu/lbm mol} \qquad (175\ 273 \text{ kJ/kg mol})$$

$$\bar{h}_{CO} = 47\ 147 \text{ Btu/lbm mol} \qquad (109\ 659 \text{ kJ/kg mol})$$

$$\bar{h}_{O_2} = 49\ 283 \text{ Btu/lbm mol} \qquad (114\ 626 \text{ kJ/kg mol})$$

The total enthalpy of the products is

$$\begin{aligned} H_p &= (0.422)(\bar{h}^\circ + \bar{h}_T - \bar{h}_{537})_{CO_2} \\ &\quad + (0.578)(\bar{h}^\circ + \bar{h}_T - \bar{h}_{537})_{CO} \\ &\quad + (0.289)(\bar{h}_T - \bar{h}_{537})_{O_2} \\ &= (0.422)(-169\ 290 + 75\ 357 - 4030) \\ &\quad + (0.578)(-47\ 540 + 47\ 147 - 3730) \\ &\quad + (0.289)(49\ 283 - 3725) \\ &= -30\ 557 \text{ Btu/mol reactant} \qquad (-71\ 072 \text{ kJ/mol}) \end{aligned}$$

From Eq. (*a*) the heat transfer is

$$Q = H_p - H_R = -30\,557 - (-169\,290) = 138\,733 \text{ Btu/lbm mol}$$

If there were no dissociation, the heating of CO_2 could be calculated from the entries in the gas tables as

$$\begin{aligned} Q &= \bar{h}_T - \bar{h}_{537} = \bar{h}_{5760} - \bar{h}_{537} \\ &= 75\,357 - 4030 \\ &= 71\,327 \text{ Btu/lbm mol} \qquad (165\,899 \quad \text{kJ/kg mol}) \end{aligned}$$

The extra energy of 138 733 − 71 327 = 67 406 Btu/lbm mol is required to dissociate the CO_2.

From this example we can see that the effects of dissociation in a combustion process would be to produce a lower combustion temperature than the adiabatic flame temperature calculated with no dissociation. When dissociation is involved, part of the heat of combustion is absorbed in the dissociation reaction(s), and thus less energy is available to heat the products to higher temperatures.

EXAMPLE 8-13

Determine the equilibrium compositions for ionization of cesium at 2000 K and 1-atm pressure.

SOLUTION The chemical equation for ionization is

$$\text{Cs} \leftrightarrows \text{Cs}^+ + e^- \tag{a}$$

In terms of stoichiometric coefficients,

$$\nu_1\text{Cs} \leftrightarrows \nu_3\text{Cs}^+ + \nu_4 e^- \tag{b}$$

and the equilibrium constant is expressed as

$$K_p = \frac{x_{\text{Cs}^+}^{\nu_3} x_{e^-}^{\nu_4}}{x_{\text{Cs}}^{\nu_1}} p_{\text{total}}^{\nu_3+\nu_4-\nu_1}$$

We have $\nu_1 = \nu_3 = \nu_4 = 1$, so that

$$K_p = \frac{x_{\text{Cs}} x_e}{x_{\text{Cs}}} \tag{c}$$

From Table 8-3 we have log K_p = 1.194 and K_p = 15.63. For the incomplete reaction we have

$$\text{Cs} \rightarrow (1 - \epsilon)\text{Cs} + \epsilon\text{Cs}^+ + \epsilon e^- \tag{d}$$

where ϵ is the degree of reaction.

The total number of moles in the equilibrium mixture is thus

$$n_{\text{total}} = (1 - \epsilon) + \epsilon + \epsilon = 1 + \epsilon$$

and the equilibrium mole fractions are

$$x_{\text{Cs}} = \frac{1 - \epsilon}{1 + \epsilon}$$

$$x_{\text{Cs}^+} = \frac{\epsilon}{1 + \epsilon}$$

$$x_{e^-} = \frac{\epsilon}{1 + \epsilon}$$

Inserting these relations into Eq. (*c*) results in

$$K_p = \frac{[\epsilon/(1 + \epsilon)][\epsilon/(1 + \epsilon)]}{(1 - \epsilon)/(1 + \epsilon)} = \frac{\epsilon^2}{1 - \epsilon^2}$$

Inserting $K_p = 15.63$ results in

$$\epsilon = 0.97$$

The equilibrium concentrations are

For Cs,

$$\frac{1 - \epsilon}{1 + \epsilon} = 0.0152$$

For Cs^+,

$$\frac{\epsilon}{1 + \epsilon} = 0.492$$

For e^-,

$$\frac{\epsilon}{1 + \epsilon} = 0.492$$

From these results it is evident that the ionization of cesium is almost complete at 2000 K.

8-11 EFFECTS OF NONREACTING GASES

In a mixture of gases it is quite common for only one constituent to be involved in a dissociation process because of the temperature level of the mixture. For such circumstances the other components of the mixture carry through the reaction unaltered; however, their presence affects the calculation of mole fractions and partial pressures for the reacting constituent and thus influences the final equilibrium concentrations. For example, one might have a mixture of 1 mol of cesium and 1 mol of nitrogen at 2000 K. Cesium is almost completely ionized at this temperature but there is hardly any dissociation of nitrogen (log $K_p = -18.092$). We would therefore have

$$Cs + N_2 \leftrightharpoons Cs^+ + e^- + N_2$$

Using the degree of reaction ϵ based on consumption of Cs,

$$Cs + N_2 \rightarrow (1 - \epsilon)Cs + \epsilon Cs^+ + \epsilon e^- + N_2$$

The total number of moles in the mixture is therefore

$$n_{\text{total}} = (1 - \epsilon) + \epsilon + \epsilon + 1 = 2 + \epsilon$$

For a mixture with 1 mol of Cs and 2 mol of N_2 the value of n_{total} would be $3 + \epsilon$, and so on. The mole fractions of the quantities used in the equilibrium constant are also changed so that

$$x_{Cs} = \frac{1 - \epsilon}{2 + \epsilon} \qquad x_{Cs^+} = \frac{\epsilon}{2 + \epsilon}$$

$$x_{e^-} = \frac{\epsilon}{2 + \epsilon}$$

We can insert these values in Eq. (*c*) of Example 8-13 to obtain

$$K_p = \frac{[\epsilon/(2 + \epsilon)][\epsilon/(2 + \epsilon)]}{(1 - \epsilon)/(2 + \epsilon)}$$

$$= \frac{\epsilon^2}{2 - \epsilon - \epsilon^2}$$

Carrying through with a numerical calculation at 2000 K, we have $K_p = 15.63$ from Example 8-13, and so a solution for ϵ in the presence of an inert gas gives

$$\epsilon = 0.98$$

There is not much difference between this result and the value of $\epsilon = 0.97$ for pure ionization of cesium because the ionization is so nearly complete in both cases. It is interesting then to make the calculation at a lower temperature, say 1400 K, to make a further comparison.

At 1400 K, $\log K_p = -1.010$ and $K_p = 0.0977$ for Cs. For pure ionization we have from Example 8-13

$$K_p = \frac{\epsilon^2}{1 - \epsilon^2}$$

and with the numerical value of K_p inserted

$$\epsilon = 0.298$$

With 1 mol of nitrogen in the mixture the value of ϵ would be $\epsilon = 0.378$. So at this lower temperature there is a marked difference in the degree of ionization of the cesium, depending on whether a nonreacting gas is present.

8-12 EQUILIBRIUM IN MULTIPLE REACTIONS

The previous section has shown how we may use equilibrium principles to establish the degree of reaction of ideal gases and ionization processes. Now let us complicate the situation by placing additional gases in the mixture so that more than one reaction may take place. One example of such multiple reactions would be the dissociation of H_2O by the two reactions

$$H_2O \leftrightharpoons H_2 + \tfrac{1}{2}O_2$$

$$H_2O \leftrightharpoons \tfrac{1}{2}H_2 + OH$$

Another example might be the combined dissociation of CO_2 and N_2:

$$CO_2 \leftrightharpoons CO + \tfrac{1}{2}O_2$$

$$\tfrac{1}{2}O_2 + \tfrac{1}{2}N_2 \leftrightharpoons NO$$

We may write for the two general reactions

$$\nu_1 A_1 + \nu_2 A_2 \leftrightharpoons \nu_3 A_3 + \nu_4 A_4 \qquad (8\text{-}52)$$

$$\nu_5 A_1 + \nu_6 A_6 \leftrightharpoons \nu_7 A_7 + \nu_8 A_8 \qquad (8\text{-}53)$$

where we are taking one reactant A_1 as common to both equations but with a possibly different stoichiometric proportion, that is, ν_1 does not necessarily equal ν_5. We shall designate Eq. (8-52) as reaction 1 and Eq. (8-53) as reaction 2, with their corresponding degrees of reaction ϵ_1 and ϵ_2. Both degrees of reaction are based on consumption of reactant A_1 in each equation. For reaction 1 we have

$$dn_1 = -\nu_1 \, d\epsilon_1$$

$$dn_2 = -\nu_2 \, d\epsilon_1$$

$$dn_3 = +\nu_3 \, d\epsilon_1$$

$$dn_4 = +\nu_4 \, d\epsilon_1$$

while for reaction 2,

$$dn_1 = -\nu_5 \, d\epsilon_2$$

$$dn_6 = -\nu_6 \, d\epsilon_2$$

$$dn_7 = +\nu_7 \, d\epsilon_2$$

$$dn_8 = +\nu_8 \, d\epsilon_2$$

For the combined reactions the total consumption of reactant A_1 is

$$dn_1 = -\nu_1 \, d\epsilon_1 - \nu_5 \, d\epsilon_2$$

while the other constituents remain independent. We may now apply Eq. (8-36) as the condition for equilibrium at constant temperature and pressure and write

$$\left(\sum \bar{g}_i \, dn_i\right)_{T,p} = 0$$

or

$$(-\bar{g}_i\nu_1 - \bar{g}_2\nu_2 + \bar{g}_3\nu_3 + \bar{g}_4\nu_4) \, d\epsilon_1 + (-\bar{g}_1\nu_5 - \bar{g}_6\nu_6 + \bar{g}_7\nu_7 + \bar{g}_8\nu_8) \, d\epsilon_2 = 0 \quad (8\text{-}54)$$

Because we are assuming an ideal-gas mixture we can use Eq. (8-41) to express the Gibbs function at the reference temperature so that we obtain two relations analogous to that for the single reaction:

$$-\mathcal{R}T \ln \left(\frac{p_3^{\nu_3} p_4^{\nu_4}}{p_1^{\nu_1} p_2^{\nu_2}}\right) = \Delta G_1^\circ \quad (8\text{-}55)$$

$$-\mathcal{R}T \ln \left(\frac{p_7^{\nu_7} p_8^{\nu_8}}{p_1^{\nu_5} p_6^{\nu_6}}\right) = \Delta G_2^\circ \quad (8\text{-}56)$$

where the free energies are

$$\Delta G_1^\circ = \nu_3 \bar{g}_3^\circ + \nu_4 \bar{g}_4^\circ - \nu_1 \bar{g}_1^\circ - \nu_2 \bar{g}_2^\circ \quad (8\text{-}57)$$

$$\Delta G_2^\circ = \nu_7 \bar{g}_7^\circ + \nu_8 \bar{g}_8^\circ - \nu_5 \bar{g}_1^\circ - \nu_6 \bar{g}_6^\circ \quad (8\text{-}58)$$

Equations (8-54) and (8-55) may be arranged in the equilibrium constant form as

$$\ln K_{p_1} = \frac{-\Delta G_1^\circ}{\mathcal{R}T} \tag{8-59}$$

$$\ln K_{p_2} = \frac{-\Delta G_2^\circ}{\mathcal{R}T} \tag{8-60}$$

where we recognize that both equilibrium constants may be expressed in terms of mole fractions of the constituents in accordance with Eq. (8-51). To obtain the equilibrium concentrations the two nonlinear equations must be solved simultaneously. For systems involving many simultaneous reactions this becomes a rather formidable task and is best accomplished with a computer.

EXAMPLE 8-14

Calculate the equilibrium concentrations for the following two reactions at 1 atm and 3000 K, starting with 1 mol of CO_2 and $\frac{1}{2}$ mol each of O_2 and N_2:

$$CO_2 \leftrightarrows CO + \tfrac{1}{2}O_2$$

$$\tfrac{1}{2}O_2 + \tfrac{1}{2}N_2 \leftrightarrows NO$$

SOLUTION The two dissociation reactions are

$$CO_2 \rightarrow (1 - \epsilon_1)CO_2 + \epsilon_1 CO + \frac{\epsilon_1}{2} O_2$$

$$\tfrac{1}{2}O_2 + \tfrac{1}{2}N_2 \rightarrow (1 - \epsilon_2)\tfrac{1}{2}O_2 + (1 - \epsilon_2)\tfrac{1}{2}N_2 + \epsilon_2 NO$$

where ϵ_1 and ϵ_2 are the two degrees of reaction. The equilibrium mixture is the sum of the right sides of these equations or

$$(1 - \epsilon_1)CO_2 + \epsilon_1 CO + \left(\frac{1}{2} - \frac{1}{2}\epsilon_2 + \frac{1}{2}\epsilon_1\right) O_2 + \left(\frac{1}{2} - \frac{1}{2}\epsilon_2\right) N_2 + \epsilon_2 NO$$

The total number of moles in the equilibrium mixture is therefore

$$n_{\text{total}} = 1 - \epsilon_1 + \epsilon_1 + \frac{1}{2} - \frac{1}{2}\epsilon_2 + \frac{1}{2}\epsilon_1 + \frac{1}{2} - \frac{1}{2}\epsilon_2 + \epsilon_2$$

$$= 2 + \frac{1}{2}\epsilon_1$$

and the mole fractions at equilibrium are

$$x_{CO_2} = \frac{1 - \epsilon_1}{2 + \epsilon_1/2} \qquad x_{CO} = \frac{\epsilon_1}{2 + \epsilon_1/2}$$

$$x_{O_2} = \frac{1 - \epsilon_2 + \epsilon_1}{4 + \epsilon_1} \qquad x_{N_2} = \frac{1 - \epsilon_2}{4 + \epsilon_1}$$

$$x_{NO} = \frac{\epsilon_2}{2 + \epsilon_1/2}$$

At 3000 K we have

$$\log K_{p_1} = -0.485 \qquad K_1 = 0.3272 \qquad (CO_2)$$

$$\log K_{p_2} = -0.913 \qquad K_{p_2} = 0.1222 \qquad (O_2, N_2)$$

Because the reaction occurs at 1 atm we may apply Eq. (8-51) to obtain

$$K_{p_1} = 0.3273 = \frac{\left(\dfrac{\epsilon_1}{2 + \epsilon_1/2}\right)^1 \left(\dfrac{1 - \epsilon_2 + \epsilon_1}{4 + \epsilon_1}\right)^{1/2}}{\left(\dfrac{1 - \epsilon_1}{2 + \epsilon_1/2}\right)^1} \tag{a}$$

and

$$K_{p_2} = 0.1222 = \frac{\left(\dfrac{\epsilon_2}{2 + \epsilon_1/2}\right)^1}{\left(\dfrac{1 - \epsilon_2 + \epsilon_1}{4 + \epsilon_1}\right)^{1/2} \left(\dfrac{1 - \epsilon_2}{4 + \epsilon_1}\right)^{1/2}} \tag{b}$$

Equations (*a*) and (*b*) must now be solved simultaneously to obtain the values of ϵ_1 and ϵ_2. The results are

$$\epsilon_1 = 0.3736 \qquad \epsilon_2 = 0.05644$$

and the resulting mole fractions at equilibrium are obtained as

$$x_{CO_2} = 0.2864 \qquad x_{CO} = 0.1708$$

$$x_{O_2} = 0.3012 \qquad x_{N_2} = 0.2157$$

$$x_{NO} = 0.0258$$

The final molal reaction is

$$CO_2 + \tfrac{1}{2}O_2 + \tfrac{1}{2}N_2 \rightarrow$$
$$0.6264CO_2 + 0.3736CO + 0.6587O_2 + 0.4717N_2 + 0.0564NO$$

8-13 THE VAN'T HOFF EQUATION

We can extend our discussion of equilibrium constants to derive a very useful relation between the enthalpy of reaction and the equilibrium constant for a constant-pressure reaction. Let us now use the symbols K_p to designate the equilibrium constant at total pressure p and $\Delta G°\,(T, p_0)$ to designate the corresponding Gibbs free energy. Then we may write Eq. (8-50) as

$$\Delta G°(T, p_0) = -\mathcal{R}T \ln K_p \tag{8-61}$$

Differentiating with respect to T, at constant p, gives

$$\frac{d(\Delta G°)}{dT} = -\mathcal{R}T \frac{1}{K_p}\frac{dK_p}{dT} - \mathcal{R} \ln K_p$$

which may be rearranged and combined with Eq. (8-61) to give

$$\frac{dK_p}{dT} = \frac{K_p}{\mathcal{R}T^2}\left[\Delta G°(T, p_0) - T\frac{d(\Delta G°)}{dT}\right] \tag{8-62}$$

We can express the Gibbs free energy as

$$\Delta G° = \Delta H° - T\,\Delta S° \tag{8-63}$$

From Eq. (6-15*b*) we know that

$$dg = -s\,dT + v\,dp$$

so for a reaction at constant pressure,

$$\frac{dg}{dT} = -s$$

and

$$\frac{d(\Delta G°)}{dT} = -\Delta S° \tag{8-64}$$

Now, substituting Eqs. (8-63) and (8-64) into (8-62) gives

$$\frac{dK_p}{dT} = \frac{K_p}{\mathcal{R}T^2}\,\Delta H°$$

or written in logarithmic form

$$\frac{d(\ln K_p)}{dT} = \frac{\Delta H°(T, p_0)}{\mathcal{R}T^2} \tag{8-65}$$

where we have now inserted p_0 again to emphasize the reference pressure basis.

Equation (8-65) is called the van't Hoff isobar equation and is quite useful in that it enables us to calculate enthalpy change in a reaction from the equilibrium compositions as determined by K_p. In a reciprocal way, energy balances on reactions can yield important information on the variation of K_p with temperature. Let us rearrange Eq. (8-65) as a function of $1/T$. Then

$$\frac{d(\ln K_p)}{d(1/T)} = \frac{-\Delta H^\circ(T, p_0)}{\mathcal{R}} \tag{8-66}$$

Equation (8-66) indicates to us that the slope of the curve in a plot of ln K_p versus $1/T$ will yield the value of $\Delta H^\circ/\mathcal{R}$.

For many reactions the value of ΔH° is very nearly constant over a rather wide range of temperatures. In such cases Eq. (8-65) can be integrated between two temperatures T_1 and T_2 to yield

$$\ln \frac{K_{p_2}}{K_{p_1}} = \frac{\Delta H^\circ}{\mathcal{R}} \left(\frac{1}{T_1} - \frac{1}{T_2} \right) \tag{8-67}$$

and if one knows the value of K_{p_1} at T_1 it is an easy task to calculate the value K_{p_2} at temperature T_2.

The van't Hoff equation can also be helpful in evaluating the endothermic or exothermic nature of a reaction. If the equilibrium constant increases with temperature, that is, $K_{p_2} > K_{p_1}$ for $T_2 > T_1$, Eq. (8-65) indicates that ΔH° must be positive and the reaction is endothermic. If K_p decreases with an increase in temperature, ΔH° will be negative and the reaction will be exothermic.

EXAMPLE 8-15

Estimate the enthalpy of reaction of CO_2 to form CO and O_2 at 2000 K.

SOLUTION We shall use the van't Hoff equation to make this estimate and assume that the enthalpy of reaction is constant over the range 1800 to 2200 K. From Table 8-3

$$T_1 = 1800 \text{ K} = 3240°\text{R} \qquad \log K_{p_1} = -3.693 \qquad K_{p_1} = 2.028 \times 10^{-4}$$

$$T_2 = 2200 \text{ K} = 3960°\text{R} \qquad \log K_{p_2} = -2.226 \qquad K_{p_2} = 1.306 \times 10^{-3}$$

Therefore

$$\log \frac{K_{p_2}}{K_{p_1}} = -2.226 - (-3.693) = 1.467$$

and

$$\frac{K_{p_2}}{K_{p_1}} = 29.309$$

We evaluate ΔH° at 2000 K, the mean between 1800 and 2200 K, and rearrange Eq. (8-67) to give

$$\Delta H^\circ = \frac{\mathcal{R} \ln (K_{p_2}/K_{p_1})}{1/T_1 - 1/T_2}$$
$$= \frac{(1545) \ln (29.309)}{(1/3240 - 1/3960)(778)}$$
$$= 119\,500 \text{ Btu/lbm mol} \qquad (277\,943 \text{ kJ/kg mol})$$

8-14 THE CHEMICAL POTENTIAL AND PHASE EQUILIBRIUM

Consider a homogeneous mixture of different constituents, the composition of which may change as a result of a change in temperature, pressure, or other intensive state variables. If there are i constituents, we may express the internal energy in functional form as

$$U = U(S, V, n_1, n_2, \ldots, n_i) \tag{8-68}$$

and a differential change in the internal energy is given by

$$dU = \left(\frac{\partial U}{\partial S}\right)_{V,n_i} dS + \left(\frac{\partial U}{\partial V}\right)_{S,n_i} dV + \sum_i \left(\frac{\partial U}{\partial n_i}\right)_{V,S,n_j} dn_i \tag{8-69}$$

For a single-component mixture or one for which the composition does not change, it is clear that $dn_i = 0$ and Eq. (8-69) reduces to the familiar relation

$$dU = T\,dS - p\,dV$$

whereupon we have

$$T = \left(\frac{\partial U}{\partial S}\right)_{V,n_i} \qquad -p = \left(\frac{\partial U}{\partial V}\right)_{S,n_i} \tag{8-70}$$

The partial derivatives in the summation of Eq. (8-69) are defined as the *chemical potential* μ_i:

$$\mu_i = \left(\frac{\partial U}{\partial n_i}\right)_{V,S,n_j} \qquad n_j \neq n_i \tag{8-71}$$

Now we may rewrite Eq. (8-69) as

$$dU = T\,dS - p\,dV + \sum_i \mu_i\,dn_i \tag{8-72}$$

Relations similar to Eq. (8-72) can be derived for other thermodynamic functions. Consider the Gibbs function for the mixture which we assume to take the functional form

$$G = G(p, T, n_1, n_2, \ldots, n_i)$$

Then

$$dG = \left(\frac{\partial G}{\partial p}\right)_{T,n_i} dp + \left(\frac{\partial G}{\partial T}\right)_{p,n_i} dT + \sum_i \left(\frac{\partial G}{\partial n_i}\right)_{p,T,n_j} dn_i \tag{8-73}$$

For a mixture of fixed composition we have, from Chap. 6 [Eq. (6-16*b*)],

$$V = \left(\frac{\partial G}{\partial p}\right)_{T,n_i} \qquad -S = \left(\frac{\partial G}{\partial T}\right)_{p,n_i} \tag{8-74}$$

so that Eq. (8-73) becomes

$$dG = V\,dp - S\,dT + \sum_i \left(\frac{\partial G}{\partial n_i}\right)_{T,p,n_j} dn_i \tag{8-75}$$

Now, for a single component,

$$\begin{aligned} dG &= dH - T\,dS - S\,dT \\ &= dU + p\,dV + V\,dp - T\,dS - S\,dT \end{aligned} \tag{8-76}$$

Equation dG as obtained from Eqs. (8-75) and (8-76) and comparing with Eq. (8-72) indicates that

$$\mu_i = \left(\frac{\partial G}{\partial n_i}\right)_{T,p,n_j} \qquad n_j \neq n_i \tag{8-77}$$

whereupon we have

$$dG = V\,dp - S\,dT + \sum_i \mu_i\,dn_i \tag{8-78}$$

In a similar manner the enthalpy and Helmholtz functions may be expressed by

$$dH = T\,dS + V\,dp + \sum_i \mu_i\,dn_i \tag{8-79}$$

$$dA = -S\,dT - p\,dV + \sum_i \mu_i\,dn_i \tag{8-80}$$

We therefore have the following equivalent relations for the chemical potential:

$$\mu_i = \left(\frac{\partial U}{\partial n_i}\right)_{V,S,n_j} = \left(\frac{\partial G}{\partial n_i}\right)_{T,p,n_j} = \left(\frac{\partial A}{\partial n_i}\right)_{T,V,n_j} \tag{8-81}$$

Now let us examine the relationships between the chemical potentials when we have an equilibrium mixture of several components which may exist in more than one phase. In general, a mixture could have many phases but in most practical circumstances only two or three might be present. For equilibrium at constant temperature and pressure we have

$$dG_{T,p} = 0 \tag{8-82}$$

The total Gibbs function is formed as the sum of the Gibbs functions for each phase, which we assume are homogeneous in order to apply our reasoning used to derive the chemical potential. For simplicity, consider a mixture of two constituents a and b in two phases 1 and 2. For a closed *total* system but one in which mass may be transferred between phases we would have

$$n_{a_1} + n_{a_2} = n_{a,\text{total}} = \text{const.}$$
$$n_{b_1} + n_{b_2} = n_{b,\text{total}} = \text{const.}$$

or, in differential form,

$$dn_{a_1} + dn_{a_2} = 0 \tag{8-83}$$

$$dn_{b_1} + dn_{b_2} = 0 \tag{8-84}$$

If both phases are homogeneous, we could write the Gibbs function for each in accordance with Eq. (8-78):

$$dG_1 = V\,dp - S\,dT + \mu_{a_1}\,dn_{a_1} + \mu_{b_1}\,dn_{b_1} \tag{8-85}$$

$$dG_2 = V\,dp - S\,dT + \mu_{a_2}\,dn_{a_2} + \mu_{b_2}\,dn_{b_2} \tag{8-86}$$

Now, for equilibrium at constant T and p

$$dG|_{T,p} = dG_1|_{T,p} + dG_2|_{T,p} = 0 \tag{8-87}$$

Combining Eqs. (8-83) and (8-87) gives

$$(\mu_{a_1} - \mu_{a_2})\,dn_{a_1} + (\mu_{b_1} - \mu_{b_2})\,dn_{b_1} = 0$$

which indicates that, for equilibrium at constant T and p

$$\mu_{a_1} = \mu_{a_2} \qquad \mu_{b_1} = \mu_{b_2} \tag{8-88}$$

We could extend the analysis to the general case of i constituents and k phases and would find that

$$\mu_{i_1} = \mu_{i_2} = \cdot \cdot \cdot = \mu_{i_k} \tag{8-89}$$

for equilibrium with all phases at the same pressure and temperature. Simply stated, *the chemical potential must have the same value in all phases for equilibrium at constant temperature and pressure.*

Now let us consider a *pure phase,* i.e., one which consists of only one constituent. At constant T and p the change in the Gibbs function from Eq. (8-78) reduces to

$$dG_{T,p} = \mu \, dn \tag{8-90}$$

which indicates that the change in the Gibbs function must come from an increase in mass of the phase (increase in dn). Because the total Gibbs function G is directly proportional to the total number of moles in the phase, we have

$$\mu = \frac{G}{n} = \bar{g} \tag{8-91}$$

or very simply, the chemical potential for a pure phase is equal to the molal Gibbs function for the phase. As a simple example of this point we can see that for equilibrium of a pure substance (like water) at constant T and p the Gibbs functions must be equal in each phase, as we have already noted in Eq. (8-32).

8-15 THE GIBBS PHASE RULE

From the foregoing discussion it should be obvious that the determination of properties of mixtures which may exist in multiple phases is a complicated affair, and even the number of properties necessary to fix the state is not easily apparent. One of the most famous relations of thermodynamics is the phase rule developed by Josiah Willard Gibbs in 1875, which we shall now discuss. We shall first consider an equilibrium mixture without the possibility of chemical reaction. Let the mixture consist of C constituents and P phases. For equilibrium at constant temperature and pressure we would write for each phase, corresponding to Eq. (8-78),

$$dG|_{T,p} = \sum_i \mu_i \, dn_i = 0 \tag{8-92}$$

We would thus have C values of μ_i but only $C - 1$ of these are independent because the equation is subject to the restriction that the total number of moles is fixed for a closed system. Additionally, we have a choice of T and p at our disposal to fix the state of the system for one phase, or

$$(C - 1) + 2 = C + 1$$

independent properties. Now, for an equation written for each phase the pressure and temperature are the same so they do not add new independent variables, and there are additional restrictions added because of the conditions expressed in Eq. (8-89). Thus, for two phases we add the restriction that

$$\mu_{i_1} = \mu_{i_2}$$

For three phases we have the *two* restrictions that

$$\mu_{i_1} = \mu_{i_2} = \mu_{i_3}$$

and the general notion is that there are $P - 1$ such restrictions for the P phases. Taking all the factors into account the total number of *degrees of freedom* F or *number of variants* of the system is

$$F = C + 1 - (P - 1) = C - P + 2 \tag{8-93}$$

This is the *phase rule of Gibbs,* and a quick application to a simple water system can illustrate its validity. In the vapor-liquid saturation region there are two phases ($P = 2$), one component ($C = 1$) so

$$F = 1 - 2 + 2 = 1$$

This means that only *one* intensive property can be independently varied to maintain the equilibrium region; saturation temperature or pressure is normally selected. For the case where there are three phases in equilibrium

$$F = 1 - 3 + 2 = 0$$

which indicates to us that there are no choices open to vary this equilibrium state. Of course, that is exactly the case at the triple point. In a single-phase region, as with a gas or vapor, a single-component substance has

$$F = 1 - 1 + 2 = 2$$

or, there are two properties which can be varied independently. This, of course, is what we know to be the case for an ideal gas or superheated vapor, and temperature and pressure are usually selected as the independent variables for defining the state of the system.

Phase Rule for Chemical Reactions

If it is possible for the different constituents to enter into chemical reactions with one another, the matter of phase equilibrium becomes even more complicated because the various constituents may be consumed or generated depending on the extent of reaction. Let N_c be the number of chemical substances which are present in the overall system. For each reaction equilibrium equation the number of independent intensive properties will be reduced by one, and if there are r reaction equations the number of components C to be used in the phase rule becomes

$$C = N_c - r \tag{8-94}$$

So, in the general context, *C is the minimum number of chemical substances which can produce the overall system.* The phase rule then becomes

$$F = (N_c - r) - P + 2 \tag{8-95}$$

EXAMPLE 8-16

As an example consider an equilibrium mixture of solid carbon C(s), O_2, CO, and CO_2 at elevated temperatures where chemical reactions may occur. There are two phases, gas and solid, so $P = 2$. The reaction equations for formation of the chemical substances are

$$C(s) + O_2(g) \rightarrow CO_2(g) \tag{a}$$

$$C(s) + \tfrac{1}{2}O_2(g) \rightarrow CO(g) \tag{b}$$

Because we have stipulated that both carbon monoxide and oxygen are present in the system, neither can be eliminated from the reaction equations as, for example, in writing

$$CO + \tfrac{1}{2}O_2 \rightarrow CO_2 \tag{c}$$

Therefore we must have at least two of these reaction equations, so $r = 2$. The value of N_c is 4, so

$$C = N_c - r = 4 - 2 = 2$$

The phase rule then yields

$$F = C - P + 2 = 2 - 2 + 2 = 2 \tag{d}$$

8-16 THE THIRD LAW OF THERMODYNAMICS

In the analysis of many chemical reactions it is necessary to fix a reference state for the entropy. We may always choose some arbitrary reference level when only one component is involved; 32°F is chosen for the conventional steam tables. On the basis of observations by Nernst, and others, Planck stated the third law of thermodynamics in 1912 as

The entropy of all perfect crystalline solids is zero at absolute zero of temperature.

A "perfect" crystal is one which is in thermodynamic equilibrium. Consequently, the third law is usually stated in a more general form as

The entropy of any pure substance in thermodynamic equilibrium approaches zero as the absolute temperature approaches zero.

The importance of the third law is clear. It furnishes a basis for calculation of the absolute entropies of substances which may then be utilized in the appropriate equations to determine the direction of chemical reactions.

There are several instances reported in the literature where calculations based on the third law are in disagreement with experiment. However, in all cases it is possible to explain the disagreement on the basis that the substance is not "pure," i.e., there may be two or more isotopes, or the presence of different molecules, or else a nonequilibrium distribution of molecules. In these cases there is more than one quantum state at absolute zero, and the entropy does not approach zero.

8-17 THE CONCEPT OF MAXIMUM WORK

As an additional topic for this chapter we consider a calculation of the maximum work which a system can perform as it changes from one equilibrium state to another. This concept is related to the notions of available energy and availability discussed in Chap. 5 but includes work interactions the system may have with its surroundings. Because we shall be interested in the maximum work *output* of a process we shall take the work as positive when it is done *by* the system. Consider a closed system which receives heat from its surroundings at the constant temperature T_0. The system does work in the amount $d'W$, part of which may be $p\,dV$ work. From the first law

$$d'W = d'Q - dU \tag{8-96}$$

From the principle of increase of entropy we have

$$dS_0 + dS_{\text{sys}} \geq 0 \tag{8-97}$$

and the change in entropy for the surroundings is given by

$$dS_0 = -\frac{d'Q}{T_0} \tag{8-98}$$

Inserting Eq. (8-98) in (8-97) and thence in (8-96) gives

$$d'W = d'Q - dU \le T_0\, dS - dU \tag{8-99}$$

where dS pertains to the system. The inequality holds for irreversible processes and the equality for reversible processes. If the overall process involving the system and its surroundings is to be reversible there must be no heat transfer across a finite temperature difference; i.e., the temperature of the system must equal the temperature of the surroundings. Thus, Eq. (8-98) becomes, for $T = T_0$,

$$d'W \le T\, dS - dU \tag{8-100}$$

where the inequality is retained to account for possible internal irreversibilities in the system. Clearly $d'W$ is maximized when the equality sign holds, so it is the reversible process which produces the maximum work. Then

$$d'W_{\max} = T\, dS - dU \tag{8-101}$$

In terms of the Helmholtz and Gibbs functions

$$d'W_{\max} = -dA - S\, dT \tag{8-102}$$

$$d'W_{\max} = -dG + V\, dp + p\, dV - S\, dT \tag{8-103}$$

For a process at constant temperature

$$(d'W_{\max})_T = (-dA)_T \tag{8-104}$$

Eq. (8-103) may be rewritten

$$d'W_{\max} - p\, dV = -dG + V\, dp - S\, dT$$

Then, for a process at constant pressure and temperature

$$(d'W_{\max} - p\, dV)_{T,p} = (-dG)_{T,p} \tag{8-105}$$

Equations (8-104) and (8-105) now furnish us with the results:

1 The maximum work *output* in a constant-temperature process is equal to the decrease in the Helmholtz function for the system.

2 The maximum work *output, exclusive of p dV work,* in a process at constant temperature and pressure is equal to the decrease in the Gibbs function for the system.

REVIEW QUESTIONS

1 What is meant by the term *stoichiometric proportions*?
2 Define the meaning of excess air.
3 For what purpose is an Orsat apparatus used?
4 What is the base, or zero level, chosen for measuring the energy level of chemical substances?
5 Define enthalpy of formation.
6 Describe endothermic and exothermic reactions.
7 How does heat of reaction differ from enthalpy of formation?
8 What is the distinguishing difference between higher heating value and lower heating value?
9 Define adiabatic flame temperature. What influence does excess air have on its value? What effect does dissociation have on its value?
10 What reference state is used to determine the standard Gibbs functions?
11 What is the criterion for equilibrium for a reaction occurring at constant temperature and pressure? At constant temperature and volume?
12 Define degree of reaction.
13 What is meant by the term *free-energy change*?
14 What is the law of mass action?
15 How is the equilibrium constant defined?
16 How is it possible for the presence of nonreacting gases to affect the equilibrium concentrations of a reacting component in a mixture?
17 How is the van't Hoff isobar used to advantage?
18 State the third law of thermodynamics.
19 Discuss the concept of reference levels for thermodynamic properties.
20 What is the phase rule of Gibbs?
21 Discuss the factors which influence the generation of irreversibility in processes involving chemical reactions.

PROBLEMS (ENGLISH UNITS)

8-1 Determine the dew-point temperature for the water vapor in the products of combustion for the following reactions at atmospheric pressure:
(*a*) Benzene with 50 percent excess air
(*b*) Pentane with stoichiometric proportions
(*c*) A fuel with a volumetric analysis of 70 percent CH_4 and 30 percent C_4H_{10} along with 100 percent excess air
(*d*) Ethyl alcohol with 20 percent excess air

8-2 A mixture of gaseous fuels is used in a certain combustion process. The volumetric analysis of the fuel is

$$CH_4 = 35 \text{ percent}$$
$$C_3H_8 = 25 \text{ percent}$$
$$C_5H_{12} = 40 \text{ percent}$$

A volumetric analysis of the dry products of combustion yields

$$CO_2 = 12 \text{ percent}$$
$$CO = 1 \text{ percent}$$
$$O_2 = 4 \text{ percent}$$
$$N_2 = 83 \text{ percent}$$

Calculate the percent theoretical air and the dew point of the products assuming combustion at atmospheric pressure.

8-3 Calculate the heat liberated per mole of fuel when gaseous propane is burned with the stoichiometric proportions of air at 77°F and 1 atm.

8-4 Using the data of Table 8-1, calculate the lower heating value for *liquid* propane at 77°F.

8-5 Calculate the enthalpy of combustion for gaseous octane at 540°F.

8-6 Calculate the energy liberated when $C_{12}H_{26}$ is burned at 77°F with theoretical oxygen and the water in the products is in the liquid state.

8-7 A certain hydrocarbon fuel is burned with excess air, and the analysis of the dry products of combustion yields the following volumetric proportions:

$$CO_2 = 10.4 \text{ percent}$$
$$CO = 1 \text{ percent}$$
$$O_2 = 5 \text{ percent}$$
$$N_2 = 83.6 \text{ percent}$$

Estimate the approximate composition of the fuel and the percent of theoretical air.

8-8 An analysis of the dry products of combustion for CH_4 yields

$$CO_2 = 10 \text{ percent}$$
$$CO = 0.7 \text{ percent}$$
$$O_2 = 2.0 \text{ percent}$$
$$N_2 = 87.3 \text{ percent}$$

Calculate the air/fuel ratio.

8-9 Consider the combustion of liquid fuels—octane, methyl alcohol, propane, and ethyl alcohol—with the theoretical quantities of air at 77°F. Compare the reactions on the basis of:
(*a*) Volume of products per unit energy liberated
(*b*) Volume of air per unit energy liberated
(*c*) Energy liberated per unit mass of fuel
Discuss these comparisons.

8-10 Consider the dissociation of diatomic oxygen. Calculate the equilibrium concentrations of O_2 and O at 6000°R for system pressures of 1, 5, and 10 atm.

8-11 CO_2 is heated at a constant pressure of 5 atm from 77°F to 2800 K. Calculate the heat transfer for the process and determine the final concentrations at the end of heating.

8-12 Assume that the human metabolic system can be approximated by a simple combustion reaction occurring at 100°F. For a total energy consumption of 3000 kcal/day what quantity of methane and air at 20°C would be required to carry on the combustion process? How much carbon dioxide would be produced?

8-13 Propane gas is burned with 50 percent excess air. The reactants enter the combustion chamber at 100°F and leave at 1200°F. Calculate the heat liberated per mole of fuel.

8-14 Liquid octane is burned with a certain quantity of excess air such that the adiabatic flame temperature is 1500°F. Calculate the percent excess air if the reactants enter the combustion chamber at 77°F.

8-15 Calculate the adiabatic flame temperature for gaseous methane burning with the theoretical air. Assume complete combustion.

8-16 Liquid octane fuel at 77°F is mixed in a combustion chamber with 400 percent theoretical air at 300°F. The products of combustion leave the chamber at 1200°F. Calculate the heat loss per mole of fuel.

8-17 Determine the adiabatic flame temperature for solid carbon burning with 200 percent theoretical air. What percent excess air would be required to produce an adiabatic flame temperature of 2000°R? Assume complete combustion.

8-18 Determine the enthalpy of combustion of gaseous methane and liquid octane at 77°F when liquid hydrogen peroxide is used as an oxidizer for the combustion reaction.

8-19 Determine the adiabatic flame temperature for hydrogen burning with 300 percent theoretical air. Assume complete combustion.

8-20 1 lbm·mol/min of butane gas is burned with 100 percent excess air in a steady-flow process at 1-atm pressure. The reactants enter the combustion chamber at 77°F. Calculate the maximum temperature for which the ducts carrying the products of combustion should be designed if the heat loss in the combustion is 30 000 Btu/min.

8-21 A radiant heater for use in large warehouse areas burns gaseous butane with 25 percent excess air. Both fuel and air enter the heater at 77°F. The products of combustion leave the device at 1240°F. How much heat is lost by the heater per pound-mass of fuel? What percent of the higher heating value does this represent?

8-22 Calculate the maximum work output as saturated liquid water changes to saturated vapor at 100 psia in (*a*) a process at constant temperature and (*b*) exclusive of $p\ dV$ work in a process at constant temperature and pressure. How do these values compare with the $p\ dV$ work done in the process?

8-23 In a home heating furnace methane is burned with 25 percent excess air, and the manufacturer claims that 70 percent of the higher heating value of the fuel is recovered as useful heat. What is the outlet temperature of the products of combustion in this circumstance if the air and fuel enter at 77°F?

8-24 In a certain power-plant combustion air enters the furnace at 250°F, and methane fuel enters at 77°F. Twenty percent excess air is used, and the products of combustion leave the furnace at 2140°F. How much energy is obtained per pound-mass of fuel?

8-25 The burner for a gas-turbine plant is designed to burn liquid propane at 77°F with air at 350°F. The outlet temperature of the products of combustion is to be 2040°F. Calculate the air/fuel ratio required and percent excess air for steady flow and no heat loss from the burner.

8-26 Some authorities claim that hydrogen produced by electrolysis of water will become a widely used fuel as more and more of our hydrocarbon fuel reserves are depleted. Calculate the higher and lower heating value for gaseous hydrogen at 77°F and 1 atm. What size tank for storage at 2000 psia and 70°F would be required to provide the same energy as a 25-gal tank of gasoline (octane)?

8-27 What adiabatic flame temperature would be obtained by burning hydrogen with 400 percent excess air?

8-28 Nitrogen oxides (that is, NO, etc.) are present in the stack discharge gas from power plants because they are "frozen" in the flow when the hot combustion gases come in rapid contact with the relatively cool surface of the boiler tubes and sufficient time is not available to perform the reverse reaction back to N_2 and O_2. Suppose the combustion gases are at 3140°F and the boiler tubes are at 1340°F. Estimate the percentage of NO present at the higher temperature that will be retained after sudden cooling to the lower temperature.

8-29 What is the maximum work output exclusive of $p\ dV$ work when CO is burned with O_2 to form CO_2 at constant pressure of 1 atm and constant temperature of 77°F?

8-30 Methane (CH_4) is burned with the stoichiometric proportion of oxygen at 1 atm and 77°F. What is the maximum work output in this process? What would be the actual work output if the process occurs under steady-flow conditions?

8-31 How many degrees of freedom are there for an equilibrium mixture of N_2, O_2, and NO at elevated temperatures?

8-32 What is the number of degrees of freedom for an equilibrium mixture of CO, CO_2, H_2, and H_2O at elevated temperatures?

8-33 What is the maximum number of homogeneous phases which can exist for (*a*) a one-component, (*b*) a two-component, and (*c*) a three-component system?

8-34 A mixture of air and water vapor can be cooled so that liquid is present along with the vapor, or at low enough temperatures a solid phase (ice) will be present along with the vapor. Apply the Gibbs phase rule to this system to determine the number of degrees of freedom for the three regimes: (*a*) air and vapor, (*b*) air, vapor, and liquid, and (*c*) air, vapor, and ice.

8-35 How many degrees of freedom are there for a mixture of N_2 and O_2 at room temperature?

PROBLEMS (METRIC UNITS)

8-1M Calculate the heat liberated per mole of fuel when gaseous propane is burned with the stoichiometric proportions of air at 25°C and 1 atm.

8-2M Using the data of Table 8-1, calculate the lower heating value for *liquid* propane at 25°C.

8-3M Calculate the enthalpy of combustion for gaseous octane at 280°C.

8-4M Calculate the energy liberated when $C_{12}H_{26}$ is burned at 25°C with theoretical oxygen and the water in the products is in the liquid state.

8-5M Propane gas is burned with 50 percent excess air. The reactants enter the combustion chamber at 40°C and leave at 650°C. Calculate the heat liberated per mole of fuel.

8-6M Calculate the air-fuel ratio for ethane (C_2H_6) burned with 25 percent excess air. Also calculate the dew point of the products at a pressure of 100 kPa.

8-7M A certain gaseous fuel has the following composition on a mol basis:

$$\begin{aligned} CH_4 &= 60 \text{ percent} \\ C_2H_6 &= 30 \text{ percent} \\ N_2 &= 4 \text{ percent} \\ CO_2 &= 6 \text{ percent} \end{aligned}$$

Calculate the air/fuel ratio when the fuel is burned with 20 percent excess air. Also calculate the dew point of the products for a pressure of 100 kPa.

8-8M Repeat Prob. 8-7M for 100 percent excess air.

8-9M Liquid octane is burned with a certain quantity of excess air such that the adiabatic flame temperature is 1100 K. Calculate the percent excess air if the reactants enter the combustion chamber at 298 K.

8-10M Calculate the adiabatic flame temperature for gaseous methane burning with the theoretical air. Assume complete combustion.

8-11M Liquid octane fuel at 298 K is mixed in a combustion chamber with 400 percent theoretical air at 425 K. The products of combustion leave the chamber at 900 K. Calculate the heat loss per mole of fuel.

8-12M Calculate the air/fuel ratio for liquid benzene (C_6H_6) when burned with 130 percent of the theoretical air. How much heat will be liberated if the reaction occurs at 25°C?

8-13M Octane (C_8H_{18}) is burned with air and an Orsat apparatus used to determine composition of the dry product of combustion as CO_2, 10 percent; CO, 2 percent; and O_2, 4 percent. Calculate the air/fuel ratio.

8-14M What temperature is necessary to dissociate diatomic hydrogen to a point that the monatomic hydrogen composes 15 percent of the total number of moles of the mixture?

8-15M 1 mol of CO is mixed with 1 mol of water vapor at 25°C and 1-atm pressure. The mixture is then heated to 1800 K. Calculate the heating required and the composition of the final mixture.

8-16M Determine the adiabatic flame temperature for solid carbon burning with 200 percent theoretical air. What percent excess air would be required to produce an adiabatic flame temperature of 1100 K? Assume complete combustion.

8-17M Determine the enthalpy of combustion of gaseous methane and liquid octane at 25°C when liquid hydrogen peroxide is used as an oxidizer for the combustion reaction.

8-18M Butane gas at the rate of 0.5 kg/min is burned with 100 percent excess air in a steady-flow process at 1-atm pressure. The reactants enter the combustion chamber at 25°C. Calculate the maximum temperature for which the ducts carrying the products of combustion should be designed if the heat loss in the combustion is 530 kW.

8-19M Consider the combustion of liquid fuels—octane, methyl alcohol, propane, and ethyl alcohol—with the theoretical quantities of air at 25°C. Compare the reactions on the basis of

(*a*) Volume of products per unit energy liberated
(*b*) Volume of air per unit energy liberated
(*c*) Energy liberated per unit mass of fuel

Discuss these comparisons.

8-20M Octane (liquid) at 25°C is burned with 30 percent excess air at a constant pressure of 100 kPa. The air enters the reaction at 400 K. Calculate the energy released for a products temperature of 800 K.

8-21M Gaseous methane (CH_4) is burned with 400 percent theoretical air at 25°C and 100 kPa. Calculate the energy released for a products temperature of 600 K. Also calculate the adiabatic flame temperature.

8-22M Consider the dissociation of diatomic oxygen. Calculate the equilibrium concentrations of O_2 and O at 3300 K for system pressures of 1, 5, and 10 atm.

8-23M CO_2 is heated at a constant pressure of 500 kPa from 298 to 2800 K. Calculate the heat transfer for the process and determine the final concentrations at the end of heating.

8-24M One mol of CO is mixed with 1 mol of water vapor at 298 K and 1-atm pressure. The mixture is then heated to 1800 K. Calculate the heating required and the composition of the final mixture.

8-25M Consider the dissociation of water vapor into H_2 and O_2. How much heat must be added to 1 mol of water vapor at 200 K to raise the temperature in a steady-flow process to 2500 K?

8-26M Calculate the equilibrium composition for the reaction

$$CO_2 \rightleftharpoons CO + \tfrac{1}{2}O_2$$

at 1 atm and 2800 K.

8-27M Repeat Prob. 8-26M for the reaction

$$H_2O(g) \rightleftharpoons H_2 + \tfrac{1}{2}O_2$$

at 1 atm and 3000 K.

8-28M Consider the ionization reaction for cesium as discussed in Example 8-13: Calculate the number density of electrons (number per cubic meter) for conditions of the example. Repeat for a total pressure of 0.1 atm and 1000 K.

8-29M In magnetohydrodynamic power-generation systems it is desirable to employ gases which will ionize at modest temperatures. One technique for producing such gases is to "seed" a gas with a substance like cesium which ionizes at a low temperature. Consider argon as the carrier gas at a total pressure of 1 atm and 1000 K. Calculate the electron number density (number per cubic meter) for cesium mass concentrations of 0.1, 0.5, and 1.0 percent. The molecular weight for cesium is 132.91. Repeat the calculation for a total pressure of 10 atm. Assume that the electrons behave like an ideal monatomic gas.

8-30M Repeat Prob. 8-29M using sodium as the seeding gas.

8-31M Gaseous butane at the rate of 2 kg mol/min is burned with the theoretical air at 25°C and 100 kPa in a steady-flow process. Calculate the energy released and irreversibility for a products temperature of 500 K and lowest available temperature of 25°C.

8-32M Carbon monoxide is burned with the theoretical air at 100 kPa and 25°C in a steady-flow process. The products leave at 100 kPa, 25°C, and the lowest available temperature is 25°C. Calculate the energy released and the irreversibility.

8-33M Calculate the adiabatic flame temperature for carbon monoxide burned with the theoretical air at 25°C and 1 atm.

8-34M Liquid propane is burned with air at 100 kPa and 25°C in a steady-flow furnace. The products leave at a temperature of 400 K. The furnace is used to heat water. If 75 percent of the HHV of the fuel is to be supplied to the water, calculate the excess air required.

8-35M Gaseous ethane is burned with 25 percent excess air at 25°C and 100 kPa in a steady-flow process. The products leave at 25°C and 100 kPa and T_0 = 25°C. Calculate the energy released and irreversibility per mol of fuel.

8-36M Some authorities claim that hydrogen produced by electrolysis of water will become a widely used fuel as more and more of our hydrocarbon fuel reserves are depleted. Calculate the higher and lower heating value for gaseous hydrogen at 298 K and 100 kPa. What size tank for storage at 14 MPa and 295 K would be required to provide the same energy as a 95-liter tank of gasoline (octane)?

8-37M Nitrogen oxides (that is, NO, etc.) are present in the stack discharge gas from power plants because they are "frozen" in the flow when the hot combustion gases come in rapid contact with the relatively cool surface of the boiler tubes and sufficient time is not available to perform the reverse reaction back to N_2 and O_2. Suppose the combustion gases are at 2000 K and the boiler tubes are at 1000 K. Estimate the percentage of NO present at the higher temperature that will be retained after sudden cooling to the lower temperature.

8-38M Calculate the equilibrium constant for the reaction $CO_2 \rightarrow CO + \frac{1}{2}O_2$ at 1000 K using heat of reaction data calculated from Table A-6 and the equilibrium constant at 298 K from Table 8-3.

8-39M How much energy is required to ionize cesium at 2400 K, at 2800 K, and at 4000 K?

8-40M Calculate the degree of ionization of cesium in a mixture of 1 mol cesium and 2 mol nitrogen at 2000 K and 10 atm.

8-41M In deriving Eq. (8-67) the heat of reaction was assumed constant over a range of temperatures. How accurate is this assumption for dissociation of CO_2 in the temperature range from 1500 to 3000 K?

8-42M Estimate the energy required to heat an equimolal mixture of nitrogen and oxygen from 298 to 3000 K.

8-43M Assume that the human metabolic system can be approximated by a simple combustion reaction occurring at 20°C. For a total energy consumption of 13 MJ/day what quantity of methane and air at 20°C would be required to carry on the combustion process? How much carbon dioxide would be produced?

8-44M Water vapor is heated at a constant pressure of 250 kPa until it is 10 percent dissociated into H_2 and O_2. At what temperature does this occur?

8-45M Calculate the temperature at which CO_2 will be 15 percent dissociated at a pressure of 300 kPa.

8-46M Calculate the energy per mol used to perform the heating in Probs. 8-44M and 8-45M if the initial temperature is 25°C.

8-47M 1 mol of hydrogen is heated in a constant-pressure process from 300 K to 3000 K at a total pressure of 50 kPa. Calculate the heating required.

8-48M Estimate the enthalpy of ionization for cesium at 1 atm and 2000 K. Repeat for sodium at 2000 K.

8-49M For a reaction at 100 kPa and 1500 K calculate the equilibrium compositions for the reaction

$$\tfrac{1}{2}O_2 + \tfrac{1}{2}N_2 \rightleftharpoons NO$$

8-50M Water vapor is heated to 2800 K. Calculate the equilibrium concentrations assuming the following two reaction equations apply:

$$H_2O(g) \rightleftharpoons H_2 + \tfrac{1}{2}O_2$$
$$H_2O(g) \rightleftharpoons OH + \tfrac{1}{2}H_2$$

8-51M In a home heating furnace methane is burned with 25 percent excess air, and the manufacturer claims that 70 percent of the higher heating value of the fuel is recovered as useful heat. What is the outlet temperature of the products of combustion in this circumstance if the air and fuel enter at 25°C?

8-52M In a certain power-plant combustion air enters the furnace at 400 K, and methane fuel enters at 298 K. Twenty percent excess air is used and the products of combustion leave the furnace at 1450 K. How much energy is obtained per kilogram of fuel?

8-53M The burner for a gas-turbine plant is designed to burn liquid propane at 298 K with air at 450 K. The outlet temperature of the products of combustion is to be 1400 K. Calculate the air/fuel ratio required and percent excess air for steady flow and no heat loss from the burner.

8-54M A radiant heater for use in large warehouse areas burns gaseous butane with 25 percent excess air. Both fuel and air enter the heater at 298 K. The products of combustion leave the device at 950 K. How much heat is lost by the heater per kilogram of fuel? What percent of the higher heating value does this represent?

8-55M Calculate the maximum work output as saturated liquid water changes to saturated vapor at 700 kPa in (*a*) a process at constant temperature and (*b*) exclusive of $p\ dV$ work in a process at constant temperature and pressure. How do these values compare with the $p\ dV$ work done in the process?

8-56M What is the maximum work output exclusive of $p\ dV$ work when CO is burned with O_2 to form CO_2 at constant pressure of 1 atm and constant temperature of 298 K?

8-57M Methane (CH_4) is burned with the stoichiometric proportion of oxygen at 1 atm and 298 K. What is the maximum work output in this process? What would be the actual work output if the process occurs under steady-flow conditions?

8-58M Calculate the energy released per kilogram of coal for the conditions in Prob. 8-66M.

8-59M A gaseous fuel contains 76 percent CH_4 and 24 percent C_2H_6 on a molal basis. Determine the air/fuel ratio for stoichiometric combustion. Also determine the higher- and lower-heating values at 1 atm, 25°C, per kilogram of fuel and per kilogram of air/fuel mixture.

8-60M Gaseous ethane, C_2H_6, is burned with 200 percent theoretical air at 1 atm. The reactants enter at 25°C and leave at 1000 K in a steady-flow process. Calculate the total energy released for a fuel flow of 15 kg/s. Make the calculation using individual component enthalpies, and also using Table A-18.

8-61M 4 kg·mol of CH_4 are burned with 9 kg mol of O_2. Calculate the percent excess of O_2 and the dew point of the products for a total pressure of 1 atm.

8-62M For the reactants in Prob. 8-60M, calculate the energy released if the reactants enter at 1 atm, 25°C, and products leave at 1 atm, 500 K in a steady-flow device. Also calculate the irreversibility for the reaction.

8-63M 3 kg·mol/h propane, C_3H_8, are burned with the theoretical amount of air at 1 atm and 25°C. The products leave at 450 K and 1 atm in a steady-flow process. Calculate the energy released in this process, the volume flow rate of reactants and products in liters/h, and the irreversibility for the reaction.

8-64M Liquid octane, C_8H_{18}, is burned adiabatically with 400 percent theoretical air at 1 atm and 25°C. Calculate the adiabatic flame temperature using (*a*) enthalpies of each compound and (*b*) Table A-18.

8-65M Carbon monoxide is burned with the theoretical oxygen at 4 atm and 25°C in a steady-flow process. Calculate the adiabatic flame temperature taking into account dissociation.

8-66M An analysis is performed on a coal-fired boiler to estimate air flow requirements for combustion and the amount of waste products which must be moved for disposal. The analysis of the coal on a mass basis is

75 percent C
7 percent H_2
7 percent O_2
2 percent N_2
4 percent H_2O
5 percent ash

An Orsat analysis of the dry products of combustion gives a volumetric analysis as 13 percent CO_2, 1.5 percent CO, 6.5 percent O_2, and 79 percent N_2. A mass analysis of the waste material indicates 21 percent C and 79 percent ash. For 1.0 kg of incoming coal, determine the volume of air required at 1 atm, 30°C, in liters/s. Also calculate the mass of refuse per kilogram of coal. If the products are discharged at 150°C, calculate their volume flow per kilogram of coal.

8-67M 1 kg·mol of CH_4 and 2.0 kg mol of O_2 at 1 atm and 25°C react in a steady-flow process to produce products of 1 mol of CO_2 and 2 mol of H_2O (vapor) at 1 atm and 25°C. Calculate the change in entropy for the reaction, the energy released, and the decrease in availability for T_0 = 25°C.

8-68M A mixture of H_2 and O_2 is allowed to react to H_2O in a rigid tank. The mass of H_2 is 1.0 kg and the mass of O_2 is 29 kg, and the initial condition of the reactants is 1 atm and 35°C. Calculate the volume of the tank. Calculate the amount of condensation if the products are cooled to 25°C.

REFERENCES

1 JANAF Thermochemical Tables, *Document PB 168–370,* Clearinghouse for Federal Scientific and Technical Information, August 1965.

2 Selected Values of Chemical Thermodynamic Properties, *NBS Technical Notes,* 270-1 and 270-2, 1955.

3 Selected Values of Physical and Thermodynamic Properties of Hydrocarbons and

Related Compounds, *API Res. Project 44,* Carnegie Press, Carnegie Institute of Technology, Pittsburgh, Pa.

4 Callen, H. B.: "Thermodynamics," John Wiley & Sons, Inc., New York, 1960.

5 Kirkwood, J. G., and I. Oppenheim: "Chemical Thermodynamics," McGraw-Hill Book Company, New York, 1961.

6 Lewis, G. N., and M. Randall, "Thermodynamics," 2d ed. (revised by K. S. Pitzer and L. Brewer), McGraw-Hill Book Company, New York, 1961.

7 Reynolds, W. C.: "Thermodynamics," 2d ed., McGraw-Hill Book Company, New York, 1968.

8 Wark, Kenneth: "Thermodynamics," 2d ed., McGraw-Hill Book Company, New York, 1971.

9 Van Wylen, G. J., and R. E. Sonntag: "Fundamentals of Classical Thermodynamics," John Wiley & Sons, Inc., New York, 1965.

10 Sears, F. W.: "Thermodynamics," 2d ed., Addison-Wesley Publishing Company, Inc., Reading, Mass., 1953.

11 Hatsopoulos, G. N., and J. H. Keenan: "Principles of General Thermodynamics," John Wiley & Sons, New York, 1965.

12 Keenan, J. H., and J. Kaye: "Gas Tables," John Wiley & Sons, Inc., New York, 1948.

13 Holman, J. P.: "Experimental Methods for Engineers," 4th ed., chap. 13, McGraw-Hill Book Company, New York, 1984.

9

POWER AND REFRIGERATION CYCLES

9-1 INTRODUCTION

One of the important objectives of engineering practice is to convert energy from one form to another—chemical energy of a fuel to heat through a combustion process, heat energy to electric energy in a steam power plant, gas turbine, internal combustion engine, or other devices. We have seen that the second law of thermodynamics serves as a guide in predicting the permissible direction of an energy-conversion process. Our objective in this chapter is to describe some of the practical arrangements employed for large-scale energy conversion and to illustrate the methods of analysis applicable to these devices. It is clear that a key item of interest is the effectiveness or efficiency of a particular energy-conversion process. We shall therefore want to establish a consistent basis for calculation of such quantities so that various processes may be compared in a meaningful manner.

The term *power cycle* in this chapter title describes a process in which devices are used that produce power continuously, i.e., a repetitive cyclic process which converts energy into mechanical or electrical work. Mechanical and electrical work are of course the most practicable forms of energy in that they can be adapted to a variety of applications.

The applicability of any power cycle is eventually related to cost and efficiency. Environmental effects associated with power production must also be taken into account, and the cost of pollution-control equipment is becoming a major factor in automotive, aircraft, and electric-power applications. In some cases environmental considerations will influence the choice of fuel alternatives: fuel oil or natural gas in favor of coal in a power plant, natural gas in favor of fuel oil in homes and apartments.

A fundamental point often overlooked in environmental considerations is that the amount of pollution in most power-production schemes is directly proportional to the amount of fuel consumed which in turn is related to the efficiency of energy conversion. The higher the efficiency is, the lower the fuel consumption, and thus the lower the total amount of pollution. What this means is that a thermodynamic analysis or design which is capable of improving efficiency will also help to minimize negative environmental side effects.

In this chapter we are primarily concerned with technical analysis; economic and environmental factors are beyond the scope of our discussion. It is necessary to bear in mind, nonetheless, that these variables must be kept in sharp focus in any practical power plant design.

9-2 GENERAL CONSIDERATIONS

The basis for a great number of power cycles is the Carnot cycle discussed in Chap. 5, because, for given temperature limits, the thermal efficiency of such a cycle is the maximum obtainable; that is,

$$\eta_t = 1 - \frac{T_L}{T_H} \tag{9-1}$$

From Eq. (9-1) we sense two basic notions:

1 The higher the temperature at which heat is added, the higher the efficiency
2 The lower the temperature at which heat is rejected, the higher the efficiency

These two notions carry over into practical power cycles. The basic idea of efficiency, it will be remembered, is

$$\eta = \frac{\text{useful effect}}{\text{energy that must be purchased}} \tag{9-2}$$

We shall be considering both reversible and actual processes in the following sections and will show the effects of irreversibility on efficiency.

A refrigeration cycle is just the opposite of a power cycle; there is a net work input and a net heat output. Compare the Carnot cycles in Fig. 9-1. The efficiency of the power cycle is given by Eq. (9-1), and the characteristic parameter for a refrigeration cycle is the coefficient of performance (COP), defined by

$$\text{COP} = \frac{\text{refrigeration effect}}{\text{work input}} \tag{9-3}$$

For the reversible Carnot refrigeration cycle this becomes

$$\text{COP} = \frac{Q_L}{W} = \frac{Q_L}{Q_H - Q_L} = \frac{1}{T_H/T_L - 1} \tag{9-4}$$

It is not uncommon for the COP to be greater than unity. This fact is the reason why it is not called an efficiency even though it is a "useful effect" divided by "energy that must be purchased."

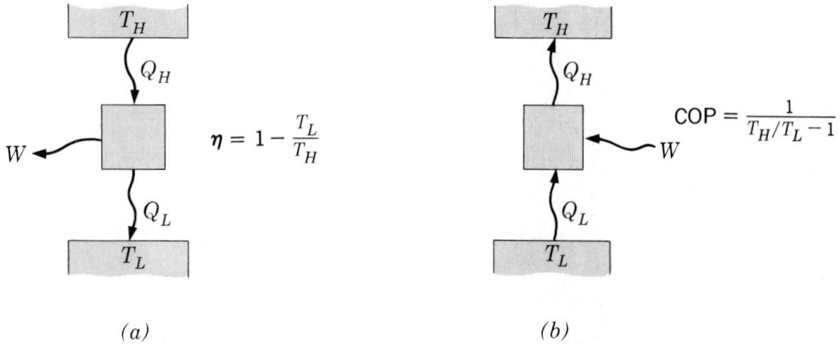

FIG. 9-1 (*a*) Carnot power cycle; (*b*) Carnot refrigerator.

9-3 VAPOR POWER CYCLES

To initiate our discussion of power cycles consider the schematic arrangement shown in Fig. 9-2*a*. Heat is supplied to the boiler where liquid is converted to vapor. The vapor is then expanded adiabatically in the turbine to produce a work output. Vapor leaving the turbine then enters the condenser where heat is removed until the vapor is condensed into the liquid state. The condensation process is the heat-rejection mechanism for the cycle. Saturated liquid is delivered to a pump where its pressure is raised to the saturation pressure corresponding to the boiler temperature, and the high-pressure liquid is delivered to the boiler where the cycle repeats itself. For a reversible cycle the turbine and pump processes are isentropic, as indicated in Fig. 9-2*b*.

FIG. 9-2 Basic Rankine vapor power cycle: (*a*) schematic; (*b*) temperature-entropy diagram.

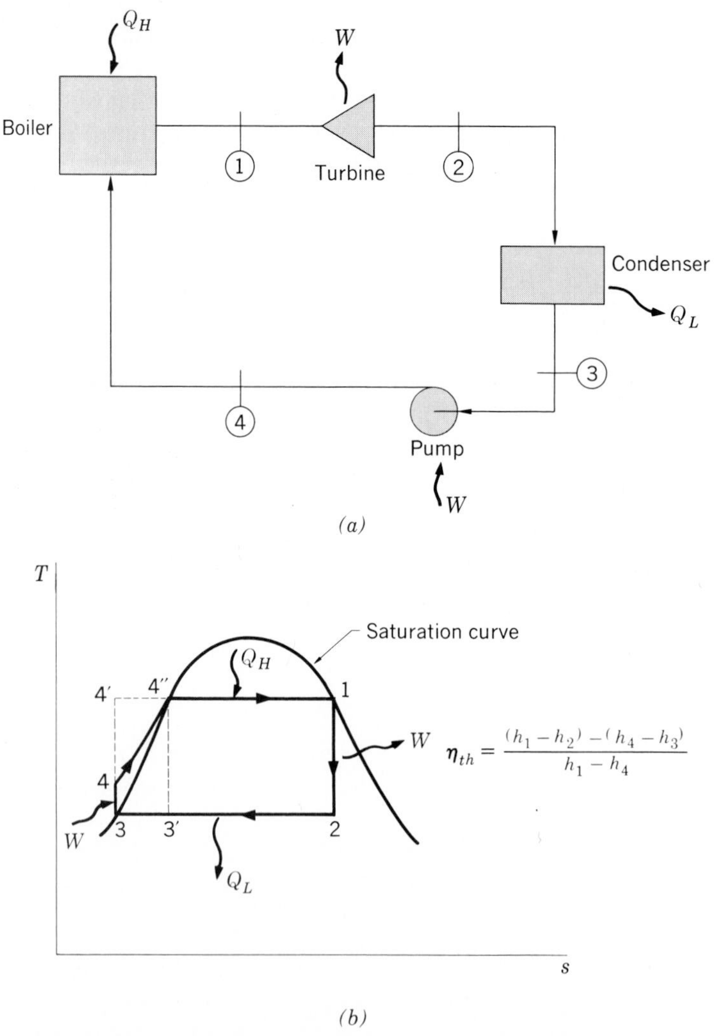

The vapor power cycle is perhaps the most widely used thermal cycle and accounts for most of the electric energy generation in the world. The basic *Rankine* cycle is shown in Fig. 9-2. We shall assume that the working fluid is steam although, in some specialized applications, other fluids might be employed. The basic Rankine cycle shown differs from a Carnot cycle in that the heat-addition process does not occur at constant temperature. If it were possible to operate the cycle either as 1-2-3′-4″ or 1-2-3-4′, a Carnot cycle would be obtained exactly. Practical difficulties prohibit these two arrangements. The process 3′-4″ is impractical because it would involve the compression of a wet mixture into a saturated liquid. It simply is not possible to design a turbine pump to accomplish this compression. The process 3-4′ would involve liquid compressions to enormous pressures, and the pumping machinery to accomplish this feat would also be highly impractical. The average temperature at which heat is added to the Rankine cycle is less than T_1 because part of the process involves heating the compressed liquid up to the saturation temperature at the high pressure. Thus the efficiency is less than would be attained by a Carnot cycle operating between T_1 and T_2.

Choice of Working Fluids

From the discussion so far we can see that the shape of the *T-s* diagram for the particular fluid governs the adaptability of that fluid to use with the Rankine cycle. It would be most desirable to have a fluid with a high critical temperature and with very steep saturated-liquid and -vapor lines so that the Rankine cycle would more closely approximate the Carnot performance. At the same time, the pressure limits required should be reasonable, and availability and cost of the fluid are also important factors to consider. The total quantity of fluid which must be circulated will depend on the enthalpy of vaporization because this governs the energy which may be delivered to the fluid at the high temperature. Taking all these factors into account, water is a good choice of working fluid for the Rankine power cycle. It is readily available at low cost, it has a very high enthalpy of vaporization (~900 Btu/lbm), it has a reasonably high critical temperature (705°F), and it is nontoxic. Corrosion problems with water are significant but can be controlled with careful engineering design.

Improvement of Efficiency

One way of improving the efficiency of the Rankine cycle is to continue the constant-pressure heating process past point 1 and into the superheat region, as shown in Fig. 9-3. Such a heating process has two advantages:

1. It increases the average temperature at which heat is added, thereby increasing the cycle efficiency.
2. It results in an operating condition which involves less moisture in the turbine expansion process. A major source of wear on turbine blades is the erosion caused by liquid droplets. The superheat process produces less moisture and thus serves to alleviate this difficulty.

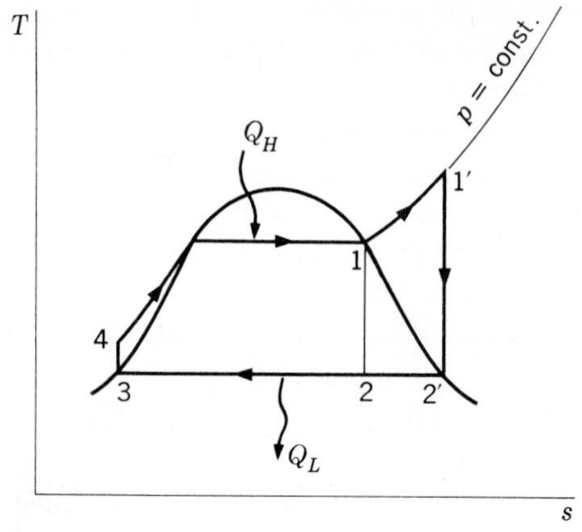

FIG. 9-3 Temperature-entropy diagram for Rankine cycle with superheat.

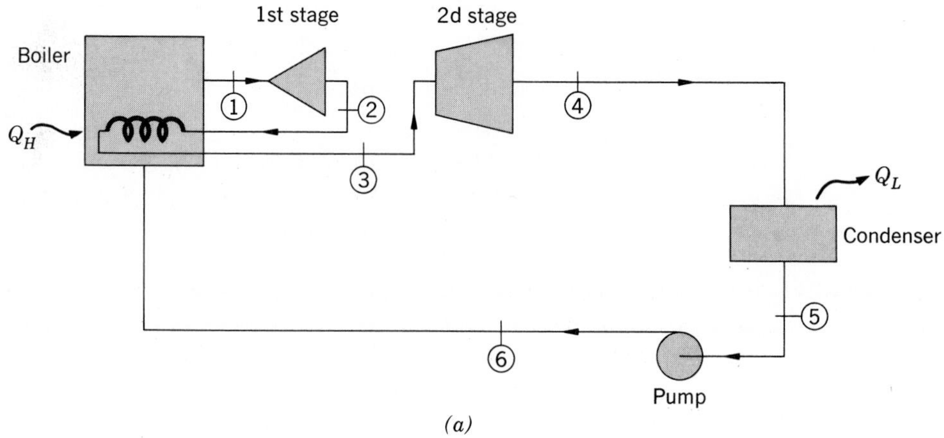

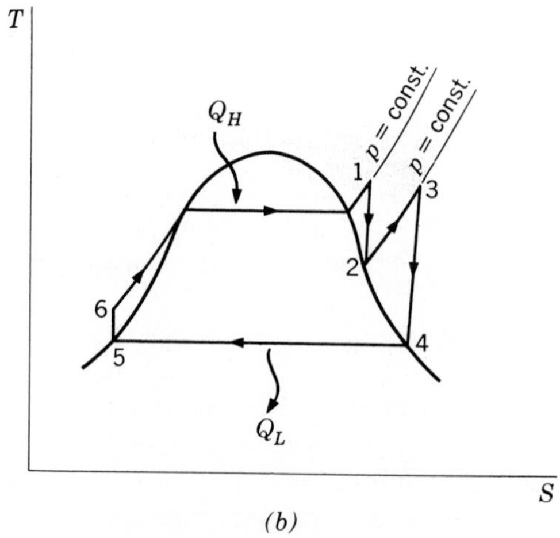

FIG. 9-4 Rankine cycle with reheat: (*a*) schematic; (*b*) temperature-entropy diagram.

Another way to improve the efficiency of the Rankine cycle is to engage in a *reheat* operation, as shown in Fig. 9-4. In this situation the steam is expanded in the first turbine stage until it approximately reaches the saturation line at p_2. Then it is returned to the boiler and reheated at constant pressure to T_3. The steam is then expanded in the second-stage turbine to p_4. The reheat arrangement has the same two advantages as the basic superheat cycle and is even more effective in preventing moisture from entering the turbine.

Heat Rate

A term in common use among power plant engineers is the *heat rate,* defined as

Heat rate = heat input to boiler (Btu) per electric energy output (kWh)

or in terms of the thermal efficiency

$$\text{Heat rate} = \frac{3413}{\eta}\ \text{Btu/kWh} \tag{9-5}$$

EXAMPLE 9-1 Rankine Cycle
A Rankine cycle operates between pressure limits of 500 and 2 psia. The temperature of the steam entering the turbine is 700°F. Calculate the thermal efficiency of the cycle assuming isentropic expansion in the turbine.

SOLUTION Figures 9-2*a* and 9-3 apply to this problem. Using either the steam tables or a Mollier diagram, we have for the steam properties

$h_1 = 1357$ (500 psia, 700°F)

$h_2 = 935$ ($s_1 = s_2$, $p_2 = 2$ psia)

$h_3 = 94.02$ (saturated liquid at 2 psia)

The pump work, $h_4 - h_3$, is computed from (see Example 4-2)

$$\begin{aligned} h_4 - h_3 &= v_f(p_4 - p_3) \\ &= \frac{(0.016\,23)(500 - 2)(144)}{778} \\ &= 1.496\ \text{Btu/lbm} \end{aligned}$$

Thus

$$h_4 = 95.52\ \text{Btu/lbm}$$

The net work output of the cycle is the work output of the turbine less the work input to the pump, or $(h_1 - h_2) - (h_4 - h_3)$. The thermal efficiency of the cycle is based on the net work output and is calculated as

$$\eta_{\text{th}} = \frac{W}{Q_H} = \frac{(h_1 - h_2) - (h_4 - h_3)}{h_1 - h_4}$$

$$= \frac{(1357 - 935) - 1.496}{1357 - 95.52}$$

$$= 33.3 \text{ percent}$$

EXAMPLE 9-2 Rankine Cycle with Reheat
A steam cycle operates between the same upper and lower pressure and temperature limits as in Example 9-1. The steam is extracted at 100 psia and reheated to 700°F. Calculate the thermal efficiency under this new arrangement.

SOLUTION The reheat cycle is shown schematically in Fig. 9-4. Using the nomenclature of this figure, the various enthalpies may be found from a Mollier diagram:

$h_1 = 1357$	(500 psia, 700°F)
$h_2 = 1194$	($s_1 = s_2$, $p_2 = 100$ psia)
$h_3 = 1379$	(100 psia, 700°F)
$h_4 = 1047$	($s_3 = s_4$, $p_4 = 2$ psia)
$h_5 = 94.02$	(saturated liquid at 2 psia)
$h_6 = 95.52$	(see Example 9-1)

The thermal efficiency is now calculated as

$$\eta_{\text{th}} = \frac{W}{Q} = \frac{(h_1 - h_2) + (h_3 - h_4) - (h_6 - h_5)}{(h_1 - h_6) + (h_3 - h_2)}$$

$$= \frac{(1357 - 1194) + (1379 - 1047) - 1.496}{(1357 - 95) + (1379 - 1194)}$$

$$= 34.2 \text{ percent}$$

Regeneration

Neither the superheat nor reheat arrangement alleviates the efficiency "loss" which results from the low-temperature heat-addition process encountered as the liquid is heated from the exit temperature at the pump up to the saturation temperature at the upper pressure limit. This problem may be solved by making use of *regeneration*. An ideal regenerative cycle is shown in Fig. 9-5. The basic

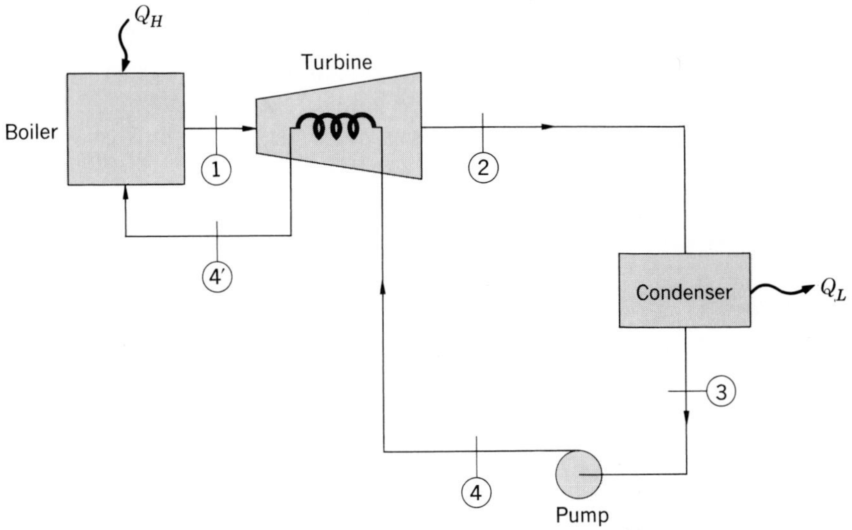

FIG. 9-5 Ideal-vapor regenerative cycle.

idea is that the high-temperature steam expanding in the turbine is employed to preheat the liquid before it enters the boiler. In this way, the heat tranferred to the overall cycle occurs at constant temperature, and the arrangement will have an efficiency equal to the Carnot value. In terms of the *T-s* diagram in Fig. 9-6, the actual cycle 1-2-3-4-4′ is equivalent to the Carnot cycle 1-2′-3′-4′. We might think of the ideal regenerative cycle as a Carnot cycle which is distorted sideways. The term *regeneration* means that heat is generated internally to the cycle by the heat-exchange process.

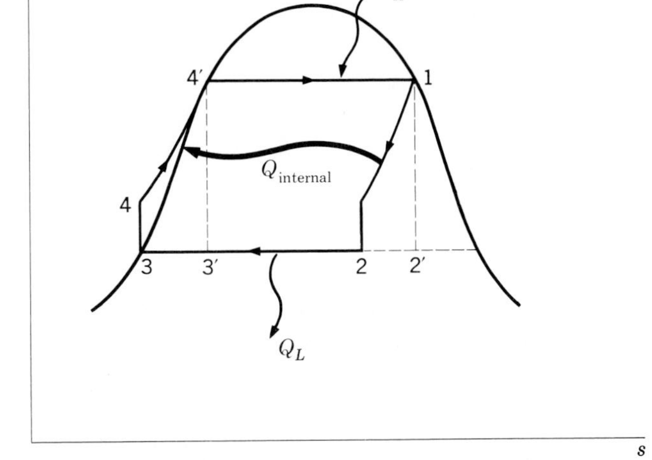

FIG. 9-6 Temperature-entropy diagram for ideal regenerative cycle.

There are practical difficulties with the ideal regenerative cycle. It is not possible to design a turbine which will serve as a power-production device and a heat exchanger as well. Accordingly, practical heat exchangers must be introduced which cause the actual regeneration cycle to have an efficiency somewhat less than the ideal value.

Feedwater Heaters

A practical regenerative cycle arrangement with one feedwater heater is shown in Fig. 9-7. We have shown the high-pressure steam expanded through one turbine stage. Then a quantity m is extracted and fed to a feedwater heater. For unit mass leaving the boiler we will have $1 - m$ mass to be expanded in the second turbine stage. In this instance the heater is an *open* type; this means that the steam extraction from the turbine is mixed directly with the incoming water from the low-pressure pump. Sometimes this is called a *direct-contact* heat exchanger.

For an ideal open feedwater heater we assume that the mixing of steam and cool water occurs at constant pressure. In terms of the cycle shown in Fig. 9-7 this means that $p_5 = p_6 = p_2$. The best we can hope to accomplish in this heater is to produce saturated liquid at point 6. We could not produce a higher energy level because a wet mixture of steam would be encountered and the high-pressure pump, designed for liquids, would be rendered inoperative. In an actual steam cycle there could be some pressure drop through the heater, and the outlet temperature would be controlled at a value somewhat below the saturation temperature. However, for the cycle analysis here, we shall assume no pressure drop and saturation temperature at exit from the heater.

A *closed* feedwater heater could also be employed, and a schematic of such a device is shown in Fig. 9-8. The term *closed* means that the incoming

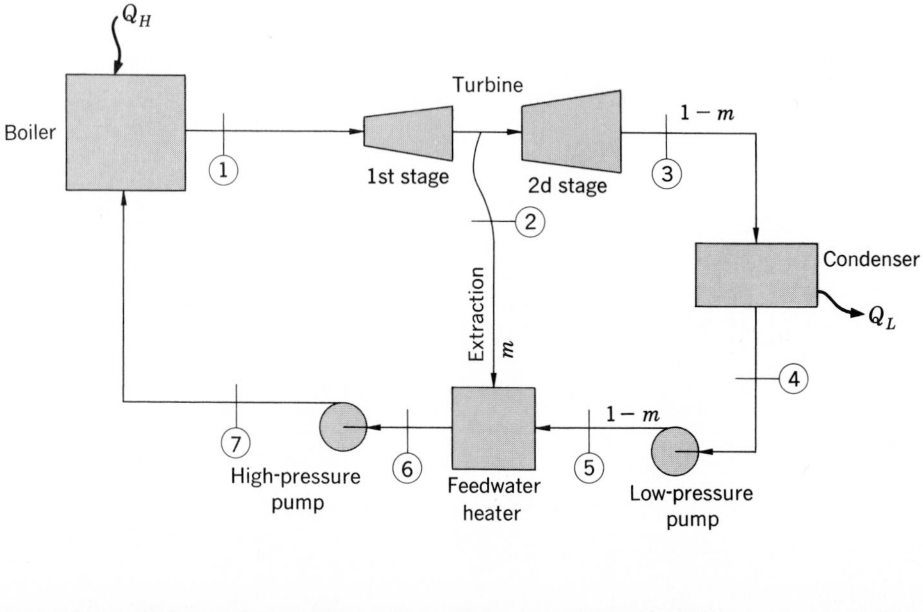

FIG. 9-7 Practical regenerative cycle with one open feedwater heater.

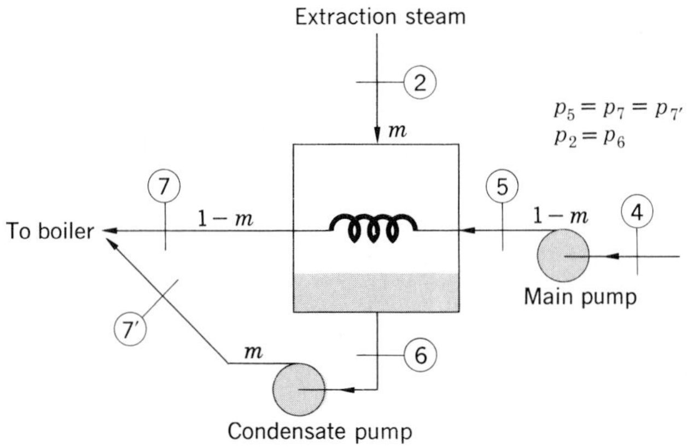

FIG. 9-8 Closed feedwater heater with condensate pump.

feedwater is not mixed with the extraction steam. A separate smaller pump is used to pump the condensate up to the boiler pressure. There is only one main feedwater pump in this instance. An alternative scheme for using a closed feedwater heater is shown in Fig. 9-9. In this example the condensate is allowed to pass through a steam trap back to the condenser hot well. A steam trap is a device with a float-operated valve which permits the passage of liquid but does not allow passage of vapor.

The closed feedwater heater is different in operation from its open counterpart. As we have said, the main pump raises the liquid water to the boiler pressure so that $p_5 = p_7 =$ boiler pressure in Fig. 9-8. The extraction steam at point 2 is at a lower pressure because of passage through the first turbine stage. It may be superheated or in the wet-mixture region, depending on the particular cycle conditions. In the closed heater the extraction steam is exposed to tubes through which flows the cool, high-pressure water. As a result, the steam

FIG. 9-9 Closed feedwater heater with condensate trapped to condenser hot well.

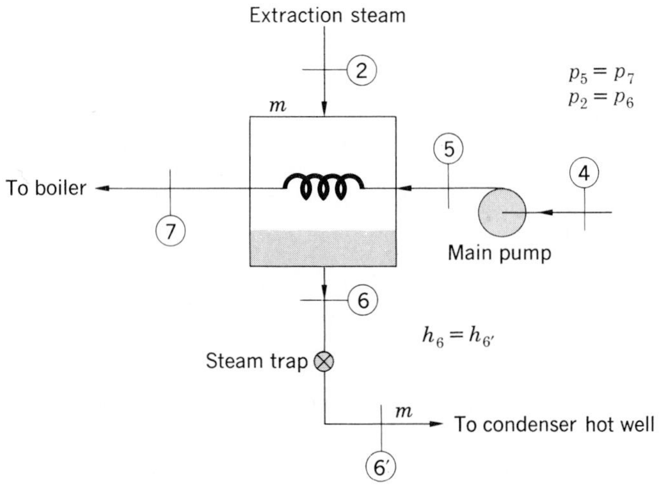

condenses on the tubes, and the energy released in the condensation process raises the temperature of the water to be delivered to the boiler. The best we can hope to accomplish in this process is to heat the high-pressure water up to the *saturation temperature corresponding to the extraction pressure;* this is the temperature at which condensation takes place. Thus, at point 7 a compressed liquid will be experienced.

We can summarize our discussion of the exit temperatures from feedwater heaters with the following table:

Type of Heater	Exit Feedwater Pressure	Exit Feedwater Temperature
Open	Extraction pressure	Saturation temperature corresponding to extraction pressure (saturated liquid)
Closed	Boiler pressure	Saturation temperature corresponding to extraction pressure (compressed liquid)

In practice, several regeneration stages may be employed for closer approximation of the ideal regenerative cycle. A schematic for a cycle using two closed feedwater heaters is shown in Fig. 9-10. For unit mass leaving the boiler $1 - m_1$ will be left for expansion in the second stage after the m_1 is extracted for the first heater. Then, when m_2 is extracted for the second heater, there will remain $1 - m_1 - m_2$ for the third turbine stage.

The number of feedwater heaters employed in an actual power plant is determined by cost considerations. In general, the increased capital expenditures for additional heaters must be more than compensated by reduced operating costs resulting from higher cycle efficiencies, if the additional expense is to be justified.

To compute the work output and efficiencies of regenerative cycles we must make energy balances on the cycle components, as in any other cycle. To perform such energy balances on the turbines and pumps requires that we determine the values of the extraction rate(s). These quantities may be obtained from energy balances on the feedwater heaters in accordance with the exit temperatures we discussed above. For example, we could write energy balances for the two feedwater heaters of Fig. 9-10 as

First heater: $$(1)h_7 + m_1h_2 = (1)h_8 + m_1h_9$$

Second heater: $$(1)h_6 + m_2h_3 = (1)h_7 + m_2h_{11}$$

Once the enthalpies are determined, m_1 and m_2 are easily calculated.

We can now see the general pattern. For each feedwater heater there is an extraction point and an unknown extraction flow rate. For each heater we can write an energy balance so that we will obtain an equation for each extraction flow. The solution of this set of equations will establish the particular extraction flows required.

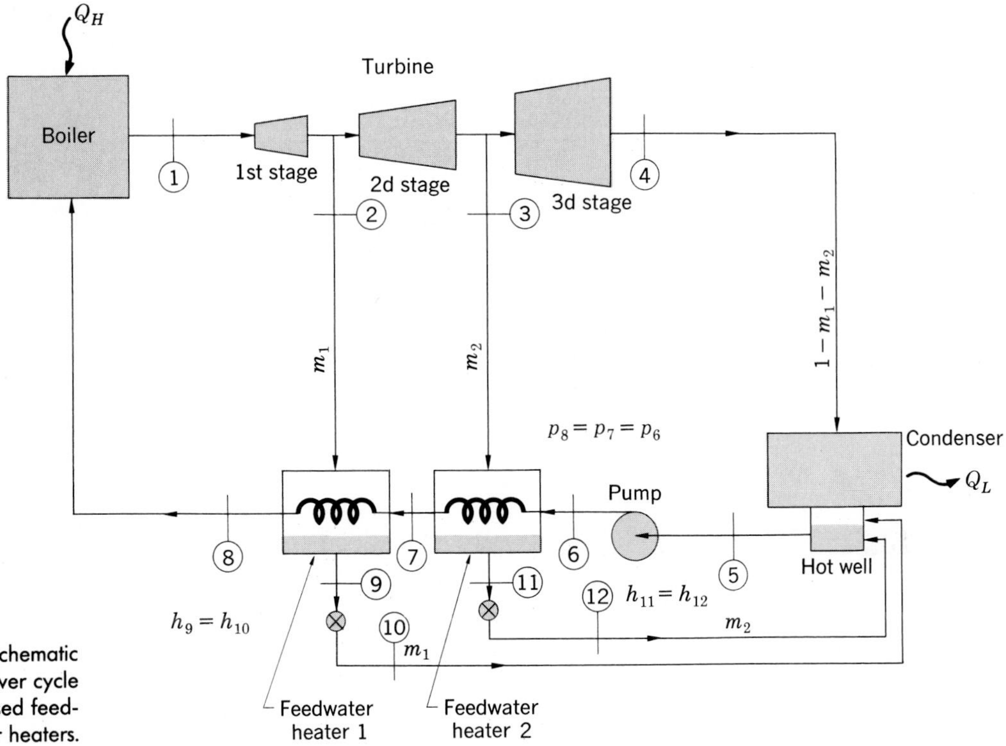

FIG. 9-10 Schematic of vapor power cycle with two closed feedwater heaters.

EXAMPLE 9-3 Regenerative Steam Cycle

A regenerative steam cycle operates between the same upper and lower pressure and temperature limits as in Example 9-1. Steam extraction occurs at 100 psia and an open feedwater heater is employed. Calculate the thermal efficiency of the cycle assuming isentropic expansion in the turbine.

SOLUTION This cycle is shown schematically in Fig. 9-7. For isentropic expansion in the turbine,

$$s_1 = s_2 = s_3$$

From a Mollier diagram we obtain

$$h_1 = 1357 \qquad (500 \text{ psia, } 700^\circ\text{F})$$

$$h_2 = 1194 \qquad (100 \text{ psia})$$

$$h_3 = 935 \qquad (2 \text{ psia})$$

$$h_4 = 94.02 \qquad (\text{saturated liquid at 2 psia})$$

The low-pressure pump raises the pressure of the condensate to 100 psia so that

$$\begin{aligned} h_5 - h_4 &= v_f(p_5 - p_4) \\ &= \frac{(0.016\ 23)(100 - 2)(144)}{778} \\ &= 0.294 \text{ Btu/lbm} \end{aligned}$$

Thus

$$\begin{aligned} h_5 &= 0.294 + 94.02 \\ &= 94.31 \text{ Btu/lbm} \end{aligned}$$

For an ideal open feedwater heater $p_2 = p_5 = p_6$ and saturated-liquid conditions are assumed to exist at point 6. Thus

$$h_6 = 298.6 \qquad \text{(saturated liquid at 100 psia)}$$

For the high-pressure pump we have

$$\begin{aligned} h_7 - h_6 &= v_f(p_7 - p_6) \\ &= \frac{(0.017\ 74)(500 - 100)(144)}{778} \\ &= 1.313 \text{ Btu/lbm} \qquad (3.0589 \text{ kJ/kg}) \end{aligned}$$

and

$$\begin{aligned} h_7 &= 1.313 + 298.6 \\ &= 299.9 \text{ Btu/lbm} \qquad (697.07 \text{ kJ/kg}) \end{aligned}$$

We are now in a position to calculate the extraction rate m with an energy balance on the feedwater heater. The energy balance on the feedwater heater is

$$mh_2 + (1 - m)(h_5) = (1)(h_6) \tag{a}$$

Inserting the proper numerical values,

$$m(1194) + (1 - m)(94.3) = 298.6$$

and

$$m = 0.186 \text{ lbm extracted/lbm total flow}$$

The thermal efficiency of the cycle is now calculated with

$$\eta_{th} = \frac{W}{Q_H}$$

$$= \frac{(1)(h_1 - h_2) + (1 - m)(h_2 - h_3) - (1 - m)(h_5 - h_4) - (1)(h_7 - h_6)}{h_1 - h_7}$$

$$= \frac{1}{1357 - 299.9}[(1)(1357 - 1194) + (1 - 0.186)(1194 - 935) - (1 - 0.186)(0.3) - (1)(1.313)]$$

$$= 35.2 \text{ percent}$$

9-4 DEVIATIONS FROM THEORETICAL VAPOR CYCLES

All of the vapor cycles discussed in the foregoing are idealized in that the expansion and compression processes are assumed to be reversible. It has been tacitly assumed that no pressure losses occur between cycle elements as a result of fluid friction. Perhaps the most significant deviation from ideal behavior occurs in the turbine expansion process. In Fig. 9-11 this deviation is illustrated on an *h-s* or Mollier diagram. Ideally, the process would occur reversibly and adiabatically as in 1-2*s*. Actually, various influences of fluid friction would cause the process to behave as in 1-2. The process would still be adiabatic, but as a result of irreversibilities, an increase in entropy would be experienced.

The turbine efficiency is defined by

$$\eta_{\text{turb}} = \frac{\text{actual work}}{\text{isentropic work}} = \frac{h_1 - h_2}{h_1 - h_{2s}} \tag{9-6}$$

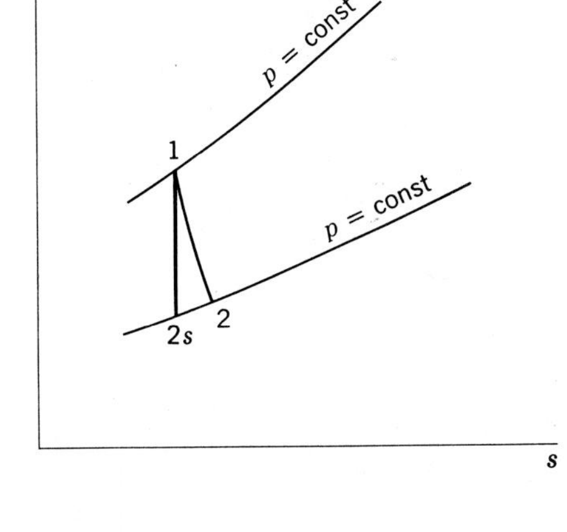

FIG. 9-11 Comparison of actual and ideal turbine expansion processes.

A similar definition is employed to take account of irreversibilities in the pumps:

$$\eta_{\text{pump}} = \frac{\text{isentropic work}}{\text{actual work}}$$

or

$$\frac{h_{4s} - h_3}{h_4 - h_3} \tag{9-7}$$

for the basic Rankine cycle. In Eq. (9-7) it is to be noted that the isentropic work for a pump is less than the actual work because the device absorbs power. In general, turbine and pump efficiencies may only be determined from experiment. The effects of pump and turbine efficiencies on the *T-s* diagram for the basic Rankine cycle are shown in Fig. 9-12.

Practical Considerations

Modern steam power plants are much more complicated than might be gathered from the above discussions. Each component of the cycle is an intricate piece of machinery which has variable performance depending on load, speed, and different pressure and temperature conditions. Elaborate control systems must be provided to ensure reliable and stable operation as well as to protect the machines from too rapid changes in loads, speeds, or other variables.

As we have seen in Chap. 8, combustion processes can be involved affairs, and this is especially true for a large steam boiler. The fuel and air flow rates

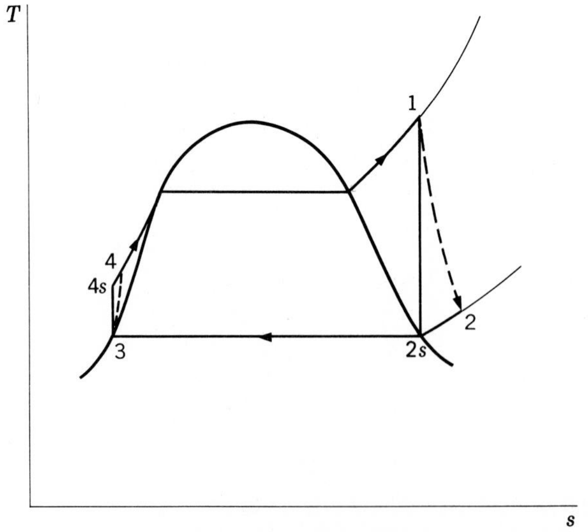

FIG. 9-12 Influence of actual turbine and pump processes on *T-s* diagram of basic Rankine cycle.

must be carefully controlled to avoid excessive temperatures which can destroy the tubes in the boiler. At the same time, environmental constraints may be placed on allowable discharges from the combustion process which can restrict the modes of operation. If coal is the fuel, the process of getting it to the boiler in satisfactory form can be quite a demanding engineering problem. It must be mined, shipped, and then stored at the plant site. It must then be pulverized to a proper size and transported to the boiler by a conveyor system. Not the least of the problems is the removal and disposal of the huge quantities of ash which accumulate after coal is burned; and, if high-sulfur coal is burned, one must use elaborate methods to remove SO_2 from the stack discharge gases. Despite all the above complications modern steam power plants operate throughout the world to produce dependable electric power at all times.

At this point the reader may have temporarily lost sight of the purpose of reheat, feedwater heaters, etc. The objective is to increase efficiency by increasing the average temperature at which heat is added to the cycle. One might ask: Why not build a hotter fire in the boiler? The answer is that metallurgical considerations limit the temperatures and/or pressures which the boiler tubes and turbine blades can tolerate. As better materials are developed, we may indeed raise the boiler and turbine temperatures and achieve a higher efficiency; still, regeneration and reheat will be employed for their contribution to efficiency gain.

We have noted that efficiency can also be increased by lowering the temperature of heat rejection; however, this temperature is essentially limited by local atmospheric conditions. Because of the large amounts of energy which must be removed from the condenser, it is usually desirable to locate a power plant adjacent to a lake or river for cooling purposes. In some cases power companies will construct a lake for this purpose when the power plant is built. In some locales lakes or rivers are not available, and the heat-rejection process is accomplished with large cooling towers, as described in Sec. 7-8. In some parts of the world, water is simply unavailable for use as a coolant, and the heat from the condenser must be dissipated directly to the surrounding air. Enormous fans and heat exchangers are required in these instances.

EXAMPLE 9-4 Effects of Turbine Efficiencies

Calculate the thermal efficiency for the reheat cycle of Example 9-2, assuming that the turbines have efficiencies of 80 percent and the expansion processes are no longer isentropic.

SOLUTION The expansion processes for the turbines are shown in schematic form on the accompanying figure. The isentropic processes are shown as dashed lines, and the actual expansion processes are indicated with the solid lines. The overall cycle schematic is indicated in Fig. 9-4. Using the definition of turbine efficiency from Eq. (9-6), we have for the first turbine stage,

$$\eta = \frac{h_1 - h_2}{h_1 - h_{2s}}$$

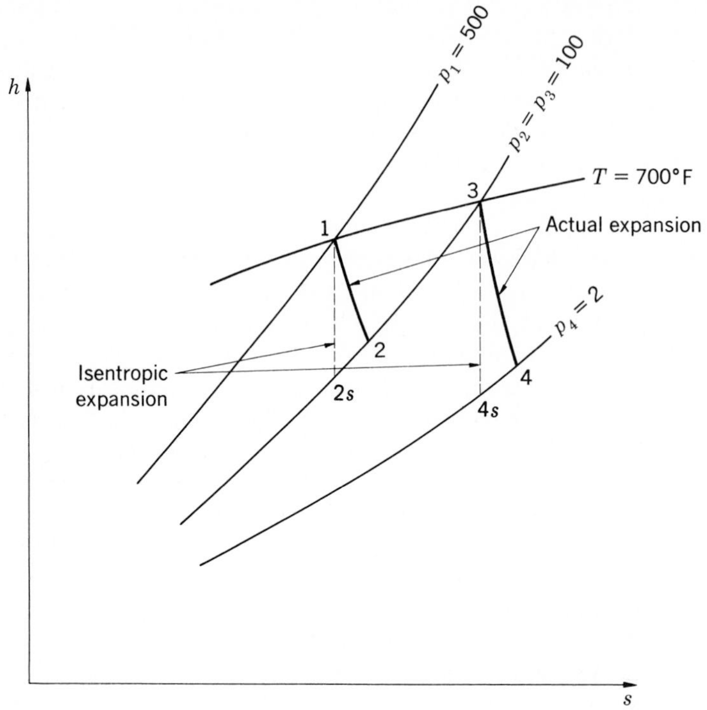

FIG. EXAMPLE 9-4

Also, for the second turbine stage,

$$\eta = \frac{h_3 - h_4}{h_3 - h_{4s}}$$

From Example 9-2,

$$\begin{aligned} h_1 &= 1357 & h_{4s} &= 1047 \\ h_{2s} &= 1194 & h_5 &= 94.02 \\ h_3 &= 1379 & h_6 &= 95.52 \end{aligned}$$

Then

$$0.8 = \frac{1357 - h_2}{1357 - 1194}$$

and

$$0.8 = \frac{1379 - h_4}{1379 - 1047}$$

so that

$$h_2 = 1227$$
$$h_4 = 1113$$

The thermal efficiency is now calculated as

$$\eta_{\text{th}} = \frac{W}{Q_H} = \frac{(h_1 - h_2) + (h_3 - h_4) - (h_6 - h_5)}{(h_1 - h_6) + (h_3 - h_{2'})}$$

$$= \frac{(1357 - 1227) + (1379 - 1113) - 1.496}{(1357 - 95.5) + (1379 - 1227)}$$

$$= 27.9 \text{ percent}$$

It should be noted that the total heat added is reduced as a result of the nonisentropic expansion in the first turbine stage. This reduction, however, is not enough to offset the reduced work output of the turbines.

Nuclear Power Cycles

Nuclear power plants which are in operation today, and those projected for the future, all use the Rankine steam cycle for conversion of heat to useful work. Instead of the combustion of fossil fuels, a nuclear reactor serves as the heat source. Because of radioactive hazards, elaborate schemes are used to prevent contamination of the main steam power cycle. Liquid sodium is frequently employed as an intermediate fluid to remove heat from the reactor and then vaporize the high-pressure water. Special piping and pumps are required for these applications to prevent leaks and any direct contact between sodium and water.

Nuclear power plants have two obvious advantages. First, they do not produce air pollution because there are no products of combustion to be dispersed in the atmosphere, and second, they alleviate demands on petroleum and natural gas fuels which may be required in home, small industrial, and automotive applications. But the nuclear plants are not without their disadvantages too. With an increase in the amount of power produced by nuclear means there is the growing problem of disposal of radioactive waste products. Burial in deep underground caverns or at sea is a possibility but needs careful evaluation to determine the eventual environmental impact.

Because of the fluids which must be circulated through the reactor and the materials which must be used to contain the fissionable substances, the practical upper temperature limit of a nuclear power plant is not as high as its fossil-fuel counterpart. As a consequence its efficiency will be lower and for a given power output more heat must be rejected to the surrounding atmosphere. Thus the heating of streams and lakes, a serious environmental problem with any power plant, is even more critical with a nuclear plant. The upper limit of efficiency for a fossil fuel plant is about 40 percent, and only about 30 percent for a nuclear plant. We can see how this reduction in efficiency affects waste heat rejection by making a simple calculation for two plants, each with 500-MW output. For the fossil fuel plant

$$Q_H = \frac{500 \text{ MW}}{0.4} = 1250 \text{ MW}$$

$$Q_L = Q_H - W = 1250 - 500 = 750 \text{ MW}$$

And for the nuclear plant

$$Q_H = \frac{500 \text{ MW}}{0.3} = 1667 \text{ MW}$$

$$Q_L = Q_H - W = 1667 - 500 = 1167 \text{ MW}$$

Thus in this case a reduction of efficiency from 40 to 30 percent increases the heat rejection from 750 to 1167 MW, or *56 percent*. So we see that there must be trade-offs made between the advantages and disadvantages of nuclear power production. However, the important point to remember is that a thermodynamic analysis of the Rankine cycle will proceed in the same fashion, whatever the application.

9-5 POWER-CYCLE ANALYSIS

From the foregoing discussion we can see that vapor power cycles can have many components, and one must be careful to make the proper energy balances. Too often, one is tempted to think of each cycle as a special case with its own peculiar analysis technique. A far better procedure, and one which will be adaptable to other types of power cycles in this chapter, is to view cycle analysis in a general way. This procedure will almost always yield correct results:

1. Draw a schematic block diagram for the cycle. Label all points in the cycle. The type of nomenclature does not matter, e.g., 1, 2, 3; a, b, c; x, y, z.
2. Draw arrows indicating heat or work flows delivered to or rejected from the cycle components.
3. Designate mass flows in, out, or through the various components.
4. Write down the known properties, states, and work or heat flows in terms of the nomenclature devised above.
5. Write down the desired results of the analysis, again, in terms of the above nomenclature.
6. Write out energy balances for works, heat flows, etc.

7 Write any energy balances which may be necessary to establish flow rates, as in a feedwater heater.
8 Determine needed properties to execute the energy balance calculations using process information (isentropic process, constant-pressure process, etc.).
9 Assemble the information and calculations to obtain the desired results.

EXAMPLE 9-5 Regeneration with Two Feedwater Heaters
A regenerative steam cycle employs two closed feedwater heaters and three turbine stages. The condensate from the first heater flows through a steam trap into the second heater, and the condensate from the second heater is pumped up to the boiler pressure. The steam enters the turbine at 10 MPa, 540°C, extraction occurs at 2 MPa and 700 kPa, and discharge from the third turbine stage at 6 kPa. The turbine efficiencies are 85 percent, and the pump efficiencies are 100 percent. Calculate the thermal efficiency and steam flow required for a total electrical output from the turbine of 20 000 kW. The schematic diagram is shown in the accompanying figure (Fig. Example 9-5*b*), along with a Mollier diagram for the turbine processes (Fig. Example 9-5*a*).

FIG. EXAMPLE 9-5*a*

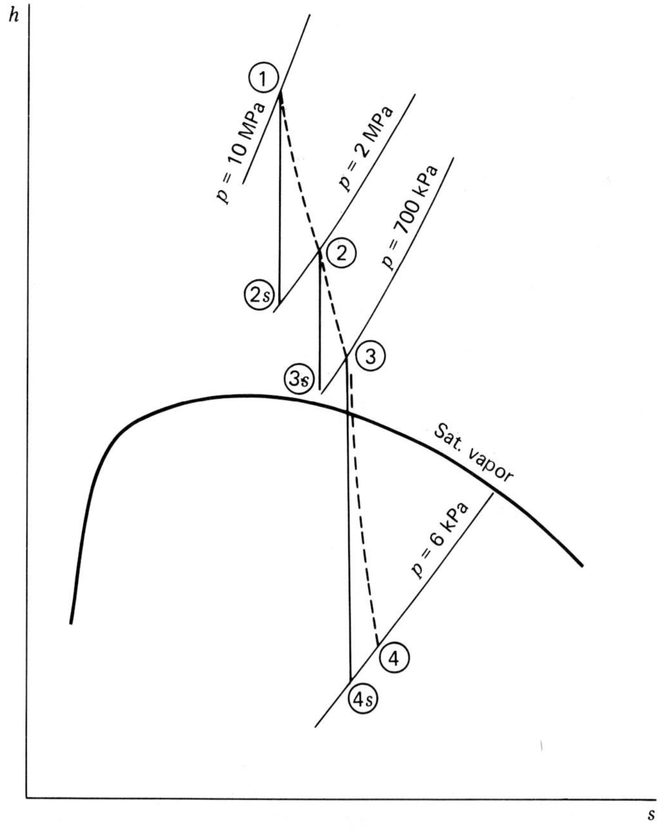

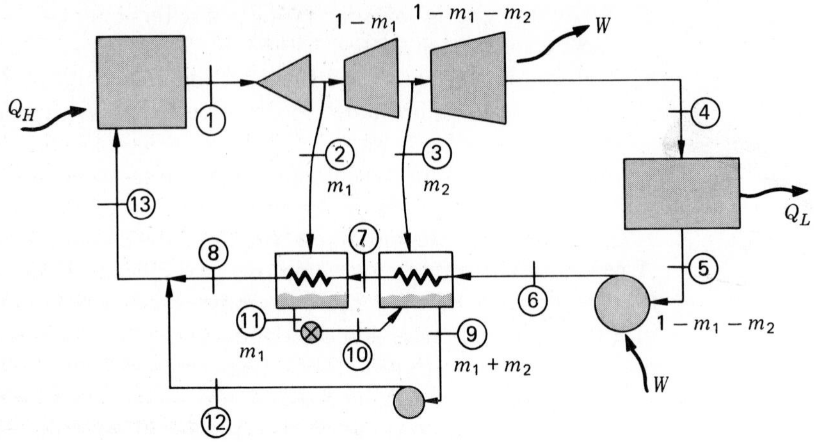

FIG. EXAMPLE 9-5*b*

SOLUTION From the problem statement

$$p_1 = p_{13} = p_{12} = p_8 = p_7 = p_6 = 10 \text{ MPa}$$
$$T_1 = 540°\text{C} \qquad p_2 = p_{11} = 2 \text{ MPa}$$
$$p_3 = p_9 = 700 \text{ kPa} \qquad p_4 = p_5 = 6 \text{ kPa}$$

From the Mollier diagram

$$h_1 = 3475 \qquad h_{2s} = 3005 \text{ kJ/kg}$$

Using the turbine efficiency,

$$0.85 = \frac{h_1 - h_2}{h_1 - h_{2s}} \qquad h_2 = 3076 \text{ kJ/kg} \tag{a}$$

Then,

$$h_{3s} = 2828$$

and

$$0.85 = \frac{h_2 - h_3}{h_2 - h_{3s}} \qquad h_3 = 2865 \text{ kJ/kg} \tag{b}$$

Now,

$$h_{4s} = 2136$$

and

$$0.85 = \frac{h_3 - h_4}{h_3 - h_{4s}} \qquad h_4 = 2245 \text{ kJ/kg} \tag{c}$$

The liquid enthalpies are now calculated:

$$\begin{aligned}
h_5 &= 151.53 \qquad (h_f \text{ at 6 kPa}) \\
h_{6s} - h_5 &= v_5(p_6 - p_5) = (1.0064 \times 10^{-3})(10\,000 - 6) \\
&= 10.06 \text{ kJ/kg} \\
h_{6s} &= 161.59 \text{ kJ/kg} \\
h_9 &= 697.22 \qquad (h_f \text{ at 700 kPa, } T = 165.0°\text{C}) \\
h_{11} &= h_{10} = 908.79 \qquad (h_f \text{ at 2 MPa, } T = 212.4°\text{C})
\end{aligned} \tag{d}$$

$$\begin{aligned}
h_{12s} - h_9 &= v_9(p_{12} - p_9) = (1.1080 \times 10^{-3})(10\,000 - 700) \\
&= 10.3 \text{ kJ/kg} \\
h_{12s} &= 707.5 \text{ kJ/kg} \\
h_7 &= 703.18 \qquad (\text{Table A-10M, 165°C, 10 MPa}) \\
h_8 &= 912.07 \qquad (\text{Table A-10M, 212.4°C, 10 MPa})
\end{aligned} \tag{e}$$

The energy balances on the heaters are

$$m_1 h_2 + (1 - m_1 - m_2)h_7 = (1 - m_1 - m_2)h_8 + m_1 h_1 \tag{f}$$

$$m_2 h_3 + (1 - m_1 - m_2)h_6 + m_1 h_{10} = (1 - m_1 - m_2)h_7 + (m_1 + m_2)h_9 \tag{g}$$

The appropriate enthalpy values may be inserted, and the equations solved simultaneously to give

$$m_1 = 0.0721 \text{ kg} \qquad m_2 = 0.1799 \text{ kg} \tag{h}$$

We are now in a position to calculate the thermal efficiency of the power cycle:

$$\eta = \frac{(h_1 - h_2) + (1 - m_1)(h_2 - h_3) + (1 - m_1 - m_2)(h_3 - h_4) - (1 - m_1 - m_2)(h_{6s} - h_5) - (m_1 + m_2)(h_{12s} - h_9)}{h_1 - h_{13}} \tag{i}$$

The enthalpy at point 13 is obtained from the energy balance:

$$\begin{aligned}
(1 - m_1 - m_2)h_8 + (m_1 + m_2)h_{12} &= (1)h_{13} \\
h_{13} &= 860.5 \text{ kJ/kg}
\end{aligned} \tag{j}$$

Inserting the enthalpy values in Eq. (*i*) we obtain the efficiency as

$$\eta = \frac{1048.4}{2614.5} = 40.1 \text{ percent} \tag{k}$$

The net work output is 1048.4 kJ/kg so the steam flow required for 20 000 kW is

$$\dot{m} = \frac{20\,000}{1048.4} = 19.08 \text{ kg/s} = 6.87 \times 10^4 \text{ kg/h} \tag{l}$$

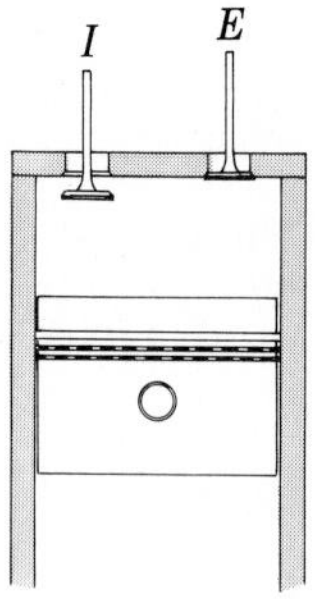

FIG. 9-13 Reciprocating air compressor.

9-6 GAS COMPRESSORS

A gas compressor may be considered as an energy-conversion device since it converts mechanical work into stored potential energy resulting from high pressure. We initiate our discussion with the reciprocating type of machine illustrated in Fig. 9-13. The ideal pressure-volume diagram for such a machine is shown in Fig. 9-14. At point 4 the piston has just completed the compression and delivery of the gas in the cylinder and the exhaust valve E has just closed. The volume of gas remaining in the cylinder at this time V_4 is called the *clearance volume*. As the piston moves back from the top-dead-center position (tdc), this volume expands until the pressure p_1 is reached. At this point the intake valve I is opened, and the gas is pulled into the cylinder by the continuing backward motion of the piston. When the piston reaches bottom-dead-center (bdc) at point 2, the intake valve is closed. The gas is then compressed from 2 to 3 and the exhaust valve is opened at point 3, thereby allowing the piston to force the gas out of the cylinder in process 3-4.

The actual compression and expansion processes in the cylinder do not follow the nice theoretical processes shown in Fig. 9-14. Intake and exhaust

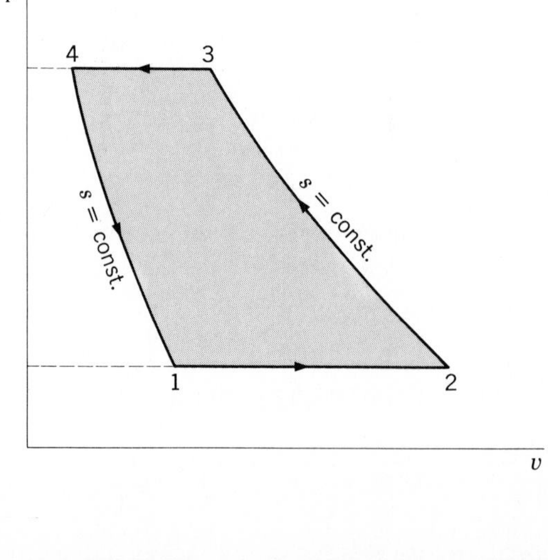

FIG. 9-14 p-v diagram for ideal reciprocating air compressor.

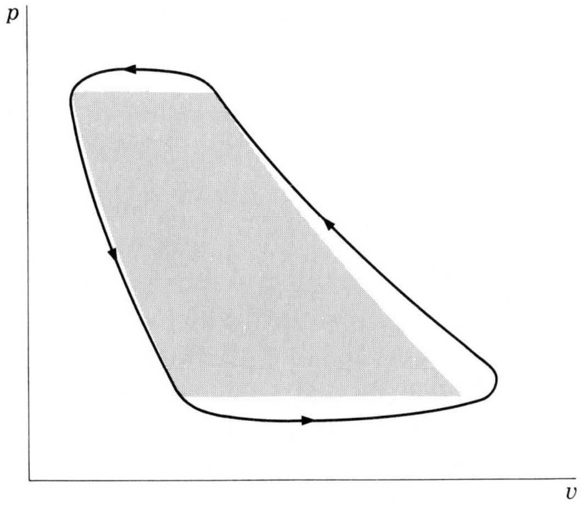

FIG. 9-15 Actual *p-v* diagram for reciprocating air compressor.

valves do not open instantaneously, and various pressure-loss and frictional effects may cause substantial deviations from the ideal processes. Figure 9-15 indicates the shape an actual *p-v* diagram might take.

Figure 9-16 illustrates the basic operation of a centrifugal compressor. Air enters the casing and is turned by the rotating-vane impeller. Because of the centrifugal force action the pressure is increased at the outer periphery. In Fig. 9-16*a* the rotating vanes are constructed in a radial fashion, while the stationary diffuser section in the casing has vanes which are curved forward in the direc-

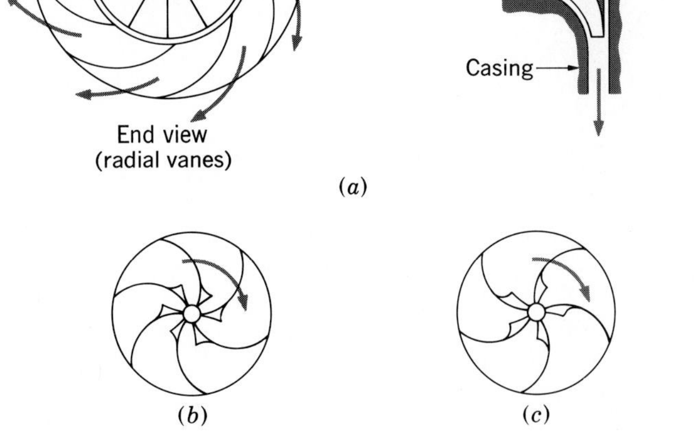

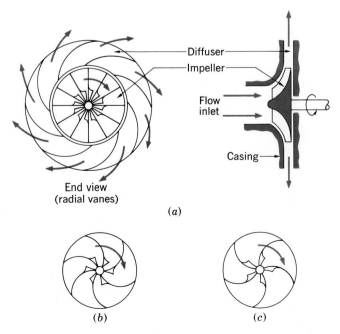

FIG. 9-16 Centrifugal compressor: (*a*) radial impeller vanes; (*b*) backward-curved vanes; (*c*) forward-curved vanes. Diagram according to Wood [11].

tion of rotation. Centrifugal compressors may also be constructed with backward- or forward-curved moving vanes, as illustrated in Fig. 9-16*b* and *c*, depending on the pressure-speed characteristics desired. Of course, a centrifugal fan or squirrel-cage blower is just a compressor to handle large flow volumes with a small pressure rise.

The operation of an axial flow compressor is illustrated in Fig. 9-17. The machine consists of a series of rotating blades (rotor) alternating with stationary blade sections (stator). A stage is composed of one rotor and one stator. The rotors are all connected to a common shaft, while the stators are fixed to the enclosing casing. The blades are special airfoil sections, and the pressure rise results from the lifting action on the blades. For this reason, the pressure rise per stage is rather small, and axial compressors always consist of many stages in series in order to accomplish any substantial compression ratio. The design of the airfoil sections in the rotor and stator is a crucial matter and will vary significantly depending on the application for the compressor; e.g., an axial compressor for use in the inlet of a high-speed jet engine would be substantially different from a compressor for low-speed air in a stationary power plant.

Gas compressors are basically steady-flow devices. In reciprocating machines a pulsating steady flow is experienced, but the work input is calculated from the steady-flow energy equation. Neglecting any cooling which may be performed on the cylinder walls, the machine is adiabatic, and the work is usually computed as the change in enthalpy for the air. For large volumes of air and modest pressure ratios, centrifugal or turbine-type compressors are nor-

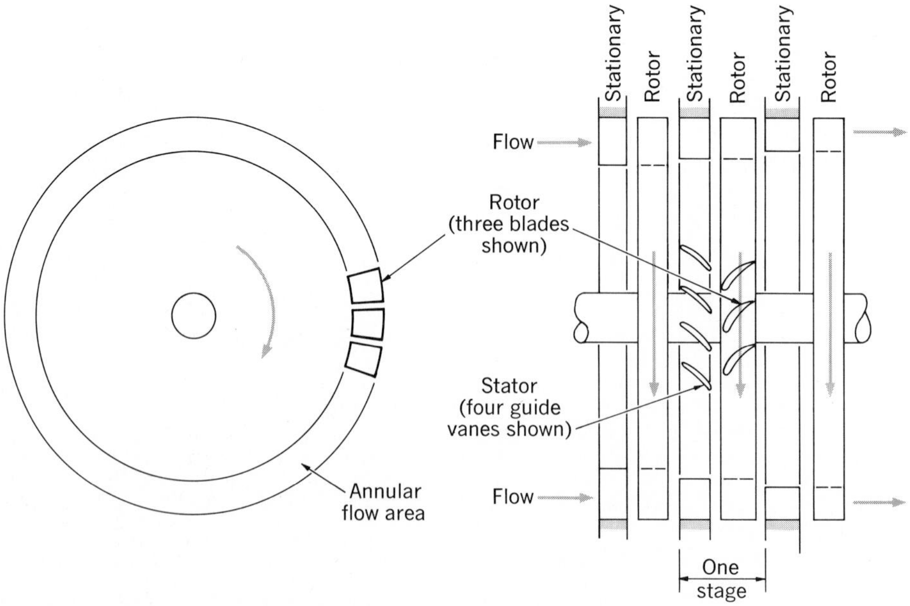

FIG. 9-17 Axial flow compressor Diagram according to Wood [11].

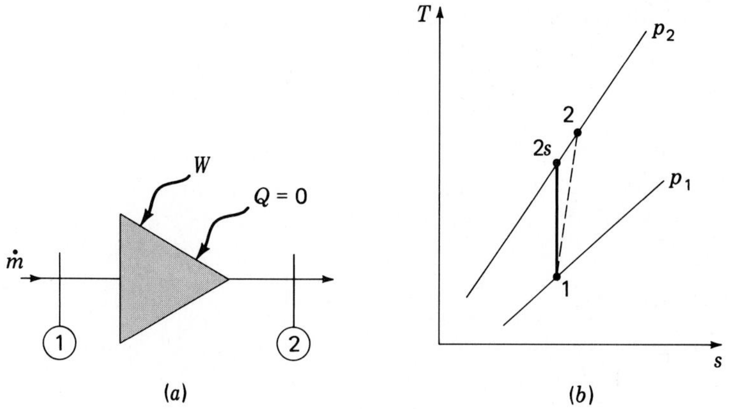

FIG. 9-18 Adiabatic steady-flow compressor: (*a*) schematic; (*b*) *T-S* diagram.

mally employed. The turbojet is one application for turbine compressors. Centrifugal compressors are widely used in large air conditioning systems.

For the steady-flow adiabatic compressor shown schematically in Fig. 9-18, the energy equation would be

$$\dot{m}h_1 + W = \dot{m}h_2$$

or, for an ideal gas,

$$\dot{W} = \dot{m}c_p(T_2 - T_1)$$

If the process is also reversible,

$$T_{2s} = T_1 \left(\frac{p_2}{p_1}\right)^{(\gamma-1)/\gamma}$$

and

$$\dot{W} = \dot{m}c_p T_1\left[\left(\frac{p_2}{p_1}\right)^{(\gamma-1)/\gamma} - 1\right] \tag{9-8}$$

Compressor efficiency is defined similar to that for a pump:

$$\eta_{\text{comp}} = \frac{\text{isentropic input work required}}{\text{actual input work}}$$

or for an ideal gas, according to the *T-s* diagram of Fig. 9-8*b*,

$$\eta_{\text{comp}} = \frac{T_{2s} - T_1}{T_2 - T_1} \tag{9-9}$$

Multistage Compression and Intercooling

If very large pressures are to be attained by the compression process, the task is very difficult to accomplish with a single piston-cylinder arrangement. First, compression to high pressures will result in prohibitively high temperatures in the cylinder; e.g., isentropic compression from 1 atm and 70°F to 1500 psia produces a temperature of 1530°F. Second, if the cylinder is large enough to handle the volume of air at the low pressure, the proportionate clearance volume at tdc will be very small, with serious leakage problems encountered around the piston rings. These difficulties can be alleviated by performing the compression in stages, as indicated in Fig. 9-19. The high-temperature problem is solved by cooling the gas between stages, and the difficulty with leakage at high pressure is minimized because the second-stage piston cylinder can be smaller since the input volume of gas to it is reduced. Of course, many stages of compression can be employed if very high outlet pressures are to be attained.

The interstage cooler is appropriately called an *intercooler*, and for the ideal arrangement of Fig. 9-19 we assume that the gas is cooled to the initial

FIG. 9-19 Intercooler arrangement for air compressor: (*a*) schematic; (*b*) temperature-entropy diagram.

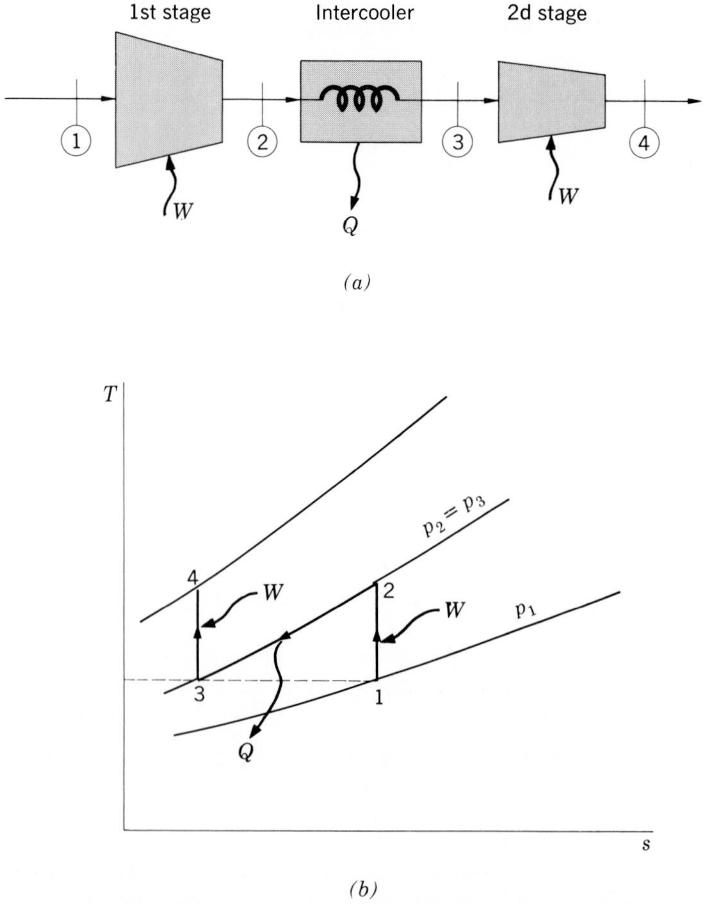

inlet temperature before admittance to the next stage. In Fig. 9-19 this means that $T_1 = T_3$ *for an ideal intercooler.*

In a multistage compressor it is desirable to select the interstage pressure so that the total work input is minimized. Using the nomenclature of Fig. 9-19 the work in each stage for isentropic compression is

$$W_{12} = h_2 - h_1 = c_p(T_2 - T_1) = c_p T_1 \left[\left(\frac{p_2}{p_1}\right)^{(\gamma-1)/\gamma} - 1\right]$$

$$W_{34} = h_4 - h_3 = c_p(T_4 - T_3) = c_p T_3 \left[\left(\frac{p_4}{p_3}\right)^{(\gamma-1)/\gamma} - 1\right]$$

The total work is the sum of these stage works. We wish to find the interstage pressure p_2 which minimizes this total work. Thus we form the derivative

$$\frac{\partial W_{\text{total}}}{\partial p_2} = 0 = \frac{\partial}{\partial p_2}\left\{c_p T_1 \left[\left(\frac{p_2}{p_1}\right)^{(\gamma-1)/\gamma} + \left(\frac{p_4}{p_2}\right)^{(\gamma-1)/\gamma} - 2\right]\right\}$$

where $p_2 = p_3$ and $T_1 = T_3$ for ideal intercooling. Performing the indicated differentiation we obtain

$$\frac{p_2}{p_1} = \frac{p_4}{p_3} \tag{9-10}$$

or, the pressure ratio is the same across each stage. But $p_2 = p_3$, so

$$\left(\frac{p_2}{p_1}\right)\left(\frac{p_4}{p_2}\right) = \left(\frac{p_2}{p_1}\right)^2$$

and

$$p_2 = (p_1 p_4)^{1/2}$$

or

$$\frac{p_2}{p_1} = \left(\frac{p_4}{p_1}\right)^{1/2}$$

We have developed this relation for a two-stage compressor, but a further analysis would show that the result applies to multiple stages in general; i.e., minimum work is obtained when the pressure ratio is the same across each stage, provided isentropic compression and ideal intercooling are experienced. For example, for a three-stage compressor,

$$\text{Single-stage pressure ratio} = (\text{overall pressure ratio})^{1/3}$$

EXAMPLE 9-6

A reciprocating compressor receives atmospheric air at 14.7 psia and 70°F and discharges at 50 psia. The clearance volume is 5 percent of the piston displacement, and the compression and expansion processes are assumed to occur isentropically. Calculate:

(*a*) The piston displacement in cubic feet per minute necessary to compress 100 ft³/min of air at inlet conditions

(*b*) The *volumetric efficiency* defined as

$$\eta_{vol} = \frac{\text{actual mass of gas compressed}}{\begin{array}{c}\text{mass of gas occupying piston displacement}\\ \text{at inlet pressure and temperature}\end{array}}$$

(*c*) The horsepower required

SOLUTION Figure 9-14 applies to this problem. The *piston displacement* is the volume swept out by the piston as it moves from tdc to bdc. Thus

$$PD = V_2 - V_4 \tag{a}$$

V_4 is the clearance volume, so that

$$V_4 = 0.05(V_2 - V_4) \tag{b}$$

The volume of free air which is pulled into the cylinder is $V_2 - V_1$, so that

$$V_2 - V_1 = 100 \text{ ft}^3/\text{min} \qquad (2832 \text{ liters/min}) \tag{c}$$

Since the compression and expansion processes are isentropic, we have

$$\frac{V_1}{V_4} = \left(\frac{p_4}{p_1}\right)^{1/\gamma} = \left(\frac{50}{14.7}\right)^{1/1.4} = 2.395 \tag{d}$$

$$\frac{V_2}{V_3} = \left(\frac{p_3}{p_2}\right)^{1/\gamma} = \left(\frac{50}{14.7}\right)^{1/1.4} = 2.395 \tag{e}$$

Combining Eqs. (*b*), (*c*), and (*d*),

$$V_4 = 5.38 \text{ ft}^3/\text{min} \qquad (152.3 \text{ liters/min})$$

$$V_1 = 12.9 \text{ ft}^3/\text{min} \qquad (365.3 \text{ liters/min})$$

$$V_2 = 112.9 \text{ ft}^3/\text{min} \qquad (3196.8 \text{ liters/min})$$

The piston displacement is thus

$$PD = V_2 - V_4 = 107.5 \text{ ft}^3/\text{min} \qquad (3044 \text{ liters/min})$$

The volumetric efficiency may be calculated as

$$\eta_{\text{vol}} = \frac{p_1(V_2 - V_1)/RT_1}{p_1(V_2 - V_4)/RT_1} = \frac{V_2 - V_1}{V_2 - V_4} = \frac{100}{107.5}$$
$$= 93 \text{ percent}$$

Of course, if there were no clearance volume, the volumetric efficiency would be 100 percent, but from a practical standpoint it is not possible to design a compressor with zero clearance. We may calculate the work by integrating the area on the p-V diagram. For a closed cycle we may either compute

$$\oint p\, dV$$

or, as is a bit easier in this case,

$$W = \oint V\, dp$$

For the isentropic processes we have

$$\begin{aligned}\int_2^3 V\, dp &= \frac{\gamma}{1-\gamma}(p_3V_3 - p_2V_2)\\ &= \frac{\gamma p_2 V_2}{1-\gamma}\left[\left(\frac{p_3}{p_2}\right)^{(\gamma-1)/\gamma} - 1\right]\\ &= \frac{(1.4)(14.7)(144)(112.9)}{1-1.4}\left[\left(\frac{50}{14.7}\right)^{0.4/1.4} - 1\right]\\ &= -3.51 \times 10^5 \text{ ft·lbf/min} \qquad (-7.93 \text{ kW})\end{aligned} \tag{f}$$

$$\begin{aligned}\int_4^1 V\, dp &= \frac{\gamma p_4 V_4}{1-\gamma}\left[\left(\frac{p_1}{p_4}\right)^{(\gamma-1)/\gamma} - 1\right]\\ &= \frac{(1.4)(50)(144)(5.38)}{1-1.4}\left[\left(\frac{14.7}{50}\right)^{0.4/1.4} - 1\right]\\ &= 4.0 \times 10^4 \text{ ft·lbf/min} \qquad (0.904 \text{ kW})\end{aligned} \tag{g}$$

For the constant-pressure processes

$$\int_3^4 V\, dp = \int_1^2 V\, dp = 0 \tag{h}$$

Thus

$$\begin{aligned}W_{\text{total}} &= -3.51 \times 10^5 + 4.0 \times 10^4\\ &= -3.11 \times 10^5 \text{ ft·lbf/min}\end{aligned}$$

or

$$W_{\text{total}} = -9.42 \text{ hp} \qquad (-7.024 \text{ kW})$$

This is the work done on the face of the piston by the gas. The work *input* to the compressor would be the negative of this, or +9.42 hp. We could also recognize that the compressor is a pulsating steady-flow device, and for an adiabatic process the rate of doing work could be calculated from

$$W = \dot{m}(h_2 - h_1) \tag{i}$$

where now subscript 1 refers to inlet and subscript 2 to outlet conditions. For ideal-gas behavior with constant specific heats

$$W = \dot{m}c_p(T_2 - T_1) \tag{j}$$

We have, for isentropic compression,

$$T_2 = T_1 \left(\frac{p_2}{p_1}\right)^{(\gamma-1)/\gamma} = 530 \left(\frac{50}{14.7}\right)^{0.286} = 752°\text{R} \qquad (418 \text{ K})$$

The mass flow rate is calculated from the volumetric flow at inlet:

$$\dot{m} = \frac{p_1 \dot{V}_1}{RT_1} = \frac{(14.7)(144)(100)}{(53.35)(530)} = 7.486 \text{ lbm/min} \qquad (3.396 \text{ kg/min})$$

The work input is then obtained from Eq. (*j*) as

$$\begin{aligned} W &= (7.486)(0.24)(752 - 530) = 399 \text{ Btu/min} \\ &= 9.42 \text{ hp} \qquad (7.024 \text{ kW}) \end{aligned}$$

in agreement with the previous calculation.

EXAMPLE 9-7 Two-Stage Compressor

A two-stage steady-flow compressor compresses air from 14.7 psia and 70°F to 100 psia. An ideal intercooler is used along with optimum staging. Calculate the horsepower required if the mass flow of air is 10 lbm/min. Also calculate the heat removed in the intercooler. Assume constant specific heats. Compare with the work required for a single-stage compressor.

SOLUTION For this problem the schematic diagrams of Fig. 9-18 apply. We have

$$T_1 = T_3 = 70°\text{F} = 530°\text{R} \qquad \text{(ideal intercooling)}$$

$$p_2 = p_3 \qquad p_1 = 14.7 \text{ psia} \qquad p_4 = 100 \text{ psia}$$

For ideal staging

$$\frac{p_2}{p_1} = \frac{p_4}{p_3}$$

or

$$p_2^2 = p_1 p_4$$
$$p_2 = 38.3 \text{ psia}$$

For isentropic compressions

$$\frac{T_2}{T_1} = \left(\frac{p_2}{p_1}\right)^{(\gamma-1)/\gamma} \qquad \frac{T_4}{T_3} = \left(\frac{p_4}{p_3}\right)^{(\gamma-1)/\gamma}$$

Thus

$$T_2 = 530\left(\frac{38.3}{14.7}\right)^{0.286}$$
$$= 697°\text{R} \qquad (387.2 \text{ K})$$

Since the pressure ratios and inlet temperatures are equal for each stage, we also have

$$T_4 = 697°\text{R}$$

The work input per pound-mass of air for the two stages is

$$\begin{aligned} W &= h_2 - h_1 + h_4 - h_3 \\ &= (0.24)(697 - 530) + (0.24)(697 - 530) \\ &= 80.2 \text{ Btu/lbm} \qquad (186.5 \text{ kJ/kg}) \end{aligned}$$

The total work is therefore, for a flow of 10 lbm/min,

$$W_{\text{total}} = 802 \text{ Btu/min} \qquad (14.1 \text{ kW})$$

Since 1 hp = 2545 Btu/h,

$$\text{hp} = \frac{(802)(60)}{2545}$$
$$= 18.9$$

The heat removed in the intercooler is

$$\begin{aligned} Q &= \dot{m}_a c_p (T_2 - T_3) \\ &= (10)(0.24)(697 - 530) \\ &= 401 \text{ Btu/min} \qquad (7.05 \text{ kW}) \end{aligned}$$

For a single-stage isentropic compression the exit temperature from the compressor would be

$$T_{4'} = T_1 \left(\frac{p_4}{p_1}\right)^{(\gamma-1)/\gamma}$$
$$= 530 \left(\frac{100}{14.7}\right)^{0.286}$$
$$= 917°\text{R} \qquad (509 \text{ K})$$

In this circumstance the work would be

$$W = \dot{m}_a(h_{4'} - h_1)$$
$$= (10)(0.24)(917 - 530)$$
$$= 928 \text{ Btu/min} \qquad (16.31 \text{ kW})$$

The advantage of the intercooler and multistaging is obvious; it results in a reduction in work of 13.5 percent (928 to 802 Btu/min). In a practical situation the reduced power requirements must be balanced against the increased cost of an intercooler-multistage arrangement.

9-7 INTERNAL COMBUSTION ENGINE CYCLES

In this section we shall discuss several power cycles which form a theoretical basis for spark-ignition and compression-ignition (diesel) engines. These cycles are employed in *reciprocating* engines. Cycles using a steady flow of fluid are discussed in connection with gas turbines in Sec. 9-12. From a practical standpoint, it is worthwhile to discuss the meaning of four-stroke and two-stroke cycles. Consider the two typical reciprocating engine configurations shown in Fig. 9-20. Let us first examine the four-cycle engine. When the piston is at tdc, the intake valve *I* opens and a charge of fuel and air is drawn into the cylinder as the piston moves downward (intake stroke). At bdc the intake valve closes and the piston moves upward to compress the mixture (compression stroke). At tdc the ignition device fires, thereby burning the fuel and forcing the piston downward again (power stroke). At bdc, the exhaust valve *E* opens and the piston moves upward to dispel the products of combustion (exhaust stroke). Finally, when the piston reaches tdc the exhaust valve closes and the intake valve opens, thereby starting the cycle once again. It is to be noted that *two* complete revolutions of the crankshaft are involved for each power stroke and firing of the ignition device.

The two-cycle engine combines functions so that a power stroke is obtained for each revolution of the crankshaft. Let us start the cycle at tdc just as the ignition device is fired. The piston moves downward in the power stroke. In the two-cycle engine the crankcase is sealed and serves as a storage reservoir for fuel-air mixture. Thus, as the piston moves downward in the power stroke,

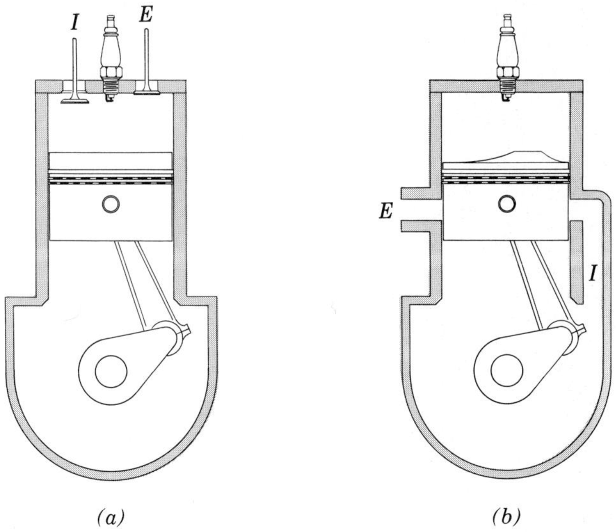

FIG. 9-20 Basic reciprocating internal-combustion-engine arrangements: (*a*) four cycle; (*b*) two cycle.

the fuel-air mixture is compressed in the crankcase. As the piston passes the exhaust and intake ports, the exhaust gases move outward and the compressed fuel-air mixture shoots into the cylinder. Thus this stroke is a combined power-exhaust-intake stroke. Upon completion of these processes, the piston moves upward again in a compression stroke until at tdc the spark plug fires to start the cycle over again.

In the two-cycle engine the unavoidable mixing of inlet fuel-air mixture with exhaust gases and even some loss of fuel out the exhaust port makes the engine generally less efficient than its four-cycle counterpart. But the fact that a power stroke is obtained with each revolution of the crankshaft means that two-cycle engines are usually smaller and lighter in weight for a given power output. For this reason we normally see two-cycle engines employed in applications where weight and size are important factors, viz., outboard motors, chain saws, and motorcycles.

There are several factors which cause problems in practical internal-combustion engines: incomplete exhaust, incomplete combustion, mixing of exhaust and intake gases, improper valve timing, too rapid combustion, too slow combustion, etc. However, a consideration of these factors is beyond the scope of our discussion.

In the following we shall investigate several theoretical cycles which may be used to describe internal-combustion-engine behavior. For purposes of analysis we shall assume that the working fluid is air, in which case the cycles are called *air-standard cycles*.

9-8 THE OTTO CYCLE

Consider first the air-standard Otto cycle shown in Fig. 9-21. The compression process is represented by 1-2, followed by a constant-volume heat addition 2-3,

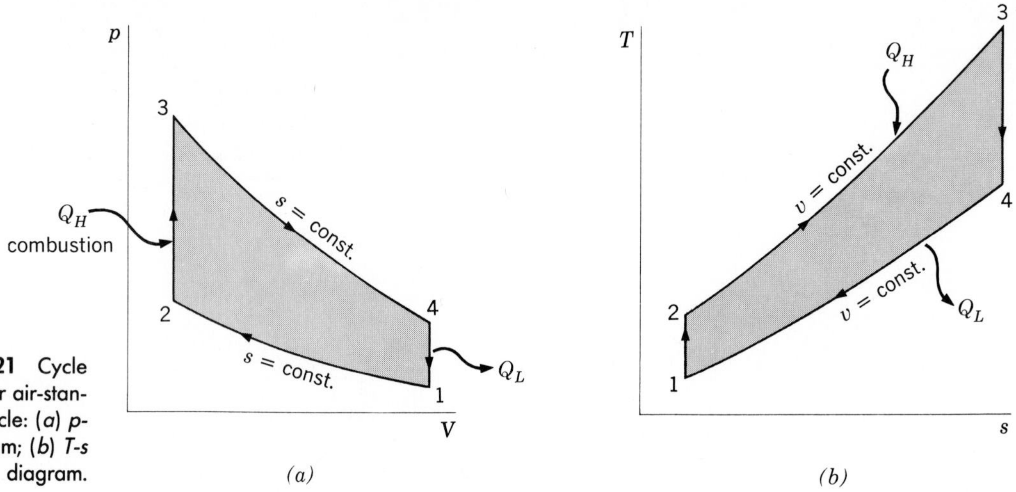

FIG. 9-21 Cycle diagrams for air-standard Otto cycle: (*a*) *p*-*V* diagram; (*b*) *T*-*s* diagram.

which is an approximation to the rapid combustion occurring in a spark-ignition engine. The power stroke is then accomplished in 3-4, followed by a constant-volume heat rejection in 4-1, which corresponds approximately to the dispelling of combustion products through the exhaust valve. The thermal efficiency of the cycle is calculated from

$$\eta = \frac{Q_H - Q_L}{Q_H} = 1 - \frac{Q_L}{Q_H} = 1 - \frac{mc_v(T_4 - T_1)}{mc_v(T_3 - T_2)} \tag{9-11}$$

In the isentropic processes

$$\frac{T_2}{T_1} = \left(\frac{v_1}{v_2}\right)^{\gamma-1} = \left(\frac{v_4}{v_3}\right)^{\gamma-1} = \frac{T_3}{T_4}$$

Making appropriate substitutions for temperatures in Eq. (9-11) the efficiency becomes

$$\eta = 1 - \frac{T_1}{T_2} = 1 - \left(\frac{v_1}{v_2}\right)^{1-\gamma}$$

The ratio v_1/v_2 is called the *compression ratio r,* so that the efficiency may be expressed as

$$\eta = 1 - r^{1-\gamma} \tag{9-12}$$

It is clear that the thermal efficiency of the Otto cycle increases with increasing compression ratio.

A higher compression ratio means a higher temperature at which heat addition is started, and we once again find that an increase in the temperature for heat addition produces an increase in efficiency. For many years automotive engine manufacturers used this principle to increase the efficiencies of their engines through successive increases in compression ratio without large increases in physical size of the engine. High compression ratios are not without disadvantage, however. The higher pressures and temperatures encountered at tdc cause very rapid burning of gasoline fuels, so that engine knock results from detonation waves in the cylinder. To cure these ills, more expensive processing of the fuel must be performed to increase the burning times and antiknock characteristics.

In particular, tetraethyl lead was used as a fuel additive for this purpose. To alleviate air pollution caused by automobiles, catalytic converters are now required on the exhaust system of most passenger vehicles. The use of lead fuel additives renders the converters inoperative so manufacturers must use unleaded gasolines, and correspondingly produce engines with lower compression ratios. This is an example of the balance which must be drawn between environmental protection and efficiency in arriving at a proper engineering design. Internal combustion engines are major sources of air pollution, and stringent efforts are made to minimize the discharge of pollutants from them. Cost of pollution-control equipment and increased fuel consumption caused by its installation are important factors which must be considered in final designs.

EXAMPLE 9-8 Efficiency of an Otto Cycle

An air-standard Otto cycle operates between temperature limits of 70 and 1500°F. Calculate the thermal efficiency of the cycle for compression ratios of (*a*) 10 : 1, (*b*) 12 : 1, and (*c*) 15 : 1. Compare with the appropriate Carnot efficiency.

SOLUTION We immediately calculate the efficiencies with Eq. (9-12) using $\gamma = 1.4$:

$$\eta = 1 - r^{1-\gamma}$$

At $r = 10$,

$$\eta = 1 - (10)^{-0.4} = 60.2 \text{ percent}$$

At $r = 12$,

$$\eta = 1 - (12)^{-0.4} = 63 \text{ percent}$$

At $r = 15$,

$$\eta = 1 - (15)^{-0.4} = 66.2 \text{ percent}$$

These efficiencies could have been read directly from Fig. 9-23. A Carnot cycle operating between the given temperature limits would have an efficiency of

$$\eta_{\text{Carnot}} = 1 - \frac{T_L}{T_H}$$
$$= 1 - \frac{530}{1960}$$
$$= 73 \text{ percent}$$

9-9 THE DIESEL CYCLE

Now consider the air-standard diesel cycle shown in Fig. 9-22. The diesel cycle is similar to the Otto cycle in that isentropic compression is assumed as well as constant-volume heat rejection of the exhaust gases from 4-1. It differs from the Otto cycle in that heat addition occurs in a constant-pressure process 2-3. This means that ignition and combustion form part of the total power stroke 2-3-4. The diesel cycle is the basis of practical compression-ignition engines. In these engines air is compressed to high pressures (point 2) and correspondingly high temperatures. A fuel with a low ignition point is then injected into the cylinder at tdc. The high temperature of compression is sufficient to ignite the fuel, and the burning continues from 2 to 3 as more fuel is injected into the cylinder. The hot products of combustion then expand in the cylinder in process 3-4. In practical diesel engines the injection process is a crucial matter, and considerable engineering effort is devoted to the design of injector mechanisms.

The efficiency of the diesel cycle is given by

$$\eta = 1 - \frac{Q_L}{Q_H} = 1 - \frac{c_v(T_4 - T_1)}{c_p(T_3 - T_2)} \tag{9-13}$$

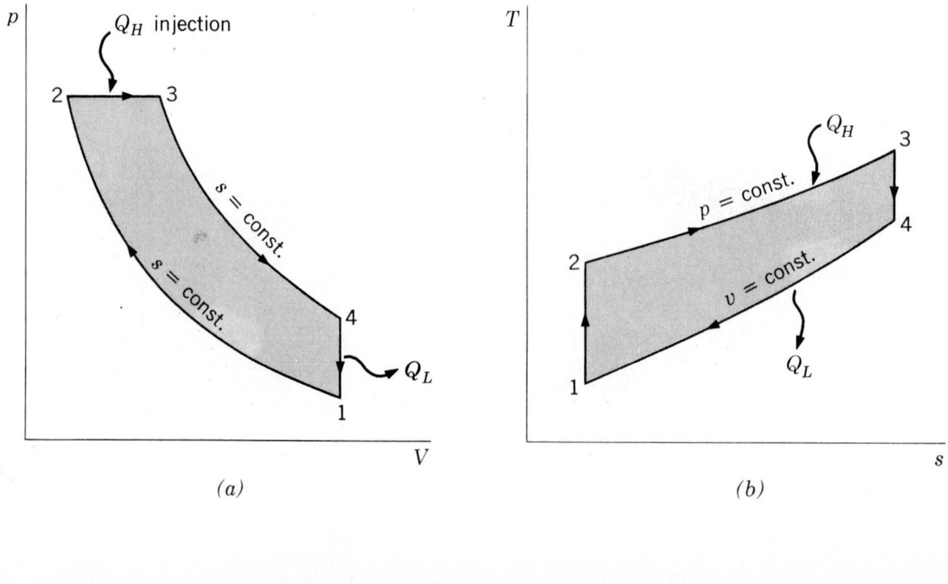

FIG. 9-22 Cycle diagrams for air-standard diesel cycle: (*a*) *p-V* diagram; (*b*) *T-s* diagram.

Using the isentropic relations

$$\frac{T_2}{T_1} = \left(\frac{v_1}{v_2}\right)^{\gamma-1} \quad \frac{T_3}{T_4} = \left(\frac{v_4}{v_3}\right)^{\gamma-1}$$

the efficiency expression becomes

$$\eta = 1 - \frac{r^{1-\gamma}(r_c^\gamma - 1)}{\gamma(r_c - 1)} \tag{9-14}$$

where $r = v_1/v_2$ is the compression ratio and $r_c = v_3/v_2$ is called the *cutoff ratio*.

Efficiencies of the Otto and diesel cycles are plotted in Fig. 9-23. It is clear that, for a given compression ratio, the Otto cycle is more efficient than the diesel cycle. The physical reason for this behavior is that the constant-volume heat-addition process results in heat being added at a higher temperature than in a constant-pressure process for the same compression ratio. This does not mean, however, that an Otto-cycle or spark-ignition engine is always to be preferred over a diesel. In practice, the diesel can operate at much higher compression ratios than a spark-ignition engine because of detonation and preignition problems encountered with compression of fuel-air mixtures to high pressures. If a fuel-air mixture is compressed to pressures which are too high, it may ignite before the piston reaches tdc and, consequently, reduce the power output of the engine. The diesel does not encounter such problems because

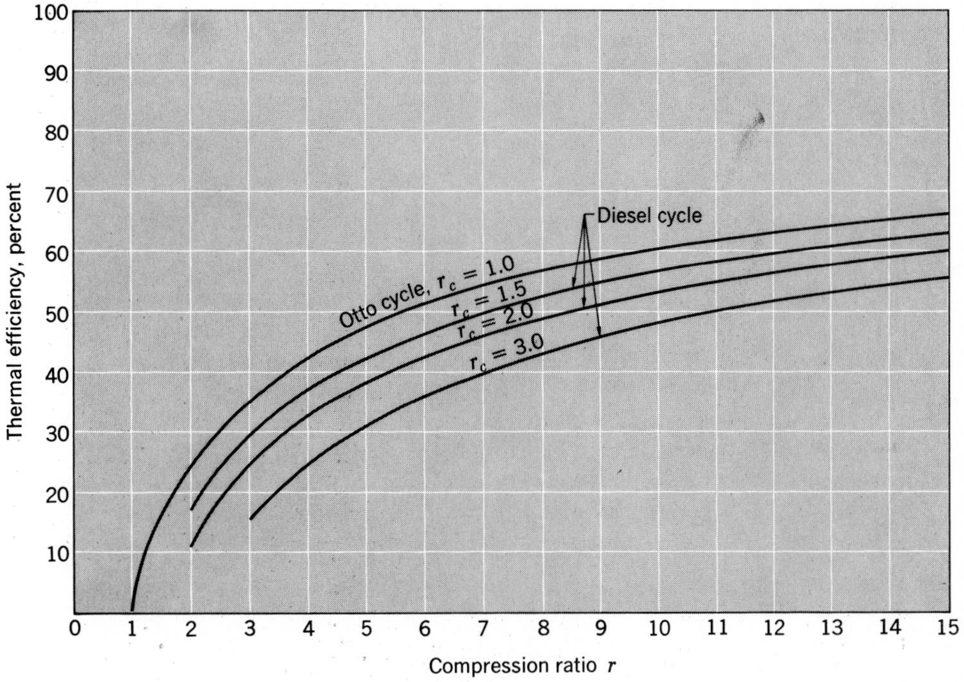

FIG. 9-23 Comparison of efficiencies of air-standard Otto and diesel cycles for $\gamma = 1.4$.

only air is compressed and fuel does not enter the cylinder until the appropriate time for combustion.

One does not select an Otto (spark-ignition) or diesel cycle strictly on the basis of the thermodynamic information given above. Consideration must be given to the specific design of each type of engine. Most diesel engines employ a carefully metered injection system which injects just the right amount of fuel into the cylinder. Of course, the time available for injection, and thus the injection pressure, is strongly dependent on the engine speed. It turns out that if optimum performance is to be achieved the diesel engine should operate over a fairly narrow speed range. In contrast, a spark-ignition engine can operate efficiently over a very wide speed range with a properly designed carburetion system. For these reasons we usually find diesel engines used in applications where fairly constant speeds are required, as in locomotives, large trucks, stationary electric generators, and the like, while spark-ignition engines are used for wide-speed ranges of personal automobiles.

We may expect that electronic microprocessor control of the injection process will be employed extensively with diesels of the future, thereby enabling an incumbent improvement in variable-speed performance.

EXAMPLE 9-9 Diesel Cycle

Calculate the work output per pound-mass for an air-standard diesel cycle with $r = 15$ and $r_c = 2.0$. The inlet conditions are 14.7 psia and 70°F. Also calculate the efficiency.

SOLUTION Figure 9-22 gives the process and cycle schematic for the diesel cycle. We have

$$T_1 = 70°\text{F} = 530°\text{R}$$

$$p_1 = 14.7 \text{ psia}$$

From the isentropic relations we have

$$\frac{T_2}{T_1} = \left(\frac{v_1}{v_2}\right)^{\gamma-1}$$

$$T_2 = 530(15)^{0.4} = 1565°\text{R} \qquad (869 \text{ K})$$

For the constant-pressure heat-addition process

$$\frac{v_3}{v_2} = \frac{T_3}{T_2}$$

so that

$$T_3 = (1565)(2.0) = 3130°\text{R} \qquad (1739 \text{ K})$$

The temperature at the end of the power stroke is determined from

$$\frac{T_4}{T_3} = \left(\frac{v_3}{v_4}\right)^{\gamma-1}$$

But $v_3 = 2v_2$, so that

$$\frac{v_3}{v_4} = \frac{2v_2}{v_4} = \frac{2v_2}{v_1} = \frac{2}{15}$$

Thus

$$T_4 = 3130\left(\frac{2}{15}\right)^{0.4}$$
$$= 1395°\text{R} \qquad (775\text{ K})$$

The heat added and rejected may now be calculated:

$$Q_H = c_p(T_3 - T_2)$$
$$= (0.24)(3130 - 1565) = 375\text{ Btu/lbm} \qquad (872.2\text{ kJ/kg})$$
$$Q_L = c_v(T_4 - T_1)$$
$$= (0.1715)(1395 - 530) = 148\text{ Btu/lbm} \qquad (344.2\text{ kJ/kg})$$

The work output per pound-mass is therefore

$$W = Q_H - Q_L = 375 - 148$$
$$= 227\text{ Btu/lbm} \qquad (528\text{ kJ/kg})$$

The efficiency is

$$\eta = \frac{W}{Q_H} = \frac{227}{375} = 60.5\text{ percent}$$

This value may be checked by referring to Fig. 9-23.

9-10 THE WANKEL ENGINE

A rotary-type internal combustion engine with increasing applications is the *Wankel engine* shown in Fig. 9-24. In this arrangement a three-lobed rotor is mounted eccentrically on the shaft which in turn is connected to a planetary gear train so that the rotor turns at one-third the crankshaft speed. The internal surface of the housing is a special trochoid shape generated by the radius R. As the rotor moves clockwise, simultaneous processes occur in the three chambers as follows: As the rotor moves by the intake port a fresh charge of fuel-air

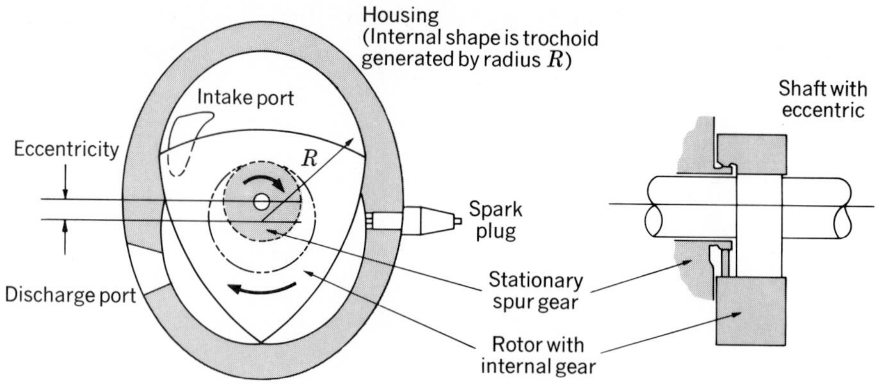

FIG. 9-24 Schematic of a Wankel rotary combustion engine. Diagram according to Wood [11].

mixture is drawn in, and with further movement the intake port is closed and the mixture is compressed. When the tip of the rotor arrives at the approximate location of the spark plug, combustion and power expansion occur in the lower right chamber. Finally, the products of combustion are forced out the discharge port in the lower left chamber. The device is very smooth in operation, with lower vibration and noise levels than its Otto cycle counterpart. It has the additional advantage of very lightweight construction for a given power output. The main practical design problems are those associated with rubbing and end seals for the rotor.

9-11 THE STIRLING CYCLE

We shall now consider a cycle which offers opportunity for high efficiency as well as reduced emissions from combustion products. It is called the *Stirling cycle* and is depicted on the p-V and T-s diagrams of Fig. 9-25.

The working fluid is a gas, and we assume reversible processes for an ideal cycle. In process 1-2 heat is added at the constant temperature T_H. Process 2-3

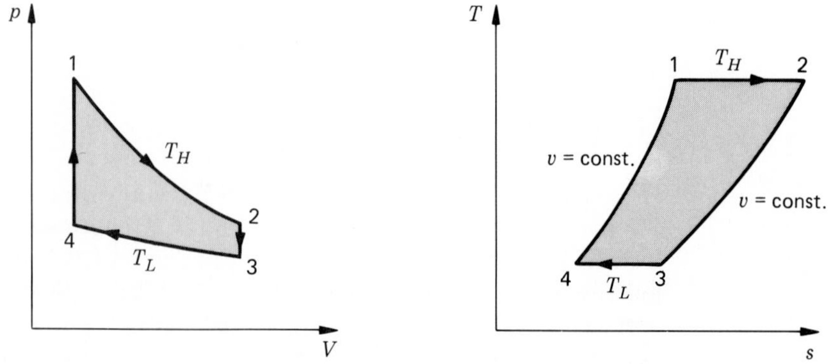

FIG. 9-25 Stirling cycle: (*a*) *p*-*V* diagram; (*b*) *T*-*s* diagram.

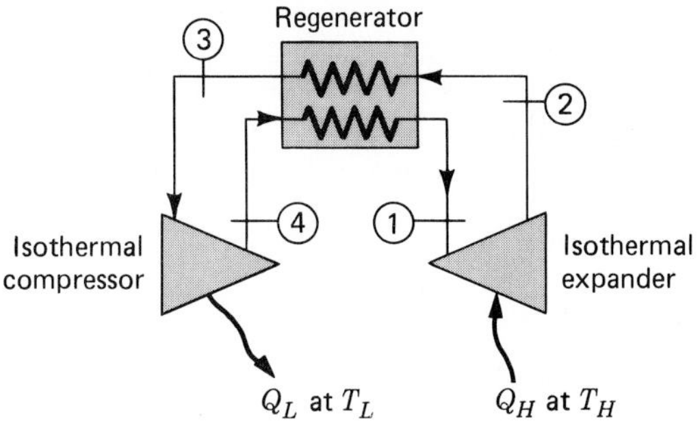

FIG. 9-26 Stirling cycle with regenerator.

is a constant-volume heat rejection; process 3-4 is a constant-temperature heat rejection at T_L, and finally, 4-1 is a constant-volume heat addition. We assume a closed system and ideal-gas behavior. Because processes 2-3 and 4-1 occur between the same temperature limits, the heat rejected in 2-3 must be numerically equal to the heat added in 4-1. This fact immediately suggests the use of regeneration.

If we can transfer the heat from 2-3 over to process 4-1, internal to the cycle, then we will have a reversible cycle with heat addition and rejection at constant temperatures. Such a cycle will have an efficiency equal to that of a Carnot cycle, or

$$\eta = 1 - \frac{T_L}{T_H} \tag{9-15}$$

The Stirling cycle with regeneration is illustrated in Fig. 9-26. The cycle can operate as an *external* combustion device and has been investigated as a means to reduce exhaust emissions in comparison with the internal combustion engine. There are problems with the cycle. To achieve acceptable power-to-weight ratios, very high pressures must be developed and the regenerator involves stringent heat-transfer design requirements. Some devices have been developed using air, helium, and hydrogen as working fluids.

9-12 GAS-TURBINE CYCLES

The thermodynamic cycle employed with gas turbines is called the *Brayton cycle* and is illustrated in Fig. 9-27. In the *open* cycle the compressor raises the pressure of the air, and heat is added at the high pressure by burning a fuel with the air. The high-temperature products of combustion are then expanded in the turbine to produce a work output. Part of the turbine work is used to drive the compressor, and the remainder is available for driving external mechanisms. The *closed* gas-turbine cycle recirculates the working fluid while heat is added at the high pressure and rejected at the low pressure. Temperature-entropy diagrams for the ideal cycles are shown in Fig. 9-28.

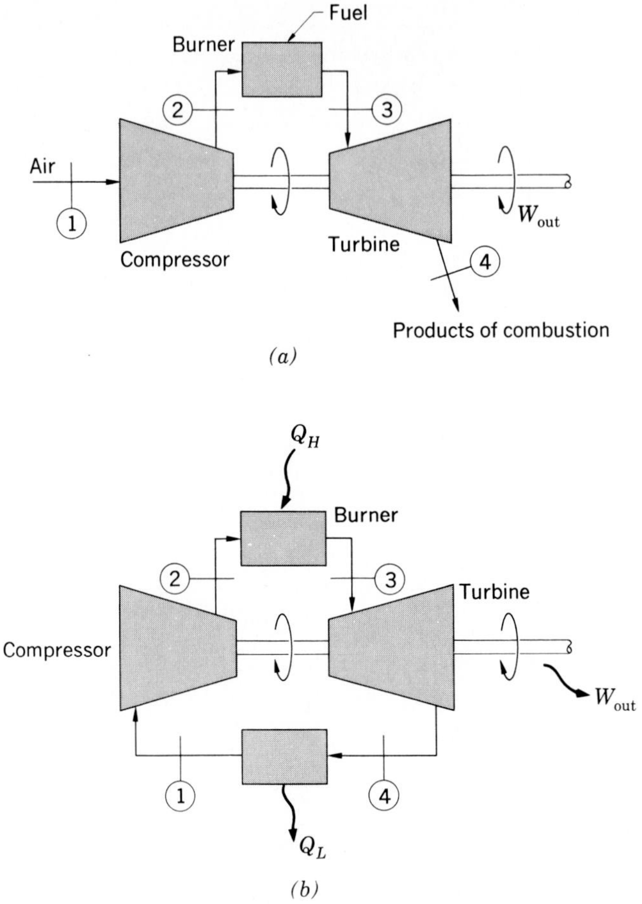

FIG. 9-27 Basic Brayton cycle for gas turbines: (*a*) open cycle; (*b*) closed cycle.

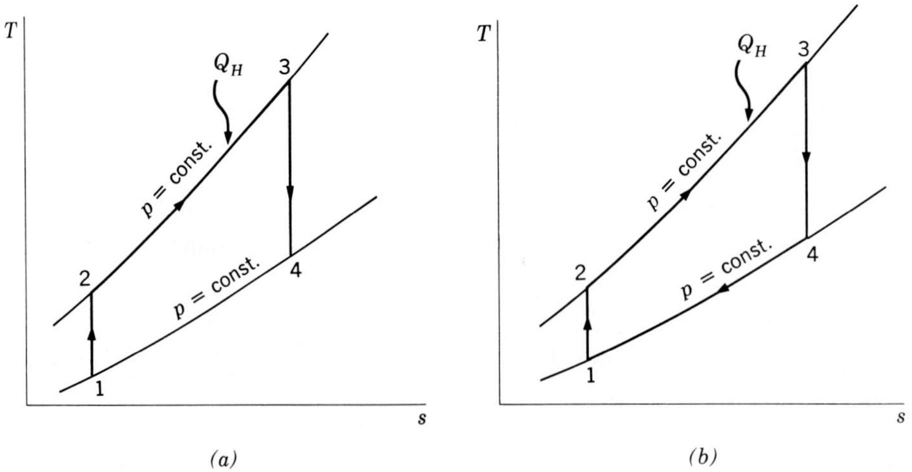

FIG. 9-28 Temperature-entropy diagram for Brayton cycles: (*a*) open cycle; (*b*) closed cycle.

In the *air-standard* Brayton cycle the analysis is performed by assuming that the working fluid is air in both the compressor and turbine. In an actual cycle the mass flow in the turbine would be somewhat greater than the mass flow in the compressor because of the added mass of the fuel. In addition, the properties of the products of combustion may be different from those of air; however, most gas turbines use a substantial amount of excess air, so that an analysis of the air-standard cycle offers reasonable estimates of the performance of actual cycles.

The gas-turbine cycle is a steady-flow device and, as such, the energy quantities may be determined with enthalpy differences. Both the compressor and turbine work processes are assumed adiabatic so the net work output of the cycle is the work output of the turbine less the work input to the compressor, or $(h_3 - h_4) - (h_2 - h_1)$. This is the shaft work which the cycle can deliver to other things. The burner is assumed to operate as a constant-pressure heating device so the net heat input is $h_3 - h_2$. Accordingly, the thermal efficiency for the ideal air-standard cycle may be calculated as

$$\eta = \frac{W_{\text{net}}}{Q_{\text{in}}} = \frac{(h_3 - h_4) - (h_2 - h_1)}{h_3 - h_2} \tag{9-16}$$

where the subscript nomenclature refers to Fig. 9-27. Expressing the enthalpy differences in terms of temperature differences and using the isentropic relations

$$\frac{p_2}{p_1} = \left(\frac{T_2}{T_1}\right)^{\gamma/(\gamma-1)} = \frac{p_3}{p_4} = \left(\frac{T_3}{T_4}\right)^{\gamma/(\gamma-1)} \tag{9-17}$$

gives for the efficiency

$$\eta = 1 - \left(\frac{p_2}{p_1}\right)^{(1-\gamma)/\gamma} \tag{9-18}$$

Regeneration

Most gas turbines operate on the open cycle and the exhaust gases from the turbine are dissipated to the surroundings. In the simple cycle discussed above, this represents a substantial loss of high-energy gas. In a *regenerator,* energy from these hot gases is used to preheat the high-pressure air from the compressor before it enters the combustion chamber, thereby requiring less fuel for the overall cycle. The arrangement for a cycle with regeneration is shown in Fig. 9-29.

It is easy to see that the highest temperature which can be achieved at point 5 is the exhaust temperature from the turbine, T_4. This means that in an ideal case the air would be heated to point 5′ (equal to T_4) and the products would be cooled to point 6′ (equal to T_2). An actual regenerator does not accomplish such an ideal heat exchange, and the efficiency of the regenerator is

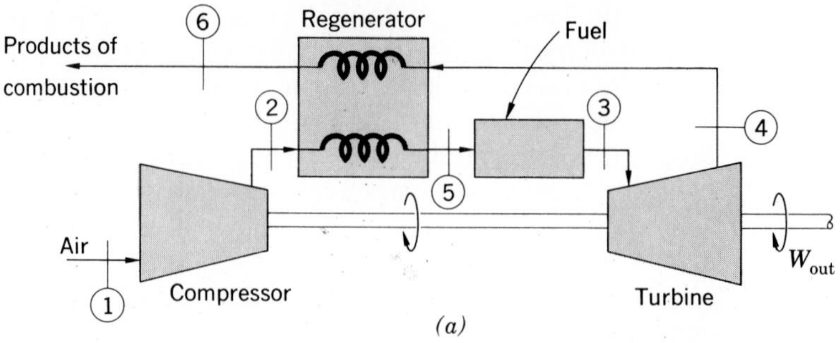

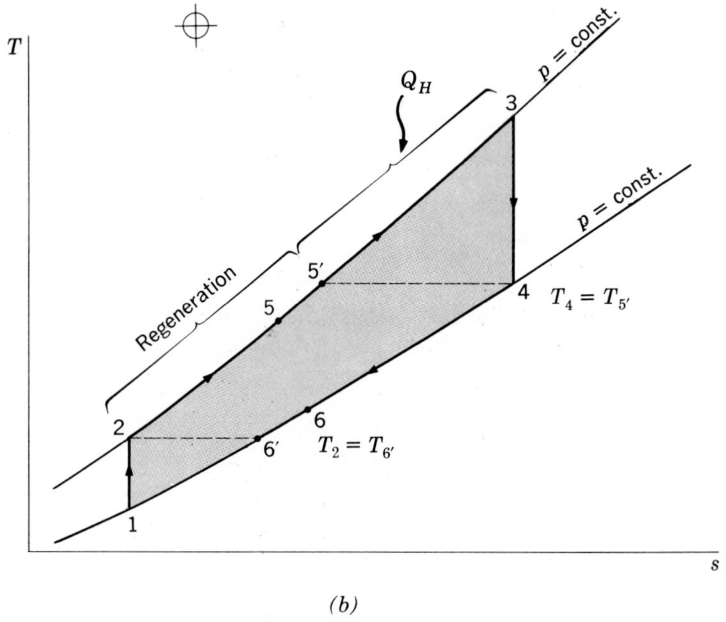

FIG. 9-29 Gas-turbine cycle with ideal regeneration: (*a*) schematic; (*b*) *T-s* diagram.

normally expressed by

$$\eta_{reg} = \frac{\text{actual heat transfer}}{\text{maximum heat transfer}} = \frac{h_5 - h_2}{h_{5'} - h_2} \tag{9-19}$$

For constant specific heats this becomes

$$\eta_{reg} = \frac{T_5 - T_2}{T_{5'} - T_2} \tag{9-20}$$

Actual gas-turbine cycles differ from theoretical cycles in that compressor and turbine efficiencies must be taken into account. The *T-s* diagram for an actual cycle would resemble the one shown in Fig. 9-30.

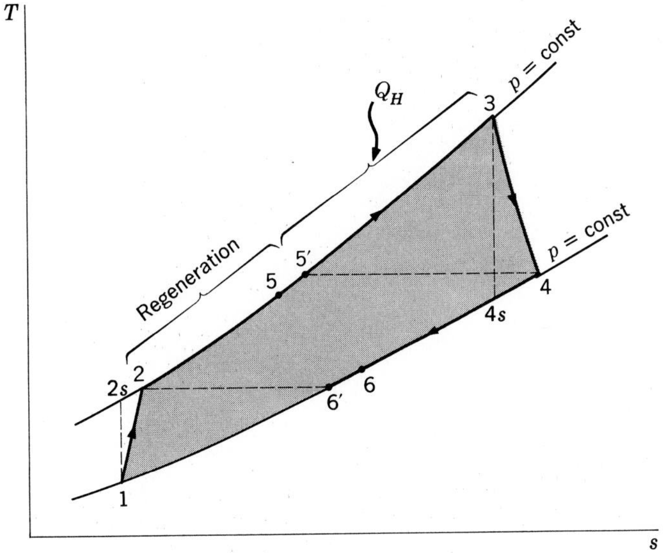

FIG. 9-30 Effect of turbine and compressor efficiencies on T-s diagram of gas-turbine cycle with regeneration.

It is possible to improve the efficiency of the gas-turbine cycle by using multiple-stage compression with intercooling as well as multiple-stage turbines with reheat. A schematic for an ideal cycle employing two stages of compression and expansion is shown in Fig. 9-31. The efficiency for this cycle would be expressed by

$$\eta = \frac{\text{net work output}}{\text{net heat input}} = \frac{(h_5 - h_6) + (h_7 - h_8) - (h_4 - h_3) - (h_2 - h_1)}{(h_5 - h_4) + (h_7 - h_6)} \tag{9-21}$$

Obviously, it is possible to incorporate a regenerator along with the intercooler and reheat arrangement.

At this point it is well to remind the reader that our discussion has been limited to air-standard cycles and some of the ideal-efficiency relations have involved the further assumption of constant specific heats. The examples are based on these assumptions. A first refinement in the analysis may be made by employing the air tables to compute the various enthalpies. Still further refinement may be made by considering actual fuel-air mixtures and the corresponding products of combustion. In this case we would employ the appropriate relations of Chap. 8 to compute flame temperature, etc. Of course, we are no longer dealing with an air-standard cycle when actual combustion processes are considered, but actual gas-cycle processes are beyond the scope of our discussion. Despite its simplicity, air-standard cycle analysis does give a very good approximation to the behavior and performance of real cycles.

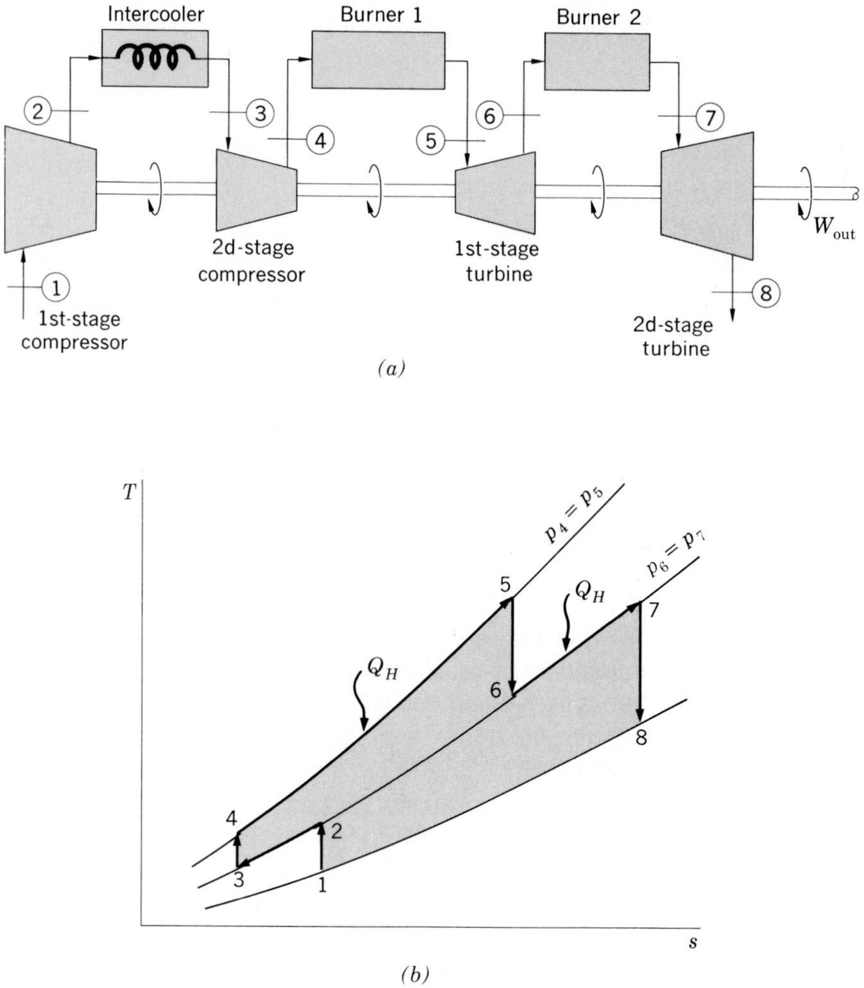

FIG. 9-31 Two-stage gas-turbine cycle with intercooling and reheat: (*a*) schematic; (*b*) *T-s* diagram.

EXAMPLE 9-10 Brayton Cycle Efficiency

An air-standard Brayton cycle receives air at 14.7 psia and 70°F. The upper pressure and temperature limits of the cycle are 60 psia and 1500°F, respectively. The compressor efficiency is 85 percent, and the turbine efficiency is 90 percent. Calculate the thermal efficiency of the cycle, assuming constant specific heats.

SOLUTION We shall employ the diagram in Fig. 9-28*a* for this problem, except that the actual compression and expansion processes are nonisentropic. From the problem statement,

$$p_1 = p_4 = 14.7 \text{ psia} \qquad T_1 = 70°\text{F} = 530°\text{R}$$

$$p_2 = p_3 = 60 \text{ psia} \qquad T_3 = 1500°\text{F} = 1960°\text{R}$$

For isentropic processes we have

$$T_2 = T_1 \left(\frac{p_2}{p_1}\right)^{(\gamma-1)/\gamma}$$

$$= 530 \left(\frac{60}{14.7}\right)^{0.286}$$

$$= 792°\text{R} \qquad (440 \text{ K})$$

$$T_4 = T_3 \left(\frac{p_4}{p_3}\right)^{(\gamma-1)/\gamma}$$

$$= 1960 \left(\frac{14.7}{60}\right)^{0.286}$$

$$= 1310°\text{R} \qquad (728 \text{ K})$$

Using the given efficiencies with constant specific heats,

$$\eta_{\text{comp}} = \frac{W_{\text{isen}}}{W_{\text{act}}} = \frac{T_{2s} - T_1}{T_2 - T_1} = 0.85$$

where T_2 is the actual exit temperature from the compressor. Inserting the proper numerical values,

$$\frac{792 - 530}{T_2 - 530} = 0.85 \qquad T_2 = 838°\text{R}$$

For the turbine,

$$\eta_{\text{turb}} = 0.90 = \frac{W_{\text{act}}}{W_{\text{isen}}} = \frac{T_3 - T_4}{T_3 - T_4}$$

$$0.90 = \frac{1960 - T_4}{1960 - 1310} \qquad T_4 = 1375°\text{R} \qquad (764 \text{ K})$$

The thermal efficiency of the cycle is now calculated as

$$\eta_{\text{th}} = \frac{W_{\text{act}}}{Q_H}$$

$$= \frac{(h_3 - h_4) - (h_2 - h_1)}{h_3 - h_2}$$

$$= \frac{(0.24)(1960 - 1375) - (0.24)(838 - 530)}{(0.24)(1960 - 838)}$$

$$= 24.7 \text{ percent}$$

EXAMPLE 9-11 Gas Turbine with Regenerator
A regenerator having an efficiency of 83 percent is installed on the gas-turbine cycle of Example 9-10. Calculate the thermal efficiency of the cycle under these conditions. Make the calculation using constant specific heats and also using the air tables.

SOLUTION We use the schematic diagram of Fig. 9-30 for this problem. From Example 9-10,

$$T_1 = 530°\text{R}$$

$$T_2 = 838°\text{R} = T_{6'}$$

$$T_3 = 1960°\text{R}$$

$$T_4 = 1375°\text{R} = T_{5'}$$

Now, from Eq. (9-21) for the regenerator efficiency,

$$\eta_{\text{reg}} = \frac{T_5 - T_2}{T_{5'} - T_2}$$

$$0.83 = \frac{T_5 - 838}{1375 - 838}$$

$$T_5 = 1284°\text{R} \qquad (713\text{ K})$$

The thermal efficiency of the cycle is now written as

$$\eta_{\text{th}} = \frac{(h_3 - h_4) - (h_2 - h_1)}{h_3 - h_5}$$

$$= \frac{(0.24)(1960 - 1375) - (0.24)(838 - 530)}{(0.24)(1960 - 1284)}$$

$$= 41 \text{ percent}$$

We now must repeat the calculations using the air tables which take into account variable specific heats. We still have

$$T_1 = 530°\text{R} \qquad T_3 = 1960°\text{R}$$

Using the air tables for the ideal isentropic process

$$\frac{p_2}{p_1} = \frac{P_{r_2}}{P_{r_1}} \qquad P_{r_1} = 1.3004 \text{ at } 530°\text{R}$$

$$P_{r_2} = \left(\frac{60}{14.7}\right)(1.3004) = 5.308$$

Therefore $T_{2s} = 790.8°\text{R}$. Also,

$$\frac{p_4}{p_3} = \frac{P_{r_4}}{P_{r_3}} \qquad P_{r_3} = 160.37 \text{ at } 1960°\text{R}$$

$$P_{r_4} = \left(\frac{14.7}{60}\right)(160.37) = 39.291$$

and $T_4 = 1388.1°\text{R}$. We now can find the corresponding enthalpies from the air tables (Table A-17):

$$h_1 = 126.67 \text{ Btu/lbm} \qquad h_{2s} = 189.57 \text{ Btu/lbm}$$

$$h_3 = 493.64 \text{ Btu/lbm} \qquad h_{4s} = 339.79 \text{ Btu/lbm}$$

Because the specific heats are no longer constant, the turbine and compressor efficiencies must be expressed in terms of enthalpies. Thus

$$\eta_{\text{comp}} = 0.85 = \frac{h_{2s} - h_1}{h_2 - h_1}$$

Solving for h_2 we obtain

$$h_2 = 200.67 \text{ Btu/lbm}$$

and for the turbine

$$\eta_{\text{turb}} = 0.90 = \frac{h_3 - h_4}{h_3 - h_{4s}}$$

so that

$$h_4 = 355.18 \text{ Btu/lbm}$$

The regenerator effectiveness is also written in terms of enthalpies:

$$\eta_{\text{neg}} = 0.83 = \frac{h_5 - h_2}{h_{5'} - h_2} \qquad \text{with } h_{5'} = h_4$$

Then,

$$\begin{aligned} h_5 &= (0.83)(355.18 - 200.67) + 200.67 \\ &= 328.31 \text{ Btu/lbm} \end{aligned}$$

The relation for the thermal efficiency is the same as before:

$$\eta_{th} = \frac{(h_3 - h_4) - (h_2 - h_1)}{h_3 - h_5}$$

$$= \frac{(493.64 - 355.18) - (200.67 - 126.67)}{493.64 - 328.31}$$

$$= 39 \text{ percent}$$

So there is a small variation in the calculation, depending on whether constant specific heats are assumed. It should be noted that the use of a regenerator can produce very striking increases in thermal efficiency for the gas-turbine cycle.

EXAMPLE 9-12 Closed-Cycle Gas Turbine

Almost all current applications of gas-turbine cycles employ the open cycle with combustion of a hydrocarbon fuel providing the heat source. Should it become possible to harness the thermonuclear reaction (nuclear fusion), new techniques may be desired to extract the energy from the reaction and convert it to useful work. It is expected that the thermonuclear reaction will produce a hot plasma gas at billions of degrees temperature, and one proposed arrangement calls for initial removal of energy from the reaction with a liquid lithium coolant, which could then be used to supply heat to a closed helium gas-turbine cycle, as shown in Fig. 9-32. Helium is chosen for its high specific heat and inert nature, and some of the design factors for such an installation are discussed in Ref. [9]. The upper temperature and pressure limits for our example cycle are 1700°F and 500 psia and the lower limits are 100°F and 200 psia. Ideal intercooling and interstaging are used in the two stages of compression along with a regenerator having an effectiveness of 91 percent. The turbine efficiency is 93 percent, and the compressor efficiencies are 88 percent. Calculate the thermal efficiency of the cycle and the mass flow of helium required per megawatt of power output. Also calculate the cooling water–flow requirements assuming a 10°F water temperature rise in both the intercooler and main cooler. Assume constant specific heats.

SOLUTION Using the nomenclature of Fig. 9-32*a*,

$$T_1 = T_3 = 100°\text{F} = 560°\text{R} \qquad T_6 = 1700°\text{F} = 2160°\text{R}$$

$$p_1 = 200 \text{ psia} = p_8 = p_7 \qquad p_4 = p_5 = p_6 = 500 \text{ psia}$$

For ideal interstaging, with $p_2 = p_3$,

$$\frac{p_2}{p_1} = \frac{p_4}{p_3}$$

$$\frac{p_2}{p_1} = \left(\frac{500}{200}\right)^{1/2} = 1.581$$

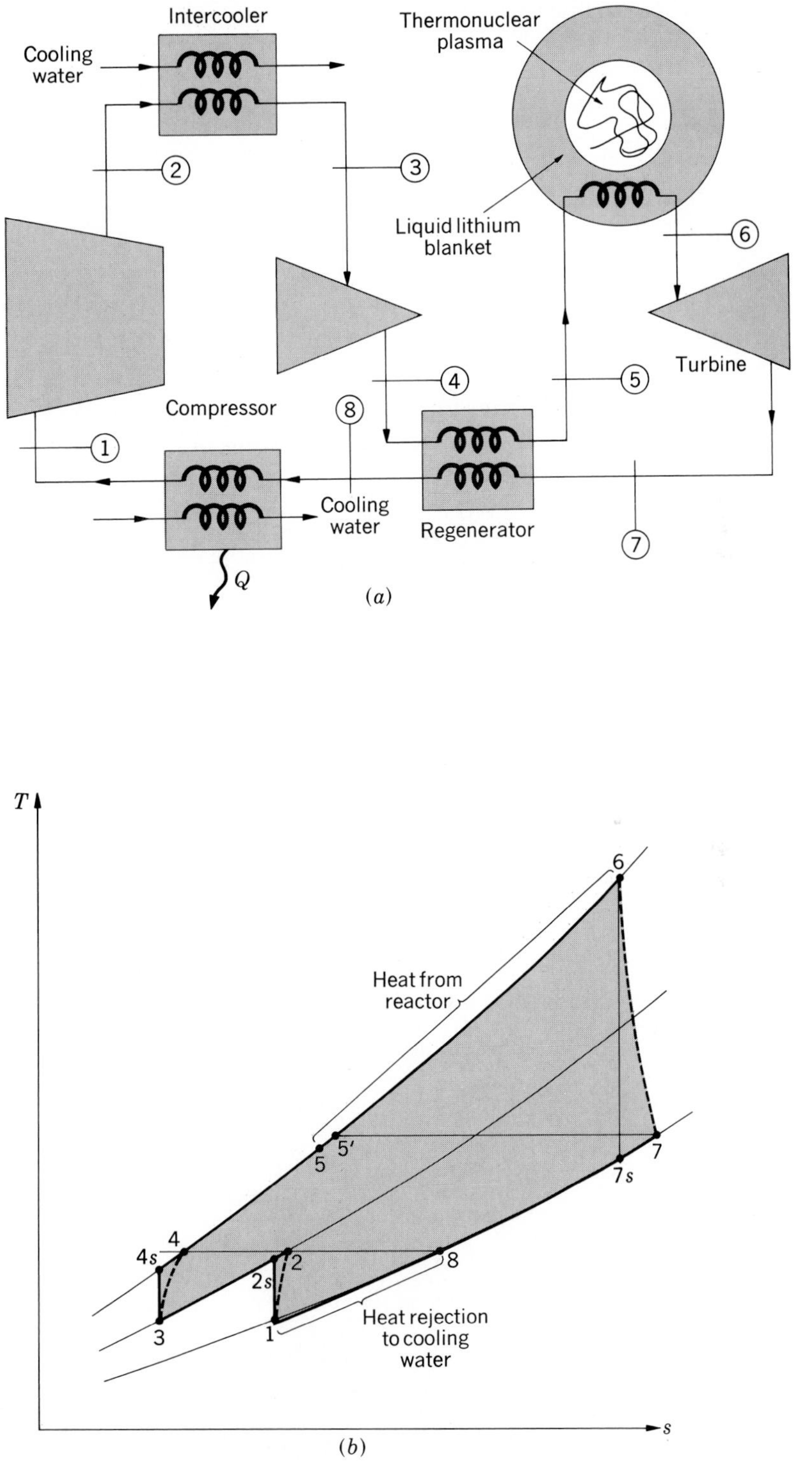

FIG. 9-32 A closed helium gas-turbine cycle for power production with a thermonuclear reactor: (*a*) schematic; (*b*) *T-s* diagram.

For helium we obtain from Table 2-2

$$c_p = 1.25 \text{ Btu/lbm}\cdot{}^\circ\text{F} \qquad \gamma = 1.66$$

For ideal interstaging the isentropic temperatures at the discharge are the same because the pressure ratios and inlet temperatures are the same. Thus

$$\frac{T_{2s}}{T_1} = \left(\frac{p_2}{p_1}\right)^{(\gamma-1)/\gamma} = (1.581)^{0.398} = 1.120$$

$$T_{2s} = (560)(1.120) = 672^\circ\text{R} = T_{4s} \qquad (373 \text{ K})$$

The T-s diagram for this cycle is shown in Fig. 9-32b. The actual discharge temperatures are determined by using the compressor efficiencies. For constant specific heats,

$$\eta_{\text{comp}} = \frac{T_{2s} - T_1}{T_2 - T_1} = 0.88$$

so that

$$T_2 = 687^\circ\text{R} = T_4 \qquad (382 \text{ K})$$

For the turbine the isentropic exit temperature is

$$T_{7s} = T_6 \left(\frac{p_7}{p_6}\right)^{(\gamma-1)/\gamma} = 2160 \left(\frac{200}{500}\right)^{0.398} = 1500^\circ\text{R} \qquad (833 \text{ K})$$

and using the turbine efficiency the actual exit temperature is obtained from

$$\eta_{\text{turb}} = \frac{T_6 - T_7}{T_6 - T_{7s}} = 0.93$$

$$T_7 = 1546^\circ\text{R} \qquad (859 \text{ K})$$

For constant specific heats the regenerator effectiveness is

$$\eta_{\text{reg}} = 0.91 = \frac{T_5 - T_4}{T_{5'} - T_4} \text{ with } T_{5'} = T_{7'}$$

so that

$$0.91 = \frac{T_5 - 687}{1546 - 687} \qquad T_5 = 1469^\circ\text{R} \qquad (816 \text{ K})$$

The overall thermal efficiency is therefore

$$\eta_{th} = \frac{\text{net work output}}{\text{heat added}} = \frac{\begin{matrix}\text{work output} \\ \text{of turbine}\end{matrix} - \begin{matrix}\text{work input} \\ \text{to compressors}\end{matrix}}{\text{heat added}}$$

$$= \frac{(h_6 - h_7) - (h_2 - h_1) - (h_4 - h_3)}{h_6 - h_5}$$

which becomes, for constant specific heats,

$$\eta_{th} = \frac{(1.25)[(2160 - 1546) - (687 - 560) - (687 - 560)]}{(1.25)(2160 - 1469)}$$

$$= \frac{450 \text{ Btu/lbm}}{863.8 \text{ Btu/lbm}} = 52.1 \text{ percent}$$

The net work output per pound-mass is 450 Btu so the required mass flow for 1 MW is

$$\dot{m}_{HE}(450) = 10^6 \text{ W} = 3.413 \times 10^6 \text{ Btu/h}$$

$$\dot{m}_{HE} = 7584 \text{ lbm/h·MW}$$

This same mass flow moves through the intercooler and main cooler. The heat given up in the intercooler is

$$Q_{int} = \dot{m}_{He}c_p(T_2 - T_3)$$
$$= (7584)(1.25)(672 - 560) = 1.062 \times 10^6 \text{ Btu/h} \qquad (311.1 \text{ kW})$$

To calculate the heat rejected in the main cooler we must first determine T_8, the helium temperature at inlet to the cooler. From the energy balance on the regenerator, assuming constant specific heats,

$$T_7 - T_8 = T_5 - T_4$$

so that

$$T_8 = 1546 - 1469 + 687 = 764°\text{R} \qquad (424 \text{ K})$$

The heat lost by the helium in the main cooler is then

$$Q_{cooler} = \dot{m}_{He}c_p(T_8 - T_1)$$
$$= (7584)(1.25)(764 - 560) = 1.934 \times 10^6 \text{ Btu/h} \qquad (566.6 \text{ kW})$$

The total heat rejection is therefore

$$Q_{tot} = (1.062 + 1.934) \times 10^6 = 2.996 \times 10^6 \text{ Btu/h} \qquad (877.8 \text{ kW})$$

which must be picked up by the cooling water. Recalling that $c_p \approx 1.0$ Btu/lbm·°F for liquid water, we have

$$Q_{tot} = \dot{m}_w c_w \, \Delta T_w = m_w(1.0)(10) = 2.996 \times 10^6 \text{ Btu/h}$$

and

$$\dot{m}_w = 2.996 \times 10^5 \text{ lbm/h·MW} \qquad (1.359 \times 10^5 \text{ kg/h·MW})$$

This is a flow of about 600 gal/min for each megawatt of power output. For these working temperatures a moderate-size power plant of 500 MW would require a cooling water flow of 300 000 gal/min. Even with the very favorable efficiency figure of 52 percent, this example illustrates the enormous cooling load placed on the environment by large power plants. In many cases the load is dissipated to the atmosphere by evaporation in a cooling tower. At normal atmospheric temperature the enthalpy of vaporization is about 1000 Btu/lbm, so the 500-MW plant would require evaporation of about

$$\frac{(2.996 \times 10^6 \text{ Btu/h·MW})(500 \text{ MW})}{1000 \text{ Btu/lbm}} = 1.498 \times 10^6 \text{ lbm/h}$$

$$(6.8 \times 10^5 \text{ liters/h})$$

This is an evaporation rate of about 3000 gal/min, quite a large amount of water. In extremely calm weather such evaporation rates may cause ground fog in the surroundings adjacent to the power plant.

From our discussions and the examples, we see that gas-turbine cycles offer the advantage of high efficiency when regeneration is employed. In many cases they also are more economical on the basis of initial capital expenditure for production of power in the range of 25 to 100 MW than competing equipment. Because they operate with high excess air they produce fewer air pollutants in the form of unburned hydrocarbons than other internal combustion cycles, and the high-temperature flow processes in the turbine help to reduce NO_x emissions.

The high efficiency of gas-turbine cycles results, of course, from the high upper-temperature limits which may be accommodated in the burner and turbine processes. As better high-temperature materials are developed for turbine blades, one may expect further increases in efficiency. Much of the development of high-temperature materials will come as a result of research on advanced turbojet engine designs for aircraft.

9-13 SECOND-LAW ANALYSIS OF POWER CYCLES

In Chap. 5 we discussed the concepts of availability and irreversibility as applied to steady-flow problems. We made further examination of these concepts as they might apply to combustion processes in Chap. 8. Let us now consider how they may apply to the exposition of this chapter, specifically to power cycles.

As we have stated several times, real processes are not reversible; we cannot have frictionless pumps, compressors, or turbines, and heat-exchange devices always involve heat transfer across a finite temperature difference with some resultant irreversibility. Ultimately, we might choose as our thermodynamic objective a minimization of the overall irreversibility generation by a particular power cycle. Generally speaking, we may expect this objective to also coincide with a condition of minimum fuel consumption and energy costs. For example, a steam-power cycle with feedwater heaters will have less total irreversibility than one without; a gas-turbine cycle with regeneration will have less irreversibility than one without; and so on. However, these facts do not tell us anything about the economic cost saving which may result. The advantage of second-law analysis of power cycles is that it may point the way to *possible* alternatives which can result in savings in fuel or equipment costs.

A second-law analysis of a power cycle is normally performed with the following sequence of operations:

1 The availability function $b = h - T_0 s$ is computed for each point in the cycle.

2 The maximum work output for each component is calculated as the decrease in the availability across the component.

3 The actual work output is compared with the maximum output, and the irreversibility of each component is computed from Eq. (5-49):

$$\dot{I} = \dot{W}_{\text{max output}} - \dot{W}_{\text{actual output}}$$

4 The overall cycle irreversibility is calculated by summing the irreversibilities for all the components.

5 Comparisons can then be drawn between different cycles, different operating conditions within a cycle, and alteration of component performance within a cycle.

EXAMPLE 9-13 Irreversibility in a Boiler

The combustion system in Example 8-10 is used to heat water from 500 psia, 200°F liquid, to superheated conditions of 500 psia, 1000°F. What is the overall irreversibility for this process? (*Note:* The high steam temperature at exit is possible because the combustion temperature can be quite high inside the boiler.)

SOLUTION The overall energy release per pound of fuel was calculated as 20 285 Btu. In accordance with the schematic diagram, this energy must go to raise the enthalpy of the water. Thus

$$q = 20\ 285 \text{ Btu} = \dot{m}_w(h_2 - h_1) \qquad (a)$$

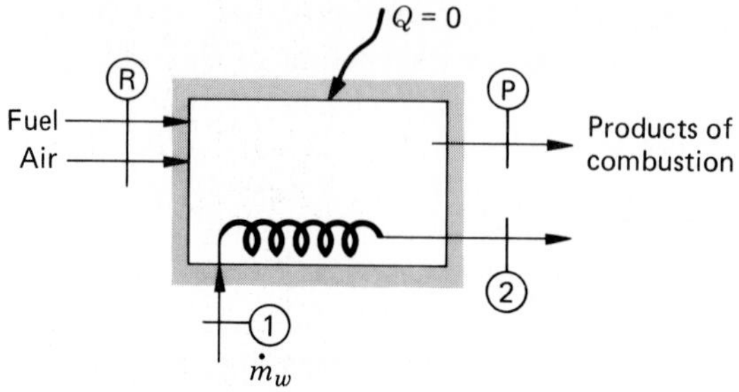

FIG. EXAMPLE 9-13

From the steam tables

$$h_1 = 167.65 \text{ Btu/lbm} \qquad \text{(compressed liquid)}$$
$$s_1 = 0.293\ 41 \text{ Btu/lbm}\cdot{}^\circ\text{R}$$
$$h_2 = 1520.7 \text{ Btu/lbm}$$
$$s_2 = 1.7471 \text{ Btu/lbm}\cdot{}^\circ\text{R}$$

Thus

$$\dot{m}_w = 14.99 \text{ lbm/lbm fuel}$$

The decrease in the availability of the water (steam) is

$$\begin{aligned} B_1 - B_2 &= \dot{m}_w[(h_1 - T_0 s_1) - (h_2 - T_0 s_2)] \\ &= 14.99[(167.65 - 1520.7) - (537)(0.292\ 41 - 1.7471)] \qquad (b) \\ &= -8573 \text{ Btu/lbm fuel} \qquad (-19\ 940 \text{ kJ/kg fuel}) \end{aligned}$$

From Example 8-10, the decrease in the availability for the combustion process is 21 516 Btu/lbm fuel, so the overall decrease for the boiler is

$$\begin{aligned} B_R - B_P + B_1 - B_2 &= 21\ 516 - 8573 = 12\ 943 \text{ Btu/lbm fuel} \\ &= 30\ 105 \text{ kJ/kg fuel} \end{aligned}$$

This is equal to the irreversibility because there is no actual work done:

$$\begin{aligned} \dot{I} &= +12\ 943 \text{ Btu/lbm fuel} \\ &= +30\ 105 \text{ kJ/kg fuel} \end{aligned}$$

EXAMPLE 9-14 Availability Analysis of a Gas-Turbine Cycle
Let us examine the cycle of Example 9-11 from the standpoint of the concepts of availability and irreversibility. We shall perform the analysis only for constant specific heats and $T_0 = 70°\text{F}$, but for two variations:

(*a*) Using the turbine, compressor, and regenerator efficiencies of Example 9-11
(*b*) Assuming 100 percent efficiencies for the cycle components

SOLUTION The availability function for steady flow was defined in Chap. 5 as

$$b = h - T_0 s \tag{a}$$

and the irreversibility for a process with unit mass was

$$\dot{I} = \dot{W}_{\substack{\text{max}\\\text{output}}} - \dot{W}_{\substack{\text{actual}\\\text{output}}} \tag{b}$$

$$= -\Delta b - \dot{W}_{\text{act}}$$

$$= -\Delta h + T_0 \Delta s - \dot{W}_{\text{act}} \tag{c}$$

For ideal gases the entropy changes may be computed with

$$\Delta s = c_p \ln\left(\frac{T_2}{T_1}\right) - R \ln\left(\frac{p_2}{p_1}\right) \tag{d}$$

(*a*) For this case, corresponding to Example 9-11, we go around the cycle and compute the needed quantities.

Compressor For the compressor,

$$h_2 - h_1 = c_p(T_2 - T_1) = (0.24)(838 - 530)$$
$$= 73.92 \text{ Btu/lbm} \qquad (165.3 \text{ kJ/kg})$$

$$s_2 - s_1 = 0.24 \ln\left(\frac{838}{530}\right) - \frac{53.35}{778} \ln\left(\frac{60}{14.7}\right)$$

$$= +0.0135 \text{ Btu/lbm·°R} \qquad (0.0565 \text{ kJ/kg·K})$$

$$b_2 - b_1 = 73.92 - (530)(0.0135) = 66.76 \text{ Btu/lbm} \qquad (155.3 \text{ kJ/kg})$$

This is the *negative* of the maximum work *output*.

The *actual* work *output* is the negative of the enthalpy change, or

$$\dot{W}_{\substack{\text{act}\\\text{output}}} = -(h_2 - h_1) = -73.92 \text{ Btu/lbm}$$

so that the irreversibility is

$$\dot{I}_{1,2} = -\Delta b_{1,2} - W_{\text{act }1,2}$$
$$= -66.76 - (-73.92) = +7.16 \text{ Btu/lbm} \qquad (16.65 \text{ kJ/kg})$$

Regenerator As the air is heated in the regenerator there is zero actual work for the constant-pressure process. We therefore calculate

$$h_5 - h_2 = (0.24)(1284 - 838) = 107.04 \text{ Btu/lbm} \qquad (248.98 \text{ kJ/kg})$$

$$s_5 - s_2 = 0.24 \ln\left(\frac{1284}{838}\right) - \frac{53.35}{778} \ln\left(\frac{60}{60}\right)$$

$$= +0.1024 \text{ Btu/lbm·°R}$$

$$b_5 - b_2 = 107.04 - (530)(0.1024) = 52.76 \text{ Btu/lbm} \qquad (122.7 \text{ kJ/kg})$$

On the other side of the regenerator the hot exhaust gases are cooled from point 4 to point 6. The temperature at point 6 is obtained from the energy balance on the regenerator,

$$h_5 - h_2 = h_4 - h_6$$

so that

$$T_6 = T_4 - T_5 + T_2 = 1375 - 1284 + 838 = 929°\text{R}$$

or, the temperature rise of the air from the compressor is equal to the temperature drop of the hot exhaust gases. We may now calculate the properties of interest:

$$h_6 - h_4 = (0.24)(929 - 1375) = -107.04 \text{ Btu/lbm}$$

$$\begin{aligned} s_6 - s_4 &= 0.24 \ln\left(\frac{929}{1375}\right) - \frac{53.35}{778} \ln\left(\frac{14.7}{14.7}\right) \\ &= -0.0941 \text{ Btu/lbm·°R} \end{aligned}$$

$$\begin{aligned} b_6 - b_4 &= -107.04 - (530)(-0.0941) \\ &= -57.16 \text{ Btu/lbm} \end{aligned}$$

Now the overall irreversibility for the regenerator is the negative of the overall increase in availability less the actual work output (zero), or

$$\begin{aligned} \dot{I}_{\text{reg}} &= -(b_5 - b_2) - (b_6 - b_4) - \dot{W}_{\text{act}} \\ &= -(+52.76) - (-57.16) - 0 \\ &= +4.40 \text{ Btu/lbm} \qquad (10.23 \text{ kJ/kg}) \end{aligned}$$

Burner In the burner the air is heated from point 5 to point 3 (for an air-standard cycle). From a practical standpoint this would be accomplished by burning a fuel, and some irreversibility would result, as described in Chap. 8. In this problem we do not have enough information to determine the irreversibility of the combustion process, so we shall just calculate the change in availability of the air:

$$h_3 - h_5 = (0.24)(1960 - 1284) = 162.24 \text{ Btu/lbm}$$

$$\begin{aligned} s_3 - s_5 &= 0.24 \ln\left(\frac{1960}{1284}\right) - \frac{53.35}{778} \ln\left(\frac{60}{60}\right) \\ &= +0.1015 \text{ Btu/lbm·°R} \end{aligned}$$

$$\begin{aligned} b_3 - b_5 &= 162.24 - (530)(+0.1015) \\ &= 108.44 \text{ Btu/lbm} \qquad (252.23 \text{ kJ/kg}) \end{aligned}$$

Turbine The calculations for the turbine are

$$h_4 - h_3 = (0.24)(1375 - 1960) = -140.4 \text{ Btu/lbm}$$

$$s_4 - s_3 = 0.24 \ln\left(\frac{1375}{1960}\right) - \frac{53.35}{778} \ln\left(\frac{14.7}{60}\right)$$
$$= +0.011\ 37 \text{ Btu/lbm}\cdot{}^\circ\text{R} \qquad (0.0476 \text{ kJ/kg}\cdot\text{K})$$

$$b_4 - b_3 = -140.4 - (530)(0.011\ 37)$$
$$= -146.43 \text{ Btu/lbm} \qquad (-340.6 \text{ kJ/kg})$$

The actual work *output* is $h_3 - h_4 = 140.4$ Btu/lbm, so the irreversibility is

$$\dot{I}_{3,4} = -\Delta b_{3,4} - W_{\text{act }3,4}$$
$$= -(-146.43) - 140.4$$
$$= +6.03 \text{ Btu/lbm} \qquad (14.03 \text{ kJ/kg})$$

The various quantities are shown in the table below to illustrate the overall behavior of the cycle.

As mentioned above, the combustion process in the burner has not been considered in this analysis. To obtain an estimate of the effect of combustion we might consult the results of Example 8-10. In that example the entrance and exit temperatures and pressures are nearly the same as those here. Summarizing that example,

Availabilities for Actual Gas-Turbine Cycle

Component	Process(es)	Δh, Btu/lbm	Δb, Btu/lbm	$\dot{W}_{max}$, Btu/lbm	$\dot{W}_{act}$, Btu/lbm	$\dot{I}$, Btu/lbm
Compressor	1-2	73.92	66.76	−66.76	−73.92	7.16
Regenerator	2-5	107.04	52.76			
	4-6	−107.04	−57.16			
	Total	0	−4.40	4.40	0	4.40
Burner	5-3	162.24	108.44	—	0	—
Turbine	3-4	−140.4	−146.43	146.43	140.4	6.03
Overall cycle excluding combustion process		95.76	24.37	84.07	66.48	17.59

Octane fuel in at 4 atm, 25°C (58.8 psia, 77°F)
Air in at 4 atm, 710 K (58.8 psia, 1278°R)
Products out at 4 atm, 1100 K (58.8 psia, 1980°R)
Irreversibility of combustion reaction = 226.31 kJ/kg reactants
= 97.29 Btu/lbm reactants

This result is very close to the change in availability calculated in the present example. Of course, the air fuel ratio in Example 8-9M is so high (94.74) that the reaction is closely represented by a calculation on the basis of air alone.

(*b*) For the ideal cycle, assuming 100 percent efficiencies in the components, we will notice a marked change in performance. Both the turbine and compressor will have zero irreversibility (because the processes are reversible), and the actual work output will be equal to the maximum possible. Using the nomenclature of Fig. 9-29, the regenerator temperature for $\eta_{reg} = 100$ percent will be

$$T_4 = T_5 = 1310°\text{R} \qquad \text{and} \qquad T_6 = T_2 = 792°\text{R}$$

We then have the following calculations:

$$\begin{aligned} h_2 - h_1 &= b_2 - b_1 = -\dot{W}_{max} = \dot{W}_{act} \qquad (\text{because } s_1 = s_2) \\ &= (0.24)(792 - 530) = 62.88 \text{ Btu/lbm} \end{aligned}$$

$$\begin{aligned} h_4 - h_3 &= b_4 - b_3 = -\dot{W}_{max} = \dot{W}_{act} \qquad (s_3 = s_4) \\ &= (0.24)(1310 - 1960) = -156 \text{ Btu/lbm} \end{aligned}$$

For the regenerator,

$$h_5 - h_2 = (0.24)(1310 - 792) = 124.32 \text{ Btu/lbm}$$

$$\begin{aligned} s_5 - s_2 &= 0.24 \ln\left(\frac{1310}{792}\right) - \frac{53.35}{778} \ln\left(\frac{60}{60}\right) \\ &= 0.1208 \text{ Btu/lbm·°R} \qquad (0.5058 \text{ kJ/kg·K}) \end{aligned}$$

$$b_5 - b_2 = 60.31 \text{ Btu/lbm} \qquad (140.3 \text{ kJ/kg})$$

$$h_6 - h_4 = (0.24)(792 - 1310) = -124.32 \text{ Btu/lbm}$$

$$\begin{aligned} s_6 - s_4 &= 0.24 \ln\left(\frac{792}{1310}\right) - \frac{53.35}{778} \ln\left(\frac{14.7}{14.7}\right) \\ &= -0.1208 \text{ Btu/lbm·°R} \end{aligned}$$

$$b_6 - b_4 = -60.31 \text{ Btu/lbm}$$

so that the overall $\Delta b = 0$ along with $\dot{W}_{act} = 0$ and

$$\dot{I}_{reg} = \dot{W}_{max} - \dot{W}_{act} = 0$$

In the burner we have

$$h_3 - h_5 = (0.24)(1960 - 1310) = 156 \text{ Btu/lbm}$$

$$\begin{aligned} s_3 - s_5 &= 0.24 \ln\left(\frac{1960}{1310}\right) - \frac{53.35}{778} \ln\left(\frac{60}{60}\right) \\ &= 0.0967 \text{ Btu/lbm} \qquad (0.4049 \text{ kJ/kg·K}) \end{aligned}$$

$$b_3 - b_5 = 104.75 \text{ Btu/lbm} \qquad (243.6 \text{ kJ/kg})$$

We may now tabulate the properties for the ideal cycle in the following table:

Availability for Ideal Gas-Turbine Cycle

Component	Process(es)	Δh, Btu/lbm	Δb, Btu/lbm	$\dot{W}_{max}$, Btu/lbm	$\dot{W}_{act}$, Btu/lbm	$\dot{I}$, Btu/lbm
Compressor	1-2	62.88	62.88	−62.88	−62.88	0
Regenerator	2-5	124.32	60.31			
	4-6	−124.32	−60.31			
	Total	0	0	0	0	0
Burner	5-3	156	104.75	—	0	—
Turbine	3-4	−156	−156	156	156	0
Overall cycle excluding combustion process		62.88	11.63	93.12	93.12	0

The thermal efficiency for this ideal cycle is

$$\eta = \frac{\dot{W}_{act}}{Q_{burner}} = \frac{93.12}{156} = 59.7 \text{ percent}$$

9-14 HYBRID CYCLES

Depending on the application, it may be advantageous to combine the gas-turbine and vapor-power cycle in a hybrid arrangement, as shown in Fig. 9-33. Instead of using a regenerator, the hot exhaust gases are used to supply heat to

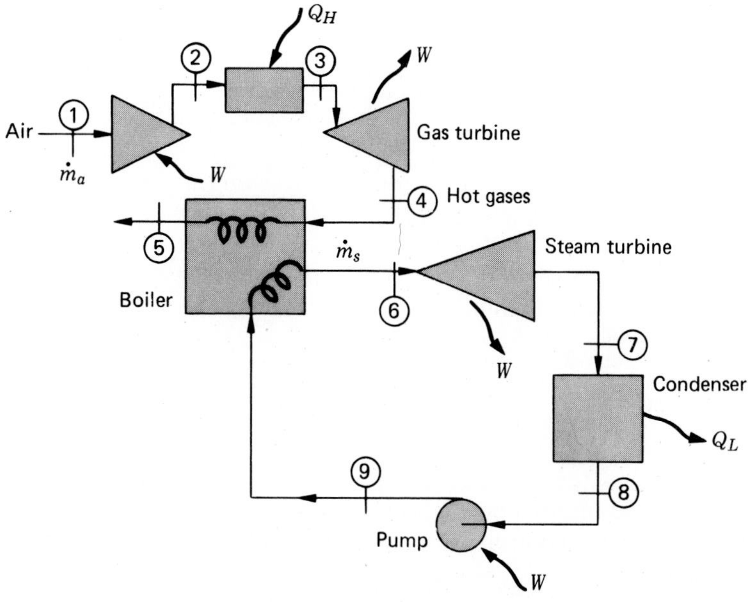

FIG. 9-33 Hybrid cycle schematic for gas-turbine–vapor-power cycle.

the boiler for driving the vapor cycle. If the exhaust gases are the only source of energy for the boiler, then we would have for the energy balance on the boiler

$$\dot{m}_a(h_4 - h_5) = \dot{m}_s(h_6 - h_9) \tag{9-22}$$

If supplemental heat Q is supplied to the boiler, then that value must be added to the left side of the equation.

The advantage of the hybrid or combined cycle is that we take advantage of high-temperature heat addition for the gas turbine as well as the low-temperature heat rejection for the vapor cycle. We have the added advantage that the products of combustion from the gas turbine are "cleaner" owing to the high excess air rates which can be used. For the cycle in Fig. 9-33 the total work output of the cycle would be

$$\begin{aligned} W_{\text{total}} &= W(\text{gas turbine}) - W(\text{air compressor}) + W(\text{steam turbine}) \\ &\qquad - W(\text{pump}) \\ &= \dot{m}_a[(h_3 - h_4) - (h_2 - h_1)] + \dot{m}_s[(h_6 - h_7) - (h_9 - h_8)] \end{aligned} \tag{9-23}$$

The relation between $\dot{m}_a$ and $\dot{m}_s$ is obtained from the boiler energy balance, Eq. (9-22). For no supplemental heat addition to the boiler the only heat added to the combined cycle *from external sources* occurs in the burner from point 2 to point 3. Thus

$$Q_H = \dot{m}_a(h_3 - h_2) \tag{9-24}$$

and the thermal efficiency is the ratio of Eq. (9-23) to Eq. (9-24).

The heat rejection occurs in the condenser from point 7 to point 8 and with the gases at point 5 so the overall heat rejection can be calculated from

$$Q_L = \dot{m}_s(h_7 - h_8) + \dot{m}_a h_5 \tag{9-25}$$

From an overall energy-balance standpoint the total work output can also be calculated from

$$W = Q_H - Q_L \tag{9-26}$$

Of course, a hybrid cycle can also include regeneration for the vapor-cycle portion and multistage compression with intercooling on the gas-cycle portion.

9-15 THE GAS TURBINE FOR JET PROPULSION

The turbojet engine which is extensively used for aircraft propulsion is a simple modification of the open gas-turbine cycle and is shown schematically in Fig.

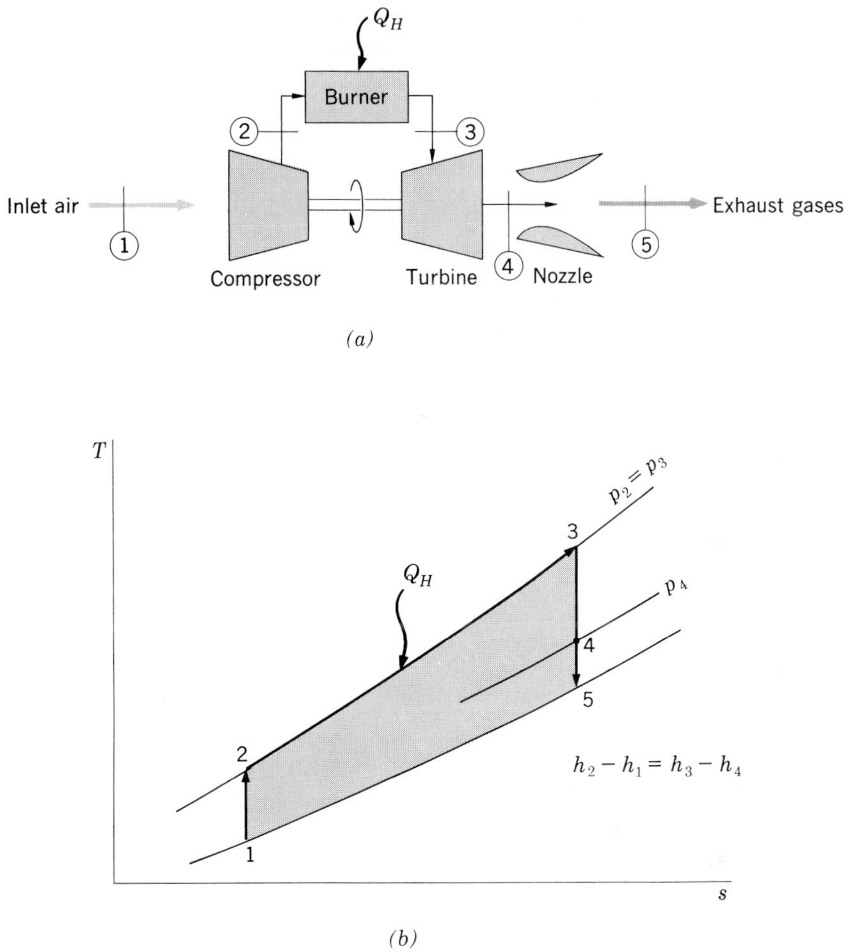

FIG. 9-34 Gas-turbine cycle for jet propulsion: (*a*) schematic; (*b*) *T-s* diagram.

9-34. The hot combustion gases are expanded in the turbine only far enough to generate work to drive the compressor. The remaining energy is then converted to high-velocity kinetic energy by expanding it in a nozzle downstream from the turbine. The jet thrust results from the difference in momentum of the entering air flow to the compressor and the high-velocity exhaust gases. This thrust is given by

$$F = \frac{\dot{m}}{g_c}(V_1 - V_5) \tag{9-27}$$

where $\dot{m}$ is the mass flow in an air-standard cycle and V_1 and V_5 are the flow velocities corresponding to the points in Fig. 9-34. We shall discuss the principles of high-speed compressible flow systems in Chap. 10 with explanations of the design factors which must be considered in establishing the configuration of the nozzle and flow exit geometries. For now we can be content with an example which illustrates the thermodynamic principles involved in the turbojet cycle.

EXAMPLE 9-15 Turbojet Engine
A turbojet engine operates between pressure limits of 5 and 50 psia. The inlet air temperature to the compressor is −40°F, and the upper temperature limit for the engine is 2000°F. Calculate the thrust for 1 lbm/s of air flow, assuming isentropic compression and expansion and an inlet velocity of 300 ft/s. Assume constant specific heats. Also calculate the heat input per pound-mass of air.

SOLUTION The schematic diagram of Fig. 9-34 is used for the solution of this problem. We have

$$T_1 = -40°\text{F} = 420°\text{R} \qquad (233\ \text{K})$$

$$T_3 = 2000°\text{F} = 2460°\text{R} \qquad (1367\ \text{K})$$

$$V_1 = 300\ \text{ft/s} \qquad (91.44\ \text{m/s})$$

$$p_1 = p_5 = 5\ \text{psia} \qquad (34.5\ \text{kPa})$$

$$p_2 = p_3 = 50\ \text{psia} \qquad (345\ \text{kPa})$$

For the isentropic compression process,

$$T_2 = T_1\left(\frac{p_2}{p_1}\right)^{(\gamma-1)/\gamma}$$

$$= 420\left(\frac{50}{5}\right)^{0.286}$$

$$= 812°\text{R} \qquad (451\ \text{K})$$

Since the compressor and turbine works are equal,

$$h_2 - h_1 = h_3 - h_4$$

or, for constant specific heats,

$$T_2 - T_1 = T_3 - T_4$$

Thus

$$T_4 = 2460 - 812 + 420 = 2068°\text{R} \qquad (1149\ \text{K})$$

For the isentropic turbine-exhaust nozzle process

$$T_5 = T_3 \left(\frac{p_5}{p_3}\right)^{(\gamma-1)/\gamma}$$
$$= 2460 \left(\frac{5}{50}\right)^{0.286}$$
$$= 1272°\text{R} \qquad (707\text{ K})$$

For the nozzle,

$$h_4 + \frac{V_4^2}{2g_c} = h_5 + \frac{V_5^2}{2g_c}$$

We shall assume that V_4 is small so that

$$V_5 = [2g_c(h_4 - h_5)]^{1/2}$$
$$= [2g_c c_p(T_4 - T_5)]^{1/2}$$
$$= [(2)(32.2)(0.24)(2068 - 1272)(778)]^{1/2}$$
$$V_5 = 3093\text{ ft/s} \qquad (943\text{ m/s})$$

The thrust is now calculated from Eq. (9-27) as

$$F = T = \dot{m}(V_1 - V_5)$$
$$= \frac{(1)(300 - 3093)}{32.2}$$
$$T = -96.8\text{ lbf}$$

The negative sign indicates that the thrust is in the opposite direction to the velocity.

The heat added per pound-mass is

$$Q_H = \dot{m}(h_3 - h_2)$$
$$= (1)(0.24)(2460 - 812)$$
$$= 395\text{ Btu/s}$$

If we assume that the turbojet moves with a velocity equal to the inlet velocity, the power developed by the thrust force is

$$P = TV_1$$
$$= (86.8)(300) = 26\,000\text{ ft·lbf/s}$$

or

$$P = 33.46 \text{ Btu/s}$$
$$= 47.3 \text{ hp} \qquad (35.3 \text{ kW})$$

These calculations are for the 1.0 lbm/s flow rate and would be proportionately higher for higher flows.

9-16 THE RAMJET

The ramjet is a device employed for aircraft propulsion at very high speeds. It operates on the same principle as the turbojet, except that a flow passage is used to convert the kinetic energy of the incoming flow into pressure. The device is shown schematically in Fig. 9-35. The incoming flow velocity is reduced in the inlet nozzle in process 1-2. Fuel is burned at constant pressure in

FIG. 9-35 The ramjet: (*a*) schematic; (*b*) *T-s* diagram.

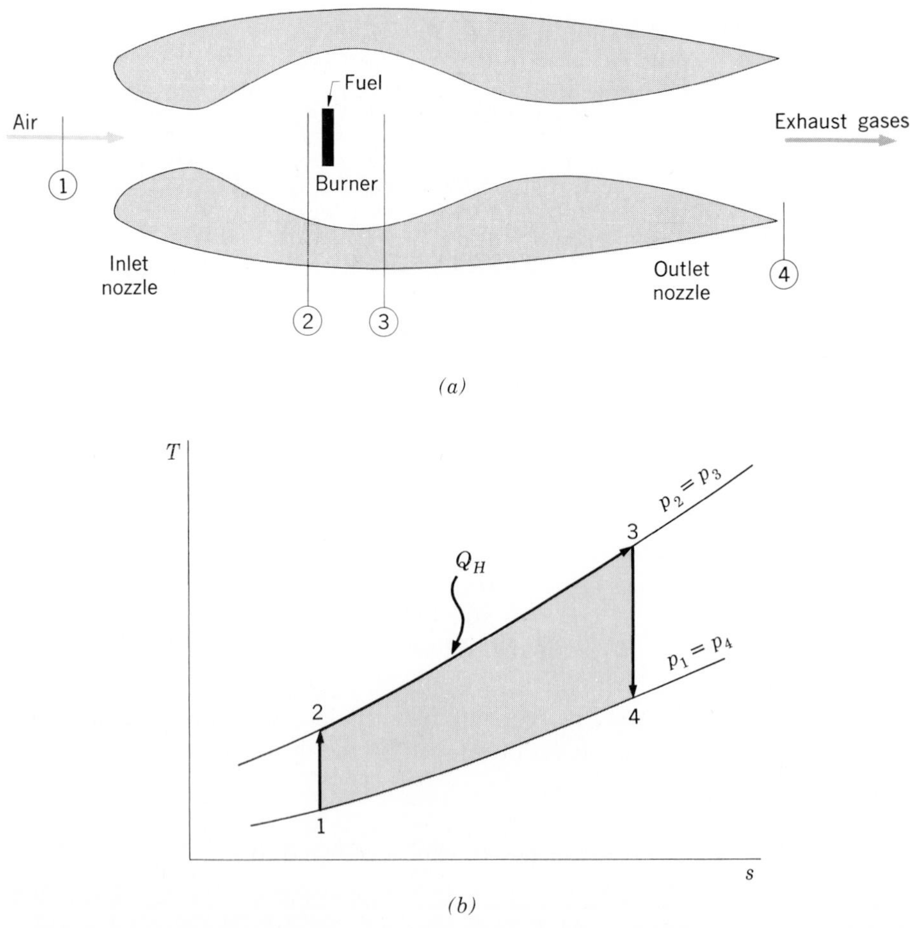

process 2-3, thereby increasing the temperature. Finally, the flow is expanded to a high velocity in the outlet nozzle. Because of the energy which has been added to the flow, the velocity at exit V_4 is greater than that at inlet. The net thrust is given as the change in momentum flux

$$T = F = \frac{\dot{m}}{g_c}(V_4 - V_1) \tag{9-28}$$

where $\dot{m}$ is the mass rate of flow. The ramjet is only feasible for high inlet-flow velocities; otherwise it would not be possible to achieve a high enough compression ratio in the inlet to counteract the losses incurred because of friction and irreversibilities in the engine.

9-17 THE ROCKET

A rocket is a simple device whereby a fuel and oxidizer are supplied to a combustion chamber (burner) at high pressure. The high-pressure products of

FIG. 9-36 The rocket: (*a*) schematic; (*b*) *T-s* diagram.

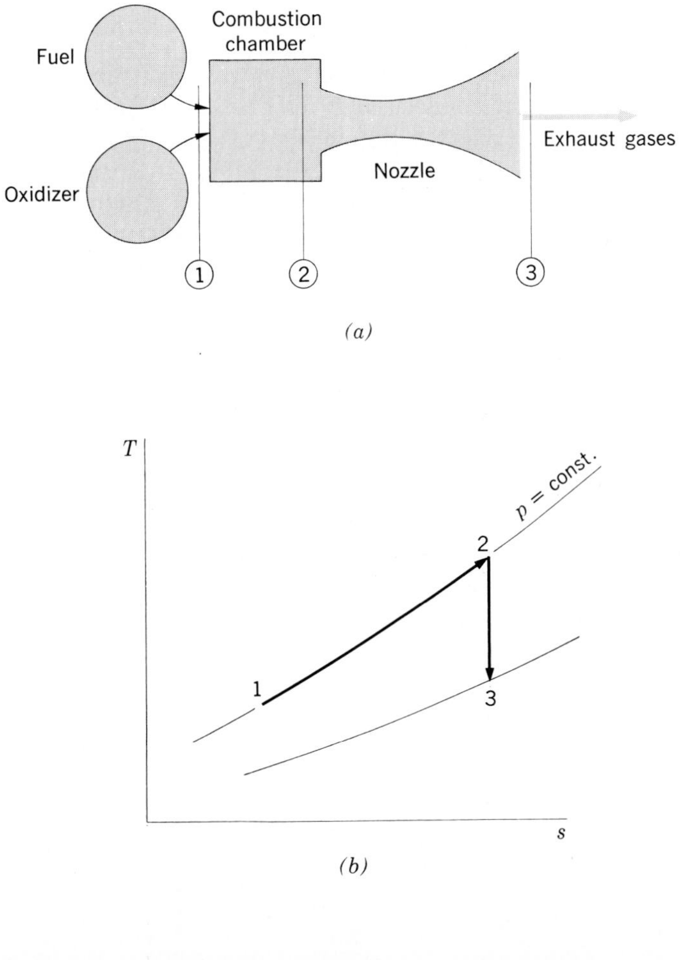

combustion are then expanded through a nozzle to high velocities, thereby producing thrust. The rocket engine is shown schematically in Fig. 9-36. Both liquid and solid propellants (fuel) may be used. The thrust is again given as the change in momentum flux, or

$$T = F = \frac{\dot{m}}{g_c} V_3 \tag{9-29}$$

Of course, the total mass of the rocket engine steadily decreases as more and more combustion products leave the exhaust nozzle.

9-18 SPECIFIC IMPULSE

A parameter frequently used to compare propulsion systems is the *specific impulse,* defined as

$$I_{sp} = \text{specific impulse} = \frac{\text{thrust}}{\text{mass flow rate}} \tag{9-30}$$

Table 9-1 gives a comparison of approximate specific impulses for the three propulsion devices described in the preceding sections. The reader is cautioned against comparing the systems on the basis of I_{sp} alone. A ramjet, for example, is completely inoperable for takeoff of an aircraft. Both the turbojet and ramjet are completely inoperable in space flight, and the rocket consumes huge amounts of expensive fuel. Each of these systems has its own place of application, but a discussion of the details involved is beyond the scope of this exposition.

9-19 VAPOR REFRIGERATION CYCLES

The vapor refrigeration cycle is a practical device which attempts to duplicate the Carnot refrigeration cycle. A schematic of the basic practical arrangement is shown in Fig. 9-37. In the ideal case compression is isentropic. Heat is added to the cycle (Q_L) by evaporating fluid at the lower temperature limit, while heat rejection (Q_H) takes place by condensing the vapor at the higher temperature limit. In the evaporator section heat is added until a saturated or slightly superheated vapor is obtained at point 1.

TABLE 9-1
Approximate Specific Impulses for Some Propulsion Devices

Turbojet	1500-2000
Chemical rocket	160-400
Ramjet	Very large when used at extreme altitudes

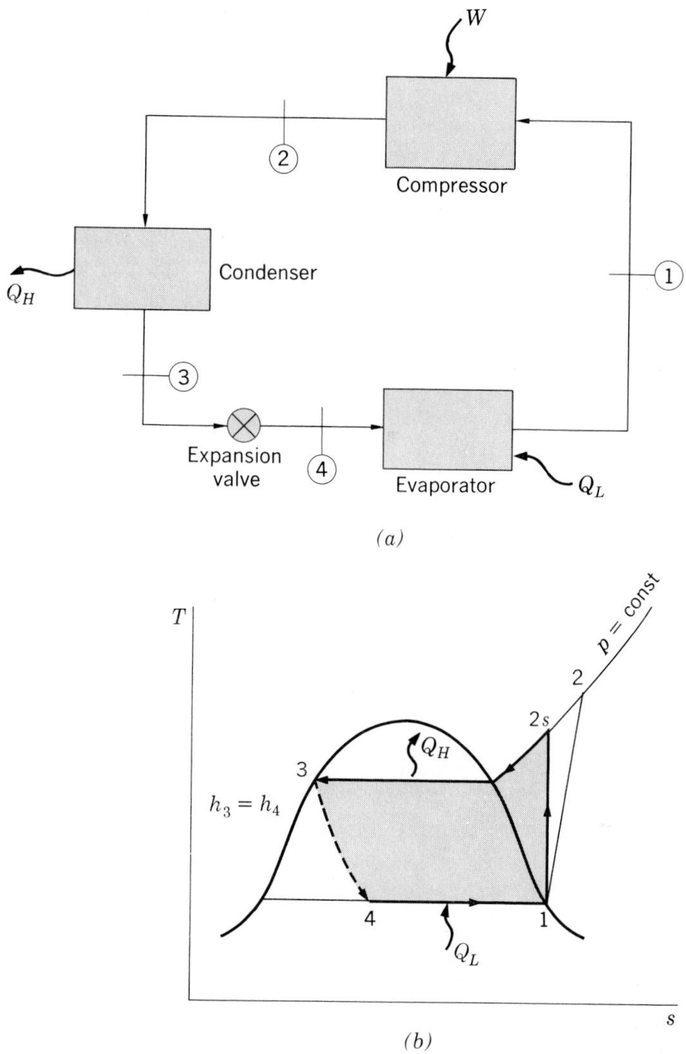

FIG. 9-37 Vapor refrigeration cycle: (*a*) schematic; (*b*) *T-s* diagram.

Condensation at the high temperature produces a saturated liquid at point 3 which must then be expanded to low pressure in order to accommodate the low-temperature evaporation process. An isentropic expansion is practically unfeasible so the simple alternative of throttling the fluid from point 3 to point 4 is selected. For small kinetic energies an adiabatic throttling process is one of constant enthalpy so $h_3 = h_4$. The process is shown as a dashed line on the *T-s* diagram because it is irreversible. In an actual cycle the compression process would also be irreversible, and the outlet from the compressor might appear as point $2'$.

To select a working fluid for refrigeration cycles the behavior of the fluid at low temperatures is an important factor. Water, for example, could not be used below 0°C, and even at moderately low temperatures a vacuum would need to be maintained in the evaporator section. Ammonia (NH_3) is widely used as a refrigerant at very low temperatures but is not normally used in air-conditioning applications because of its toxic nature should leaks occur in the piping or evaporator systems. The fluorinated hydrocarbons (Freons) are most often used in air-conditioning work because of their low cost, inert behavior, and nontoxic effect on humans. They have an additional advantage in that they have a strong affinity for oil and thus can serve as self-lubricating agents in the compressor. Both ammonia and some of the Freons have saturation pressures above 1 atm even at low temperatures, thereby alleviating any necessity to maintain vacuums. Properties of ammonia and dichlorodifluoromethane (Freon 12) are given in the Appendix for calculation purposes.

The coefficient of performance is

$$\text{COP} = \frac{\text{refrigeration effect}}{\text{work supplied}} = \frac{Q_L}{W} \tag{9-31}$$

$$\text{COP} = \frac{h_1 - h_4}{h_2 - h_1}$$

where the subscripts refer to the points in Fig. 9-37. A common unit used in practice to describe the refrigeration effect is the *ton,* defined as the heat rate to melt a ton of ice in a 24-h period. This definition reduces to

$$1 \text{ ton} = 12\,000 \text{ Btu/h} = 3.516 \text{ kW} \tag{9-32}$$

The following example illustrates the overall analysis of a refrigeration cycle.

EXAMPLE 9-16 Freon Refrigerator

An ideal-vapor refrigeration cycle uses Freon 12 as the working fluid. The saturation temperature in the condenser is 120°F, and the evaporator temperature is 30°F. Calculate the coefficient of performance and the horsepower input necessary to produce a cooling effect of 5 tons.

SOLUTION The schematic diagram of Fig. 9-37 applies to this problem. We may consult the Freon 12 property tables in the Appendix to obtain

$h_1 = 80.419$ Btu/lbm (saturated vapor at 30°F)

$s_1 = 0.16648$ Btu/lbm·°R $= s_{2s}$

$h_3 = h_4 = 36.013$ Btu/lbm (saturated liquid at 120°F)

$p_3 = p_2 = 172.35$ psia (saturation pressure at 120°F)

Since $s_2 = s_1$ we may consult the superheat tables to obtain

$h_{2s} = 91.5$ Btu/lbm (212.8 kJ/kg) at $p = 172$ psia and $s = 0.16648$

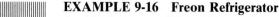

We may now calculate the COP from Eq. (9-31) as

$$\mathrm{COP} = \frac{h_1 - h_4}{h_2 - h_1} = \frac{80.4 - 36.0}{91.5 - 80.4}$$

$$\mathrm{COP} = 4.0$$

The work input required per pound-mass of fluid is

$$W = h_2 - h_1 = 11.1 \text{ Btu/lbm} \qquad (25.82 \text{ kJ/kg})$$

The total flow rate of fluid required may be obtained from

$$Q_L = \dot{m}(h_1 - h_4)$$

$$(5)(12\,000) = \dot{m}(80.4 - 36.0)$$

$$\dot{m} = 1350 \text{ lbm/h} \qquad (612.4 \text{ kg/h})$$

The total work required is therefore

$$W_{\text{total}} = \dot{m}(h_2 - h_1) = (1350)(11.1) = 15\,000 \text{ Btu/h}$$

$$W = 5.9 \text{ hp} \qquad (4.39 \text{ kW})$$

9-20 THE HEAT PUMP

In viewing the vapor refrigeration cycle we could have thought of the device as a mechanism to "pump" heat uphill on the temperature scale and given it the name "heat pump." We used the term *refrigeration* cycle because the cooling was the useful effect we were trying to achieve. The same device may be used to accomplish another objective, that of *delivering heat* Q_H at the condenser. Thus, one might employ the same machine for both summer cooling and winter heating in residences and smaller industrial or commercial buildings. In the heating mode it is normal practice to define the coefficient of performance as

$$\mathrm{COP}_H = \frac{\text{useful effect}}{\text{energy that must be purchased}} = \frac{\text{heating effect}}{\text{work input}} = \frac{Q_H}{W} = \frac{1}{1 - Q_L/Q_H} \tag{9-33}$$

For a Carnot heat pump the corresponding COP_H would be

$$COP_H(\text{Carnot}) = \frac{1}{1 - T_L/T_H} \tag{9-34}$$

The heat pump has an obvious advantage over electric resistance heating in that one obtains the "multiplier effect" through the COP; for example, a unit with a cooling COP of 3.0 would have

$$3.0 = \frac{Q_L}{W} \qquad Q_L = 3W$$

so that

$$Q_H = Q_L + W = 4.0W$$

and

$$COP_H = 4.0$$

Thus we would only have to *purchase* one unit of energy for each *four* units of heat delivered. This is not getting something for nothing! One must make a capital investment in the machine in order to have a future savings in electric costs.

The heat pump may or may not offer economic advantages over natural gas- or oil-fired heating systems. Final cost comparisons are strongly dependent on climate. In moderate climates the heat pump is usually competitive with these other modes of heating, while it is not in very cold climates. Capital costs for heat pump installations are about 20 to 25 percent greater than for a comparable air conditioner and natural gas furnace.

We may expect that future heat pump designs will make extensive use of solar energy as a heat source on the evaporator. By significantly raising the evaporator temperature above ambient outside temperature it is possible to make dramatic improvements in the overall coefficient of performance.

EXAMPLE 9-17 Carnot Heat Pump
A Carnot heat pump operates between an outside temperature of −15°C (5°F) and inside room heating unit temperature of 38°C (100.4°F). Calculate the percent of improvement in performance if a solar energy collector is installed so that the low-temperature limit is raised to 10°C (50°F)

SOLUTION At the first condition we have

$$T_L = -15°\text{C} = 258 \text{ K}$$

$$T_H = 38°\text{C} = 311 \text{ K}$$

so that the heating coefficient of performance is

$$\text{COP}_H = \frac{1}{1 - T_L/T_H} = \frac{1}{1 - 258/311}$$
$$= 5.868$$

When the low temperature is raised to 10°C we have

$$T_L = 10°\text{C} = 283 \text{ K}$$

$$T_H = 38°\text{C} = 311 \text{ K}$$

And

$$\text{COP}_H = \frac{1}{1 - 283/311} = 11.107$$

or an increase of 89 percent. Thought of another way, by raising the low side temperature from −15 to 10°C, we can produce the same heating output with a reduction in power input of 47 percent. Naturally, real heat pumps will not give the high values of COP illustrated in this example; however, one may expect about the same *percentage* improvements when the evaporator (low side) temperature is raised.

EXAMPLE 9-18 Cost Comparisons: Heat Pump vs. Fossil-Fuel Heater Freon 12 is employed as the working fluid in a heat pump with an evaporator temperature (outdoor temperature) of 40°F and a condenser temperature (indoor temperature) of 120°F. The compressor efficiency is 85 percent. The heating requirements are 100 000 Btu/h. If electricity costs $0.053/kWh, compare the cost of operating the heat pump with a fossil-fuel heater in which fuel costs $3.25 per 10^6 Btu. Assume the efficiency of the fossil-fuel heater is 70 percent; i.e., only 70 percent of the fuel energy is delivered to the indoor space to be heated.

SOLUTION The heat pump is a vapor refrigeration device and the schematic is shown in Fig. 9-37. The enthalpies are determined as

$h_1 = 81.436$ Btu/lbm (saturated vapor at 40°F, Table A-15)

$s_1 = 0.16586$ Btu/lbm·°R $= s_2$

$h_3 = h_4 = 36.013$ Btu/lbm (saturated liquid at 120°F)

$p_3 = p_2 = 172.35$ psia (saturation pressure at 120°F)

Using $s_1 = s_2$, we consult the superheat tables (Table A-16) at 172.35 psia and obtain by interpolation

$$h_{2s} = 90.64 \text{ Btu/lbm}$$

Because of the compressor efficiency we must write

$$0.85 = \frac{h_{2s} - h_1}{h_2 - h_1} = \frac{90.64 - 81.436}{h_2 - 81.436} \qquad (a)$$

$$h_2 = 92.264 \text{ Btu/lbm}$$

The heating capacity is $Q_H = 100\ 000$ Btu/h so

$$Q_H = \dot{m}(h_2 - h_3)$$

and the flow rate of Freon is calculated as

$$\dot{m} = \frac{100\ 000}{92.264 - 36.013} = 1778 \text{ lbm/h} \qquad (b)$$

The corresponding work input is

$$W = \dot{m}(h_2 - h_1) = (1778)(92.264 - 81.436)$$
$$= 19\ 252 \text{ Btu/h} = 5.64 \text{ kW} \qquad (c)$$

At \$0.053/kWh (\$15.53/10^6 Btu) the corresponding cost is

$$C_{HP} = (0.053)(5.64) = \$0.2989/\text{h}$$

The unit heating cost is

$$C_{HP}/10^6 \text{ Btu} = \frac{\$0.2989/\text{h}}{0.1 \times 10^6 \text{ Btu/h}} = \$2.989/10^6 \text{ Btu} \qquad (d)$$

For the fossil-fuel heater

$$E_{\text{fuel}}(0.7) = 100\ 000 \text{ Btu/h}$$
$$E_{\text{fuel}} = 142\ 860 \text{ Btu/h}$$

and, at \$3.25/10^6 Btu the cost is

$$C_F = (\$3.25/10^6 \text{ Btu})(0.142\ 86 \times 10^6 \text{ Btu/h})$$
$$= \$0.4643/\text{h}$$

The cost per 10^6 Btu *delivered to the inside* is

$$C_F/10^6 \text{ Btu} = \frac{0.4643}{0.1} = \$4.643/10^6 \text{ Btu} \qquad (e)$$

Thus we find that the unit heating costs for the heat pump are about 35 percent cheaper than for the fossil-fuel heater, even though the unit cost of energy *purchased* is much higher for the heat pump (\$15.53/10⁶ Btu for the heat pump versus \$3.25/10⁶ Btu for fuel). This results of course from the coefficient of performance effect. As a matter of interest we can calculate the heating coefficient of performance as

$$\mathrm{COP}_H = \frac{Q_H}{W} = \frac{100\,000}{19\,252} = 5.194 \qquad (f)$$

9-21 ABSORPTION REFRIGERATION CYCLES

In many plants and buildings a ready supply of steam or hot flue gases are available as a by-product of some process or as unused capacity of a heating system which is inoperative in off-seasons. In such cases there is a strong economic motivation to make use of the heat supply for other purposes. The absorption refrigeration system provides a vehicle for making use of waste heat, either for air conditioning in the summer or for an appropriate refrigeration storage system.

It is clear that a pressure differential must be involved in a refrigeration system to accomplish a temperature difference for heat addition and rejection. In the vapor cycle discussed in the preceding section the pressure differential is accomplished with the vapor compressor. Such a device requires a relatively large power input per unit mass of fluid circulated. A liquid pump would not require such a large amount of power. The absorption system represents a scheme whereby the vapor compressor is replaced by a liquid pump and a suitable heater. The basic absorption system is shown in Fig. 9-38. For the sake of discussion we shall assume that the refrigerant is ammonia, NH_3; this is the

FIG. 9-38 Schematic of an absorption refrigeration system.

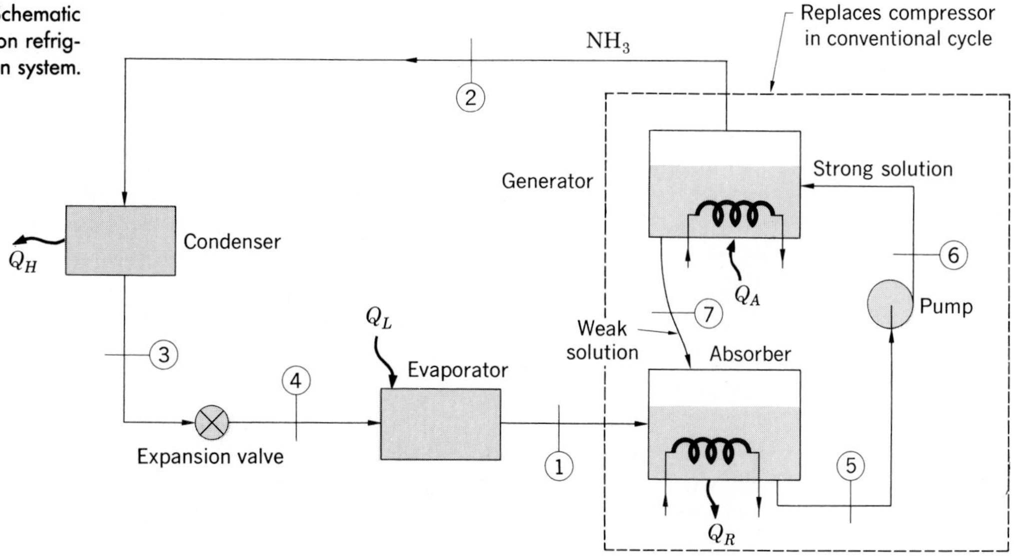

most widely used refrigerant in absorption systems. The condenser, evaporator, and expansion valve are the same as in the vapor compression cycle, so we are primarily concerned with the elements replacing the compressor—the absorber, pump, and generator.

The basic principle involved is that water absorbs ammonia and liberates heat in the process, provided the temperature is held constant. Conversely, when heat is added to a strong solution of NH_4OH, it will break down into NH_3 vapor and H_2O. In the absorber, NH_3 vapor is bubbled through H_2O and heat Q_R is removed to maintain the temperature constant. The strong liquid solution is then taken off at point 5 and pumped to a higher pressure at point 6. The strong solution is then placed in the generator where heat Q_A is added to boil off the NH_3 at the elevated pressure. The remaining weak solution of liquid is then returned to the absorber to acquire a fresh charge of NH_3.

In the overall absorption cycle the work input to the pump is very small, and the major source of energy which must be supplied from the outside is Q_A for the generator. It is to be noted, however, that the absorption cycle requires substantially more heat-rejection facilities than the vapor compression cycle because of the necessity to dissipate the heat of absorption Q_R in addition to the quantity Q_H. Consideration must be given to such matters in an overall economic analysis of a system to be used in practice.

Another pair of fluids which may be used in absorption systems for air-conditioning applications where evaporator temperatures are not too low is the combination of water as the refrigerant and lithium bromide as an absorber. With water as the refrigerant, a vacuum must be maintained in the evaporator to achieve low temperatures; an absolute pressure of 0.12 psia would have to be maintained to achieve an evaporator temperature of 40°F. Such vacuums are normally maintained by steam ejectors.

The schematic for a practical LiBr–H_2O system is shown in Fig. 9-39. Both the evaporator and absorber operate on the low-pressure side of the system and thus are enclosed in the common shell with a connection for purging to maintain vacuum conditions. The condenser and generator operate on the high-pressure side and so are similarly enclosed in a common shell. This system is normally used in large cooling systems which require cooling rates above 10^6 Btu/h, so a substantial amount of cooling water must be supplied to remove the heat from the absorber and condenser. Steam is commonly employed to supply energy to the generator, and the evaporator is used to produce chilled water.

Now let us trace through the processes involved. At point 6 water vapor in the lower shell is being absorbed in the weak LiBr solution arriving at point 5. After the absorption, and accompanying heat release to the cooling water, the strong LiBr–H_2O solution at point 1 is pumped through pump 1 to the generator. The three-way valve and heat exchanger will be discussed in a moment. The strong solution enters the bottom of the upper shell where it is heated with steam coils, thereby causing water vapor to come out of solution at point 3. This vapor is subsequently condensed in the pan at the top of the upper shell. The weak LiBr solution is then returned to the lower shell at points 4 and 5. The pressure difference between the two shells is not large because the temperature

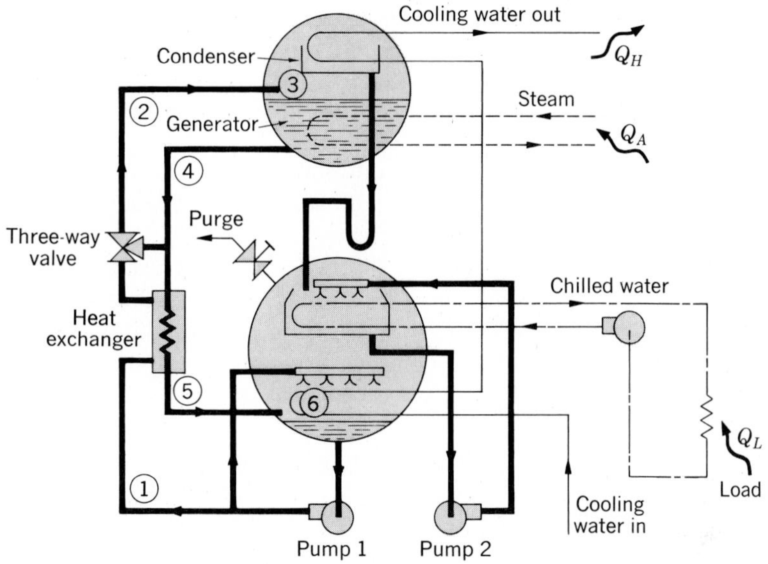

FIG. 9-39 Schematic of a practical lithium-bromide–water absorption refrigeration system. (*Carrier Corporation.*)

for condensation is modest, say 85°F, where the saturation pressure is 0.6 psia. For an evaporator temperature of 40°F the pressure difference would be 0.6 − 0.12 = 0.48 psia. Cooling is accomplished when the liquid water flashes to wet vapor in the lower shell, and the U arrangement in the liquid water line can maintain the pressure differential between the two shells.

Low pressure is maintained in the lower shell by the rapid absorption of water vapor in the LiBr solution and by the purge valve, when necessary. Pump 2 is provided to keep a fresh spray of cool water over the chilled water coils to enhance the heat-transfer rates. A similar spray system is provided in the bypass of pump 1 to enhance the absorption process at point 6.

Steam heating is normally supplied at temperatures of 210 to 220°F, so the solution at point 4 is quite warm. The solution at point 1 is much cooler, and the heat exchanger shown improves overall performance in a manner similar to the regenerators used in power cycles. The high-temperature liquid at 4 is used to preheat the cool fluid from 1 to 2 so that less heat must be added in the generator section. There is also a saving in cooling water requirements for the absorber because the solution entering at 5 is cooler than it would have been without the heat exchanger. The three-way valve serves as a load-control device to increase or decrease liquid flow between the two shells in accordance with the chilled water cooling requirements.

REVIEW QUESTIONS

1 What are the primary factors which influence thermal efficiency of a power cycle or coefficient of performance of a refrigeration cycle?

2 How is the thermal efficiency of a power cycle related to pollution generated by the cycle?
3 Why is it advantageous to employ superheat in the basic Rankine cycle?
4 Describe some important factors in the selection of a working fluid for a vapor power cycle.
5 How does a regenerator improve the performance of a vapor power cycle?
6 Distinguish between closed and open feedwater heaters.
7 Define *turbine efficiency*.
8 Why does a lower thermal efficiency result in more "thermal pollution"?
9 What advantages are offered by intercoolers in air-compression machines?
10 What is the general principle which is applied to selection of interstage pressure(s) in compressors with ideal intercooling?
11 Define *volumetric efficiency*.
12 Why is it possible to analyze an air compressor as a steady-flow device even though the compression occurs in the closed confines of a piston-cylinder arrangement?
13 Distinguish between the Otto and diesel cycles.
14 What is an air-standard cycle?
15 Compare the spark-ignition and compression-ignition cycles on the basis of their respective advantages and disadvantages.
16 How does compression ratio affect thermal efficiency of an internal combustion engine?
17 Describe some advantages of gas-turbine cycles for power production.
18 How does a turbojet work?
19 Define *regenerator efficiency*.
20 How does a ramjet work?
21 What is coefficient of performance?
22 What is a *ton*?
23 How does an absorption refrigeration system work?
24 What are the main factors affecting the coefficient of performance of a refrigeration machine?
25 What is a *heat pump*?
26 What is a *Stirling cycle*? How is it possible for it to have a thermal efficiency approaching the Carnot value?
27 What are some advantages of hybrid cycles combining the gas-turbine with the vapor-power cycle?

PROBLEMS (ENGLISH UNITS)

9-1 In a Rankine cycle steam enters the single-stage turbine at 1000 psia and 800°F. Heat is rejected in the condenser at 80°F. Calculate the thermal efficiency of the cycle. Repeat the calculation for reheat of the steam at 500 psia back to 800°F.

9-2 A certain Rankine cycle operates with two reheat stages. Steam enters the turbine at 1500 psia and 1200°F. Reheat to 1200°F occurs at 1000 and

500 psia. The lower-pressure limit of the cycle is 1.0 psia. Each turbine stage has an efficiency of 90 percent. Calculate the thermal efficiency of the cycle and compare with the Carnot efficiency between the same temperature limits.

9-3 A small steam turbine is designed to operate with 200 psia of steam. The turbine is to be employed in a simple Rankine cycle which rejects heat at 90°F. To what temperature must the high-pressure steam be heated to ensure that the moisture content does not exceed 5 percent in the turbine? Calculate the thermal efficiency of the cycle under these conditions. Assume isentropic expansion in the turbine.

9-4 Repeat Prob. 9-3 for a turbine efficiency of 90 percent.

9-5 A regenerative steam cycle is to operate between 1000 psia, 800°F, and 2 psia. One open feedwater heater is to be employed. Isentropic expansion occurs in the turbine, and the extraction pressure is chosen such that the enthalpy drop is the same in each of the two stages. Calculate the thermal efficiency of the cycle.

9-6 Repeat Prob. 9-5 using a closed feedwater heater with the condensate trapped back to the condenser hot well.

9-7 A steam power plant operates with three open feedwater heaters and reheat after each turbine stage. Draw a schematic of the cycle, label all appropriate points, and write an expression for the thermal efficiency. Also write all necessary energy balances.

9-8 A steam power cycle employs three turbine stages with two extractions to closed feedwater heaters and reheat after each extraction. Condensate is pumped to the boiler. Draw a schematic diagram and write an expression for the thermal efficiency of the cycle. Also write any necessary energy balances.

9-9 A steam power plant uses two feedwater heaters. The first extraction is fed to an open feedwater heater, and the remaining steam is reheated before entering the second turbine stage. The second extraction is fed to a closed feedwater heater with the condensate pumped up to the boiler pressure. There is no reheat after the second extraction. Draw a schematic diagram for the cycle, label it, and write an expression for the thermal efficiency of the cycle. Also write all necessary energy balances.

9-10 A steam power plant uses two extractions and two open feedwater heaters. Reheat is employed after both extractions. Draw a schematic for the cycle, label it, and write an expression for the thermal efficiency of the cycle. Also write all necessary energy balances.

9-11 A reciprocating air compressor receives air at 14 psia and 100°F and discharges at 45 psia. The clearance volume is 5 percent of the piston displacement. The compressor has a 4-in diameter bore and a 5-in stroke and rotates at 500 rpm. Calculate the volume of air delivered at inlet conditions, the volumetric efficiency, and the horsepower required for (*a*) isentropic compression and (*b*) polytropic compression with $n = 1.30$.

9-12 A two-stage machine is used to compress air from 14.7 psia, 70°F, to 500 psia. The compressor handles 110 ft^3/min at inlet conditions, and ideal interstage pressure is used. Each stage is adiabatic but has an efficiency of 70 percent. An ideal intercooler is employed between the stages. Calculate the horsepower required if the machine operates as a steady-flow device. What horsepower would be required if a single-stage machine were used?

9-13 A regenerative steam cycle operates between 1500 psia, 1200°F, and 2 psia. Two closed feedwater heaters are used with extraction at 800 and 200 psia. The condensate is trapped back to the condenser hot well. Calculate the thermal efficiency of the cycle, assuming (*a*) isentropic expansion in the turbine and (*b*) a turbine efficiency of 85 percent for each of the three stages.

9-14 A combined reheat-regenerative steam cycle has the following operating conditions:

Steam inlet to first turbine stage: 1000 psia, 1000°F

Condenser pressure: 2 psia

Extraction pressure: 100 psia

After a portion of the steam is extracted for use in an open feedwater heater, the remainder is reheated to 1000°F and then expanded in the second turbine stage. The turbine efficiency for both stages is 85 percent, and the high-pressure pump efficiency is 60 percent. Calculate the thermal efficiency of the cycle.

9-15 A steam power plant operates with one closed feedwater heater. Steam enters the turbine at 1000 psia and 1200°F and extraction occurs at 300 psia. The low-temperature limit of the cycle is 100°F. Condensate from the heater is trapped to the condenser. Calculate the thermal efficiency of the cycle.

9-16 A steam power cycle employs one open feedwater heater. Entrance conditions to the turbine are 1120°F and 1600 psia, and the condenser temperature is 100°F. Each turbine stage has an efficiency of 87 percent. Calculate the thermal efficiency of the cycle. Neglect pump work. Extraction occurs at 140 psia.

9-17 A Rankine cycle has turbine inlet conditions of 1200 psia and 950°F and a condenser temperature of 100°F. The turbine efficiency is 87 percent. Calculate the thermal efficiency of the cycle.

9-18 A steam cycle operates with inlet steam to the turbine at 800 psia and 1200°F and a condenser pressure of 2 psia. Extraction occurs at 300 psia with one closed feedwater heater, the condensate from which is pumped up to the boiler pressure. The efficiencies of the turbine stages are 90 percent. Calculate the thermal efficiency of the cycle. Neglect pump work.

9-19 A steam cycle is designed to produce a power output of 100 MW. Inlet to the turbine is at 1000 psia and 1200°F, and one closed feedwater heater is used with extraction at 100 psia. The condenser temperature is 100°F. Calculate the heat added in the boiler assuming isentropic expansion and neglecting pump work. Condensate is trapped to the condenser.

9-20 A large three-stage air compressor receives inlet air at 14.7 psia and 70°F for a discharge pressure of 900 psia. The efficiency of each stage is 92 percent. Calculate the power input for an air flow rate of 15 lbm/min. Assume ideal interstaging.

9-21 The compression ratio for an air-standard diesel cycle is 19 : 1, and the heat input per pound-mass of air is 400 Btu. The conditions at the start of compression are 80°F and 14.5 psia. Calculate the temperature and pressure at all points in the cycle, the cutoff ratio, and the thermal efficiency. Compare with the efficiency of a Carnot cycle operating between the same temperature limits.

9-22 An air-standard diesel cycle has a compression ratio of 17 : 1 with conditions at the beginning of the compression stroke of 14.0 psia and 50°F. The maximum temperature in the cycle is 3500°R. Calculate the temperature and pressure at each point in the cycle, the cutoff ratio, and the thermal efficiency. Compare with the efficiency of a Carnot cycle operating between the same temperature limits.

9-23 Calculate the thermal efficiency of an air-standard Otto cycle having the same upper pressure limit as the diesel cycle of Prob. 9-21.

9-24 Calculate the thermal efficiency of an Otto cycle having the same upper temperature limit as the diesel cycle of Prob. 9-22.

9-25 What conclusions can be drawn from the results of Probs. 9-23 and 9-24?

9-26 An air-standard Brayton cycle receives air at 14 psia and 60°F. The pressure ratio is 4 : 1, and the upper temperature limit of the cycle is 1800°F. The compressor and turbine efficiencies are 85 percent. Calculate the thermal efficiency of the cycle and the net work output per lbm of air.

9-27 A three-stage compressor is used to compress air from 14.7 psia, 70°F, to 1000 psia. Calculate the work required per lbm of air compressed, assuming ideal interstaging and intercooling. Assume that the compression processes occur isentropically.

9-28 An air-standard Otto cycle has a compression ratio of 9 : 1. At the start of the compression stroke the conditions are 14.0 psia and 90°F. The maximum pressure in the cycle is 560 psia. Determine the maximum temperature in the cycle and the thermal efficiency. Compare this efficiency with that of a Carnot cycle operating between the same temperature limits.

9-29 An air-standard Otto cycle operates such that the compression and expansion processes follow polytropic relations with $n = 1.30$. Calculate the thermal efficiency for a cycle with a compression ratio of 8 : 1, inlet conditions of 14 psia, 80°F, and a maximum temperature limit of 2000°F.

9-30 An air-standard Otto cycle operates with a compression ratio of 10 : 1 and conditions at the start of the compression stroke of 14.5 psia, 100°F.

During the combustion process the heat addition rate is 300 Btu/lbm of air. Calculate the temperature and pressure at all points in the cycle and the thermal efficiency, assuming (*a*) constant specific heats and (*b*) variable specific heats based on the air tables. Compare with the efficiency of a Carnot cycle operating between the same temperature limits.

9-31 Repeat Prob. 9-26, using air tables which take into account variable specific heats.

9-32 An ideal regenerator is employed with the Brayton cycle of Prob. 9-26. Calculate the thermal efficiency under these conditions.

9-33 A gas-turbine cycle uses two stages of compression and expansion along with an ideal intercooler for the compression stages. Inlet conditions are 14.7 psia and 70°F, and the maximum pressure is 150 psia. The temperature at the inlet to each turbine stage is 1700°F. Calculate the thermal efficiency for the cycle. Assume isentropic processes in the compression and turbine stages.

9-34 A regenerator having an efficiency of 80 percent is installed on the two-stage cycle of Prob. 9-33. Calculate the thermal efficiency under these conditions and compare with the efficiency of a Carnot cycle operating between the same temperature limits.

9-35 Rework Prob. 9-34, using compressor and turbine efficiencies of 80 and 85 percent, respectively, for each stage.

9-36 A gas turbine cycle employs a two-stage compressor with ideal interstaging and ideal intercooling. The overall pressure ratio is 6.3. Inlet air is at 14 psia and 100°F, and the upper temperature limit of the cycle is 1750°F. Only one turbine stage is used and the turbine efficiency is 88 percent. The compressors are isentropic. A regenerator is also used and it has an efficiency of 78 percent. Calculate the thermal efficiency of the cycle.

9-37 A gas turbine cycle operates with a pressure ratio of 6.0 and inlet conditions of 10 psia and 50°F. The upper temperature limit is 1850°F. The compressor efficiency is 86 percent, and the turbine efficiency is 83 percent. Calculate the thermal efficiency of the cycle.

9-38 A gas turbine cycle has air input conditions of 120°F and 1 atm. The pressure ratio is 6.0. The upper temperature limit is 1900°F and the turbine efficiency is 90 percent. The compressor is isentropic. A regenerator with an efficiency of 82 percent is also employed. Calculate the thermal efficiency of the cycle.

9-39 A gas turbine cycle has two stages of compression with ideal interstaging and intercooling and one turbine stage. A regenerator with an efficiency of 90 percent is also used. Both the compressor and turbine are isentropic. Calculate the thermal efficiency of the cycle if the overall pressure ratio is 6.7, inlet conditions are 14 psia and 75°F, and the upper temperature is 1750°F.

9-40 A jet engine has an overall pressure ratio of 5.0, inlet conditions of 7 psia and −40°F, and an upper temperature limit of 1950°F. Compression and expansion are isentropic. Calculate the mass flow rate necessary to produce a net thrust force of 10 000 lbf if the inlet velocity to the compressor is 600 ft/s.

9-41 A gas turbine is used for jet propulsion, with inlet conditions of 5 psia and −40°F. The overall pressure ratio is 5.3 and the upper temperature limit of the engine is 2000°F. Assuming isentropic compression and expansion, calculate the flow rate necessary to produce a thrust force of 15 000 lbf with an inlet velocity to the engine of 200 ft/s.

9-42 One way of making use of the energy contained in the hot exhaust gases of a gas-turbine cycle is to employ them for heating steam in a Rankine cycle. Consider the gas-turbine cycle of Prob. 9-33. The hot exhaust gases are used to supply heat to a Rankine cycle operating between 500 and 2 psia. The upper temperature in the steam cycle is 700°F, and isentropic expansion is assumed in the turbine. The exhaust gases leave the steam heater at 500°F. Calculate:
(*a*) The circulation of steam per lbm of air in the gas-turbine cycle
(*b*) The net work output of the combination gas-turbine-steam cycle per lbm of air
(*c*) The thermal efficiency of the combined cycle

9-43 Rework Prob. 9-42, assuming steam reheat to 700°F at 100 psia.

9-44 A turbojet engine takes in air at 14.5 psia and 70°F while in a stationary position on the ground. The pressure ratio in the compressor is 5.0, and the upper temperature limit in the cycle is 1800°F. The compressor and turbine efficiencies are 80 and 85 percent, respectively, and the expansion in the exhaust nozzle may be assumed isentropic. Calculate the thrust and heat input per lbm/s of air, neglecting the inlet velocity.

9-45 Rework Prob. 9-44 for flight at an altitude where the inlet conditions are 3 psia, −60°F, and 800 ft/s.

9-46 An ideal-vapor refrigeration cycle uses Freon 12 as the working fluid. The unit is used for an air-conditioning application such that the evaporator temperature is 45°F and the condenser saturation temperature is 90°F. Calculate the coefficient of performance and the horsepower required per ton of cooling.

9-47 A low-temperature refrigeration system uses ammonia as the working fluid. The evaporator temperature is −35°F, and the upper pressure limit of the cycle is 180.6 psia. The compressor efficiency is 80 percent. Calculate the COP and the horsepower required per ton of cooling. Compare these figures with the values for a Carnot refrigerator operating between the same temperature limits.

9-48 Suppose chilled water is available for cooling the condenser of the cycle in Prob. 9-47 such that the upper pressure limit of the cycle is reduced to 80.00 psia. Calculate the COP and kilowatts input per ton under these new conditions.

9-49 Water is used as the refrigerant in a vapor refrigeration cycle. The evaporator temperature is 45°F and the condenser temperature is 100°F.

The compressor efficiency is 80 percent. Calculate the coefficient of performance of the cycle.

9-50 Despite the fact that it must operate in a vacuum, a novel refrigeration cycle is designed to use water as the working fluid. The evaporator temperature is 40°F and the condenser temperature is 100°F. Calculate the coefficient of performance which may be achieved if the compressor is isentropic.

9-51 A large building complex employs an ammonia air-conditioning system which must produce a cooling rate of 2000 tons. The evaporator operates at 40°F, and a cooling tower is used to maintain the condenser at 80°F. The compressor efficiency is 83 percent. Calculate the work input and the flow rate of ammonia required.

9-52 A refrigeration system for a cold storage plant uses ammonia as the working fluid. The evaporator temperature is −20°F and the condenser temperature is 90°F. The compressor efficiency is 85 percent. Calculate the work required to produce 75 tons of cooling.

9-53 An ammonia refrigeration system is designed to operate with an evaporator temperature of 40°F and a condenser pressure of 180.6 psia. Calculate the work input in kilowatts to produce a refrigeration effect of 1250 tons. Assume isentropic compression.

9-54 In a certain heat-pump application Freon 12 is used as the working fluid, and a heating rate of 100 000 Btu/h is to be supplied to the interior of a building. It may be assumed that the condenser temperature is 110°F for this case. The evaporator outside the building must operate at 20°F in the winter. Calculate the power input for this unit in Btu/h and hp, assuming isentropic compression.

9-55 Freon 12 is employed as the working fluid in a refrigeration device which operates with an evaporator temperature of −10°F and a condenser temperature of 100°F. The compressor efficiency is 85 percent, and the total cooling performed by the machine is 25 tons. Calculate the power required to drive the compressor.

9-56 A refrigeration cycle uses Freon 12 as the working fluid and has an evaporator temperature of 40°F. The condenser temperature is 130°F and the compressor efficiency is 78 percent. Calculate the work input to produce a cooling effect of 12 tons.

9-57 In a certain refrigeration system for low-temperature applications a two-stage operation is desirable which employs a Freon 12 system in combination with an ammonia system. The ammonia operates between a low-temperature limit of −40°F and an upper temperature of 30°F. The condenser for the ammonia cycle (at 30°F) is also the evaporator for the Freon 12 cycle which then rejects heat in its own condenser at 100°F. The combined ammonia condenser and Freon evaporator is insulated so that no heat is gained from the surroundings, and all the heat rejected by the ammonia goes to evaporate the Freon. The overall system is to produce a refrigeration effect of 5 tons at the −40°F condition. Assuming isentropic compression processes, calculate the mass flows of each fluid required, the power input to each compressor, and the coefficient of performance for the overall system.

9-58 A certain absorption air-conditioning system uses water as the working fluid and lithium bromide as the absorber. The evaporator temperature for the cycle is 40°F, and the condenser saturation temperature is 100°F. Calculate the cooling effect per lbm of water.

9-59 An automobile air conditioner uses a Freon 12 refrigeration system. The evaporator temperature at high-speed driving conditions may be as low as 20°F while the saturation temperature in the condenser may be as high as 140°F. Calculate the horsepower required to provide 3 tons of cooling with a compressor efficiency of 80 percent. What is the coefficient of performance under these conditions?

9-60 A regenerative steam cycle employs one closed feedwater heater and two turbine stages. Steam enters the turbine at 800 psia and 700°F, and extraction occurs at 100 psia. Discharge pressure is 1 psia. The efficiency of each turbine stage is 85 percent, and a condensate pump pumps the condensate from the heater up to the boiler pressure. The pump efficiencies are 100 percent. Calculate the thermal efficiency of the cycle and the pump horsepowers required if the total electric output from the turbine is 20 000 kW.

9-61 A gas-turbine cycle receives air at 14.7 psia and 70°F and operates with an upper pressure limit of 250 psia. Two compression states are employed with ideal intercooling and the interstage pressure selected accordingly. Two turbine stages are also used with reheat between stages. A regenerator having an efficiency of 85 percent is connected to the discharge from the second turbine and the high-pressure air entering the first burner. Compressor efficiencies are 88 percent, and turbine efficiencies are 90 percent. The upper temperature limit of the cycle is 1900°F. Calculate the thermal efficiency of the cycle and the mass flow of air required for an electrical output of 10 000 kW. Assume air as the working fluid with constant specific heats. If a fuel having a heating value of 20 000 Btu/lbm is available at a price of \$0.075/lbm, calculate the cost of electric energy generated in ¢/kWh.

9-62 A regenerative steam cycle employs two closed feedwater heaters and three turbine stages. The condensate from the first heater flows through a steam trap into the second heater, and the condensate from the second heater is pumped up to the boiler pressure. The steam enters the turbine at 1500 psia, 1000°F, and extraction occurs at 300 and 100 psia, and discharge from the third turbine stage is at 1 psia. The turbine efficiencies are 85 percent, and the pump efficiencies are 100 percent. Calculate the thermal efficiency and steam flow required for a total electrical output from the turbine of 20 000 kW.

9-63 In some parts of the western United States geothermal energy, i.e., underground hot water, may be an economical source of power. Suppose that the hot water is available at 200°F and the surroundings temperature for dissipation of heat is 85°F. A special Rankine vapor cycle is devised to operate between these limits using Freon 12 as the working fluid and a maximum pressure limit of 250 psia. The turbine efficiency is 85 percent. Assuming that the hot water temperature drops to 140°F as it is giving up heat to the Freon in the boiler, calculate the water flow rate required for a power output of 25 MW.

9-64 A regenerative steam cycle operates with steam inlet to the turbine at 1000 psia and 1200°F and a condenser temperature of 100°F. One open feedwater heater is to be employed with the steam extraction pressure selected to provide the maximum increase in cycle efficiency. Select several extraction pressures and examine their effect on thermal efficiency. Make the calculations for (*a*) isentropic expansion and (*b*) turbine stage efficiencies of 87 percent.

9-65 A small gas-turbine cycle is to be designed for automotive applications with the following data:

Maximum temperature:	1800°F
Turbine efficiency:	90 percent
Compressor efficiency:	85 percent
Regenerator efficiency:	70 percent
Inlet air temperature:	120°F, 14.7 psia
Power output:	300 hp
Pressure ratio:	3.5 : 1

Assuming an air-standard cycle, calculate the thermal efficiency of the cycle and mass flow of air required. If a fuel is used with a heating value of 140 000 Btu/gal, calculate the fuel consumption rate at the maximum power output. Make the calculations assuming constant specific heats and also using the air tables.

9-66 A large diesel engine is to produce a power output of 5000 hp and operates with a compression ratio of 18 : 1. The cutoff ratio is 2.0, and isentropic compression and expansion may be assumed. Assuming an air/fuel ratio of 16 : 1 and a fuel with a heating value of 20 000 Btu/lbm, calculate the fuel consumption rate necessary to produce the indicated power output. Make the calculations on the basis of an air-standard cycle.

9-67 A "heat pump" is a reversed refrigeration machine which may be used for heating in cold weather. The condenser becomes the heating unit inside the building, while the evaporator receives heat from the cold air outdoors. Normally a reciprocating compressor provides the work input to power the cycle. Consider a Freon 12 unit which has the following specifications:

Condenser temperature:	100°F
Evaporator temperature:	20°F
Heating required:	125 000 Btu/h at 100°F

Assuming isentropic compression, calculate the work input to produce the indicated heating and the mass flow rate of Freon 12. If electric energy costs \$0.042/kWh, what is the cost of operation of the device? How does this compare with the cost of performing the same heating with fuel oil costing \$4.00 per million Btu?

9-68 The heat pump of Prob. 9-67 is to be operated as an air-conditioning unit in the summer with the compressor providing the same mass flow rate capabilities, but the condenser rejects heat to the outdoors at 120°F (an air-cooled unit) while the evaporator functions at 55°F indoors. Calculate the cooling rate in tons for these conditions and the required horsepower input. Also calculate the electric operating cost of the unit for these conditions.

9-69 A closed gas-turbine cycle using helium as the working fluid is to be designed to accomplish the same power output as the cycle in Prob. 9-65. The same conditions may be assumed to apply except that the helium inlet conditions to the compressor are 300°F and 150 psia. Calculate the thermal efficiency of this cycle and the mass flow of helium required assuming constant specific heats. Also calculate the heat which must be dissipated to the surroundings.

9-70 To operate at high-compression ratios spark-ignition engines (Otto cycles) require special tetraethyl lead additives to the fuel which environmentalists claim are serious health hazards. To reduce these hazards some engine designs employ lower compression ratios. Estimate the percentage increase in fuel consumption which would result from lowering the compression ratio from 10.75 : 1 to a value of 8.5 : 1.

9-71 A steady-flow device is designed to compress helium from 1.5 atm, 25°C, to 100 atm in three stages. Ideal intercooling is employed with optimum staging, and the stage efficiencies are 90 percent. Calculate the work input in kW to compress 10 kg/min.

9-72 A reciprocating air compressor has a volumetric efficiency of 95 percent and is designed to compress 500 ft^3/min of air measured at inlet conditions of 1 atm and 70°F. The discharge pressure is 75 psia, and isentropic compression and expansion may be assumed. Calculate the clearance volume as a percentage of piston displacement and the horsepower input required.

9-73 A closed-cycle gas-turbine plant using helium as the working fluid is to be designed for a marine installation where heat may be rejected to seawater. An overall pressure ratio of 6 : 1 is used with two stages of compression and ideal intercooling to 70°F. Inlet conditions to the first stage are 70°F and 150 psia, and the upper temperature limit of the cycle is 2000°F. Only a single turbine stage is employed with a regenerator having an efficiency of 80 percent. Both compressor and turbine efficiencies are 90 percent. Calculate the thermal efficiency of the cycle and the mass flow of helium necessary to produce a 25 000-hp output. What is the total amount of heat which must be dissipated to the seawater, including that removed in the intercooler?

9-74 A nuclear power plant has an upper temperature limit of 850°F because of metallurgical considerations in the reactor vessel. In one particular installation the following design information is supplied:

Upper pressure limit:	1500 psia
Feedwater heaters:	Two, closed, condensate from first trapped to second, condensate from second pumped to boiler
Extraction pressures:	500 and 150 psia
Reheat:	Reheat to 850°F after each extraction
Condenser temperature:	100°F
Turbine efficiencies:	90 percent
Power output:	500 000 kW

Calculate the mass flow of steam required in the boiler, the thermal efficiency of the cycle, and the waste heat which must be dissipated in the condenser.

PROBLEMS (METRIC UNITS)

9-1M In a Rankine cycle steam enters the single-stage turbine at 8.0 MPa and 440°C. Heat is rejected in the condenser at 25°C. Calculate the thermal efficiency of the cycle. Repeat the calculation for reheat of steam at 4.0 MPa back to 440°C.

9-2M A certain Rankine cycle operates with two reheat stages. Steam enters the turbine at 10 MPa and 640°C. Reheat to 640°C occurs at 8.0 MPa and 4.0 MPa. The lower pressure limit of the cycle is 6.0 kPa. Each turbine stage has an efficiency of 90 percent. Calculate the thermal efficiency of the cycle and compare with the Carnot efficiency between the same temperature limits.

9-3M A small steam turbine is designed to operate with 1.5 MPa of steam. The turbine is to be employed in a simple Rankine cycle which rejects heat at 30°C. To what temperature must the high-pressure steam be heated to ensure that the moisture content does not exceed 5 percent in the turbine? Calculate the thermal efficiency of the cycle under these conditions. Assume isentropic expansion in the turbine.

9-4M A regenerative steam cycle is to operate between 8 MPa, 440°C, and 10 kPa. One open feedwater heater is to be employed. Isentropic expansion occurs in the turbine, and the extraction pressure is chosen such that the enthalpy drop is the same in each of the two stages. Calculate the thermal efficiency of the cycle.

9-5M A steam power plant employs two open feedwater heaters and reheat after the first turbine stage. Draw a schematic diagram for the cycle, label it, and write an expression for the thermal efficiency of the cycle in accordance with your labeling. Also write all necessary energy-balance expressions.

9-6M A steam power cycle uses two extractions. The first extraction is fed to an open feedwater heater and the remaining steam is reheated. The second extraction is fed to a closed feedwater heater, with no reheat and with the condensate pumped to the boiler pressure. Draw a schematic diagram, label it, and write an expression for the thermal efficiency of the cycle. Also write all necessary energy balances.

9-7M A steam power plant employs two closed feedwater heaters with reheat after each extraction. Draw a schematic diagram for the cycle, label it, and write an expression for the thermal efficiency of the cycle. Also write all necessary energy balances. Make your own choice as to the disposal of the condensate from the two heaters.

9-8M Steam enters the turbine of a power plant at 580°C and 8.0 MPa and expands isentropically. Steam extraction is taken at 800 kPa for use in a closed feedwater heater. The condenser temperature is 40°C. Calculate the steam flow required for a power output of 100 MW. Neglect pump work. Condensate is pumped to the boiler pressure.

9-9M A Rankine steam cycle has an upper pressure limit of 5.0 MPa and a lower pressure limit of 20 kPa. The temperature at inlet to the turbine is 500°C and the turbine efficiency is 85 percent. Calculate the thermal efficiency of the cycle.

9-10M A regenerative steam cycle employs one closed feedwater heater and two turbine stages. Steam enters the turbine at 6.0 MPa and 370°C, and extraction occurs at 700 kPa. Discharge pressure is 6 kPa. The efficiency of each turbine stage is 85 percent, and a condensate pump pumps the condensate from the heater up to boiler pressure. The pump efficiencies are 100 percent. Calculate the thermal efficiency of the cycle and the pump horsepowers required if the total electrical output from the turbine is 20 000 kW.

9-11M A gas-turbine cycle receives air at 1 atm and 20°C and operates with an upper pressure limit of 1.7 MPa. Two compression states are employed with ideal intercooling, and the interstage pressure is selected accordingly. Two turbine stages are also used with reheat between stages. A regenerator having an efficiency of 85 percent is connected to the discharge from the second turbine and the high-pressure air entering the first burner. Compressor efficiencies are 88 percent and turbine efficiencies are 90 percent. The upper temperature limit of the cycle is 1050°C. Calculate the thermal efficiency of the cycle and the mass flow of air required for an electrical output of 10 000 kW. Assume air as the working fluid with constant specific heats. If a fuel having a heating value of 44 000 kJ/kg is available at a price of \$0.165/kg, calculate the cost of electric energy generated in cents per kilowatt hour.

9-12M A regenerative steam cycle employs two closed feedwater heaters and three turbine stages. The condensate from the first heater flows through a steam trap into the second heater, and the condensate from the second heater is pumped up to boiler pressure. The steam enters the turbine at 10 MPa, 560°C, extraction occurs at 2 MPa and 700 kPa, and discharge from the third turbine stage is at 6 kPa. The turbine efficiencies are 85 percent, and the pump efficiencies are 100 percent. Calculate the thermal efficiency and steam flow required for a total electrical output from the turbine of 20 000 kW.

9-13M A regenerative steam cycle operates with steam inlet to the turbine at 8.0 MPa and 640°C and a condenser temperature of 35°C. One open feedwater heater is to be employed with the steam-extraction pressure selected to provide the maximum increase in cycle efficiency. Select several extraction pressures and examine their effect on thermal efficiency. Make the calculations for (*a*) isentropic expansion and (*b*) turbine stage efficiencies of 87 percent.

9-14M An ideal regenerator is employed with the Brayton cycle of Prob. 9-45M. Calculate the thermal efficiency under these conditions.

9-15M A gas-turbine cycle uses two stages of compression and expansion along with an ideal intercooler for the compression stages. Inlet conditions are 1 atm and 20°C, and the maximum pressure is 1.0 MPa. The temperature at the inlet to each turbine stage is 925°C. Calculate the thermal efficiency for the cycle. Assume isentropic processes in the compression and turbine stages.

9-16M A regenerator having an efficiency of 80 percent is installed on the two-stage cycle of Prob. 9-15M. Calculate the thermal efficiency under these conditions and compare with the efficiency of a Carnot cycle operating between the same temperature limits.

9-17M Rework Prob. 9-16M using compressor and turbine efficiencies of 80 and 85 percent, respectively, for each stage.

9-18M One way of making use of the energy contained in the hot exhaust gases of a gas-turbine cycle is to employ them for heating steam in a Rankine cycle. Consider the gas-turbine cycle of Prob. 9-15M. The hot exhaust gases are used to supply heat to a Rankine cycle operating between 4.0 MPa and 10 kPa. The upper temperature in the steam cycle is 370°C, and isentropic expansion is assumed in the turbine. The exhaust gases leave the steam heater at 260°C. Calculate:

(*a*) The circulation of steam per kgm of air in the gas-turbine cycle

(*b*) The net work output of the combination gas-turbine-steam cycle per kgm of air

(*c*) The thermal efficiency of the combined cycle

9-19M Rework Prob. 9-18M, assuming steam reheat to 370°C at 700 kPa.

9-20M A turbojet engine takes in air at 100 kPa and 20°C while in a stationary position on the ground. The pressure ratio in the compressor is 5 : 1, and the upper temperature limit in the cycle is 980°C. The compressor and turbine efficiencies are 80 and 85 percent, respectively, and the expansion in the exhaust nozzle may be assumed isentropic. Calculate the thrust and heat input per kgm of air, neglecting the inlet velocity.

9-21M Rework Prob. 9-20M for flight at an altitude where the inlet conditions are 20 kPa, −50°C, and 275 m/s.

9-22M A gas turbine cycle employs two compressor stages with intercooler and one turbine stage. The discharge from the turbine is fed to the boiler of a Rankine cycle with one closed feedwater heater. Draw a schematic of the cycle, write all necessary energy balances, and write an expression for the thermal efficiency of the hybrid cycle in terms of the way you label the diagram.

9-23M A gas turbine cycle has a pressure ratio of 5 : 5 and an upper temperature limit of 1900°F. Inlet air is at 85 kPa and 25°C. The turbine efficiency is 90 percent, and the compressor efficiency is 85 percent. Calculate the thermal efficiency of the cycle and the mass flow of air required for a total power output of 12 MW.

9-24M A small gas-turbine cycle is to be designed for automotive applications with the following data:

Maximum temperature:	1000°C
Turbine efficiency:	90 percent
Compressor efficiency:	85 percent
Regenerator efficiency:	70 percent
Inlet air temperature:	50°C, 1 atm
Power output:	225 kW
Pressure ratio:	3.5 : 1

Assuming an air-standard cycle, calculate the thermal efficiency of the cycle and mass flow of air required. If a fuel is used with a heating value of 36 000 kJ/liter, calculate the fuel consumption rate at the maximum power output. Make the calculations assuming constant specific heats and also using the air tables.

9-25M A closed gas-turbine cycle using helium as the working fluid is to be designed to accomplish the same power output as the cycle in Prob. 9-24M. The same conditions may be assumed to apply except that the helium inlet conditions to the compressor are 150°C and 1.0 MPa. Calculate the thermal efficiency of this cycle and the mass flow of helium required assuming constant specific heats. Also calculate the heat which must be dissipated to the surroundings.

9-26M A reciprocating air compressor has a volumetric efficiency of 95 percent and is designed to compress 235 liters/s of air measured at inlet conditions of 1 atm and 20°C. The discharge pressure is 500 kPa, and isentropic compression and expansion may be assumed. Calculate the clearance volume as a percentage of piston displacement and the horsepower input required.

9-27M A reciprocating air compressor receives air at 95 kPa and 38°C and discharges at 300 kPa. The clearance volume is 5 percent of the piston displacement. The compressor has a 10-cm diameter bore, a 12.5-cm stroke, and rotates at 500 rpm. Calculate the volume of air delivered at inlet conditions, the volumetric efficiency, and the power required for (*a*) isentropic compression and (*b*) polytropic compression with $n = 1.30$.

9-28M A two-stage machine is used to compress air from 1 atm, 20°C, to 3.5 MPa. The compressor handles 50 liters/s at inlet conditions, and ideal interstage pressure is used. Each stage is adiabatic but has an efficiency of 70 percent. An ideal intercooler is employed between the stages. Calculate the power required if the machine operates as a steady-flow device. What power would be required if a single-stage machine were used?

9-29M A three-stage compressor is used to compress air from 1 atm, 20°C, to 8.0 MPa. Calculate the work required per kilogram of air compressed, assuming ideal interstaging and intercooling. Assume that the compression processes occur isentropically.

9-30M An air-standard Otto cycle has a compression ratio of 9 : 1. At the start of the compression stroke the conditions are 95 kPa and 30°C. The maximum pressure in the cycle is 3.8 MPa. Determine the maximum temperature in the cycle and the thermal efficiency. Compare this efficiency with that of a Carnot cycle operating between the same temperature limits.

9-31M An air-standard Otto cycle operates such that the compression and expansion processes follow polytropic relations with $n = 1.30$. Calculate the thermal efficiency for a cycle with a compression ratio of 8 : 1, inlet conditions of 95 kPa, 25°C, and a maximum temperature limit of 1100°C.

9-32M A three-stage compressor is used to compress helium from 95 kPa and

25°C to 1.5 MPa. Ideal interstaging is used, and the efficiency of each stage is 83 percent. Calculate the work input for a flow rate of 1 lbm/s.

9-33M A gas turbine used for a jet engine has an overall pressure ratio of 6.3 and inlet conditions of −40°C and 200 m/s. The upper temperature limit is 1300 K and compressor and turbine are isentropic. Calculate the thrust for an air flow rate of 100 kg/s.

9-34M A gas-turbine cycle has a pressure ratio of 6.3. The compressor efficiency is 85 percent, and the turbine efficiency is 93 percent. The regenerator efficiency is 80 percent, and the turbine inlet temperature is 1250°C. Inlet conditions to the compressor are 90 kPa and 10°C. Calculate the thermal efficiency of the cycle.

9-35M A gas-turbine cycle employs two compressor stages with intercooler and one turbine stage. The discharge from the turbine is fed to the boiler of a Rankine cycle with one open feedwater heater. Draw a schematic of the cycle, label it, write all necessary energy balances, and write an expression for the thermal efficiency of the hybrid cycle.

9-36M A gas-turbine cycle uses two stages of compression with ideal intercooling and interstaging. Only one turbine stage is used. Air inlet conditions are 10°C and 1100°C. The overall pressure ratio is 6.2. A regenerator with an efficiency of 85 percent is also employed. Assuming isentropic processes in the compressor and turbine, calculate the thermal efficiency of the cycle and the mass flow of air required to produce a work output of 18 MW.

9-37M An air-standard Otto cycle operates with a compression ratio of 10 : 1 and conditions at the start of the compression stroke of 100 kPa, 35°C. During the combustion process the heat-addition rate is 650 kJ/kg of air. Calculate the temperature and pressure at all points in the cycle and the thermal efficiency, assuming (*a*) constant specific heats and (*b*) variable specific heats based on the air tables. Compare with the efficiency of a Carnot cycle operating between the same temperature limits.

9-38M The compression ratio for an air-standard diesel cycle is 19 : 1, and the heat input per pound-mass of air is 800 kJ/kg. The conditions at the start of compression are 25°C and 100 kPa. Calculate the temperature and pressure at all points in the cycle, the cutoff ratio, and the thermal efficiency. Compare with the efficiency of a Carnot cycle operating between the same temperature limits.

9-39M An air-standard diesel cycle has a compression ratio of 17 : 1 with conditions at the beginning of the compression stroke of 95 kPa and 10°C. The maximum temperature in the cycle is 1900 K. Calculate the temperature and pressure at each point in the cycle, the cutoff ratio, and the thermal efficiency. Compare with the efficiency of a Carnot cycle operating between the same temperature limits.

9-40M Calculate the thermal efficiency of an air-standard Otto cycle having the same upper-pressure limit as the diesel cycle of Prob. 9-38M.

9-41M An automobile manufacturer produces two engines, each having a displacement of 5.7 liters. One engine operates as a diesel with a compression ratio of 22 and a cutoff ratio of 1.5. The other is a spark-ignition engine with a compression ratio of 8.0. The claim is made that the diesel will consume about 20 percent less fuel per kilometer traveled than the spark-ignition engine. Is this a reasonable claim?

9-42M In some locales it is possible to use hot pressurized water from geothermal sources for the production of power. In one application water is available from the ground at 20 MPa and 300°C. This water is throttled to a pressure of 1.5 MPa to produce a wet mixture of steam, and this mixture is subsequently passed through a device to remove the liquid droplets (separator) so that saturated vapor at 1.5 MPa is available to enter a steam turbine. Discharge from the turbine is at 1 atm. Calculate the flow of groundwater needed to produce a 10 000-kW power output from the turbine. Assume isentropic flow in the turbine.

9-43M Calculate the thermal efficiency of an Otto cycle having the same upper temperature limit as the diesel cycle of Prob. 9-39M.

9-44M What conclusions can be drawn from the results of Probs. 9-40M and 9-43M.

9-45M An air-standard Brayton cycle receives air at 95 kPa and 15°C. The pressure ratio is 4 : 1, and the upper temperature limit of the cycle is 980°C. The compressor and turbine efficiencies are 85 percent. Calculate the thermal efficiency of the cycle and the net work output per kilogram of air.

9-46M A certain Freon 12 heat pump is designed to heat 460 liters/s of air at 1 atm from 5°C to 30°C. The evaporator temperature (outside) is 20°F, and the condenser pressure is 140 psia. For isentropic compression calculate the coefficient of performance, work input, and cost per hour for heating with an electric power cost of \$0.07/kWh.

9-47M A Carnot refrigerator has a coefficient of performance of 5.4 when the low temperature is −20°C. The change in entropy for heat addition and rejection is 0.40 kJ/K. Determine the refrigeration effect and work input.

9-48M A certain refrigeration device is designed to produce cooling of 1 ton with a coefficient of performance of 4.0. Determine the power input in kilowatts. Suppose the same device is used for heating (heat pump). What would be its *heating* COP?

9-49M A reversed Carnot cycle is to be employed for heating (heat pump) in which the heating requirements are 6000 kJ/min. The heat added to the cycle is 4500 kJ/min. What power input in kilowatts is required?

9-50M A jet aircraft travels at a speed of 270 m/s and the velocity of the exhaust gases is 600 m/s. A net thrust of 18 kN is produced. Calculate the mass flow of air required.

9-51M Inlet conditions for Prob. 9-50M are 30 kPa, 0°C, and the exhaust conditions are 30 kPa and 670°C. Compute the overall irreversibility for the aircraft assuming $T_0 = 20°C$.

9-52M The work input to a reversed Carnot cycle is 5.0 kW. What is the refrigeration effect if the coefficient of performance is 4.0? What would be the coefficient of performance if the unit were used for heating? What is the upper temperature if the lower temperature is 0°C?

9-53M Air having an initial velocity of 20 m/s is compressed from 1 atm, 25°C, to 300 kPa. The final velocity is 80 m/s. Calculate the inlet volume flow if 1900-kW power is required and the process is reversible.

9-54M The air flow in Prob. 9-53M is compressed from the same inlet conditions to 300 kPa in an irreversible manner, and the discharge temperature is 160°C. Calculate the power input.

9-55M An air-standard gas turbine is designed to produce a power output of 3.7 MW with air inlet at 1 atm, 30°C, and a pressure ratio of 10.0. Inlet to the turbine is at 840°C, and the compressor and turbine efficiencies are both 85 percent. Calculate the air flow required, compressor and turbine works, and cycle thermal efficiency.

9-56M Repeat Prob. 9-55M for a regenerator installed with an efficiency of 90 percent. Compare overall cycle irreversibilities for the two cycles. For the problem, assume that the turbine discharge is cooled at constant pressure to 30°C.

9-57M A closed-cycle gas-turbine plant using helium as the working fluid is to be designed for a marine installation where heat may be rejected to seawater. An overall pressure ratio of 6.0 is used with two stages of compression and ideal intercooling to 20°C. Inlet conditions to the first stage are 20°C and 1.0 MPa, and the upper temperature limit of the cycle is 1100°C. Only a single turbine stage is employed with a regenerator having an efficiency of 80 percent. Both compressor and turbine efficiencies are 90 percent. Calculate the thermal efficiency of the cycle and the mass flow of helium necessary to produce a 18 700-kW output. What is the total amount of heat which must be dissipated to the seawater, including that removed in the intercooler?

9-58M A large diesel engine is to produce a power output of 3700 kW and operates with a compression ratio of 18 : 1. The cutoff ratio is 2.0, and isentropic compression and expansion may be assumed. Assuming an air/fuel ratio of 16 : 1 and a fuel with a heating value of 44 000 kJ/kg, calculate the fuel consumption rate necessary to produce the indicated power output. Make the calculations on the basis of an air-standard cycle.

9-59M An ammonia refrigeration machine is designed to produce 25 tons of cooling at −20°F. Outlet from the compressor is at 220 psia and 360°F, and the throttling (expansion) valve receives the liquid at 100°F. Calculate the compressor efficiency, flow rate of ammonia, and coefficient of performance.

9-60M A gas turbine employs one compression and two turbine stages. The high-pressure turbine produces just enough work to drive the compressor, while the low-pressure turbine produces the net cycle output. Inlet air is at 1 atm, 25°C, and the overall pressure ratio is 6.5. Inlet temperature to each turbine is 800°C. Isentropic compression and expansion may be assumed, and a regenerator with an efficiency of 90 percent is employed. If a fuel with heating value of 40 MJ/kg is used, calculate the thermal efficiency and fuel flow for a power output of 10 MW.

9-61M A small gas turbine is designed to operate on a closed cycle using argon as the working fluid. The compressor and turbine efficiencies are 88 percent and 82 percent, respectively. The gas enters the compressor at 400 kPa, 40°C, and the compression ratio is 10.0. The temperature at entry to the turbine is 1100°C. Calculate the argon flow rate for a total power output of 6.0 MW. Also calculate the irreversibility for the cycle with $T_0 = 20°C$.

9-62M A closed-cycle gas turbine uses argon as the working fluid with the following conditions:

Entry to compressor:	60°C, 475 kPa
Exit from compressor:	375°C, 1.9 MPa
Entry to turbine:	1100°C
Exit from turbine:	620°C

Calculate the work output per unit mass flow and the compressor and turbine efficiencies. Also calculate the irreversibility for the cycle with $T_0 = 20°C$.

9-63M A regenerative steam cycle operates between 10 MPa, 640°C, and 10 kPa. Two closed feedwater heaters are used with extraction at 500 kPa and 1.5 MPa. The condensate is trapped back to the condenser hot well. Calculate the thermal efficiency of the cycle, assuming (*a*) isentropic expansion in the turbine and (*b*) a turbine efficiency of 85 percent for each of the three stages.

9-64M A combined reheat-regenerative cycle has the following operating conditions:

Steam inlet to first turbine stage:	8 MPa, 560°C
Condenser pressure:	10 kPa
Extraction pressure:	700 kPa

After a portion of the steam is extracted for use in an open feedwater heater, the remainder is reheated to 560°C and then expanded in the second turbine stage. The turbine efficiency for both stages is 85 percent, and the high-pressure pump efficiency is 60 percent. Calculate the thermal efficiency of the cycle.

9-65M A nuclear power plant has an upper temperature limit of 450°C because of metallurgical considerations in the reactor vessel. In one particular installation the following design information is supplied:

Upper pressure limit:	10.0 MPa
Feedwater heaters:	Two, closed, condensate from first trapped to second, condensate from second pumped to boiler
Extraction pressures:	3.5 MPa and 1.0 MPa
Reheat:	Reheat to 450°C after each extraction
Condenser temperature:	35°C
Turbine efficiencies:	90 percent
Power output:	500 000 kW

Calculate the mass flow of steam required in the boiler, the thermal efficiency of the cycle, and the waste heat which must be dissipated in the condenser.

9-66M Measurements are performed on an actual regenerative-reheat steam power cycle. One closed feedwater heater is employed with condensate pumped to the boiler. The following data are obtained:

Rated power output:	50 000 kW
Steam entrance to turbine:	p = 9.0 MPa, T = 490°C
Extraction properties:	p = 900 kPa, T = 210°C
Fraction extracted:	24 percent
Steam entrance to second turbine stage after reheat:	p = 850 kPa, T = 490°C
Condenser temperature:	30°C
Discharge temperature from feedwater heater:	175°C
Turbine efficiency for second stage:	78 percent

For these data calculate the steam flow required and the thermal efficiency of the cycle.

9-67M In modern steam power plants it is common practice to use supercritical pressures in the boiler. Consider an ideal cycle with steam leaving the boiler at 40 MPa and 800°C. Five closed feedwater heaters are employed with extractions at 8.0 MPa, 5.0 MPa, 1.5 MPa, 800 kPa, and 100 kPa. Discharge from the turbine is at 8 kPa. Assume an isentropic turbine and the condensate from each heater trapped to the next heater, and that the condensate from the last heater is pumped to the boiler. Calculate the thermal efficiency of the cycle. What power output may be obtained with a flow rate of 500 000 kg/h?

9-68M An ideal reheat cycle employs two stages of reheat. Initially, steam enters the turbine at 20.0 MPa and 540°C and is reheated back to 540°C at pressures of 4.0 and 1.0 MPa. Discharge from the turbine is at 20 kPa. Calculate the thermal efficiency of the cycle and the power output for a steam flow of 10 kg/s.

9-69M Proposals have been made to use a Rankine cycle for automotive applications with Freon 12 as the working fluid. Suppose a power output of 38 kW is desired and that the Freon 12 enters an isentropic turbine at 500 psia and discharges as a saturated vapor at 100°F. Calculate the thermal efficiency of the cycle and the flow rate of Freon 12 required for the rated output.

9-70M One proposal for power production would obtain methane gas from off-shore geothermal sources to fuel a hybrid gas-turbine steam power cycle. The plant is to be installed on an offshore platform so that heat rejection can be to the ocean. The following conditions are assumed to apply:

Inlet to compressor:	1 atm, 20°C
Compression ratio:	10 : 1
Temperature at inlet to gas turbine:	1400 K
Inlet to steam turbine:	6.0 MPa, 320°C
Condensing temperature for steam:	15°C
Compressor efficiency:	87 percent
Gas-turbine efficiency:	90 percent
Steam-turbine efficiency:	92 percent
Gas-exit temperature from boiler and water preheater:	100°C

Calculate the thermal efficiency for the hybrid cycle, the flow rate of air and steam required for a power output of 500 MW, and the heat rejected to the ocean for this power level. If the fuel supplies 38 kJ/liter at 1 atm, 20°C, calculate the flow rate required at 10 atm, 50°C, to produce the required power output.

9-71M Rework Prob. 9-70M assuming two stages of compression and expansion with ideal interstaging and intercooling. Assume the temperature at inlet to each turbine stage is 1400 K.

9-72M Steam enters the turbine of a Rankine cycle at a pressure of 3.0 MPa and discharges at a pressure of 30 kPa. Calculate the thermal efficiency of the cycle for three inlet temperatures: (*a*) saturated vapor, (*b*) 320°C, and (*c*) 700°C.

9-73M A steam power plant used two open feedwater heaters with reheat after each extraction. Draw a schematic diagram for the cycle, label it appropriately, and write an expression for the thermal efficiency of the cycle. Also write all necessary energy balances.

REFERENCES

1 Gaffert, G. A.: "Steam Power Stations," 4th ed., McGraw-Hill Book Company, New York, 1952.

2 Hill, P. G., and C. R. Peterson: "Mechanics and Thermodynamics of Propulsion," Addison-Wesley Publishing Company, Inc., Reading, Mass., 1965.

3 Jones, J. B., and G. A. Hawkins: "Engineering Thermodynamics," 2d ed., John Wiley & Sons, Inc., New York, 1985.

4 Obert, E. F., and R. A. Gaggioli: "Thermodynamics," McGraw-Hill Book Company, New York, 1963.

5 Doolittle, J. S.: "Thermodynamics for Engineers," 2d ed., International Textbook Company, Scranton, Pa., 1964.

6 Stoecker, W. F.: "Refrigeration and Air Conditioning," McGraw-Hill Book Company, New York, 1958.

7 Barron, Randall: "Cryogenic Systems," McGraw-Hill Book Company, New York, 1966.

8 Culp, A. W.: "Principles of Energy Conversion," McGraw-Hill Book Company, New York, 1979.

9 Fraas, A. P.: Problems in Coupling a Gas Turbine to a Thermonuclear Reactor, *ASME,* 72-GT-98, May 1972.

10 Cole, D. E.: The Wankel Engine, *Scientific American,* vol. 227, pp. 14–23, August 1972.

11 Wood, B. D.: "Applications of Thermodynamics," Addison-Wesley Publishing Company, Inc., Reading, Mass., 1969.

10

THERMO-DYNAMICS OF COM-PRESSIBLE FLOW

10-1 INTRODUCTION

We have already applied thermodynamic principles to a number of flow processes and analyzed the behavior of certain open systems with the control-volume concept and appropriate mass and energy balances. In this chapter we wish to extend the analysis to situations which involve the very-high-speed flow of a compressible fluid—in most cases, air. In such systems the concepts of the Mach number and shock waves become important. Our discussion will be confined to one-dimensional compressible flows, i.e., those flows which involve only a single velocity direction.

10-2 THE STAGNATION STATE

When a fluid is brought to rest from some velocity we say that it has attained a *stagnation state*. The manner in which the slowing-down process is accomplished obviously influences the final stagnation state which will be experienced. The reversible adiabatic, or isentropic, stagnation state will be of great interest later. For such a stagnation process the steady-flow energy equation is

$$h + \frac{V^2}{2g_c} = h_0 \tag{10-1}$$

where h_0 designates the enthalpy of the fluid after it is brought to rest and h is the fluid enthalpy when it has the velocity V. If the fluid is brought to rest adiabatically but *irreversibly*, the stagnation enthalpy would still be the same because Eq. (10-1) is an energy balance and does not depend on the reversibility of the process. However, the final stagnation *state* would not be the same because an increase in entropy would occur in the irreversible process. The two processes are illustrated in Fig. 10-1 on an h-s diagram. The main difference between the two is that the isentropic stagnation pressure p_0 is higher than p', the value which might be attained in some actual stagnation process. Although both stagnation states involve zero velocity, we shall reserve the zero subscript to designate the isentropic stagnation state.

The pressure in the flow which would be measured by an observer moving along with the fluid is usually termed the *static pressure* and is the pressure which determines the thermodynamic state of the moving fluid. The choice of the word *static* is somewhat unfortunate because it implies a pressure of the fluid after it is brought to rest, but, as we have already discussed, it is the *stagnation* properties that are measured at zero velocity. To keep the two types of pressure straight the reader is urged to recall the first sentence of this paragraph. Anytime we use the term *static* applied to a thermodynamic property it will mean a value measured by an observer moving at the local flow velocity. Extended discussions of this matter may be found in the references at the end of the chapter.

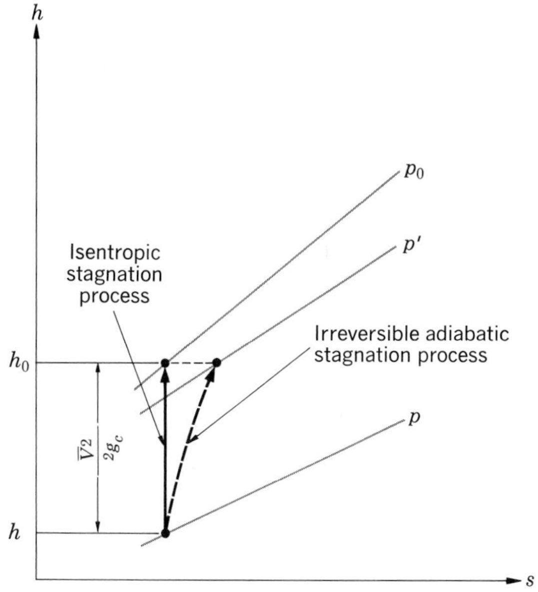

FIG. 10-1 Stagnation states.

10-3 SONIC VELOCITY AND THE MACH NUMBER

The velocity at which a small pressure disturbance will propagate in a fluid is called the velocity of sound and designated by the symbol c. We shall now derive an expression for calculating this velocity for an ideal gas in terms of the thermodynamic state of the gas. Consider the tube shown in Fig. 10-2. In (*a*) we are viewing the wave front as a stationary observer, watching the infinitesimal wave move by at a velocity c. Across the front there are small changes in the thermodynamic properties, as illustrated. In Fig. 10-2*b*, an alternate viewpoint is taken of moving along with the wave front and observing the flow in and out of the control surface enclosing the front. We assume that the flow through the front is adiabatic so a small change in enthalpy dh will cause a change in velocity dV because of energy conservation.

Now let us make a mass and energy balance on the control volume surrounding the wave front in Fig. 10-2*b*. From the mass-continuity relation of Chap. 4 we have

$$\rho A c = (\rho + d\rho)A(c - dV) \qquad (10\text{-}2)$$

or

$$c\, d\rho - \rho\, dV = 0 \qquad (10\text{-}3)$$

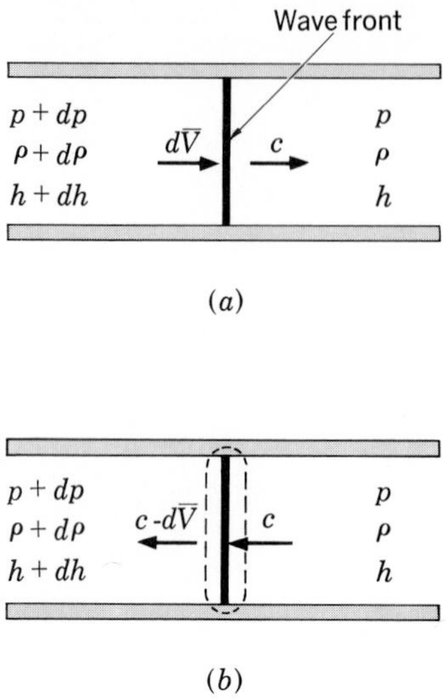

FIG. 10-2 Sonic wave: (*a*) stationary observer; (*b*) observer moving with wave front.

where A is the flow cross-sectional area. We know that

$$T\,ds = dh - v\,dp = dh - \frac{dp}{\rho} \tag{10-4}$$

If the process is reversible as well as adiabatic, Eq. (10-4) becomes

$$0 = dh - \frac{dp}{\rho} \tag{10-5}$$

The energy balance on the wave front is

$$h + \frac{c^2}{2g_c} = (h + dh) + \frac{(c - dV)^2}{2g_c}$$

or, clearing terms,

$$dh - \frac{c\,dV}{g_c} = 0 \tag{10-6}$$

Combining Eqs. (10-5) and (10-6) gives

$$\frac{dp}{\rho} - \frac{c\,dV}{g_c} = 0 \tag{10-7}$$

Now, combining Eqs. (10-3) and (10-7) gives

$$\frac{c^2}{g_c} = \frac{dp}{d\rho}$$

Because we have assumed an isentropic process, this should be written

$$\frac{c^2}{g_c} = \left(\frac{\partial p}{\partial \rho}\right)_s \tag{10-8}$$

Equation (10-8) now provides a mechanism for calculating the speed of sound in a fluid medium. For an ideal gas further simplification is possible. We have shown that the isentropic process for an ideal gas can be represented by

$$pv^\gamma = \text{const.} \qquad \text{or} \qquad p\left(\frac{1}{\rho}\right)^\gamma = \text{const.} \tag{10-9}$$

We can therefore perform a differentiation of Eq. (10-9) to obtain

$$\frac{dp}{d\rho} = \left(\frac{\partial p}{\partial \rho}\right)_s = \frac{\gamma p}{\rho} \tag{10-10}$$

Now, combining Eqs. (10-8) and (10-10) we obtain for an ideal gas

$$c^2 = \frac{\gamma p g_c}{\rho}$$

But $p/\rho = RT$ for the ideal gas so that

$$c^2 = \gamma g_c RT \tag{10-11}$$

$$c = \sqrt{\gamma g_c RT} \tag{10-12}$$

Thus the speed of sound in an ideal gas is a function of temperature alone. Our discussion in Sec. 1-16 identified temperature with mean molecular kinetic energy, which means that $v_{\text{rms}} \sim T^{1/2}$. It is all the more interesting to note that the velocity of sound is also proportional to $T^{1/2}$. It can be shown that various transport processes are related to the mean molecular speed available to convey energy, momentum, and mass. It is not surprising, therefore, to find that the rate at which we can propagate a small pressure disturbance, i.e., the velocity of sound, follows the same temperature dependence. From a molecular viewpoint, of course, it is molecular collisions that must propagate the pressure disturbance, and the collision rates are proportional to mean molecular speed.

The Mach number M is a dimensionless ratio which we shall find useful in later analytical developments and is defined as

$$M = \frac{V}{c} \tag{10-13}$$

where V is the flow velocity. Different flow regions are designated in terms of values of the Mach number as supersonic for $M > 1$, subsonic for $M < 1$, and sonic flow for $M = 1$.

EXAMPLE 10-1 Isentropic Stagnation
What temperature and pressure will air attain if it is stagnated isentropically from $M = 4$, $T = -60°\text{F}$, and $p = 0.1$ atm?

SOLUTION The speed of sound at the given conditions is calculated from Eq. (10-12):

$$\begin{aligned} c &= (\gamma g_c R T)^{1/2} = [(1.4)(32.2)(53.35)(460 - 60)]^{1/2} \\ &= 980.0 \text{ ft/s} = 299 \text{ m/s} \end{aligned}$$

For $M = 4$ the corresponding flow velocity is

$$V = (4)(980.0) = 3923 \text{ ft/s} = 1196 \text{ m/s}$$

Using the energy balance for the stagnation process

$$h + \frac{V^2}{2g_c} = h_0$$

or

$$c_p(T_0 - T) = \frac{V^2}{2g_c}$$

$$\begin{aligned} T_0 &= -60 + \frac{(3923)^2}{(2)(0.24)(32.2)(778)} \\ &= -60 + 1279 = 1219°\text{F} = 660°\text{C} \end{aligned}$$

The final stagnation pressure is obtained from the isentropic relation:

$$\frac{p_0}{p} = \left(\frac{T_0}{T}\right)^{\gamma/(\gamma-1)}$$

$$p_0 = 0.1 \left(\frac{1679}{400}\right)^{3.5} = 15.15 \text{ atm} = 2227 \text{ psia}$$

EXAMPLE 10-2 Speed of Sound
Calculate the speed of sound in nitrogen at 200°C.

SOLUTION We shall use SI units for this calculation. The gas constant is

$$R = \frac{\mathcal{R}}{M} = \frac{8315}{28} = 297 \text{ J/kg·K}$$

From Table 2-2 $\gamma = 1.40$ for nitrogen, so

$$\begin{aligned} c &= (\gamma g_c RT)^{1/2} \\ &= [(1.4)(1.0\ \text{kg}\cdot\text{m/N}\cdot\text{s}^2)(297\ \text{J/kg}\cdot\text{K})(200 + 273)]^{1/2} \\ &= 443.5\ \text{m/s} = 1455\ \text{ft/s} \end{aligned}$$

10-4 STEADY, ONE-DIMENSIONAL ISENTROPIC FLOW

Let us now derive some useful relations for a general one-dimensional isentropic flow in steady state. Such a flow system may be approximated in a number of practical cases, most often when an ideal gas flows in a smooth channel, nozzle, or diffuser. At high velocities the sonic velocity and Mach number become important parameters which determine how the flow will react to changes in pressure, flow cross-sectional area, etc. To analyze the problem we have four general notions at our disposal: the energy relation for the flow, the mass-continuity equation, an equation representing the thermodynamic equation of state, and finally a process representation which, in this case, is obtained from our previous development for the sonic velocity.

For isentropic steady flow the energy equation is

$$h + \frac{V^2}{2g_c} = \text{const.}$$

or in differential form,

$$dh + \frac{V\,dV}{g_c} = 0 \tag{10-14}$$

The mass-continuity relation is

$$\rho A V = \dot{m} = \text{const.}$$

or again, in differential form,

$$AV\,d\rho + \rho A\,dV + \rho V\,dA = 0$$

or

$$\frac{d\rho}{\rho} + \frac{dV}{V} + \frac{dA}{A} = 0 \tag{10-15}$$

Our thermodynamic property relation for the isentropic case is

$$T\,ds = dh - v\,dp = dh - \frac{dp}{\rho} = 0 \tag{10-16}$$

If we combine the energy and property relations [Eqs. (10-14) and (10-16)], we obtain

$$dh = \frac{dp}{\rho} = \frac{-V\,dV}{g_c} \tag{10-17}$$

or

$$\frac{V\,dV}{g_c} + \frac{1}{\rho}\,dp = 0 \tag{10-18}$$

Equation (10-18) is the differential form of *Bernoulli's equation* and indicates to us that as the velocity increases ($dV > 0$) the pressure must decrease ($dp < 0$). In a physical sense this means that the pressure force serves to accelerate the flow.

Because the flow is isentropic we can obtain from our equation for the speed of sound,

$$\left(\frac{\partial p}{\partial \rho}\right)_s = \frac{dp}{d\rho} = \frac{c^2}{g_c}$$

or

$$dp = \frac{c^2}{g_c}\,d\rho \tag{10-19}$$

We may now combine the differential continuity relation, Eq. (10-15), the property relation, Eq. (10-16), and Eq. (10-19) to obtain

$$\frac{dA}{A} = \frac{dp}{\rho V^2/g_c}\left(1 - \frac{V^2}{c^2}\right) = \frac{dp}{\rho V^2/g_c}(1 - M^2) \tag{10-20}$$

Equation (10-20) provides us with the basic information to determine the shape which a channel must assume for subsonic or supersonic isentropic flow. These shapes are illustrated in Fig. 10-3. Consider a subsonic flow case where we wish to *increase* the velocity (a nozzle). From Eq. (10-18) $dp < 0$ for $dV > 0$. For $M < 1$ and $dV > 0$, the right side of Eq. (10-20) is therefore negative and dA must be negative, so we must decrease the flow area to cause an increase in velocity, as illustrated in Fig. 10-3*c*. For $M > 1$ just the opposite is the case because the right side of Eq. (10-20) becomes positive for $dV > 0$, and we must therefore *increase* the flow area to increase the velocity in supersonic flow, as illustrated in Fig. 10-3*b*. To effect a decrease in velocity (a diffuser) requires the opposite changes in flow area, as shown in Fig. 10-3*a* and *d* for $M < 1$ and $M > 1$, respectively.

When $M = 1$ the flow area attains its minimum value and we say that a *throat* condition is attained. A nozzle which is to accelerate the fluid from

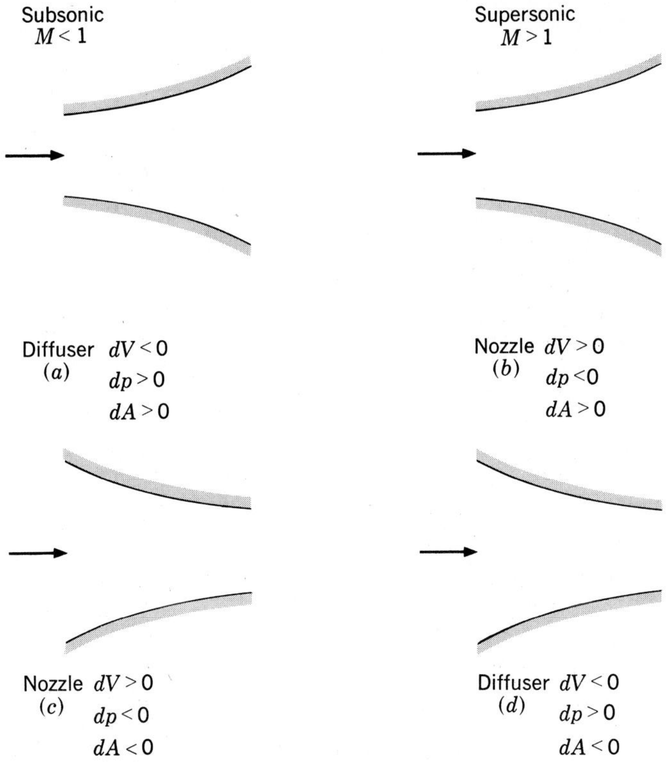

FIG. 10-3 Subsonic and supersonic nozzles and diffusers.

subsonic to supersonic conditions therefore must have first a converging and then a diverging section, as shown in Fig. 10-4. Such a flow channel is appropriately called a *converging-diverging nozzle,* and $M = 1$ is experienced at the minimum flow area, or throat.

We can now express the temperature variation in the above isentropic flow situations through the use of the energy equation

$$h_0 - h = c_p(T_0 - T) = \frac{V^2}{2g_c} \tag{10-21}$$

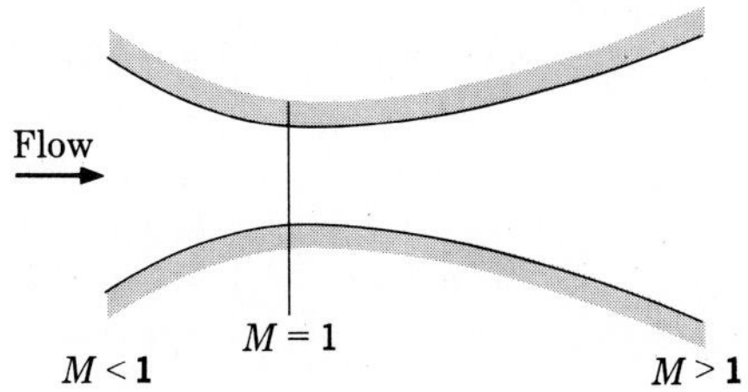

FIG. 10-4 Converging-diverging supersonic nozzle.

Noting that

$$c_p = \frac{\gamma R}{\gamma - 1} \qquad \text{and} \qquad c^2 = \gamma g_c RT$$

Eq. (10-21) can be written

$$\frac{V^2}{c^2} = M^2 = 2g_c \frac{\gamma RT}{(\gamma - 1)(\gamma g_c RT)} \left(\frac{T_0}{T} - 1\right)$$

Clearing terms,

$$\frac{T_0}{T} = 1 + \frac{\gamma - 1}{2} M^2 \tag{10-22}$$

For isentropic flow the pressure and density are obtained from

$$\frac{p_0}{p} = \left(\frac{T_0}{T}\right)^{\gamma/(\gamma-1)} = \left(1 + \frac{\gamma - 1}{2} M^2\right)^{\gamma/(\gamma-1)} \tag{10-23}$$

$$\frac{\rho_0}{\rho} = \left(\frac{T_0}{T}\right)^{1/(\gamma-1)} = \left(1 + \frac{\gamma - 1}{2} M^2\right)^{1/(\gamma-1)} \tag{10-24}$$

The flow properties at the throat are designated with a superscript asterisk (*) and correspond to $M = 1$. Therefore

$$\frac{T^*}{T_0} = \frac{2}{\gamma + 1} \tag{10-25}$$

$$\frac{p^*}{p_0} = \left(\frac{2}{\gamma + 1}\right)^{\gamma/(\gamma-1)} \tag{10-26}$$

$$\frac{\rho^*}{\rho_0} = \left(\frac{2}{\gamma + 1}\right)^{1/(\gamma-1)} \tag{10-27}$$

It is common practice to refer to the throat properties as *critical* properties (distinguished from the critical properties of Chap. 6) and the above ratios are *critical ratios*. Their values for several values of γ are indicated in Table 10-1. Values of the various temperature, pressure, and density ratios are tabulated in Table A-19 of the Appendix.

For steady flow the mass flow rate is constant and may be calculated from

$$\begin{aligned} \dot{m} &= \rho AV = \frac{p}{RT} VA \\ &= \frac{p}{RT} M\sqrt{\gamma g_c RT}\, A \end{aligned} \tag{10-28}$$

TABLE 10-1
Critical Pressure, Temperature, and Density Ratio for Isentropic Flow of an Ideal Gas

	$\gamma = 1.2$	$\gamma = 1.3$	$\gamma = 1.4$	$\gamma = 1.67$
$\frac{p^*}{p_0}$	0.5644	0.5457	0.5283	0.4867
$\frac{T^*}{T_0}$	0.9091	0.8696	0.8333	0.7491
$\frac{\rho^*}{\rho_0}$	0.6209	0.6276	0.6340	0.6497

Using the relation for T_0/T [Eq. (10-22)], this becomes

$$\frac{\dot{m}}{A} = \frac{pM}{\sqrt{T_0}}\sqrt{\frac{\gamma g_c}{R}}\left(1 + \frac{\gamma - 1}{2}M^2\right)^{1/2} \tag{10-29}$$

Substituting for p in terms of p_0 from Eq. (10-23) gives

$$\frac{\dot{m}}{A} = \frac{p_0 M}{\sqrt{T_0}}\sqrt{\frac{\gamma g_c}{R}}\frac{1}{\left(1 + \frac{\gamma - 1}{2}M^2\right)^{(\gamma+1)/2(\gamma-1)}} \tag{10-30}$$

At the throat $M = 1$ and $A = A^*$ so that

$$\frac{\dot{m}}{A} = p_0\sqrt{\frac{\gamma g_c}{RT_0}}\left(\frac{2}{\gamma + 1}\right)^{(\gamma+1)/2(\gamma-1)} \tag{10-31}$$

This yields the interesting result that the total mass flow in the nozzle is a function of only the stagnation properties and the throat area, which leads us to the concept of *choked flows*.

For a given set of stagnation properties there is a maximum flow which can be forced through the nozzle as governed by the throat area. No matter what one does downstream from the throat in the way of decreasing the pressure, or increasing the flow area, the flow rate will remain the same if sonic conditions are established at the throat. If the flow has not been accelerated enough to establish $M = 1$ at the minimum flow area, then choked flow conditions are never established and the above reasoning does not apply.

The overall area variation through the nozzle may be determined by combining Eqs. (10-30) and (10-31) to give

$$\frac{A}{A^*} = \frac{1}{M}\left[\left(\frac{2}{\gamma + 1}\right)\left(1 + \frac{\gamma - 1}{2}M^2\right)\right]^{(\gamma+1)/2(\gamma-1)} \tag{10-32}$$

Values of A/A^* are also given in Table A-19 of Appendix A.

EXAMPLE 10-3 Converging Nozzle for Flow Measurement
A sonic nozzle (converging section only) is sometimes used as a flow-measurement device because for $M = 1$ at the exit the flow rate is dependent only on the upstream stagnation pressure and temperature. What is the maximum air flow rate for such a device with exit area of 0.5 cm in diameter and stagnation properties of 700 kPa and 500 K?

SOLUTION For this calculation we can make use of Eq. (10-31) to calculate the flow rate with

$$A^* = \frac{\pi(0.005)^2}{4} = 1.963 \times 10^{-5}\ \text{m}^2$$

$$p_0 = 700\ \text{kPa} \qquad T_0 = 500\ \text{K}$$

Then, substituting in Eq. (10-31),

$$\dot{m} = (1.963 \times 10^{-5})(700 \times 10^3)\left[\frac{(1.4)(1.0)}{(287)(500)}\right]^{1/2} \times \left(\frac{2}{1.4 + 1}\right)^{(1.4+1)/(2)(1.4-1)}$$

$$= 0.0248\ \text{kg/s}$$

EXAMPLE 10-4 Converging-Diverging Nozzle
A small converging-diverging nozzle having an exit area of 4.0 cm² is to be designed to produce an exit flow of air at atmospheric pressure and $M = 2.5$. The upstream stagnation temperature is 125°C. Calculate (*a*) the stagnation pressure, (*b*) the exit temperature, (*c*) the throat area, and (*d*) the mass flow rate.

SOLUTION For this calculation we could make use of the appropriate equations which have been derived, but it is much simpler to employ the tables in Appendix A for $\gamma = 1.4$. Designating the exit with the subscript e we have the given data:

$$A_e = 4.0\ \text{cm}^2 \qquad M_e = 2.5 \qquad T_0 = 125°\text{C} = 398\ \text{K} \qquad p_e = 1\ \text{atm}$$

Entering Table A-19 at $M = 2.5$ we find

$$\frac{A_e}{A^*} = 2.6367 \qquad \frac{p_e}{p_0} = 0.05853 \qquad \frac{T_e}{T_0} = 0.44444$$

The stagnation pressure is therefore

$$p_0 = \frac{1}{0.058\ 53} = 17.09\ \text{atm} = 1.73\ \text{MPa}$$

and the exit temperature is

$$T_e = (398)(0.444\,44) = 176 \text{ K} = -96°\text{C} \qquad (-141°\text{F})$$

The throat area is

$$A^* = \frac{4.0}{2.6367} = 1.517 \text{ cm}^2$$

There are several formulas available to us for calculating the mass flow rate. We shall use Eq. (10-29) applied at exit conditions. Because we are working in SI units we recall that $g_c = 1.0$ kg·m/N·s².

For air

$$R = \frac{\mathcal{R}}{28.97} = \frac{8315}{28.97} = 287 \text{ J/kg K}$$

Substituting in Eq. (10-29),

$$\begin{aligned} m &= \frac{(4.0 \times 10^{-4}\, m^2)(1.0132 \times 10^5 \text{ N/m}^2)(2.5)}{(398)^{1/2}} \\ &\quad \times \left[\frac{(1.4)(1.0 \text{ kg·m/N·s}^2)}{287 \text{ J/kg·K}}\right]^{1/2}\left[1 + \frac{(1.4 - 1)(2.5)^2}{2}\right]^{1/2} \\ &= 0.532 \text{ kg/s} \qquad (1.171 \text{ lbm/s}) \end{aligned}$$

We would obtain the same answer if we had applied Eq. (10-31) to the throat conditions.

EXAMPLE 10-5

For the nozzle of Example 10-4 what would be the flow area, temperature, pressure, and velocity at the section where $M = 1.5$?

SOLUTION This calculation is very simple because we have already determined the throat area and stagnation properties. We simply enter Table A-19 at $M = 1.5$ and read

$$\frac{A}{A^*} = 1.1762 \qquad \frac{p}{p_0} = 0.2724 \qquad \frac{T}{T_0} = 0.689\,65$$

so that

$$A = (1.1762)(1.517) = 1.784 \text{ cm}^2$$

$$p = (0.2764)(17.09) = 4.724 \text{ atm}$$

$$T = (0.689\,65)(398) = 274 \text{ K} = 1°\text{C}$$

To calculate the flow velocity we first compute the speed of sound at this location using Eq. (10-12):

$$c = (\gamma g_c RT)^{1/2} = [(1.4)(1.0)(287)(274)]^{1/2} = 331.8 \text{ m/s}$$

For $M = 1.5$ the flow velocity is therefore

$$V = (1.5)(331.8) = 498 \text{ m/s} = 1517 \text{ ft/s}$$

10-5 MOMENTUM EQUATION FOR A CONTROL VOLUME

Because most of our discussion of flow processes will be concerned with control volumes (open systems), it will be useful to express Newton's momentum principle in a form which is most easily adapted to the control-volume analysis technique. Momentum is a vector quantity and a properly general analysis must be done in three dimensions; however, because we are concerned with one-dimensional flow systems in this chapter we can simplify our procedure somewhat by just analyzing the x component of momentum.

Newton's second law of motion states that the force acting on a body is proportional to the rate of change of momentum. For the x direction

$$F_x \sim \frac{d(mV_x)}{d\tau}$$

or, with the proportionality constant,

$$F_x = \frac{1}{g_c}\frac{d(mV_x)}{d\tau} \tag{10-33}$$

Equation (10-33) applies to a fixed mass. Now consider the rather general control volume shown in Fig. 10-5. In Fig. 10-5*a* the volume is indicated at time τ with fluid moving in and out of the control surface defining the control-volume region. Inside the control volume is a certain mass of fluid having a certain velocity and corresponding momentum. Now note the situation in Fig. 10-5*b* at time $\tau + \Delta\tau$. The control surface is indicated with the dashed line and is in the same spatial location as before, but the mass, which was totally enclosed by the control surface, has now moved out of the control volume somewhat. We shall

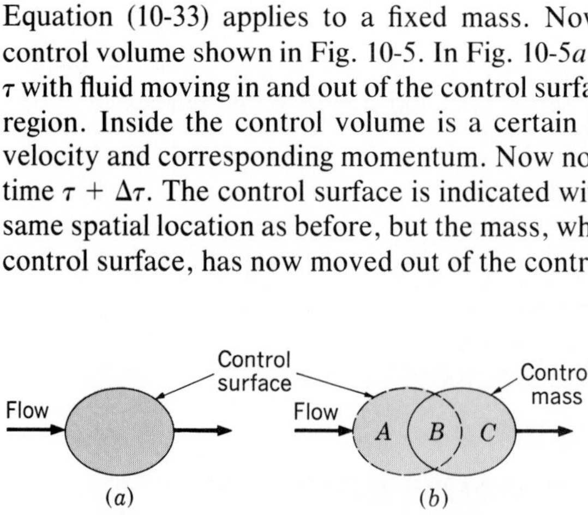

FIG. 10-5 Control volume for momentum analysis: (*a*) at time τ; (*b*) at time $\tau + \alpha\tau$.

refer to this quantity as the *control mass*. Now let us apply Eq. (10-33) to this control mass, using only forces and momenta in the x direction. Expressing the derivative in limit form,

$$\sum F_x = \lim_{\Delta\tau \to 0} \frac{1}{g_c} \frac{(mV_x)_{\tau+\Delta\tau} - (mV_x)_\tau}{\Delta\tau} \tag{10-34}$$

At the time $\tau + \Delta\tau$ the momentum of the control mass can be expressed in terms of the momenta in regions A, B, and C as

$$(mV_x)_{\tau+\Delta\tau} = (mV_x)_B + (mV_x)_C \tag{10-35}$$

Designating the control volume with the subscript σ

$$(mV_x)_{A+B} = (mV_x)_\sigma = (mV_x)_A + (mV_x)_B \tag{10-36}$$

Solving for $(mV_x)_B$ from Eq. (10-36) at time $\tau + \Delta\tau$ and substituting in Eq. (10-35) gives

$$(mV_x)_{\tau+\Delta\tau} = (mV_x)_c - (mV_x)_A + (mV_x)_{\sigma,\tau+\Delta\tau} \tag{10-37}$$

At time τ the control mass is contained entirely in the control volume so that

$$(mV_x)_\tau = (mV_x)_{\sigma,\tau}$$

Substituting the various expressions back into Eq. (10-34) gives

$$\sum F_x = \frac{1}{g_c} \lim_{\Delta\tau \to 0} \left[\frac{(mV_x)_c}{\Delta\tau} - \frac{(mV_x)_A}{\Delta\tau} + \frac{(mV_x)_{\sigma,\tau+\Delta\tau} - (mV_x)_{\sigma,\tau}}{\Delta\tau} \right]$$

As $\Delta\tau \to 0$, m_A and m_C approach δm_A and δm_C, and the ratios $\delta m_A/\delta\tau$ and $\delta m_C/\Delta\tau$ become the mass flow rates entering and leaving the control volume. The third term in the limit becomes the rate of accumulation of momentum in the control volume, and the overall expression becomes

$$\sum F_x = \frac{1}{g_c} \left[(\dot{m}V_x)_e - (\dot{m}V_x)_i + \frac{d(mV_x)_\sigma}{d\tau} \right] \tag{10-38}$$

where now the subscripts i and e have been used to designate inlet and exit streams, respectively. The term $\dot{m}V$ is called the *momentum flux*, and an equation like (10-38) could be written for both the y and z directions. Equation (10-38) may be expressed in words as

Sum of forces acting on control volume
= momentum flux out − momentum flux in + rate of accumulation of momentum in control volume

In steady state there will be no accumulation within the control volume. For steady state and only one stream entering and leaving the control volume we obtain for the three force components

$$\sum F_x = \frac{\dot{m}}{g_c}(V_{e_x} - V_{i_x})$$

$$\sum F_y = \frac{\dot{m}}{g_c}(V_{e_y} - V_{i_y}) \tag{10-39}$$

$$\sum F_z = \frac{\dot{m}}{g_c}(V_{e_z} - V_{i_z})$$

EXAMPLE 10-6 Force and Momentum Flux

A large tank is mounted on rollers, as shown, and a water jet discharges water into the tank at the rate of 8 kg/s with a velocity of 30 m/s. Calculate the force F_x necessary (*a*) to hold the tank stationary and (*b*) to allow it to move to the left with a velocity of 1 m/s.

SOLUTION This is an unsteady-state problem because mass, and therefore momentum, is accumulating in the tank with time. We select the control surface shown in the dashed line so that it always encloses the tank. The *x*-direction momentum equation is

$$F_x = \frac{1}{g_c}\left[(\dot{m}V_x)_e - (\dot{m}V_x)_i + \frac{d(mV_x)_\sigma}{d\tau}\right] \tag{a}$$

We have only an inlet stream so $(mV_x)_e = 0$, and choosing to the right as positive gives

$$(\dot{m}V_x)_i = (8 \text{ kg/s})(-30 \cos 30° \text{ m/s})$$
$$= -207.8 \text{ kg·m/s}^2$$

FIG. EXAMPLE 10-6

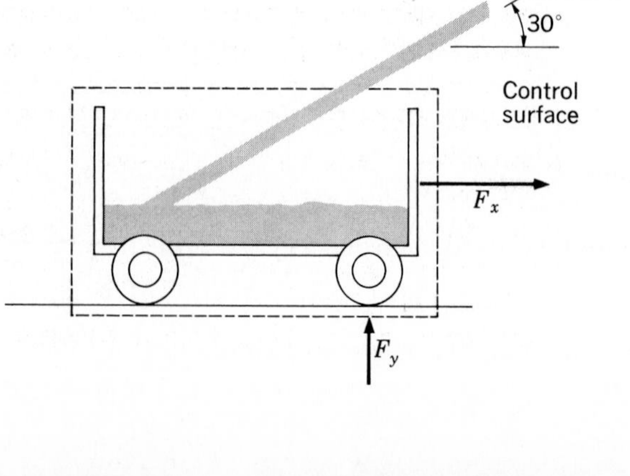

For part (*a*) of the problem the tank is at rest and $V_{x,\sigma} = 0$, so there is no accumulation of momentum even though there is an accumulation of mass. In this case then, Eq. (*a*) becomes

$$F_x = \frac{1}{1.0\ \text{kg}\cdot\text{m/N}\cdot\text{s}^2}[0 - (-207.8\ \text{kg}\cdot\text{m/s}^2) + 0]$$
$$= +207.8\ \text{N}$$

The positive sign indicates that the force is to the right because this was the positive direction for the coordinate system. In case (*b*) the tank is moving to the left, and there is a rate of increase in momentum in the control volume amounting to

$$\frac{d(mV_x)_\sigma}{d\tau} = (8\ \text{kg/s})(-1\ \text{m/s}) = -8\ \text{kg}\cdot\text{m/s}^2 \qquad (b)$$

The -1 m/s is the velocity of the water inside the control volume. The value of the force for case (*b*) is therefore

$$F_x = \frac{1}{1.0}[0 - (-207.8) + (-8)]$$
$$= 199.8\ \text{N}$$

10-6 SHOCK WAVES IN AN IDEAL GAS

We have already seen that a small pressure disturbance is propagated in a fluid at the speed of sound. We might say that this is the speed of movement of an infinitesimal wave in the fluid. When stronger waves are encountered such that there are rapid and sharp changes in fluid properties in a very small region of the flow we say that a *shock wave* has developed, and we now wish to analyze the conditions which must prevail if such waves are to exist. We shall limit our discussion to *normal* shock waves, or waves that are perpendicular to the flow velocity.

Consider the one-dimensional flow channel shown in Fig. 10-6. We assume the flow is adiabatic and consider the control volume bounded by the dashed line containing the shock wave which is *stationary* in the channel. Upstream from the shock the properties are designated with the subscript x, while the subscript y is used to designate the downstream properties. There is no heat or work added so the steady-flow energy equation is

$$h_{0x} = h_x + \frac{V_x^2}{2g_c} = h_y + \frac{V_y^2}{2g_c} = h_{0y} \qquad (10\text{-}40)$$

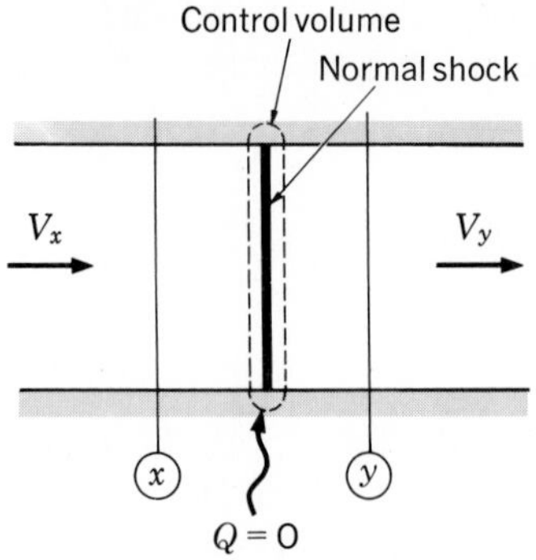

FIG. 10-6 Flow across a normal shock wave.

The mass continuity relation is

$$\frac{\dot{m}}{A} = \rho_x V_x = \rho_y V_y \tag{10-41}$$

and the momentum equation for the control volume is obtained by setting the net pressure force equal to the increase in momentum flux through the control volume:

$$A(p_x - p_y) = \frac{\dot{m}}{g_c}(V_y - V_x) \tag{10-42}$$

We have assumed the process to be adiabatic, but there is no assurance that it is reversible so the second law of thermodynamics requires that

$$s_y - s_x \geqslant 0 \tag{10-43}$$

If the energy and continuity equations (10-40) and (10-41) are combined and the combination expressed in appropriate form, it is possible to plot the resulting relation on an enthalpy-entropy diagram. Such a plot is called a *Fanno line* and is illustrated in Fig. 10-7. A combination of the momentum and continuity equations (10-41) and (10-42) can also be plotted on an h-s diagram and is called a *Rayleigh line*. The intersections of these two lines at points x and y represent the simultaneous solution of the three equations and therefore must represent the states x and y before and after the shock wave. The second-law relation of Eq. (10-43) requires that $s_y > s_x$ so this fact establishes the specific designation of the points on the diagram.

A more detailed analysis of the Fanno and Rayleigh line relations would show that

1 The maximum entropy on both curves (points a and b) corresponds to $M = 1$.
2 Above point a or b on a particular line the flow is subsonic, $M < 1$.
3 Below point a or b on a particular line the flow is supersonic, $M > 1$.

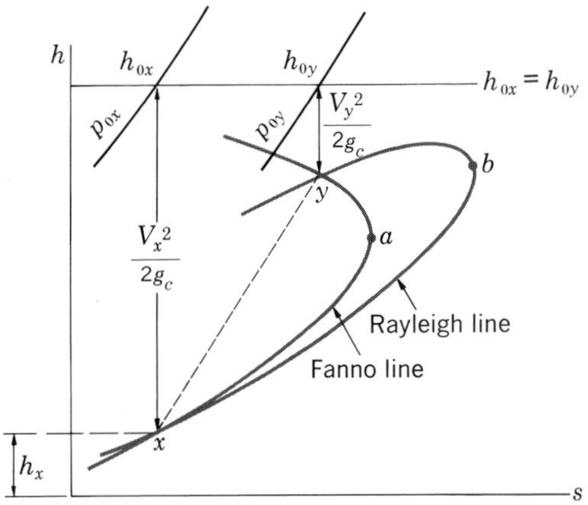

FIG. 10-7 Fanno and Rayleigh lines for evaluation of normal shock waves.

These considerations are illustrated in Fig. 10-7, and the net result is that if a normal shock wave is to exist the flow must change from supersonic to subsonic conditions across the shock. In the limit, if V_x is just sonic ($M = 1$), we will have an infinitesimal shock wave, or the sonic wave discussed before. The *strength* of the shock is usually measured in terms of the magnitude of the change in the Mach number across it. The adiabatic flow assumption dictates that the stagnation enthalpy remain constant across the shock, and for an ideal gas with constant specific heats this means

$$T_{0x} = T_{0y} \tag{10-44}$$

or the stagnation temperature remains constant regardless of the strength of the shock.

Although we have assumed an abrupt or almost step change in fluid properties across the shock for the analysis so far, we know that, in fact, there must be some finite distance required for communication of energy and momentum between gas molecules. This distance is called the *mean free path* of the molecules. At normal atmospheric pressure the mean free path is very small ($\sim 10^{-7}$ m) and shock waves at these pressures are indeed very thin. At much lower pressures, as encountered in very high altitude flight, the mean free path is much larger and shock waves encountered in those circumstances may have substantial thicknesses. For purposes of our analysis here we will assume that the pressures are always high enough that the shock waves can be assumed very thin. From Eq. (10-22) we have

$$\frac{T_{0x}}{T_x} = 1 + \frac{\gamma - 1}{2} M_x^2$$

$$\frac{T_{0y}}{T_y} = 1 + \frac{\gamma - 1}{2} M_y^2$$

Dividing these equations,

$$\frac{T_y}{T_x} = \frac{1 + [(\gamma - 1)/2]\, M_x^2}{1 + [(\gamma - 1)/2]\, M_y^2} \tag{10-45}$$

Using the continuity relation, Eq. (10-41), along with

$$\rho_x = \frac{p_x}{RT_x} \qquad \text{and} \qquad \rho_y = \frac{p_y}{RT_y}$$

gives

$$\frac{T_y}{T_x} = \frac{p_y V_y}{p_x V_x} \tag{10-46}$$

But $V = cM$ and $c = \sqrt{\gamma g_c RT}$, so that Eq. (10-46) may be expressed as

$$\frac{T_y}{T_x} = \frac{p_y M_y \sqrt{T_y}}{p_x M_x \sqrt{T_x}} = \left(\frac{p_y}{p_x}\right)^2 \left(\frac{M_y}{M_x}\right)^2 \tag{10-47}$$

Equations (10-45) and (10-47) may now be combined to give the pressure ratio across the shock as

$$\frac{p_y}{p_x} = \frac{M_x \sqrt{1 + [(\gamma - 1)/2]\, M_x^2}}{M_y \sqrt{1 + [(\gamma - 1)/2]\, M_y^2}} \tag{10-48}$$

Because Eq. (10-48) results from a combination of the energy and continuity equations, it is a representation of the Fanno line expressed in terms of pressure and Mach number. A similar relation for the Rayleigh line in terms of these same variables may be obtained by combining the momentum and continuity equations. We have

$$p_x - p_y = \frac{\dot{m}}{Ag_c}(V_y - V_x)$$

But $\dot{m} = \rho_x A V_x = \rho_y A V_y$ so that

$$p_x - p_y = \frac{\rho_y V_y^2 - \rho_x V_x^2}{g_c}$$

Introducing $V = cM$, $\rho = p/RT$, and $c = \sqrt{\gamma g_c RT}$ gives

$$p_x g_c + \frac{p_x M_x^2}{RT_x}(\gamma g_c RT_x) = p_y g_c + \frac{p_y M_y^2}{RT_y}(\gamma g_c RT_y)$$

Collecting terms,

$$\frac{p_y}{p_x} = \frac{1 + \gamma M_x^2}{1 + \gamma M_y^2} \tag{10-49}$$

The combination of Eqs. (10-48) and (10-49) represents the intersections of the Fanno and Rayleigh lines and gives

$$M_y^2 = \frac{M_x^2 + [2/(\gamma - 1)]}{[2\gamma/(\gamma - 1)]\, M_x^2 - 1} \tag{10-50}$$

Given the upstream Mach number M_x, the downstream Mach number M_y is uniquely defined. Table A-20 gives a convenient tabulation of the properties across a normal shock wave using the upstream Mach number as the primary variable. We have already mentioned that the stagnation temperature remains constant across the shock wave because of the adiabatic flow condition. The stagnation *pressure*, however, always *decreases* across the shock because of the requirement that the entropy must increase. We can see this rather easily by writing the relation for entropy change between the isentropic stagnation states before and after the shock.

$$s_{0y} - s_{0x} = c_p \ln \frac{T_{0y}}{T_{0x}} - R \ln \frac{p_{0y}}{p_{0x}} \geqslant 0$$

With the requirement that $T_{0y} = T_{0x}$, it is clear that p_{0y} must always be less than p_{0x} in order to produce an increase in entropy. We note in Table A-20 that the values of p_{0y}/p_{0x} are indeed always less than unity.

10-7 SHOCK WAVES IN CONVERGENT-DIVERGENT PASSAGES

We have already discussed isentropic flow in a converging-diverging passage and pointed out that a variety of flow conditions can be established, depending on the particular flow-area configuration and stagnation properties available to drive the flow. But the previous discussion considered only the ideal isentropic flow. Under suitable combinations of back pressure and nozzle design it is possible to produce shock-wave phenomena in the passage, and it is these phenomena we now wish to discuss.

Consider the converging-diverging nozzle shown in Fig. 10-8 connected to the stagnation-chamber supply pressure p_0 and discharging into another chamber where we designate the pressure as p_b and call it the *back pressure*. We assume that the stagnation pressure is sufficiently high to produce sonic flow conditions at the throat and supersonic flow in the downstream diverging section. Now let us examine the effect of changes in the back pressure on the flow in the nozzle.

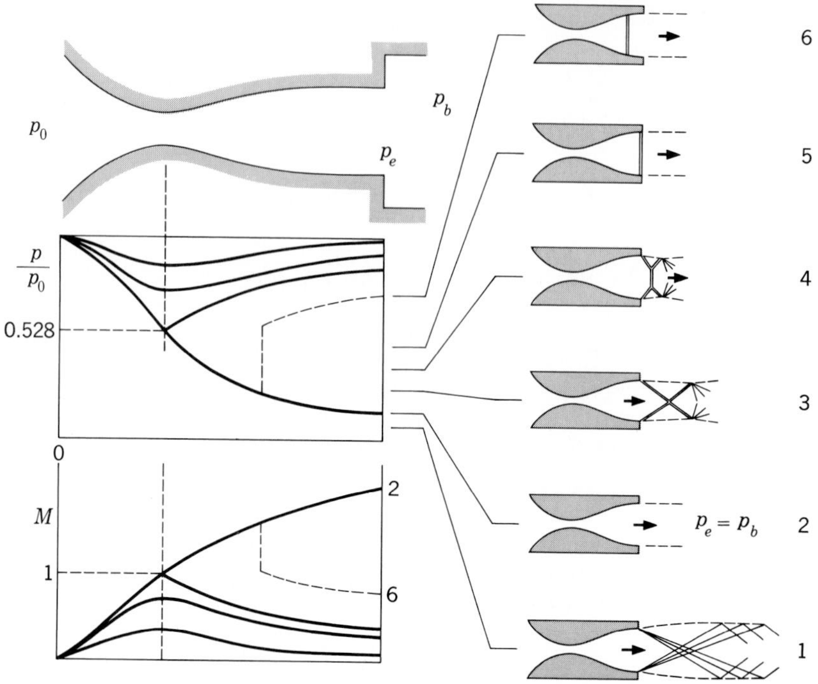

FIG. 10-8 Influence of back pressure on flow in a supersonic nozzle.

As long as the back pressure remains equal to or less than the exit pressure corresponding to the exit area there is no effect on the flow inside the nozzle. There may, however, be complicated fan-type expansion waves downstream from the exit for $p_b < p_e$, as shown at point 1. As the back pressure is increased above p_e, progressively stronger "oblique" shock waves are formed downstream from the exit, illustrated as points 3 and 4 in Fig. 10-8. Still, the flow inside the nozzle is unaffected. Point 5 illustrates the condition where the back pressure is just high enough to create a normal shock wave at the exit, that is, p_e/p_b at that point corresponds to the value for a normal shock with an upstream Mach number of M_e. Further increase in the back pressure causes the normal shock to move back into the nozzle. Downstream from the shock the flow is subsonic so the diverging section will cause a *decrease* in flow velocity and an increase in pressure, as illustrated in point 6. Eventually, an increase in the back pressure can cause the shock to move all the way back to the throat at which point it will disappear and the flow in the diverging section becomes totally subsonic. Still further increase in the back pressure causes subsonic flow in the entire nozzle and a progressive reduction in the mass flow rate. The examples illustrate the calculation procedures for nozzle-shock phenomena.

EXAMPLE 10-7 Shock at Nozzle Exit
To what value must the back pressure be raised to just produce a normal shock wave at the exit of the nozzle of Example 10-4?

SOLUTION This calculation is most easily performed by using the normal shock relations presented in Table A-20 of Appendix A. For this case $M_x = 2.5$ just before entering the shock wave at the exit. The pressure at that point is 1 atm as given in Example 10-4. Entering Table A-20 at $M_x = 2.5$ we find

$$M_y = 0.512\,99 \qquad \frac{p_y}{p_x} = 7.125 \qquad \frac{p_{0y}}{p_x} = 8.5262$$

The pressure in the flow just downstream from the shock is therefore

$$p_y = (7.125)(1.0) = 7.125 \text{ atm}$$

The back pressure in a discharge chamber like that of Fig. 10-8 is the stagnation pressure downstream from the shock, which is

$$p_{0y} = (8.5262)(1.0) = 8.5262 \text{ atm}$$

EXAMPLE 10-8 Shock in Diffuser Section

Suppose the back pressure were increased on the nozzle of Example 10-4 until a normal shock wave was formed in the diffuser section where $M_x = 1.5$. What back pressure would be necessary to accomplish this, and what would be the resulting exit temperature and velocity?

SOLUTION This problem is complicated further because the flow will be reduced to subsonic downstream from the shock and will then slow down further because of the area increase. The schematic indicates the nomenclature to be employed in the calculations. We can use the results of Examples 10-4 and 10-5 to obtain

$$A^* = 1.517 \text{ cm}^2 \qquad A_x = A_y = 1.784 \text{ cm}^2 \qquad p_x = 4.724 \text{ atm}$$

$$T_x = 274 \text{ K} \qquad M_x = 1.5 \qquad A_e = 4.0 \text{ cm}^2 \qquad T_{0x} = T_{0y} = 398 \text{ K}$$

Entering Table A-20 of the Appendix at $M_x = 1.5$ we find

$$M_y = 0.701\,99 \qquad \frac{p_y}{p_x} = 2.4583 \qquad \frac{T_y}{T_x} = 1.3202 \qquad \frac{p_{0y}}{p_x} = 3.4133$$

FIG. EXAMPLE 10-8

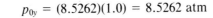

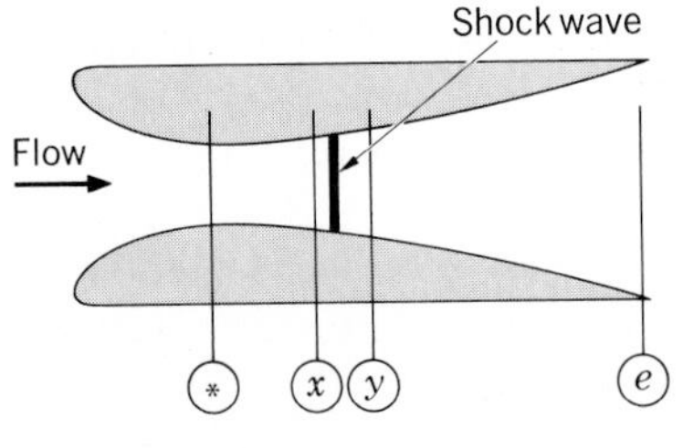

The back pressure in the discharge chamber is p_{0y} so

$$p_{0y} = (3.4133)(4.724) = 16.12 \text{ atm}$$

To calculate the exit temperature and flow velocity we note that the flow is isentropic as it moves from y to e, so we can use Table A-19 for our calculations. However, the effective throat area for the downstream-shock flow is *not* the same as that calculated previously; i.e., it is not 1.517 cm² because the downstream-shock flow is not the same isentropic flow as the upstream-shock flow. This results, of course, from the fact that the entropy increases across the shock.

We shall designate the effective throat in the downstream-shock region as $A^{*\prime}$ and determine it by entering Table A-19 at $M = 0.701\,99$. Then

$$\frac{A_y}{A^{*\prime}} = 1.093\,25$$

and

$$A^{*\prime} = \frac{1.784}{1.093\,25} = 1.632 \text{ cm}^2$$

Our actual exit area is 4.0 cm² so that

$$\frac{A_e}{A^{*\prime}} = \frac{4.0}{1.632} = 2.451$$

We can now determine the exit-flow conditions by entering Table A-19 at a value of A/A^* of 2.451 for *subsonic* flow. By interpolation the results are

$$\frac{T_e}{T_0} = 0.986\,70 \qquad M_e = 0.255$$

The exit temperature is therefore

$$T_c = (0.986\,70)(398) = 393 \text{ K} = 120°\text{C}$$

The speed of sound at exit is calculated from Eq. (10-12):

$$c_e = [(1.4)(1.0)(287)(393)]^{1/2} = 397 \text{ m/s}$$

so the actual flow velocity is

$$V_e = c_e M_e = (397)(0.255) = 101.2 \text{ m/s} = 308.6 \text{ ft/s}$$

We should remark that the mass flow rate in the nozzle is not altered by the presence of the shock because sonic conditions are still present at the throat and we still have so-called choked-flow conditions.

EXAMPLE 10-9 Nuclear Ramjet
A nuclear ramjet can be idealized as shown in the accompanying sketch. The inlet flow is at $M = 4.0$, $T = -50°C$, and $p = 0.05$ atm. A normal shock "stands" at the inlet, and the flow is accelerated to $M = 0.7$ at section 2 whereupon heat is added from the nuclear reactor at constant pressure and velocity until the temperature is raised to 1200°C at point 3. The flow is subsequently expanded to $p = 0.05$ atm at the exit. Calculate the reactor heat required and thrust produced for an inlet area of 1 m^2.

SOLUTION To evaluate the thrust force on the ramjet we must be able to evaluate the momentum fluxes at inlet and exit as well as the pressure forces at those points. As we shall see, the momentum forces will make a major contribution to the overall thrust. Using the x and y subscripts to designate upstream and downstream conditions on the shock we have, using Table A-20,

$$M_x = 4.0 \qquad M_y = 0.434\,96 \qquad \frac{p_{0y}}{p_x} = 21.068$$

$$\frac{T_y}{T_x} = 4.0469 \qquad T_x = -50°C = 223 \text{ K} \qquad p_x = 0.05 \text{ atm}$$

Using Table A-19 at $M = 4.0$,

$$\frac{T_x}{T_0} = 0.238\,10 \qquad (a)$$

and

$$T_{0x} = T_{0y} = \frac{223}{0.238\,10} = 937 \text{ K} = 664°C = T_{02}$$

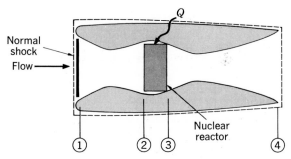

FIG. EXAMPLE 10-9

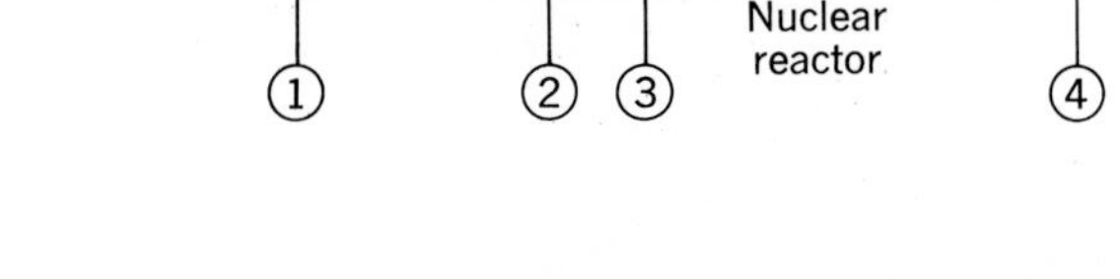

Also, $p_{0y} = p_{02} = p_{03} = p_{04} = (21.068)(0.05) = 1.0534$ atm. Entering Table A-19 again with $M_2 = 0.7$ gives

$$\frac{A_2}{A_2^*} = 1.094\,37 \qquad \frac{p_2}{p_{02}} = 0.720\,92 \qquad \frac{T_2}{T_{02}} = 0.910\,75$$

whereupon

$$T_2 = (0.910\,75)(937) = 853 \text{ K} = 580°\text{C} \tag{b}$$

We are given that

$$p_3 = p_2 \qquad T_3 = 1200°\text{C} = 1473 \text{ K} \qquad V_2 = V_3$$

Because the velocities are equal,

$$T_3 - T_2 = T_{03} - T_{02} \tag{c}$$

or

$$T_{03} = 937 + (1200 - 580) = 1557 \text{ K} = 1284°\text{C} = T_{04}$$

and

$$\frac{T_3}{T_{03}} = \frac{1473}{1557} = 0.946\,05 \tag{d}$$

We now have a new isentropic flow from point 3 to point 4 which we can calculate using relation (*d*) and Table A-19. Entering the table at the value given in (*d*) gives

$$M_3 = 0.5324 \qquad \frac{p_3}{p_{03}} = 0.823\,90$$

We have

$$p_2 = p_3 = (1.0534)(0.720\,92) = 0.759\,42 \text{ atm} \tag{e}$$

and

$$p_{03} = \frac{0.759\,42}{0.823\,90} = 0.921\,73 \text{ atm} = p_{04}$$

Now, at section 4 $p = 0.05$ atm and

$$\frac{p_4}{p_{04}} = \frac{0.05}{0.921\,73} = 0.054\,25 \tag{f}$$

We may now once again enter Table A-19 using the value from (f) with the result

$$M_4 = 2.551 \qquad \frac{T_4}{T_{04}} = 0.434\,63$$

$$\frac{A_4}{A_4^*} = 2.780$$

so that

$$T_4 = (0.434\,63)(1557) = 676 \text{ K} = 403°\text{C}$$

$$c_4 = [(1.4)(1.0)(287)(676)]^{1/2} = 521 \text{ m/s}$$

and

$$V_4 = c_4 M_4 = (521)(2.551) = 1330 \text{ m/s}$$

At the inlet, $T_{1,x} = 223$ K so that

$$c_{1,x} = [(1.4)(1.0)(287)(223)]^{1/2} = 229 \text{ m/s}$$

and

$$V_{1x} = c_{1x} M_1 = (299)(4.0) = 1197 \text{ m/s}$$

The mass flow may be calculated at entrance conditions. First, the density is evaluated as

$$\rho_1 = \frac{p_1}{RT_1} = \frac{(0.05)(1.0132 \times 10^5)}{(287)(223)} = 0.0792 \text{ kg/m}^3$$

Then

$$\dot{m} = \rho_1 A_1 V_1 = (0.0792 \text{ kg/m}^3)(1 \text{ m}^2)(1197 \text{ m/s}) = 94.80 \text{ kg/s}$$

At exit the density is

$$\rho_4 = \frac{p_4}{RT_4} = \frac{(0.05)(1.0132 \times 10^5)}{(287)(676)} = 0.0261 \text{ kg/m}^3$$

and the exit area can be determined from the mass-continuity relation

$$\dot{m} = \rho_4 A_4 V_4$$

$$A_4 = \frac{94.80}{(0.0261)(1330)} = 2.731 \text{ m}^2$$

The force balance is now made on the control volume bounded by the dashed line with the thrust force T on the fluid assumed to the right and the positive force direction taken to the right. The thrust on the ramjet structure would be to the left:

$$\sum F_x = \frac{\dot{m}}{g_c}(V_e - V_i)$$

$$T + p_1A_1 - p_4A_4 = \frac{\dot{m}}{g_c}(V_4 - V_1) \tag{g}$$

We now have all the numerical values for insertion in Eq. (g) to solve for the thrust:

$$T = -(0.05)(1.0132 \times 10^5)(1 - 2.731) + \frac{94.80}{1.0}(1330 - 1197)$$

$$= 8769 + 12\,608 = 21\,377 \text{ N} = 4805 \text{ lbf}$$

The heat which must be supplied by the reactor is

$$Q = \dot{m}(h_3 - h_2) = \dot{m}c_p(T_3 - T_2) \tag{h}$$

The specific heat for air is

$$c_p = 1005.8 \text{ J/kg}\cdot°\text{C}$$

so that

$$Q = (94.80)(1005.8)(1200 - 580)$$
$$= 59.06 \text{ MW}$$
$$= 2.016 \times 10^8 \text{ Btu/h}$$

As a matter of interest we might calculate the rate of doing work by the thrust force moving at the velocity V_1.

$$W = TV_1 = (21\,377 \text{ N})(1197 \text{ m/s})$$
$$= 2.559 \times 10^7 \text{ W}$$

The energy-conversion efficiency for the ramjet is therefore

$$\text{Efficiency} = \frac{W}{Q} = \frac{2.559 \times 10^7}{5.906 \times 10^7} = 43.3 \text{ percent}$$

This is a rather high value and results from the fact that the heat is added at a very high temperature (1200°C).

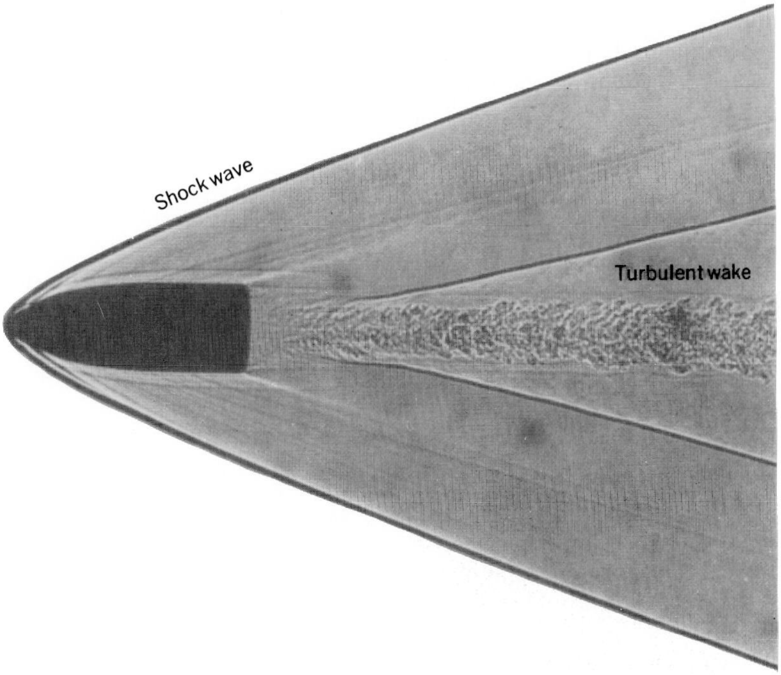

FIG. 10-9 Photograph of flight of a 0.22-caliber Remington "swift" shell; velocity = 4100 ft/s. (*Courtesy ARO Inc., Arnold Air Force Base, Tenn.*)

We have already seen from the many flow regimes which may develop in the downstream section of a nozzle that shock waves are not always the simple "normal" type. Shock waves can form at an oblique orientation to the flow velocity and may assume curved shapes under certain circumstances. In Fig. 10-9 we have a photograph of the "bow shock" produced by a high-speed bullet, and a similar wave would be formed around the nose of a high-speed aircraft. The flow just behind the most forward shock position can be calculated as that downstream from a normal shock wave. As the Mach number is increased, the trailing shock wave is swept backward at progressively steeper angles. The sweep angle is approximated by the so-called Mach cone shown in Fig. 10-10.

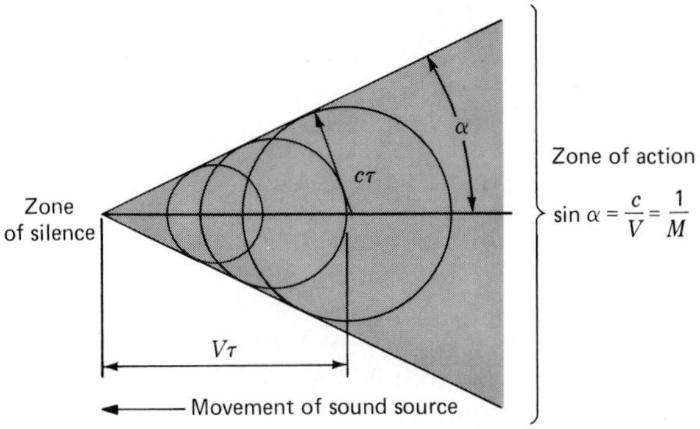

FIG. 10-10 Mach cone and zone of silence.

In front of the Mach cone is a zone of silence because disturbances can only propagate at the speed of sound. For the source movement of distance $V\tau$ in time τ the sound only moves $c\tau$, or the edge of the cone. Behind the cone, disturbances are heard. Evidently, the cone angle is given by

$$\sin \alpha = \frac{c}{V} = \frac{1}{M}$$

For $M < 1$ disturbances can propagate ahead of the source as well as behind. This means that a supersonic aircraft may have passed well overhead before an observer on the ground hears the trailing bow shock moving along with it.

As a simple example consider an aircraft at an altitude of 50 000 ft traveling at $M = 2$. The Mach cone angle is

$$a = \sin^{-1} \frac{1}{M} = \sin^{-1} \frac{1}{2} = 30^\circ$$

When the edge of the cone arrives at the ground the aircraft has moved past the ground observer a distance x where

$$\frac{50\ 000}{x} = \tan \alpha$$

or

$$x = 86\ 602 \text{ ft} = 16.4 \text{ mi} \qquad (26.4 \text{ km})$$

10-8 NOZZLE FLOW OF A VAPOR

We have already learned that a vapor does not necessarily behave as an ideal gas, particularly when the thermodynamic state approaches the saturation region. The calculation of high-speed vapor flows in nozzles therefore must be performed differently than in the case of an ideal gas with constant specific heats. Consider the *h-s* diagram shown in Fig. 10-11 where steam is to be expanded isentropically from state 1 to state 3 in a nozzle. As the saturation line is encountered at point 2, one would expect condensation to begin, but this is not always the case. If the flow velocity is high enough (as it would be in a supersonic nozzle), sufficient time may not be available to form nucleation of droplets and accomplish the heat-transfer process associated with condensation. Once the flow moves far enough into the saturation region, large enough droplets will be formed to sustain the condensation process and a "condensation shock" will be experienced. The delay of condensation produces a situation which is appropriately called *supersaturation*.

The condensation phenomenon described above is usually described in thermodynamic terms as a problem in *metastable equilibrium*. In a strict sense

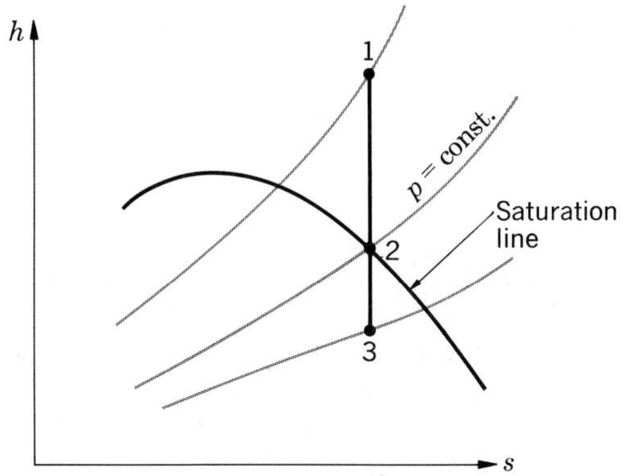

FIG. 10-11 Enthalpy-entropy diagram for high-speed flow of a vapor.

this means a circumstance in which the thermodynamic equilibrium state remains the same for small displacements, but then moves to a new equilibrium position for large displacements. In the above example this means that large enough droplets must be formed to effect the change. From a microscopic viewpoint things are very complicated and involve the capture of vapor molecules by the liquid surface of a droplet, droplet growth rates, and their general transport in the flow stream.

As a result of the phenomenon of supersaturation, vapor flow in a nozzle may be calculated with more precision than might otherwise be possible by assuming a constant value of γ over a rather wide range of temperatures and pressures. For steam $\gamma = 1.3$ is a good approximation. The critical pressure ratio is then

$$\frac{p^*}{p_0} = \left(\frac{2}{\gamma + 1}\right)^{\gamma/(\gamma-1)} = 0.545 \tag{10-51}$$

If the steam enters the nozzle as a saturated vapor the critical ratio is, to a good approximation, 0.577, or that corresponding to a value of $\gamma = 1.14$.

EXAMPLE 10-10 Steam Flow in a Nozzle
Steam expands isentropically in a nozzle from stagnation conditions of 100 psia and 500°F to 14.7 psia. The mass flow rate is 1 lbm/s. Calculate the exit area required using the steam tables and the throat area using the constant-property assumptions pertaining to Eq. (10-51).

SOLUTION Designating exit properties with a subscript e, we have

$$T_0 = 500°\text{F} = 960°\text{R} \qquad p_e = 14.7 \text{ psia}$$

$$p_0 = 100 \text{ psia} \qquad \dot{m} = 1 \text{ lbm/s}$$

From the steam tables (Table A-9)

$$s_0 = s_e = 1.7085 \text{ Btu/lbm}\cdot{}^\circ\text{R}$$
$$h_0 = 1279.1 \text{ Btu/lbm} \qquad (a)$$

At the exit pressure of 14.7 psia we find that the flow is in the saturation region with values from Table A-8 of

$$s_f = 0.3121 \qquad s_{fg} = 1.4446 \text{ Btu/lbm}\cdot{}^\circ\text{R}$$
$$h_f = 180.15 \qquad h_{fg} = 970.4 \text{ Btu/lbm}$$
$$v_f = 0.016\,72 \qquad v_g = 26.80 \text{ ft}^3/\text{lbm}$$

Using Eq. (*a*) we calculate the exit quality from

$$s_e = 1.7085 = s_f + xs_{fg} = 0.3121 + x(1.4446)$$
$$x = 0.9666$$

The resulting exit enthalpy and specific volume are therefore

$$h_e = 180.15 + (0.9666)(970.4) = 1118.1 \text{ Btu/lbm}$$
$$v_e = 0.016\,72 + (0.9666)(26.8 - 0.016\,72) = 25.91 \text{ ft}^3/\text{lbm}$$

The exit velocity can be calculated from the steady-flow energy equation for this problem:

$$h_e + \frac{V_e}{2g_c} = h_0 \qquad (b)$$

With the numerical values inserted

$$V_e = [(2)(32.2)(778)(1279.1 - 1118.1)]^{1/2}$$
$$= 2840 \text{ ft/s} \qquad (931.8 \text{ m/s})$$

The mass rate of flow can be expressed as

$$\dot{m} = \rho AV = \frac{AV}{v} \qquad (c)$$

At exit conditions this gives a flow area of

$$A_e = \frac{(1 \text{ lbm/s})(25.91 \text{ ft}^3/\text{lbm})}{2840 \text{ ft/s}} = 9.123 \times 10^{-3} \text{ ft}^2 = 1.314 \text{ in}^2$$

Using the constant-property assumptions of Eq. (10-51), the throat area is calculated using a value of $\gamma = 1.3$. This value may then be applied to Eq. (10-31) with the gas constant for water vapor taken as

$$R = \frac{1545}{18} = 85.8 \text{ ft·lbf/lbm·°R}$$

Using Eq. (10-31),

$$\frac{\dot{m}}{A^*} = (100)(144)\left[\frac{(1.3)(32.2)}{(85.8)(960)}\right]^{1/2}\left(\frac{2}{1.3+1}\right)^{(1.3+1)/2(1.3-1)}$$
$$= 190 \text{ lbm/s·ft}^2$$

With $\dot{m} = 1$ lbm/s the throat area required is

$$A^* = \frac{1}{190} = 5.263 \times 10^{-3} \text{ ft}^2 = 0.758 \text{ in}^2 \qquad (4.89 \text{ cm}^2)$$

10-9 NOZZLE AND DIFFUSER COEFFICIENTS

Up to this point we have considered one-dimensional adiabatic flow systems where compressibility effects are important. With the exception of situations involving shock-wave phenomena the flows were also considered reversible, or isentropic, for the ideal case. The adiabatic assumption is a reasonable one for actual nozzle and diffuser flow situations, but there are several factors which cause the actual flow to be irreversible. Fluid friction is always present at the wall and will cause the overall flow to be irreversible. The exact configuration of a nozzle strongly influences the nature of the frictional effects because of the viscous boundary-layer buildup on the wall. A detailed treatment of these effects is beyond the scope of our discussion but we can indicate their diverse nature by noting the two situations in Fig. 10-12. In (*a*) the nozzle contour is smooth and gradual, and if the flow is to be supersonic, smooth acceleration to $M = 1$ at the throat and then $M > 1$ in the diffusing section is experienced with a viscous boundary layer at the wall, as shown.

In Fig. 10-12*b* the flow is again accelerated smoothly to $M = 1$ at the throat, but the area change in the diffuser section downstream is much more rapid than in Fig. 10-12*a*. As a result, the high-speed flow may not be able to follow the wall and may become "detached" or "separated" with a highly turbulent mixing region at the wall. Not surprisingly, the nature of the fluid friction effects in Figs. 10-12*b* can be quite different from those in Fig. 10-12*a*. Fundamental analyses of the nozzle flow systems can sometimes predict the frictional effects, and the results are usually expressed in terms of certain

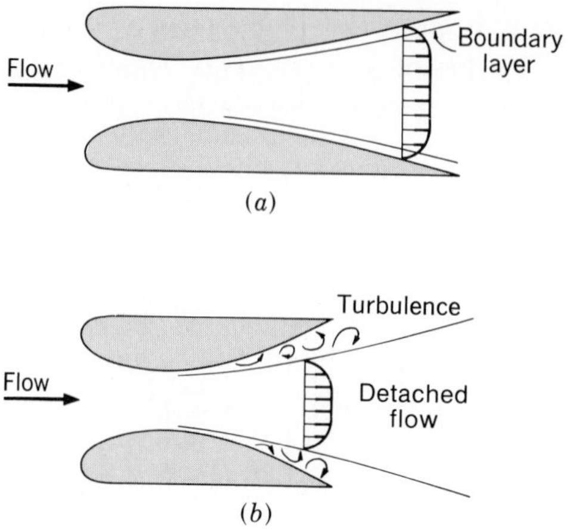

FIG. 10-12 Effect of nozzle contour on flow: (*a*) gradual expansion; (*b*) too rapid expansion and separated flow.

performance coefficients which can be compared with experiments. Regardless of the success of the analysis, experimental data can be used to determine the values of the coefficients for design purposes, and it is the more important coefficients which we now wish to discuss.

The *nozzle efficiency* η_N is defined by

$$\eta_N = \frac{\text{actual KE at exit}}{\text{KE at exit with isentropic flow to same exit pressure}} \tag{10-52}$$

Efficiencies range between 90 and 99 percent with larger nozzles having the higher efficiencies because the viscous boundary layer occupies a relatively smaller portion of the flow, the larger the flow area.

If the kinetic energy at inlet is small, then the inlet enthalpy is very nearly h_0, and the nozzle efficiency can be expressed by

$$\eta_N = \frac{h_0 - h_e}{h_0 - h_{es}} \tag{10-53}$$

where h_e is the actual exit enthalpy and h_{es} is the isentropic exit enthalpy for expansion to the actual exit pressure.

The *nozzle velocity coefficient* C_V is defined in a similar way:

$$C_V = \frac{\text{actual velocity at exit}}{\text{velocity at exit for isentropic flow to same exit pressure}} \tag{10-54}$$

The *coefficient of discharge* C_D is defined as

$$C_D = \frac{\text{actual mass rate of flow}}{\text{mass rate of flow for isentropic flow}} \tag{10-55}$$

In some cases the design objective for the flow channel is to slow the flow down and increase the pressure. For these instances it is frequently advantageous to define a *diffuser pressure recovery factor* C_P as

$$C_P = \frac{\text{actual pressure rise}}{\text{isentropic pressure rise}} \tag{10-56}$$

An alternate method for describing diffuser performance is to define a *diffuser efficiency* η_D as

$$\eta_D = \frac{\begin{array}{c}\text{isentropic enthalpy rise for discharge at}\\ \text{actual exit stagnation pressure}\end{array}}{\text{inlet kinetic energy}} \tag{10-57}$$

The reasoning behind this definition is not immediately apparent but may be clarified by referring to Fig. 10-13. In this diagram the inlet kinetic energy is

$$\frac{V_i^2}{2g_c} = h_{0i} - h_i$$

while the kinetic energy at exit is

$$\frac{V_e^2}{2g_c} = h_{0e} - h_e$$

The numerator of Eq. (10-57) represents the maximum kinetic energy which could be recovered for discharge at the actual exit stagnation pressure. So, in terms of the nomenclature of Fig. 10-13 we obtain

$$\eta_D = \frac{h_{es} - h_i}{h_{0i} - h_i} \tag{10-58}$$

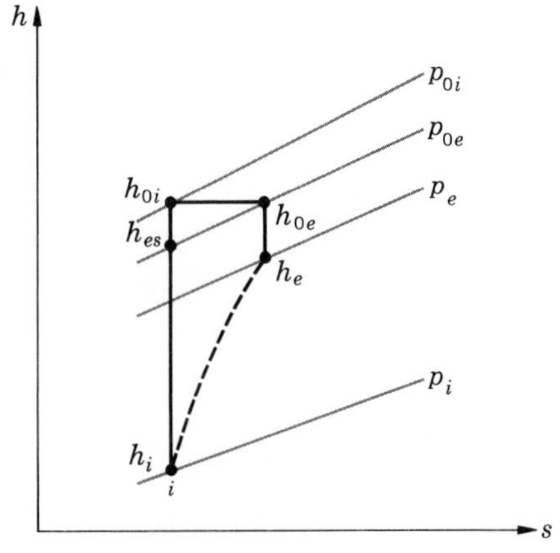

FIG. 10-13 Parameters for evaluating diffuser efficiency.

EXAMPLE 10-11 Diffuser Pressure Recovery
A certain diffuser having an efficiency of 90 percent is used to reduce the velocity of an air stream initially at 150 m/s, 300 K, and 100 kPa down to 60 m/s. Calculate the ratio of exit area to inlet area to accomplish this reduction. What is the value of the pressure-recovery factor for this diffuser?

SOLUTION From the steady-flow energy equation,

$$h_{0i} - h_i = \frac{V_1^2}{2g_c} = \frac{(150)^2}{(2)(1.0)(1000)} = 11.25 \text{ kJ/kg} \qquad (a)$$

Using the nomenclature of Fig. 10-13 and Eq. (10-58) gives

$$h_{es} - h_i = (0.9)(11.25) = 10.125 \text{ kJ/kg} \qquad (b)$$

we have $T_i = 300$ K so that, from Eqs. (*a*) and (*b*),

$$T_{0i} = \frac{11.25}{1.005} + 300 = 311.2 \text{ K} = 38.2°\text{C} = T_{0e}$$
$$T_{es} = \frac{10.125}{1.005} + 300 = 310.1 \text{ K} = 37.1°\text{C} \qquad (c)$$

The actual exit temperature can be obtained from

$$T_e = T_{0e} - \frac{1}{c_p}\frac{V_e^2}{2g_c}$$
$$= 311.2 - \frac{(60)^2}{(1.005)(2)(1000)} = 309.4 \text{ K} = 36.4°\text{C}$$

Now, observing Fig. 10-13 we can follow the isentropic line from point *i* to point *es* to calculate the exit stagnation pressure:

$$\frac{p_{0e}}{p_i} = \frac{p_{es}}{p_i} = \left(\frac{T_{es}}{T_i}\right)^{\gamma/(\gamma-1)}$$
$$p_{0e} = (100)\left(\frac{310.1}{300}\right)^{1.4/0.4} = 112.3 \text{ kPa} \qquad (d)$$

In a similar manner we can determine the actual exit pressure by going from point 0_e to point *e*:

$$\frac{p_e}{p_{0_e}} = \left(\frac{T_e}{T_{0_e}}\right)^{\gamma/(\gamma-1)}$$
$$p_e = (112.3)\left(\frac{309.4}{311.2}\right)^{3.5} = 110.04 \text{ kPa} \qquad (e)$$

The area ratio is determined from the mass continuity relation

$$\dot{m} = \rho_i A_i V_i = \rho_e A_e V_e \tag{f}$$

But $\rho = p/RT$, so that

$$\frac{A_e}{A_i} = \frac{p_i V_i T_e}{p_e V_e T_i} = \frac{(100)(150)(309.4)}{(110.04)(60)(300)} = 2.343$$

To calculate the pressure-recovery factor we must first determine the isentropic pressure rise. The inlet stagnation pressure is calculated from

$$\frac{p_{0_i}}{p_i} = \left(\frac{T_{0_i}}{T_i}\right)^{\gamma/(\gamma-1)}$$

$$p_{0_i} = (100)\left(\frac{311.2}{300}\right)^{3.5} = 113.7 \text{ kPa} \tag{g}$$

The pressure-recovery factor is calculated from Eq. (10-56):

$$C_p = \frac{p_e - p_i}{p_{0_i} - p_i} = \frac{110.04 - 100}{113.7 - 100} = 0.733$$

REVIEW QUESTIONS

1 What is meant by the term *isentropic stagnation state*?
2 Why is the stagnation pressure reached in an irreversible adiabatic process lower than that reached in an isentropic process?
3 What is meant by the term *static pressure*?
4 Define the Mach number. Why is it important?
5 Give a word definition of the speed of sound.
6 Explain, in a physical sense, why the speed of sound in an ideal gas is a function of temperature alone.
7 What is Bernoulli's equation?
8 Explain the meaning of the term *choked flow*.
9 Why must the flow area be increased to cause an increase in flow velocity in the supersonic section of a nozzle?
10 Explain in a physical sense why the pressure must always decrease for an increase in velocity in an isentropic nozzle, regardless of subsonic or supersonic flow conditions.
11 What is meant by the term *momentum flux*?
12 Why cannot Newton's second law of motion be applied directly to a control volume (open system)?
13 Describe a shock wave in terms which could be understood by a layperson.

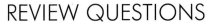

14 Why are we justified in assuming that the flow across a shock wave is adiabatic?
15 What is the approximate thickness of a shock wave?
16 What are the Fanno and Rayleigh lines?
17 Why must the flow always proceed from supersonic to subsonic conditions when moving across a shock wave?
18 For an ideal gas why does the stagnation temperature remain constant across a shock wave?
19 Why does the presence of a shock wave in the diverging section of a converging-diverging nozzle not affect the mass flow rate through the nozzle?

PROBLEMS

10-1 Calculate the stagnation temperature and pressure for flow of
(*a*) Helium at 2000 ft/s, 100°F, 1 psia
(*b*) Air at $M = 5$, -40°F, 0.01 atm
(*c*) Hydrogen at 100 m/s, 100°C, 0.5 atm
(*d*) Steam at 1500 ft/s, 120°F, 1 psia

10-2 Calculate the Mach number or flow velocity for each condition in Prob. 10-1. Assume $\gamma = 1.3$ for steam.

10-3 A ballistic missile may reenter the atmosphere at very high velocities of the order of $M = 10$. For a location where $T = -60$°F and $p = 0.03$ atm estimate the stagnation temperature which will be present on the front of the missile assuming ideal-gas behavior. If a bow shock envelops the nose cone, estimate the pressure which will exist at the stagnation point.

10-4 Consider a normal shock wave as a slowing-down mechanism for supersonic flow. For $M_x = 1.5$ calculate the value of a "diffuser efficiency" as defined by Eq. (10-57).

10-5 A converging-diverging nozzle is designed to operate with inlet stagnation conditions of 150 psia and 200°F with air discharging at $M = 2.5$. Assuming the stagnation conditions remain constant, to what value must the exit pressure be raised to just produce subsonic flow in the entire nozzle?

10-6 A converging-diverging nozzle is designed to produce an exit flow of air at $M = 4.0$ and 0.1 atm. The stagnation temperature is 120°F. Calculate the throat area and mass flow for an exit area of 1.0 in^2.

10-7 To what value must the back pressure be raised to just cause a normal shock to stand in the exit of the nozzle of Prob. 10-6?

10-8 Calculate the speed of sound for air, hydrogen, helium, carbon dioxide, and propane at 400°F.

10-9 Calculate the speed of sound in water vapor at 500 psia and 1000°F using the steam tables.

10-10 What percent reduction in flow rate would result from lowering the stagnation pressure in Prob. 10-5 to 100 psia?

10-11 A conveyor belt moves with a velocity of 10 ft/s and receives material from directly above at the rate of 900 lbm/min. What horsepower motor is required to drive the belt neglecting frictional losses in pulleys, drives, etc.?

10-12 A novel motorboat design employs a turbine pump inside the boat which sucks water through a front inlet and discharges it at the rear of the boat at a higher velocity. For a particular design it is determined that 500 lbf thrust is required to propel the boat at 30 mph. The inlet flow area is 1 ft^2, and the water may be assumed to enter at the boat speed. What exit flow area is required considering all the thrust to result from increased fluid momentum? What horsepower engine is required to drive the boat?

10-13 If the local free-stream temperature is 0°F for the flow around the bullet of Fig. 10-9, calculate the stagnation temperature and pressure on the nose of the bullet.

10-14 A diffuser is used to reduce the velocity of an air stream from $M = 0.8$, $p = 14.7$ psia, and $T = 70$°F down to a value of 200 ft/s. Calculate the ratio of exit to inlet area if the diffuser efficiency is 85 percent. Also calculate the pressure-recovery factor. How much heat is added per lbm of fluid as it flows through the diffuser?

10-15 Air enters a converging section at 200 ft/s, 50 psia, and 150°F. The exit and inlet areas are 4 and 6 in^2, respectively. Calculate the exit velocity, temperature, and pressure for isentropic flow.

10-16 Suppose an air flow stream at $M = 2.5$, $p = 0.1$ atm, and $T = 10$°F is imposed on the inlet of the converging-diverging nozzle of Prob. 10-30M and that a normal shock wave develops at that point. What would be the exit flow conditions in this circumstance?

10-17 A tank filled with water and having a volume of 150 ft^3 is placed on wheels whereby it can move freely on a level surface. An engine-driven pump mounted on the tank discharges water at the rate of 50 lbm/s through a nozzle having an exit area of 1.0 in^2. What is the net thrust force produced when the tank is moving at a speed of 20 ft/s opposite to the direction of the nozzle discharge velocity?

10-18 Very low temperatures can be produced by expanding high-pressure air in a supersonic nozzle. What ratio of exit to throat area is required to produce −60°F air from high-pressure air at 70°F and 20 atm?

10-19 There are design limitations placed on high-speed aircraft because of temperatures which the skin materials can withstand at high speeds. For a given speed the maximum temperature of the aircraft is the stagnation temperature. What is the maximum speed for flight at an altitude where $T = -50$°F, $p = 0.1$ atm, and there is a maximum allowable skin temperature of 700°F?

10-20 A small supersonic wind tunnel is to be designed so that it will have a test section with $M = 2.0$, $T = -50$°F, and $p = 2$ psia. The flow cross-sectional area is to be 1.0 ft^2, and the air flow is to be established with a high-pressure tank discharging through a converging-diverging nozzle section. What stagnation properties would be necessary in the tank, and what mass flow of air would be required? Also calculate the throat area of the nozzle.

10-21 Suppose the steam nozzle of Prob. 10-23 has a velocity coefficient of 0.95. What would be the mass flow in this circumstance and the coefficient of discharge?

10-22 A small toy rocket is charged with air at 2000 psia and 110°F. The air is allowed to discharge through a converging-diverging nozzle having a throat area of 1.0 mm^2 and an exit area of 3.0 mm^2. Discharge takes place to atmospheric pressure. Calculate the net thrust force when the rocket is in a stationary position.

10-23 Steam expands isentropically in a nozzle from 500 psia, 600°F, to 200 psia. The inlet velocity is small. Calculate the exit velocity and exit area for a flow rate of 2 lbm/s.

10-24 Nitrogen expands isentropically from 150 psia and 300°F to atmospheric pressure. What throat and exit areas would be required for a mass flow of 0.1 lbm/s?

10-25 A converging-diverging nozzle having an exit-to-throat area ratio of 2.40 is connected to a large tank containing air at 200 psia and 70°F. What back pressure must be exerted to just produce a normal shock at the exit of the nozzle?

PROBLEMS (METRIC UNITS)

10-1M Calculate the stagnation temperature and pressure for flow of

(*a*) Helium at 610 m/s, 35°C, 7 kPa
(*b*) Air at $M = 5$, −40°C, 0.01 atm
(*c*) Hydrogen at 1000 m/s, 100°C, 0.5 atm
(*d*) Steam at 460 m/s, 49°C, 0.5 atm

10-2M A ballistic missile may reenter the atmosphere at very high velocities of the order of $M = 10$. For a location where $T = -50$°C and $p = 0.03$ atm, estimate the stagnation temperature which will be present on the front of the missile assuming ideal-gas behavior. If a bow shock envelops the nose cone, estimate the pressure which will exist at the stagnation point.

10-3M A converging-diverging nozzle is designed to operate with inlet stagnation conditions of 1.0 MPa and 100°C, with air discharging at $M = 2.5$. Assuming the stagnation conditions remain constant, to what value must the exit pressure be raised to just produce subsonic flow in the entire nozzle?

10-4M A converging-diverging nozzle is designed to produce an exit flow of air at $M = 4.0$ and 1.0 atm. The stagnation temperature is 50°C. Calculate the throat area and mass flow for an exit area of 6.5 cm^2.

10-5M Calculate the speed of sound for air, hydrogen, helium, carbon dioxide, and propane at 200°C.

10-6M Calculate the speed of sound in water vapor at 3.5 MPa and 280°C using the steam tables.

10-7M Air at 900 kPa and 30°C expands isentropically in a nozzle to a pressure of 170 kPa. The inlet velocity is very low. Calculate the temperature and velocity at exit.

10-8M Air at 1.0 kPa and 60°C is decelerated in an isentropic diffuser from $M = 3.0$ to $M = 1.0$ at exit from the diffuser. The flow rate is 12.0 kg/s. Determine the temperature and pressure at exit and the inlet and exit flow areas.

10-9M What percent reduction in flow rate would result from lowering the stagnation pressure in Prob. 10-3M to 700 kPa?

10-10M A conveyor belt moves with a velocity of 3.0 m/s and receives material from directly above at the rate of 33 kg/s. What power motor is required to drive the belt neglecting frictional losses in pulleys, drives, etc.?

10-11M A novel motorboat design employs a turbine pump inside the boat which sucks water through a front inlet and discharges it at the rear of the boat at a higher velocity. For a particular design it is determined that 2.2 kN thrust is required to propel the boat at 13 m/s. The inlet flow area is 0.1 m^2, and the water may be assumed to enter at the boat speed. What exit flow area is required considering all the thrust to result from increased fluid momentum? What power engine is required to drive the boat?

10-12M Very low temperatures can be produced by expanding high-pressure air in a supersonic nozzle. What ratio of exit-to-throat area is required to produce −50°C air from high-pressure air at 20°C and 20 atm?

10-13M Calculate the stagnation temperature and pressure for steam flowing at 6.0 MPa and 440°C with a velocity of 150 m/s.

10-14M In a certain jet aircraft the stagnation properties of the air are measured as $p_0 = 8.5$ MPa and $T_0 = 875$°C for a free steam pressure of 35 kPa. Calculate the speed of the aircraft and the Mach number.

10-15M A small sonic flow nozzle meters the flow of nitrogen at 5 atm and 50°C. What flow rate will the nozzle measure for an exit area of 4.0 mm^2 and a discharge at atmospheric pressure?

10-16M What would the flow rate be for the nozzle of Prob. 10-15M using air and helium for the same temperature and pressure?

10-17M A high-temperature gas stream of helium is to be accelerated in a converging nozzle from $T = 1000$°C, $p = 3$ atm, and $V = 100$ m/s to $M = 0.9$. Assuming isentropic flow, calculate the ratio of exit to inlet area and the exit temperature and pressure.

10-18M A cylinder of high-pressure air discharges through a small converging nozzle having an exit area of 5 mm^2. The conditions in the cylinder are 100 atm and 30°C and may be assumed to remain constant during the discharge process. What thrust force results, assuming isentropic flow and discharge into a large room at 1-atm pressure? What thrust force would result if the nozzle efficiency were 97 percent?

10-19M There are design limitations placed on high-speed aircraft because of the temperatures which the skin materials can withstand at high speeds. For a given speed the maximum temperature of the aircraft is the stagnation temperature. What is the maximum speed for flight at an altitude where $T = -45°C$, $p = 0.1$ atm, and there is a maximum allowable skin temperature of 370°C?

10-20M What minimum stagnation properties would be required to produce flow at $M = 3.0$, $p = 1$ atm, and $T = 20°C$?

10-21M A tank filled with water and having a volume of 4200 liters is placed on wheels whereby it can move freely on a level surface. An engine-driven pump mounted on the tank discharges water at the rate of 23 kg/s through a nozzle having an exit area of 6.5 cm^2. What is the net thrust force produced when the tank is moving at a speed of 6 m/s opposite to the direction of the nozzle discharge velocity?

10-22M Steam expands isentropically in a nozzle from 3.5 MPa, 320°C, to 1.5 MPa. The inlet velocity is small. Calculate the exit velocity and exit area for a flow rate of 0.9 kg/s.

10-23M Nitrogen expands isentropically from 1 MPa and 150°C to atmospheric pressure. What throat and exit areas would be required for a mass flow of 0.05 kg/s?

10-24M A converging-diverging nozzle having an exit-to-throat area ratio of 2.40 is connected to a large tank containing air at 1.3 MPa and 20°C. What back pressure must be exerted to just produce a normal shock at the exit of the nozzle?

10-25M Suppose the steam nozzle of Prob. 10-22M has a velocity coefficient of 0.95. What would be the mass flow in this circumstance and the coefficient of discharge?

10-26M A small supersonic wind tunnel is to be designed so that it will have a test section with $M = 2.0$, $T = -45°C$, and $p = 14$ kPa. The flow cross-sectional area is to be 0.1 m^2, and the air flow is to be established with a high-pressure tank discharging through a converging-diverging nozzle section. What stagnation properties would be necessary in the tank, and what mass flow of air would be required? Also calculate the throat area of the nozzle.

10-27M If the local free-stream temperature is $-18°C$ for the flow around the bullet of Fig. 10-9, calculate the stagnation temperature and pressure on the nose of the bullet.

10-28M A diffuser is used to reduce the velocity of an air stream from $M = 0.8$, $p = 1$ atm, and $T = 20°C$ down to a value of 60 m/s. Calculate the ratio of exit to inlet area if the diffuser efficiency is 85 percent. Also calculate the pressure-recovery factor. How much heat is added per kilogram of fluid as it flows through the diffuser?

10-29M Air enters a converging section at 60 m/s, 340 kPa, and 65°C. The exit and inlet areas are 4 and 6 cm^2, respectively. Calculate the exit velocity, temperature, and pressure for isentropic flow.

10-30M A converging-diverging nozzle for use with air is constructed so that the inlet area is 8.0 cm^2, the minimum area is 4.0 cm^2, and the exit area is also 8.0 cm^2. What Mach number must be present at the inlet section to produce sonic conditions at the minimum area? What would be the exit Mach number under these conditions? If the inlet temperature and pressure are 25°C and 5 atm, what is the mass flow rate for sonic conditions at the minimum area?

10-31M What are the stagnation temperature and pressure for the sonic flow of Prob. 10-30M?

10-32M Air is contained in a 0.5-m^3 tank at 5 atm and 25°C and is allowed to discharge through a small converging nozzle having an exit area of 1.0 cm^2. Discharge occurs into a large room at 1 atm, and the tank is perfectly insulated. How long will it take the discharge process to lower the pressure in the tank to 3 atm?

10-33M A normal shock wave is established in helium at a point where M = 3.0, and the free-stream pressure and temperature are 0.2 atm and −10°C, respectively. Calculate the stagnation pressure downstream from the shock and the downstream-flow velocity. Compare these results with the values for air undergoing a shock at the same upstream conditions.

10-34M A small toy rocket is charged with air at 13.5 MPa and 45°C. The air is allowed to discharge through a converging-diverging nozzle having a throat area of 1.0 mm^2 and an exit area of 3.0 mm^2. Discharge takes place to atmospheric pressure. Calculate the net thrust force when the rocket is in a stationary position.

10-35M Calculate the maximum velocity which may theoretically be obtained by expanding air from stagnation conditions of 10 atm and 100°C.

10-36M Suppose the converging-diverging nozzle of Example 10-4 has an inlet area of 8.0 cm^2 and is placed in a flow stream at M = 3.0, p = 0.5 atm, and T = 0°C, and that a normal shock forms at the inlet. What will be the exit temperature, pressure, and velocity under these conditions? What would be the mass flow rate under these conditions, and the net force exerted on the fluid as it moves from just upstream of the shock to the exit of the nozzle?

REFERENCES

1 Liepmann, H. W., and A. Roshko: "Elements of Gas Dynamics," John Wiley & Sons, Inc., New York, 1957.
2 Shapiro, A. H.: "The Dynamics and Thermodynamics of Compressible Flow," The Ronald Press Company, New York, 1954.
3 Thompson, P. A.: "Compressible Fluid Dynamics," McGraw-Hill Book Company, New York, 1972.
4 Chapman, A. J., and W. F. Walker: "Introductory Gas Dynamics," Holt, Rinehart and Winston, Inc., New York, 1971.

11

ELEMENTS OF HEAT TRANSFER

11-1 INTRODUCTION

We have seen in previous chapters that thermodynamics may be used for energy balances with a variety of physical situations. Some examples are

1 An energy balance is performed on a feedwater heater to determine the amount of steam required to produce the high-temperature water.
2 An energy balance is performed on an air-conditioning coil to determine the flow rate of refrigerant necessary to cool a given quantity of air.
3 An energy balance is performed to determine the amount of heat necessary to vaporize a given quantity of water.

Despite our success in performing such energy balances, and others, we have never indicated any techniques for determining the *size* of the heat exchangers which may be required for a particular heating or cooling application. Consider another simple process:

A steel block is heated in an oven to 200°C and then removed and allowed to cool to 50°C in room air. A thermodynamic energy analysis may be used to determine the energy lost by the steel block in the cooling process but it cannot answer the question, How *long* does the process take? The science of heat transfer can provide the answer and can also be used to determine the sizes of heat exchangers for applications like those outlined above.

In this chapter we shall provide an abbreviated treatment of the subject of heat transfer. First we will discuss the principles of the three modes of heat transfer: conduction, convection, and radiation. Then we will discuss various applications of the principles to the design of heat exchangers.

There are three modes of heat transfer: conduction, convection, and radiation. *Conduction* results from the transport of energy by molecular motion in gases and liquids and by a combination of lattice vibration and electron transport in solids. In general, good electrical conductors are also good heat conductors.

Convection heat transfer occurs because of the motion of a fluid past a heated surface—the faster the motion, the greater the heat transfer. The convection heat transfer is usually assumed to be proportional to the surface area in contact with the fluid and the difference in temperature of the surface and fluid. Thus,

$$q_{\text{conv}} = hA(T_{\text{surf}} - T_{\text{fluid}}) \qquad \text{W} \tag{11-1}$$

where h is called the *convection heat transfer coefficient,* which is a strong function of both fluid properties and fluid velocity. h has the units of W/m$^2\cdot$°C when q is in W and ΔT is in °C. Sometimes it is convenient to think in terms of a convection resistance $R_{\text{conv}} = 1/hA$ so that

$$q_{\text{conv}} = \frac{T_{\text{surf}} - T_{\text{fluid}}}{R_{\text{conv}}} \tag{11-2}$$

Radiation heat transfer is the result of electromagnetic radiation emitted by a surface because of the temperature of the surface. This differs from other forms of electromagnetic radiation such as radio, television, x-rays, and gamma rays which are not related to temperature. In the following sections we shall examine calculation techniques for the different modes in an elementary way. The reader should recognize, however, that this is an abbreviated presentation of the subject, and more detailed and elaborate treatments are available in the end-of-chapter references.

11-2 STEADY-STATE CONDUCTION HEAT TRANSFER

Conduction is described by *Fourier's law,* which states that the heat transfer rate q (watts) is proportional to the temperature gradient in the direction of heat flow and the area perpendicular to the heat flow (see Fig. 11-1), or

$$q_x = -kA \frac{\partial T}{\partial x} \tag{11-3a}$$

where k is the proportionality constant, which is defined as the *thermal conductivity*. The minus sign is inserted because heat must flow downhill on the temperature scale; that is, q_x is positive when $\partial T/\partial x$ is negative. The units for thermal conductivity are W/m·°C or Btu/h·ft·°F.

Some typical values of k are shown in Table 11-1. Values of the thermal conductivities for a number of metals and thermal insulators are given in Tables B-1 and B-2 in Appendix B.

For the simple system shown in Fig. 11-2 the conduction is obtained by integrating Eq. (11-3a) to give

$$q = kA \frac{T_1 - T_2}{\Delta x} = \frac{T_1 - T_2}{\Delta x/kA} \tag{11-3b}$$

FIG. 11-1 Sketch showing direction of heat flow.

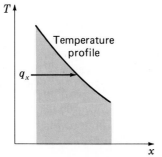

TABLE 11-1 Thermal Conductivity of Various Materials at 0°C

Material	Thermal Conductivity k W/m·°C	Thermal Conductivity k Btu/h·ft·°F
Metals		
Silver (pure)	410	237
Copper (pure)	385	223
Aluminum (pure)	202	117
Nickel (pure)	93	54
Iron (pure)	73	42
Carbon steel, 1% C	43	25
Lead (pure)	35	20.3
Chrome-nickel steel (18% Cr, 8% Ni)	16.3	9.4
Nonmetallic solids		
Quartz, parallel to axis	41.6	24
Magnesite	4.15	2.4
Marble	2.08–2.94	1.2–1.7
Sandstone	1.83	1.06
Glass, window	0.78	0.45
Maple or oak	0.17	0.096
Sawdust	0.059	0.034
Glass wool	0.038	0.022
Liquids		
Mercury	8.21	4.74
Water	0.556	0.327
Ammonia	0.540	0.312
Lubricating oil, SAE 50	0.147	0.085
Freon 12, CCl_2F_2	0.073	0.042
Gases		
Hydrogen	0.175	0.101
Helium	0.141	0.081
Air	0.024	0.0139
Water vapor (saturated)	0.0206	0.0119
Carbon dioxide	0.0146	0.008 44

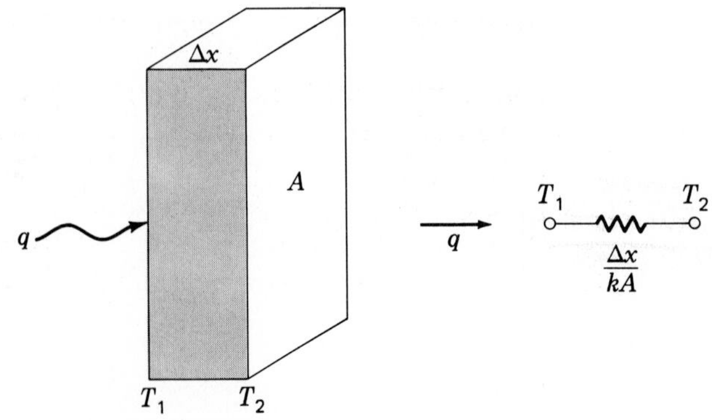

FIG. 11-2 One-dimensional heat flow in a wall and resistance analog.

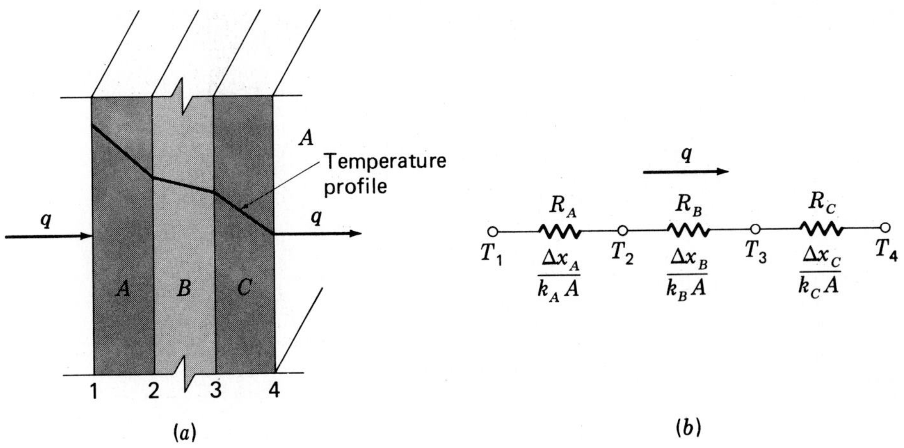

FIG. 11-3 One-dimensional heat transfer through a composite wall and electrical analog.

and an electric resistance analogy may be constructed as shown.

$$q = \frac{T_1 - T_2}{R} \tag{11-4}$$

with $R = \Delta x/kA$. Conduction through a three-layer wall, as shown in Fig. 11-3, could thus be described by

$$q = \frac{T_1 - T_2}{R_A} = \frac{T_2 - T_3}{R_B} = \frac{T_3 - T_4}{R_C}$$

or

$$q = \frac{T_1 - T_4}{(\Delta x/kA)_A + (\Delta x/kA)_B + (\Delta x/kA)_C} \tag{11-5}$$

For the hollow-cylinder system shown in Fig. 11-4 the area for heat transfer in the radial direction is $2\pi rL$, so Fourier's law becomes

$$q_r = -k2\pi rL \frac{dT}{dr} \tag{11-6a}$$

With the temperatures specified at the inside and outside radii Eq. (11-5) is integrated to give

$$q = \frac{T_i - T_o}{[\ln (r_o/r_i)]/2\pi kL} \tag{11-6b}$$

and the denominator can be taken as a thermal resistance, as shown in Fig. 11-4. As in the case of the multiple-plane wall of Fig. 11-3, multiple-layer cylin-

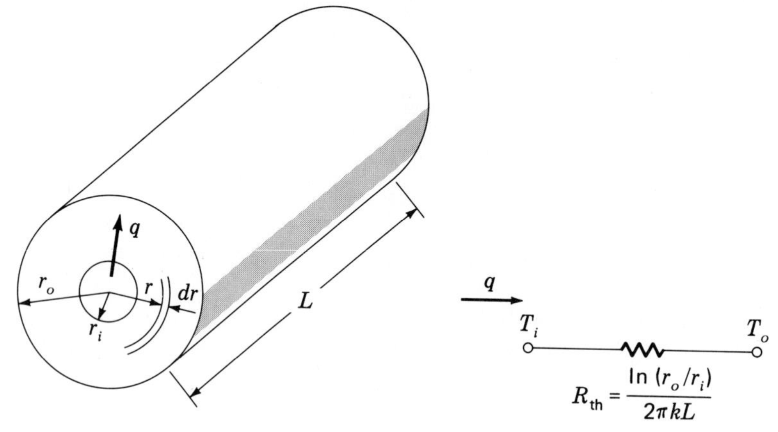

FIG. 11-4 One-dimensional heat flow through a hollow cylinder and electrical analog.

$$R_{th} = \frac{\ln (r_o/r_i)}{2\pi kL}$$

drical systems can be constructed, as shown in Fig. 11-5. Similar relations may also be derived for spherical systems.

The above analysis pertains to one-dimensional systems or those where the temperature is a function of one space coordinate. There are also two-dimensional systems where the temperature is a function of two coordinates. For some simple cases involving only two temperature limits the heat transfer can be expressed in the form

$$q = kS\, T_{overall} \tag{11-7}$$

where S is a *conduction shape factor*, which may be determined by both analytical and empirical means. Table 11-2 lists equations which may be employed for the calculation of S in a number of circumstances.

The examples illustrate applications of the conduction relations.

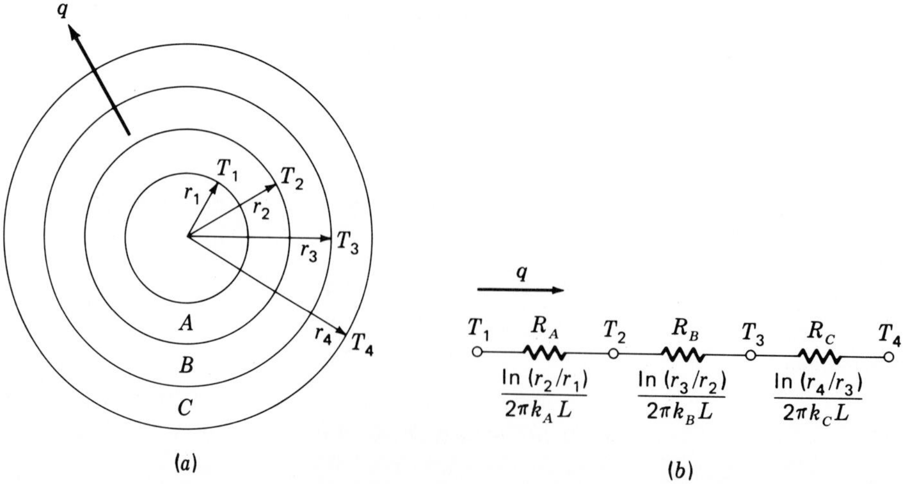

FIG. 11-5 One-dimensional heat flow through multiple cylindrical sections and electrical analog.

$$\frac{\ln (r_2/r_1)}{2\pi k_A L} \qquad \frac{\ln (r_3/r_2)}{2\pi k_B L} \qquad \frac{\ln (r_4/r_3)}{2\pi k_C L}$$

(a)

(b)

TABLE 11-2
Conduction Shape Factors, Summarized from Refs. [8, 9]

Physical System	Schematic	Shape Factor	Restrictions
Isothermal cylinder of radius r buried in semi-infinite medium having isothermal surface	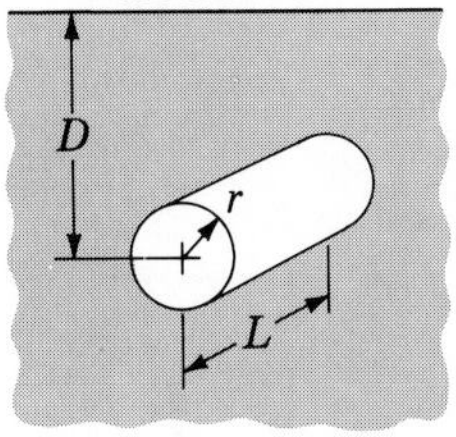	$\dfrac{2\pi L}{\cosh^{-1}(D/r)}$	$L \gg r$
		$\dfrac{2\pi L}{\ln(2D/r)}$	$L \gg r$ $D > 3r$
		$\dfrac{2\pi L}{\ln\dfrac{L}{r}\left\{1 - \dfrac{\ln[L/(2D)]}{\ln(L/r)}\right\}}$	$D \gg r$ $L \gg D$
Isothermal sphere of radius r buried in infinite medium	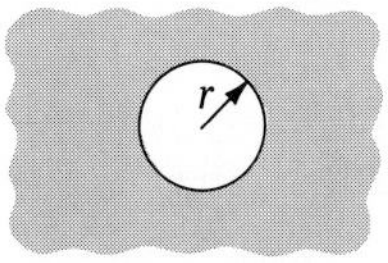	$4\pi r$	
Isothermal sphere of radius r buried in semi-infinite medium having isothermal surface	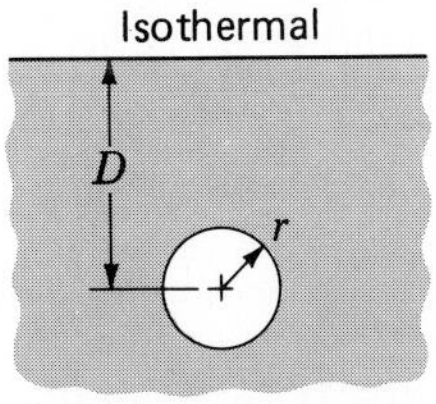	$\dfrac{4\pi r}{1 - r/2D}$	
Conduction between two isothermal cylinders buried in infinite medium	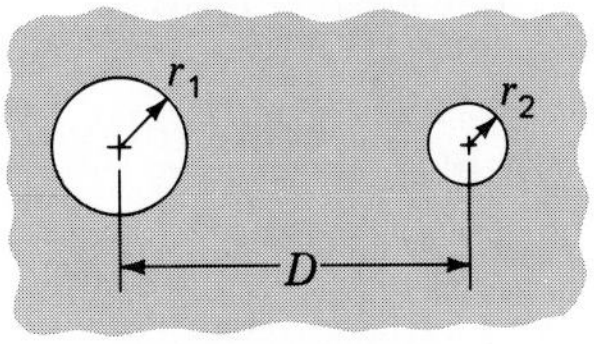	$\dfrac{2\pi L}{\cosh^{-1}\left(\dfrac{D^2 - r_1^2 - r_2^2}{2r_1r_2}\right)}$	$L \gg r$ $L \gg D$
Thin horizontal disk buried in semi-infinite medium with isothermal surface	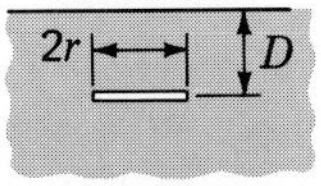	$4r$	$D = 0$
		$8r$	$D \gg 2r$
Hemisphere buried in semi-infinite medium	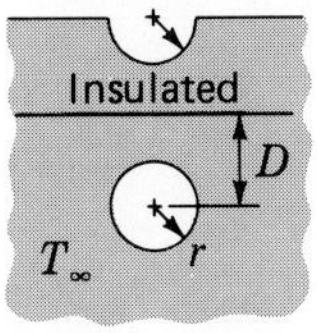	$2\pi r$	
Isothermal sphere buried in semi-infinite medium with insulated surface		$\dfrac{4\pi r}{1 + r/2D}$	
Two isothermal spheres buried in infinite medium	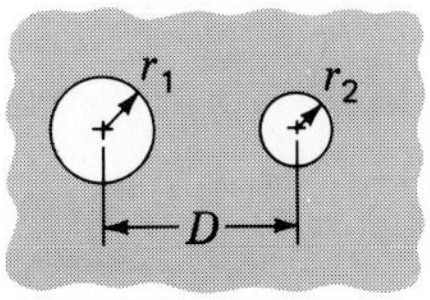	$\dfrac{4\pi}{\dfrac{r_2}{r_1}\left[1 - \dfrac{(r_1/D)^4}{1 - (r_2/D)^2}\right] - \dfrac{2r_2}{D}}$	$D > 5r$

TABLE 11-2 (*Continued*)
Conduction Shape Factors

Physical System	Schematic	Shape Factor	Restrictions
Thin rectangular plate of length L, buried in semi-infinite medium having isothermal surface	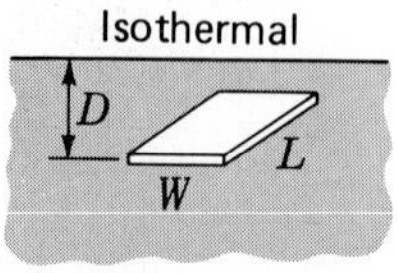	$\dfrac{\pi W}{\ln(4W/L)}$	$D = 0$
		$\dfrac{2\pi W}{\ln(4W/L)}$	$D \gg W$

EXAMPLE 11-1

An exterior wall of a house may be approximated by a 4-in layer of common brick ($k = 0.7$ W/m·°C) followed by a 1.5-in layer of gypsum plaster ($k = 0.48$ W/m·°C). What thickness of loosely packed rock-wool insulation ($k = 0.065$ W/m·°C) should be added to reduce the heat loss (or gain) through the wall by 80 percent?

SOLUTION The overall heat loss will be given by

$$q = \frac{\Delta T}{\Sigma R_{\text{th}}}$$

Because the heat loss with the rock-wool insulation will be only 20 percent (80 percent reduction) of that before insulation

$$\frac{q \text{ with insulation}}{q \text{ without insulation}} = 0.2 = \frac{\Sigma R_{\text{th}} \text{ without insulation}}{\Sigma R_{\text{th}} \text{ with insulation}}$$

We have for the brick and plaster for unit area

$$R_b = \frac{\Delta x}{k} = \frac{(4)(0.0254)}{0.7} = 0.145 \text{ m}^2\cdot{}^\circ\text{C/W}$$

$$R_p = \frac{\Delta x}{k} = \frac{(1.5)(0.0254)}{0.48} = 0.079 \text{ m}^2\cdot{}^\circ\text{C/W}$$

so that the thermal resistance without insulation is

$$R = 0.145 + 0.079 = 0.224 \text{ m}^2\cdot{}^\circ\text{C/W}$$

Then

$$R \text{ with insulation} = \frac{0.224}{0.2} = 1.122 \text{ m}^2\cdot{}^\circ\text{C/W}$$

and this represents the sum of our previous value and the resistance for the rock wool:

$$1.122 = 0.224 + R_{rw}$$

$$R_{rw} = 0.898 = \frac{\Delta x}{k} = \frac{\Delta x}{0.065}$$

so that

$$\Delta x_{rw} = 0.0584 \text{ m} = 2.30 \text{ in}$$

EXAMPLE 11-2

A thick-walled tube of stainless steel (18% Cr, 8% Ni, $k = 19$ W/m · °C) with 2-cm inner diameter (ID) and 4-cm outer diameter (OD) is covered with a 3-cm layer of asbestos insulation ($k = 0.2$ W/m·°C). If the inside-wall temperature of the pipe is maintained at 600°C and the outside of the insulation at 100°C, calculate the heat loss per meter of length.

SOLUTION The accompanying figure shows the thermal network for this problem. The heat flow is given by

$$\frac{q}{L} = \frac{2\pi(T_1 - T_2)}{\ln (r_2/r_1)/k_s + \ln (r_3/r_2)/k_a} = \frac{2\pi(600 - 100)}{(\ln 2)/19 + (\ln 5/2)/0.2} = 680 \text{ W/m}$$

EXAMPLE 11-3

An isothermal cylinder having a surface temperature of 50°C and a diameter of 15 cm is buried in the earth at a position where the depth to centerline is 30 cm and the thermal conductivity of the soil is 1.7 W/m·°C. The surface temperature of the earth is 20°C. Calculate the heat loss by the cylinder per unit length.

FIG. EXAMPLE 11-2

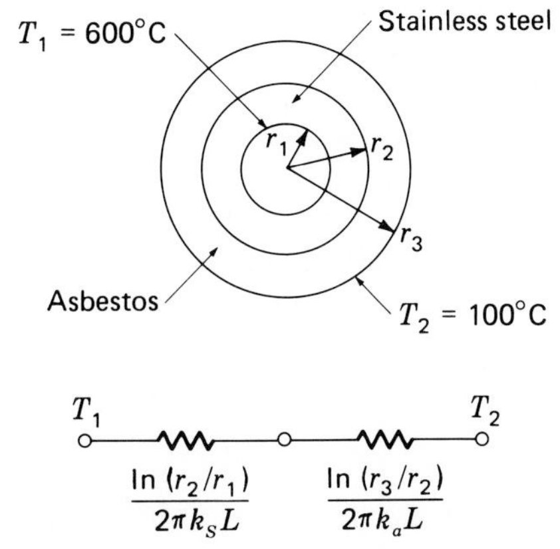

SOLUTION This is a two-dimensional problem with $D/r = 30/15 = 2.0$ using the nomenclature of Table 11-2. We thus use

$$q = kS\,\Delta T$$

to calculate the heat transfer and

$$S = \frac{2\pi L}{\cosh^{-1}(D/r)}$$

from Table 11-2. We may note that

$$\cosh^{-1}x = \ln\,(x \pm \sqrt{x^2 - 1})$$

so that

$$\cosh^{-1}\frac{D}{r} = \cosh^{-1}\frac{30}{15} = 1.317$$

and, for unit length

$$S = \frac{2}{1.317} = 4.77$$

and

$$q = kS\,\Delta T = (1.7)(4.77)(50 - 20) = 24.3 \text{ W/m}$$

11-3 TRANSIENT CONDUCTION

Up to now we have considered steady-state conduction where the temperatures do not vary with time. Obviously, there are many problems where one must be concerned with transient conditions as the temperature changes with time. We shall examine only one simple case involving cooling of a solid in a convection environment. For this analysis we consider a thermal "lump," shown in Fig. 11-6, which has a low internal conduction resistance compared with the convection resistance at the surface. This means that there will be small temperature gradients in the solid material, so that at any instant of time the material remains essentially uniform in temperature. Then, the heat lost by convection will equal the decrease in internal energy of the solid or

$$hA(T - T_\infty) = -\rho c V \frac{dT}{d\tau} \tag{11-8}$$

where T is the temperature of the solid, T_∞ is the stream temperature, A is the surface area for convection, ρ is the density, c is the specific heat, and V is the

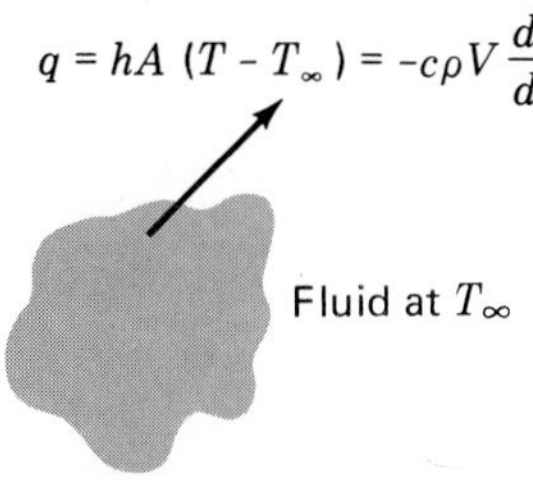

FIG. 11-6 Nomenclature for single-lump heat-capacity analysis.

volume. If the solid is initially at T_0, we have

$$T = T_0 \qquad \text{at} \qquad \tau = 0$$

and integration of Eq. (11-8) yields

$$\frac{T - T_\infty}{T_0 - T_\infty} = e^{-(hA/\rho cV)\tau} \tag{11-9}$$

As we mentioned before, this lumped-capacity analysis can apply when the conduction resistance is small compared to the convection resistance. In practice this normally applies when

$$\frac{h(V/A)}{k} < 0.1 \tag{11-10}$$

Eq. (11-9) may also be expressed in terms of a thermal resistance for convection, $R_{th} = 1/hA$, and a thermal capacitance, $C_{th} = \rho cV$, so that

$$\frac{hA}{\rho cV} = \frac{1}{R_{th}C_{th}}$$

and the system behaves like an electric capacitance discharging through a resistor.

EXAMPLE 11-4

A steel ball ($c = 0.46$ kJ/kg·°C, $k = 35$ W/m·°C) 5.0 cm in diameter and initially at a uniform temperature of 450°C is suddenly placed in a controlled environment in which the temperature is maintained at 100°C. The convection heat-transfer coefficient is 10 W/m²·°C. Calculate the time required for the ball to attain a temperature of 150°C.

SOLUTION We anticipate that the lumped-capacity method will apply because of the low value of h and high value of k. We can check by using Eq. (11-10):

$$\frac{h(V/A)}{k} = \frac{(10)[(4/3)\pi(0.025)^3]}{4\pi(0.025)^2(35)} = 0.0023 < 0.1$$

so we may use Eq. (11-9). We have

$$T = 150°\text{C} \qquad \rho = 7800 \text{ kg/m}^3 \text{ (486 lbm/ft}^3\text{)}$$

$$T_\infty = 100°\text{C} \qquad h = 10 \text{ W/m}^2\cdot°\text{C (1.76 Btu/h}\cdot\text{ft}^2\cdot°\text{F)}$$

$$T_0 = 450°\text{C} \qquad c = 460 \text{ J/kg}\cdot°\text{C (0.11 Btu/lbm}\cdot°\text{F)}$$

$$\frac{hA}{\rho c V} = \frac{(10)4\pi(0.025)^2}{(7800)(460)(4\pi/3)(0.025)^3} = 3.344 \times 10^{-4}\ s^{-1}$$

$$\frac{T - T_\infty}{T_0 - T_\infty} = e^{-(hA/\rho c V)\tau}$$

$$\frac{150 - 100}{450 - 100} = e^{-3.344\times 10^{-4}\tau}$$

$$\tau = 5819 \text{ s} = 1.62 \text{ h}$$

11-4 CONVECTION HEAT TRANSFER

We have noted that convection heat transfer results from motion of a fluid over a heated surface. The energy transfer depends on the *rate* at which the fluid can convect, or transport, heat away. For continuum flow, the velocity of the fluid layer near the heat-transfer surface is zero so all the heat transfer for convection occurs by conduction through this stationary layer. In terms of Fig. 11-7 we have

$$q_{\text{conv}} = q \text{ (conduction at surface layer)}$$

$$= -kA \left.\frac{dT}{dy}\right|_{y=0}$$

where k is the thermal conductivity of the fluid. The region near the plate where substantial velocity gradients occur is called a *boundary layer*.

There are three main types of convection heat transfer:

1 Forced convection, where the fluid is forced across the surface as shown in Figs. 11-7 and 11-8*a*, or through a tube or channel, as shown in Fig. 11-8*b*.

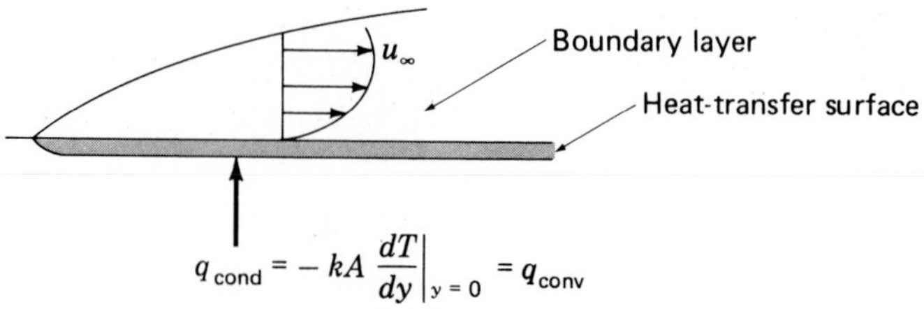

FIG. 11-7 Convection heat transfer from flat plate.

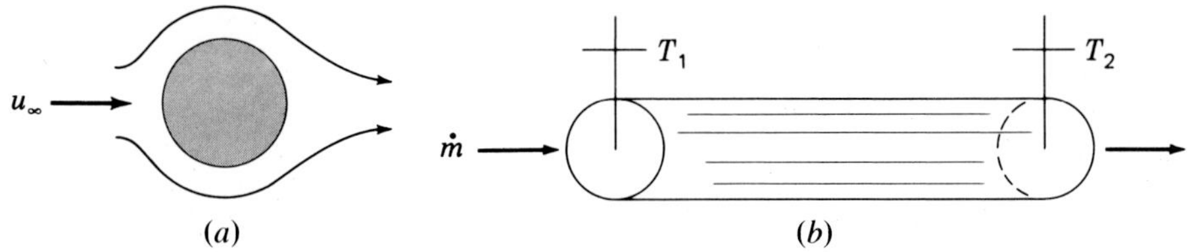

FIG. 11-8 (*a*) Cross-flow over a cylinder; (*b*) forced convection in a tube.

2 Free, or natural convection, over surfaces as shown in Fig. 11-9. Free convection results from the fact that the density of a fluid is reduced near a heated surface and therefore will rise because of buoyancy effects. The strength of the buoyancy effect is greater the larger the temperature difference between the heated surface and the fluid.

3 Change-of-phase convection heat transfer as observed in boiling or condensation phenomena. Mechanisms of these processes are very complicated and the reader should consult the references for further information.

As we mentioned earlier, it is common practice in convection heat transfer to define a *heat transfer coefficient h* such that

$$q_{\text{conv}} = hA(T_{\text{surface}} - T_{\text{fluid}}) \tag{11-11}$$

where h is in W/m$^2\cdot$°C and A is the *surface area* for convection heat transfer. Equation (11-11) is sometimes called *Newton's law of cooling*.

In forced convection over external surfaces T_{fluid} is the so-called *free-stream* temperature T_∞, or a temperature far removed from the surface. For flow in tubes or channels T_{fluid} is the *bulk*, or energy-average, temperature. For those thermodynamic balances we have performed on devices like feedwater heaters and air-conditioning cooling coils the stated "inlet" and "exit" temperatures are in fact bulk temperatures.

Convection heat transfer depends on fluid properties and they in turn depend on temperature. In many practical problems the temperature dependence is not strong, but is nevertheless present. For most external forced- and free-convection problems satisfactory calculations can be performed by evalu-

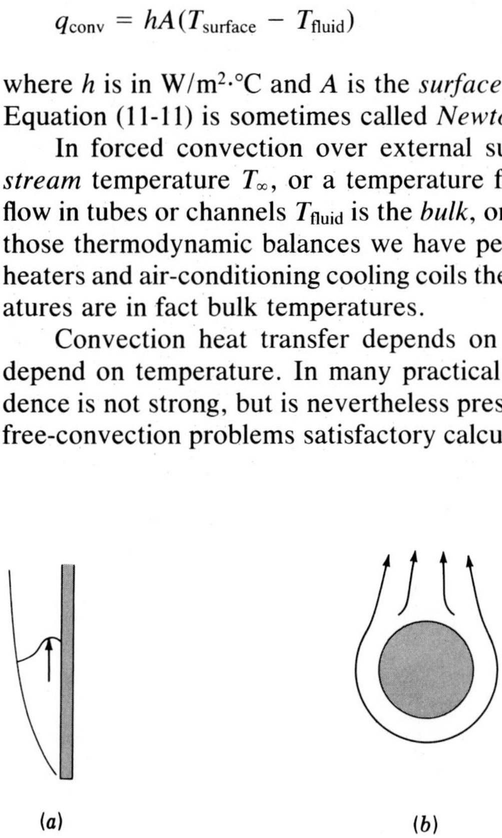

FIG. 11-9 Free convection on (*a*) vertical plate and (*b*) horizontal cylinder.

ating properties at the so-called film temperature T_f defined by

$$T_f = \frac{T_{\text{surface}} + T_{\text{free stream}}}{2} \tag{11-12}$$

while for channel flows an average bulk temperature is usually correct:

$$T_b = \frac{T_{b\text{ inlet}} + T_{b\text{ outlet}}}{2} \tag{11-13}$$

Over a period of decades, researchers and practitioners have been able to develop many analytical and empirical relations for the heat-transfer coefficient in a variety of situations. In almost all cases the results can be expressed in terms of dimensionless variables which take the following forms:

$$\text{Nu} = f(\text{Re}, \text{Pr}) \qquad \text{for forced convection} \tag{11-14}$$

and

$$\text{Nu} = f(\text{Gr}, \text{Pr}) \qquad \text{for free convection} \tag{11-15}$$

where the terms are as defined in Tables 11-3 and 11-4.

In general, convection heat transfer depends on flow system behavior, which is characterized by the Reynolds number in forced convection and by the Grashof number in free convection. These parameters determine whether the flow is laminar or turbulent. The Prandtl number determines the relative rates of transport of momentum and energy in the fluid. Expressions for these heat-transfer relations are summarized in Tables 11-5 and 11-6.

As we have seen before, convection heat transfer may also be expressed in terms of a thermal resistance concept as

$$q = hA(T_{\text{surface}} - T_{\text{fluid}}) = \frac{T_{\text{surface}} - T_{\text{fluid}}}{1/hA} \tag{11-16}$$

For the plane wall exposed to two convection environments T_A and T_B in Fig. 11-10, we would have

$$q = \frac{T_A - T_1}{1/h_A A} = \frac{T_1 - T_2}{\Delta x/kA} = \frac{T_2 - T_B}{1/h_B A}$$
$$= \frac{T_A - T_B}{1/h_A A + \Delta x/kA + 1/h_B A} \tag{11-17}$$

and we see that we have three resistances in series: two convection and one conduction. A similar approach could be taken for a hollow cylinder, as shown

TABLE 11-3 Dimensionless Groups Used in Convection Heat Transfer

Name	Symbol	Relation
Grashof number	Gr	$\mathrm{Gr}_x = \dfrac{g\beta\rho^2(T_w - T_\infty)x^3}{\mu^2}$
		Gr_L = same, with $x = L$
		$\mathrm{Gr}_d = \dfrac{g\beta\rho^2(T_w - T_\infty)d^3}{\mu^2}$
Nusselt number	Nu	$\mathrm{Nu}_x = \dfrac{h_x x}{k}$
		$\mathrm{Nu}_d = \dfrac{hd}{k}$
Nusselt number, average	$\overline{\mathrm{Nu}}$	$\overline{\mathrm{Nu}}_L = \dfrac{\bar{h}L}{k}$
		$\overline{\mathrm{Nu}}_d = \dfrac{\bar{h}d}{k}$
Reynolds number	Re	$\mathrm{Re}_x = \dfrac{\rho u_\infty x}{\mu} = \dfrac{u_\infty x}{\nu}$
For flow over flat plate		$\mathrm{Re}_L = \dfrac{\rho u_\infty L}{\mu} = \dfrac{u_\infty L}{\nu}$
For flow across cylinders		$\mathrm{Re}_d = \dfrac{\rho u_\infty d}{\mu} = \dfrac{u_\infty d}{\nu}$
For tube flow		$\mathrm{Re}_d = \dfrac{\rho u_m d}{\mu} = \dfrac{u_m d}{\nu}$
Prandtl number	Pr	$\mathrm{Pr} = \dfrac{c_p\mu}{k} = \dfrac{\nu}{\alpha}$

FIG. 11-10 Overall heat transfer through a plane wall.

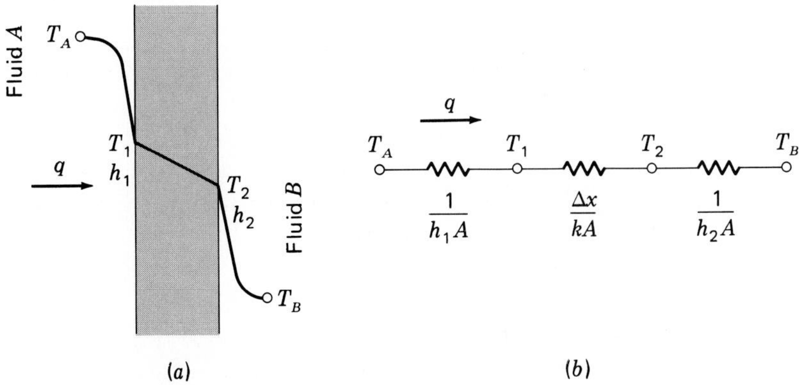

TABLE 11-4 Properties and Parameters Used in Dimensionless Groups

Symbol	Property or Parameter	Units
A	Surface area for heat transfer	m^2 or ft^2
A_c	Flow cross-section area in tube flow, $\pi d^2/4$	m^2 or ft^2
d	Diameter of cylinder or tube	m or ft
g	Acceleration of gravity	m/s^2 or ft/s^2
h_x	Local heat-transfer coefficient at distance x from leading edge of plate	W/m·°C or Btu/h·ft^2·°F
$\bar{h}$	Average heat-transfer coefficient over entire plate or cylinder, or inside tube	W/m^2·°C or Btu/h·ft^2·°F
k	Thermal conductivity	W/m·°C or Btu/h·ft·°F
L	Length of flat plate	m or ft
$\dot{m}$	Mass rate of flow	kg/s or lbm/h
T_b	Bulk temperature in tube flow	°C or °F
$\bar{T}_b$	Mean bulk temperature in tube flow	°C or °F
$T_f = \dfrac{T_\infty + T_w}{2}$	Film temperature	°C or °F
T_∞	Free-stream temperature	°C or °F
T_w	Wall or surface temperature	°C or °F
u_∞	Free-stream velocity across plate or cylinder	m/s or ft/s
u_m	Mean flow velocity in tube, defined by $\dot{m} = \rho u_m A_c$	m/s or ft/s
x	Distance from leading edge of plate	m or ft
$\beta = \dfrac{1}{v}\left(\dfrac{\partial v}{\partial T}\right)_p$	Volume coefficient of expansion (1/abs. temp. for ideal gas)	K^{-1} or $°R^{-1}$
μ	Dynamic viscosity	kg/m·s or lbm/h·ft
$\nu = \dfrac{\mu}{\rho}$	Kinematic viscosity	m^2/s or ft^2/s
ρ	Density	kg/m^3 or lbm/ft^3

TABLE 11-5

Convection Heat-Transfer Correlations for Forced Convection (See Tables 11-3 and 11-4 for Definitions of Terms)

Properties Evaluated at

Flow in Circular Tube

$q = \bar{h}A(T_w - \bar{T}_b) = \dot{m}c_p\Delta T_b$

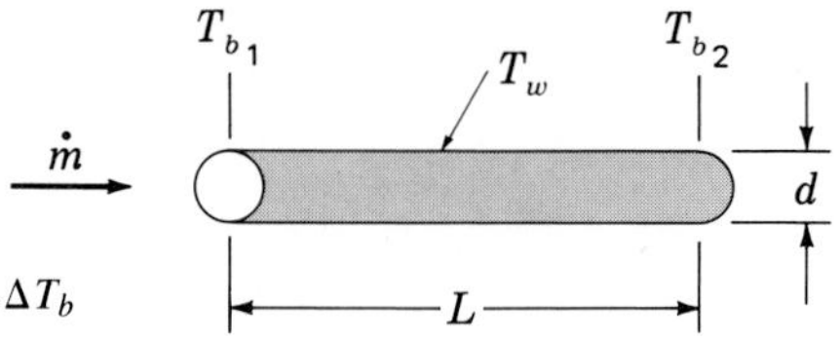

$\Delta T_b = T_{b_2} - T_{b_1}$

(a)

Fully developed laminar flow: $\mathrm{Re}_d = \dfrac{\rho u_m d}{\mu} < 2000 \qquad \dfrac{L/d}{\mathrm{Re}_d \mathrm{Pr}} > 0.05$

$\overline{\mathrm{Nu}}_d = 4.364$ for constant wall heat flux — $\bar{T}_b$

$\overline{\mathrm{Nu}}_d = 3.66$ for constant wall temperature — $\bar{T}_b$

Fully developed turbulent flow: $\mathrm{Re}_d = \dfrac{\rho u_m d}{\mu} > 2000$

$\mathrm{Nu}_d = 0.023\,\mathrm{Re}_d^{0.8}\,\mathrm{Pr}^n$ $\quad n = 0.4$ for heating; $n = 0.3$ for cooling — $\bar{T}_b$

Flow over Flat Plate, Constant Wall Temperature

$q = \bar{h}A(T_w - T_\infty)$

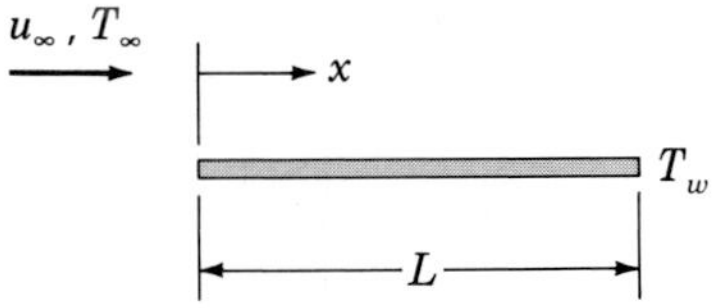

Laminar flow: $\mathrm{Re}_L < 5 \times 10^5$

$\mathrm{Nu}_x = 0.332\mathrm{Re}_x^{1/2}\,\mathrm{Pr}$ — T_f

$\overline{\mathrm{Nu}}_L = 0.664\mathrm{Re}_L^{1/2}\,\mathrm{Pr}$

(b)

Turbulent flow: $\mathrm{Re}_L > 5 \times 10^5$

$\mathrm{Nu}_x = 0.0296\mathrm{Re}_x^{0.8}\,\mathrm{Pr}^{1/3}$ — T_f

$\overline{\mathrm{Nu}}_L = (0.037\mathrm{Re}_L^{0.8} - 850)\mathrm{Pr}^{1/3}$

Flow across Cylinder, Constant Wall Temperature

$q = \bar{h}A(T_w - T_\infty)$ — T_f

$\overline{\mathrm{Nu}}_d = C\,\mathrm{Re}_d^n\,\mathrm{Pr}^{1/3}$

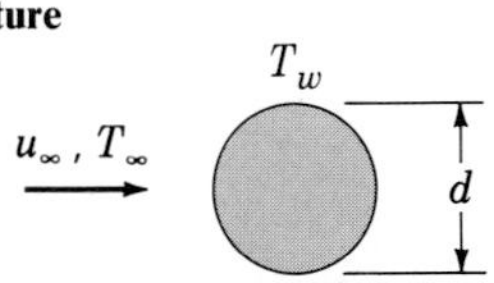

Re_d	C	n
0.4–4	0.989	0.330
4–40	0.911	0.385
40–4000	0.683	0.466
4000–40000	0.193	0.618
40000–400000	0.0266	0.805

(c)

Flow across Sphere, Constant Wall Temperature

Gases: $\overline{\mathrm{Nu}}_d = 0.37\mathrm{Re}_d^{0.6}$ — T_f

Water and oil: $\overline{\mathrm{Nu}}_d = (1.2 + 0.53\mathrm{Re}_d^{0.54})\mathrm{Pr}^{0.3}$ — T_f

Overbars on $\overline{\mathrm{Nu}}$ and $\bar{h}$ indicate average value over entire heat transfer surface

TABLE 11-6 Convection Heat-Transfer Correlations for Free Convection, Constant Surface Temperature

General relation: $\overline{\mathrm{Nu}} = C(\mathrm{GrPr})^m$ with properties evaluated at film temperature	GrPr	C	m
Vertical planes and cylinders			
	10^4–10^9	0.59	1/4
	10^9–10^{13}	0.10	1/3
Horizontal cylinders			
	10^{-10}–10^{-2}	0.675	0.058
	10^{-2}–10^2	1.02	0.148
	10^2–10^4	0.85	0.188
	10^4–10^9	0.53	1/4
	10^9–10^{12}	0.13	1/3
Upper surface of heated plates or lower surface of cooled plates	2×10^4–8×10^6	0.54	1/4
Upper surface of heated plates or lower surface of cooled plates	8×10^6–10^{11}	0.15	1/3
Lower surface of heated plates or upper surface of cooled plates	10^5–10^{11}	0.58	1/5

in the network of Fig. 11-11 and the overall heat transfer would be

$$q = \frac{T_A - T_B}{1/h_iA_i + [\ln (r_o/r_i)]/2\pi kL + 1/h_oA_o} \tag{11-18}$$

for inner and outer fluid temperatures of T_A and T_B.

Calculation Procedure for Convection Heat Transfer

Although the collection of formulas may appear rather formidable, the calculation of convection heat-transfer coefficients is rather straightforward when these steps are followed:

1 Establish whether the problem is free or forced convection.
2 Establish the geometry.
3 Determine the flow regime by calculating the Reynolds number if forced convection or the Grashof-Prandtl number product if free convection. Be sure to use the proper temperature to evaluate fluid properties. This is

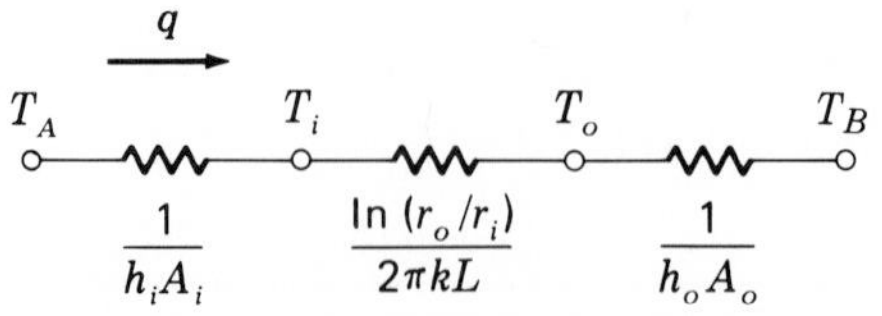

FIG. 11-11 Resistance analogy for hollow cylinder with convection boundaries.

usually film temperature for external flows and mean bulk temperature for internal flows.

4 Select the appropriate equations from Tables 11-5 and 11-6 in accordance with the above three steps and calculate the value of h, the convection heat-transfer coefficient.

5 If required, calculate the heat-transfer rate q or surface area from Eq. (11-11).

Physical properties for air and water to be used in heat-transfer calculations are given in Tables B-3 and B-4 of Appendix B. The following examples illustrate the concepts for calculation of convection heat transfer. Most of the convection heat-transfer relations presented here, and others, are available in a computer software package by Kraus [7].

EXAMPLE 11-5

Air at 1 atm and 300 K flows across a 5-cm-diameter cylinder at a velocity of 50 m/s. The cylinder is maintained at a constant temperature of 400 K. Calculate the heat lost per unit length of the cylinder.

SOLUTION We consult Table 11-5 and find that the heat-transfer coefficient may be calculated from the relation

$$\overline{\mathrm{Nu}}_d = \frac{\bar{h}d}{k} = C\,\mathrm{Re}_d^{\,n}\,\mathrm{Pr}^{1/3}$$

with properties evaluated at the film temperature T_f:

$$T_f = \frac{T_w + T_\infty}{2} = \frac{400 + 300}{2} = 350\ \mathrm{K}$$

From Table B-3 the properties of air at 350 K are

$$\nu = 20.76 \times 10^{-6}\ \mathrm{m^2/s} \qquad k = 0.03003\ \mathrm{W/m\cdot{}^\circ C} \qquad \mathrm{Pr} = 0.697$$

so that

$$\mathrm{Re}_d = \frac{u_\infty d}{\nu} = \frac{(50)(0.05)}{20.76 \times 10^{-6}} = 1.204 \times 10^5$$

Again consulting Table 11-5, we find

$$C = 0.0266 \qquad n = 0.805$$

so we have

$$\frac{\bar{h}d}{k} = (0.0266)(1.204 \times 10^5)^{0.805}(0.697)^{1/3} = 290.1$$

and

$$\bar{h} = \frac{(290.1)(0.03003)}{0.05} = 174 \text{ W/m}^2\cdot{}^\circ\text{C}$$

For unit length the surface area of the cylinder is

$$A = \pi d = \pi(0.05) = 0.1571 \text{ m}^2/\text{m}$$

so that the heat transfer is

$$q = (hA(T_w - T_\infty) = (174)(0.1571)(400 - 300) = 2733 \text{ W/m}$$

EXAMPLE 11-6

Water is to be heated from 50 to 70°F in a 2-cm-diameter tube maintained at 120°F. The flow rate is 11 kg/min. Calculate the length of tube required to accomplish this heating.

SOLUTION Consulting Table 11-5, we find that properties are evaluated at average bulk temperature for this type of problem. Thus

$$T_b = \frac{50 + 70}{2} = 60^\circ\text{F}$$

and from Table B-4 the properties of water are

$$c_p = 4186 \text{ kJ/kg}\cdot{}^\circ\text{C}$$

$$\rho = 999 \text{ kg/m}^3$$

$$\mu = 1.12 \times 10^{-3} \text{ kg/m}\cdot\text{s}$$

$$k = 0.595 \text{ W/m}\cdot{}^\circ\text{C}$$

$$\text{Pr} = 7.88$$

The flow regime is determined by calculating the Reynolds number:

$$\text{Re}_d = \frac{\rho u_m d}{\mu}$$

The mean flow velocity is determined from

$$\dot{m} = \rho \frac{\pi d^2}{4} u_m$$

$$u_m = \frac{11/60}{(999)\pi(0.02)^2/4} = 0.584 \text{ m/s}$$

and

$$\mathrm{Re}_d = \frac{(999)(0.584)(0.02)}{1.12 \times 10^{-3}} = 10421 > 2000$$

Thus the flow is turbulent and we use

$$\overline{\mathrm{Nu}}_d = \frac{\bar{h}d}{k} = 0.023\ \mathrm{Re}_d^{0.8}\ \mathrm{Pr}^{0.4}$$

to calculate $\bar{h}$:

$$\overline{\mathrm{Nu}}_d = (0.023)(10421)^{0.8}(7.88)^{0.4} = 86.03$$

and

$$\bar{h} = \frac{(86.03)(0.595)}{0.02} = 2560\ \mathrm{W/m^2 \cdot {}^\circ C}$$

The total heat transfer is

$$q = \dot{m}c_p\ \Delta T_b = \left(\frac{11}{60}\right)(4186)(70 - 50)\left(\frac{5}{9}\right)$$
$$= 8527\ \mathrm{W}$$

and this is also equal to

$$q = \bar{h}A(T_w - T_b)$$

where the surface area for convection is $A = \pi dL$. Thus

$$q = 8527\ \mathrm{W} = (2560)(\pi)(0.02)L(120 - 60)\left(\frac{5}{9}\right)$$

and

$$L = 1.59\ \mathrm{m}$$

EXAMPLE 11-7

Air at 1 atm and 300 K flows across a 50-by-50-cm-square plate at a velocity of 50 m/s. The plate is maintained at a constant temperature of 400 K. Calculate the heat lost by the plate.

SOLUTION Consulting Table 11-5, we find that properties for this geometry are evaluated at the film temperature. For air at 1 atm we obtain from Table B-3

$$T_f = \frac{T_w + T_\infty}{2} = \frac{400 + 300}{2} = 350\ \mathrm{K}$$

and

$$\nu = 20.76 \times 10^{-6}\ \text{m}^2/\text{s} \qquad k = 0.03003\ \text{W/m}\cdot{}^\circ\text{C} \qquad \text{Pr} = 0.697$$

We now must evaluate the Reynolds number to determine the flow regime and subsequently the correlation to use for calculation of the convection heat-transfer coefficient:

$$\text{Re}_L = \frac{u_\infty L}{\nu} = \frac{(50)(0.5)}{20.76 \times 10^{-6}} = 1.204 \times 10^6$$

Thus from Table 11-5 the relation for determining h is

$$\overline{\text{Nu}}_L = \frac{\bar{h}L}{k} = (0.037\text{Re}_L^{0.8} - 850)\text{Pr}^{1/3}$$

$$= [(0.037)(1.204 \times 10^6)^{0.8} - 850](0.697)^{1/3}$$

$$= 1648$$

and

$$\bar{h} = \frac{(1648)(0.03003)}{0.5} = 99\ \text{W/m}^2\cdot{}^\circ\text{C}$$

The surface area for convection is $A = 0.50 \times 0.50 = 0.25\ \text{m}^2$, so

$$q = hA(T_w - T_\infty) = (99)(0.25)(400 - 200) = 2475\ \text{W}$$

EXAMPLE 11-8

A horizontal cylinder 6 cm in diameter is exposed to stagnant room air at 1 atm and 300 K. The surface of the cylinder is maintained at a constant temperature of 400 K. Calculate the heat lost by the cylinder per unit length.

SOLUTION This is a *free* convection problem because there is no forced movement of the air across the heat-transfer surface. Consulting Table 11-6, we find that the fluid properties are evaluated at the film temperature:

$$T_f = \frac{T_w + T_\infty}{2} = \frac{400 + 300}{2} = 350\ \text{K}$$

From Table B-3, the properties of air at 1 atm are

$$\rho = 0.998\ \text{kg/m}^3$$

$$\mu = 2.075 \times 10^{-5}\ \text{kg/m}\cdot\text{s}$$

$$k = 0.03003\ \text{W/m}\cdot{}^\circ\text{C}$$

$$\text{Pr} = 0.697$$

$$\beta = \frac{1}{T_f} = \frac{1}{350} = 0.00286$$

The flow regime is determined by evaluating the Grashof-Prandtl number product:

$$\text{Gr}_d \text{ Pr} = \frac{g\beta\rho^2(T_w - T_\infty)d^3}{\mu^2} \text{ Pr}$$

$$= \frac{(9.8)(0.00286)(0.998)^2(400 - 300)(0.06)^3}{(2.075 \times 10^{-5})^2} 0.697$$

$$= 9.76 \times 10^5$$

From Table 11-6 we find

$$\overline{\text{Nu}}_d = C(\text{GrPr})^n$$

and $C = 0.53$, $m = 1/4$. Thus

$$\overline{\text{Nu}}_d = (0.53)(9.76 \times 10^5)^{1/4} = 16.66$$

and

$$\bar{h} = \frac{(16.66)(0.03003)}{0.06} = 8.34 \text{ W/m}^2\cdot{}^\circ\text{C}$$

The surface area per unit length is $A = \pi d = \pi(0.06) = 0.1884 \text{ m}^2\text{/m}$ so the heat transfer is

$$q = \bar{h}A(T_w - T_\infty) = (8.34)(0.1884)(400 - 300) = 157 \text{ W/m}$$

11-5 RADIATION HEAT TRANSFER

As we have mentioned before, thermal radiation is electromagnetic radiation which is emitted by a body as a result of its temperature. The basic physical principle which governs such radiation is the Stefan-Boltzmann law, which states that the energy emitted by an ideal radiator, or *blackbody,* per unit area and per unit time is proportional to absolute temperature to the fourth power. Thus

$$\frac{q}{A} = \sigma T^4 = E_b \qquad \text{W/m}^2 \tag{11-19}$$

where the proportionality factor σ is defined as the *Stefan-Boltzmann constant* and has the value

$$\begin{aligned}\sigma &= 5.669 \times 10^{-8}\ \mathrm{W/m^2 \cdot K^4} \\ &= 0.1714 \times 10^{-8}\ \mathrm{Btu/h \cdot ft^2 \cdot {}^\circ R^4}\end{aligned} \tag{11-20}$$

E_b is called the *blackbody emissive power*. A blackbody is one that absorbs all radiation incident upon it.

Thermal radiation extends over a spectral range of wavelengths from about 0.1 to 100 μm and the spectral distribution is described by the Planck blackbody distribution formula:

$$E_{b\lambda} = \frac{C_1 \lambda^{-5}}{e^{C_2/\lambda T} - 1} \tag{11-21}$$

where
λ = wavelength, μm
T = temperature, K
$C_1 = 3.743 \times 10^8\ \mathrm{W \cdot \mu m^4/m^2}$
$C_2 = 1.4387 \times 10^4\ \mathrm{\mu m \cdot K}$

$E_{b\lambda}$ is called the *monochromatic* blackbody emissive power and has the units of $\mathrm{W/m^2 \cdot \mu m}$.

Real surfaces emit less radiant energy than a blackbody, and a property called the *emissivity* ϵ is defined as

$$\epsilon = \frac{E}{E_b} \tag{11-22}$$

where E is the emissive power of the real surface. A *gray body* is defined such that the monochromatic emissivity is constant over all wavelengths:

$$\epsilon_\lambda = \frac{E_\lambda}{E_{b\lambda}} = \text{constant for gray body}$$

Real surfaces are not gray either but may have a rather jagged emissive power distribution as indicated in Fig. 11-12. Even so, the gray body assumption is a sufficiently good approximation for most practical radiation problems.

When radiation strikes a surface as shown in Fig. 11-13 a fraction ρ is reflected, a fraction α is absorbed, and a fraction τ transmitted. These fractions are called respectively the *reflectivity, absorptivity,* and *transmissivity*. It can be shown (see Ref. [1]) that, for a gray body,

$$\epsilon = \alpha \tag{11-23}$$

This is called *Kirchhoff*'s identity. The actual radiant energy transfer between the two surfaces 1 and 2 shown in Fig. 11-14 depends upon the two surface

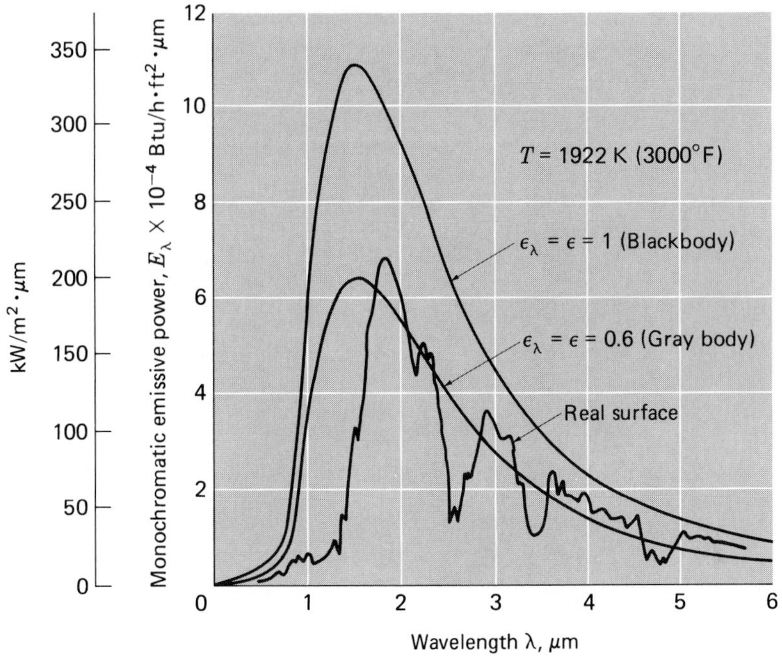

FIG. 11-12 Comparison of ideal blackbody, gray body, and real surface.

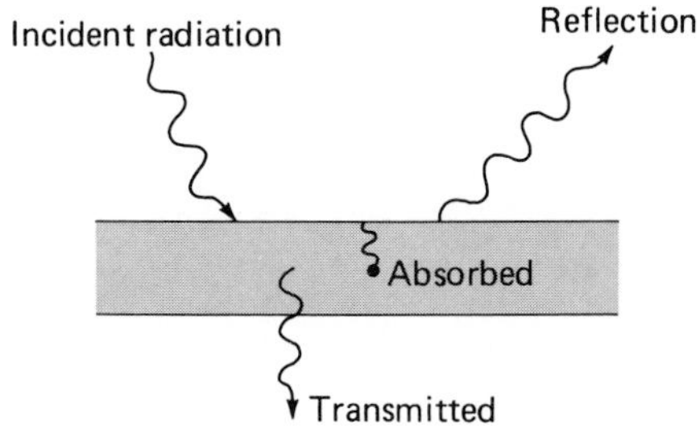

FIG. 11-13 Sketch showing effects of incident radiation at surface.

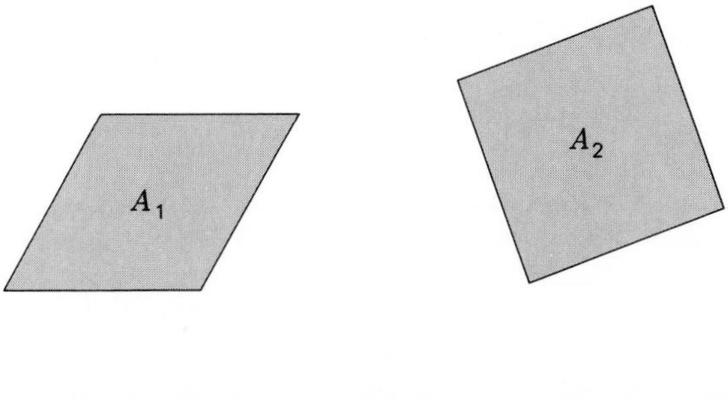

FIG. 11-14 Effect of geometric orientation on radiation exchange.

temperatures, the surface emissivities, and the geometric orientation in space. We shall define a *radiation shape factor* F_{12} as the fraction of energy leaving surface 1 which arrives at surface 2. Similarly, F_{21} is the fraction leaving 2 which arrives at 1. F_{12} and F_{21} are pure functions of geometry.

It can be shown (see Ref. [1]) that a reciprocity relation exists such that

$$A_1F_{12} = A_2F_{21} \tag{11-24}$$

or for general surfaces i and j,

$$A_iF_{ij} = A_jF_{ji} \tag{11-25}$$

Charts of F_{12} for three geometries are shown in Figs. 11-15, 11-16, and 11-17.

Analysis of Radiation Exchange

Two new terms are now defined as

Radiosity $= J =$ total energy leaving a surface per unit area and per unit time (sum of emitted and reflected energies)

Irradiation $= G =$ total energy incident on a surface per unit area and per unit time

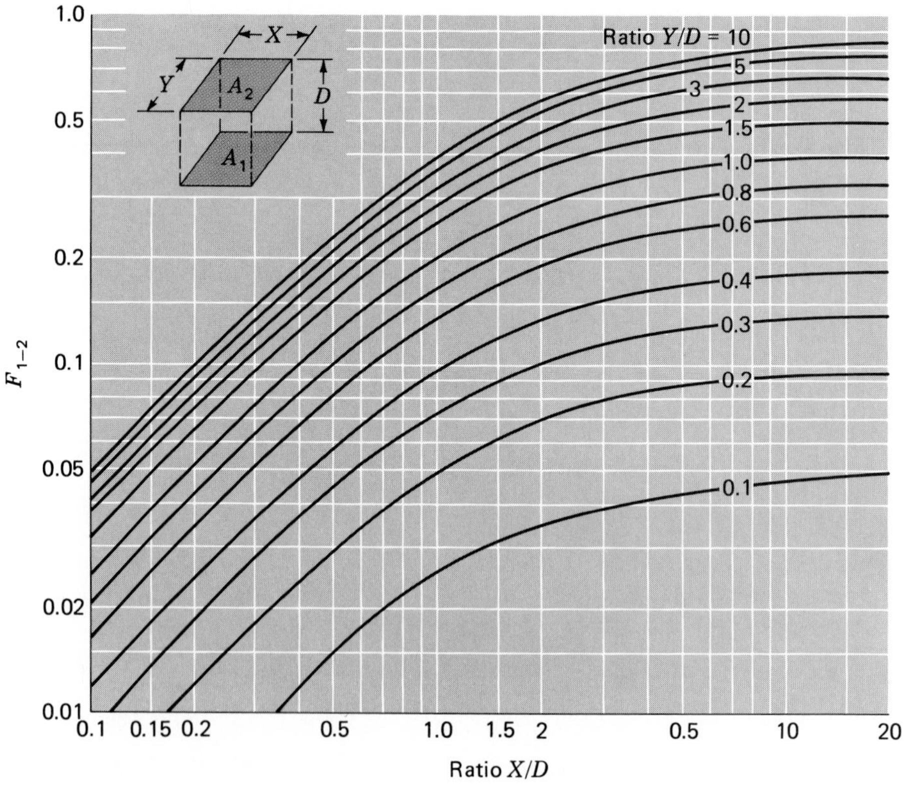

FIG. 11-15 Radiation shape factor for radiation between parallel rectangles.

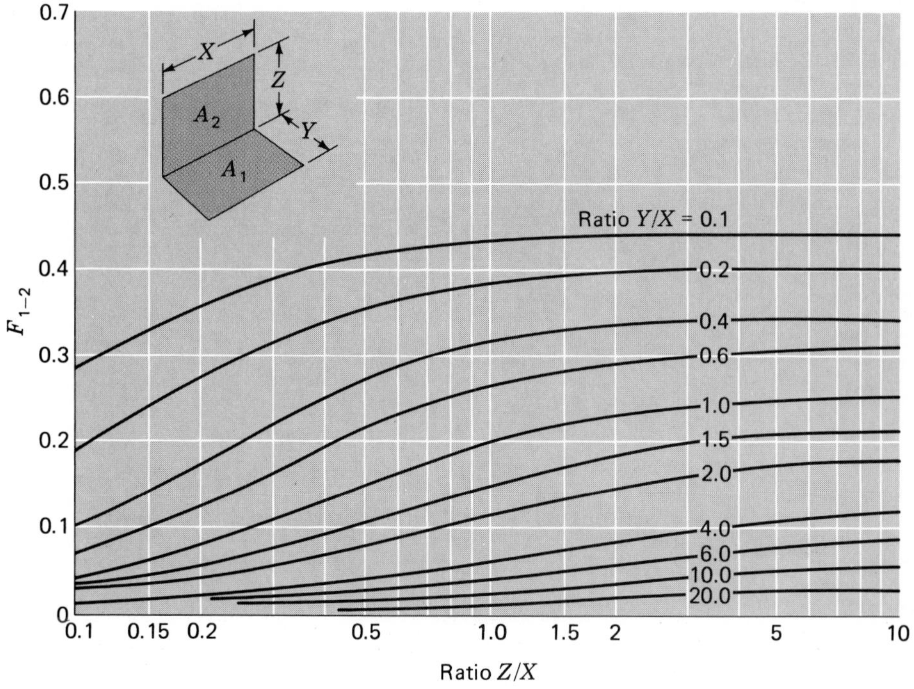

FIG. 11-16 Radiation shape factor for radiation between perpendicular rectangles with a common edge.

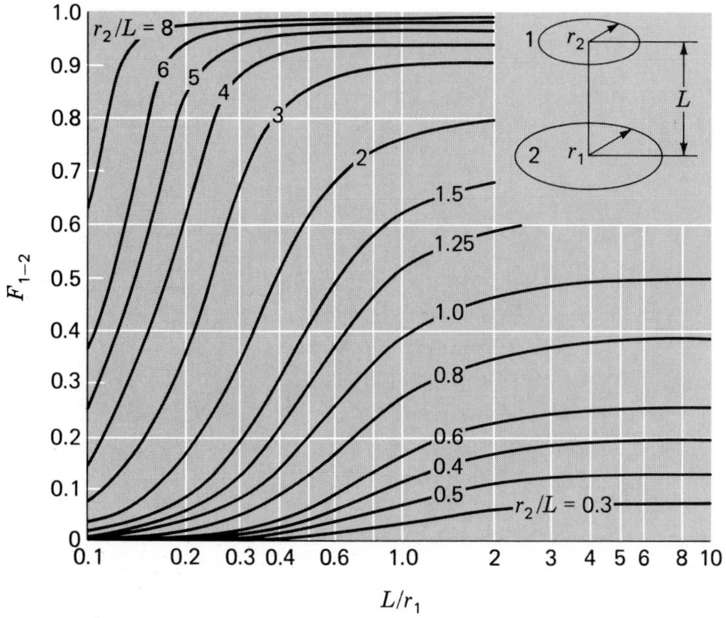

FIG. 11-17 Radiation shape factor for radiation between two parallel concentric disks.

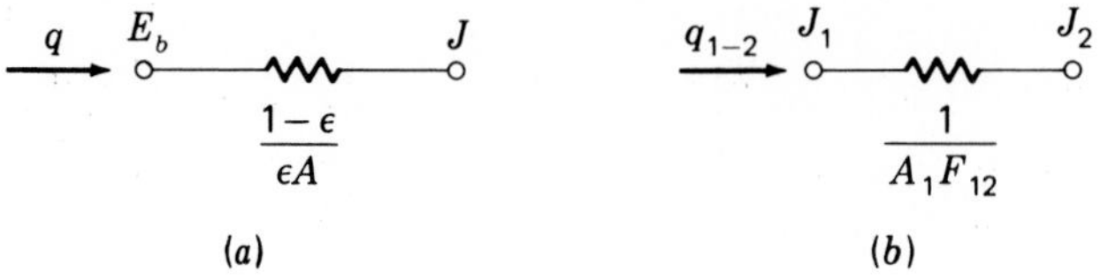

FIG. 11-18 Resistance analogy for radiation: (*a*) surface resistance, (*b*) space resistance.

In analytical terms for an opaque surface with $\tau = 0$,

$$J = \epsilon E_b + \rho G = \epsilon E_b + (1 - \alpha)G$$

The net radiation heat *leaving* the surface is

$$\frac{q}{A} = J - G$$

Setting $\alpha = \epsilon$, and with algebraic manipulation,

$$q = \frac{E_b - J}{(1 - \epsilon)/\epsilon A} \tag{11-26}$$

and a resistance analogy for this equation is shown in Fig. 11-18*a*. Of the energy leaving surface 1 the amount arriving at 2 will be

$$q_{1\to 2} = J_1 A_1 F_{12}$$

and of that energy leaving 2 the amount that arrives at 1 is

$$q_{2\to 1} = J_2 A_2 F_{21}$$

The net exchange is

$$q_{12\ \text{net}} = J_1 A_1 F_{12} - J_2 A_2 F_{21} = \frac{J_1 - J_2}{1/A_1 F_{12}} \tag{11-27}$$

where the reciprocity relation $A_1F_{12} = A_2F_{21}$ has been used. The resistance analogy for Eq. (11-27) is shown in Fig. 11-18*b*.

The *radiation network method* consists of assembling the above resistance elements and then solving for the current flows as heat transfers.

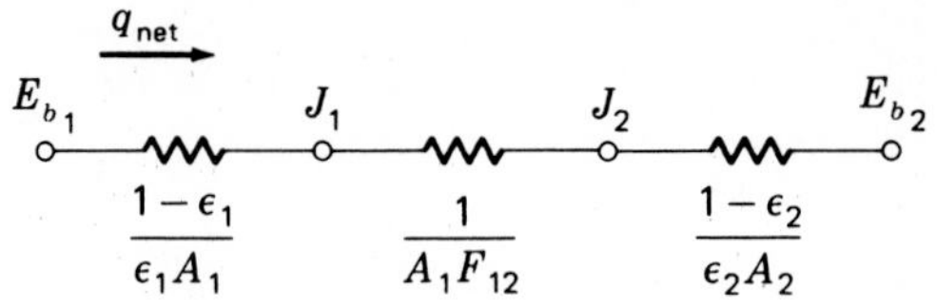

FIG. 11-19 Radiation network for two surfaces which see each other and nothing else.

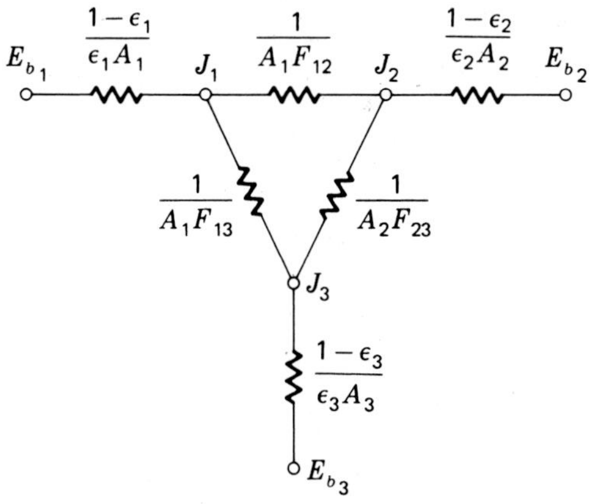

FIG. 11-20 Radiation network for three surfaces which see each other and nothing else.

Networks for two- and three-surface problems are shown in Figs. 11-19 and 11-20. The general technique is to attach a surface resistance to each E_b and J, and a space resistance between the J's.

For the special case of a surface 1 surrounded by a very large surface 2, the resistor $(1 - \epsilon_2)/\epsilon_2 A_2$ in Fig. 11-19 is very small compared with the other two resistors. In addition, if surface 1 is convex $F_{12} = 1.0$ and the overall heat transfer from 1 to 2 is

$$q = \frac{E_{b_1} - E_{b_2}}{[(1 - \epsilon_1)/\epsilon_1 A_1] + (1/(A_1) + 0} \qquad (11\text{-}28)$$
$$= A_1\epsilon_1(E_{b_1} - E_{b_2}) = A_1\epsilon_1\sigma(T_1^4 - T_2^4)$$

EXAMPLE 11-9

Two long concentric cylinders having diameters of 5 and 10 cm exchange heat by radiation. The inner cylinder has $T_1 = 800$ K and $\epsilon_1 = 0.65$ while the outer cylinder has $T_2 = 400$ K and $\epsilon_2 = 0.4$. Calculate the net heat transfer between the two cylinders per unit length.

FIG. EXAMPLE 11-9

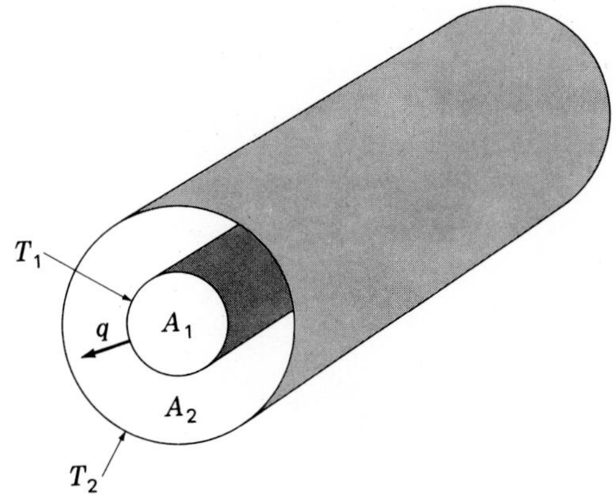

SOLUTION The schematic for the problem is shown in the accompanying sketch and the radiation network in Fig. 11-19. This is a two-body problem because the cylinders are very long and little radiation leaks out the ends. We thus compute the resistors for the network per unit length and the blackbody emissive powers.

All of the radiation leaving the inner cylinder arrives at the outer cylinder, so $F_{12} = 1.0$. Then, for unit length

$$A_1 = \pi d_1 = \pi(0.05) = 0.157 \qquad A_2 = \pi(0.1) = 0.314$$

$$\frac{1-\epsilon_1}{\epsilon_1 A_1} = \frac{1-0.65}{(0.65)(0.157)} = 3.43 \qquad \frac{1-\epsilon_2}{\epsilon_2 A_2} = \frac{1-0.4}{(0.4)(0.314)} = 4.78$$

$$\frac{1}{A_1 F_{12}} = \frac{1}{(0.157)(1.0)} = 6.38$$

$$E_{b_1} = \sigma T_1^4 = (5.669 \times 10^{-8})(800)^4 = 23220 \text{ W/m}^2$$

$$E_{b_2} = \sigma T_2^4 = (5.669 \times 10^{-4})(400)^4 = 1451 \text{ W/m}^2$$

The total heat transfer is now obtained from the network of Fig. 11-19 as

$$q = \frac{E_{b_1} - E_{b_2}}{\Sigma R} = \frac{23220 - 1451}{3.43 + 6.37 + 4.78} = 1493 \text{ W/m length}$$

EXAMPLE 11-10 Hot Plates in a Room

Two parallel plates 0.5 by 1.0 m are spaced 0.5 m apart. One plate is maintained at 1000°C and the other at 500°C. The emissivities of the plates are 0.2 and 0.5, respectively. The plates are located in a very large room, the walls of which are maintained at 27°C. The plates exchange heat with each other and with the room, but only the plate surfaces facing each other are to be considered in the analysis. Find the net transfer to each plate and to the room.

SOLUTION This is a three-body problem, the two plates and the room, so the radiation network is shown in Fig. 11-20. From the data of the problem

$$T_1 = 1000°\text{C} = 1273 \text{ K} \qquad A_1 = A_2 = 0.5 \text{ m}^2$$

$$T_2 = 500°\text{C} = 773 \text{ K} \qquad \epsilon_1 = 0.2$$

$$T_3 = 27°\text{C} = 300 \text{ K} \qquad \epsilon_2 = 0.5$$

FIG. EXAMPLE 11-10

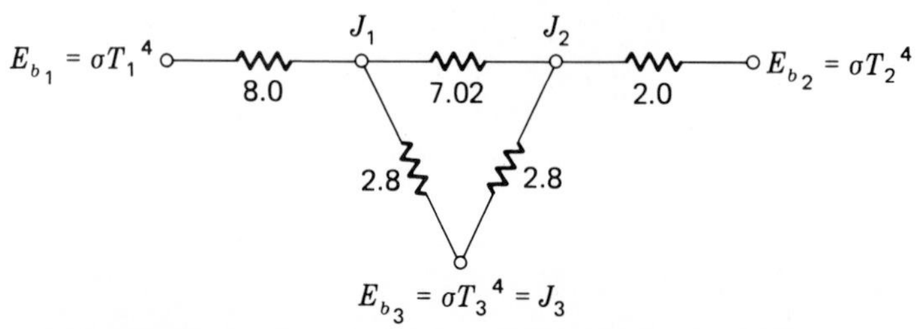

Because the area of the room A_3 is very large, the resistance $(1 - \epsilon_3)/\epsilon_3 A_3$ may be taken as zero and we obtain $E_{b_3} = J_3$. The shape factor is obtained from Fig. 11-15:

$$F_{12} = 0.285 = F_{21}$$
$$F_{13} = 1 - F_{12} = 0.715$$
$$F_{23} = 1 - F_{21} = 0.715$$

The resistances in the network are calculated as

$$\frac{1 - \epsilon_1}{\epsilon_1 A_1} = \frac{1 - 0.2}{(0.2)(0.5)} = 8.0 \qquad \frac{1 - \epsilon_2}{\epsilon_2 A_2} = \frac{1 - 0.5}{(0.5)(0.5)} = 2.0$$

$$\frac{1}{A_1 F_{12}} = \frac{1}{(0.5)(0.285)} = 7.018 \qquad \frac{1}{A_1 F_{13}} = \frac{1}{(0.5)(0.715)} = 2.797$$

$$\frac{1}{A_2 F_{23}} = \frac{1}{(0.5)(0.715)} = 2.797$$

Taking the resistance $(1 - \epsilon_3)/\epsilon_3 A_3$ as zero, we have the network as shown. To calculate the heat flows at each surface we must determine the radiosities J_1 and J_2. The network is solved by setting the sum of the heat currents entering nodes J_1 and J_2 to zero:

$$\textit{Node } J_1\text{:} \quad \frac{E_{b_1} - J_1}{8.0} + \frac{J_2 - J_1}{7.018} + \frac{E_{b_3} - J_1}{2.797} = 0 \qquad (a)$$

$$\textit{Node } J_2\text{:} \quad \frac{J_1 - J_2}{7.018} + \frac{E_{b_3} - J_2}{2.797} + \frac{E_{b_2} - J_2}{2.0} = 0 \qquad (b)$$

Now

$$E_{b_1} = \sigma T_1^4 = 148.87 \text{ kW/m}^2 \qquad (47190 \text{ Btu/h}\cdot\text{ft}^2)$$
$$E_{b_2} = \sigma T_2^4 = 20.241 \text{ kW/m}^2 \qquad (6416 \text{ Btu/h}\cdot\text{ft}^2)$$
$$E_{b_3} = \sigma T_3^4 = 0.4592 \text{ kW/m}^2 \qquad (145.6 \text{ Btu/h}\cdot\text{ft}^2)$$

Inserting the values of E_{b_1}, E_{b_2}, and E_{b_3} into Eqs. (a) and (b), we have two equations and two unknowns J_1 and J_2 which may be solved simultaneously to give

$$J_1 = 33.469 \text{ kW/m}^2 \qquad J_2 = 15.054 \text{ kW/m}^2$$

The total heat lost by plate 1 is

$$q_1 = \frac{E_{b_1} - J_1}{(1 - \epsilon_1)/\epsilon_1 A_1} = \frac{148.87 - 33.469}{8.0} = 14.425 \text{ kW}$$

and the total heat lost by plate 2 is

$$q_2 = \frac{E_{b_2} - J_2}{(1 - \epsilon_2)/\epsilon_2 A_2} = \frac{20.241 - 15.054}{2.0} = 2.594 \text{ kW}$$

The total heat received by the room is

$$q_3 = \frac{J_1 - J_3}{1/A_1F_{13}} + \frac{J_2 - J_3}{1/A_2F_{23}}$$

$$= \frac{33.469 - 0.4592}{2.797} + \frac{15.054 - 0.4592}{2.797} = 17.020 \text{ kW} \qquad [58070 \text{ Btu/h}]$$

From an overall-balance standpoint we must have

$$q_3 = q_1 + q_2$$

because the net energy lost by both plates must be absorbed by the room.

EXAMPLE 11-11 Surface in Radiant Balance

Two rectangles 50 by 50 cm are placed perpendicularly with a common edge. One surface has $T_1 = 1000$ K, $\epsilon_1 = 0.6$, while the other surface is insulated and in radiant balance with a large surrounding room at 300 K. Determine the temperature of the insulated surface and the heat lost by the surface at 1000 K.

SOLUTION The radiation network is shown in the accompanying figure where surface 3 is the room and surface 2 is the insulated surface. Note that

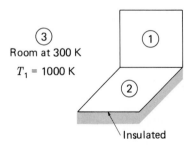

FIG. EXAMPLE 11-11

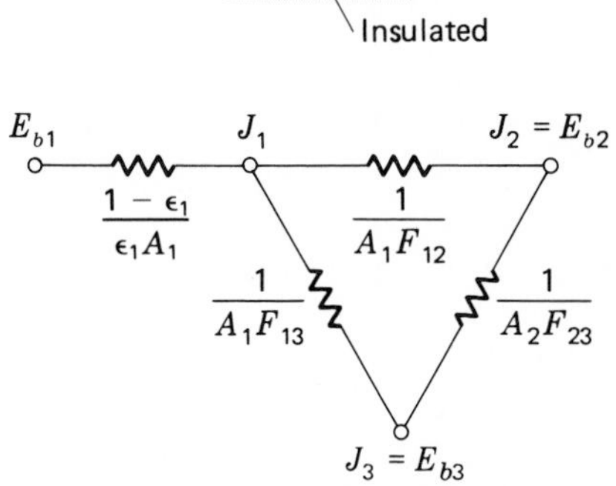

$J_3 = E_{b_3}$ because the room is large and $(1 - \epsilon_3)/\epsilon_3 A_3$ approaches zero. Because surface 2 is insulated it has zero heat transfer and $J_2 = E_{b_2}$. J_2 "floats" in the network and is determined from the overall radiant balance. From Fig. 11-16 the shape factors are

$$F_{12} = 0.2 = F_{21}$$

Because $F_{11} = 0$ and $F_{22} = 0$, we have

$$F_{12} + F_{13} = 1.0 \qquad \text{and} \qquad F_{13} = 1 - 0.2 = 0.8 = F_{23}$$

$$A_1 = A_2 = (0.5)^2 = 0.25 \text{ m}^2$$

The resistances are

$$\frac{1 - \epsilon_1}{\epsilon_1 A_1} = \frac{0.4}{(0.6)(0.25)} = 2.667$$

$$\frac{1}{A_1 F_{13}} = \frac{1}{A_2 F_{23}} = \frac{1}{(0.25)(0.8)} = 5.0$$

$$\frac{1}{A_1 F_{12}} = \frac{1}{(0.25)(0.2)} = 20.0$$

We also have

$$E_{b_1} = (5.669 \times 10^{-8})(1000)^4 = 5.669 \times 10^4 \text{ W/m}^2$$

$$J_3 = E_{b_3} = (5.669 \times 10^{-8})(300)^4 = 459.2 \text{ W/m}^2$$

The overall circuit is a series-parallel arrangement and the heat transfer is

$$q = \frac{E_{b_1} - E_{b_3}}{R_{\text{equiv}}}$$

We have

$$R_{\text{equiv}} = 2.667 + \frac{1}{1/5 + 1/(20 + 5)} = 6.833$$

and

$$q = \frac{56\,690 - 459.2}{6.833} = 8.229 \text{ kW} \qquad (28\,086 \text{ Btu/h})$$

This heat transfer can also be written

$$q = \frac{E_{b_1} - J_1}{(1 - \epsilon_1)/\epsilon_1 A_1}$$

Inserting the values we obtain

$$J_1 = 34\ 745\ \text{W/m}^2$$

The value of J_2 is determined from proportioning the resistances between J_1 and J_3, so that

$$\frac{J_1 - J_2}{20} = \frac{J_1 - J_3}{20 + 5}$$

and

$$J_2 = 7316 = E_{b_2} = \sigma T_2^4$$

Finally, we obtain the temperature as

$$T_2 = \left(\frac{7316}{5.669 \times 10^{-8}}\right)^{1/4} = 599.4\ \text{K} \qquad (619°\text{F})$$

11-6 FINS

Most readers will have seen some type of heat-transfer surface with fins attached to increase the area for convection and thus increase the heat-transfer rate. One type of fin is the straight fin of rectangular profile shown in Fig. 11-21. By combining a conduction and convection energy balance on the fin it is possible to derive an expression for the temperature distribution at any x position as [see Ref. (1)]

$$\frac{T - T_\infty}{T_0 - T_\infty} = \frac{\cosh[m(L_c - x)]}{\cosh(mL_c)} \tag{11-29}$$

where T_0 is the temperature at the base of the fin, $L_c = L + t/2$, L is the fin length, t is the fin thickness, T_∞ the free-stream temperature, h is the heat

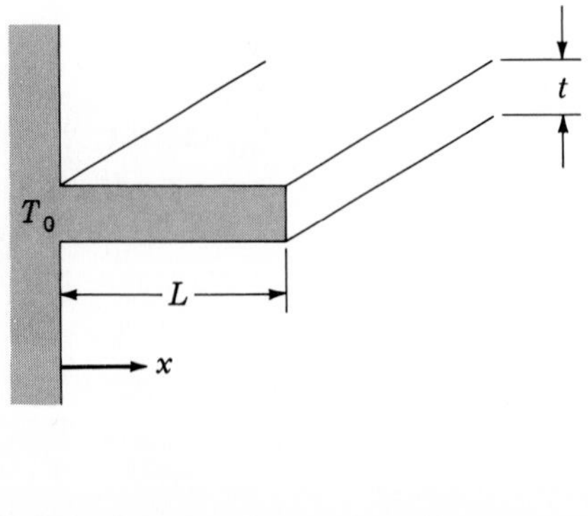

FIG. 11-21 Straight fin of rectangular profile.

transfer coefficient, k is the thermal conductivity of the fin, and

$$m = (2h/kt)^{1/2} \tag{11-30}$$

A fin efficiency may be defined as

$$\text{Fin efficiency} = \frac{\text{actual heat transfer}}{\text{heat which would be transferred if entire fin area were at base temperature}} = \eta_f \tag{11-31}$$

For the rectangular fin the fin efficiency may be calculated from

$$\eta_f = \frac{\tanh(mL_c)}{mL_c} \tag{11-32}$$

The hyperbolic functions are defined as

$$\cosh x = \frac{e^x + e^{-x}}{2} \qquad \tanh x = \frac{e^x - e^{-x}}{e^x + e^{-x}}$$

Analytical expressions for circular fins may also be derived but are very complicated, so the results are given here in graphical form in Fig. 11-22. The top curve for $r_{2c}/r_1 = 1.0$ is also the efficiency for the straight fin of rectangular

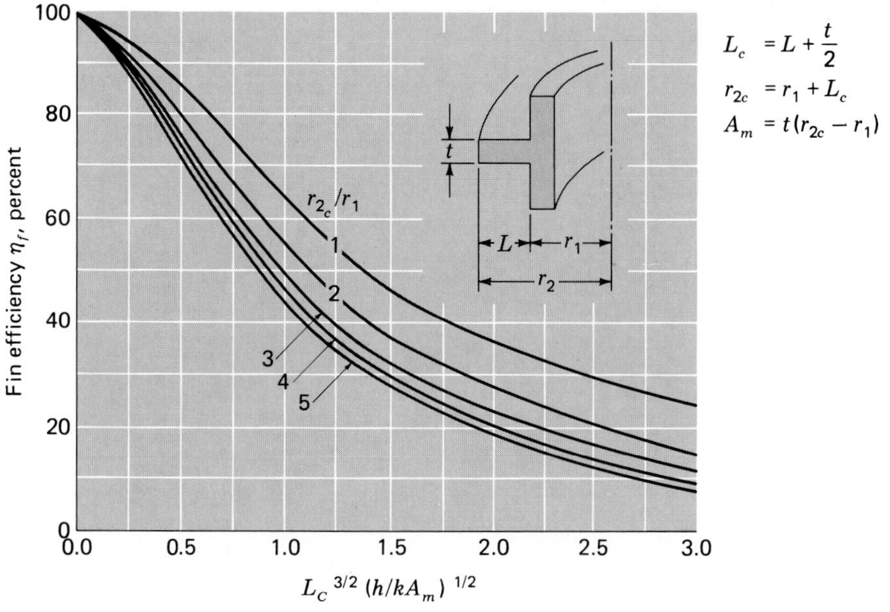

FIG. 11-22 Efficiencies of circumferential fins of rectangular profile.

profile. From the definition of η_f, the actual heat transfer is thus

$$q_{\text{act}} = \eta_f \times h \times (\text{surface area for convection}) \times (T_0 - T_\infty) \tag{11-33}$$

EXAMPLE 11-12

An aluminum fin ($k = 200$ W/m·°C) 3.0 mm thick and 7.5 cm long protrudes from a wall, as shown in Fig. 11-21. The base is maintained at 300°C, and the ambient temperature is 50°C with $h = 10$ W/m²·°C. Calculate the heat loss from the fin per unit depth of material.

SOLUTION We have

$$L_c = L + \frac{t}{2} = 7.5 + 0.15 = 7.65 \text{ cm} \qquad (3.01 \text{ in})$$

$$m = \frac{2h}{kt}$$

or

$$m = \left[\frac{(2)(10)}{(200)(3 \times 10^{-3})}\right]^{1/2} = 5.774$$

From Eq. (11-32) the fin efficiency is

$$\eta_f = \frac{\tanh(mL_c)}{mL_c}$$

and

$$mL_c = (5.774)(0.0765) = 0.4417$$

so that

$$\eta_f = \frac{0.4151}{0.4417} = 0.94$$

For unit depth the maximum heat transfer is

$$\begin{aligned} q_{\max} &= h \times \text{surface area} \times (T_0 - T_\infty) \\ &= (10)(2)(0.0765)(300 - 50) = 382.5 \text{ W/m} \end{aligned}$$

The actual heat transfer is then

$$q_{\text{act}} = \eta_f q_{\max} = (0.94)(382.5) = 360 \text{ W/m}$$

EXAMPLE 11-13

Aluminum fins 1.5 cm wide and 1.0 mm thick are placed on a 2.5-cm-diameter tube to dissipate the heat. The tube surface temperatures is 170°C and the ambient fluid temperature is 25°C. Calculate the heat loss per fin for $h = 130$ W/m²·°C. Assume $k = 200$ W/m·°C for aluminum.

SOLUTION For this example we can compute the heat transfer by using the fin-efficiency curves in Fig. 11-22. The parameters needed are

$$L_c = L + \frac{t}{2} = 1.5 + 0.05 = 1.55 \text{ cm}$$

$$r_1 = \frac{2.5}{2} = 1.25 \text{ cm}$$

$$r_{2_c} = r_1 + L_c = 1.25 + 1.55 = 2.80 \text{ cm}$$

$$\frac{r_{2_c}}{r_1} = \frac{2.80}{1.25} = 2.24$$

$$A_m = t(r_{2_c} - r_1) = (0.001)(2.8 - 1.25)(10^{-2}) = 1.55 \times 10^{-5} \text{ m}^2$$

$$L_c^{3/2}\left(\frac{h}{kA_m}\right)^{1/2} = (0.0155)^{3/2}\left[\frac{130}{(200)(1.55 \times 10^{-5})}\right]^{1/2} = 0.396$$

From Fig. 11-22 $\eta_f = 82$ percent. The heat which would be transferred if the entire fin were at the base temperature is (both sides of fin exchanging heat)

$$\begin{aligned} q_{max} &= 2\pi(r_{2_c}^2 - r_1^2)h(T_0 - T_\infty) \\ &= 2\pi(2.8^2 - 1.25^2)(10^{-4})(130)(170 - 25) \\ &= 74.35 \text{ W} \qquad (253.7 \text{ Btu/h}) \end{aligned}$$

The actual heat transfer is then the product of the heat flow and the fin efficiency:

$$q_{act} = (0.82)(74.35) = 60.97 \text{ W} \qquad (208 \text{ Btu/h})$$

11-7 HEAT EXCHANGERS

When discussing power and refrigeration cycles we have mentioned heat exchangers such as feedwater heaters and air-conditioning cooling coils. Let us first consider a basic type of heat exchanger: the double-pipe exchanger shown in Fig. 11-23 along with the thermal resistance network. If fluid A is the hotter fluid, the heat transfer is given by Eq. (11-18) which is repeated here for convenience:

$$q = \frac{T_A - T_B}{1/h_iA_i + [\ln(r_o/r_i)/2\pi kL] + 1/h_oA_o} \tag{11-34}$$

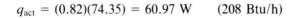

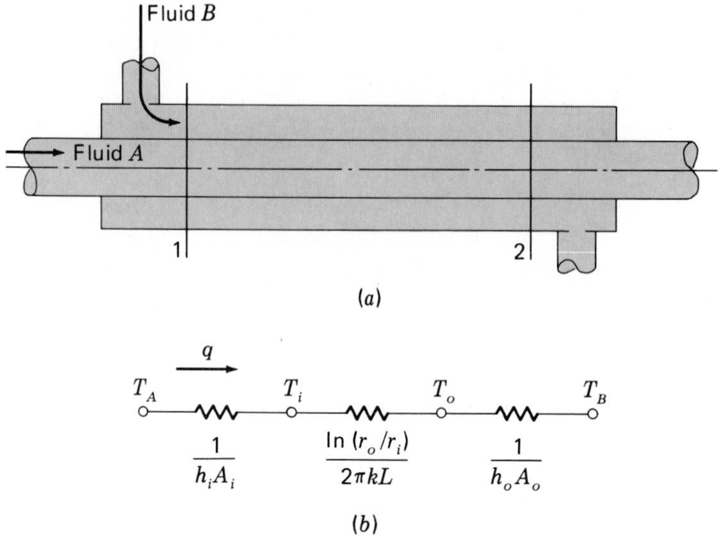

FIG. 11-23 Double-pipe heat exchanger: (a) schematic; (b) thermal-resistance network for overall heat transfer.

and it is customary to define an *overall* heat-transfer coefficient U as

$$q = UA(T_A - T_B) \tag{11-35}$$

The product UA is the inverse of the denominator in Eq. (11-34), and some approximate values of U are given in Table 11-7.

TABLE 11-7
Approximate Values of Overall Heat-Transfer Coefficients

Physical Situation	**U** Btu/h·ft²·°F	**U** W/m²·°C
Brick exterior wall, plaster interior, uninsulated	0.45	2.55
Frame exterior wall, plaster interior:		
Uninsulated	0.25	1.42
With rock-wool insulation	0.07	0.4
Plate-glass window	1.10	6.2
Double plate-glass window	0.40	2.3
Steam condenser	200–1000	1100–5600
Feedwater heater	200–1500	1100–8500
Freon-12 condenser with water coolant	50–150	280–850
Water-to-water heat exchanger	150–300	850–1700
Finned-tube heat exchanger, water in tubes, air across tubes	5–10	25–55
Water-to-oil heat exchanger	20–60	110–350
Steam to light fuel oil	30–60	170–340
Steam to heavy fuel oil	10–30	56–170
Steam to kerosene or gasoline	50–200	280–1140
Finned-tube heat exchanger, steam in tubes, air over tubes	5–50	28–280
Ammonia condenser, water in tubes	150–250	850–1400
Alcohol condenser, water in tubes	45–120	255–680
Gas-to-gas heat exchanger	2–8	10–40

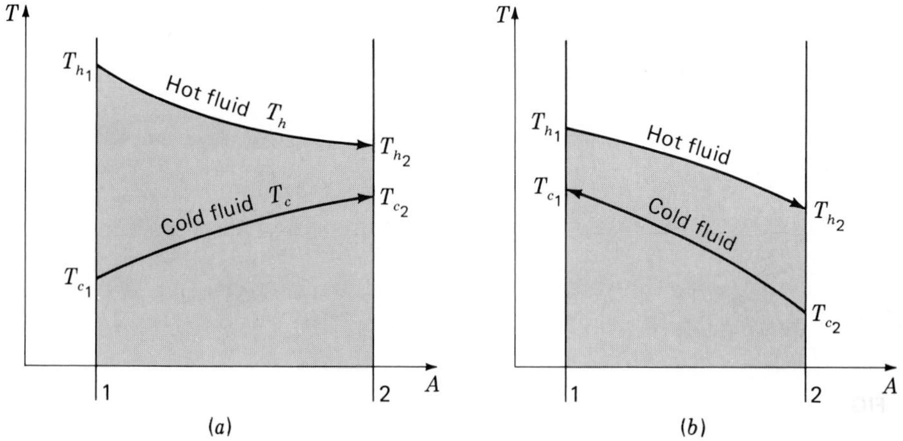

FIG. 11-24 Temperature profiles for parallel flow and counterflow in double-pipe heat exchanger.

The double-pipe heat exchanger may be operated in the parallel-flow mode or counterflow mode, as indicated in Fig. 11-24. Obviously, entirely different temperature profiles result and the central question is: What average temperature difference shall be used in calculating the heat transfer in Eq. (11-35)? An analysis shows that for the double-pipe exchanger the correct value is the log-mean-temperature difference, LMTD, defined by

$$\text{LMTD} = \Delta T_m = \frac{(T_{h_2} - T_{c_2}) - (T_{h_1} - T_{c_1})}{\ln [(T_{h_2} - T_{c_2})/(T_{h_1} - T_{c_1})]} \tag{11-36}$$

where the nomenclature is indicated in Figs. 11-23 and 11-24. Then the total heat transfer for the exchanger is

$$q = UA\ \Delta T_m \tag{11-37}$$

FIG. 11-25 Cross-flow heat exchanger, both fluids unmixed.

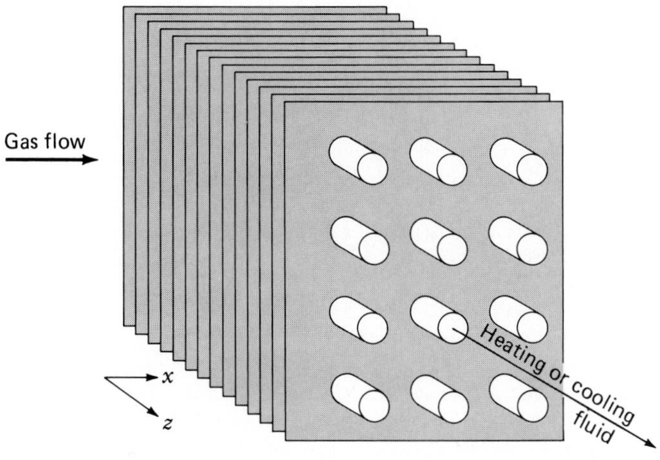

Another type of heat exchanger is the finned tube, or cross-flow, exchanger shown in Fig. 11-25. This is the type of exchanger normally employed in air-conditioning cooling coils.

An alternate method of analyzing heat exchangers hinges on defining a *heat-exchanger effectiveness* ϵ as

$$\epsilon = \frac{\text{actual heat transfer}}{\text{maximum possible heat transfer}} \tag{11-38}$$

For the parallel-flow double-pipe exchanger an energy balance yields

$$q = \dot{m}_h c_h (T_{h_1} - T_{h_2}) = \dot{m}_c C_c (T_{c_2} - T_{c_1})$$

The maximum possible heat transfer is expressed as

$$q_{\max} = (\dot{m}c)_{\min}(T_{h_{\text{inlet}}} - T_{c_{\text{inlet}}}) \tag{11-39}$$

FIG. 11-26 Effectiveness for parallel-flow exchanger performance.

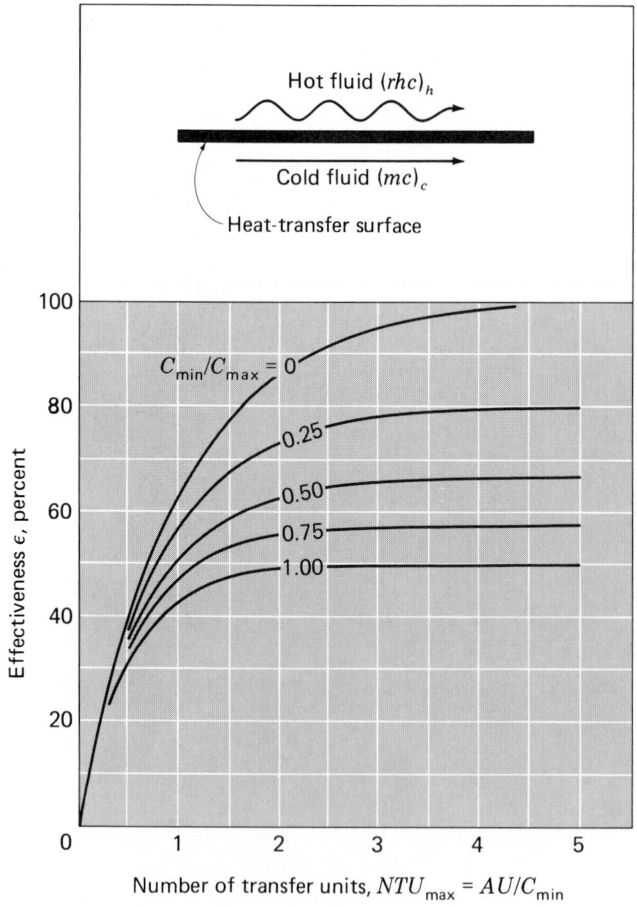

which would be

$$q_{max} = (\dot{m}c)_{min}(T_{h_1} - T_{c_1})$$

for the parallel-flow exchanger and

$$q_{max} = (\dot{m}c)_{min}(T_{h_1} - T_{c_2})$$

for the counterflow exchanger.

Designating $\dot{m}c = C$, analysis shows that the effectiveness can be expressed in the functional form

$$\epsilon = fcn\left(NTU, \frac{C_{min}}{C_{max}}\right) \tag{11-40}$$

FIG. 11-27 Effectiveness for counterflow exchanger performance.

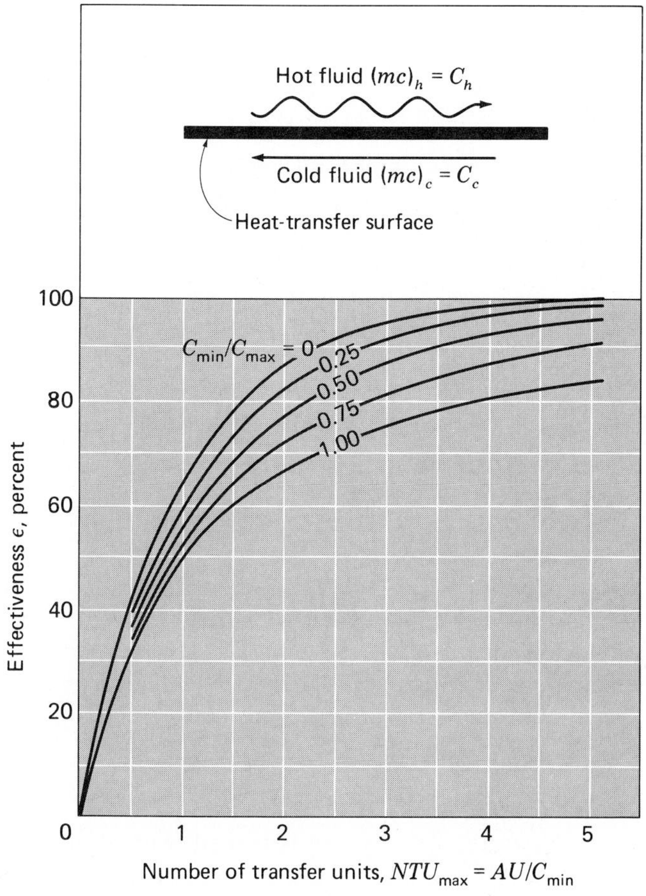

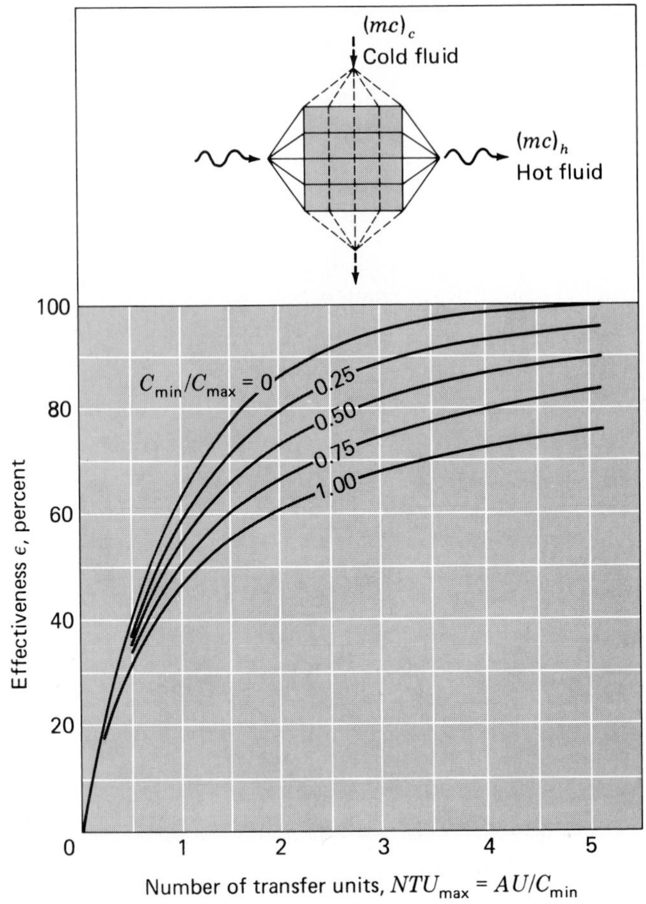

FIG. 11-28 Effectiveness for cross-flow exchanger with fluids unmixed.

where

$$NTU = \frac{UA}{C_{min}}$$

NTU is called the *number of transfer units* because it is related to the size or area of the heat exchanger. Plots of the effectiveness as a function of these variables are shown in Figs. 11-26 to 11-28 for the parallel-flow and counterflow double-pipe exchangers and the cross-flow finned-tube exchanger. The examples illustrate the application of the above relations to calculation of heat-exchanger performance.

EXAMPLE 11-14

Water at the rate of 68 kg/min is heated from 35 to 75°C by an oil having a specific heat of 1.9 kJ/kg·°C. The fluids are used in a counterflow double-pipe heat exchanger, and the oil enters the exchanger at 110°C and leaves at 75°C. The overall heat-transfer coefficient is 320 W/m$^2\cdot$°C. Calculate the heat-exchanger area.

SOLUTION The total heat transfer is determined from the energy absorbed by the water:

$$q = \dot{m}_w c_w \, \Delta T_w = (68)(4180)(75 - 35) = 11.37 \text{ MJ/min} \qquad (a)$$
$$= 189.5 \text{ kW} \qquad (6.47 \times 10^5 \text{ Btu/h})$$

Since all the fluid temperatures are known, the LMTD can be calculated by using the temperature scheme in Fig. 11-24:

$$\Delta T_m = \frac{(110 - 75) - (75 - 35)}{\ln [(110 - 75)/(75 - 35)]} = 37.44\text{°C} \qquad (b)$$

Then, since $q = UA \, \Delta T_m$,

$$A = \frac{1.895 \times 10^5}{(320)(37.44)} = 15.82 \text{ m}^2 \qquad (170 \text{ ft}^2)$$

EXAMPLE 11-15

The heat exchanger of Example 11-14 is used for heating water as described in the example. Using the same entering-fluid temperatures calculate the exit water temperature when only 40 kg/min of water is heated but the same quantity of oil is used. Also calculate the total heat transfer under these new conditions.

SOLUTION The flow rate of oil is calculated from the energy balance for the original problem:

$$\dot{m}_h c_h \, \Delta T_h = \dot{m}_c c_c \, \Delta T_c \qquad (a)$$

$$\dot{m}_h = \frac{(68)(4180)(75 - 35)}{(1900)(110 - 75)} = 170.97 \text{ kg/min}$$

The capacity rates for the new conditions are now calculated as

$$\dot{m}_h c_h = \frac{170.97}{60} 1900 = 5414 \text{ W/°C}$$

$$\dot{m}_c c_c = \frac{40}{60} 4180 = 2787 \text{ W/°C}$$

so that the water (cold fluid) is the minimum fluid, and

$$\frac{C_{\text{min}}}{C_{\text{max}}} = \frac{2787}{5414} = 0.515$$

$$NTU_{\text{max}} = \frac{UA}{C_{\text{min}}} = \frac{(320)(15.82)}{2787} = 1.816 \qquad (b)$$

where the area of 15.82 m² is taken from Example 11-14. From Fig. 11-27 the effectiveness is

$$\epsilon = 0.744$$

and because the cold fluid is the minimum, we can write

$$\epsilon = \frac{\Delta T_{\text{cold}}}{\Delta T_{\text{max}}} = \frac{\Delta T_{\text{cold}}}{110 - 35} = 0.744 \qquad (c)$$

$$\Delta T_{\text{cold}} = 55.8°\text{C}$$

and the exit water temperature is

$$T_{w,\text{exit}} = 35 + 55.8 = 90.8°\text{C}$$

The total heat transfer under the new flow conditions is calculated as

$$q = \dot{m}_c c_c \, \Delta T_c = \frac{40}{60}(4180)(55.8) = 155.5 \text{ kW} \qquad (5.29 \times 10^5 \text{ Btu/h}) \qquad (d)$$

Notice that although the flow rate has been reduced by 41 percent (68 to 40 kg/min), the heat transfer is reduced by only 18 percent (189.5 to 155.5 kW) because the exchanger is more effective at the lower flow rate.

REVIEW QUESTIONS

1 How does the subject of *heat transfer* differ from the subject of *thermodynamics*?
2 What is Fourier's law of heat conduction?
3 Define thermal conductivity.
4 Discuss the thermal resistance analogy for conduction and convection.
5 What is the conduction-shape factor?
6 What is meant by a *lumped-capacity* analysis?
7 Discuss the physical mechanism of convection.
8 Define the convection heat-transfer coefficient.
9 Distinguish between *free* and *forced* convection.

10 What is Newton's law of cooling?
11 Define film and bulk temperatures.
12 Define Nusselt, Reynolds, and Grashof numbers.
13 What is the Stefan-Boltzmann law?
14 Define emissivity.
15 What is a gray body?
16 What is the radiation network method?
17 Define fin efficiency.
18 What is meant by the *overall heat transfer coefficient*?
19 Write an expression for the LMTD of a double-pipe heat exchanger.
20 Define heat exchanger effectiveness.

PROBLEMS

11-1 Find the heat transfer per unit area through the composite wall sketched. Assume one-dimensional heat flow.

11-2 One side of a copper block 5 cm thick is maintained at 260°C. The other side is covered with a layer of fiber glass 2.5 cm thick. The outside of the fiber glass is maintained at 38°C, and the total heat flow through the copper–fiber-glass combination is 44 kW. What is the area of the slab?

11-3 A steel pipe with a 5-cm outer diameter is covered with a 6.4-mm asbestos insulation ($k = 0.096$ Btu/h·ft·°F) followed by a 2.5-cm layer of fiber-glass insulation ($k = 0.028$ Btu/h·ft·°F). The pipe-wall temperature is 315°C, and the outside insulation temperature is 38°C. Calculate the interface temperature between the asbestos and fiber glass.

11-4 A wall 2 cm thick is to be constructed from material which has an average thermal conductivity of 1.3 W/m·°C. The wall is to be insulated with material having an average thermal conductivity of 0.35 W/m·°C so that the heat loss per square meter will not exceed 1830 W. Assuming that the inner and outer surface temperatures of the insu-

FIG. P11-1

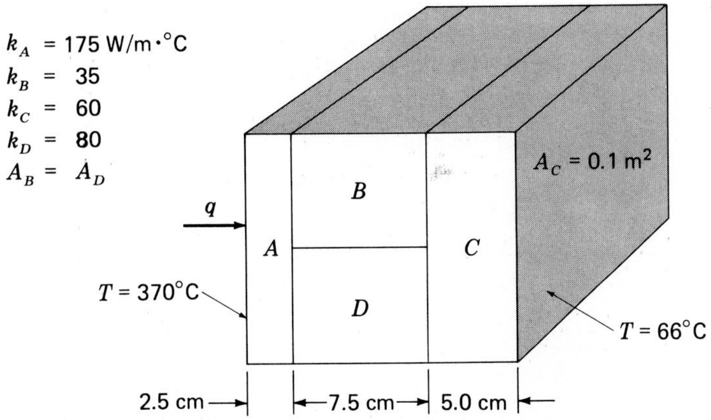

lated wall are 1300 and 30°C, calculate the thickness of insulation required.

11-5 A temperature difference of 85°C is impressed across a fiber-glass layer of 13-cm thickness. The thermal conductivity of the fiber glass is 0.035 W/m·°C. Compute the heat transferred through the material per hour per unit area.

11-6 The temperatures on the faces of a plane wall 15 cm thick are 370 and 93°C. The wall is constructed of a special glass with the following properties: $k = 0.78$ W/m·°C, $\rho = 2700$ kg/m^3, $c_p = 0.84$ kJ/kg·°C. What is the heat flow through the wall at steady-state conditions?

11-7 A certain superinsulation material having a thermal conductivity of 2×10^{-4} W/m·°C is used to insulate a tank of liquid nitrogen that is maintained at −320°F; 85.8 Btu is required to vaporize each pound-mass of nitrogen at this temperature. Assuming that the tank is a sphere having an inner diameter (ID) of 2 ft, estimate the amount of nitrogen vaporized per day for an insulation thickness of 1.0 in and an ambient temperature of 70°F. Assume that the outer temperature of the insulation is 70°F.

11-8 A 50-cm-diameter pipeline in the Arctic carries hot oil at 30°C and is exposed to a surrounding temperature of −20°C. A special powder insulation 5 cm thick surrounds the pipe and has a thermal conductivity of 7 mW/m·°C. The convection heat-transfer coefficient on the outside of the pipe is 12 W/m^2·°C. Estimate the energy loss from the pipe per meter of length.

11-9 One side of a plane wall is maintained at 100°C while the other side is exposed to a convection environment having $T = 10$°C and $h = 10$ W/m^2·°C. The wall has $k = 1.6$ W/m·°C and is 40 cm thick. Calculate the heat-transfer rate through the wall.

11-10 One side of a copper block 4 cm thick is maintained at 200°C. The other side is covered with a layer of fiber glass 2.5 cm thick. The outside of the fiber glass is maintained at 90°C, and the total heat flow through the composite slab is 300 W. What is the area of the slab?

11-11 A wall is constructed of a section of stainless steel ($k = 16$ W/m·°C) 4.0 mm thick with identical layers of plastic on both sides of the steel. The overall heat-transfer coefficient, considering convection on both sides of the plastic, is 200 W/m^2·°C. If the overall temperature difference across the arrangement is 100°C, calculate the temperature difference across the stainless steel.

11-12 A hot steam pipe having an inside surface temperature of 300°C has an inside diameter of 8 cm and a wall thickness of 5.5 mm. It is covered with a 9-cm layer of insulation having $k = 50$ W/m·°C, followed by a 4-cm layer of insulation having $k = 0.35$ W/m·°C. The outside temperature of the insulation is 30°C. Calculate the heat lost per meter of length. Assume $k = 47$ W/m·°C for the pipe.

11-13 A heavy-wall tube of Monel, 2.5-cm ID and 5-cm OD, is covered with a

2.5-cm layer of glass wool. The inside tube temperature is 260°C, and the temperature at the outside of the insulation is 38°C. How much heat is lost per foot of length? Take k = 11 Btu/h·ft·°F for Monel.

11-14 Two long cylinders 7.5 and 2.5 cm in diameter are completely surrounded by a medium with k = 1.4 W/m·°C. The distance between centers is 10 cm, and the cylinders are maintained at 200 and 35°C. Calculate the heat-transfer rate per unit length.

11-15 A 1-m-diameter sphere maintained at 35°C is buried in the earth at a place where k = 1.7 W/m·°C. The depth to the centerline is 2.4 m, and the earth surface temperature is 4°C. Calculate the heat lost by the sphere.

11-16 A copper sphere 4.0 cm in diameter is maintained at 70°C and submerged in a large earth region where k = 1.3 W/m·°C. The temperature at a large distance from the sphere is 12°C. Calculate the heat lost by the sphere.

11-17 Two long, eccentric cylinders having diameters of 15 and 4 cm, respectively, are maintained at 100 and 20°C and separated by a material with k = 3.0 W/m·°C. The distance between centers is 4.5 cm. Calculate the heat transfer per unit length between the cylinders.

11-18 Two pipes are buried in the earth and maintained at temperatures of 300 and 125°C. The diameters are 8 and 16 cm, and the distance between centers is 40 cm. Calculate the heat-transfer rate per unit length if the thermal conductivity of earth at this location is 0.7 W/·°C.

11-19 A hot sphere having a diameter of 1.5 m is maintained at 300°C and buried in a material with k = 1.2 W/m·°C and outside surface temperature of 30°C. The depth of the centerline of the sphere is 3.75 m. Calculate the heat loss.

11-20 A piece of aluminum weighing 5.5 kg and initially at a temperature of 290°C is suddenly immersed in a fluid at 15°C. The convection heat-transfer coefficient is 58 W/m^2·°C. Taking the aluminum as a sphere having the same weight as that given, estimate the time required to cool the aluminum to 90°C, using the lumped-capacity method of analysis.

11-21 A copper sphere having a diameter of 3.0 cm is initially at a uniform temperature of 50°C. It is suddenly exposed to an airstream at 10°C with h = 15 W/m^2·°C. How long does it take the sphere temperature to drop to 25°C?

11-22 Air at standard conditions of 1 atm and 20°C flows over a flat plate at 30 m/s. The plate is 60 cm square and is maintained at 90°C. Calculate the heat transfer from the plate.

11-23 Air at 20°C and 14 kPa flows at a velocity of 150 m/s past a flat plate 1 m long which is maintained at a constant temperature of 150°C. What is the average heat-transfer rate per unit area of plate?

11-24 Air at 90°C and atmospheric pressure flows over a horizontal flat plate at 60 m/s. The plate is 60 cm square and is maintained at a uniform temperature of 10°C. What is the total heat transfer?

11-25 Calculate the heat transfer from a 30-cm-square plate over which air flows at 35°C and 14 kPa. The plate temperature is 250°C, and the free-stream velocity is 6 m/s.

11-26 Air at 27°C and 1 atm blows over a 4.0-m-square flat plate at a velocity of 40 m/s. The plate temperature is 70°C. Calculate the total heat transfer.

11-27 Air at 1 atm and 27°C blows across a large concrete surface 15 m wide maintained at 55°C. The flow velocity is 4.5 m/s. Calculate the convection heat loss from the surface.

11-28 Water flows across a 15-cm-square plate at a velocity of 10 m/s. The plate is maintained at a constant temperature of 120°F and the free-stream temperature is 60°F. Calculate the heat lost by the plate.

11-29 Repeat Prob. 11-28 for flow across a 7.5-cm-diameter cylinder.

11-30 Water at 70°F flows across a 1-ft-square flat plate at a velocity of 20 ft/s. The plate is maintained at a constant temperature of 130°F. Calculate the heat lost by the plate.

11-31 Water at the rate of 1 kg/s is forced through a tube with a 2.5-cm ID. The inlet water temperature is 15°C, and the outlet water temperature is 50°C. The tube wall temperature is 14°C higher than the water temperature all along the length of the tube. What is the length of the tube?

11-32 Water at the rate of 3 kg/s is heated from 5 to 15°C by passing it through a 5-cm-ID copper tube. The tube wall temperature is maintained at 90°C. What is the length of the tube?

11-33 Water at the rate of 0.8 kg/s at 93°C is forced through a 5-cm-ID copper tube at a suitable velocity. The wall thickness is 0.8 mm. Air at 15°C and atmospheric pressure is forced over the outside of the tube at a velocity of 15 m/s in a direction normal to the axis of the tube. What is the heat loss per meter of length of the tube?

11-34 A heated cylinder at 150°C and 2.5 cm in diameter is placed in an atmospheric airstream at 1 atm and 38°C. The air velocity is 30 m/s. Calculate the heat loss per meter of length for the cylinder.

11-35 Assuming that a human can be approximated by a cylinder 1 ft in diameter and 6 ft high with a surface temperature of 75°F, calculate the heat the person would lose while standing in a 30-mi/h wind whose temperature is 30°F.

11-36 Water flows over a 3-mm-diameter sphere at 6 m/s. The free-stream temperature is 38°C, and the sphere is maintained at 93°C. Calculate the heat-transfer rate.

11-37 A 0.13-mm-diameter wire is exposed to an airstream at −30°C and 54 kPa. The flow velocity is 230 m/s. The wire is electrically heated and is 12.5 mm long. Calculate the electric power necessary to maintain the wire surface temperature at 175°C.

11-38 A pipeline in the Arctic carries hot oil at 50°C. A strong Arctic wind blows across the 50-cm-diameter pipe at a velocity of 13 m/s and a temperature of −35°C. Estimate the heat loss per meter of pipe length.

11-39 Calculate the heat-transfer rate per unit length for flow over a 0.025-

mm-diameter cylinder maintained at 65°C. Perform the calculation for (*a*) air at 20°C and 1 atm and (*b*) water at 20°C; $u_\infty = 6$ m/s.

11-40 Water enters a 3-mm-diameter tube at 21°C and leaves at 32°C. The flow rate is such that the Reynolds number is 600. The tube length is 10 cm and is maintained at a constant temperature of 60°C. Calculate the water flow rate.

11-41 Water flows in a 2-cm-diameter tube at an average flow velocity of 8 m/s. If the water enters at 20°C and leaves at 30°C and the tube length is 10 m, estimate the average wall temperature necessary to effect the required heat transfer.

11-42 Assume that one-half the heat transfer from a cylinder in cross-flow occurs on the front half of the cylinder. On this assumption, compare the heat transfer from a cylinder in cross-flow with the heat transfer from a flat plate having a length equal to the distance from the stagnation point on the cylinder. Discuss this comparison.

11-43 A 10-cm length of platinum wire 0.4 mm in diameter is placed horizontally in a container of water at 38°C and is electrically heated so that the surface temperature is maintained at 93°C. Calculate the heat lost by the wire.

11-44 Water at the rate of 0.8 kg/s at 90°C flows through a steel pipe with 2.5-cm ID and 3-cm OD. The outside surface temperature of the pipe is 85°C, and the temperature of the surrounding air is 20°C. The room pressure is 1 atm, and the pipe is 15 m long. How much heat is lost by free convection to the room?

11-45 A horizontal pipe 7.5 cm in diameter is located in a room where atmospheric air is at 20°C. The surface temperature of the pipe is 240°C. Calculate the free-convection heat loss per meter of pipe.

11-46 A horizontal 1.25-cm-OD tube is heated to a surface temperature of 250°C and exposed to air at room temperature of 20°C and 1 atm. What is the free-convection heat transfer per unit length of tube?

11-47 Assuming that a human may be approximated by a vertical cylinder 1 ft in diameter and 6 ft tall, estimate the free-convection heat loss for a surface temperature of 75°F in ambient air at 68°F.

11-48 A large circular duct 3.0 m in diameter carries hot gases at 250°C. The outside of the duct is exposed to room air at 1 atm and 20°C. Estimate the heat loss per unit length of the duct.

11-49 A 15-cm-square vertical plate is maintained at 100°F and submerged in liquid water at 60°F. Calculate the free-convection heat loss from the plate.

11-50 Repeat Prob. 11-49 for a 7.5-cm-diameter horizontal cylinder.

11-51 A 10-cm length of platinum wire 0.4 mm in diameter is placed horizontally in a container of water at 38°C and is electrically heated so that the surface temperature is maintained at 93°C. Calculate the heat lost by the wire.

11-52 A 1-m-square vertical plate is maintained at 120°F and exposed to room air at 70°F. Calculate the heat lost by the plate.

11-53 A 1-m-square vertical plate is heated to 400°C and placed in room air at 25°C. Calculate the heat loss from one side of the plate.

11-54 A horizontal 1.25-cm-OD tube is heated to a surface temperature of 250°C and exposed to air at room temperature of 20°C and 1 atm. What is the free-convection heat transfer per unit length of tube?

11-55 Two infinite black plates at 500 and 100°C exchange heat by radiation. Calculate the heat-transfer rate per unit area. If another perfectly black plate is placed between the 500 and 100°C plates, by how much is the heat transfer reduced? What is the temperature of the center plate?

11-56 A room 3-by-3-by-3 m has one side wall maintained at 260°C; the floor is maintained at 90°C. The other four surfaces are perfectly insulated. Assume that all surfaces are black. Calculate the net heat transfer between the hot wall and the cool floor.

11-57 A square room 3-by-3 m has a floor heated to 25°C, a ceiling at 13°C, and walls that are assumed perfectly insulated. The height of the room is 2.5 m. The emissivity of all surfaces is 0.8. Using the network method, find the net interchange between floor and ceiling and the wall temperature.

11-58 It is desired to transmit energy from one spaceship to another. A 1.5-m-square plate is available on each ship to accomplish this. The ships are guided so that the plates are parallel and 30 cm apart. One plate is maintained at 800°C and the other at 280°C. The emissivities are 0.5 and 0.8, respectively. Find (*a*) the net heat transferred between the spaceships in watts and (*b*) the total heat lost by the hot plate in watts. Assume that outer space is a blackbody at 0 K.

11-59 Two perfectly black parallel planes 1.2-by-1.2 m are separated by a distance of 1.2 m. One plane is maintained at 550°C and the other at 250°C. The planes are located in a large room whose walls are at 20°C. What is the net heat transfer between the planes?

11-60 Two parallel planes 90-by-60 cm are separated by a distance of 60 cm. One plane is maintained at a temperature of 550°C and has an emissivity of 0.6. The other plane is insulated. The planes are placed in a large room which is maintained at 10°C. Calculate the temperature of the insulated plane and the energy lost by the heated plane.

11-61 Three infinite parallel plates are arranged as shown. Plate 1 is maintained at 1200 K, and plate 3 is maintained at 60 K; $\epsilon_1 = 0.2$, $\epsilon_2 = 0.5$, and $\epsilon_3 = 0.8$. Plate 2 receives no heat from external sources. What is the temperature of plate 2?

FIG. P11-61

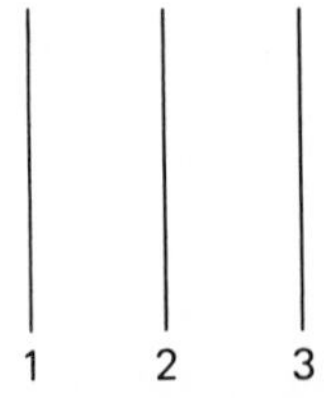

11-62 Two parallel concentric disks have $d_1 = 10$ cm, $d_2 = 5$ cm and are spaced 10 cm apart. Determine F_{12} and F_{21}.

11-63 Two parallel disks 30 cm in diameter are separated by a distance of 5 cm in a large room at 20°C. One disk contains an electric heater that produces a constant heat flux of 100 kW/m^2 and $\epsilon = 0.9$ on the surface facing the other disk. Its back surface is insulated. The other disk has $\epsilon = 0.5$ on both sides and is in radiant balance with the other disk and room. Calculate the temperatures of both disks.

11-64 A long duct has an equilateral triangle shape as shown. The surface conditions are $T_1 = 1100$ K, $\epsilon_1 = 0.6$, $T_2 = 2100$ K, $\epsilon_2 = 0.8$, $(q/A)_3 = 1000$ W/m^2, $\epsilon_3 = 0.7$. Calculate the heat fluxes for surfaces 1 and 2 and the temperature of surface 3.

11-65 A long pipe 5 cm in diameter passes through a room and is exposed to air at atmospheric pressure and temperature of 20°C. The pipe surface temperature is 93°C. Assuming that the emissivity of the pipe is 0.6, calculate the radiation heat loss per foot of length of pipe.

11-66 Two parallel disks, 60 cm in diameter, are separated by a distance of 15 cm and completely enclosed by a large room at 30°C. The properties of the surfaces are $T_1 = 540$°C, $\epsilon_1 = 0.7$, $T_2 = 300$°C, $\epsilon_2 = 0.5$. What is the net radiant heat transfer with each surface? (Do not include back-side exchange, only that from the surfaces facing each other.)

11-67 Two parallel disks having diameters of 50 cm are separated by a distance of 10 cm. One disk also has a 20-cm-diameter hole cut in the center. Find the shape factor from this disk to the one without the hole.

11-68 A straight rectangular fin has a length of 2.0 cm and a thickness of 1.5 mm. The thermal conductivity is 55 W/m·°C, and it is exposed to a convection environment at 20°C and $h = 500$ W/m^2·°C. Calculate the maximum possible heat loss for a base temperature of 200°C. What is the actual heat loss?

11-69 An aluminum fin 1.6 mm thick is placed on a circular tube with 2.5-cm OD. The fin is 6.4 mm long. The tube wall is maintained at 150°C, the environment temperature is 15°C, and the convection heat-transfer coefficient is 23 W/m^2·°C. Calculate the heat lost by the fin.

11-70 A 2.5-cm-diameter tube has circumferential fins of rectangular profile spaced at 9.5-mm increments along its length. The fins are constructed of aluminum and are 0.8 mm thick and 12.5 mm long. The tube wall temperature is maintained at 200°C, and the environment temperature is 93°C. The heat-transfer coefficient is 110 W/m^2·°C. Calculate the heat loss from the tube per meter of length.

FIG. P11-64

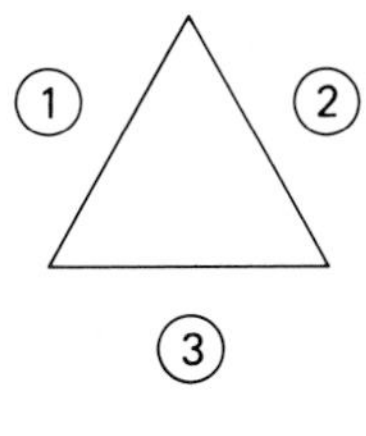

11-71 A circumferential fin of rectangular cross section surrounds a 2.5-cm-diameter tube. The length of the fin is 6.4 mm, and the thickness is 3.2 mm. The fin is constructed of mild steel. If air blows over the fin so that a heat-transfer coefficient of 28 $W/m^2 \cdot °C$ is experienced and the temperatures of the base and air are 260 and 93°C, respectively, calculate the heat transfer from the fin.

11-72 A straight rectangular fin 2.5 cm thick and 15 cm long is constructed of steel and placed on the outside of a wall maintained at 200°C. The environment temperature is 15°C, and the heat-transfer coefficient for convection is 17 $W/m^2 \cdot °C$. Calculate the heat lost from the fin per unit depth.

11-73 An aluminum fin 1.6 mm thick surrounds a tube 2.5 cm in diameter. The length of the fin is 12.5 mm. The tube wall temperature is 200°C, and the environment temperature is 20°C. The heat-transfer coefficient is 60 $W/m^2 \cdot °C$. What is the heat lost by the fin?

11-74 A certain internal combustion engine is air-cooled and has a cylinder constructed of cast iron (k = 35 Btu/h·ft·°F). The fins on the cylinder have a length of $\frac{5}{8}$ in and thickness of $\frac{1}{8}$ in. The convection coefficient is 12 $Btu/h \cdot ft^2 \cdot °F$. The cylinder diameter is 4 in. Calculate the heat loss per fin for a base temperature of 450°F and environment temperature of 100°F.

11-75 A rectangular fin has a length of 2.5 cm and thickness of 1.1 mm. The thermal conductivity is 55 W/m·°C. The fin is exposed to a convection environment at 20°C and h = 500 $W/m^2 \cdot °C$. Calculate the heat loss for a base temperature of 125°C.

11-76 Copper fins with a thickness of 1.0 mm are installed on a 2.5-cm-diameter tube. The length of each fin is 12 mm. The tube temperature is 250°C and the fins are exposed to air at 30°C with a convection heat-transfer coefficient of 120 $W/m^2 \cdot °C$. Calculate the heat lost by each fin.

11-77 Hot exhaust gases are used in a finned-tube cross-flow heat exchanger to heat 2.5 kg/s of water from 35 to 85°C. The gases (c_p = 1.09 kJ/kg·°C) enter at 200 and leave at 93°C. The overall heat-transfer coefficient is 180 $W/m^2 \cdot °C$. Calculate the area of the heat exchanger using (*a*) the LMTD approach and (*b*) the effectiveness-*NTU* method.

11-78 For the exchanger in Prob. 11-77 the water flow rate is reduced by half, while the gas flow rate is maintained constant along with the fluid inlet temperatures. Calculate the percentage reduction in heat transfer as a result of this reduced flow rate. Assume that the overall heat-transfer coefficient remains the same.

11-79 It is desired to heat 230 kg/h of water from 35 to 93°C with oil (c_p = 2.1 kJ/kg·°C) having an initial temperature of 175°C. The mass flow of oil is also 230 kg/h. Two double-pipe heat exchangers are available:

Exchanger 1: U = 570 $W/m^2 \cdot °C$ A = 0.47 m^2

Exchanger 2: U = 370 $W/m^2 \cdot °C$ A = 0.94 m^2

Which exchanger should be used?

11-80 Water at the rate of 230 kg/h at 35°C is available for use as a coolant in a double-pipe heat exchanger whose total surface area is 1.4 m^2. The water is to be used to cool oil (c_p = 2.1 kJ/kg·°C) from an initial temperature of 120°C. Because of other circumstances, an exit water temperature greater than 99°C cannot be allowed. The exit temperature of the oil must not be below 60°C. The overall heat-transfer coefficient is 280 W/m^2·°C. Estimate the maximum flow rate of oil which may be cooled, assuming the flow rate of water is fixed at 230 kg/h.

11-81 Hot water enters a counterflow heat exchanger at 99°C. It is used to heat a cool stream of water from 4 to 32°C. The flow rate of the cool stream is 1.3 kg/s, and the flow rate of the hot stream is 2.6 kg/s. The overall heat-transfer coefficient is 830 W/m^2·°C. What is the area of the heat exchanger? Calculate the effectiveness of the heat exchanger.

11-82 Water at 80°C enters a counterflow heat exchanger. It leaves at 35°C. The water is used to heat an oil from 25 to 48°C. What is the effectiveness of the heat exchanger?

11-83 Water at 90°C enters a double-pipe heat exchanger and leaves at 55°C. It is used to heat a certain oil from 25 to 50°C. Calculate the effectiveness of the heat exchanger.

11-84 A counterflow double-pipe heat exchanger is to be used to heat 0.6 kg/s of water from 35 to 90°C with an oil flow of 0.9 kg/s. The oil has a specific heat of 2.1 kJ/kg·°C and enters the heat exchanger at a temperature of 175°C. The overall heat-transfer coefficient is 425 W/m^2·°C. Calculate the area of the heat exchanger and the effectiveness.

11-85 A cross-flow finned-tube heat exchanger uses hot water to heat an appropriate quantity of air from 15 to 25°C. The water enters the heat exchanger at 70°C and leaves at 40°C, and the total heat-transfer rate is to be 29 kW. The overall heat-transfer coefficient is 45 W/m^2·°C. Calculate the area of the heat exchanger.

11-86 Calculate the heat-transfer rate for the exchanger in Prob. 11-85 when the water flow rate is reduced to half that of the design value.

11-87 A small steam condenser is designed to condense 0.76 kg/min of steam at 83 kN/m^2 with cooling water at 10°C. The exit water temperature is not to exceed 57°C. The overall heat-transfer coefficient is 3400 W/m^2·°C. Calculate the area required for a double-pipe heat exchanger.

11-88 Suppose the inlet water temperature in the exchanger of Prob. 11-87 is raised to 30°C. What percentage increase in flow rate would be necessary to maintain the same rate of condensation?

11-89 A counterflow double-pipe heat exchanger is used to heat water from 25 to 50°C by cooling an oil from 100 to 65°C. The exchanger is designed for a total heat transfer of 29 kW with an overall heat-transfer coefficient of 340 W/m^2·°C. Calculate the surface area of the exchanger.

11-90 An air-to-air heat recovery unit uses a cross-flow exchanger with both fluids unmixed and an air flow of 0.5 kg/s on both sides. The hot air enters at 40°C, while the cool air enters at 20°C. Calculate the exit temperatures for $U = 40$ W/m$^2\cdot$°C and a total exchanger area of 20 m^2.

11-91 Suppose both flow rates in Prob. 11-90 were cut in half. What would be the exit temperatures in this case, assuming no change in U? What if the flow rates were doubled?

11-92 Hot water at 90°C is used in the tubes of a finned-tube heat exchanger. Air flows across the fins and enters at 1 atm, 30°C, with a flow rate of 65 kg/min. The overall heat-transfer coefficient is 52 W/m$^2\cdot$°C, and the exit air temperature is to be 45°C. Calculate the exit water temperature if the total area is 8.0 m^2.

REFERENCES

1 Holman, J. P.: "Heat Transfer," 6th ed., McGraw-Hill Book Company, New York, 1986.
2 Schneider, P. J.: "Conduction Heat Transfer," Addison-Wesley Publishing Company, Inc., Reading Mass., 1955.
3 Kern, D. Q., and A. D. Kraus: "Extended Surface Heat Transfer," McGraw-Hill Book Company, New York, 1972.
4 Sparrow, E. M., and R. D. Cess: "Radiation Heat Transfer," Wadsworth Publishing Company, Belmont, Calif., 1966.
5 Eckert, E. R. G., and R. M. Drake: "Heat and Mass Transfer," 2d ed., McGraw-Hill Book Company, New York, 1959.
6 Kays, W. M., and A. L. London: "Compact Heat Exchangers," 3d ed., McGraw-Hill Book Company, New York, 1982.
7 Kraus, A. D.: "Heat Transfer Software," McGraw-Hill Book Company, New York, 1987.
8 Rudenberg, R.: Die Ausbreitung der Luft-und Erdfelder und Hochspannungsleitungen, besonders bei Erd-und Kurzschlussen, *Electrotech. Z.*, vol. 46, p. 1342, 1925.
9 Andrews, R. V.: Solving Conductive Heat Transfer Problems with Electrical Analogue Shape Factors, *Chem. Eng. Prog.*, vol. 51, no. 2, p. 67, 1955.

12

PRINCIPLES OF STATISTICAL THERMO-DYNAMICS

12-1 INTRODUCTION

Up to this point we have been primarily concerned with macroscopic thermodynamics and the conservation-of-energy principle. Some preliminary remarks concerning microscopic thermodynamics were given in the introductory chapter, but we have not yet shown how these developments are related to the overall subject matter of thermodynamics. Some experimental consequences of the second law of thermodynamics were also stated, but their relationship to the microscopic behavior of matter has not been established.

The objective of this chapter is to establish some mathematical and physical modeling concepts for microscopic thermodynamics which may be invoked in Chap. 13, which is concerned with specific applications. At this point the reader is cautioned against expecting final results from the exposition of this chapter. As we shall show, microscopic analysis can yield fruitful results only when combined with the natural macroscopic laws of thermodynamics. For this reason, the macroscopic analysis of Chaps. 5 and 6 are used to provide a proper link between the microscopic models of this chapter and the behavior of matter as observed in finite engineering systems.

12-2 PROBABILITY

For our purposes, probability will be related to the fraction of times a certain phenomenon is observed in a large number of occurrences. An unweighted coin will land on heads one-half of the time when flipped a large number of times, an unweighted die will land on each side one-sixth of the time in a large number of rolls, the probability of drawing a *particular* card from a deck of cards is 1/52, and so on. Later on we shall be interested in counting the number of ways various objects and groups of objects may be arranged, subject to certain restrictions. This counting process is related to probability because the probability of occurrence of a certain event is the number of ways that event can occur divided by the total number of possible events. For example, consider the toss of two dice. We might want to know the probability that a 7 will be obtained. There are six possible ways to obtain a 7 and there are 36 possible arrangements. The probability of getting a 7 is thus 6/36, or 1/6. If we know the probability that separate events will occur, the probability that either of the events will occur is the sum of the individual probabilities for the events. The probability for a double 6 to occur on the toss of two dice is 1/36 (there is only one way this can occur), so that the probability of getting *either* a double 6 or 7 would be

$$\frac{6}{36} + \frac{1}{36} = \frac{7}{36}$$

If several *independent* events occur at the same time such that each event has an independent probability p_i, the probability that *all* events will occur simultaneously is given as the product of the probabilities for the individual events. Thus

$$p = \prod_i p_i \tag{12-1}$$

where the Π sign denotes a product. A probability of unity corresponds to certainty.

It will be of interest to calculate the number of ways of arranging objects in sequence. Let us assume that we have N distinguishable objects and we want to calculate the number of different ways the objects may be arranged in sequence. For the first object in the sequence there are N choices, for the second object $N - 1$ choices, $N - 2$ choices for the third object, and so on. The total number of possible sequences is therefore

$$N(N - 1)(N - 2) \cdots N \cdots (2)(1) = N! \tag{12-2}$$

It is important to note that all of these possible sequences would be identical if the objects were indistinguishable. Various sequences of a set of identical coins, for example, would all be the same. A sequence of playing cards, on the other hand, would have several possibilities.

Our later discussions will be concerned with the number of ways microscopic particles may be arranged among various energy states, subject to specific restrictions such as conservation of energy and mass. Once we are able to predict the number of particles which occupy each energy state, then we may calculate the total energy of all particles by performing the summation

$$E = \Sigma\, n_i \epsilon_i$$

where ϵ_i is an energy level and n_i is the number of particles occupying this energy level. The allowable energy states which a particle may attain are determined from quantum mechanics.

12-3 PHASE SPACE

In the introductory chapter an analogy was drawn between mechanics and thermodynamics to the extent that the free body was likened to the system concept and thermodynamic properties were shown to be analogous to space and velocity (or momentum) coordinates in mechanics. It has been noted more than once that the observable thermodynamic properties, i.e., the macroscopic properties, are dependent on the microscopic state of the system. If we were able to say exactly what was happening at the microscopic level, then, presumably, we would be able to calculate the macroscopic thermodynamic properties.

Phase space is a *six-dimensional* space formed by the three cartesian space coordinates x, y, and z, and the three momentum coordinates p_x, p_y, and

p_z, where p denotes the momentum vector:

$$p_x = mv_x$$

$$p_y = mv_y$$

$$p_z = mv_z$$

There is really nothing new about phase space. It is a concept used in mechanics for the solution of even simple impulse-momentum problems. We are merely relating the fact that it is necessary to specify these six coordinates for a precise description of the dynamical state of a particle. We choose to call the region formed by these coordinates a *phase space*. The fact that a six-dimensional space is not readily accessible to intuitive visualization does not alter the basic concept. If we can establish the location of each microscopic particle in phase space, then we are able to specify the precise thermodynamic state of the system.

12-4 QUANTUM CONSIDERATIONS

It is well known that classical mechanics must be supplemented and modified by quantum theory when operating on a microscopic scale. The purpose of this brief discussion of quantum theory is to lay the foundation for a realistic physical model to describe the properties and behavior of matter on a microscopic scale and to show how the use of this physical model will result in calculation of macroscopic thermodynamic properties which are in agreement with experiment. Our objective is to use the principles of quantum theory to establish a physical model to be used for thermodynamic analysis. We shall, therefore, not go into a historical development of the subject but refer the reader to Ref. [1] for additional background information.

The basic premise of quantum theory is that energy occurs only in discrete quanta and that matter has a wave nature as well as a particle nature. The characteristic increment of energy for electromagnetic radiation is

$$E = h\nu \tag{12-3}$$

where ν is the frequency of the radiation and h is Planck's constant, with the value of 6.625×10^{-34} J·s. Energy transfer at the microscopic level occurs as a result of the transition from one quantum state to another. In this transition process a photon having the energy given by Eq. (12-3) is frequently emitted or absorbed, depending on the direction of the energy transfer. As a matter of experimental evidence, the energy transitions occur in discrete increments and thus the quantum concepts are accepted as basic physical postulates.

Electromagnetic radiation travels at the speed of light c and its wavelength λ is defined as

$$c = \lambda\nu \tag{12-4}$$

According to the special theory of relativity, energy is related to mass with

$$E = mc^2 \tag{12-5}$$

The "mass" of a photon is therefore identified with

$$m = \frac{E}{c^2} = \frac{h\nu}{c^2} \tag{12-6}$$

and the momentum of a photon is calculated as

$$p = mc = \frac{h\nu}{c} = \frac{h}{\lambda} \tag{12-7}$$

All physical principles are based on experiment. This is true of quantum theory as well as of macroscopic energy-conservation principles; however, an additional principle limits the precision with which measurements may be performed at the microscopic level. The *Heisenberg uncertainty principle* limits the specification of momentum and incremental displacement according to

$$\Delta p_x \, \Delta x \sim h \tag{12-8}$$

That is, the closer we measure the location of a particle, the less precise will be our measurement of its momentum. The net result of the uncertainty principle is that the best we can hope to do is to locate particles in small volumes of phase space having the dimensions of h^3. This is analogous to having several boxes all the same size with balls placed in the boxes. The lid is closed on each box so that we are unable to specify the exact location of each ball in the boxes. We are only allowed to specify the *number* of balls in each box.

At first glance, we might say that the uncertainty principle introduces no serious measurement problem because the allowable uncertainties are so small. It does not, for macroscopic systems; but with the very small dimensions and momenta involved in microscopic analysis, it represents a significant limitation.

Quantum theory postulates that particles as well as electromagnetic radiation have energies and momenta which are quantized and have wave properties. From Eq. (12-7), the characteristic wavelength† is expressed in terms of momentum by

$$\lambda = \frac{h}{p} \tag{12-9}$$

† This parameter is frequently termed the *de Broglie wavelength*.

In general, we may expect that particles may only have speeds and momenta which result in integral numbers of characteristic wavelengths. This means that if the characteristic wavelength is very small compared with a characteristic dimension of the particle or some characteristic displacement encountered in the motion of the particle, then quantum effects are not too important, since a large number of wavelengths is available for the particle motion and a continuous displacement spectrum is observed. We may see the nature of this argument by examining two simple cases.

Consider first the motion of a 1-kg mass at a velocity of 1 m/s. The characteristic wavelength for this motion is

$$\lambda = \frac{h}{p} = \frac{h}{mv} = \frac{6.625 \times 10^{-34}}{(1)(1)} = 6.625 \times 10^{-34} \text{ m}$$

Certainly there is no great restriction in saying that the motion of this mass may occur only in increments of 6×10^{-34} m. We may, therefore, conclude that quantum considerations are not important in this motion—as we knew from the start.

Now consider the motion of an electron having a mass of 9.106×10^{-31} kg at a velocity of 10^6 m/s. The characteristic wavelength in this case is (neglecting relativistic effects)

$$\lambda = \frac{h}{mv} = \frac{6.625 \times 10^{-34}}{(9.106 \times 10^{-31})10^6} = 7.28 \times 10^{-10} \text{ m}$$

Although 7.28×10^{-10} m may seem a very small distance, it is quite large for the motion of microscopic particles. For comparison purposes, it may be noted that the radius of an oxygen molecule is about 2×10^{-10} m. The aforementioned wavelength is, therefore, rather large for electron motion, and we might expect that quantum effects could exert a significant influence on electron motion.

The purpose of this brief discussion has been solely to show that quantum effects can be significant when considering the motion, energy, and momentum distributions of microscopic particles. A development of precise calculation techniques for quantum-mechanical energy distributions is beyond the scope of our discussion.

12-5 DEGENERACY

We have already recognized that energy at the microscopic level is quantized. In addition to the notion of discrete energy levels we also need to recognize that a particular energy level may occur in more than one way and still satisfy the principles of quantum mechanics involved. For a given quantum energy level ϵ_i we shall designate the number of quantum *states* having the energy ϵ_i with the symbol g_i. If more than one quantum state with the energy ϵ_i is possible, then the system of particles is said to be *degenerate*. At the moment we have no specific indication as to the possible values of g_i for different energy levels, but

1 energy unit (eu) = minimum quantum increment											
$\epsilon_i = 0$			$\epsilon_i = 1$ eu			$\epsilon_i = 2$ eu			$\epsilon_i = 3$ eu		
KE_x	KE_y	KE_z	KE_x	KE_y	KE_z	KE_x	KE_y	KE_z	KE_x	KE_y	KE_z
0	0	0	1	0	0	1	1	0	1	1	1
			0	1	0	0	1	1	1	2	0
			0	0	1	1	0	1	1	0	2
						2	0	0	2	1	0
						0	2	0	0	1	2
						0	0	2	2	0	1
									0	2	1
									3	0	0
									0	3	0
									0	0	3
$g_i = 1$			$g_i = 3$			$g_i = 6$			$g_i = 10$		

FIG. 12-1 Illustration of degeneracy concept for simple translational kinetic-energy model.

we may obtain a physical understanding of the degeneracy idea by considering a single particle which may have three translational degrees of freedom. Let us choose some energy unit which is the minimum increment allowed according to quantum considerations and examine the possible number of ways by which four energy levels could be attained. The situations of interest are depicted in Fig. 12-1. This simple tabulation shows that the number of possible quantum states increases very rapidly with an increase in the energy level. We may remark that the minimum quantum increment of energy may not always be the same for all energy modes. For example, a diatomic gas molecule may involve translational, rotational, and vibrational energy modes, as illustrated in Fig. 12-2. The characteristic quantum energy increment for rotational energy may

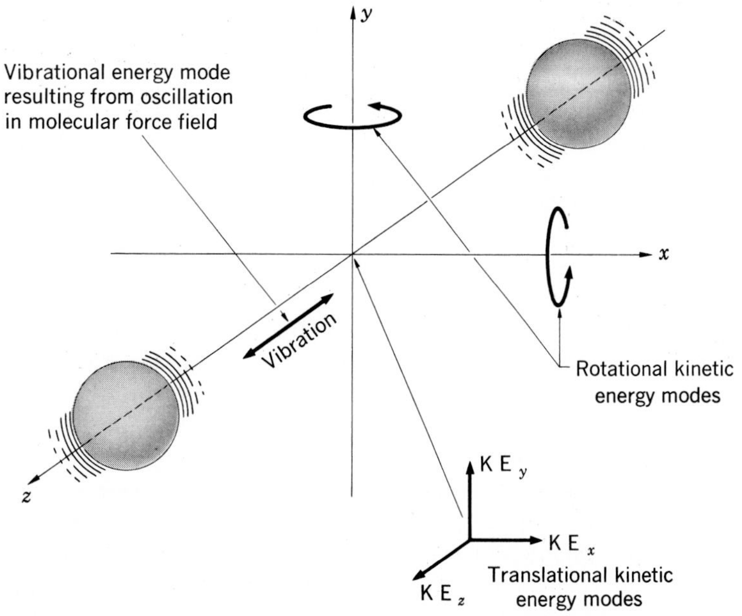

FIG. 12-2 Energy modes for a diatomic molecule.

not be the same as for the translational modes, and we shall find it necessary to take this into account in our later calculations. Accordingly, the term *degeneracy* may be applied to several energy modes.

It is interesting to note at this point that for ideal gases at moderately high temperatures (room temperature is moderately high), the number of available energy states is very large compared with the number of particles which are to occupy these states. Such a gas is highly *degenerate* because there are many states which have the same energy level.

The *nondegenerate* system is one for which $g_i = 1$, that is, each energy level may be attained in only one way.

12-6 MICROSTATES, MACROSTATES, AND THERMODYNAMIC PROBABILITY

For the moment let us consider a monatomic gas the molecules of which act as perfect billiard balls. We could presumably calculate the internal energy of the gas by taking the sum of the kinetic energies of all the molecules. The pressure of the gas could be expressed in terms of the momentum exchange of the gas molecules with a containing wall. To specify the *microstate* of such a system, we would have to specify the exact location and velocity of *each* particle in phase space, i.e., we would have to specify the location and velocity of every particle in the system. Such a detailed specification, however, is unnecessary for purposes of calculating the macroscopically observable thermodynamic properties; that is, to calculate the energy, we need to know only the *number* of particles which have a certain kinetic energy (or velocity). More specifically, we need to know the *number* of particles in each element of volume (a six-dimensional element) of phase space, and it is unnecessary to specify *which* particle is in each volume element. When the number of particles in each element of phase space is specified without regard to which particles are involved, we say that a *macrostate* has been determined. We shall have more to say about these concepts later, but the reader should realize that a determination of the macrostate of a system will be of central concern since it is related to observable thermodynamic properties.

Before we can analyze the physical models discussed above some mathematical preliminaries are required. These preliminaries are directed toward a calculation of the number of permutations and combinations which may be achieved by arranging balls in boxes. Later the balls will represent particles and the boxes will represent energy levels and quantum states which the particles may occupy.

Our eventual objective is to determine the macrostate which is most probable and then to calculate the energy corresponding to this state. The most probable macrtostate will be the one which has the largest number of possible microstates, i.e., the one that can occur in the largest number of ways. In the following calculations we shall determine a quantity Ω called the *thermody-*

namic probability, which is the number of microstates corresponding to a macrostate which gives a total of 7 on the throw. There are six ways to throw a bility.'' Again, using the throw of two dice as an example we might examine the macrostate which gives a total of 7 on the throw. There are six ways to throw a 7 so there are six microstates for the designated macrostate. In this case we would say that Ω is 6. The value of Ω for a throw of 11 would be two, etc. The important point to remember is that the number of arrangements is recognized to be proportional to probability, so that a study of the probability of observing certain arrangements may be made by examining the behavior of Ω.

The three mathematical cases corresponding to our later physical models are as follows:

Case 1 The number of ways a total of N *distinguishable* balls may be arranged in a set of large boxes with N_1 balls in the first box, N_2 balls in the second box, N_i balls in the ith box, and so on, with no restriction on the number of balls in each box (except that the number obviously cannot be greater than N, the total number of balls). Inside each ith large box there are g_i smaller boxes which are distinguishable and into which the N_i balls may be distributed. The number of small boxes g_i may be greater or less than the number of balls N_i which will be distributed among them, and the number of small boxes in each large box is not necessarily the same. This situation is depicted in Fig. 12-3.

Case 2 The same as case 1, except that the balls are *indistinguishable*.

Case 3 The same as case 2, except that a restriction is imposed on the system such that no more than *one* ball may be placed in each *small* box.

FIG. 12-3 Balls and boxes for calculation of number of arrangements.

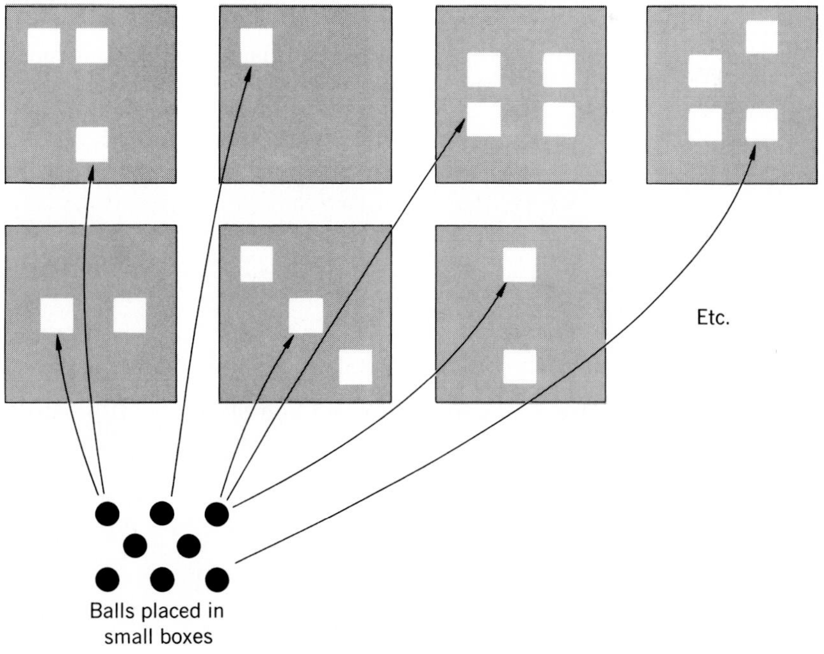

In these models the large boxes will correspond to energy levels, the small boxes to the number of quantum states in each energy level, or degeneracy, and the balls are the particles.

Case 1 may be calculated as follows. The *total* number of ways N distinguishable balls may be arranged in sequence is $N!$. The number of ways of arranging the balls inside the large boxes is $N_1!$ in the first box, $N_2!$ in the second box, and so on. Thus, if we are to determine the number of ways of arranging the balls in the various large boxes, we must divide out the number of arrangements in each box by forming the quotient

$$\frac{N!}{N_1!N_2!N_3! \cdots N!}$$

or

$$\Omega = \frac{N!}{\prod_i N_i!} \tag{12-10}$$

where N_i designates the number of balls in the ith large box. We must also consider the number of possible ways the N_i balls may be arranged in the g_i small boxes contained in each large box. For each ball we have g_i choices, so that for all balls in the ith large box we have the following number of possibilities:

$$g_i \cdot g_i \cdot g_i \cdots = g_i^{N_i} \tag{12-11}$$

We thus have the result that there are $g_i^{N_i}$ possible ways to arrange the N_i balls in the g_i small boxes inside the ith large box. This means that Eq. (12-10) must be modified in the following way to account for all possible arrangements:

$$\Omega = \frac{N!}{\prod_i N_i!} g_1^{N_1} \cdot g_2^{N_2} \cdot g_3^{N_3} \cdots g_i^{N_i}$$

or

$$\Omega = N! \prod_i \frac{g_i^{N_i}}{N_i!} \tag{12-12}$$

Case 2 is somewhat more complicated. In this instance the balls are assumed to be indistinguishable so that the permutations of a number of balls inside a box do not alter the basic arrangement. There are g_i small boxes of the ith type and a total of N_i indistinguishable balls distributed among these boxes. Let us consider first the number of ways of arranging N_i indistinguishable balls in the g_i distinguishable boxes. Suppose that we start placing the N_i balls in the g_i small boxes one at a time, and *assume for the moment* that the balls are distinguishable. To show the nature of the counting process we shall designate the small boxes with numbers 1, 2, 3, etc., and the balls with the letters A, B, C, etc. For two boxes and two balls the possible arrangements are

1 A, 2 B	2 B, 1 A
1 AB, 2 0	2 0, 1 AB
1 BA 2 0	2 0, 1 BA
1 B, 2 A	2 A, 1 B
1 0, 2 AB	2 AB, 1 0
1 0, 2 BA	2 BA, 1 0

where the zero indicates zero balls in the box. If we were to expand the listing with larger numbers of balls and boxes we would find that each possible arrangement could be designated by a sequence beginning with a number (a box) and including all the letters (balls) and other numbers. The total number of sequences of g_i and N_i distinguishable objects is

$$(g_i + N_i)! = (g_i + N_i)(g_i + N_i - 1)!$$

and the number always beginning with g_i (a box) is

$$g_i(g_i + N_i - 1)! \tag{12-13}$$

In terms of our balls and boxes this is the number of ways of arranging N_i distinguishable balls in g_i distinguishable boxes. Because the balls are actually indistinguishable, such pairs of sequences as

1 A, 2 B	1 B, 2 A
1 AB, 2 0	1 BA, 2 0

are actually the same. Now consider the two pairs:

1 AB, 2 0	2 0, 1 AB
1 B, 2 A	2 A, 1 B

In terms of a microstate of the system the arrangements in each pair are identical. It makes no difference how the small boxes (energy states) are arranged; the number of particles in each is what matters. We can take care of the indistinguishability of the balls and unimportance of ordering of small boxes by dividing the relation in Eq. (12-13) by $N_i!$ and $g_i!$, respectively, so that

$$\Omega_i = \frac{g_i(g_i + N_i - 1)!}{g_i!N_i!} = \frac{g_i(g_i + N_i - 1)!}{g_i(g_i - 1)!N_i!}$$

or

$$\Omega_i = \frac{(g_i + N_i - 1)!}{(g_i - 1)!N_i!} \tag{12-14}$$

The total thermodynamic probability is computed by taking the product of all such functions over all the "large" boxes:

$$\Omega = \prod_i \Omega_i = \prod_i \frac{(g_i + N_i - 1)!}{(g_i - 1)!N_i!} \tag{12-15}$$

In this equation we are making use of the product relation in Eq. (12-1) relating the total probability to the probability of independent events.

It is of interest to note that Eq. (12-15) simplifies to the following relation when $g_i \gg 1$:

$$\Omega = \prod_i \frac{(g_i + N_i)!}{g_i!N_i!} \tag{12-16}$$

We should note here that although individual quantum states are distinguishable (the g_i), they behave as if they were indistinguishable for determining the thermodynamic probability. Summarizing, we can see how this reasoning operates with the 1, 2, A, B sequence discussed above ($N_i = 2$, $g_i = 2$). For the four distinguishable objects the total number of sequences is $4! = 24$, and the number of sequences starting with a number is $g(g + N - 1) = (2)(2 + 2 - 1)! = 12$. If the balls are indistinguishable we divide by 2! and obtain only six arrangements, and if the ordering of the boxes is not important we divide by another 2! and get only three different arrangements for computing the macrostate. These three arrangements are one ball in each box and both balls in either box.

Case 3 obviously involves the restriction that $N_i \leq g_i$; otherwise there would have to be more than one ball in each small box. Suppose we start placing the N_i balls in the g_i small boxes, one at a time. For the first ball there would be g_i open boxes, $g_i - 1$ for the second ball, $g_i - 2$ for the third ball, etc., so that the possible number of arrangements in the ith large box, if the balls were indistinguishable, would be

$$g_i(g_i - 1)(g_i - 2) \cdots [g_i - (N_i - 1)] = \frac{g_i!}{(g_i - N_i)!}$$

Since the balls are indistinguishable, we must divide by $N_i!$ to obtain the desired number of arrangements. Thus

$$\Omega_i = \frac{g_i!}{N_i!(g_i - N_i)!} \tag{12-17}$$

The possible number of arrangements for all the ith boxes is therefore

$$\Omega = \prod_i \Omega_i = \prod_i \frac{g_i!}{N_i!(g_i - N_i)!} \tag{12-18}$$

The following examples illustrate the application of these formulas for calculation purposes. Subsequent sections will show how they are related to a microscopic physical model.

EXAMPLE 12-1
Calculate the number of ways arranging six distinguishable balls in four boxes so that $N_1 = 1$, $N_2 = 1$, $N_3 = 3$, $N_4 = 1$.

SOLUTION Equation (12-10) applies for this case, so that

$$\Omega = \frac{N!}{\prod_i N_i!} = \frac{6!}{1!1!3!1!} = 120$$

EXAMPLE 12-2
Calculate the number of ways of arranging six indistinguishable balls in four distinguishable boxes.

SOLUTION For this problem we use Eq. (12-13) and take out the distinguishability of the balls by dividing by $N_i!$ so that

$$\Omega = \frac{g(g + N - 1)!}{N!} = \frac{4(4 + 6 - 1)!}{6!} = 2016$$

If the boxes were indistinguishable, i.e., their ordering were unimportant, Eq. (12-14) would apply and we would have

$$\Omega = \frac{(g + N - 1)!}{(g - 1)N!} = \frac{(4 + 6 - 1)!}{(4 - 1)!6!} = 84$$

EXAMPLE 12-3
Calculate the number of ways of arranging six indistinguishable balls in eight distinguishable boxes with no more than one ball per box.

SOLUTION Equation (12-17) applies, so that

$$\Omega = \frac{g!}{N!(g - N)!} = \frac{8!}{6!(8 - 6)!} = 28$$

12-7 PHYSICAL MODELS

Let us now formulate several possible physical models for the microscopic behavior of matter. We shall, for the moment, consider only particulate matter (no electromagnetic radiation) and will limit our attention to isolated systems.

Four fundamental assumptions are made for all the models:

1 The total energy of the system remains constant.
2 The total number of particles remains constant.
3 A sufficiently large number of particles is involved so that their behavior may be inferred from a statistical analysis. In other words, we assume the sample is large enough that statistical principles will apply.
4 All microstates are equally probable. In other words, a particle is just as likely to be in one volume element of phase space as in any other volume element.

The first two assumptions are necessary because the energy and mass of an isolated system remain constant, while the last two assumptions govern the way we attack the problem in a statistical sense.

Three possible physical models are proposed:

1 *Maxwell-Boltzmann (MB) model.* The particles are distinguishable and distributed among various quantum energy levels designated by the subscript i. Thus there are N_i particles having an energy ϵ_i in the ith energy level. The values of the energies ϵ_i are quantized, and there are several ways that a particle may attain the energy ϵ_i. For example, a particle having only translational kinetic energy could have the energy distributed several different ways among each translational mode. Three possibilities for a particle having a total kinetic energy of six units could be

$$\begin{Bmatrix} KE_x \\ KE_y \\ KE_x \end{Bmatrix} = \begin{Bmatrix} 1 \\ 2 \\ 3 \end{Bmatrix} \text{ or } \begin{Bmatrix} 3 \\ 0 \\ 3 \end{Bmatrix} \text{ or } \begin{Bmatrix} 2 \\ 1 \\ 3 \end{Bmatrix}$$

Thus, for generality, we must assume that each group of particles (in the ith energy level) could be composed of several quantum *states* g_i, all having the same energy ϵ_i. There are no restrictions on the number of particles which may occupy each energy level ϵ_i or on the number of quantum states within each energy level.
2 *Bose-Einstein (BE) model.* Here the physical situation is the same as the MB model except that the particles are *indistinguishable*. Still, there is no restriction on the number of particles which may occupy each energy level or on the various quantum states g_i.
3 *Fermi-Dirac (FD) model.* This model is the same as the BE model except that no more than one particle is allowed to occupy each quantum state. This restriction requires that $g_i \geqslant N_i$.

These three physical models serve to analyze a great body of microscopic phenomena. All three models assume that the particular quantum energy level may be attained in several different ways. The MB model assumes that each particle is different in some way, i.e., distinguishable, whereas the BE and FD models assume that they are indistinguishable. The characteristics of the three models are summarized in Table 12-1.

TABLE 12-1
Summary of Characteristics of Statistical Models

Statistics	Type of Particles	Energy Quantized	No. of Particles per Quantum State
Maxwell-Boltzmann (MB)	Distinguishable	Yes	Any number
Bose-Einstein (BE)	Indistinguishable	Yes	Any number
Fermi-Dirac (FD)	Indistinguishable	Yes	One

12-8 PURPOSE OF STATISTICAL ANALYSIS

At this time we shall not be concerned with the type of system (gas, liquid, or solid) for which each of the foregoing physical models applies. Our purpose is to obtain the equilibrium energy distribution for each of the physical models subject to constant-total-energy and number-of-particles restrictions. In other words, we want to determine the number of particles which will occupy each energy state (level), which, in turn, will give the number of particles in each energy level under the most probable conditions; i.e., we seek N_i as a function of ϵ_i. The most probable distribution is called the *equilibrium distribution*.

We are interested in the equilibrium distribution because this distribution governs the macroscopic properties of systems normally observed experimentally. Thus our analysis is being directed toward an eventual tie-in with macroscopic thermodynamics—as was promised in the opening remarks of this chapter.

We shall use the term *most probable* to designate the state which is attainable through the largest number of permutations. Since the objective of the analysis is to determine the number distribution of particles in various energy states, we are, in fact, seeking a specification of the most probable *macrostate* of the system. This macrostate may then be related to the macroscopically observable thermodynamic properties of the system. Once the energy distributions are obtained, we shall be able to apply the information to specific physical systems. Most of the applications of the statistical material will be presented in Chap. 13.

It should be fairly obvious that the three physical models discussed in the preceding section correspond *exactly* to the mathematical statistics models concerning balls and boxes discussed in Sec. 12-6. The thermodynamic probability or number of arrangements Ω is then the quantity we wish to maximize, subject to the constant-energy and number-of-particles restrictions.

12-9 STIRLING'S APPROXIMATION

In subsequent sections we shall be dealing with factorials of large numbers. In particular we shall be interested in evaluating

$$\ln x! \qquad \text{for } x \gg 1$$

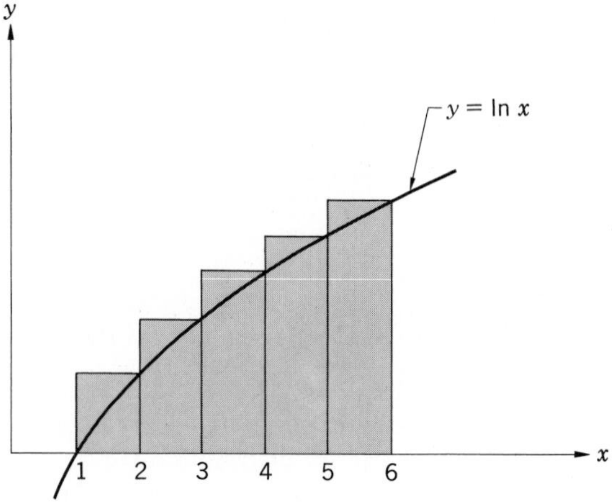

FIG. 12-4 Stirling's approximation.

We may write

$$\ln x! = \ln 2 + \ln 3 + \cdot \cdot \cdot + \ln x = \sum_{i=1}^{n} \ln x_i \tag{12-19}$$

This sum is approximated by the area under the curve shown in Fig. 12-4. The summation in Eq. (12-19) is the sum of the small rectangular areas shown in the figure. When x becomes very large, the sum of these rectangular areas is closely approximated by the area under the curve $y = \ln x$. Thus we may write

$$\ln x! \approx \int_1^x \ln x \, dx \qquad \text{for } x \gg 1$$

Evaluating the integral,

$$\ln x! \approx x \ln x - x - 1$$

Finally, for large x we may neglect the 1, so that

$$\ln x! \approx x \ln x - x \qquad \text{for } x \gg 1 \tag{12-20}$$

Equation (12-20) is called Stirling's approximation for the factorial.

12-10 BOSE-EINSTEIN STATISTICS

We shall now obtain the equilibrium distribution for the BE physical model. We have

$$\Omega = \prod_i \frac{(N_i + g_i - 1)!}{N_i!(g_i - 1)!} \tag{12-21}$$

and want to maximize Ω subject to

$$N = \sum_i N_i \qquad \text{constant number of particles} \tag{12-22}$$

$$U = \sum_i \epsilon_i N_i \qquad \text{constant total internal energy} \tag{12-23}$$

Here the symbol ϵ_i designates the energy of each particle occupying the ith energy level.

We now want to maximize Ω with respect to the number of particles N_i having each energy ϵ_i. Thus we seek the distribution of particle energies which is most probable and, therefore, the one which we may call the equilibrium energy distribution. If Ω is maximized, so is $\ln \Omega$. It will be more convenient to work with the logarithm so we write

$$\ln \Omega = \sum_i [\ln (N_i + g_i - 1)! - \ln N_i! - \ln (g_i - 1)!]$$

If both g_i and N_i are large compared with unity, ones may be neglected and Stirling's approximation may be applied to obtain

$$\ln \Omega = \sum_i [(N_i + g_i) \ln (N_i + g_i) - (N_i + g_i) - N_i \ln N_i + N_i - g_i \ln g_i + g_i] \tag{12-24}$$

The maximization process is denoted by setting

$$\frac{\partial (\ln \Omega)}{\partial N_i} = 0 \tag{12-25}$$

This condition means that the variation in $\ln \Omega$ is zero for a small variation in N_i from the equilibrium distribution or

$$\delta(\ln \Omega) = 0 = \frac{\partial (\ln \Omega)}{\partial N_i} \delta N_i$$

Using Eq. (12-24), this condition becomes

$$\sum_i \ln \frac{N_i + g_i}{N_i} \delta N_i = 0 \tag{12-26}$$

The constant-energy condition is

$$\delta U = 0 = \frac{\partial U}{\partial N_i} \delta N_i$$

or

$$\sum_i \epsilon_i \, \delta N_i = 0 \tag{12-27}$$

Finally, the particle-conservation condition is

$$\delta N = 0 = \frac{\partial N}{\partial N_i} \, \delta N_i$$

or

$$\sum_i \delta N_i = 0 \tag{12-28}$$

We thus have three conditions to satisfy in order to establish the maximum probability. If we were not bound by the constant energy and number of particles restrictions, then the number of particles in each energy state N_i could all be considered as independent variables. In this case we would only need to apply Eq. (12-26) and obtain

$$\ln \frac{N_i + g_i}{N_i} = 0$$

as the maximizing condition. Clearly, as soon as the total number of particles is specified, not all the N_i's are independent. Similarly, the energy-conservation condition imposes an additional restriction on the independence of the N_i's. The particular restrictions are governed by the particular total energy of the system and the particular total number of particles. Suppose Eq. (12-27) is multiplied by a constant $-\beta$, which is a function of the total system energy, and Eq. (12-28) is multiplied by another constant, $-\alpha$, which is a function of the total number of particles. If the resulting equations are added to Eq. (12-26), we obtain

$$\sum_i \left[\ln \left(1 + \frac{g_i}{N_i} \right) - \beta \epsilon_i - \alpha \right] \delta N_i = 0 \tag{12-29}$$

In effect, the constants β and α have incorporated the energy and number-of-particles restriction into Eq. (12-26) so that the N_i's may now be treated as independent. We thus obtain the condition

$$\ln \left(1 + \frac{g_i}{N_i} \right) - \beta \epsilon_i - \alpha = 0$$

or

$$\frac{N_i}{g_i} = \frac{1}{A e^{\beta \epsilon_i} - 1} \tag{12-30}$$

where now A has been substituted for e^{α}. The constants α and β are called *undetermined multipliers* and the technique of maximizing Ω is called the *Lagrange multiplier method*. The constants α and β play a role similar to constants of integration in the solution of differential equations in that their values must be determined from appropriate physical boundary conditions. We shall determine the values of these constants in subsequent sections.

12-11 FERMI-DIRAC STATISTICS

The FD distribution is determined in an analogous way to the BE model. The thermodynamic probability distribution is given by Eq. (12-18), and we want to maximize ln Ω subject to the restrictions of constant total energy and constant total number of particles. We have, after applying Stirling's approximation to ln Ω,

$$\delta(\ln \Omega) = \frac{\partial(\ln \Omega)}{\partial N_i}\,\delta N_i = \sum_i \ln \frac{g_i - N_i}{N_i}\,\delta N_i = 0 \tag{12-31}$$

$$\delta U = \frac{\partial U}{\partial N_i}\,\delta N_i = \sum_i \epsilon_i\,\delta N_i = 0 \tag{12-32}$$

$$\delta N = \sum_i \delta N_i = 0 \tag{12-33}$$

Using the undetermined-multiplier technique as before, we obtain

$$\sum_i \left[\ln\left(\frac{g_i}{N_i} - 1\right) - \alpha - \beta\epsilon_i\right] \delta N_i = 0 \tag{12-34}$$

There results

$$\frac{N_i}{g_i} = \frac{1}{Ae^{\beta\epsilon_i} + 1} \tag{12-35}$$

For the MB model we would obtain

$$\frac{N_i}{g_i} = \frac{1}{Ae^{\beta\epsilon_i}} \tag{12-36}$$

Equations (12-30), (12-35), and (12-36) are very similar in that they differ only in the way the factor of one appears in the denominator. Consider the physical circumstance where $N_i \ll g_i$; i.e., the number of particles is much smaller than the available number of quantum states for each energy level. In this case the factor of 1 in the denominator of Eqs. (12-30) and (12-35) is very small compared with the $Ae^{\beta\epsilon_i}$ term, and both the BE and FD distributions approach Eq. (12-36), the MB model. This limiting case is important because it will enable us to analyze indistinguishable particles with the simpler MB distribution for cases where $N_i \ll g_i$.

12-12 CLASSICAL MAXWELL-BOLTZMANN MODEL

The quantum nature of the statistical model can be removed in the following way. We examine the MB model for the case where $g_i = 1$ for all energy levels, and the energy is assumed to have a continuous distribution so that no quantum restrictions are placed on the energy levels. In this instance each energy level can be slightly different from others and there is no need to be concerned about different "kinds" of energy states, so we take $g_i = 1$ for each energy level. There is, in effect, an infinite number of possible energy states for such a continuous distribution. Such a model is called a *classical* statistical model because it ignores the possibility of different types of quantum states having the same energy. In this case the number of particles N_i having the energy ϵ_i is simply

$$N_i = \frac{1}{Ae^{\beta\epsilon_i}} \tag{12-37}$$

There is a problem with the MB model. Certain types of microscopic particles just are not distinguishable. It is not possible, for example, to tell one oxygen molecule from another or one electron from another. It will be seen later that this fact makes the MB model unsuitable for analyzing some substances, except when employed as a limiting case of the BE or FD statistics.

12-13 THE EQUILIBRIUM DISTRIBUTION

The foregoing analyses have served to establish a most probable distribution of particles among various energy states subject to restrictions of constant energy and number of particles. We have yet to determine the values of the constants A and β, but the functional form of the distribution has been established. This distribution represents, in fact, the most probable *macrostate*. It defines the energy distribution which should be observed most often. It is of interest to know whether the distribution is just slightly more probable than some other distribution or whether it is overwhelmingly the most probable occurrence. We choose to call this most probable distribution the *equilibrium macrostate* of the system since it is the one we would expect to observe after the system is left alone for a while. If it is overwhelmingly the most probable occurrence, then it should be directly related to the *macroscopic* thermodynamic properties which are normally observed. For simplicity, let us consider the MB distribution. We have

$$\Omega = N! \prod_i \frac{g_i^{N}}{N_i!}$$

Forming the logarithm and using Stirling's approximation,

$$\ln \Omega = N \ln N + \sum_i N_i \ln g_i - \sum_i N_i \ln N_i \tag{12-38}$$

The most probable distribution is obtained by inserting Eq. (12-36) into Eq. (12-38). Thus

$$\ln \Omega_{\max} = N \ln N + \sum_i N_i(\ln A + \beta\epsilon_i) \tag{12-39}$$

We now wish to examine the effect of a change in N_i on Ω. In particular, we want to compare $\Omega_{\max} + \delta\Omega$ with $\Omega_{\max}$, where $\delta\Omega$ is the deviation from $\Omega_{\max}$ resulting from some deviation δN_i from the most probable distribution. We can write

$$\ln (\Omega + \delta\Omega) = N \ln N + \sum_i (N_i + \delta N_i) \ln g_i - \sum_i (N_i + \delta N_i) \ln (N_i + \delta N_i) \tag{12-40}$$

Subtracting Eq. (12-38) from Eq. (12-40),

$$\ln \frac{\Omega + \delta\Omega}{\Omega} = \sum_i \ln g_i\, \delta N_i - \sum_i N_i \ln \left(1 + \frac{\delta N_i}{N_i}\right) - \sum_i \ln (N_i + \delta N_i)\, \delta N_i \tag{12-41}$$

One of the conditions for $\Omega_{\max}$ is $\delta(\ln \Omega) = 0$ or

$$\sum_i (\ln g_i - \ln N_i)\, \delta N_i = 0 \tag{12-42}$$

Subtracting Eq. (12-42) from Eq. (12-41),

$$\ln \frac{\Omega_{\max} + \delta\Omega}{\Omega_{\max}} = - \sum_i N_i \ln \left(1 + \frac{\delta N_i}{N_i}\right) - \sum_i \ln \left(1 + \frac{\delta N_i}{N_i}\right) \delta N_i \tag{12-43}$$

With some algebraic manipulation Eq. (12-43) can be reduced to the following expression under the condition that δN_i be small compared with N_i:

$$\ln \frac{\Omega_{\max} + \delta\Omega}{\Omega_{\max}} \approx -\frac{1}{2} \sum_i \frac{(\delta N_i)^2}{N_i} \tag{12-44}$$

This relation provides a means for comparing the probability of a distribution deviating by an amount $\delta\Omega$ from the equilibrium distribution as a result of a change δN_i in the occupation of the ith energy level. As an example, let us consider just two energy cells both having the same energy and containing 10^{20} total particles between them. Since the energy of each cell is the same, the most probable distribution of particles is one of equal numbers of particles in each cell, which we take as 5×10^{19}. Let us investigate the likelihood of a deviation from this most probable distribution. Suppose that only 0.01 percent of the particles were to change cells. We have

$$N_1 = N_2 = 5 \times 10^{19}$$
$$\delta N_1 = -\delta N_2 = (0.0001)(5 \times 10^{19}) = 5 \times 10^{15}$$

Forming the sum in Eq. (12-44) gives

$$\begin{aligned}\ln \frac{\Omega_{\max} + \delta\Omega}{\Omega_{\max}} &= -\frac{1}{2} \sum_{i=1}^{2} \frac{(\delta N_i)^2}{N_i} \\ &= -\frac{1}{2}\left[\frac{(5 \times 10^{15})^2}{5 \times 10^{19}} + \frac{(-5 \times 10^{15})^2}{5 \times 10^{19}}\right] \\ &= -5 \times 10^{11}\end{aligned}$$

or

$$\frac{\Omega_{\max} + \delta\Omega}{\Omega_{\max}} \approx e^{-5\times 10^{11}}$$

Thus we find that the probability of even this slight deviation from the most probable state is very small indeed.

We could carry this discussion further, but it should already be evident to the reader that the equilibrium distributions given in Eqs. (12-35) and (12-36) are not just the most probable ones—they are overwhelmingly so. Furthermore, they are so predominant that we would not expect to observe macroscopic properties which deviate from those calculated with these distributions, except in very special circumstances.

As mentioned previously, our analytical efforts are directed toward a calculation of macroscopic thermodynamic properties from microscopic considerations. We should mention, however, that there are important areas of fluctuation theory and transport phenomena where deviations from the most probable equilibrium distribution are very significant, but these subjects are beyond the scope of our discussion.

12-14 MICROSCOPIC INTERPRETATION OF HEAT AND WORK

We have already seen that the first law of thermodynamics can be written as

$$dU = d'Q + d'W \tag{12-45}$$

for a closed system. From Eq. (12-23) the change in internal energy can be written

$$dU = \sum_i \epsilon_i \, dN_i + \sum_i N_i \, d\epsilon_i \tag{12-46}$$

Now consider a simple system of noninteracting particles whose only work interaction occurs in the form of a compression (volume change) work mode. It can be shown from quantum mechanics that the allowable translational energy states of such a system are proportional to $V^{-2/3}$ where V is the system volume. When V is fixed the allowable energy states are fixed. We can therefore take $\epsilon_i = f(V)$ and rewrite Eq. (12-46) as

$$dU = \sum_i \epsilon_i \, dN_i + \sum_i N_i \left(\frac{d\epsilon_i}{dV}\right) dV \tag{12-47}$$

If we apply this equation to a quasistatic process for a simple compressible substance, we see that for zero volume change the work is zero and the internal energy must increase because of the heat interactions, so we identify

$$d'Q = \sum_i \epsilon_i \, dN_i \tag{12-48}$$

For zero heat transfer, but with work addition at the boundary, we interpret

$$d'W = \sum_i N_i \left(\frac{d\epsilon_i}{dV}\right) dV = \sum_i N_i \, d\epsilon_i \tag{12-49}$$

From this discussion we are able to make the following microscopic appraisal of the quasistatic process. A constant-volume heat addition is evidenced by a change in the population of the energy states (N_i) but the allowable energy levels are not changed; heat addition causes an increase in the population of higher-energy states and a corresponding decrease in population of lower-energy states. In contrast, an adiabatic-compression process causes a shift up-

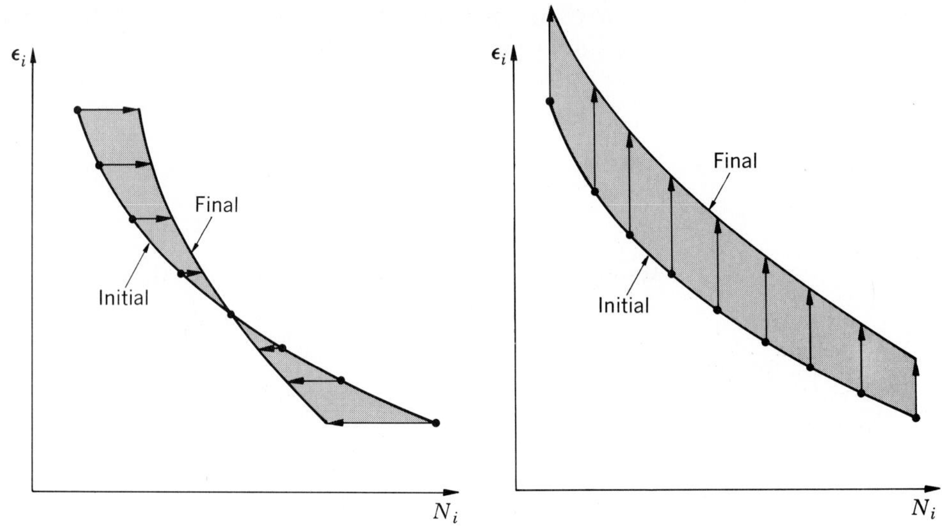

FIG. 12-5 (Left) Constant-volume heat addition causes shift in populations; (right) adiabatic compression causes shift in energy levels.

ward in the energy levels ϵ_i while the population of each level remains the same. These two types of processes are illustrated in Fig. 12-5.

12-15 ENTROPY

Because the most probable distribution is so predominant, we shall specify Ω_{max} as a macroscopic property of the system. For convenience we choose to work with ln Ω_{max} and write

$$S = k \ln \Omega_{max} \tag{12-50}$$

where now k is some constant to be determined later and the property S is a *macroscopic equilibrium property* called the *entropy*.

For a large number of particles the states represented by the values of Ω other than Ω_{max} are much less likely than Ω_{max} but still represent *possible* physical states of the system. Accordingly, we could define the entropy for these possible states as

$$S = k \ln \Omega \tag{12-51}$$

A comparison of Eqs. (12-50) and (12-51) indicates that, as an *isolated* (constant energy and number of particles) system approaches its most likely state, its entropy approaches a maximum value. Furthermore, it is highly improbable that the entropy would ever decrease from its maximum because of the overwhelming nature of Ω_{max}. Note that such a decrease is *not* impossible, only highly improbable. The behavior of the property entropy is thus quite different from the internal-energy property for an isolated system. We accept the premise that it is *impossible* for the internal energy to change in such a system. This premise, of course, results from the first law of thermodynamics.

Another way of stating the foregoing remarks is to say that in any change involving an isolated system the entropy will most likely increase. Written in analytical form,

$$\Delta S_{\text{isolated}} \geqslant 0 \qquad \text{most probable} \tag{12-52}$$

12-16 THE SECOND LAW OF THERMODYNAMICS

In Sec. 5-8 we obtained a macroscopic definition for entropy based on the impossibility of building a perpetual motion machine of the second kind. In reality, the definition of entropy for microscopic thermodynamics adopted above has been chosen to match with macroscopic concepts. As we have seen, an isolated system will always tend toward a condition of maximum entropy, at which point equilibrium will be established. Equation (12-52) furnishes the same information of a "most-probable" basis, but because for most practical problems the equilibrium distribution is so overwhelmingly probable, we may expect the second law to be "certain." While some purists may argue that there may be special occurrences when the second law is just "most probable," such events are not likely to be observed in engineering practice, and the second law may be taken as absolute.

12-17 THE PARTITION FUNCTION

According to Eq. (12-36) we may write the following relation for the MB model:

$$N_i = \frac{1}{A} g_i e^{-\beta\epsilon_i} \tag{12-53}$$

The quantity g_i is the number of quantum states having an energy ϵ_i. If we assume a classical model such that a continuous distribution of energy is available then the number of particles having an energy ϵ_i is given by Eq. (12-37):

$$N_i = \frac{1}{A} e^{-\beta\epsilon_i} \qquad g_i = 1 \text{ for classical model}$$

Now

$$N = \sum_i N_i = \frac{1}{A} \sum_i e^{-\beta\epsilon_i} \tag{12-54}$$

The summation on the right side of this equation is defined as the *partition function*, or *sum of state*† Z.

$$Z = \sum_i e^{-\beta\epsilon_i} \tag{12-55}$$

It follows that we may rewrite Eq. (12-37) as

$$\frac{N_i}{N} = \frac{1}{Z} e^{-\beta\epsilon_i} \tag{12-56}$$

The partition function is just what its name implies: a function related to the specific way particles are "partitioned" among the various energy states accessible to them.

The foregoing discussion is restricted in that we considered a simplified model, i.e., one for which $g_i = 1$. Let us adopt a general definition of the partition function in accordance with Eq. (12-55), but require that the summation be taken over all energy *states* accessible to the particles. Then, using the subscript j to designate energy states and retaining the i subscript to denote energy levels, we obtain

$$Z = \sum_i e^{-\beta\epsilon_j} = \sum_i g_i e^{-\beta\epsilon_i} \tag{12-57}$$

This expression results because there are g_i energy states having the value ϵ_i. We have already called g_i the degeneracy of the particular energy level. Other writers may prefer to call g_i the *statistical weight,* thus saying that certain energy levels may be more heavily "weighted" than others.

By adopting the definition of partition function given by Eq. (12-57) we may then examine the MB distribution without the restriction that $g_i = 1$. Thus

$$N = \sum_i N_i = \sum \frac{g_i e^{-\beta\epsilon_i}}{A} = \frac{Z}{A}$$

or

$$\frac{N_i}{N} = \frac{1}{Z} g_i e^{-\beta\epsilon_i} \tag{12-58}$$

EXAMPLE 12-4

For the four energy levels illustrated in Fig. 12-1 compute the value of the partition function assuming that the particles obey MB statistics and $\beta = 1.0$ per energy unit. Subsequently calculate the fractional populations of the four energy levels.

---

† Translation of German *Zustandssumme*.

SOLUTION To calculate the partition function, we make use of Eq. (12-57) and the information from Fig. 12-1:

$$Z = \sum_i g_i e^{-\beta\epsilon_i} = g_0 e^{-\beta\epsilon_0} + g_1 e^{-\beta\epsilon_1} + g_2 e^{-\beta\epsilon_2} + g_3 e^{-\beta\epsilon_3}$$

$$= (1)e^0 + (3)e^{-1} + (6)e^{-2} + (10)e^{-3}$$

$$= 1.0 + 1.104 + 0.813 + 0.498$$

$$Z = 3.415$$

The fractional populations of the four energy levels are calculated with Eq. (12-58):

$$\frac{N_i}{N} = \frac{1}{Z} g_i e^{-\beta\epsilon_i}$$

$$\frac{N_0}{N} = \frac{1.0}{3.415} = 0.293$$

$$\frac{N_1}{N} = \frac{1.104}{3.415} = 0.323$$

$$\frac{N_2}{N} = \frac{0.813}{3.415} = 0.238$$

$$\frac{N_3}{N} = \frac{0.498}{3.415} = 0.146$$

For this example it appears that the second energy level ($\epsilon_i = 1$ energy unit) is the most heavily populated. The populations are depicted in Fig. Example 12-4.

FIG. EXAMPLE 12-4

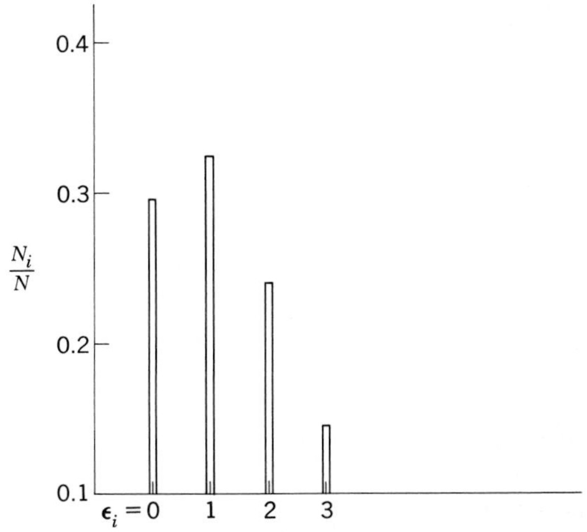

12-18 THE CONSTANTS A AND β

We have shown above how the constant A may be evaluated for the MB distribution [Eq. (12-58)]. The evaluation of A for the BE and FD distributions is not so simple and depends on the physical system (gas, liquid, or solid) to which these statistics are applied. We shall defer our evaluation for these cases until Chap. 13, where we apply the statistical distributions to some specific circumstances.

To evaluate the constant β we need to utilize the second law of thermodynamics and some general thermodynamic property relations; however, we can at least state the result at this point. It will be shown that

$$\beta = \frac{1}{kT} \tag{12-59}$$

where k is the gas constant per molecule, or *Boltzmann's constant,*

$$k = 1.3807 \times 10^{-23} \text{ J/molecule·K} \tag{12-60}$$

and T is the absolute temperature. It will also be shown that Boltzmann's constant is the value used in Eq. (12-50) to define the entropy. The value of β is given by Eq. (12-59) for all three of the statistical distribution discussed previously. The Boltzmann constant is given in terms of the universal gas constant as

$$k = \frac{\mathcal{R}}{N_0} \tag{12-61}$$

where as before, N_0 is Avogadro's number.

REVIEW QUESTIONS

1. What is phase space?
2. What is the de Broglie wavelength?
3. Why are quantum considerations unimportant for analysis of macroscopic systems?
4. What is meant by the term *degeneracy*?
5. Distinguish between quantum *state* and energy *level.*
6. Distinguish between the terms *microstate* and *macrostate.*
7. How is the concept of thermodynamic probability used to predict the macroscopic behavior of a system?
8. What are the assumptions pertaining to the MB, BE, and FD statistical models?
9. Describe heat and work addition to a simple compressible substance in terms of microscopic phenomena.
10. How is entropy defined?
11. Define the partition function.

12 What is meant by the term *equilibrium distribution*?
13 How is entropy related to thermodynamic probability?
14 What is a property?
15 How is the second law related to microscopic thermodynamic analysis?
16 How is the Boltzmann constant related to the universal gas constant?

PROBLEMS

12-1 Calculate the number of ways of arranging three distinguishable balls in two boxes so that $N_1 = 1$, $N_2 = 2$.

12-2 Calculate the number of ways of arranging eight indistinguishable balls in four indistinguishable boxes.

12-3 Calculate the number of ways of arranging 10 indistinguishable balls in 12 boxes with no more than one ball per box.

12-4 Calculate the thermodynamic probability for arranging five particles among the energy levels and states shown in Fig. 12-1 with $N_0 = 1$, $N_1 = 2$, $N_3 = 1$. Make the calculation for all three physical models.

12-5 Repeat Prob. 12-4, but with $N_0 = 1$, $N_1 = 1$, $N_2 = 2$, $N_3 = 1$.

12-6 Determine the degeneracy for a translational energy level of four energy units when the minimum quantum increment is two energy units. Repeat for an energy level of six energy units.

12-7 Determine the degeneracy for a translational energy level of four energy units when the minimum quantum increment is one energy unit.

12-8 Consider a system of particles for which the energy levels are quantized according to

$$\epsilon_i = (i + 1)i \qquad \text{energy units (eu)} \qquad i = 0, 1, 2, \ldots$$

and the degeneracy of each level is

$$g_i = 2i + 1$$

Assuming that the particles obey MB statistics and that $\beta = 1.0\ (\text{eu})^{-1}$, calculate the partition function for a group of particles occupying (*a*) the first four energy levels; (*b*) the first five energy levels. Subsequently calculate the fractional populations of the energy levels.

12-9 Consider a system of particles for which the energy levels are quantized according to the following for $\beta = 1.0\ (\text{eu})^{-1}$:

$$\epsilon_i = (i + \tfrac{1}{2})\ \text{eu} \qquad i = 0, 1, 2, \ldots$$

and the energy levels are nondegenerate. Calculate and plot the fractional populations of the first 10 energy levels assuming that the particles obey MB statistics.

12-10 Four indistinguishable particles are to be placed in two energy levels, each of which has a degeneracy of three. Evaluate the thermodynamic probability for all possible arrangements. What is the most probable arrangement?

12-11 Calculate the thermodynamic probability for six indistinguishable balls to be placed in three boxes with
(*a*) $N_1 = 1, N_2 = 4, N_3 = 1$
(*b*) $N_1 = N_2 = N_3 = 2$
(*c*) $N_1 = 1, N_2 = 3, N_3 = 2$

12-12 Calculate the maximum thermodynamic probability for 100 distinguishable particles distributed among the four energy levels illustrated in Fig. 12-1. Establish the number of particles in each level for this equilibrium distribution. Estimate the relative decrease in the thermodynamic probability resulting from a switch of one particle from the equilibrium distribution [$\beta = 1.0\ (\text{eu})^{-1}$].

12-13 Recalculate the maximum probability for Prob. 12-12 assuming that the particles are indistinguishable.

12-14 Repeat Prob. 12-12 for the first four energy levels as described in Prob. 12-8. Repeat the solution assuming that the particles are indistinguishable.

REFERENCES

1 Kaplan, I.: "Nuclear Physics," 2d ed., Addison-Wesley Publishing Company, Inc., Reading, Mass., 1963.
2 Allis, W. P., and M. A. Herlin: "Thermodynamics and Statistical Mechanics," McGraw-Hill Book Company, New York, 1952.
3 Callen, H. B.: "Thermodynamics," John Wiley & Sons, Inc., New York, 1960.
4 Crawford, F. H.: "Heat, Thermodynamics, and Statistical Physics," Harcourt, Brace & World, New York, 1963.
5 Lee, J. F., F. W. Sears, and D. L. Turcotte: "Statistical Thermodynamics," Addison-Wesley Publishing Company, Inc., Reading, Mass., 1963.
6 Lewis, G. N., and M. Randall: "Thermodynamics," 2d ed. (revised by K. S. Pitzer and L. Brewer), McGraw-Hill Book Company, New York, 1961.
7 Mayer, J. E., and M. G. Mayer: "Statistical Mechanics," John Wiley & Sons, Inc., New York, 1940.
8 Reynolds, W. C.: "Thermodynamics," 2d ed., McGraw-Hill Book Company, New York, 1968.
9 Schrödinger, E.: "Statistical Thermodynamics," Cambridge University Press, Cambridge, 1952.
10 Sears, F. W.: "An Introduction to Thermodynamics, The Kinetic Theory of Gases, and Statistical Mechanics," Addison-Wesley Publishing Company, Inc., Reading, Mass., 1950.
11 Sommerfeld, A.: "Thermodynamics and Statistical Mechanics," Academic Press, Inc., New York, 1956.
12 Tolman, C.: "The Principles of Statistical Mechanics," Oxford University Press, London, 1938.
13 Hatsopoulos, G. N., and J. H. Keenan: "Principles of General Thermodynamics," John Wiley & Sons, Inc., New York, 1965.

13

APPLICATIONS OF STATISTICAL THERMO-DYNAMICS

13-1 INTRODUCTION

The basic statistical presentations of Chap. 12 along with macroscopic discussions of the second law of thermodynamics have indicated possible links between macroscopically observable thermodynamic phenomena and the underlying molecular behavior of matter. Statistical implications of this correspondence have already been noted. Our purpose in this chapter is to carry through with the analysis in order to develop calculation techniques for thermodynamic properties based on a microscopic model. It turns out that the macroscopic definition of entropy is needed to introduce the temperature concept into a microscopic analysis. This fact makes sense because temperature, as measured macroscopically, is a collective concept, that is, it represents the total effect of a collection of microscopic particles on some measuring device. It would be foolish then for us to speak of the temperature of a *single* molecule—we must consider a statistically large sample of particles. This means that we need a macroscopic concept of temperature to be able to link microscopic thermodynamic analysis with calculations of equilibrium thermodynamic properties.

By judicious use of the temperature concept, it will be possible for us to describe statistically several types of thermodynamic systems (gases, solids, thermal radiation) and calculate their properties. It is well to caution the reader at this point that the statistical approach followed here is not the only one available and, in many respects, is quite restrictive in scope. In a later section of this chapter we shall mention some features of the *ensemble* method, which furnishes powerful techniques for analysis of very complicated thermodynamic systems.

13-2 THE CONSTANTS β AND k

The development of general property relations in Chap. 6 furnishes an analytical basis for determination of the energy constant β that appears in all of the statistical distributions of Chap 12. We shall now establish the relationship between β and k, the entropy constant, from

$$S = k \ln \Omega \tag{13-1}$$

In a later section we shall show that k is the Boltzmann constant.

Consider the Bose-Einstein (BE) energy distribution:

$$\frac{N_i}{g_i} = \frac{1}{Ae^{\beta\epsilon_i} - 1} \tag{13-2}$$

The logarithm of the thermodynamic probability for this distribution was given by

$$\ln \Omega = \Sigma[(N_i + g_i) \ln (N_i + g_i) - g_i \ln g_i - N_i \ln N_i] \tag{13-3}$$

when Stirling's approximation is employed. Rearranging terms and using Eq. (13-1), the entropy is written as

$$S = k \sum \left(g_i \ln \frac{N_i + g_i}{g_i} + N_i \ln \frac{N_i + g_i}{N_i} \right) \tag{13-4}$$

One of the intermediate steps in the derivation of the Maxwell relations [Eq. (6-16*d*)] yields

$$\frac{1}{T} = \left(\frac{\partial S}{\partial U}\right)_V \tag{13-5}$$

Equation (13-4) expresses the entropy for a fixed volume (isolated system), so that we easily perform the differentiation and obtain

$$\left(\frac{\partial S}{\partial U}\right)_V = k \sum g_i \frac{g_i}{N_i + g_i} \frac{1}{g_i} \left(\frac{\partial N_i}{\partial U}\right)_V + N_i \frac{N_i}{N_i + g_i} g_i \frac{-1}{N_i^2} \left(\frac{\partial N_i}{\partial U}\right)_V + \ln \frac{N_i + g_i}{N_i} \left(\frac{\partial N_i}{\partial U}\right)_V \tag{13-6}$$

In this differentiation g_i is considered constant since it does not vary with the total energy of the system. Canceling terms in Eq. (13-6) gives

$$\left(\frac{\partial S}{\partial U}\right)_V = k \sum \ln \frac{N_i + g_i}{N_i} \left(\frac{\partial N_i}{\partial U}\right)_V \tag{13-7}$$

From the equilibrium distribution of Eq. (13-2)

$$\ln \frac{N_i + g_i}{N_i} = \ln A + \beta \epsilon_i \tag{13-8}$$

so that

$$\left(\frac{\partial S}{\partial U}\right)_V = k \ln A \sum \left(\frac{\partial N_i}{\partial U}\right)_V + k\beta \sum \epsilon_i \left(\frac{\partial N_i}{\partial U}\right)_V \tag{13-9}$$

The constant total energy and number of particles restrictions may be differentiated to give

$$U = \Sigma \epsilon_i N_i$$
$$\left(\frac{\partial U}{\partial U}\right)_V = 1 = \sum \epsilon_i \left(\frac{\partial N_i}{\partial U}\right)_V \tag{13-10}$$

$$N = \Sigma N_i$$
$$\left(\frac{\partial N}{\partial U}\right)_V = 0 = \sum \left(\frac{\partial N_i}{\partial U}\right)_V \tag{13-11}$$

Inserting Eqs. (13-5), (13-10), and (13-11) into Eq. (13-9) gives

$$\frac{1}{T} = k\beta$$

or

$$\beta = \frac{1}{kT} \tag{13-12}$$

It may be shown that this same relation also applies for the Fermi-Dirac (FD) and Maxwell-Boltzmann (MB) distributions. As mentioned previously, we have yet to show that k is the Boltzmann constant.

13-3 PROPERTIES AND PARTITION FUNCTIONS

Let us assume that a collection of microscopic particles is available which obeys MB statistics or BE statistics in the limit where $g_i \gg N_i$. In this case we are able to define a partition function by

$$Z = \Sigma g_i e^{-\beta \epsilon_i} \tag{13-13}$$

and the equilibrium distribution becomes

$$N_i = \frac{N}{Z} g_i e^{-\beta \epsilon_i} \tag{13-14}$$

The entropy may be written as follows for $g_i \gg N_i$, using Eq. (13-4):

$$S = k \ln \Omega_{\max} = k \sum g_i \frac{N_i}{g_i} + N_i \ln \frac{g_i}{N_i}$$

$$= Nk + k \sum N_i \ln \frac{g_i}{N_i}$$

Making use of the equilibrium distribution, Eq. (13-14),

$$S = Nk + k \sum N_i \left(\ln \frac{Z}{N} + \beta \epsilon_i\right)$$

The total energy is $U = \Sigma \epsilon_i N_i$ and $\beta = 1/kT$ so that

$$S = Nk + Nk \ln \frac{Z}{N} + \frac{U}{T} \tag{13-15}$$

We thus have expressed the entropy in terms of the partition function and the internal energy of the system.

For an isolated system (constant total energy and number of particles), the partition function depends only on the temperature, since g_i and ϵ_i are established by quantum mechanical considerations. The other variable in the partition function is $\beta = 1/kT$, and the rate of change of the partition function with temperature is

$$\begin{aligned}\left(\frac{\partial Z}{\partial T}\right)_V &= \sum - g_i\epsilon_i e^{-\beta\epsilon_i}\left(\frac{\partial \beta}{\partial T}\right)_V \\ &= \frac{1}{kT^2}\sum g_i\epsilon_i e^{-\beta\epsilon_i}\end{aligned} \tag{13-16}$$

The total internal energy may now be expressed as

$$U = \sum \epsilon_i N_i = \frac{N}{Z}\sum g_i\epsilon_i e^{-\beta\epsilon_i} \tag{13-17}$$

Comparing Eq. (13-16) with (13-17), we obtain

$$U = \frac{NkT^2}{Z}\left(\frac{\partial Z}{\partial T}\right)_V = NkT^2\left(\frac{\partial \ln Z}{\partial T}\right)_V \tag{13-18}$$

Equation (13-15) for the entropy may now be rewritten as

$$S = Nk\left[1 + \ln\frac{Z}{N} + T\left(\frac{\partial \ln Z}{\partial T}\right)_V\right] \tag{13-19}$$

The differentiation is performed at constant volume because we are considering an isolated system with a constant number of particles.

We now have the important result that both the entropy and internal energy of the system for $g_i \gg N_i$ can be expressed in terms of a single property Z, the value of which may be calculated from microscopic considerations. Other thermodynamic properties may also be expressed in terms of the partition function, e.g., the Helmholtz function

$$A = U - TS = -NkT\left(1 + \ln\frac{Z}{N}\right) \tag{13-20}$$

According to Eq. (6-16a), the pressure is expressible as

$$p = -\left(\frac{\partial A}{\partial V}\right)_T$$

Performing the indicated differentiation on Eq. (13-20) gives

$$p = NkT\left(\frac{\partial \ln Z}{\partial V}\right)_T \tag{13-21}$$

The enthalpy and Gibbs function are now written immediately as

$$H = U + pV = NkT\left[T\left(\frac{\partial \ln Z}{\partial T}\right)_V + V\left(\frac{\partial \ln Z}{\partial V}\right)_T\right] \tag{13-22}$$

$$G = A + pV = -NkT\left[1 + \ln\frac{Z}{N} - V\left(\frac{\partial \ln Z}{\partial V}\right)_T\right] \tag{13-23}$$

We now have related the partition function to all macroscopic thermodynamic properties of interest for simple substances.

The reader may find it interesting to muse over the developments relating the partition function to macroscopic thermodynamic properties. As the connecting link between these properties and microscopic analysis the partition function and its calculation form a substantial part of statistical thermodynamic analysis. To aid in this reflection we may mention a few points of interest:

1 To link microscopic analysis with the macroscopic world we must introduce the concept of temperature. This is done through Eq. (13-5), which is obtained from purely macroscopic considerations.
2 It is necessary to identify the behavior of the thermodynamic probability with the behavior of a quantity called the entropy. Otherwise the temperature definition could not be applied.
3 To define temperature, only systems in equilibrium may be considered. Otherwise it is difficult to attach a physical meaning to the temperature obtained from $\beta = 1/kT$.

13-4 THE IDEAL MONATOMIC GAS

As the first example of an application of statistical thermodynamic analysis, let us consider a simple monatomic gas whose molecules are so widely spaced that molecular force fields do not exert a significant influence on collision processes or distribution of energy. For this system we assume that the molecules represent point masses, so that their energy may be expressed as translational kinetic energy. The molecules are assumed to be indistinguishable and the translational kinetic energies are quantized. We therefore choose the BE statistical model,

$$\frac{N_i}{g_i} = \frac{1}{Ae^{\beta\epsilon_i} - 1} \tag{13-24}$$

If the gas is at a high temperature (room temperature is sufficiently high), there will be many more quantum states available than there are particles to fill them,† and the equilibrium distribution is given by

$$\frac{N_i}{N} = \frac{1}{Z} g_i e^{-\beta\epsilon_i} \tag{13-25}$$

Our problem is thus one of determining the values of g_i and ϵ_i so that the partition function may be calculated. Once the partition function is obtained, other thermodynamic properties may be evaluated.

The energies ϵ_i may be expressed by

$$\epsilon_i = \frac{1}{2} m(v_x^2 + v_y^2 + v_z^2) \tag{13-26}$$

where v_x, v_y, and v_z are the velocity components of the particles. Now consider an elemental volume in phase space, $dx\, dy\, dz\, dp_x\, dp_y\, dp_z$, where $p = mv$ is the particle momentum. According to the uncertainty principle, the accuracy of specification of momentum and displacement is limited by

$$\Delta x\, \Delta p_x \sim h$$

This means that the smallest discernible volume in phase space is h^3. The number of these small volumes in some larger volume element represents the degeneracy or the number of ways a certain energy level may occur. Thus we take

$$g_i = \frac{dx\, dy\, dz\, dp_x\, dp_y\, dp_z}{h^3} \tag{13-27}$$

In this analysis we assume that the volume element in phase space is large enough to contain a large number of particles so that the statistics will apply. Using Eqs. (13-25) and (13-27), the expression for the partition function is obtained as

$$Z = \sum \frac{m^3}{h^3} \exp\left[\frac{-m\beta}{2}(v_x^2 + v_y^2 + v_z^2)\right] dx\, dy\, dz\, dv_x\, dv_y\, dv_z \tag{13-28}$$

where $p_x = mv_x$, etc., have been substituted for the momenta. The sum in Eq. (13-28) must be extended over six dimensions. If the energy states are sufficiently close-spaced, and we assume they are for gases at high temperatures,

† See discussion of Sec. 12-7.

the sum may be replaced by an integral. Thus

$$Z = \int\int\int\int\int\int \frac{m^3}{h^3} \exp\left[\frac{-m\beta}{2}(v_x^2 + v_y^2 + v_z^2)\right] dx\, dy\, dz\, dv_x\, dv_y\, dv_z$$

The triple integral over the three space coordinates is simply the total system volume V, and the exponential integrals involving the velocity components are all identical, so that the expression for the partition function becomes

$$\begin{aligned} Z &= \frac{m^3V}{h^3}\left[\int_{-\infty}^{\infty} \exp\left(\frac{-\beta m v_x^2}{2}\, dv_x\right)\right]^3 \\ &= \frac{m^3V}{h^3}\left[2\int_0^{\infty} \exp\left(\frac{-\beta m v_x^2}{2}\, dv_x\right)\right]^3 \end{aligned} \tag{13-29}$$

The upper limit of this integral obviously has no physical meaning, but so few particles have very high velocities that no appreciable error is involved. In other words, the exponential function approaches zero so rapidly for large values of v_x that the integral is a suitable representation of the true physical situation.

The integral in Eq. (13-29) may be evaluated to give

$$Z = \frac{m^3V}{h^3}\left(\frac{2\pi}{m\beta}\right)^{3/2}$$

Letting $\beta = 1/kT$ gives

$$Z = V\left(\frac{2\pi mkT}{h^2}\right)^{3/2} \tag{13-30}$$

Inserting Eqs. (13-30) and (13-27) into Eq. (13-24) results in the energy or velocity distribution

$$\frac{d^6N}{N} = \frac{m^3}{V}\left(\frac{1}{2\pi mkT}\right)^{3/2} e^{-mv^2/2kT}\, dx\, dy\, dz\, dv_x\, dv_y\, dv_z \tag{13-31}$$

The symbol d^6N represents the number of particles in a six-dimensional differential element, as indicated in the equation. The symbol d^3N will be used to designate the number of particles in a three-dimensional differential element.

In this model of a monatomic gas the gas is assumed to fill uniformly the system volume so that the molecular density (molecules per unit volume) is constant. Thus Eq. (13-31) may be integrated over the system volume to give

$$\frac{d^3N}{N} = \left(\frac{m}{2\pi kT}\right)^{3/2} e^{-mv^2/2kT}\, dv_x\, dv_y\, dv_z \tag{13-32}$$

Equation (13-32) is normally called the Maxwell-Boltzmann velocity distribution because the limiting case of BE statistics ($g_i \gg N_i$) corresponds to that obtained from the MB model.

The internal energy of the monatomic gas may now be evaluated with Eq. (13-18). We have, from Eq. (13-30),

$$\ln Z = \ln \left[V \left(\frac{2\pi mk}{h^2} \right)^{3/2} \right] + \frac{3}{2} \ln T$$

$$\left(\frac{\partial \ln Z}{\partial T} \right)_V = \frac{3}{2} \frac{1}{T}$$

and

$$U = \frac{3}{2} NkT \tag{13-33}$$

Equation (13-33) is a very important result because it shows that the internal energy for this ideal monatomic gas is a function only of temperature.

We may derive the equation of state for the ideal monatomic gas very quickly by referring back to Eq. (13-21):

$$p = NkT \left(\frac{\partial \ln Z}{\partial V} \right)_T$$

or

$$p = NkT \frac{1}{V}$$

Thus

$$pV = NkT \tag{13-34}$$

This relation is identical to the perfect gas relation of Chap. 1 [Eq. (1-30)] if we identify

$$Nk = \eta \mathfrak{R} \tag{13-35}$$

where η is the number of moles and $\mathfrak{R}$ is the universal gas constant. The total number of molecules N is

$$N = \eta N_0$$

where N_0 is the number of molecules per mole, or Avogadro's number. Then from Eq. (13-35) we have the result that

$$k = \frac{\mathcal{R}}{N_0} \tag{13-36}$$

where k is called the *gas constant per molecule*, or the *Boltzmann constant*, having the numerical value 1.3807×10^{-23} J/K·molecule.

Inserting the relations for internal energy and the partition function into Eq. (13-15) results in an expression for the entropy of the ideal monatomic gas.

$$S = Nk\left(\frac{5}{2} + \frac{3}{2}\ln 2\pi mk - \ln h^3 + \ln V + \frac{3}{2}\ln T - \ln N\right) \tag{13-37}$$

Specific heats for the ideal monatomic gas may be obtained immediately with

$$C_V = \left(\frac{\partial U}{\partial T}\right)_V = \frac{3}{2}Nk = \frac{3}{2}\eta\mathcal{R}$$

or, on a molal basis,

$$\bar{c}_v = \frac{3}{2}\mathcal{R} \tag{13-38}$$

Then

$$\bar{c}_p = \bar{c}_v + \mathcal{R} = \frac{5}{2}\mathcal{R} \tag{13-39}$$

These specific-heat relations are in excellent agreement with experiment for monatomic gases under pressure and temperature conditions which satisfy the original assumptions in the analysis, i.e., sufficiently low density so that the molecules are far apart and molecular force fields do not strongly influence molecular energy distributions.

13-5 DIATOMIC AND POLYATOMIC GASES

The major difference between monatomic and polyatomic gases is that the latter have rotational and vibrational energy states as well as translational energies based on the motion of the center of mass of the molecule. Still, if the molecules are sufficiently far apart, i.e., the gas density is low enough, the interaction force fields between molecules will not exert a significant influence on the energy distribution of the total system. As before, we assume $g_i \gg N_i$ and employ the limiting case of the BE statistics to calculate the thermodynamic properties. The problem reduces to that of determining the various energy levels and degeneracies to evaluate the system partition function. We assume that the total energy of a molecule may be expressed by the sum

$$\epsilon_{\text{total}} = \epsilon_{\text{trans}} + \epsilon_{\text{rot}} + \epsilon_{\text{vib}}$$

The partition function is then written as

$$Z = \Sigma(g_{\text{trans}}e^{-\beta\epsilon_{\text{trans}}})(g_{\text{rot}}e^{-\beta\epsilon_{\text{rot}}})(g_{\text{vib}}e^{-\beta\epsilon_{\text{vib}}})$$

if it is possible to assume that each energy mode is independent of the others. Then the sums may be taken separately and

$$Z = Z_{\text{trans}}Z_{\text{rot}}Z_{\text{vib}}$$

where

$$Z_{\text{trans}} = \Sigma g_{\text{trans}}e^{-\beta\epsilon_{\text{trans}}}$$
$$Z_{\text{rot}} = \Sigma g_{\text{rot}}e^{-\beta\epsilon_{\text{rot}}}$$
$$Z_{\text{vib}} = \Sigma g_{\text{vib}}e^{-\beta\epsilon_{\text{vib}}}$$

The assumption that various energy modes are independent is not strictly correct. Consider a diatomic dumbbell molecule, as shown in Fig. 13-1. The molecule has a translational energy expressed in terms of the motion of the center of mass, and a rotational energy which may be expressed in terms of the

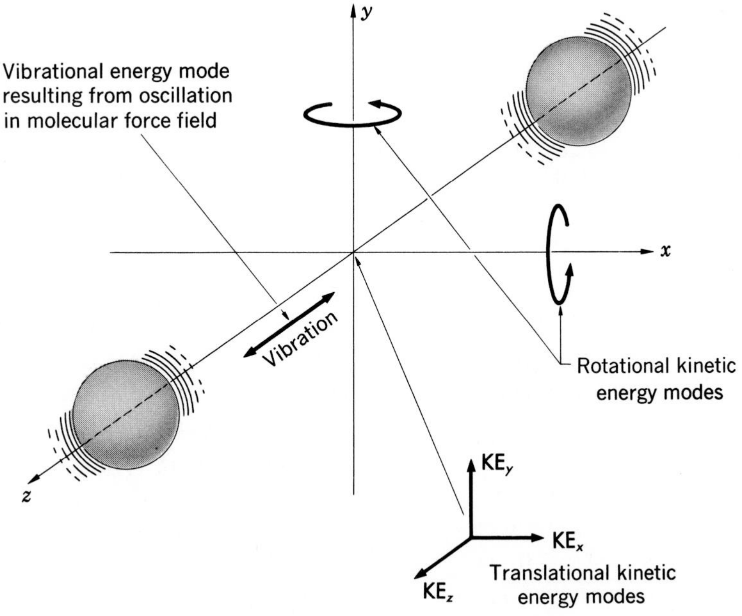

FIG. 13-1 Energy modes of a diatomic dumbbell molecule.

moment of inertia of the molecule about the center of mass. The vibrational energy results from vibration of the atoms in the force field acting between their centers. However, when these atoms vibrate, there can be a corresponding change in the moment of inertia of the molecule and, consequently, a change in the rotational energy. This is not usually an important consideration, except at high temperatures, and requires an analysis much too complicated for our study. At sufficiently high temperatures we should also consider changes in electron energy levels within the atoms.

For a diatomic dumbbell molecule like that shown in Fig. 13-1 a quantum mechanical analysis shows that the allowable rotational energy levels are given by

$$\epsilon_j = j(j+1)\frac{h^2}{8\pi^2 I} \qquad j = 0, 1, 2, \ldots \tag{13-40}$$

where I is the moment of inertia of the molecule about the center of mass. The degeneracy of the rotational energy levels can be shown to be

$$g_{\text{rot}} = 2j + 1 \tag{13-41}$$

We have used the index j to avoid confusion with the index i in the translational partition function. The rotational partition function is now written as

$$Z_{\text{rot}} = \sum_{j=0,1,2,\ldots} (2j+1)\exp\left[-j(j+1)\frac{h^2}{8\pi^2 IkT}\right] \tag{13-42}$$

We now define the *characteristic temperature for rotation*, Θ_{rot}, as

$$\Theta_{\text{rot}} = \frac{h^2}{8\pi^2 Ik} \tag{13-43}$$

so that the expression for the partition function becomes

$$Z_{\text{rot}} = \sum_{j=0,1,2,\ldots} (2j+1)\exp\left[-j(j+1)\frac{\Theta_{\text{rot}}}{T}\right] \tag{13-44}$$

Let us examine the behavior of Eq. (13-44) as a function of the ratio Θ_{rot}/T. For very low temperatures compared with Θ_{rot}, that is, $T \ll \Theta_{\text{rot}}$, the exponential function is very small and, hence, the value of Z_{rot} is very small. When T is of the same order as Θ_{rot}, the partition function undergoes rapid changes with T. For $T/\Theta_{\text{rot}} > 1$, Eq. (13-44) may be expanded to give

$$Z_{\text{rot}} = \frac{T}{\Theta_{\text{rot}}}\left[1 + \frac{1}{3}\left(\frac{\Theta_{\text{rot}}}{T}\right) + \frac{1}{15}\left(\frac{\Theta_{\text{rot}}}{T}\right)^2 + \cdots\right] \tag{13-45}$$

As the temperature becomes very large compared with Θ_{rot}, the higher-order terms become negligible and

$$Z_{rot} = \frac{T}{\Theta_{rot}} \qquad \text{for } \frac{T}{\Theta_{rot}} \gg 1 \tag{13-46}$$

We may now employ Eq. (13-46) to evaluate the rotational contribution to internal energy at high temperatures compared with Θ_{rot}. There results

$$U_{rot} = NkT \qquad \text{for } \frac{T}{\Theta_{rot}} \gg 1 \tag{13-47}$$

The molal specific heat for rotation then becomes

$$\bar{c}_{v,rot} = \mathcal{R} \qquad \text{for } \frac{T}{\Theta_{rot}} \gg 1 \tag{13-48}$$

To compute the specific heat for conditions other than the limiting one given by Eq. (13-48), it is necessary to evaluate the partition function from Eq. (13-44). The results of such a calculation are shown in Fig. 13-2. As expected, the rotational contributions to specific heat are small for $T/\Theta_{rot} \ll 1$ and approach a constant value of $\mathcal{R}$ at high temperatures compared with the characteristic temperature.

To compute the vibrational partition function an analysis similar to the previous one is performed. From quantum mechanics it is found that the allowable vibrational energy levels for a diatomic molecule are given by

$$\epsilon_{vib} = \left(n + \frac{1}{2}\right)h\nu \qquad n = 0, 1, 2, \ldots \tag{13-49}$$

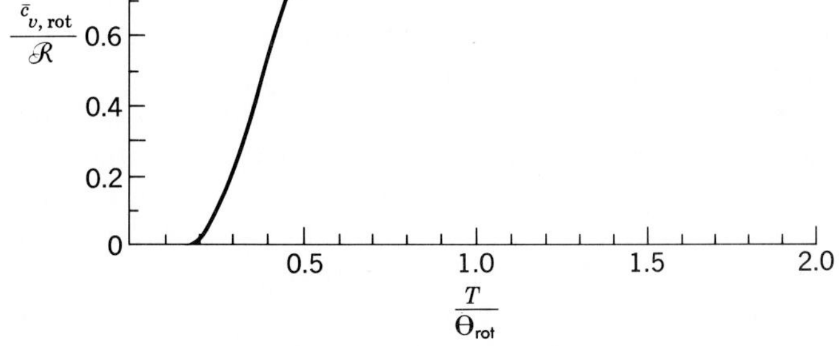

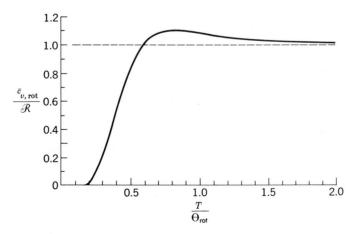

FIG. 13-2 Rotational specific heat for a diatomic gas.

where ν is a characteristic frequency of vibration. It can be shown that each vibrational energy level may occur only one way so that the degeneracy is unity (or the vibrational energy states are nondegenerate). Equation (13-49) is obtained by assuming that the vibrating diatomic molecule behaves as a one-dimensional simple harmonic oscillator and it is well to remark that this assumption may not be valid at very high temperatures.

The vibrational partition function is now computed with $g_{\text{vib}} = 1.0$:

$$Z_{\text{vib}} = \sum_{n=0}^{\infty} \exp\left[-\frac{(n+1/2)h\nu}{kT}\right] \tag{13-50}$$

This sum may fortunately be expressed in closed form. Defining the *characteristic temperature for vibration,* Θ_{vib}, as

$$\Theta_{\text{vib}} = \frac{h\nu}{k} \tag{13-51}$$

the final relation for the partition function is

$$Z_{\text{vib}} = \frac{e^{-\Theta_{\text{vib}}/2T}}{1 - e^{-\Theta_{\text{vib}}/T}} \tag{13-52}$$

Using Eq. (13-18), the vibrational contribution to internal energy may be obtained as

$$U_{\text{vib}} = Nk\Theta_{\text{vib}}\left(\frac{1}{2} + \frac{1}{e^{\Theta_{\text{vib}}/T} - 1}\right) \tag{13-53}$$

The vibrational contribution to the molal specific heat then becomes

$$\bar{c}_{v,\text{vib}} = \left(\frac{\partial \bar{u}_{\text{vib}}}{\partial T}\right)_V = \mathcal{R}\left(\frac{\Theta_{\text{vib}}}{T}\right)^2 \frac{e^{\Theta_{\text{vib}}/T}}{(e^{\Theta_{\text{vib}}/T} - 1)^2} \tag{13-54}$$

Figure 13-3 gives a graphical display of the behavior of $\bar{u}_{\text{vib}}$ and $\bar{c}_{v,\text{vib}}$ as a function of the ratio T/Θ_{vib}. The behavior of specific heat is similar to that obtained for the rotational contribution. At very low temperatures compared with Θ_{vib} the vibrational contribution to specific heat is negligible. At high temperatures it approaches a constant value of $\mathcal{R}$. The behavior of vibrational internal energy, however, is quite different from that of rotational internal energy. An inspection of Eq. (13-40) shows that the lowest rotational energy state is zero. This is not the case for vibration as may be seen from Eq. (13-49). Therefore, as temperature is reduced it is not surprising to find that the vibrational internal energy does not approach zero.

A summary of values of Θ_{rot} and Θ_{vib} for various gases is given in Table 13-1. It may be noted that the characteristic temperatures for rotation are

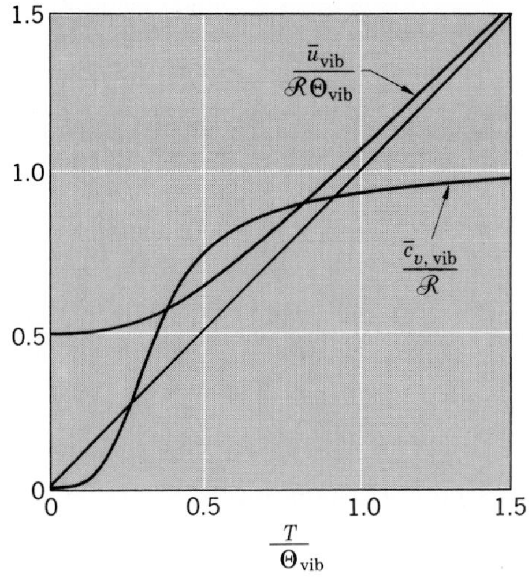

FIG. 13-3 Vibrational specific heat and internal energy for a diatomic gas.

substantially lower than those for vibration. This means that essentially all of the rotational energy modes for a diatomic molecule will be excited before there is any appreciable change in the vibrational energy.

We are now in a position to predict the behavior of the total specific heat of a diatomic gas. Since the energies are additive, we may write

$$\bar{c}_v = \bar{c}_{v,\text{trans}} + \bar{c}_{v,\text{rot}} + \bar{c}_{v,\text{vib}} \tag{13-55}$$

At low temperatures compared with Θ_{rot}, the essential contribution to specific heat results from translational energy modes, so that we would expect

$$\bar{c}_v = \frac{3}{2}\mathcal{R} \qquad T \ll \Theta_{\text{rot}}$$

TABLE 13-1 Characteristic Temperatures for Rotation and Vibration for Various Gases

Substance	Θ_{vib}, K	Θ_{rot}, K
H_2	6140	85.5
OH	5360	27.5
HCl	4300	15.3
CH	4100	20.7
N_2	3340	2.86
HBr	3700	12.1
HI	3200	9.0
CO	3120	2.77
NO	2740	2.47
O_2	2260	2.09
Cl_2	810	0.347
Br_2	470	0.117
Na_2	230	0.224
K_2	140	0.081

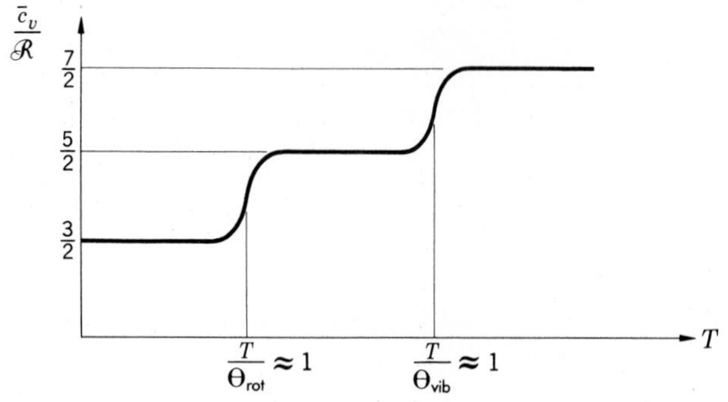

FIG. 13-4 Approximate behavior of total specific heat of a diatomic gas.

As the temperature is raised, rotational energy modes are excited and the specific heat is increased corresponding to the behavior shown in Fig. 13-4. Once the temperature is raised substantially above Θ_{rot}, but still much below Θ_{vib}, the specific heat approaches

$$\bar{c}_v = \frac{3}{2}\mathcal{R} + \mathcal{R} = \frac{5}{2}\mathcal{R} \qquad \Theta_{rot} \ll T \ll \Theta_{vib}$$

Finally at much higher temperatures, vibrational energy states are excited and the specific heat is increased further until it approaches (see also Fig. 13-4)

$$\bar{c}_v = \frac{3}{2}\mathcal{R} + \mathcal{R} + \mathcal{R} = \frac{7}{2}\mathcal{R} \qquad \Theta_{vib} \ll T$$

Constant-pressure specific heats may be computed from

$$\bar{c}_p = \bar{c}_v + \mathcal{R} \tag{13-56}$$

as long as ideal-gas behavior is observed.

This rather involved analysis may be put in quick perspective by examining the specific heat behavior of a familiar gas like N_2 at room temperature (~300 K). According to Table 13-1 the rotational energy levels are fully excited while there is essentially no excitation of vibrational energy states. Thus

$$\bar{c}_v = \frac{5}{2}\mathcal{R}$$

$$\bar{c}_p = \frac{7}{2}\mathcal{R}$$

The ratio of specific heats is

$$\gamma = \frac{\bar{c}_p}{\bar{c}_v} = \frac{7}{5} = 1.40$$

TABLE 13-2
Comparison of Calculated Specific Heats of Gases with Experiment for $\Theta_{rot} \ll T \ll \Theta_{vib}$

	g		$\frac{\bar{c}_p}{\mathcal{R}}$		$\frac{\bar{c}_v}{\mathcal{R}}$		$\frac{\bar{c}_p - \bar{c}_v}{\mathcal{R}}$	
Gas	**Calc.**	**Exp.**	**Calc.**	**Exp.**	**Calc.**	**Exp.**	**Calc.**	**Exp.**
A	1.667	1.67	2.500	2.52	1.500	1.509	1.000	1.008
He		1.659		2.52		1.519		1.001
CO	1.400	1.404	3.500	3.50	2.500	2.488	1.000	1.005
H_2		1.410		3.42		2.438		0.999
N_2		1.404		3.50		2.448		1.005
O_2		1.401		3.52		2.504		1.004

Table 13-2 gives a convenient comparison of specific-heat values calculated with the foregoing relations and values obtained experimentally. The agreement is excellent.

For temperatures in the range of $T \approx \Theta_{rot}$ hydrogen is an exception to the above analysis, because nuclear spin makes a contribution to its specific heat. A detailed analysis is beyond the scope of our discussion, but we may note that there are two types of hydrogen present in this temperature regime: *para*hydrogen with antisymmetrical nuclear spin states, and *ortho*hydrogen with symmetrical nuclear spin states. Experiments indicate that for $T \approx \Theta_{rot}$ the actual hydrogen is a mixture of three parts ortho- to one part parahydrogen, and that rotational specific heats based on Fig. 13-2 will be too high. For further information the interested reader should consult Ref. [4].

EXAMPLE 13-1 Rotational Energy Levels
Calculate the values of the first three rotational energy levels for nitrogen.

SOLUTION The rotational energy levels are given by Eq. (13-40):

$$\epsilon_j = j(j+1)\frac{h^2}{8\pi^2 I} \qquad (a)$$

Using Table 13-1 the characteristic temperature for rotation is

$$\Theta_{rot} = 2.86\ \text{K} = \frac{h^2}{8\pi^2 Ik} \qquad (b)$$

so that

$$\frac{h^2}{8\pi^2 I} = (2.86)(1.38 \times 10^{-23}) = 3.95 \times 10^{-23}\ \text{J/molecule}$$

The first three nonzero energy levels are therefore obtained from Eq. (*a*) as

$$\epsilon_1 = (1)(1 + 1)(3.95 \times 10^{-23}) = 7.9 \times 10^{-23} \text{ J/molecule}$$

$$\epsilon_2 = (2)(2 + 1)(3.95 \times 10^{-23}) = 2.37 \times 10^{-22} \text{ J/molecule}$$

$$\epsilon_3 = (3)(3 + 1)(3.95 \times 10^{-23}) = 4.74 \times 10^{-22} \text{ J/molecule}$$

EXAMPLE 13-2
Calculate the specific heat of nitrogen (N_2) at 2000 K.

SOLUTION Consulting Table 13-1, we find that $T \gg \Theta_{\text{rot}}$, so the rotational energy states are fully excited. The vibrational states are only partially excited so we compute

$$\frac{T}{\Theta_{\text{vib}}} = \frac{2000}{3340} = 0.599$$

From Fig. 13-3 we have

$$\bar{c}_{v,\text{vib}} = 0.85\mathcal{R}$$

and the total specific heat becomes

$$\bar{c}_{v,\text{total}} = \frac{3}{2}\mathcal{R} + \mathcal{R} + 0.85\mathcal{R}$$
$$= 3.35\mathcal{R}$$

Polyatomic molecules are much more complicated than the simple diatomic model considered in the foregoing, although the basic notion for evaluating partition functions and specific heats is the same. In particular, much more complicated rotational and vibrational energy modes are encountered and require a more tedious evaluation procedure for the partition function.

13-6 EQUIPARTITION OF ENERGY

Our analysis of monatomic and diatomic gases has yielded rather interesting information on the specific heat behavior of ideal gases. In this analysis, several energy modes were allowed, and we shall say that for each energy mode there is a *degree of freedom*. There are three translational energy modes, so there are three translational degrees of freedom. In the diatomic gas the molecule can rotate in three directions; however, the moment of inertia about the axis through the centers of the two atoms is negligibly small compared with the moments of inertia about the other two cartesian axes passing through the center of mass. Thus we say that the molecule has two important rotational energy modes. The vibrational energy modes of a diatomic molecule result from oscillation along the axis joining the two atoms. To specify this vibrational energy we would need to specify both the kinetic energy of the oscillating atoms and their potential energy resulting from position in the force field. Consequently, we say that there are two vibrational degrees of freedom.

A quick inspection of the preceding analysis shows that for each *fully excited* degree of freedom a contribution of 1/2 Nk to the specific heat is observed. This is the essential content of the *equipartition-of-energy principle*. Note that this contribution to specific heat is independent of the mass of the particles.

The equipartition-of-energy principle is of limited value because we can never be sure when it applies without consulting tabular information such as that shown in Table 13-1. As somewhat general analytical specification of the principle may be given in the following way. Suppose that the energy associated with a degree of freedom is expressible as a quadratic function of the coordinate specifying the particle position, velocity, etc.,

$$\epsilon_x = ax^2$$

Furthermore, let us assume the distribution among energy states is given by the MB exponential formula:

$$dN_x = e^{-\beta\epsilon_x}\, dx$$

where we are assuming a continuous distribution of energy states. The average energy associated with the degree of freedom is therefore

$$\bar{\epsilon}_x = \frac{\int \epsilon_x\, dN_x}{\int dN_x} = \frac{\int_0^\infty ax^2\, e^{-ax^2/kT}\, dx}{\int_0^\infty e^{-ax^2/kT}\, dx} \tag{13-57}$$

$$\bar{\epsilon}_x = \frac{1}{2}kT$$

This is the average energy per particle per degree of freedom. The total energy associated with the x mode would be 1/2 NkT, and the specific heat contribution would be 1/2 Nk.

There are several restrictive assumptions in the simple analysis of the preceding paragraphs. A quadratic energy function, a continuous distribution of energy, and the MB type of exponential distribution formula are all necessary to obtain the 1/2 kT result. The translational energy modes for a monatomic gas would fulfill these conditions because of the v_x^2 energy dependence and an almost continuous distribution of energy states. If rotational energy modes behaved in a "classical" manner such that $\epsilon_{rot} \sim 1/2\, I\omega^2$, where ω is the angular velocity, we should expect equipartition to hold in this case also. As already observed, however, this behavior is experienced only at temperatures which are high compared with Θ_{rot}.

13-7 THE SOLID STATE

Our previous discussions have considered the application of statistical thermodynamics to systems where the molecular spacing is very large compared with the size of the molecules. Because of this wide spacing, it was reasonable to assume that the molecules were free to move as independent entities, with their energy essentially unaffected by the force fields of neighbor molecules. The solid state represents an entirely different circumstance. In this instance the atoms are so closely spaced that their force fields essentially lock them in place, so that they are only free to oscillate about some equilibrium position under the influence of the force fields of neighboring atoms. One of the problems of a statistical analysis of the solid state is to be able to describe satisfactorily the contribution of this oscillatory motion to the internal energy and specific heat of the solid.

If the solid is an electric conductor, an appreciable number of electrons are relatively free to move about and transport electric charge. These free electrons are usually termed the *electron gas*. Since the electrons have motion and kinetic energy, they would be expected to contribute to the total internal energy and specific heat of the solid. An additional part of the statistical analysis of the solid state is, therefore, addressed to a specification of the energy distribution in the electron gas for a solid. This information is then combined with that concerning the lattice vibration energy to predict the overall energy behavior of the solid. For electric insulators it is expected that the electron-gas contribution to the internal energy of the solid would be small.

Let us first consider the energy contribution resulting from lattice vibrations. The first reasonably satisfactory analysis of this problem was performed by Einstein, who assumed that the forces controlling the vibration of each atom produced a simple harmonic oscillator which could be analyzed with quantum mechanical methods to determine the allowable energy levels. The allowable energy levels are given by the same relation employed for analysis of vibrational energies in diatomic molecules:

$$\epsilon = \left(n + \frac{1}{2}\right)h\nu \qquad n = 0, 1, 2, \ldots$$

where ν is the frequency of the oscillator. It is assumed that only one quantum state per energy level is allowed so the partition function is given as before by

$$Z = \sum_{n=0}^{\infty} \exp\left[-\left(n + \frac{1}{2}\right)\frac{h\nu}{kT}\right] = \frac{e^{-\Theta_E/2T}}{1 - e^{-\Theta_E/T}} \tag{13-58}$$

where now $\Theta_E = h\nu/k$ is called the *Einstein temperature* of the solid.

The total number of atoms in the solid is taken as N; however, it is expected that there will be three harmonic oscillators for each atom (one for each space coordinate), so that the total number of oscillators to be considered is $3N$. The internal energy is thus written as

$$U = 3NkT^2 \left(\frac{\partial \ln Z}{\partial T}\right)_V \tag{13-59}$$

Combining this relation with Eq. (13-58) gives

$$U = \frac{3}{2} NkT \frac{\Theta_E}{T} + 3NkT \frac{\Theta_E/T}{e^{\Theta_E/T} - 1} \tag{13-60}$$

The molal constant-volume specific heat is thus

$$\bar{c}_v = 3\mathcal{R} \left(\frac{\Theta_E}{T}\right)^2 \frac{e^{\Theta_E/T}}{(e^{\Theta_E/T} - 1)^2} \tag{13-61}$$

This relation for specific heat is plotted in Fig. 13-5 as a function of T/Θ_E. At very high temperatures the specific heat approaches a constant value of $3\mathcal{R}$. This value is exactly what would be expected from a simple equipartition of energy appraisal of the problem. The atoms have three directions of motion, and there are two vibrational energy modes for each coordinate, so that a total of six degrees of freedom are expected. Each degree of freedom contributes 1/2 $\mathcal{R}$ to the specific heat; thus the total of $3\mathcal{R}$ is quite reasonable. The shape of the curve in Fig. 13-5 is in general agreement with experimental data for specific heats of solids but predicts values for $\bar{c}_v$ which are too low at low temperatures.

The main fault in the Einstein analysis is that all atoms in a solid simply do not oscillate with the same frequency as is assumed for the calculation. This

FIG. 13-5 Specific heat of a solid according to Einstein theory compared with experimental values.

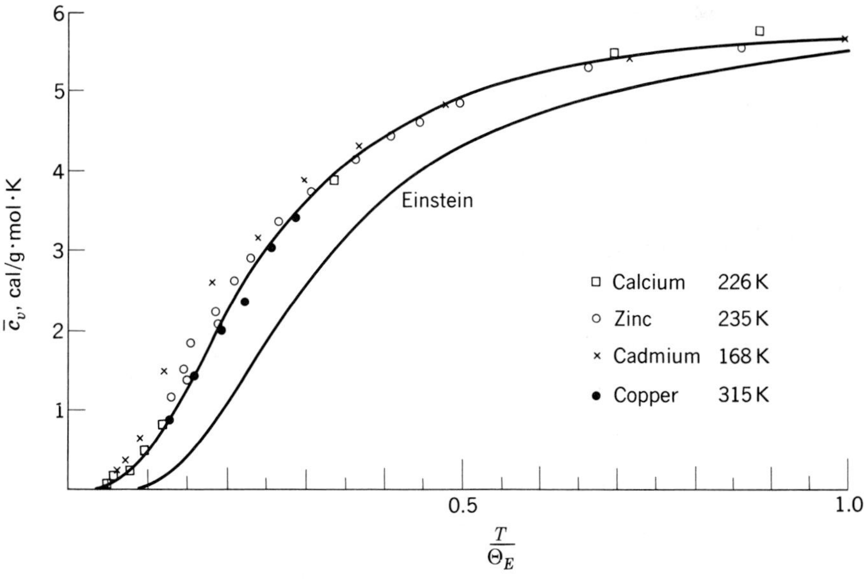

difficulty was alleviated by Debye who assumed a continuous distribution of frequencies which could be computed from elastic-wave theory for isotropic media. We shall not present the rather cumbersome analysis but only state the result. For the internal energy Debye obtained

$$U - U_0 = \frac{9NkT}{(\Theta_D/T)^3} \int_0^{\Theta_D/T} \frac{x^3}{e^x - 1}\,dx \tag{13-62}$$

where U_0 is the internal energy of the lattice at $T = 0$. This expression may be differentiated to obtain the specific heat as

$$\bar{c}_v = 3\mathcal{R}\left[\frac{12}{(\Theta_D/T)^3}\int_0^{\Theta_D/T}\frac{x^3\,dx}{e^x - 1} - \frac{3(\Theta_D/T)}{e^{\Theta_D/T} - 1}\right] \tag{13-63}$$

In these expressions Θ_D is called the *Debye temperature* and is given by

$$\Theta_D = \frac{h\nu_m}{k} \tag{13-64}$$

where ν_m is the maximum frequency encountered in the lattice vibrations. The integrals in Eqs. (13-62) and (13-63) must be evaluated numerically and results of such an evaluation are shown in Fig. 13-6. This curve is in excellent agreement with experiment, as indicated by the experimental data points. Values of the Debye temperature Θ_D for a number of solid substances are given in Table 13-3.

For high temperatures compared with Θ_D the specific heat approaches $3\mathcal{R}$. For very low temperatures compared with Θ_D the exponential term of the

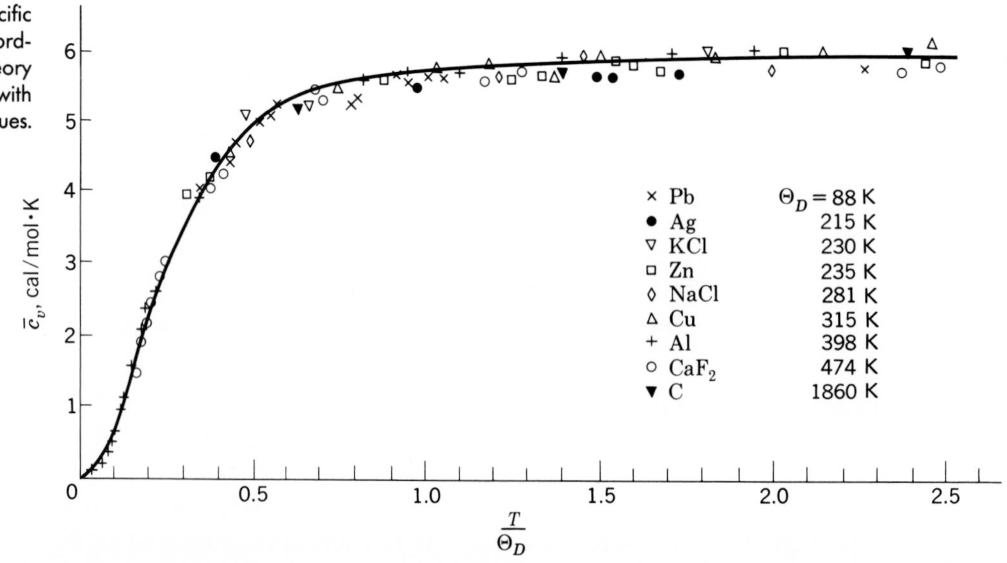

FIG. 13-6 Specific heat of a solid according to Debye theory and comparison with experimental values.

TABLE 13-3
Debye Temperatures for Various Solids

Substance	Range, K	Θ_D	Substance	Range, K	Θ_D
Lead	14–573	88	Silvium (KCl)	23–550	230
Thallium	23–301	96	Zinc	33–673	235
Mercury	31–232	97	Rock salt (NaCl)	25–664	281
Iodine	22–298	106	Copper	14–773	315
Cadmium	50–380	168	Aluminum	19–773	398
Sodium	50–240	172	Iron	32–95	453
Potassium bromide	79–417	177	Fluorspar (CaF_2)	17–328	474
Silver	35–873	215	Pyrites (FeS_2)	22–57	645
Calcium	22–62	226	Diamond	30–1169	1860

denominators in Eq. (13-63) dominates, and $\bar{c}_v$ is given approximately as

$$\begin{aligned}\bar{c}_v &\approx 3\mathcal{R}\left[\frac{12}{(\Theta_D/T)^3}\int_0^{\Theta_D/T}\frac{x^3\,dx}{e^x}\right]\\ &= \frac{12}{5}\mathcal{R}\frac{\pi^4}{(\Theta_D/T)^3} = 464.4\left(\frac{T}{\Theta_D}\right)^3 \text{ cal/g·mol·K} \qquad T \ll \Theta_D\end{aligned} \tag{13-65}$$

since $\mathcal{R} = 1.987$ cal/g·mol·K. Equation (13-65) is generally valid for $T < \Theta_D/10$.

EXAMPLE 13-3
Calculate the specific heat of aluminum at 200 K.

SOLUTION From Table 13-3 we have $\Theta_D = 398$ K for aluminum. We compute

$$\frac{T}{\Theta_D} = \frac{200}{398} = 0.502$$

Consulting Fig. 13-6, we find

$$\bar{c}_v = 4.9 \text{ cal/g·mol·K} = 20.5 \text{ kJ/kg·mol·K}$$

EXAMPLE 13-4
Calculate the specific heat of copper at 10 K.

SOLUTION From Table 13-3 $\Theta_D = 315$ K for copper so that $T \ll \Theta_D$. We may therefore use Eq. (13-65) to compute the specific heat:

$$\begin{aligned}\bar{c}_v &= 464.4\left(\frac{T}{\Theta_D}\right)^3\\ &= 464.4\left(\frac{10}{315}\right)^3\\ &= 0.014\,86 \text{ cal/g·mol·K} = 0.0622 \text{ kJ/kg·mol·K}\end{aligned}$$

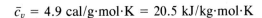

13-8 THE ELECTRON GAS

It will be recalled that our initial discussion of solids indicated that the free electrons in a metal, or electron gas, might contribute to the total energy of the solid and, hence, to the constant-volume specific heat. Assuming that these electrons do, indeed, behave like an ideal monatomic gas, then their contribution to the specific heat should be about 3/2 $\mathcal{R}$, as calculated in Sec. 13-4. But this apparently is not the case, because specific heats of both conductors and nonconductors are predicted quite nicely by the Debye theory alone. This theory, of course, does not consider the electrons in any way. In other words, the electrons apparently contribute nothing to the specific heat of the solid. This is a matter of serious concern, and we shall now see how the apparent anomaly is resolved.

The electrons in a solid are not really "free" in the same sense as the atoms of a monatomic gas even though they can move about. They are still bound to the atoms in the solid even if in a very loose fashion. Their allowable energy levels are influenced by *all* the atoms in the solid, not by just one atom as in the case of electrons attached to a gas atom. The basic statistical postulate concerning the electron gas is the *Pauli exclusion principle* which states that no more than two electrons may occupy the same quantum state. The factor of 2 arises because electrons may have a right- or left-handed spin. Note that the principle applies to quantum *state*, not energy *level*. It should be fairly easy to see that we are encountering a situation where the Fermi-Dirac statistics apply, i.e., a circumstance where only one particle may occupy each quantum state. (In Sec. 12-6 this meant no more than one ball per small box.) All we need to do is to introduce a factor of 2 to account for spin and proceed with the analysis. Still, we assume that the uncertainty principle applies so that the degeneracy is given by

$$g_i = \frac{2}{h^3}\, dx\, dy\, dz\, dp_x\, dp_y\, dp_z \tag{13-66}$$

The FD distribution was

$$\frac{N_i}{g_i} = \frac{1}{Ae^{\beta\epsilon} + 1} \tag{13-67}$$

Let us now assume that there is a very large number of energy levels available to the electrons in a solid, so that the energy spectrum is almost continuous. Then, using Eqs. (13-66) and (13-67), we obtain

$$d^2N = \frac{2/h^3}{Ae^{\beta\epsilon} + 1}\, dx\, dy\, dz\, dp_x\, dp_y\, dp_z$$

Integrating over the system volume V gives

$$d^3N = \frac{2V/h^3}{Ae^{\beta\epsilon} + 1}\, dp_x\, dp_y\, dp_z \tag{13-68}$$

We are now faced with the problem of evaluating the constant A. Let us make the substitution

$$A = e^{-\beta\mu_0} \tag{13-69}$$

where μ_0 is a new constant. The FD distribution corresponding to Eq. (13-67) then becomes

$$\frac{N_i}{g_i} = \frac{1}{e^{(\epsilon-\mu_0)/kT} + 1} \tag{13-70}$$

A plot of this relation is indicated in Fig. 13-7 for different temperatures. This plot reveals a rather remarkable result. At $T = 0$ all the energy states are completely filled up to an energy of μ_0. Above this energy none of the energy states is occupied. For higher temperatures (we have not yet said how high), the energy distribution drops off so that there are some energy states below μ_0 which are not completely filled and some above μ_0 which may be occupied. Still, the curves are approximately symmetric about $\epsilon = \mu_0$ so that for all these temperatures the total energy might be assumed to be approximately that observed at $T = 0$. The maximum energy of the electrons at $T = 0$ is μ_0 and this energy is called the *Fermi level*.

We now need to obtain an expression for μ_0. All of the electron energy is assumed to be translational kinetic energy, so that

$$\epsilon = \frac{1}{2} m_e v^2 = \frac{p^2}{2m_e} \tag{13-71}$$

where p is the momentum. The physical meaning of the plot in Fig. 13-7 is that energy space is uniformly populated at $T = 0$ up to an energy of μ_0. Above this energy there is no population. For this maximum energy there corresponds a maximum momentum

$$p_m^2 = 2m_e\mu_0 \tag{13-72}$$

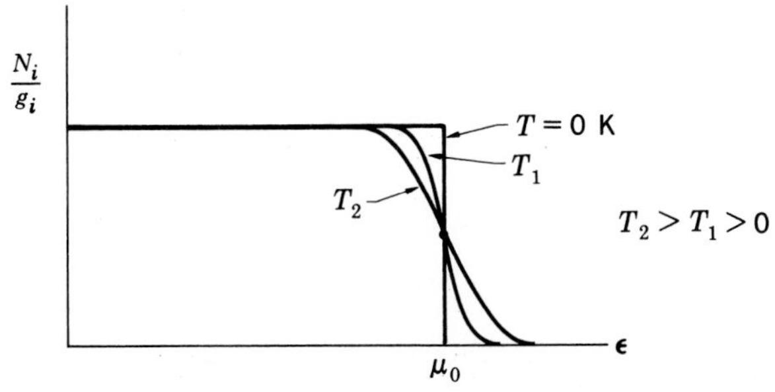

FIG. 13-7 Fermi-Dirac distribution as a function of temperature.

Now the number of electrons per unit volume of momentum space at $T = 0$ is given by Eq. (13-68) as

$$\rho_{e_0} = \frac{2V}{h^3}$$

This number is a constant over all the applicable portion of momentum space, so to obtain the total number of electrons we multiply this electron density ρ_{e_0} by the total volume of momentum space up to p_m. This volume is that of a sphere of radius p_m:

$$N_{e_0} = \frac{2V}{h^3}\left(\frac{4}{3}\pi p_m^3\right) \tag{13-73}$$

Substituting Eq. (13-73) into Eq. (13-72), we obtain

$$\mu_0 = \frac{h^2}{8m_e}\left(\frac{3N_{e_0}}{\pi V}\right)^{2/3} \tag{13-74}$$

We now have an expression for the contant A at $T = 0$ and if we assume that this expression applies at higher temperatures, we may proceed to obtain more specific information about electron velocity distributions and contributions to specific heat. At temperatures above absolute zero it has been shown by Sommerfeld [9] that Eq. (13-69) may be used for the constant A if a parameter μ is substituted for μ_0, where

$$\mu = \mu_0\left[1 - \frac{\pi^2}{12}\left(\frac{kT}{\mu_0}\right)^2 + \cdots\right] \tag{13-75}$$

We shall assume that all directions of electron motion are equally probable. The number density of electrons is thus uniform in a thin spherical shell of velocity space $4\pi v^2\, dv$. This density is obtained from Eq. (13-68) as

$$\frac{d^3N}{dv_x\, dv_y\, dv_z} = \frac{2m^3V/h^3}{Ae^{\beta\epsilon} + 1}$$

Using $\epsilon = 1/2\ mv^2$, the value for A, and multiplying by the volume of the spherical shell gives the velocity distribution as

$$dN_v = \frac{8\pi m^3 V}{h^3} = \frac{v^2\, dv}{\exp\left[\left(\frac{1}{2}mv^2 - \mu\right)\Big/kT\right] + 1} \tag{13-76}$$

This relation may be rearranged to give the energy distribution as

$$dN_\epsilon = 4\pi V\left(\frac{2m}{h^2}\right)^{3/2} \frac{\epsilon^{1/2}\, d\epsilon}{e^{(\epsilon-\mu)/kT} + 1} \tag{13-77}$$

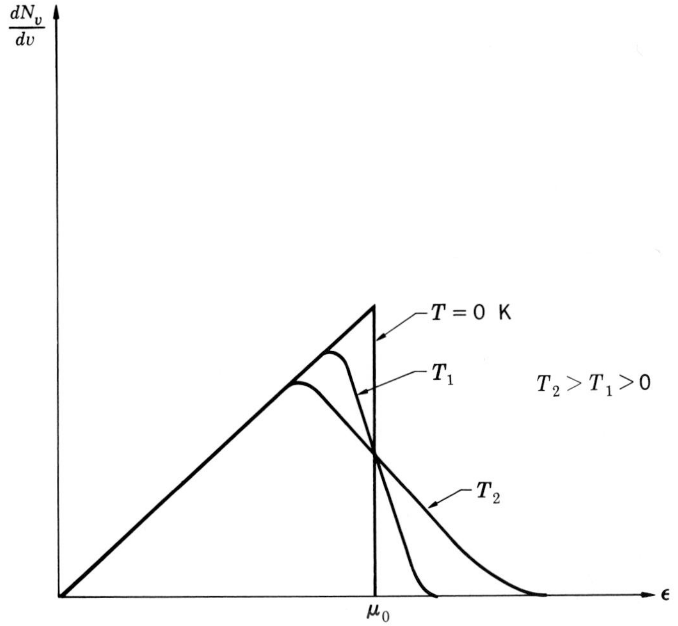

FIG. 13-8 Velocity distribution for electron gas.

In Eqs. (13-76) and (13-77) the factor

$$f_e = \frac{1}{e^{(\epsilon-\mu)/kT} + 1} \tag{13-78}$$

represents the fraction of available energy states which are occupied by electrons since in the FD model no more than one particle is allowed in each quantum state. We shall designate f_e as the *Fermi function*.

Figures 13-8 and 13-9 show plots of electron velocity and energy distributions as a function of temperature. At $T = 0$ ϵ is always less than μ_0 and the exponential term of Eq. (13-78) goes to zero, so that the energy distribution is

FIG. 13-9 Energy distribution for electron gas.

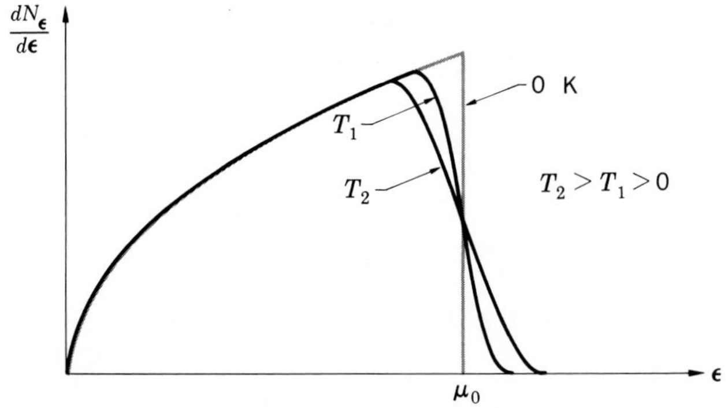

given by

$$dN_\epsilon = 4\pi V \left(\frac{2m}{h^2}\right)^{3/2} \epsilon^{1/2}\, d\epsilon \qquad \text{at } T = 0 \tag{13-79}$$

The *average* electron energy at $T = 0$ is

$$\bar{\epsilon}_0 = \frac{\int_0^{\mu_0} \epsilon\, dN_\epsilon}{\int_0^{\mu_0} dN_\epsilon} = \frac{3}{5}\mu_0 \tag{13-80}$$

The upper limit for these integrals is μ_0 since this is the maximum electron energy at $T = 0$. At temperatures above absolute zero the average electron energy must be computed from

$$\bar{\epsilon} = \frac{\int_0^{\infty} \epsilon\, dN_\epsilon}{\int_0^{\infty} dN_\epsilon} \tag{13-81}$$

since now there is no upper limit on the electron energy levels. The integrations in Eq. (13-81) may be performed by substituting the energy distribution from Eq. (13-77) along with the Sommerfeld relation for μ from Eq. (13-75). The result is

$$\bar{\epsilon} = \frac{3}{5}\mu_0 \left[1 + \frac{5}{12}\pi^2 \left(\frac{kT}{\mu_0}\right)^2 + \cdots\right] \tag{13-82}$$

The internal energy is, then,

$$U = N\bar{\epsilon} = \frac{3}{5} N\mu_0 + \frac{1}{4}\frac{\pi^2 N k^2}{\mu_0} T^2 \tag{13-83}$$

where N is the total number of electrons. The molal specific heat is then

$$\bar{c}_v = \left(\frac{\partial \bar{u}}{\partial T}\right)_v = \frac{\pi^2}{2}\frac{kT}{\mu_0}\mathcal{R} \tag{13-84}$$

If the electrons behaved as an ideal monatomic gas, we would obtain $\bar{c}_v = 3/2\ \mathcal{R}$, a much larger value.

The foregoing analysis explains why it is unnecessary to consider the electron gas when computing specific heats of solids; their contribution is very small since $kT \ll \mu_0$ for all ordinary temperatures. Another way of stating this conclusion is that the temperature-dependent part of the electron internal energy in Eq. (13-83) is small compared with the total electron energy. This total energy however, is quite large since $\mu_0 \gg kT$, as is illustrated in Example 13-5.

EXAMPLE 13-5 Fermi Level for Copper
Estimate the Fermi level for copper and compute the equivalent ideal-gas temperature for the free electrons based on their average energy at $T = 0$.

SOLUTION We shall assume that copper has one free electron per atom so that

$$\frac{N_{e_0}}{V} = \frac{N_0}{M}\frac{m}{V} = \frac{N_0}{M}\rho$$

where M is the molecular weight and ρ is the density. For copper, $M = 63.57$ and $\rho = 8.94\ \text{g/cm}^3$ so that

$$\frac{N_{e_0}}{V} = \frac{(6.025 \times 10^{23})(8.94)}{63.57}$$
$$= 8.47 \times 10^{22}\ \text{electrons/cm}^3$$

We now use Eq. (13-74) to compute μ_0:

$$\mu_0 = \frac{h^2}{8m_e}\left(\frac{3N_{e_0}}{\pi V}\right)^{2/3}$$
$$= \frac{(6.624 \times 10^{-27})^2}{(8)(9.1 \times 10^{-23})}\left[\frac{(3)(8.47 \times 10^{22})}{\pi}\right]^{2/3}$$
$$= 1.13 \times 10^{-11}\ \text{erg} = 7.07\ \text{eV}$$

The average electron energy at $T = 0$ is given by Eq. (13-80) as

$$\bar{\epsilon}_0 = \frac{3}{5}\mu_0$$
$$= (0.6)(1.13 \times 10^{-11}) = 6.78 \times 10^{-12}\ \text{erg}$$
$$= 6.78 \times 10^{-19}\ \text{J}$$

To compute the equivalent ideal-gas temperature for the free electrons we set

$$\bar{\epsilon}_0 = \frac{3}{2}kT_{\text{equiv}}$$

Solving for the equivalent temperature, we obtain

$$T_{\text{equiv}} = \frac{2\bar{\epsilon}_0}{3k}$$
$$= \frac{(2)(6.78 \times 10^{-19})}{(3)(1.38 \times 10^{-23})}$$
$$= 32\ 800\ \text{K}$$

EXAMPLE 13-6 Electron Contribution to Specific Heat
Calculate the electron contribution to specific heat of copper at 300 K (room temperature).

SOLUTION This calculation may be performed with Eq. (13-84), and the value of μ_0 is obtained in Example 13-5. Thus

$$\begin{aligned}\bar{c}_{v,\text{elec}} &= \frac{\pi^2}{2}\frac{kT}{\mu_0}\mathcal{R} \\ &= \frac{\pi^2(1.38 \times 10^{-23})(300)}{(2)(1.13 \times 10^{-18})}\mathcal{R} \\ &= 0.0181\mathcal{R}\end{aligned}$$

From this calculation we see that the electron contribution to specific heat is indeed very small.

Our brief discussion of the solid state has yielded several important bits of information.

1 Lattice vibration furnishes the primary contribution to specific heats of solids.
2 Specific heats of solids approach zero as the absolute temperature approaches zero.
3 The internal energy of solids does not approach zero as $T \to 0$. Not only is there potential energy stored in the force fields binding the atoms but also a considerable energy in electron kinetic energy. In fact, the kinetic energy of the electrons is much higher than that of an ordinary gas at room temperature.
4 Absolute zero of temperature does not represent a state where atomic and electronic motion stops.
5 The "free" electrons in a solid do not behave like a simple monatomic gas in that their specific heat may not be computed from the simple equipartition of energy principle.

13-9 CONDUCTORS AND SEMICONDUCTORS

Our analysis of the solid state up to this point has consisted of a determination of specific heats based on the Debye theory and on a demonstration that the electron gas makes no substantial contribution to specific heat. The analysis of the electron gas is based upon what might be called a *single-band model;* i.e., electron energies are primarily limited to the band between the zero energy level and the Fermi level. This is not an entirely realistic picture of a solid. A single atom may contain electrons in various energy levels which are governed by appropriate quantum mechanical relations. When a large number of atoms are forced into close proximity to one another, as in a solid, the electrons interact very strongly and occupy broadbands of allowable energy states instead of the nicely separated discrete values observed in a single atom. It can be shown by a rather involved quantum mechanical analysis that the overall result of the interaction process is to produce *allowable* and *forbidden energy bands* within the solid. An allowable energy band is one which may be occupied by electrons whereas a forbidden band is one which may not be occupied by electrons.

The lowest-level electron band is the *valence band* and results from interactions of the valence (outer shell) electrons in the single atoms. The next allowable band is called the *conduction band*. In general, the forbidden band separating these two allowable bands is very wide (that is, $\gg kT$). There are other possible allowable bands, but these first two are of most interest.

The Fermi-Dirac distribution function gives the distribution of electrons among various *allowable* energy states in the solid. At $T = 0$, all electrons would occupy the valence band. The fraction of electrons which may be raised to the conduction band depends on T and the width of the forbidden band.

If the valence band is completely filled the solid may not conduct unless sufficient energy is applied to move the electrons to make the jump and the material will act as an *insulator*.

If the forbidden band is narrow and the conduction band is empty then it will be much easier for the electrons to cross from the valence band into the conduction band. In this instance the material is called a *semiconductor*.

If the conduction band is partially filled, then relatively little energy is required to raise the electrons to a slightly higher potential within the conduction band, i.e., to an unoccupied state at some higher energy level. In this case electron motion is most easily influenced by an electric field, and we say that the material behaves as a *conductor*. Figures 13-10 and 13-11 illustrate the behavior described in the foregoing.

When an electron jumps the gap from the valence band to the conduction band, the vacant state created in the valence band is called a *hole*. The hole

FIG. 13-10 Band theory of (*a*) insulators and (*b*) semiconductors.

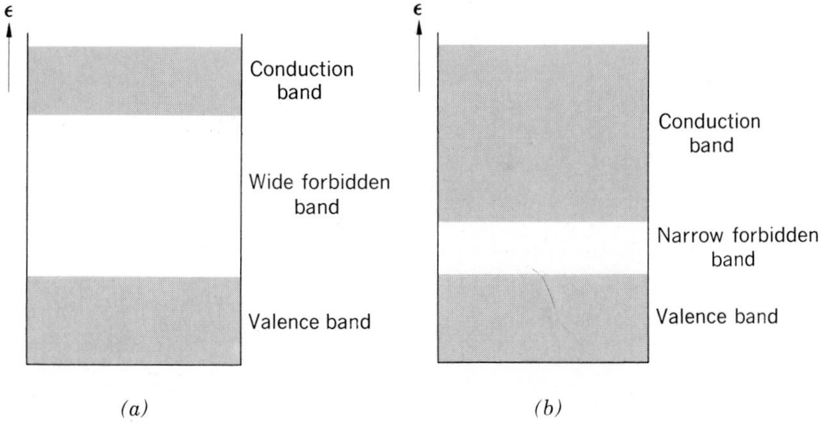

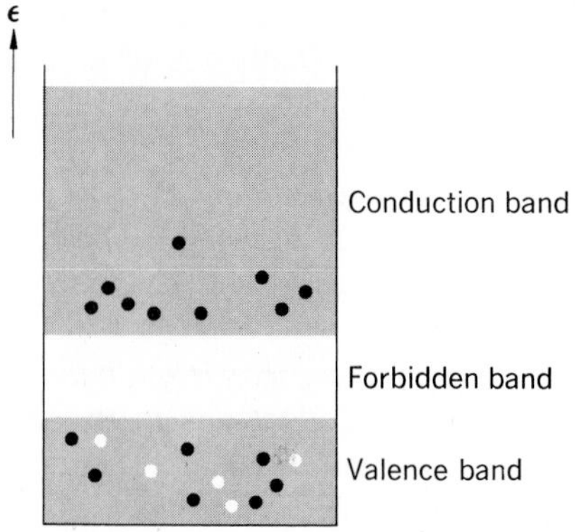

FIG. 13-11 Band theory of conductors.

may remain vacant or be filled by another electron from some other energy site within the valence band. The number of electrons which move from the valence band to the conduction band increases with temperature and, since all electrons which occupy the conduction band come from the valence band, we assume that the number of electrons in the conduction band must equal the number of holes in the valence band.

If conduction in the material results only from electron movement from valence to conduction band it is called an *intrinsic semiconductor*. The performance characteristics of semiconductors may be changed appreciably by adding small amounts of impurities to the solid, which act as *acceptors* or *donors* of electrons in the forbidden gap, as illustrated in Fig. 13-12. An acceptor is an empty band at an energy level somewhat above the valence band. Since it provides additional "holes," such an impurity semiconductor is called a *p-type semiconductor*, where the *p* signifies *positive*. A donor is a filled band at an energy level somewhat below the conduction band. This type of impurity semiconductor provides additional negative charges (electrons), so it is called an *n-type semiconductor*.

The Fermi level may be related to the forbidden bandwidth in the following way. For convenience we take the energy level as zero at the top of the valence

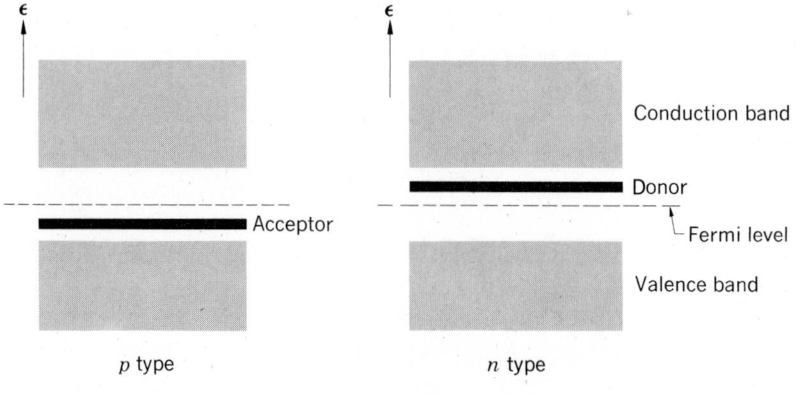

FIG. 13-12 Impurity semiconductors.

band and equal to ϵ_g at the bottom of the conduction band. It is assumed that every electron which makes the jump into the conduction band leaves a hole in the valence band. Thus

$$N_h \big|_{\text{valence}} = N_e \big|_{\text{conduction}} \tag{13-85}$$

where N_h represents the number of holes and N_e the number of electrons. The energy scale for this model is illustrated in Fig. 13-13. Because of the forbidden-band concept we are forced to require that the number of available quantum states for the electrons must be zero in this band. We then assume a FD distribution outside the forbidden band. This means that we must employ the following expression for g_i in the conduction band:

$$g_i \big|_{\text{conduction band}} = 4\pi V \left(\frac{2m_e}{h^2}\right)^{3/2} (\epsilon - \epsilon_g)^{1/2} \tag{13-86}$$

In other words, the allowable states start to build up from ϵ_g in the same way they built up from $\epsilon = 0$ for the single-band model. We still employ

$$f_e = \frac{1}{e^{(\epsilon-\mu)/kT} + 1} \tag{13-87}$$

for the FD energy distribution function.

The number of electrons in the conduction band is thus

$$N_e = 4\pi V \left(\frac{2m_e}{h^2}\right)^{3/2} \int_{\epsilon_g}^{\infty} \frac{(\epsilon - \epsilon_g)^{1/2}\, d\epsilon}{e^{(\epsilon-\mu)/kT} + 1} \tag{13-88}$$

Since the forbidden bandwidth is usually much greater than kT, it is expected that $(\epsilon - \mu) \gg kT$ in the conduction band, so that the one in the denominator of

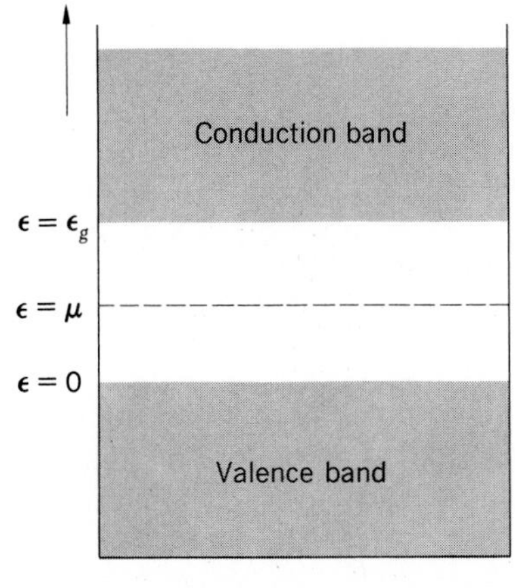

FIG. 13-13 Energy scale for band analysis.

Eq. (13-88) may be neglected and the integral may be evaluated with an appropriate variable substitution to give

$$N_e = V\left(\frac{2\pi m_e kT}{h^2}\right)^{3/2} e^{\mu/kT}\, e^{-\epsilon_g/kT} \tag{13-89}$$

The distribution function for the holes in the valence band is

$$\begin{aligned} f_h &= 1 - f_e = 1 - \frac{1}{e^{(\epsilon-\mu)/kT} + 1} \\ &= \frac{e^{-(\mu-\epsilon)/kT}}{e^{-(\mu-\epsilon)/kT} + 1} \end{aligned} \tag{13-90}$$

In the valence band it is expected that $(\mu - \epsilon) \gg kT$ so that the exponential in the denominator of Eq. (13-90) may be neglected. Then

$$f_h = e^{-(\mu-\epsilon)/kT}$$

The number of holes in the valence band is then obtained by forming the integral

$$N_h = 4\pi V\left(\frac{2m_h}{h^2}\right)^{3/2} \int_{-\infty}^{0} \epsilon^{1/2} e^{-(\mu-\epsilon)/kT}\, d\epsilon \tag{13-91}$$

where m_h is an effective mass of the hole. The integral may be evaluated to give

$$N_h = V\left(\frac{2\pi m_h kT}{h^2}\right)^{3/2} e^{-\mu/kT} \tag{13-92}$$

Equating Eqs. (13-89) and (13-92) gives

$$\mu = \frac{1}{2}\epsilon_g + \frac{3}{4} kT \ln \frac{m_h}{m_e} \tag{13-93}$$

or if the masses of the electrons and holes are equal, we obtain

$$\mu = \frac{1}{2}\epsilon_g \tag{13-93a}$$

and the Fermi level lies exactly in the center of the forbidden band. This is an important result and is of practical value in explaining thermoelectric energy-conversion devices. Some typical bandwidths for semiconductors are given in Table 13-4.

TABLE 13-4 Typical Forbidden Bandwidths for Semiconductors

Material	e_g, eV
Ge	0.78
Se	2.0
Si	1.21
Te	0.33
Cu_2O	1.0
InAs	0.5
InSb	2.3
UO_2	0.2
ZnS	3.6

13-10 THE PHOTON GAS (BLACKBODY RADIATION)

As an interesting application of quantum statistics, we consider a quantity of thermal radiation inside an insulated cavity and seek an expression for the energy distribution for such radiation. Such a system may be called a *photon gas* since it consists of energy quanta of magnitude

$$\epsilon_i = h\nu_i$$

where ν_i represents the frequency of the photons. The momentum of the photons is

$$p_i = \frac{h\nu_i}{c} = \frac{h}{\lambda_i} = \frac{\epsilon_i}{c} \tag{13-94}$$

The photons represent "particles" which occupy various quantum states, but the isolated collection does not consist of a constant number of particles. As with conventional particle systems, we seek the number density of quanta which occupy each energy state ϵ_i. The photons are indistinguishable and there is no restriciton on the number of available quantum states so we employ the BE statistical model:

$$\Omega_{\text{BE}} = \prod_i \frac{(g_i + N_i - 1)!}{N_i!(g_i - 1)!} \tag{13-95}$$

Forming the maximizing condition,

$$\delta(\ln \Omega) = 0$$

along with the constant energy condition,

$$\delta U = 0 = \Sigma \epsilon_i \delta N_i$$

gives

$$\frac{N_i}{g_i} = \frac{1}{e^{\beta\epsilon_i} - 1} \tag{13-96}$$

where it is to be noted that we have not imposed a constant-number-of-particles condition on the distribution, as in Sec. 13-4, since photons may be continuously created and destroyed.

Let us assume that the walls of the enclosure are at the temperature T and the total volume is V. The number of allowable quantum states g_i in each energy level ϵ_i is calculated by dividing the elemental volume in phase space having this energy by h^3, the smallest discernible volume in phase space. We cannot say "where" the photon is located in the cartesian space of the container, except to say it is in the volume V. On the other hand, the energy level ϵ_i is linked to momentum through Eq. (13-94) so that we choose as our elemental volume in phase space a spherical shell

$$V \times 4\pi p_i^2 \, dp_i$$

and calculate the number of quantum states in this volume by

$$g_i = \frac{4\pi V p_i^2 \, dp_i}{h^3} \tag{13-97}$$

This relation must be multiplied by a factor of 2 to account for the fact that light may be right or left circularly polarized. Combining Eq. (13-97) with Eq. (13-96) gives

$$N_i = \frac{8V\pi p_i^2 \, dp_i}{h^3(e^{\beta\epsilon_i} - 1)} \tag{13-98}$$

It is to be noted once again that N_i represents the number of photons in the volume V having the momentum between p_i and $p_i + dp_i$. This, in turn, is the number having an energy between ϵ_i and $\epsilon_i + d\epsilon$. Equation (13-98) may be expressed in terms of frequency by noting from Eq. (13-94) that

$$p_i^2 = \frac{h^2\nu_i^2}{c^2} \qquad dp_i = \frac{h d\nu_i}{c}$$

so that

$$N_i = \frac{8V\pi\nu_i^2 \, d\nu_i}{c^3(e^{\beta h\nu_i} - 1)} \tag{13-98a}$$

The number of photons per unit volume and per unit frequency is thus

$$\frac{dN}{V\,d\nu} = \frac{N_i}{V\,d\nu_i} = \frac{8\pi\nu^2}{c^3(e^{\beta h\nu} - 1)}$$

where now we have assumed that an essentially continuous spectrum of frequencies is available so that ν_i is replaced by ν. The energy of each photon is $h\nu$, so that the energy per unit volume and per unit frequency is

$$u_\nu = \frac{8\pi h\nu^3}{c^3(e^{\beta h\nu} - 1)} \tag{13-99}$$

It turns out that $\beta = 1/kT$ when a condition like Eq. (13-5) is applied as before. Making use of $\nu = c/\lambda$ and $d\nu = -(c/\lambda^2)\,d\lambda$, Eq. (13-99) may then be written in terms of wavelength as

$$u_\lambda = \frac{8\pi hc\lambda^{-5}}{e^{hc/\lambda kT} - 1} \tag{13-99a}$$

where now u_λ is the energy per unit volume and per unit wavelength.

It can be shown that the radiant energy emitted from a surface in the enclosure per unit area per unit time and per unit wavelength $E_{b\lambda}$ is related to u_λ by

$$E_{b\lambda} = \frac{u_\lambda c}{4}$$

so that

$$E_{b\lambda} = \frac{2\pi hc^2\lambda^{-5}}{e^{hc/\lambda kT} - 1}$$
$$E_{b\lambda} = \frac{c_1\lambda^{-5}}{e^{c_2/\lambda T} - 1} \tag{13-100}$$

Equation (13-100) is called the *Planck blackbody radiation law* and forms a base for calculations of many radiation heat-transfer phenomena; a plot of this relation for several temperatures is given in Fig. 13-14. Integration of Eq. (13-100) over all possible wavelengths results in

$$E_b = \int_0^\infty E_{b\lambda}\,d\lambda = \int_0^\infty \frac{c_1\lambda^{-5}\,d\lambda}{e^{c_2/\lambda T} - 1} = \sigma T^4 \tag{13-101}$$

where σ is a new constant called the *Stefan-Boltzmann constant*. Equation (13-101) expresses the total blackbody radiation leaving a surface per unit area and per unit time. E_b is usually called the *emissive power* of an ideal radiator. It describes only the *thermal* radiation emitted by a body. Thermal radiation is distinguished from other types of electromagnetic radiation by the fact that it depends on temperature.

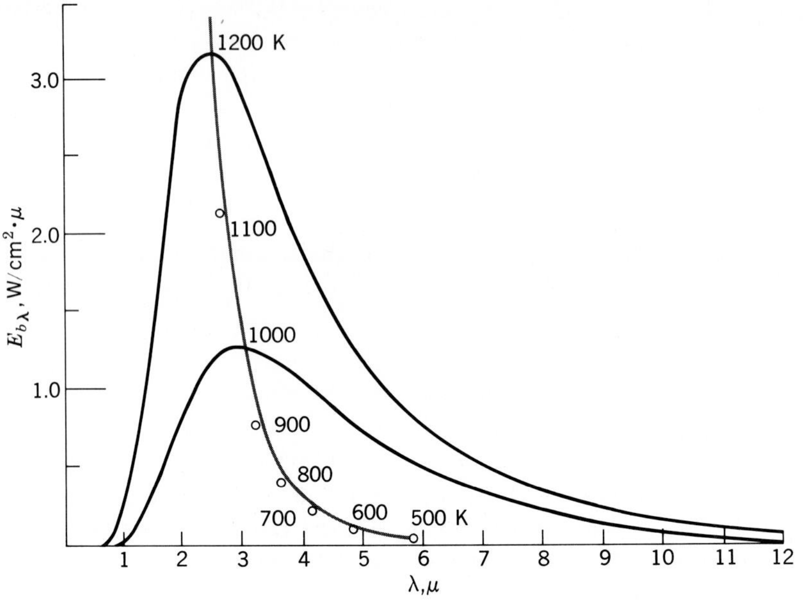

FIG. 13-14 Planck blackbody radiation distribution.

Some of the appropriate constants for use in the radiation formulas are

$$\sigma = 5.669 \times 10^{-8}\ \mathrm{W/m^2{\cdot}K^4}$$
$$= 1.714 \times 10^{-9}\ \mathrm{Btu/h{\cdot}ft^2{\cdot}{}^\circ R^4}$$

$$c_1 = 3.743 \times 10^{-8}\ \mathrm{W{\cdot}\mu m^4/m^2}$$
$$= 1.187 \times 10^{8}\ \mathrm{Btu{\cdot}\mu m^4/ft^2{\cdot}h}$$

$$c_2 = 1.4387 \times 10^{4}\ \mu\mathrm{m{\cdot}K}$$
$$= 2.5896 \times 10^{4}\ \mu\mathrm{m{}^\circ R}$$

where μm designates the unit of micrometers (formerly microns; 1 μm = 10^{-6} m).

EXAMPLE 13-7

Calculate the total energy emitted by an ideal radiator at 1000 and 2000 K.

SOLUTION This calculation is made immediately with Eq. (13-101). At T = 1000 K,

$$E_b = \sigma T^4 = (5.669 \times 10^{-8}(1000)^4$$
$$= 56.69\ \mathrm{kW/m^2}$$
$$= 5.669\ \mathrm{W/cm^2}$$

At T = 2000 K,

$$E_b = (5.669 \times 10^{-8})(2000)^4$$
$$= 907\ \mathrm{kW/m^2}$$

13-11 ENSEMBLES AND INTERACTING SYSTEMS

In the basic statistical presentations of Chap. 12 and in the applications of these statistics throughout this chapter we have been concerned with particles which do not interact with one another. To be sure, the vibration of atoms in a solid are dependent on the behavior of their neighbors. But the *statistical* portion of the analysis still considered the particles as independent, and a quantum mechanical specification of allowable energy states was relied upon to take interactions into account. For the systems considered so far an isolated assembly of particles has been assumed. For the isolated assembly the total energy is constant as well as the total number of particles.

In the real world there are many situations where particle interaction exerts a significant influence on the distribution of particles among available energy states. Another way of saying this is that the number of particles in a particular element of volume in phase space is not unlimited but may depend on the interaction forces between particles as well as on the size of the particles. The solid state is an obvious example. High-pressure gases and liquids are others. In these instances we are forced to take the interactions into account. These interesting and complicated effects are normally treated by *ensemble theory*, a brief outline of which we shall give in this section.

A *canonical ensemble* is depicted in Fig. 13-15. The large system is supposed to be a collection of many smaller subsystems, each having a fixed number of particles N and a fixed volume V. The ensemble is the collection of all these η subsystems. The total energy of the ensemble is assumed to be constant, but the energy of each subsystem may vary as the particles comprising this subsystem move about in the total system. Each of the η subsystems is assumed to be distinguishable. Note that the subsystems may exchange energy with one another, even though the total energy of the ensemble remains constant. To specify the allowable energy states of the subsystems we would need to consider particle interactions as well as other factors. Presumably, these allowable energy states are calculable. If the number of subsystems having the energy U_i is η_i, then the total average energy of the ensemble is

$$U = \sum_i \eta_i U_i \tag{13-102}$$

FIG. 13-15 Canonical ensemble.

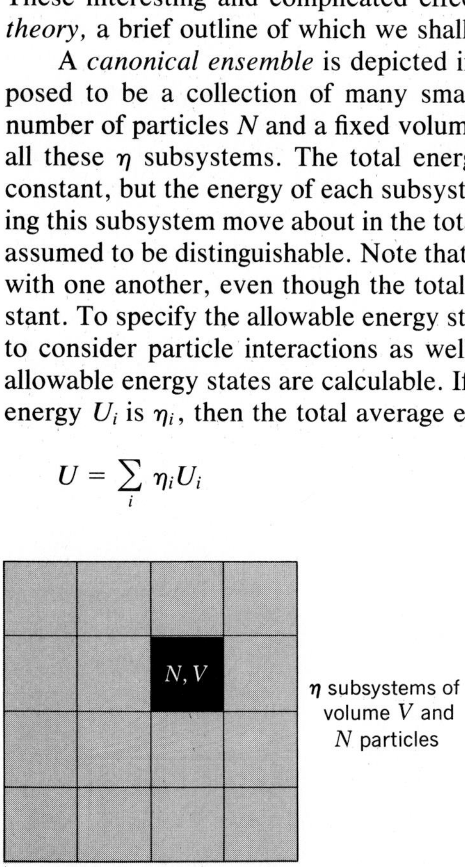

The total number of subsystems is η so that

$$\eta = \sum_i \eta_i \tag{13-103}$$

We are interested in the number of ways of arranging the subsystems among the various η_i states. This was calculated in Eq. (12-10) as

$$\Omega = \frac{\eta!}{\Pi \eta_i!} \tag{13-104}$$

Of course, the relations in Eqs. (13-102) to (13-104) bear a striking resemblance to the basic MB relations of Chap. 12. The point to remember is that the ensemble artifice allows an inclusion of interaction energies in this simplified distribution.

A *grand canonical ensemble* is depicted in Fig. 13-16. In this type of ensemble we remove the restriction of a constant number of particles in each subsystem. Now we have η subsystems of volume V, all of which are distinguishable. The total energy and number of particles of the ensemble are assumed constant but the energy and number of particles in each subsystem may vary. In this way we are able to study systems of particles which may exchange both energy and mass. The inclusion of a variable number of particles in the analysis complicates things considerably; in particular, multiple summations must be employed. This complication is justified however, by the power supplied for analysis of more involved physical systems.

A *microcanonical ensemble* is one which consists of a group of subsystems which have constant N, V, *and* energy.

We shall not be able to expand our discussion of ensembles further except to say that a detailed study of ensemble methods is essential to a more advanced study of statistical thermodynamics. The interested reader may consult the references at the end of the chapter for more information on this subject.

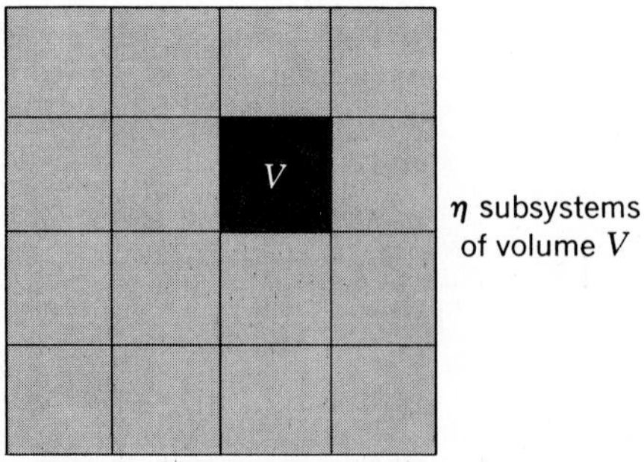

FIG. 13-16 Grand canonical ensemble.

REVIEW QUESTIONS

1 What is a partition function?
2 How is the macroscopic concept of temperature connected with microscopic thermodynamic models?
3 What is meant by an ideal gas in a physical sense?
4 How does rotational energy contribute to the specific heats of ideal gases?
5 What is the equipartition of energy principle?
6 Discuss the relative contributions of lattice vibrations and electron energies to the specific heats of solids.
7 In many secondary school science programs (and some colleges) there is a strong implication that absolute zero of temperature means zero energy. Discuss this concept.
8 What is a *hole*?
9 Distinguish between conductors, semiconductors, and electrical insulators in a microscopic sense.
10 What is blackbody radiation?
11 What is entropy?
12 What is degeneracy?
13 How is entropy related to the partition function?
14 What is meant by characteristic rotational and vibrational temperatures?
15 What is a *degree of freedom*?
16 Distinguish between the Einstein and Debye models of a solid.
17 What are free electrons?
18 What is the Fermi level?
19 What is a photon?
20 What is an ensemble?
21 Distinguish between the MB, BE, and FD models of energy distributions.
22 How do electrons contribute to specific heat of a solid?
23 Discuss the effect of impurities on the behavior of semiconductors.
24 What are meant by *p*- and *n*-type semiconductors?
25 What is a photon gas?
26 What is meant be radiation energy density?
27 What is the Stefan-Boltzmann law?

PROBLEMS

13-1 Estimate the specific heats of the following gases:
(*a*) CO at 50, 100, 300, 3000 K
(*b*) O_2 at 100, 300, 3000 K
(*c*) N_2 at 100, 300, 3000 K

13-2 Estimate the specific heats of the following solids:
(*a*) Sodium at 40, 100, 200 K
(*b*) Silver at 50, 100, 300, 500 K

(*c*) Copper at 50, 100, 300, 700 K
(*d*) Lead at 20, 100, 300, 500 K

13-3 Compute the first four terms of the series of Eq. (13-44) and calculate the corresponding value of rotational specific heat for

(*a*) $\frac{T}{\Theta_{\text{rot}}} = 0.2$

(*b*) $\frac{T}{\Theta_{\text{rot}}} = 0.5$

(*c*) $\frac{T}{\Theta_{\text{rot}}} = 0.8$

(*d*) $\frac{T}{\Theta_{\text{rot}}} = 1.0$

Compare the results of these first four terms of the series with the curve in Fig. 13-2.

13-4 Estimate the Fermi level for silver and for tungsten.

13-5 Plot the Fermi function versus $\epsilon - \mu$ for
(*a*) $T = 0$ K
(*b*) $T = 50$ K
(*c*) $T = 100$ K
(*d*) $T = 300$ K
(*e*) $T = 500$ K
Discuss this plot.

13-6 Plot the Fermi funciton versus ϵ/μ for the valence and conduction bands of a semiconductor in which $\mu = \frac{1}{2}\epsilon_g$ for the following temperatures:
(*a*) $T = 0$ K
(*b*) $T = 50$ K
(*c*) $T = 100$ K
(*d*) $T = 300$ K
(*e*) $T = 500$ K
Make the plots over the range $-5 < \epsilon/\mu < +5$, where the zero level is taken at the top of the valence band. For this calculation take the forbidden bandwidth as 1.0 eV. Discuss the plots.

13-7 Why is it that free electrons do not contribute substantially to the specific heat of solids?

13-8 Four indistinguishable particles are to be placed in two energy levels, each of which has a degeneracy of three. Evaluate the thermodynamic probability for all possible arrangements. What is the most probable arrangement?

13-9 Determine the entropy for H_2 at 50, 300, and 3000 K.

13-10 Derive an expression for enthalpy in terms of the partition function.

13-11 Calculate the relative populations in the first three rotational energy levels for N_2 at 30 and 2000 K.

13-12 Calculate the relative populations of the first three vibrational energy levels for N_2 at 2000 and 3500 K.

13-13 Construct plot of u and $\bar{c}_v$ for N_2 versus temperature from 50 to 7000 K.

13-14 Calculate the specific heat of diamond at 30 and 1000 K.

13-15 For what temperature range(s) will the equipartition-of-energy principle apply for (*a*) N_2, (*b*) O_2, (*c*) H_2, with an error of less than ± 10 percent for specific heat?

13-16 Estimate the moment of inertia for (*a*) O_2, (*b*) N_2, (*c*) H_2 molecules.

13-17 Calculate the maximum free-electron velocity in copper at $T = 0$.

13-18 Calculate, for silver at 35 K, the percent contribution to specific heat by the electron gas.

13-19 Compare the total energy of the electron gas in silver at 0 K with the energy required to raise the temperature of silver from 35 to 55 K.

13-20 Estimate the characteristic maximum vibrational frequency for the lattice of (*a*) copper, (*b*) lead, (*c*) diamond.

13-21 Calculate the radiant emission from an ideal radiator at (*a*) 100 K, (*b*) 1000 K, (*c*) 5000 K.

13-22 Calculate the energy density of radiation per unit wavelength at 1 μm and 5 μm for each of the cases in Prob. 13-21.

13-23 Show, by a suitable differentiation process, that a maximum point exists in the Planck blackbody radiation formula, and that this point is given by

$$(\lambda T)_{\max} = 5215.6\ \mu\text{m}\cdot{}^\circ\text{R}$$

This relation is called *Wien's displacement law*.

REFERENCES

1 Allis, W. P., and M. A. Herlin: "Thermodynamics and Statistical Mechanics," McGraw-Hill Book Company, New York, 1952.

2 Crawford, F. H.: "Heat, Thermodynamics and Statistical Physics," Harcourt, Brace & World, New York, 1963.

3 Hill, L.: "Statistical Mechanics," McGraw-Hill Book Company, New York, 1956.

4 Lee, J. F., F. W. Sears, and D. L. Turcotte: "Statistical Thermodynamics," Addison-Wesley Publishing Company, Inc., Reading, Mass., 1963.

5 Mayer, J. E., and M. G. Mayer: "Statistical Mechanics," John Wiley & Sons, Inc., New York, 1940.

6 Reynolds, W. C.: "Thermodynamics," 2d ed., McGraw-Hill Book Company, New York, 1968.

7 Schrödinger, E.: "Statistical Thermodynamics," Cambridge University Press, Cambridge, 1952.

8 Sears, F. W.: "An Introduction to Thermodynamics, The Kinetic Theory of Gases, and Statistical Mechanics," Addison-Wesley Publishing Company, Inc., Reading, Mass., 1950.

9 Sommerfeld, A.: "Thermodynamics and Statistical Mechanics," Academic Press, Inc., New York, 1956.

10 Tolman, C.: "The Principles of Statistical Mechanics," Oxford University Press, London, 1938.
11 Tribus, M.: "Thermostatics and Thermodynamics," D. Van Nostrand Company, Inc., Princeton, N.J., 1961.
12 Hatsopoulos, G. N., and J. H. Keenan: "Principles of General Thermodynamics," John Wiley & Sons, Inc., New York, 1965.
13 Fay, J. A.: "Molecular Thermodynamics," Addison-Wesley Publishing Company, Inc., Princeton, N.J., 1961.
14 Knuth, E. S.: "Introduction to Statistical Thermodynamics," McGraw-Hill Book Company, New York, 1966.

A

CONVERSION FACTORS AND THERMODYNAMIC PROPERTIES

TABLE A-1
IMPORTANT PHYSICAL CONSTANTS

Avogadro's number	N_0	$= 6.022\ 045 \times 10^{26}$ molecules/kg mol
Universal gas constant	$\mathcal{R}$	= 1545.35 ft·lbf/lbm·mol·°R
		= 8314.41 J/kg mol·K
		= 1.986 Btu/lbm·mol·°R
		= 1.986 kcal/kg mol·K
Planck's constant	h	$= 6.626\ 176 \times 10^{-34}$ J
Boltzmann's constant	k	$= 1.380\ 662 \times 10^{-23}$ J/molecule·K
		$= 8.6173 \times 10^{-5}$ eV/molecule·K
Speed of light in vacuum	c	$= 2.997\ 925 \times 10^{8}$ m/s
Standard gravitational acceleration	g	= 32.174 ft/s^2
		= 9.806 65 m/s^2
Electron mass	m_e	$= 9.1095 \times 10^{-31}$ kg
Charge on the electron	e	$= 1.602\ 189 \times 10^{-19}$ C
Stefan-Boltzmann constant	σ	$= 0.1714 \times 10^{-8}$ Btu/h·ft^2·R^4
		$= 5.670\ 32 \times 10^{-8}$ W/m^2·K^4

TABLE A-2
CONVERSION FACTORS

Length	1 cm	= 0.3937 in = 10^4 μm = 10^8 A
	1 in	= 2.540 cm
	1 ft	= 0.3048 m
Mass	1 lbm	= 0.453 592 37 kg
	1 slug	= 32.174 lbm
Force	1 N	= 10^5 dyn
	1 lbf	= 444 822 dyn = 4.44822 N
	1 kgf	= 9.806 65 N
		= 1.0 kilopond
Pressure	1 Pa	= 1 N/m^2
	1 lbf/in^2	= 2.036 inHg at 32°F = 6894.76 Pa
	1 inHg	= 33 864 dyn/cm^2 = 0.0334 atm
		= 0.491 lbf/in^2
	1 atm	= 14.695 95 lbf/in^2 = 760 mmHg at 32°F
		= 29.92 inHg at 32°F = 2116.21 lbf/ft^2
		= 1.013 25 × 10^5 Pa
	1 kgf/cm^2	= 9.806 65 × 10^4 Pa
	1 bar	= 10^5 N/m^2 = 0.986 92 atm
Volume	1 liter	= 0.2642 gal = 0.0353 ft^3 = 61.03 in^3
	1 gal	= 231 in^3
	1 ft^3	= 28.3168 liters = 7.4805 gal = 0.028 316 8 m^3
	1 in^3	= 16.387 cm^3
Energy	1 Btu	= 778.16 ft·lbf = 252.16 cal = 1055.04 J
	1 ft·lbf	= 1.3558 J
	1 erg	= 1 dyn·cm
	1 J	= 1 N·m = 10^7 ergs
	1 cal	= 4.1854 J (thermochemical)
Power	1 W	= 1 J/s = 860.13 cal/h = 3.413 Btu/h
	1 hp	= 746 W (electric) = 550 ft·lbf/s (mechanical)
		= 2545 Btu/h
Temperature	°F	= 1.8°C + 32
	°R	= °F + 459.67
	K	= °C + 273.15
	°R	= 1.8 K
Miscellaneous	1 Btu/lbm·°F	= 4186.8 J/kg·°C
	1 Btu/lbm	= 0 5559 cal/g
		= 2326 J/kg
	1 ft^3/lbm	= 0.062 427 m^3/kg

TABLE A-3
BASIC AND SUPPLEMENTAL SI UNITS

Quantity	Unit	Symbol
Basic Units		
Length	meter	m
Mass	kilogram	kg
Time	second	s
Electric current	ampere	A
Temperature	kelvin	K
Luminous intensity	candela	cd
Supplemental Units		
Plane angle	radian	rad
Solid angle	steradian	sr

TABLE A-4
DERIVED SI UNITS

Quantity	Name(s) of Unit	Unit Symbol or Abbreviation, Where Differing from Basic Form	Unit Expressed in Terms of Basic or Supplementary Units
Area	square meter		m^2
Volume	cubic meter		m^3
Frequency	hertz, cycle per second	Hz	s^{-1}
Density, concentration	kilogram per cubic meter		kg/m^3
Velocity	meter per second		m/s
Angular velocity	radian per second		rad/s
Acceleration	meter per second squared		m/s^2
Angular acceleration	radian per second squared		rad/s^2
Volumetric flow rate	cubic meter per second		m^3/s
Force	newton	N	$kg{\cdot}m/s^2$
Surface tension	newton per meter, joule per square meter	N/m, J/m^2	kg/s^2
Pressure	newton per square meter, pascal	N/m^2, Pa	$kg/m{\cdot}s^2$
Viscosity, dynamic	newton-second per square meter, poiseuville	$N{\cdot}s/m^2$, Pl	$kg/m{\cdot}s$
Viscosity, kinematic; diffusivity; mass conductivity	meter squared per second		m^2/s
Work, torque, energy, quantity of heat	joule, newton-meter, watt-second	J, $N{\cdot}m$, $W{\cdot}s$	$kg{\cdot}m^2/s^2$
Power, heat flux	watt, joule per second	W, J/s	$kg{\cdot}m^2/s^3$
Heat flux density	watt per square meter	W/m^2	kg/s^3
Volumetric heat release rate	watt per cubic meter	W/m^3	$kg/m{\cdot}s^3$
Heat-transfer coefficient	watt per square meter-degree	$W/m^2{\cdot}deg$	$kg/s^3{\cdot}deg$
Latent heat, enthalpy (specific)	joule per kilogram	J/kg	m^2/s^2
Heat capacity (specific)	joule per kilogram-degree	$J/kg{\cdot}deg$	$m^2/s^2{\cdot}deg$
Capacity rate	watt per degree	W/deg	$kg{\cdot}m^2/s^3{\cdot}deg$
Thermal conductivity	watt per meter-degree	$W/m{\cdot}deg$, $J{\cdot}m/s{\cdot}m^2{\cdot}deg$	$kg{\cdot}m/s^3{\cdot}deg$
Mass flux, mass flow rate	kilogram per second		kg/s
Mass flux density, mass flow rate per unit area	kilogram per square meter-second		$kg/m^2{\cdot}s$
Mass-transfer coefficient	meter per second		m/s
Quantity of electricity	coulomb	C	$A{\cdot}s$
Electromotive force	volt	V, W/A	$kg{\cdot}m^2/A{\cdot}s^3$
Electric resistance	ohm	Ω, V/A	$kg{\cdot}m^2/A^2{\cdot}s^3$
Electric conductivity	ampere per volt meter	$A/V{\cdot}m$	$A^2{\cdot}s^3/kg{\cdot}m^3$
Electric capacitance	farad	F, $A{\cdot}s/V$	$A^3{\cdot}s^4/kg{\cdot}m^2$
Magnetic flux	weber	Wb, $V{\cdot}s$	$kg{\cdot}m^2/A{\cdot}s^2$
Inductance	henry	H, $V{\cdot}s/A$	$kg{\cdot}m^2/A^2{\cdot}s^2$
Magnetic permeability	henry per meter	H/m	$kg{\cdot}m/A^2{\cdot}s^2$
Magnetic flux density	tesla, weber per square meter	T, Wb/m^2	$kg/A{\cdot}s^2$
Magnetic field strength	ampere per meter		A/m
Magnetomotive force	ampere		A
Luminous flux	lumen	lm	$cd{\cdot}sr$
Luminance	candela per square meter		cd/m^2
Illumination	lux, lumen per square meter	lx, lm/m^2	$cd{\cdot}sr/m^2$

TABLE A-5
CRITICAL CONSTANTS

Substance	Symbol	M	T_c, °K	p_c, atm	v_c, cm³/g mol (1 cm³/g mol = 10^{-3} m³/kg mol)	v_c, ft³/lbm mol	z_c
Acetylene	C_2H_2	26.038	309.5	61.6	113	1.81	0.274
Air		28.967	132.41	37.25	93.25	1.49	0.320
Ammonia	NH_3	17.032	405.4	111.3	72.5	1.16	0.243
Argon	A	39 944	150.72	47.996	75	1.20	0.291
Benzene	C_6H_6	78.114	562.6	48.6	260	4.17	0.274
n-Butane	C_4H_{10}	58.124	425.17	37.47	255	4.08	0.274
Isobutane	C_4H_{10}	58.124	408.14	36.00	263	4.21	0.283
1-Butene	C_4H_8	56.108	419.6	39.7	240	3.84	0.277
Carbon dioxide	CO_2	44.011	304.20	72.90	94	1.51	0.275
Carbon monoxide	CO	28.011	132.91	34.529	93	4.42	0.294
Carbon tetrachloride	CCl_4	153 839	556.4	45.0	276	4.42	0.272
n-Deuterium	D_2	4.029	38.43	16.421			
Dodecane	$C_{12}H_{26}$	170.340	659	17.9		11.5	0.237
Ethane	C_2H_6	30.070	305.43	48.20	148	2.37	0.285
Ethyl ether	$C_4H_{10}O$	74.124	467.8	35.6	282.9	4.53	0.262
Ethylene	C_2H_4	28.054	283.06	50.50	124	1.99	0.270
Freon, F-12	CCl_2F_2	120.925	385.16	40.6	217	3.47	0.279
Helium	He	4.003	5.19	2.26	58	0.929	0.308
n-Heptane	C_7H_{16}	100.205	540.17	27.00	426	6.82	0.260
n-Hexane	C_6H_{14}	86.178	507.9	29.94	368	5.89	0.264
Hydrogen	H_2	2.016	33.24	12.797	65	1.04	0.304
Hydrogen sulfide	H_2S	34.082	373.7	88.8	98	1.57	0.284
Mercury	Hg	200.610					
Methane	CH_4	16.043	190.7	45.8	99	1.59	0.290
Methyl fluoride	CH_3F	34.035	317.71	58.0	113	1.81	0.251
Neon	Ne	20.183	44.39	26.86	41.7	0.668	0.308
Nitric oxide	NO	30.008	179.2	65.0	58	0.929	0.256
Nitrogen	N_2	28.016	126.2	33.54	90	0.144	0.291
Octane	C_8H_{18}	114.232	569.4	24.64	486	7.77	0.256
Oxygen	O_2	32.000	154.78	50.14	74	1.19	0.292
n-Pentane	C_5H_{12}	72.151	469.78	33.31	311	4.98	0.269
Isopentane	C_5H_{12}	72.151	461.0	32.92	308	4.93	0.268
Propane	C_3H_8	44.097	370.01	42.1	200	3.20	0.277
Propylene	C_3H_6	42.081	365.1	45.40	181	2.90	0.274
Sulfur dioxide	SO_2	64.066	430.7	77.8	122	1.95	0.269
Water	H_2O	18.016	647.27	218.167	56	0.897	0.230
Xenon	Xe	131.300	289.81	58.0	118.8	1.90	0.290

TABLE A-6
(Continued)

Substance	Formula	Molecular Weight	$\bar{h}_f^\circ$		$\bar{g}_f^\circ$		$\bar{s}^\circ$	
			Btu/lbm·mol	kJ/kg·mol	Btu/lbm·mol	kJ/kg·mol	Btu/lbm·mol·°R	kJ/kg·mol·K
Carbon (solid)	C(s)	12.01	0	0	0	0	1.36	5.74
Carbon (gas)	C(g)	12.01	+309 060	+718 400	+289 520	+672 980	37.76	157.99
Hydrogen	$H_2(g)$	2.018	0	0	0	0	31.21	130.57
Oxygen	$O_2(g)$	32.00	0	0	0	0	49.00	205.03
Carbon monoxide	CO(g)	28.01	−47 540	−110 530	−59 010	−137 150	47.21	197.56
Carbon dioxide	$CO_2(g)$	44.01	−169 300	−393 520	−169 680	−394 360	51.07	213.64
Water	$H_2O(g)$	18.02	−104 040	−241 820	−98 350	−228 590	45.11	188.72
Water	$H_2O(l)$	18.02	−122 970	−285 830	−102 040	−237 180	16.71	69.92
Hydrogen peroxide	$H_2O_2(g)$	34.02	−58 640	−136 310	−45 430	−105 600	55.60	232.63
Ammonia	$NH_3(g)$	17.04	−19 750	−46 190	−7 140	−16 590	45.97	192.33
Methane	$CH_4(g)$	16.04	−32 210	−74 850	−21 860	−50 790	44.49	186.16
Acetylene	$C_2H_2(g)$	26.04	+97 540	+226 730	+90 000	+209 170	48.00	200.85
Ethylene	$C_2H_4(g)$	28.05	+22 490	+52 280	+29 306	+68 120	52.54	219.83
Ethane	$C_2H_6(g)$	30.07	−36 420	−84 680	−14 150	−32 890	54.85	229.49
Propylene	$C_3H_6(g)$	42.08	+8 790	+20 410	+26 980	+62 720	63.80	266.94
Propane	$C_3H_8(g)$	44.09	−44 680	−103 850	−10 105	−23 490	64.51	269.91
n-Butane	$C_4H_{10}(g)$	58.12	−54 270	−126 150	−6 760	−15 710	74.11	310.12
n-Pentane	$C_5H_{12}(g)$	72.15	−63 000	−146 440	−3 600	−8 360	83.34	348.69
n-Hexane	$C_6H_{14}(g)$	86.17	−71 930	−167 200	−126	−293	92.83	388.40
n-Heptane	$C_7H_{16}(g)$	100.20	−80 800	−187 820	−3 490	−8 110	102.24	427.77
n-Octane	$C_8H_{18}(g)$	114.22	−89 680	−208 450	+7 110	+16 530	111.55	466.73
Benzene	$C_6H_6(g)$	78.11	+35 680	+82 930	+55 780	+129 660	64.34	269.20
Methyl alcohol	$CH_3OH(g)$	32.05	−86 540	−200 670	−69 700	−162 000	57.29	239.70
Ethyl alcohol	$C_2H_5OH(g)$	46.07	−101 230	−235 310	−72 520	−168 570	67.54	282.59
Sulfur dioxide	$SO_2(g)$	64.07	−127 730	−296 910	−129 220	−300 370	59.40	248.53

TABLE A-7

PROPERTIES OF WATER: SATURATION TEMPERATURE TABLE (ENGLISH UNITS)

v in ft³/lbm; h and u in Btu/lbm; s in Btu/lbm·°R

		Specific Volume		Internal Energy		Enthalpy			Entropy			
Temp. T, °F	Press. p, psia	Sat. Liquid v_f	Sat. Vapor v_g	Sat. Liquid u_f	Sat. Vapor u_g	Sat. Liquid h_f	Evap. h_{fg}	Sat. Vapor h_g	Sat. Liquid s_f	Evap. s_{fg}	Sat. Vapor s_g	Temp. T, °F
32	0.0886	0.016 02	3305	−0.01	1021.2	−0.01	1075.4	1075.4	−0.000 03	2.1870	2.1870	32
35	0.0999	0.016 02	2948	2.99	1022.2	3.00	1073.7	1076.7	0.006 07	2.1704	2.1764	35
40	0.1217	0.016 02	2445	8.02	1023.9	8.02	1070.9	1078.9	0.016 17	2.1430	2.1592	40
45	0.1475	0.016 02	2037	13.04	1025.5	13.04	1068.1	1081.1	0.026 18	2.1162	2.1423	45
50	0.1780	0.016 02	1704	18.06	1027.2	18.06	1065.2	1083.3	0.036 07	2.0899	2.1259	50
55	0.2140	0.016 03	1431	23.07	1028.8	23.07	1062.4	1085.5	0.045 86	2.0641	2.1099	55
60	0.2563	0.016 04	1207	28.08	1030.4	28.08	1059.6	1087.7	0.055 55	2.0388	2.0943	60
65	0.3057	0.016 04	1022	33.09	1032.1	33.09	1056.8	1089.9	0.065 14	2.0140	2.1791	65
70	0.3632	0.016 05	867.7	38.09	1033.7	38.09	1054.0	1092.0	0.074 63	1.9896	2.0642	70
75	0.4300	0.016 06	739.7	43.09	1035.4	43.09	1051.1	1094.2	0.084 02	1.9657	2.0497	75
80	0.5073	0.016 07	632.8	48.08	1037.0	48.09	1048.3	1096.4	0.093 32	1.9423	2.0356	80
85	0.5964	0.016 09	543.1	53.08	1038.6	53.08	1045.5	1098.6	0.1025	1.9193	2.0218	85
90	0.6988	0.016 10	467.7	58.07	1040.2	58.07	1042.7	1100.7	0.1117	1.8966	2.0083	90
95	0.8162	0.016 11	404.0	63.06	1041.9	63.06	1039.8	1102.9	0.1207	1.8744	1.9951	95
100	0.9503	0.016 13	350.0	68.04	1043.5	68.05	1037.0	1105.0	0.1296	1.8526	1.9822	100
110	1.276	0.016 17	265.1	78.02	1046.7	78.02	1031.3	1109.3	0.1473	1.8101	1.9574	110
120	1.695	0.016 21	203.0	87.99	1049.9	88.00	1025.5	1113.5	0.1647	1.7690	1.9336	120
130	2.225	0.016 25	157.2	97.97	1053.0	97.98	1019.8	1117.8	0.1817	1.7292	1.9109	130
140	2.892	0.016 29	122.9	107.95	1056.2	107.96	1014.0	1121.9	0.1985	1.6907	1.8892	140
150	3.722	0.016 34	97.0	117.95	1059.3	117.96	1008.1	1126.1	0.2150	1.6533	1.8684	150
160	4.745	0.016 40	77.2	127.94	1062.3	127.96	1002.2	1130.1	0.2313	1.6171	1.8484	160
170	5.996	0.016 45	62.0	137.95	1065.4	137.97	996.2	1134.2	0.2473	1.5819	1.8293	170
180	7.515	0.016 51	50.2	147.97	1068.3	147.99	990.2	1138.2	0.2631	1.5478	1.8109	180
190	9.343	0.016 57	41.0	158.00	1071.3	158.03	984.1	1142.1	0.2787	1.5146	1.7932	190
200	11.529	0.016 63	33.6	168.04	1074.2	168.07	977.9	1145.9	0.2940	1.4822	1.7762	200
210	14.13	0.016 70	27.82	178.1	1077.0	178.1	971.6	1149.7	0.3091	1.4508	1.7599	210
212	14.70	0.016 72	26.80	180.1	1077.6	180.2	970.3	1150.5	0.3121	1.4446	1.7567	212
220	17.19	0.016 77	23.15	188.2	1079.8	188.2	965.3	1153.5	0.3241	1.4201	1.7441	220
230	20.78	0.016 85	19.39	198.3	1082.6	198.3	958.8	1157.1	0.3388	1.3901	1.7289	230
240	24.97	0.016 92	16.33	208.4	1085.3	208.4	952.3	1160.7	0.3534	1.3609	1.7143	240
250	29.82	0.017 00	13.83	218.5	1087.9	218.6	945.6	1164.2	0.3677	1.3324	1.7001	250
260	35.42	0.017 08	11.77	228.6	1090.5	228.8	938.8	1167.6	0.3819	1.3044	1.6864	260
270	41.85	0.017 17	10.07	238.8	1093.0	239.0	932.0	1170.9	0.3960	1.2771	1.6731	270
280	49.18	0.017 26	8.65	249.0	1095.4	249.2	924.9	1174.1	0.4099	1.2504	1.6602	280
290	57.53	0.017 35	7.47	259.3	1097.7	259.4	917.8	1177.2	0.4236	1.2241	1.6477	290

TABLE A-7
(Continued)

v in ft^3/lbm; h and u in Btu/lbm; s in Btu/lbm·°R

		Specific Volume		Internal Energy		Enthalpy			Entropy			
Temp. T, °F	Press. p, psia	Sat. Liquid v_f	Sat. Vapor v_g	Sat. Liquid u_f	Sat. Vapor u_g	Sat. Liquid h_f	Evap. h_{fg}	Sat. Vapor h_g	Sat. Liquid s_f	Evap. s_{fg}	Sat. Vapor s_g	Temp. T, °F
300	66.98	0.017 45	6.472	269.5	1100.0	269.7	910.4	1180.2	0.4372	1.1984	1.6356	300
310	77.64	0.017 55	5.632	279.8	1102.1	280.1	903.0	1183.0	0.4507	1.1731	1.6238	310
320	89.60	0.017 65	4.919	290.1	1104.2	290.4	895.3	1185.8	0.4640	1.1483	1.6123	320
330	103.00	0.017 76	4.312	300.5	1106.2	300.8	887.5	1188.4	0.4772	1.1238	1.6010	330
340	117.93	0.017 87	3.792	310.9	1108.0	311.3	879.5	1190.8	0.4903	1.0997	1.5901	340
350	134.53	0.017 99	3.346	321.4	1109.8	321.8	871.3	1193.1	0.5033	1.0760	1.5793	350
360	152.92	0.018 11	2.961	331.8	1111.4	332.4	862.9	1195.2	0.5162	1.0526	1.5688	360
370	173.23	0.018 23	2.628	342.4	1112.9	343.0	854.2	1197.2	0.5289	1.0295	1.5585	370
380	195.60	0.018 36	2.339	353.0	1114.3	353.6	845.4	1199.0	0.5416	1.0067	1.5483	380
390	220.2	0.018 50	2.087	363.6	1115.6	364.3	836.2	1200.6	0.5542	0.9841	1.5383	390
400	247.1	0.018 64	1.866	374.3	1116.6	375.1	826.8	1202.0	0.5667	0.9617	1.5284	400
410	276.5	0.018 78	1.673	385.0	1117.6	386.0	817.2	1203.1	0.5792	0.9395	1.5187	410
420	308.5	0.018 94	1.502	395.8	1118.3	396.9	807.2	1204.1	0.5915	0.9175	1.5091	420
430	343.3	0.019 09	1.352	406.7	1118.9	407.9	796.9	1204.8	0.6038	0.8957	1.4995	430
440	381.2	0.019 26	1.219	417.6	1119.3	419.0	786.3	1205.3	0.6161	0.8740	1.4900	440
450	422.1	0.019 43	1.101 1	428.6	1119.5	430.2	775.4	1205.6	0.6282	0.8523	1.4806	450
460	466.3	0.019 61	0.996 1	439.7	1119.6	441.4	764.1	1205.5	0.6404	0.8308	1.4712	460
470	514.1	0.019 80	0.902 5	450.9	1119.4	452.8	752.4	1205.2	0.6525	0.8093	1.4618	470
480	565.5	0.020 00	0.818 7	462.2	1118.9	464.3	740.3	1204.6	0.6646	0.7878	1.4524	480
490	620.7	0.020 21	0.743 6	473.6	1118.3	475.9	727.8	1203.7	0.6767	0.7663	1.4430	490
500	680.0	0.020 43	0.676 1	485.1	1117.4	487.7	714.8	1202.5	0.6888	0.7448	1.4335	500
520	811.4	0.020 91	0.560 5	508.5	1114.8	511.7	687.3	1198.9	0.7130	0.7015	1.4145	520
540	961.5	0.021 45	0.465 8	532.6	1111.0	536.4	657.5	1193.8	0.7374	0.6576	1.3950	540
560	1131.8	0.022 07	0.387 7	548.4	1105.8	562.0	625.0	1187.0	0.7620	0.6129	1.3749	560
580	1324.3	0.022 78	0.322 5	583.1	1098.9	588.6	589.3	1178.0	0.7872	0.5668	1.3540	580
600	1541.0	0.023 63	0.267 7	609.9	1090.0	616.7	549.7	1166.4	0.8130	0.5187	1.3317	600
620	1784.4	0.024 65	0.220 9	638.3	1078.5	646.4	505.0	1151.4	0.8398	0.4677	1.3075	620
640	2057.1	0.025 93	0.180 5	668.7	1063.2	678.6	453.4	1131.9	0.8681	0.4122	1.2803	640
660	2362	0.027 67	0.144 6	702.3	1042.3	714.4	391.1	1105.5	0.8990	0.3493	1.2483	660
680	2705	0.030 32	0.111 3	741.7	1011.0	756.9	309.8	1066.7	0.9350	0.2718	1.2068	680
700	3090	0.036 66	0.074 4	801.7	947.7	822.7	167.5	990.2	0.9902	0.1444	1.1346	700
705.4	3204	0.050 53	0.050 53	872.6	872.6	902.5	0	902.5	1.0580	0	1.0580	705.4

Source: Abridged from Keenan, J. H., F. G. Keyes, P. G. Hill, and J. G. Moore, "Steam Tables," John Wiley & Sons, Inc., New York, 1969.

TABLE A-7M

PROPERTIES OF WATER: SATURATION TEMPERATURE TABLE (SI UNITS)

v in cm^3/g, 1 cm^3/g = 10^{-3} m^3/kg; h and u in kJ/kg; s in kJ/kg·K; p in bars, 1 bar = 10^5 Pa

		Specific Volume		Internal Energy		Enthalpy			Entropy	
Temp. T, °C	Press. p, bars	Sat. Liquid v_f	Sat. Vapor v_g	Sat. Liquid u_f	Sat. Vapor u_g	Sat. Liquid h_f	Evap. h_{fg}	Sat. Vapor h_g	Sat. Liquid s_f	Sat. Vapor s_g
0	0.006 11	1.0002	206 278	−0.03	2375.4	−0.02	2501.4	2501.3	−0.0001	9.1565
5	0.008 72	1.0001	147 120	20.97	2382.3	20.98	2489.6	2510.6	0.0761	9.0257
10	0.012 28	1.0004	106 379	42.00	2389.2	42.01	2477.7	2519.8	0.1510	8.9008
15	0.017 05	1.0009	77 926	62.99	2396.1	62.99	2465.9	2528.9	0.2245	8.7814
20	0.023 39	1.0018	57 791	83.95	2402.9	83.96	2454.1	2538.1	0.2966	8.6672
25	0.031 69	1.0029	43 360	104.88	2409.8	104.89	2442.3	2547.2	0.3674	8.5580
30	0.042 46	1.0043	32 894	125.78	2416.6	125.79	2430.5	2556.3	0.4369	8.4533
35	0.056 28	1.0060	25 216	146.67	2423.4	146.68	2418.6	2565.3	0.5053	8.3531
40	0.073 84	1.0078	19 523	167.56	2430.1	167.57	2406.7	2574.3	0.5725	8.2570
45	0.095 93	1.0099	15 258	188.44	2436.8	188.45	2394.8	2583.2	0.6387	8.1648
50	0.1235	1.0121	12 032	209.32	2443.5	209.33	2382.7	2592.1	0.7038	8.0763
55	0.1576	1.0146	9568	230.21	2450.1	230.23	2370.7	2600.9	0.7679	7.9913
60	0.1994	1.0172	7671	251.11	2456.6	251.13	2358.5	2609.6	0.8312	7.9096
65	0.2503	1.0199	6197	272.02	2463.1	272.06	2346.2	2618.3	0.8935	7.8310
70	0.3119	1.0228	5042	292.95	2469.6	292.98	2333.8	2626.8	0.9549	7.7553
75	0.3858	1.0259	4131	313.90	2475.9	313.93	2321.4	2635.3	1.0155	7.6824
80	0.4739	1.0291	3407	334.86	2482.2	334.91	2308.8	2643.7	1.0753	7.6122
85	0.5783	1.0325	2828	355.84	2488.4	355.90	2296.0	2651.9	1.1343	7.5445
90	0.7014	1.0360	2361	376.85	2494.5	376.92	2283.2	2660.1	1.1925	7.4791
95	0.8455	1.0397	1982	397.88	2500.6	397.96	2270.2	2668.1	1.2500	7.4159
100	1.014	1.0435	1673.	418.94	2506.5	419.04	2257.0	2676.1	1.3069	7.3549
110	1.433	1.0516	1210.	461.14	2518.1	461.30	2230.2	2691.5	1.4185	7.2387
120	1.985	1.0603	891.9	503.50	2529.3	503.71	2202.6	2706.3	1.5276	7.1296
130	2.701	1.0697	668.5	546.02	2539.9	546.31	2174.2	2720.5	1.6344	7.0269
140	3.613	1.0797	508.9	588.74	2550.0	589.13	2144.7	2733.9	1.7391	6.9299
150	4.758	1.0905	392.8	631.68	2559.5	632.20	2114.3	2746.5	1.8418	6.8379
160	6.178	1.1020	307.1	674.86	2568.4	675.55	2082.6	2758.1	1.9427	6.7502
170	7.917	1.1143	242.8	718.33	2576.5	719.21	2049.5	2768.7	2.0419	6.6663
180	10.02	1.1274	194.1	762.09	2583.7	763.22	2015.0	2778.2	2.1396	6.5857
190	12.54	1.1414	156.5	806.19	2590.0	807.62	1978.8	2786.4	2.2359	6.5079
200	15.54	1.1565	127.4	850.65	2595.3	852.45	1940.7	2793.2	2.3309	6.4323
210	19.06	1.1726	104.4	895.53	2599.5	897.76	1900.7	2798.5	2.4248	6.3585
220	23.18	1.1900	86.19	940.87	2602.4	943.62	1858.5	2802.1	2.5178	6.2861
230	27.95	1.2088	71.58	986.74	2603.9	990.12	1813.8	2804.0	2.6099	6.2146
240	33.44	1.2291	59.76	1033.2	2604.0	1037.3	1766.5	2803.8	2.7015	6.1437
250	39.73	1.2512	50.13	1080.4	2602.4	1085.4	1716.2	2801.5	2.7927	6.0730
260	46.88	1.2755	42.21	1128.4	2599.0	1134.4	1662.5	2796.9	2.8838	6.0019
270	54.99	1.3023	35.64	1177.4	2593.7	1184.5	1605.2	2789.7	2.9751	5.9301
280	64.12	1.3321	30.17	1227.5	2586.1	1236.0	1543.6	2779.6	3.0668	5.8571
290	74.36	1.3656	25.57	1278.9	2576.0	1289.1	1477.1	2766.2	3.1594	5.7821
300	85.81	1.4036	21.67	1332.0	2563.0	1344.0	1404.9	2749.0	3.2534	5.7045
320	112.7	1.4988	15.49	1444.6	2525.5	1461.5	1238.6	2700.1	3.4480	5.5362
340	145.9	1.6379	10.80	1570.3	2464.6	1594.2	1027.9	2622.0	3.6594	5.3357
360	186.5	1.8925	6.945	1752.2	2351.1	1760.5	720.5	2481.0	3.9147	5.0526
374.14	220.9	3.155	3.155	2029.6	2029.6	2099.3	0	2099.3	4.4298	4.4298

Source: Abridged from Keenan, J. H., F. G. Keyes, P. G. Hill, and J. G. Moore, "Steam Tables," John Wiley & Sons, Inc., New York, 1969.

TABLE A-8

PROPERTIES OF WATER: SATURATION PRESSURE TABLE (ENGLISH UNITS)

v in ft³/lbm; h and u in Btu/lbm; s in Btu/lbm·°R

		Specific Volume		Internal Energy		Enthalpy			Entropy			
Abs. Press. p, psia	Temp. t, °F	Sat. Liquid v_f	Sat. Vapor v_g	Sat. Liquid u_f	Sat. Vapor u_g	Sat. Liquid h_f	Evap. h_{fg}	Sat. Vapor h_g	Sat. Liquid s_f	Evap. s_{fg}	Sat. Vapor s_g	Abs. Press. p, psia
0.4	72.84	0.016 06	792.0	40.94	1034.7	40.94	1052.3	1093.3	0.0800	1.9760	2.0559	0.4
0.6	85.19	0.016 09	540.0	53.26	1038.7	53.27	1045.4	1098.6	0.1029	1.9184	2.0213	0.6
0.8	94.35	0.016 11	411.7	62.41	1041.7	62.41	1040.2	1102.6	0.1195	1.8773	1.9968	0.8
1.0	101.70	0.016 14	333.6	69.74	1044.0	69.74	1036.0	1105.8	0.1327	1.8453	1.9779	1.0
1.2	107.88	0.016 16	280.9	75.90	1046.0	75.90	1032.5	1108.4	0.1436	1.8190	1.9626	1.2
1.5	115.65	0.016 19	227.7	83.65	1048.5	83.65	1028.0	1111.7	0.1571	1.7867	1.9438	1.5
2.0	126.04	0.016 23	173.75	94.02	1051.8	94.02	1022.1	1116.1	0.1750	1.7448	1.9198	2.0
3.0	141.43	0.016 30	118.72	109.38	1056.6	109.39	1013.1	1122.5	0.2009	1.6852	1.8861	3.0
4.0	152.93	0.016 36	90.64	120.88	1060.2	120.89	1006.4	1127.3	0.2198	1.6426	1.8624	4.0
5.0	162.21	0.016 41	73.53	130.15	1063.0	130.17	1000.9	1131.0	0.2349	1.6093	1.8441	5.0
6.0	170.03	0.016 45	61.98	137.98	1065.4	138.00	996.2	1134.2	0.2474	1.5819	1.8292	6.0
7.0	176.82	0.016 49	53.65	144.78	1067.4	144.80	992.1	1136.9	0.2581	1.5585	1.8167	7.0
8.0	182.84	0.016 53	47.35	150.81	1069.2	150.84	988.4	1139.3	0.2675	1.5383	1.8058	8.0
9.0	188.26	0.016 56	42.41	156.25	1070.8	156.27	985.1	1141.4	0.2760	1.5203	1.7963	9.0
10	193.19	0.016 59	38.42	161.20	1072.2	161.23	982.1	1143.3	0.2836	1.5041	1.7877	10
14.696	211.99	0.016 72	26.80	180.10	1077.6	180.15	970.4	1150.5	0.3121	1.4446	1.7567	14.696
15	213.03	0.016 72	26.29	181.14	1077.9	181.19	969.7	1150.9	0.3137	1.4414	1.7551	15
20	227.96	0.016 83	20.09	196.19	1082.0	196.26	960.1	1156.4	0.3358	1.3962	1.7320	20
25	240.08	0.016 92	16.31	208.44	1085.3	208.52	952.2	1160.7	0.3535	1.3607	1.7142	25
30	250.34	0.017 00	13.75	218.84	1088.0	218.93	945.4	1164.3	0.3682	1.3314	1.6996	30
35	259.30	0.017 08	11.90	227.93	1090.3	228.04	939.3	1167.4	0.3809	1.3064	1.6873	35
40	267.26	0.017 15	10.50	236.03	1092.3	236.16	933.8	1170.0	0.3921	1.2845	1.6767	40
45	274.46	0.017 21	9.40	243.37	1094.0	243.51	928.8	1172.3	0.4022	1.2651	1.6673	45
50	281.03	0.017 27	8.52	250.08	1095.6	250.24	924.2	1174.4	0.4113	1.2476	1.6589	50
55	287.10	0.017 33	7.79	256.28	1097.0	256.46	919.9	1176.3	0.4196	1.2317	1.6513	55
60	292.73	0.017 38	7.177	262.1	1098.3	262.2	915.8	1178.0	0.4273	1.2170	1.6443	60
65	298.00	0.017 43	6.647	267.5	1099.5	267.7	911.9	1179.6	0.4345	1.2035	1.6380	65
70	302.96	0.017 48	6.209	272.6	1100.6	272.8	908.3	1181.0	0.4412	1.1909	1.6321	70
75	307.63	0.017 52	5.818	277.4	1101.6	277.6	904.8	1182.4	0.4475	1.1790	1.6265	75
80	312.07	0.017 57	5.474	282.0	1102.6	282.2	901.4	1183.6	0.4534	1.1679	1.6213	80
85	316.29	0.017 61	5.170	286.3	1103.5	286.6	898.2	1184.8	0.4591	1.1574	1.6165	85
90	320.31	0.017 66	4.898	290.5	1104.3	290.8	895.1	1185.9	0.4644	1.1475	1.6119	90
95	324.16	0.017 70	4.654	294.5	1105.0	294.8	892.1	1186.9	0.4695	1.1380	1.6075	95
100	327.86	0.017 74	4.434	298.3	1105.8	298.6	889.2	1187.8	0.4744	1.1290	1.6034	100
110	334.82	0.017 81	4.051	305.5	1107.1	305.9	883.7	1189.6	0.4836	1.1122	1.5958	110

120	341.30	0.017 89	3.730	312.3	1108.3	312.7	878.5	1191.1	0.4920	1.0966	1.5886	120
130	347.37	0.017 96	3.457	318.6	1109.4	319.0	873.5	1192.5	0.4999	1.0822	1.5821	130
140	353.08	0.018 02	3.221	324.6	1110.3	325.1	868.7	1193.8	0.5073	1.0688	1.5761	140
150	358.48	0.018 09	3.016	330.2	1111.2	330.8	864.2	1194.9	0.5142	1.0562	1.5704	150
160	363.60	0.018 15	2.836	335.6	1112.0	336.2	859.8	1196.0	0.5208	1.0443	1.5651	160
170	368.47	0.018 21	2.676	340.8	1112.7	341.3	855.6	1196.9	0.5270	1.0330	1.5600	170
180	373.13	0.018 27	2.533	345.7	1113.4	346.3	851.5	1197.8	0.5329	1.0223	1.5552	180
190	377.59	0.018 33	2.405	350.4	1114.0	351.0	847.5	1198.6	0.5386	1.0122	1.5508	190
200	381.86	0.018 39	2.289	354.9	1114.6	355.6	843.7	1199.3	0.5440	1.0025	1.5465	200
250	401.04	0.018 65	1.845	375.4	1116.7	376.2	825.8	1202.1	0.5680	0.9594	1.5274	250
300	417.43	0.018 90	1.544	393.0	1118.2	394.1	809.8	1203.9	0.5883	0.9232	1.5115	300
350	431.82	0.019 12	1.327	408.7	1119.0	409.9	795.0	1204.9	0.6060	0.8917	1.4977	350
400	444.70	0.019 34	1.162	422.8	1119.5	424.2	781.2	1205.5	0.6218	0.8638	1.4856	400
450	456.39	0.019 55	1.033	435.7	1119.6	437.4	768.2	1205.6	0.6360	0.8385	1.4745	450
500	467.13	0.019 75	0.928	447.7	1119.4	449.5	755.8	1205.3	0.6490	0.8154	1.4644	500
550	477.07	0.019 94	0.842	458.9	1119.1	460.9	743.9	1204.8	0.6611	0.7941	1.4551	550
600	486.33	0.020 13	0.770	469.4	1118.6	471.7	732.4	1204.1	0.6723	0.7742	1.4464	600
700	503.23	0.020 51	0.656	488.9	1117.0	491.5	710.5	1202.0	0.6927	0.7378	1.4305	700
800	518.36	0.020 87	0.569	506.6	1115.0	509.7	689.6	1199.3	0.7110	0.7050	1.4160	800
900	532.12	0.021 23	0.501	523.0	1112.6	526.6	669.5	1196.0	0.7277	0.6750	1.4027	900
1000	544.75	0.021 59	0.446	538.4	1109.9	542.4	650.0	1192.4	0.7432	0.6471	1.3903	1000
1100	556.45	0.021 95	0.401	552.9	1106.8	557.4	631.0	1188.3	0.7576	0.6209	1.3786	1100
1200	567.37	0.022 32	0.362	566.7	1103.5	571.7	612.3	1183.9	0.7712	0.5961	1.3673	1200
1300	577.60	0.022 69	0.330	579.9	1099.8	585.4	593.8	1179.2	0.7841	0.5724	1.3565	1300
1400	587.25	0.023 07	0.302	592.7	1096.0	598.6	575.5	1174.1	0.7964	0.5497	1.3461	1400
1500	596.39	0.023 46	0.277	605.0	1091.8	611.5	557.2	1168.7	0.8082	0.5276	1.3359	1500
1600	605.06	0.023 86	0.255	616.9	1087.4	624.0	538.9	1162.9	0.8196	0.5062	1.3258	1600
1700	613.32	0.024 28	0.236	628.6	1082.7	636.2	520.6	1156.9	0.8307	0.4852	1.3159	1700
1800	621.21	0.024 72	0.218	640.0	1077.7	648.3	502.1	1150.4	0.8414	0.4645	1.3060	1800
1900	628.76	0.025 17	0.203	651.3	1072.3	660.1	483.4	1143.5	0.8519	0.4441	1.2961	1900
2000	636.00	0.025 65	0.188	662.4	1066.6	671.9	464.4	1136.3	0.8623	0.4238	1.2861	2000
2250	652.90	0.026 98	0.157	689.9	1050.6	701.1	414.8	1115.9	0.8876	0.3728	1.2604	2250
2500	668.31	0.028 60	0.131	717.7	1031.0	730.9	360.5	1091.4	0.9131	0.3196	1.2327	2500
2750	682.46	0.030 77	0.107	747.3	1005.9	763.0	297.4	1060.4	0.9401	0.2604	1.2005	2750
3000	695.52	0.034 31	0.084	783.4	968.8	802.5	213.0	1015.5	0.9732	0.1843	1.1575	3000
3203.6	705.44	0.050 53	0.0505	872.6	872.6	902.5	0	902.5	1.0580	0	1.0580	3203.6

Source: Abridged from Keenan, J. H., F. G. Keyes, P. G. Hill, and J. G. Moore, "Steam Tables," John Wiley & Sons, Inc., New York, 1969.

TABLE A-8M

PROPERTIES OF WATER: SATURATION PRESSURE TABLE (SI UNITS)

v in cm³/g, 1 cm³/g = 10^{-3} m³/kg; h and u in kJ/kg; s in kJ/kg·K; p in bars, 1 bar = 10^5 Pa

		Specific Volume		Internal Energy		Enthalpy			Entropy	
Press. p, bars	Temp. T, °C	Sat. Liquid v_f	Sat. Vapor v_g	Sat. Liquid u_f	Sat. Vapor u_g	Sat. Liquid h_f	Evap. h_{fg}	Sat. Vapor h_g	Sat. Liquid s_f	Sat. Vapor s_g
0.040	28.96	1.0040	34 800.	121.45	2415.2	121.46	2432.9	2554.4	0.4226	8.4746
0.060	36.16	1.0064	23 729.	151.53	2425.0	151.53	2415.9	2567.4	0.5210	8.3304
0.080	41.51	1.0084	18 103.	173.87	2432.2	173.88	2403.1	2577.0	0.5926	8.2287
0.10	45.81	1.0102	14 674.	191.82	2437.9	191.83	2392.8	2584.7	0.6493	8.1502
0.20	60.06	1.0172	7649.	251.38	2456.7	251.40	2358.3	2609.7	0.8320	7.9085
0.30	69.10	1.0223	5229.	289.20	2468.4	289.23	2336.1	2625.3	0.9439	7.7686
0.40	75.87	1.0265	3993.	317.53	2477.0	317.58	2319.2	2636.8	1.0259	7.6700
0.50	81.33	1.0300	3240.	340.44	2483.9	340.49	2305.4	2645.9	1.0910	7.5939
0.60	85.94	1.0331	2732.	359.79	2489.6	359.86	2293.6	2653.5	1.1453	7.5320
0.70	89.95	1.0360	2365.	376.63	2494.5	376.70	2283.3	2660.0	1.1919	7.4797
0.80	93.50	1.0380	2087.	391.58	2498.8	391.66	2274.1	2665.8	1.2329	7.4346
0.90	96.71	1.0410	1869.	405.06	2502.6	405.15	2265.7	2670.9	1.2695	7.3949
1.00	99.63	1.0432	1694.	417.36	2506.1	417.46	2258.0	2675.5	1.3026	7.3594
1.50	111.4	1.0528	1159.	466.94	2519.7	467.11	2226.5	2693.6	1.4336	7.2233
2.00	120.2	1.0605	885.7	504.49	2529.5	504.70	2201.9	2706.7	1.5301	7.1271
2.50	127.4	1.0672	718.7	535.10	2537.2	535.37	2181.5	2716.9	1.6072	7.0527
3.00	133.6	1.0732	605.8	561.15	2543.6	561.47	2163.8	2725.3	1.6718	6.9919
3.50	138.9	1.0786	524.3	583.95	2548.9	584.33	2148.1	2732.4	1.7275	6.9405
4.00	143.6	1.0836	462.5	604.31	2553.6	604.74	2133.8	2738.6	1.7766	6.8959
4.50	147.9	1.0882	414.0	622.77	2557.6	623.25	2120.7	2743.9	1.8207	6.8565
5.00	151.9	1.0926	374.9	639.68	2561.2	640.23	2108.5	2748.7	1.8607	6.8213
6.00	158.9	1.1006	315.7	669.90	2567.4	670.56	2086.3	2756.8	1.9312	6.7600
7.00	165.0	1.1080	272.9	696.44	2572.5	697.22	2066.3	2763.5	1.9922	6.7080
8.00	170.4	1.1148	240.4	720.22	2576.8	721.11	2048.0	2769.1	2.0462	6.6628
9.00	175.4	1.1212	215.0	741.83	2580.5	742.83	2031.1	2773.9	2.0946	6.6226
10.0	179.9	1.1273	194.4	761.68	2583.6	762.81	2015.3	2778.1	2.1387	6.5863
15.0	198.3	1.1539	131.8	843.16	2594.5	844.89	1947.3	2792.2	2.3150	6.4448
20.0	212.4	1.1767	99.63	906.44	2600.3	908.79	1890.7	2799.5	2.4474	6.3409
25.0	224.0	1.1973	79.98	959.11	2603.1	962.11	1841.0	2803.1	2.5547	6.2575
30.0	233.9	1.2165	66.68	1004.8	2604.1	1008.4	1795.7	2804.2	2.6457	6.1869
35.0	242.6	1.2347	57.07	1045.4	2603.7	1049.8	1753.7	2803.4	2.7253	6.1253
40.0	250.4	1.2522	49.78	1082.3	2602.3	1087.3	1714.1	2801.4	2.7964	6.0701
45.0	257.5	1.2692	44.06	1116.2	2600.1	1121.9	1676.4	2798.3	2.8610	6.0199
50.0	264.0	1.2859	39.44	1147.8	2597.1	1154.2	1640.1	2794.3	2.9202	5.9734
60.0	275.6	1.3187	32.44	1205.4	2589.7	1213.4	1571.0	2784.3	3.0267	5.8892
70.0	285.9	1.3513	27.37	1257.6	2580.5	1267.0	1505.1	2772.1	3.1211	5.8133
80.0	295.1	1.3842	23.52	1305.6	2569.8	1316.6	1441.3	2758.0	3.2068	5.7432
90.0	303.4	1.4178	20.48	1350.5	2557.8	1363.3	1378.9	2742.1	3.2858	5.6772
100.0	311.1	1.4524	18.03	1393.0	2544.4	1407.6	1317.1	2724.7	3.3596	5.6141
110.0	318.2	1.4886	15.99	1433.7	2529.8	1450.1	1255.5	2705.6	3.4295	5.5527
120.0	324.8	1.5267	14.26	1473.0	2513.7	1491.3	1193.6	2684.9	3.4962	5.4924
130.0	330.9	1.5671	12.78	1511.1	2496.1	1531.5	1130.7	2662.2	3.5606	5.4323
140.0	336.8	1.6107	11.49	1548.6	2476.8	1571.1	1066.5	2637.6	3.6232	5.3717
150.0	342.2	1.6581	10.34	1585.6	2455.5	1610.5	1000.0	2610.5	3.6848	5.3098
160.0	347.4	1.7107	9.306	1622.7	2431.7	1650.1	930.6	2580.6	3.7461	5.2455
170.0	352.4	1.7702	8.364	1660.2	2405.0	1690.3	856.9	2547.2	3.8079	5.1777
180.0	357.1	1.8397	7.489	1698.9	2374.3	1732.0	777.1	2509.1	3.8715	5.1044
190.0	361.5	1.9243	6.657	1739.9	2338.1	1776.5	688.0	2464.5	3.9388	5.0228
200.0	365.8	2.036	5.834	1785.6	2293.0	1826.3	583.4	2409.7	4.0139	4.9269
220.9	374.1	3.155	3.155	2029.6	2029.6	2099.3	0	2099.3	4.4298	4.4298

Source: Abridged from Keenan, J. H., F. G. Keyes, P. G. Hill, and J. G. Moore, "Steam Tables," John Wiley & Sons, Inc., New York, 1969.

TABLE A-9

PROPERTIES OF WATER: SUPERHEATED VAPOR (ENGLISH UNITS)

v in ft³/lbm; *h* and *u* in Btu/lbm; *s* in Btu/lbm·°R

Temp., °F	*v*	*u*	*h*	*s*	*v*	*u*	*h*	*s*
	1 psia (101.7°F)				**5 psia (162.2°F)**			
Sat.	333.6	1044.0	1105.8	1.9779	73.53	1063.0	1131.0	1.8441
150	362.6	1060.4	1127.5	2.0151				
200	392.5	1077.5	1150.1	2.0508	78.15	1076.0	1148.6	1.8715
250	422.4	1094.7	1172.8	2.0839	84.21	1093.8	1171.7	1.9052
300	452.3	1112.0	1195.7	2.1150	90.24	1111.3	1194.8	1.9367
400	511.9	1147.0	1241.8	2.1720	102.24	1146.6	1241.2	1.9941
500	571.5	1182.8	1288.5	2.2235	114.20	1182.5	1288.2	2.0458
600	631.1	1219.3	1336.1	2.2706	126.15	1219.1	1335.8	2.0930
700	690.7	1256.7	1384.5	2.3142	138.08	1256.5	1384.3	2.1367
800	750.3	1294.4	1433.7	2.3550	150.01	1294.7	1433.5	2.1775
900	809.9	1333.9	1483.8	2.3932	161.94	1333.8	1483.7	2.2158
1000	869.5	1373.9	1534.8	2.4294	173.86	1373.9	1534.7	2.2520
	10 psia (193.2°F)				**14.7 psia (212.0°F)**			
Sat.	38.42	1072.2	1143.3	1.7877	26.80	1077.6	1150.5	1.7567
200	38.85	1074.7	1146.6	1.7927				
250	41.95	1092.6	1170.2	1.8272	28.42	1091.5	1168.8	1.7832
300	44.99	1110.4	1193.7	1.8592	30.52	1109.6	1192.6	1.8157
400	51.03	1146.1	1240.5	1.9171	34.67	1145.6	1239.9	1.8741
500	57.04	1182.2	1287.7	1.9690	38.77	1181.8	1287.3	1.9263
600	63.03	1218.9	1335.5	2.0164	42.86	1218.6	1335.2	1.9737
700	69.01	1256.3	1384.0	2.0601	46.93	1256.1	1383.8	2.0175
800	74.98	1294.6	1433.3	2.1009	51.00	1294.4	1433.1	2.0584
900	80.95	1333.7	1483.5	2.1393	55.07	1333.6	1483.4	2.0967
1000	86.91	1373.8	1534.6	2.1755	59.13	1373.7	1534.5	2.1330
1100	92.88	1414.7	1586.6	2.2099	63.19	1414.6	1586.4	2.1674
	20 psia (228.0°F)				**40 psia (267.3°F)**			
Sat.	20.09	1082.0	1156.4	1.7320	10.50	1093.3	1170.0	1.6767
250	20.79	1090.3	1167.2	1.7475				
300	22.36	1108.7	1191.5	1.7805	11.04	1105.1	1186.8	1.6993
350	23.90	1126.9	1215.4	1.8110	11.84	1124.2	1211.8	1.7312
400	25.43	1145.1	1239.2	1.8395	12.62	1143.0	1236.4	1.7606
500	28.46	1181.5	1286.8	1.8919	14.16	1180.1	1284.9	1.8140
600	31.47	1218.4	1334.8	1.9395	15.69	1217.3	1333.4	1.8621
700	34.47	1255.9	1383.5	1.9834	17.20	1255.1	1382.4	1.9063
800	37.46	1294.3	1432.9	2.0243	18.70	1293.7	1432.1	1.9474
900	40.45	1333.5	1483.2	2.0627	20.20	1333.0	1482.5	1.9859
1000	43.44	1373.5	1534.3	2.0989	21.70	1373.1	1533.8	2.0223
1100	46.42	1414.5	1586.3	2.1334	23.20	1414.2	1585.9	2.0568

Source: Abridged from Kennan, J. H., F. G. Keyes, P. G. Hill, and J. G. Moore, "Steam Tables," John Wiley & Sons, Inc., New York, 1969.

TABLE A-9
(Continued)

Temp., °F	v	u	h	s	v	u	h	s
	60 psia (292.7°F)				**80 psia (312.1°F)**			
Sat.	7.17	1098.3	1178.0	1.6444	5.47	1102.6	1183.6	1.6214
300	7.26	1101.3	1181.9	1.6496				
350	7.82	1121.4	1208.2	1.6830	5.80	1118.5	1204.3	1.6476
400	8.35	1140.8	1233.5	1.7134	6.22	1138.5	1230.6	1.6790
500	9.40	1178.6	1283.0	1.7678	7.02	1177.2	1281.1	1.7346
600	10.43	1216.3	1332.1	1.8165	7.79	1215.3	1330.7	1.7838
700	11.44	1254.4	1381.4	1.8609	8.56	1253.6	1380.3	1.8285
800	12.45	1293.0	1431.2	1.9022	9.32	1292.4	1430.4	1.8700
900	13.45	1332.5	1481.8	1.9408	10.08	1332.0	1481.2	1.9087
1000	14.45	1372.7	1533.2	1.9773	10.83	1372.3	1532.6	1.9453
1100	15.45	1413.8	1585.4	2.0119	11.58	1413.5	1584.9	1.9799
1200	16.45	1455.8	1638.5	2.0448	12.33	1455.5	1638.1	2.0130
	100 psia (327.8°F)				**120 psia (341.3°F)**			
Sat.	4.434	1105.8	1187.8	1.6034	3.730	1108.3	1191.1	1.5886
350	4.592	1115.4	1200.4	1.6191	3.783	1112.2	1196.2	1.5950
400	4.934	1136.2	1227.5	1.6517	4.079	1133.8	1224.4	1.6288
450	5.265	1156.2	1253.6	1.6812	4.360	1154.3	1251.2	1.6590
500	5.587	1175.7	1279.1	1.7085	4.633	1174.2	1277.1	1.6868
600	6.216	1214.2	1329.3	1.7582	5.164	1213.2	1327.8	1.7371
700	6.834	1252.8	1379.2	1.8033	5.682	1252.0	1378.2	1.7825
800	7.445	1291.8	1429.6	1.8449	6.195	1291.2	1428.7	1.8243
900	8.053	1331.5	1480.5	1.8838	6.703	1330.9	1479.8	1.8633
1000	8.657	1371.9	1532.1	1.9204	7.208	1371.5	1531.5	1.9000
1100	9.260	1413.1	1584.5	1.9551	7.711	1412.8	1584.0	1.9348
1200	9.861	1455.2	1637.7	1.9882	8.213	1454.9	1637.3	1.9679
	140 psia (353.1°F)				**160 psia (363.6°F)**			
Sat.	3.221	1110.3	1193.8	1.5761	2.836	1112.0	1196.0	1.5651
400	3.466	1131.4	1221.2	1.6088	3.007	1128.8	1217.8	1.5911
450	3.713	1152.4	1248.6	1.6399	3.228	1150.5	1246.1	1.6230
500	3.952	1172.7	1275.1	1.6682	3.440	1171.2	1273.0	1.6518
550	4.184	1192.5	1300.9	1.6945	3.646	1191.3	1299.2	1.6785
600	4.412	1212.1	1326.4	1.7191	3.848	1211.1	1325.0	1.7034
700	4.860	1251.2	1377.1	1.7648	4.243	1250.4	1376.0	1.7494
800	5.301	1290.5	1427.9	1.8068	4.631	1289.9	1427.0	1.7916
900	5.739	1330.4	1479.1	1.8459	5.015	1329.9	1478.4	1.8308
1000	6.173	1371.0	1531.0	1.8827	5.397	1370.6	1530.4	1.8677
1100	6.605	1412.4	1583.6	1.9176	5.776	1412.1	1583.1	1.9026
1200	7.036	1454.6	1636.9	1.9507	6.154	1454.3	1636.5	1.9358

TABLE A-9
(Continued)

Temp., °F	v	u	h	s	v	u	h	s
	180 psia (373.1°F)				**200 psia (381.8°F)**			
Sat.	2.533	1113.4	1197.8	1.5553	2.289	1114.6	1199.3	1.5464
400	2.648	1126.2	1214.4	1.5749	2.361	1123.5	1210.8	1.5600
450	2.850	1148.5	1243.4	1.6078	2.548	1146.4	1240.7	1.5938
500	3.042	1169.6	1270.9	1.6372	2.724	1168.0	1268.8	1.6239
550	3.228	1190.0	1297.5	1.6642	2.893	1188.7	1295.7	1.6512
600	3.409	1210.0	1323.5	1.6893	3.058	1208.9	1322.1	1.6767
700	3.763	1249.6	1374.9	1.7357	3.379	1248.8	1373.8	1.7234
800	4.110	1289.3	1426.2	1.7781	3.693	1288.6	1425.3	1.7660
900	4.453	1329.4	1477.7	1.8174	4.003	1328.9	1477.1	1.8055
1000	4.793	1370.2	1529.8	1.8545	4.310	1369.8	1529.3	1.8425
1100	5.131	1411.7	1582.6	1.8894	4.615	1411.4	1582.2	1.8776
1200	5.467	1454.0	1636.1	1.9227	4.918	1453.7	1635.7	1.9109
	250 psia (401.0°F)				**300 psia (417.4°F)**			
Sat.	1.845	1116.7	1202.1	1.5274	1.544	1118.2	1203.9	1.5115
450	2.002	1141.1	1233.7	1.5632	1.636	1135.4	1226.2	1.5365
500	2.150	1163.8	1263.3	1.5948	1.766	1159.5	1257.5	1.5701
550	2.290	1185.3	1291.3	1.6233	1.888	1181.9	1286.7	1.5997
600	2.426	1206.1	1318.3	1.6494	2.004	1203.2	1314.5	1.6266
700	2.688	1246.7	1371.1	1.6970	2.227	1244.0	1368.3	1.6751
800	2.943	1287.0	1423.2	1.7301	2.442	1285.4	1421.0	1.7187
900	3.193	1327.6	1475.3	1.7799	2.653	1326.3	1473.6	1.7589
1000	3.440	1368.7	1527.9	1.8172	2.860	1367.7	1526.5	1.7964
1100	3.685	1410.5	1581.0	1.8524	3.066	1409.6	1579.8	1.8317
1200	3.929	1453.0	1634.8	1.8858	3.270	1452.2	1633.8	1.8653
1300	4.172	1496.3	1689.3	1.9177	3.473	1495.6	1688.4	1.8973
	350 psia (431.8°F)				**400 psia (444.7°F)**			
Sat.	1.327	1119.0	1204.9	1.4978	1.162	1119.5	1205.5	1.4856
450	1.373	1129.2	1218.2	1.5125	1.175	1122.6	1209.6	1.4901
500	1.491	1154.9	1251.5	1.5482	1.284	1150.1	1245.2	1.5282
550	1.600	1178.3	1281.9	1.5790	1.383	1174.6	1277.0	1.5605
600	1.703	1200.3	1310.6	1.6068	1.476	1197.3	1306.6	1.5892
700	1.898	1242.5	1365.4	1.6562	1.650	1240.4	1362.5	1.6397
800	2.085	1283.8	1418.8	1.7004	1.816	1282.1	1416.6	1.6844
900	2.267	1325.0	1471.8	1.7409	1.978	1323.7	1470.1	1.7252
1000	2.446	1366.6	1525.0	1.7787	2.136	1365.5	1523.6	1.7632
1100	2.624	1408.7	1578.6	1.8142	2.292	1407.8	1577.4	1.7989
1200	2.799	1451.5	1632.8	1.8478	2.446	1450.7	1631.8	1.8327
1300	2.974	1495.0	1687.6	1.8799	2.599	1494.3	1686.8	1.8648

TABLE A-9
(Continued)

Temp., °F	*v*	*u*	*h*	*s*	*v*	*u*	*h*	*s*
	450 psia (456.4°F)				**500 psia (467.1°F)**			
Sat.	1.033	1119.6	1205.6	1.4746	0.928	1119.4	1205.3	1.4645
500	1.123	1145.1	1238.5	1.5097	0.992	1139.7	1231.5	1.4923
550	1.215	1170.7	1271.9	1.5436	1.079	1166.7	1266.6	1.5279
600	1.300	1194.3	1302.5	1.5732	1.158	1191.1	1298.3	1.5585
700	1.458	1238.2	1359.6	1.6248	1.304	1236.0	1356.7	1.6112
800	1.608	1280.5	1414.4	1.6701	1.441	1278.8	1412.1	1.6571
900	1.752	1322.4	1468.3	1.7113	1.572	1321.0	1466.5	1.6987
1000	1.894	1364.4	1522.2	1.7495	1.701	1363.3	1520.7	1.7471
1100	2.034	1406.9	1576.3	1.7853	1.827	1406.0	1575.1	1.7731
1200	2.172	1450.0	1630.8	1.8192	1.952	1449.2	1629.8	1.8072
1300	2.308	1493.7	1685.9	1.8515	2.075	1493.1	1685.1	1.8395
1400	2.444	1538.1	1741.7	1.8823	2.198	1537.6	1741.0	1.8704
	600 psia (486.3°F)				**700 psia (503.2°F)**			
Sat.	0.770	1118.6	1204.1	1.4464	0.656	1117.0	1202.0	1.4305
500	0.795	1128.0	1216.2	1.4592				
550	0.875	1158.2	1255.4	1.4990	0.728	1149.0	1243.2	1.4723
600	0.946	1184.5	1289.5	1.5320	0.793	1177.5	1280.2	1.5081
700	1.073	1231.5	1350.6	1.5872	0.907	1226.9	1344.4	1.5661
800	1.190	1275.4	1407.6	1.6343	1.011	1272.0	1402.9	1.6145
900	1.302	1318.4	1462.9	1.6766	1.109	1315.6	1459.3	1.6576
1000	1.411	1361.2	1517.8	1.7155	1.204	1358.9	1514.9	1.6970
1100	1.517	1404.2	1572.7	1.7519	1.296	1402.4	1570.2	1.7337
1200	1.622	1447.7	1627.8	1.7861	1.387	1446.2	1625.8	1.7682
1300	1.726	1491.7	1683.4	1.8186	1.476	1490.4	1681.7	1.8009
1400	1.829	1536.5	1739.5	1.8497	1.565	1535.3	1738.1	1.8321
	800 psia (518.3°F)				**900 psia (532.1°F)**			
Sat.	0.569	1115.0	1199.3	1.4160	0.501	1112.6	1196.0	1.4027
550	0.615	1138.8	1229.9	1.4469	0.527	1127.5	1215.2	1.4219
600	0.677	1170.1	1270.4	1.4861	0.587	1162.2	1260.0	1.4652
650	0.732	1197.2	1305.6	1.5186	0.639	1191.1	1297.5	1.4999
700	0.783	1222.1	1338.0	1.5471	0.686	1217.1	1331.4	1.5297
800	0.876	1268.5	1398.2	1.5969	0.772	1264.9	1393.4	1.5810
900	0.964	1312.9	1455.6	1.6408	0.851	1310.1	1451.9	1.6257
1000	1.048	1356.7	1511.9	1.6807	0.927	1354.5	1508.9	1.6662
1100	1.130	1400.5	1567.8	1.7178	1.001	1398.7	1565.4	1.7036
1200	1.210	1444.6	1623.8	1.7526	1.073	1443.0	1621.7	1.7386
1300	1.289	1489.1	1680.0	1.7854	1.144	1487.8	1687.3	1.7717
1400	1.367	1534.2	1736.6	1.8167	1.214	1533.0	1735.1	1.8031

TABLE A-9
(Continued)

Temp., °F	v	u	h	s	v	u	h	s
	1000 psia (544.7°F)				**1200 psia (567.4°F)**			
Sat.	0.446	1109.0	1192.4	1.3903	0.362	1103.5	1183.9	1.3673
600	0.514	1153.7	1248.8	1.4450	0.402	1134.4	1223.6	1.4054
650	0.564	1184.7	1289.1	1.4822	0.450	1170.9	1270.8	1.4490
700	0.608	1212.0	1324.6	1.5135	0.491	1201.3	1310.2	1.4837
800	0.688	1261.2	1388.5	1.5665	0.562	1253.7	1378.4	1.5402
900	0.761	1307.3	1448.1	1.6120	0.626	1301.5	1440.4	1.5876
1000	0.831	1352.2	1505.9	1.6530	0.685	1347.5	1499.7	1.6297
1100	0.898	1396.8	1562.9	1.6908	0.743	1393.0	1557.9	1.6682
1200	0.963	1441.5	1619.7	1.7261	0.798	1438.3	1615.5	1.7040
1300	1.027	1486.5	1676.5	1.7593	0.853	1483.8	1673.1	1.7377
1400	1.091	1531.9	1733.7	1.7909	0.906	1529.6	1730.7	1.7696
1600	1.215	1624.4	1849.3	1.8499	1.011	1622.6	1847.1	1.8290
	1400 psia (587.2°F)				**1600 psia (605.1°F)**			
Sat.	0.302	1096.0	1174.1	1.3461	0.255	1087.4	1162.9	1.3258
600	0.318	1110.9	1193.1	1.3641				
650	0.367	1155.5	1250.5	1.4171	0.303	1137.8	1227.4	1.3852
700	0.406	1189.6	1294.8	1.4562	0.342	1177.0	1278.1	1.4299
800	0.471	1245.8	1367.9	1.5168	0.403	1237.7	1357.0	1.4953
900	0.529	1295.6	1432.5	1.5661	0.456	1289.5	1424.4	1.5468
1000	0.582	1342.8	1493.5	1.6094	0.504	1338.0	1487.1	1.5913
1100	0.632	1389.1	1552.8	1.6487	0.549	1385.2	1547.7	1.6315
1200	0.681	1435.1	1611.4	1.6851	0.592	1431.8	1607.1	1.6684
1300	0.728	1481.1	1669.6	1.7192	0.634	1478.3	1666.1	1.7029
1400	0.774	1527.2	1727.8	1.7513	0.675	1524.9	1724.8	1.7354
1600	0.865	1620.8	1844.8	1.8111	0.755	1619.0	1842.6	1.7955
	1800 psia (621.2°F)				**2000 psia (636.0°F)**			
Sat.	0.218	1077.7	1150.4	1.3060	0.188	1066.6	1136.3	1.2861
650	0.251	1117.0	1200.4	1.3517	0.206	1091.1	1167.2	1.3141
700	0.291	1163.1	1259.9	1.4042	0.249	1147.7	1239.8	1.3782
750	0.322	1198.6	1305.9	1.4430	0.280	1187.3	1291.1	1.4216
800	0.350	1229.1	1345.7	1.4753	0.307	1220.1	1333.8	1.4562
900	0.399	1283.2	1416.1	1.5291	0.353	1276.8	1407.6	1.5126
1000	0.443	1333.1	1480.7	1.5749	0.395	1328.1	1474.1	1.5598
1100	0.484	1381.2	1542.5	1.6159	0.433	1377.2	1537.2	1.6017
1200	0.524	1428.5	1602.9	1.6534	0.469	1425.2	1598.6	1.6398
1300	0.561	1475.5	1662.5	1.6883	0.503	1472.7	1659.0	1.6751
1400	0.598	1522.5	1721.8	1.7211	0.537	1520.2	1718.8	1.7082
1600	0.670	1617.2	1840.4	1.7817	0.602	1615.4	1838.2	1.7692

TABLE A-9
(Continued)

Temp., °F	*v*	*u*	*h*	*s*	*v*	*u*	*h*	*s*
	2500 psia (668.3°F)				**3000 psia (695.5°F)**			
Sat.	0.1306	1031.0	1091.4	1.2327	0.0840	968.8	1015.5	1.1575
700	0.1684	1098.7	1176.6	1.3073	0.0977	1003.9	1058.1	1.1944
750	0.2030	1155.2	1249.1	1.3686	0.1483	1114.7	1197.1	1.3122
800	0.2291	1195.7	1301.7	1.4112	0.1757	1167.6	1265.2	1.3675
900	0.2712	1259.9	1385.4	1.4752	0.2160	1241.8	1361.7	1.4414
1000	0.3069	1315.2	1457.2	1.5262	0.2485	1301.7	1439.6	1.4967
1100	0.3393	1366.8	1523.8	1.5704	0.2772	1356.2	1510.1	1.5434
1200	0.3696	1416.7	1587.7	1.6101	0.3086	1408.0	1576.6	1.5848
1300	0.3984	1465.7	1650.0	1.6465	0.3285	1458.5	1640.9	1.6224
1400	0.4261	1514.2	1711.3	1.6804	0.3524	1508.1	1703.7	1.6571
1500	0.4531	1562.5	1772.1	1.7123	0.3754	1557.3	1765.7	1.6896
1600	0.4795	1610.8	1832.6	1.7424	0.3978	1606.3	1827.1	1.7201
	3500 psia				**4000 psia**			
650	0.0249	663.5	679.7	0.8630	0.0245	657.7	675.8	0.8574
700	0.0306	759.5	779.3	0.9506	0.0287	742.1	763.4	0.9345
750	0.1046	1058.4	1126.1	1.2440	0.0633	960.7	1007.5	1.1395
800	0.1363	1134.7	1223.0	1.3226	0.1052	1095.0	1172.9	1.2740
900	0.1763	1222.4	1336.5	1.4096	0.1462	1201.5	1309.7	1.3789
1000	0.2066	1287.6	1421.4	1.4699	0.1752	1272.9	1402.6	1.4449
1100	0.2328	1345.2	1496.0	1.5193	0.1995	1333.9	1481.6	1.4973
1200	0.2566	1399.2	1565.3	1.5624	0.2213	1390.1	1553.9	1.5423
1300	0.2787	1451.1	1631.7	1.6012	0.2414	1443.7	1622.4	1.5823
1400	0.2997	1501.9	1696.1	1.6368	0.2603	1495.7	1688.4	1.6188
1500	0.3199	1552.0	1759.2	1.6699	0.2784	1546.7	1752.8	1.6526
1600	0.3395	1601.7	1821.6	1.7010	0.2959	1597.1	1816.1	1.6841
	4400 psia				**4800 psia**			
650	0.0242	653.6	673.3	0.8535	0.0237	649.8	671.0	0.8499
700	0.0278	732.7	755.3	0.9257	0.0271	725.1	749.1	0.9187
750	0.0415	870.8	904.6	1.0513	0.0352	832.6	863.9	1.0154
800	0.0844	1056.5	1125.3	1.2306	0.0668	1011.2	1070.5	1.1827
900	0.1270	1183.7	1287.1	1.3548	0.1109	1164.8	1263.4	1.3310
1000	0.1552	1260.8	1387.2	1.4260	0.1385	1248.3	1317.4	1.4078
1100	0.1784	1324.7	1469.9	1.4809	0.1608	1315.3	1458.1	1.4653
1200	0.1989	1382.8	1544.7	1.5274	0.1802	1375.4	1535.4	1.5133
1300	0.2176	1437.7	1614.9	1.5685	0.1979	1431.7	1607.4	1.5555
1400	0.2352	1490.7	1682.3	1.6057	0.2143	1485.7	1676.1	1.5934
1500	0.2520	1542.7	1747.6	1.6399	0.2300	1538.2	1742.5	1.6282
1600	0.2681	1593.4	1811.7	1.6718	0.2450	1589.8	1807.4	1.6605

TABLE A-9M

PROPERTIES OF WATER: SUPERHEATED VAPOR (SI UNITS)

v in cm^3/g, $1\ cm^3/g = 10^{-3}\ m^3/kg$; h and u in kJ/kg; s in kJ/kg·K

Temp., °C	v	u	h	s	v	u	h	s
	6 kPa (36.16°C)				**35 kPa (72.69°C)**			
Sat.	23739	2425.0	2546.4	8.3304	4526.	2473.0	2631.4	7.7158
80	27132	2487.3	2650.1	8.5804	4625.	2483.7	2645.6	7.7564
120	30219	2544.7	2726.0	8.7840	5163.	2542.4	2723.1	7.9644
160	33302	2602.7	2802.5	8.9693	5696.	2601.2	2800.6	8.1519
200	36383	2661.4	2879.7	9.1398	6228.	2660.4	2878.4	8.3237
240	39462	2721.0	2957.8	9.2982	6758.	2720.3	2956.8	8.4828
280	42540	2781.5	3036.8	9.4464	7287.	2780.9	3036.0	8.6314
320	45618	2843.0	3116.7	9.5859	7815.	2842.5	3116.1	8.7712
360	48696	2905.5	3197.7	9.7180	8344.	2905.1	3197.1	8.9034
400	51774	2969.0	3279.6	9.8435	8872.	2968.6	3279.2	9.0291
440	54851	3033.5	3362.6	9.9633	9400.	3033.2	3362.2	9.1490
500	59467	3132.3	3489.1	10.134	10192.	3132.1	3488.8	9.3194
	70 kPa (89.95°C)				**100 kPa (99.63°C)**			
Sat.	2365.	2494.5	2660.0	7.4797	1694.	2506.1	2675.5	7.3594
100	2434.	2509.7	2680.0	7.5341	1696.	2506.7	2676.2	7.3614
120	2571.	2539.7	2719.6	7.6375	1793.	2537.3	2716.6	7.4668
160	2841.	2599.4	2798.2	7.8279	1984.	2597.8	2796.2	7.6597
200	3108.	2659.1	2876.7	8.0012	2172.	2658.1	2875.3	7.8343
240	3374.	2719.3	2955.5	8.1611	2359.	2718.5	2954.5	7.9949
280	3640.	2780.2	3035.0	8.3162	2546.	2779.6	3034.2	8.1445
320	3905.	2842.0	3115.3	8.4504	2732.	2841.5	3114.6	8.2849
360	4170.	2904.6	3196.5	8.5828	2917.	2904.2	3195.9	8.4175
400	4434.	2968.2	3278.6	8.7086	3103.	2967.9	3278.2	8.5435
440	4698.	3032.9	3361.8	8.8286	3288.	3032.6	3361.4	8.6636
500	5095.	3131.8	3488.5	8.9991	3565.	3131.6	3488.1	8.8342
	150 kPa (111.37°C)				**300 kPa (133.55°C)**			
Sat.	1159.	2519.7	2693.6	7.2233	606.	2543.6	2725.3	6.9919
120	1188.	2533.3	2711.4	7.2693				
160	1317.	2595.2	2792.8	7.4665	651.	2587.1	2782.3	7.1276
200	1444.	2656.2	2872.9	7.6433	716.	2650.7	2865.5	7.3115
240	1570.	2717.2	2952.7	7.8052	781.	2713.1	2947.3	7.4774
280	1695.	2778.6	3032.8	7.9555	844.	2775.4	3028.6	7.6299
320	1819.	2840.6	3113.5	8.0964	907.	2838.1	3110.1	7.7722
360	1943.	2903.5	3195.0	8.2293	969.	2901.4	3192.2	7.9061
400	2067.	2967.3	3277.4	8.3555	1032.	2965.6	3275.0	8.0330
440	2191.	3032.1	3360.7	8.4757	1094.	3030.6	3358.7	8.1538
500	2376.	3131.2	3487.6	8.6466	1187.	3130.0	3486.0	8.3251
600	2685.	3301.7	3704.3	8.9101	1341.	3300.8	3703.2	8.5892

Source: Abridged from Keenan, J. H., F. G. Keyes, P. G. Hill, and J. G. Moore, "Steam Tables," John Wiley & Sons, Inc., New York, 1969.

TABLE A-9M
(Continued)

Temp., °C	v	u	h	s	v	u	h	s
	500 kPa (151.86°C)				**700 kPa (164.97°C)**			
Sat.	374.9	2561.2	2748.7	6.8213	272.9	2572.5	2763.5	6.7080
180	404.5	2609.7	2812.0	6.9656	284.7	2599.8	2799.1	6.7880
200	424.9	2642.9	2855.4	7.0592	299.9	2634.8	2844.8	6.8865
240	464.6	2707.6	2939.9	7.2307	329.2	2701.8	2932.2	7.0641
280	503.4	2771.2	3022.9	7.3865	357.4	2766.9	3017.1	7.2233
320	541.6	2834.7	3105.6	7.5308	385.2	2831.3	3100.9	7.3697
360	579.6	2898.7	3188.4	7.6660	412.6	2895.8	3184.7	7.5063
400	617.3	2963.2	3271.9	7.7938	439.7	2960.9	3268.7	7.6350
440	654.8	3028.6	3356.0	7.9152	466.7	3026.6	3353.3	7.7571
500	710.9	3128.4	3483.9	8.0873	507.0	3126.8	3481.7	7.9299
600	804.1	3299.6	3701.7	8.3522	573.8	3298.5	3700.2	8.1956
700	896.9	3477.5	3925.9	8.5952	640.3	3476.6	3924.8	8.4391
	1.0 MPa (179.91°C)				**1.5 MPa (198.32°C)**			
Sat.	194.4	2583.6	2778.1	6.5865	131.8	2594.5	2792.2	6.4448
200	206.0	2621.9	2827.9	6.6940	132.5	2598.1	2796.8	6.4546
240	227.5	2692.9	2920.4	6.8817	148.3	2676.9	2899.3	6.6628
280	248.0	2760.2	3008.2	7.0465	162.7	2748.6	2992.7	6.8381
320	267.8	2826.1	3093.9	7.1962	176.5	2817.1	3081.9	6.9938
360	287.3	2891.6	3178.9	7.3349	189.9	2884.4	3169.2	7.1363
400	306.6	2957.3	3263.9	7.4651	203.0	2951.3	3255.8	7.2690
440	325.7	3023.6	3349.3	7.5883	216.0	3018.5	3342.5	7.3940
500	354.1	3124.4	3478.5	7.7622	235.2	3120.3	3473.1	7.5698
540	372.9	3192.6	3565.6	7.8720	247.8	3189.1	3560.9	7.6805
600	401.1	3296.8	3697.9	8.0290	266.8	3293.9	3694.0	7.8385
640	419.8	3367.4	3787.2	8.1290	279.3	3364.8	3783.8	7.9391
	2.0 MPa (212.42°C)				**3.0 MPa (233.90°C)**			
Sat.	99.6	2600.3	2799.5	6.3409	66.7	2604.1	2804.2	6.1869
240	108.5	2659.6	2876.5	6.4952	68.2	2619.7	2824.3	6.2265
280	120.0	2736.4	2976.4	6.6828	77.1	2709.9	2941.3	6.4462
320	130.8	2807.9	3069.5	6.8452	85.0	2788.4	3043.4	6.6245
360	141.1	2877.0	3159.3	6.9917	92.3	2861.7	3138.7	6.7801
400	151.2	2945.2	3247.6	7.1271	99.4	2932.8	3230.9	6.9212
440	161.1	3013.4	3335.5	7.2540	106.2	3002.9	3321.5	7.0520
500	175.7	3116.2	3467.6	7.4317	116.2	3108.0	3456.5	7.2338
540	185.3	3185.6	3556.1	7.5434	122.7	3178.4	3546.6	7.3474
600	199.6	3290.9	3690.1	7.7024	132.4	3285.0	3682.3	7.5085
640	209.1	3362.2	3780.4	7.8035	138.8	3357.0	3773.5	7.6106
700	223.2	3470.9	3917.4	7.9487	148.4	3466.5	3911.7	7.7571

TABLE A-9M
(Continued)

Temp., °C	*v*	*u*	*h*	*s*	*v*	*u*	*h*	*s*
	4.0 MPa (250.40°C)				**6.0 MPa (275.64°C)**			
Sat.	49.78	2602.3	2801.4	6.0701	32.44	2589.7	2784.3	5.8892
280	55.46	2680.0	2901.8	6.2568	33.17	2605.2	2804.2	5.9252
320	61.99	2767.4	3015.4	6.4553	38.76	2720.0	2952.6	6.1846
360	67.88	2845.7	3117.2	6.6215	43.31	2811.2	3071.1	6.3782
400	73.41	2919.9	3213.6	6.7690	47.39	2892.9	3177.2	6.5408
440	78.72	2992.2	3307.1	6.9041	51.22	2970.0	3277.3	6.6853
500	86.43	3099.5	3445.3	7.0901	56.65	3082.2	3422.2	6.8803
540	91.45	3171.1	3536.9	7.2056	60.15	3156.1	3517.0	6.9999
600	98.85	3279.1	3674.4	7.3688	65.25	3266.9	3658.4	7.1677
640	103.7	3351.8	3766.6	7.4720	68.59	3341.0	3752.6	7.2731
700	111.0	3462.1	3905.9	7.6198	73.52	3453.1	3894.1	7.4234
740	115.7	3536.6	3999.6	7.7141	76.77	3528.3	3989.2	7.5190
	8.0 MPa (295.06°C)				**10.0 MPa (311.06°C)**			
Sat.	23.52	2569.8	2758.0	5.7432	18.03	2544.4	2724.7	5.6141
320	26.82	2662.7	2877.2	5.9489	19.25	2588.8	2781.3	5.7103
360	30.89	2772.7	3019.8	6.1819	23.31	2729.1	2962.1	6.0060
400	34.32	2863.8	3138.3	6.3634	26.41	2832.4	3096.5	6.2120
440	37.42	2946.7	3246.1	6.5190	29.11	2922.1	3213.2	6.3805
480	40.34	3025.7	3348.4	6.6586	31.60	3005.4	3321.4	6.5282
520	43.13	3102.7	3447.7	6.7871	33.94	3085.6	3425.1	6.6622
560	45.82	3178.7	3545.3	6.9072	36.19	3164.1	3526.0	6.7864
600	48.45	3254.4	3642.0	7.0206	38.37	3241.7	3625.3	6.9029
640	51.02	3330.1	3738.3	7.1283	40.48	3318.9	3723.7	7.0131
700	54.81	3443.9	3882.4	7.2812	43.58	3434.7	3870.5	7.1687
740	57.29	3520.4	3978.7	7.3782	45.60	3512.1	3968.1	7.2670
	12.0 MPa (324.75°C)				**14.0 MPa (336.75°C)**			
Sat.	14.26	2513.7	2684.9	5.4924	11.49	2476.8	2637.6	5.3717
360	18.11	2678.4	2895.7	5.8361	14.22	2617.4	2816.5	5.6602
400	21.08	2798.3	3051.3	6.0747	17.22	2760.9	3001.9	5.9448
440	23.55	2896.1	3178.7	6.2586	19.54	2868.6	3142.2	6.1474
480	25.76	2984.4	3293.5	6.4154	21.57	2962.5	3264.5	6.3143
520	27.81	3068.0	3401.8	6.5555	23.43	3049.8	3377.8	6.4610
560	29.77	3149.0	3506.2	6.6840	25.17	3133.6	3486.0	6.5941
600	31.64	3228.7	3608.3	6.8037	26.83	3215.4	3591.1	6.7172
640	33.45	3307.5	3709.0	6.9164	28.43	3296.0	3694.1	6.8326
700	36.10	3425.2	3858.4	7.0749	30.75	3415.7	3846.2	6.9939
740	37.81	3503.7	3957.4	7.1746	32.25	3495.2	3946.7	7.0952

TABLE A-9M
(Continued)

Temp., °C	*v*	*u*	*h*	*s*	*v*	*u*	*h*	*s*
	16.0 MPa (347.44°C)				**18.0 MPa (357.06°C)**			
Sat.	9.31	2431.7	2580.6	5.2455	7.49	2374.3	2509.1	5.1044
360	11.05	2539.0	2715.8	5.4614	8.09	2418.9	2564.5	5.1922
400	14.26	2719.4	2947.6	5.8175	11.90	2672.8	2887.0	5.6887
440	16.52	2839.4	3103.7	6.0429	14.14	2808.2	3062.8	5.9428
480	18.42	2939.7	3234.4	6.2215	15.96	2915.9	3203.2	6.1345
520	20.13	3031.1	3353.3	6.3752	17.57	3011.8	3378.0	6.2960
560	21.72	3117.8	3465.4	6.5132	19.04	3101.7	3444.4	6.4392
600	23.23	3201.8	3573.5	6.6399	20.42	3188.0	3555.6	6.5696
640	24.67	3284.2	3678.9	6.7580	21.74	3272.3	3663.6	6.6905
700	26.74	3406.0	3833.9	6.9224	23.62	3396.3	3821.5	6.8580
740	28.08	3486.7	3935.9	7.0251	24.83	3478.0	3925.0	6.9623
	20.0 MPa (365.81°C)				**24.0 MPa**			
Sat.	5.83	2293.0	2409.7	4.9269				
400	9.94	2619.3	2818.1	5.5540	6.73	2477.8	2639.4	5.2393
440	12.22	2774.9	3019.4	5.8450	9.29	2700.6	2923.4	5.6506
480	13.99	2891.2	3170.8	6.0518	11.00	2838.3	3102.3	5.8950
520	15.51	2992.0	3302.2	6.2218	12.41	2950.5	3248.5	6.0842
560	16.89	3085.2	3423.0	6.3705	13.66	3051.1	3379.0	6.2448
600	18.18	3174.0	3537.6	6.5048	14.81	3145.2	3500.7	6.3875
640	19.40	3260.2	3648.1	6.6286	15.88	3235.5	3616.7	6.5174
700	21.13	3386.4	3809.0	6.7993	17.39	3366.4	3783.8	6.6947
740	22.24	3469.3	3914.1	6.9052	18.35	3451.7	3892.1	6.8038
800	23.85	3592.7	4069.7	7.0544	19.74	3578.0	4051.6	6.9567
	28.0 MPa				**32.0 MPa**			
400	3.83	2223.5	2330.7	4.7494	2.36	1980.4	2055.9	4.3239
440	7.12	2613.2	2812.6	5.4494	5.44	2509.0	2683.0	5.2327
480	8.85	2780.8	3028.5	5.7446	7.22	2718.1	2949.2	5.5968
520	10.20	2906.8	3192.3	5.9566	8.53	2860.7	3133.7	5.8357
560	11.36	3015.7	3333.7	6.1307	9.63	2979.0	3287.2	6.0246
600	12.41	3115.6	3463.0	6.2823	10.61	3085.3	3424.6	6.1858
640	13.38	3210.3	3584.8	6.4187	11.50	3184.5	3552.5	6.3290
700	14.73	3346.1	3758.4	6.6029	12.73	3325.4	3732.8	6.5203
740	15.58	3433.9	3870.0	6.7153	13.50	3415.9	3847.8	6.6361
800	16.80	3563.1	4033.4	6.8720	14.60	3548.0	4015.1	6.7966
900	18.73	3774.3	4298.8	7.1084	16.33	3762.7	4285.1	7.0372

TABLE A-10
PROPERTIES OF WATER: COMPRESSED LIQUID TABLE (ENGLISH UNITS)

v in ft³/lbm; h and u in Btu/lbm; s in Btu/lbm·°R

Temp., °F	500 psia (T_{sat} = 467.1°F)				1000 psia (T_{sat} = 544.7°F)			
	v	u	h	s	v	u	h	s
32	0.015994	0.00	1.49	0.00000	0.015967	0.03	2.99	0.00005
50	0.015998	18.02	19.50	0.03599	0.015972	17.99	20.94	0.03592
100	0.016106	67.87	69.36	0.12932	0.016082	67.70	70.68	0.12901
150	0.016318	117.66	119.17	0.21457	0.016293	117.38	120.40	0.21410
200	0.016608	167.65	169.19	0.29341	0.016580	167.26	170.32	0.29281
300	0.017416	268.92	270.53	0.43641	0.017379	268.24	271.46	0.43552
400	0.018608	373.68	375.40	0.56604	0.018550	372.55	375.98	0.56472
Sat.	0.019748	447.70	449.53	0.64904	0.021591	538.39	542.38	0.74320
	1500 psia (T_{sat} = 596.4°F)				**2000 psia (T_{sat} = 636.0°F)**			
32	0.015939	0.05	4.47	0.00007	0.015912	0.06	5.95	0.00008
50	0.015946	17.95	22.38	0.03584	0.015920	17.91	23.81	0.03575
100	0.016058	67.53	71.99	0.12870	0.016034	67.37	73.30	0.12839
150	0.016268	117.10	121.62	0.21364	0.016244	116.83	122.84	0.21318
200	0.016554	166.87	171.46	0.29221	0.016527	166.49	172.60	0.29162
300	0.017343	267.58	272.39	0.43463	0.017308	266.93	273.33	0.43376
400	0.018493	371.45	376.59	0.56343	0.018439	370.38	377.21	0.56216
500	0.02024	481.8	487.4	0.6853	0.02014	479.8	487.3	0.6832
Sat.	0.02346	605.0	611.5	0.8082	0.02565	662.4	671.9	0.8623
	3000 psia (T_{sat} = 695.5°F)				**4000 psia**			
32	0.015859	0.09	8.90	0.00009	0.015807	0.10	11.80	0.00005
50	0.015870	17.84	26.65	0.03555	0.015821	17.76	29.47	0.03534
100	0.015987	67.04	75.91	0.12777	0.015942	66.72	78.52	0.12714
150	0.016196	116.30	125.29	0.21226	0.016150	115.77	127.73	0.21136
200	0.016476	165.74	174.89	0.29046	0.016425	165.02	177.18	0.28931
300	0.017240	265.66	275.23	0.43205	0.017174	264.43	277.15	0.43038
400	0.018334	368.32	378.50	0.55970	0.018235	366.35	379.85	0.55734
500	0.019944	476.2	487.3	0.6794	0.019766	472.9	487.5	0.6758
Sat.	0.034310	783.5	802.5	0.9732				

Source: Abridged from Keenan, J. H., F. G. Keyes, P. G. Hill, and J. G. Moore, "Steam Tables," John Wiley & Sons, Inc., New York, 1969.

TABLE A-10M

PROPERTIES OF WATER: COMPRESSED LIQUID TABLE (SI UNITS)

v in cm³/g, 1 cm³/g = 10^{-3} m³/kg; *h* and *u* in kJ/kg; *s* in kJ/kg·K

Temp., C	*v*	*u*	*h*	*s*	*v*	*u*	*h*	*s*
	2.5 MPa (223.99°C)				**5.0 MPa (263.99°C)**			
20	1.0006	83.80	86.30	0.2961	0.9995	83.65	88.65	0.2956
40	1.0067	167.25	169.77	0.5715	1.0056	166.95	171.97	0.5705
80	1.0280	334.29	336.86	1.0737	1.0268	333.72	338.85	1.0720
120	1.0590	502.68	505.33	1.5255	1.0576	501.80	507.09	1.5233
160	1.1006	673.90	676.65	1.9404	1.0988	672.62	678.12	1.9375
200	1.1555	849.9	852.8	2.3294	1.1530	848.1	848.1	2.3255
220	1.1898	940.7	943.7	2.5174	1.1866	938.4	944.4	2.5128
Sat.	1.1973	959.1	962.1	2.5546	1.2859	1147.8	1154.2	2.9202
	7.5 MPa (290.59°C)				**10.0 MPa (311.06°C)**			
20	0.9984	83.50	90.99	0.2950	0.9972	83.36	93.33	0.2945
40	1.0045	166.64	174.18	0.5696	1.0034	166.35	176.38	0.5686
80	1.0256	333.15	340.84	1.0704	1.0245	332.59	342.83	1.0688
100	1.0397	416.81	424.62	1.3011	1.0385	416.12	426.50	1.2992
140	1.0752	585.72	593.78	1.7317	1.0737	584.68	595.42	1.7292
180	1.1219	758.13	766.55	2.1308	1.1199	756.65	767.84	2.1275
220	1.8135	936.2	945.1	2.5083	1.1805	934.1	945.9	2.5039
260	1.2696	1124.4	1134.0	2.8763	1.2645	1121.1	1133.7	2.8699
Sat.	1.3677	1282.0	1292.2	3.1649	1.4524	1393.0	1407.6	3.3596
	15.0 MPa (342.24°C)				**20.0 MPa (365.81°C)**			
20	0.9950	83.06	97.99	0.2934	0.9928	82.77	102.62	0.2923
40	1.0013	165.76	180.78	0.5666	0.9992	165.17	185.16	0.5646
100	1.0361	414.75	430.28	1.2955	1.0337	413.39	434.06	1.2917
180	1.1159	753.76	770.50	2.1210	1.1120	750.95	773.20	2.1147
220	1.1748	929.9	947.5	2.4953	1.1693	925.9	949.3	2.4870
260	1.2550	1114.6	1133.4	2.8576	1.2462	1108.6	1133.5	2.8459
300	1.3770	1316.6	1337.3	3.2260	1.3596	1306.1	1333.3	3.2071
Sat.	1.6581	1585.6	1610.5	3.6848	2.036	1785.6	1826.3	4.0139
	25.0 MPa				**30.0 MPa**			
20	0.9907	82.47	107.24	0.2911	0.9886	82.17	111.84	0.2899
40	0.9971	164.60	189.52	0.5626	0.9951	164.04	193.89	0.5607
100	1.0313	412.08	437.85	1.2881	1.0290	410.78	441.66	1.2844
200	1.1344	834.5	862.8	2.2961	1.1302	831.4	865.3	2.2893
300	1.3442	1296.6	1330.2	3.1900	1.3304	1287.9	1327.8	3.1741

Source: Abridged from Keenan, J. H., F. G. Keyes, P. G. Hill, and J. G. Moore, "Steam Tables," John Wiley & Sons, Inc., New York, 1969.

TABLE A-11

PROPERTIES OF WATER: COMPRESSED-LIQUID DEVIATIONS FROM SATURATED LIQUID (ENGLISH UNITS)

Abs. Press., lbf/in² (sat. temp.)	Saturated Liquid	Temperature, °F 32	100	200	300	400	500
		p 0.088 54 v_f 0.016 022 h_f 0 s_f 0	p 0.9492 v_f 0.016 132 h_f 67.97 s_f 0.129 48	p 11.526 v_f 0.016 634 h_f 167.99 s_f 0.293 82	p 67.013 v_f 0.017 449 h_f 269.59 s_f 0.436 94	p 247.31 v_f 0.018 639 h_f 374.97 s_f 0.566 38	p 680.8 v_f 0.020 432 h_f 487.82 s_f 0.688 71
200 (381.79)	$(v - v_f) \times 10^5$	−1.2	−1.2	−0.4	−0.9		
	$h - h_f$	+0.57	+0.53	+0.50	+0.35		
	$(s - s_f) \times 10^3$	0	−0.08	−0.20	−0.20		
400 (444.59)	$(v - v_f) \times 10^5$	−2.2	−2.2	−1.4	−1.9	−1.9	
	$h - h_f$	+1.17	+1.06	+0.96	+0.72	+0.28	
	$(s - s_f) \times 10^3$	0	−0.18	−0.32	−0.34	−0.08	
800 (518.23)	$(v - v_f) \times 10^5$	−4.2	−4.2	−4.4	−4.9	−6.9	−2.2
	$h - h_f$	+2.38	+2.12	+1.87	+1.46	+0.74	+0.04
	$(s - s_f) \times 10^3$	+0.1	−0.48	−0.82	−1.14	−1.18	−0.21
1000 (544.61)	$(v - v_f) \times 10^5$	−5.2	−5.2	−5.4	−6.9	−8.9	−7.2
	$h - h_f$	+2.98	+2.64	+2.32	+1.83	+0.97	−0.10
	$(s - s_f) \times 10^3$	+0.1	−0.58	−1.02	−1.44	−1.68	−1.11
1500 (586.23)	$(v - v_f) \times 10^5$	−8.2	−7.2	−8.4	−9.9	−14.9	−18.2
	$h - h_f$	+4.48	+3.96	+3.46	+2.77	+1.57	−0.19
	$(s - s_f) \times 10^3$	+0.1	−0.88	−1.62	−2.34	−2.98	−3.21

Source: Differential values calculated from tabulated data for compressed water in "Thermodynamics and Transport Properties of Stream," American Society of Mechanical Engineers, New York, 1967.

TABLE A-12
PROPERTIES OF SATURATED AMMONIA: TEMPERATURE TABLE (ENGLISH UNITS)

		Specific Volume, ft³/lbm		Enthalpy, Btu/lbm			Entropy, Btu/lbm·°R		
Temp. t, °F	Press. p, psia	Sat. Liquid v_f	Sat. Vapor v_g	Sat. Liquid h_f	Evap. h_{fg}	Sat. Vapor h_g	Sat. Liquid s_f	Evap. s_{fg}	Sat. Vapor s_g
−60	5.55	0.022 78	44.73	−21.2	610.8	589.6	−0.0517	1.5286	1.4769
−50	7.67	0.022 99	33.08	−10.6	604.3	593.7	−0.0256	1.4753	1.4497
−40	10.41	0.023 22	24.86	0.0	597.6	597.6	0.0000	1.4242	1.4242
−30	13.90	0.023 45	18.97	10.7	590.7	601.4	0.0250	1.3751	1.4001
−20	18.30	0.023 69	14.68	21.4	583.6	605.0	0.0497	1.3277	1.3774
−10	23.74	0.023 93	11.50	32.1	576.4	608.5	0.0738	1.2820	1.3558
0	30.42	0.024 19	9.116	42.9	568.9	611.8	0.0975	2.2377	1.3352
5	34.27	0.024 32	8.150	48.3	565.0	613.3	0.1092	1.2161	1.3253
10	38.51	0.024 46	7.304	53.8	561.1	614.9	0.1208	1.1949	1.3157
20	48.21	0.024 74	5.910	64.7	553.1	617.8	0.1437	1.1532	1.2969
30	59.74	0.025 03	4.825	75.7	544.8	620.5	0.1663	1.1127	1.2790
40	73.32	0.025 33	3.971	86.8	536.2	623.0	0.1885	1.0733	1.2618
50	89.19	0.025 64	3.294	97.9	527.3	625.2	0.2105	1.0348	1.2453
60	107.6	0.025 97	2.751	109.2	518.1	627.3	0.2322	0.9972	1.2294
70	128.8	0.026 32	2.312	120.5	508.6	629.1	0.2537	0.9603	1.2140
80	153.0	0.026 68	1.955	132.0	498.7	630.7	0.2749	0.9242	1.1991
86	169.2	0.026 91	1.772	138.9	492.6	631.5	0.2875	0.9029	1.1904
90	180.6	0.027 07	1.661	143.5	488.5	632.0	0.2958	0.8888	1.1846
100	211.9	0.027 47	1.419	155.2	477.8	633.0	0.3166	0.8539	1.1705
110	247.0	0.027 90	1.217	167.0	466.7	633.7	0.3372	0.8194	1.1566
120	286.4	0.028 36	1.047	179.0	455.0	634.0	0.3576	0.7851	1.1427

Source: Data from Tables of Thermodynamic Properties of Ammonia, *Natl. Bur. Std. Circ.* 142, 1945.

TABLE A-13
PROPERTIES OF SATURATED AMMONIA: PRESSURE TABLE (ENGLISH UNITS)

		Specific Volume, ft^3/lbm		Enthalpy, Btu/lbm			Entropy, Btu/lbm·R		
Press. p, psia	Temp. t, °F	Sat. Liquid v_f	Sat. Vapor v_g	Sat. Liquid h_f	Evap. h_{fg}	Sat. Vapor h_g	Sat. Liquid s_f	Evap. s_{fg}	Sat. Vapor s_g
5	−63.11	0.022 71	49.31	−24.5	612.8	588.3	−0.0599	1.5456	1.4857
10	−41.34	0.023 19	25.81	−1.4	598.5	597.1	−0.0034	1.4310	1.4276
15	−27.29	0.023 51	17.67	13.6	588.8	602.4	0.0318	1.3620	1.3938
20	−16.64	0.023 77	13.50	25.0	581.2	606.2	0.0578	1.3122	1.3700
30	0.57	0.024 17	9.236	42.3	569.3	611.6	0.0962	1.2402	1.3364
40	11.66	0.024 51	7.047	55.6	559.8	615.4	0.1246	1.1879	1.3125
50	21.67	0.024 79	5.710	66.5	551.7	618.2	0.1475	1.1464	1.2939
60	30.21	0.025 04	4.805	75.9	544.6	620.5	0.1668	1.1119	1.2787
80	44.40	0.025 46	3.655	91.7	532.3	624.0	0.1982	1.0563	1.2545
100	56.05	0.025 84	2.952	104.7	521.8	626.5	0.2237	1.0119	1.2356
120	66.02	0.026 18	2.476	116.0	512.4	628.4	0.2452	0.9749	1.2201
140	74.79	0.026 49	2.132	126.0	503.9	629.9	0.2638	0.9430	1.2068
170	86.29	0.026 92	1.764	139.3	492.3	631.6	0.2881	0.9019	1.1900
200	96.34	0.027 32	1.502	150.9	481.8	632.7	0.3090	0.8666	1.1756
230	105.30	0.027 70	1.307	161.4	472.0	633.4	0.3275	0.8356	1.1631
260	113.42	0.028 06	1.155	171.1	462.8	633.9	0.3441	0.8077	1.1518

Source: Data from Tables of Thermodynamic Properties of Ammonia, *Natl. Bur. Std. (U.S.) Circ.* 142, 1945.

TABLE A-14
PROPERTIES OF SUPERHEATED AMMONIA (ENGLISH UNITS)

Abs. Press., psia (sat. temp.)		Temperature, °F											
		0	20	40	60	80	100	120	140	160	180	200	220
	v	28.58	29.90	31.20	32.49	33.78	35.07	36.35	37.62	38.90	40.17	41.45	
10	h	618.9	629.1	639.3	649.5	659.7	670.0	680.3	690.6	701.1	711.6	722.2	
(−41.34)	s	1.477	1.499	1.520	1.540	1.559	1.578	1.596	1.614	1.631	1.647	1.664	
	v	18.92	19.82	20.70	21.58	22.44	23.31	24.17	25.03	25.88	26.74	27.59	
15	h	617.2	627.8	638.2	648.5	658.9	669.2	679.6	690.0	700.5	711.1	721.7	
(−27.29)	s	1.427	1.450	1.471	1.491	1.511	1.529	1.548	1.566	1.583	1.599	1.616	
	v	14.09	14.78	15.45	16.12	16.78	17.43	18.08	18.73	19.37	20.02	20.66	21.3
20	h	615.5	626.4	637.0	647.5	658.0	668.5	678.9	689.4	700.0	710.6	721.2	732.0
(−16.64)	s	1.391	1.414	1.436	1.456	1.476	1.495	1.513	1.531	1.549	1.565	1.582	1.598
	v	11.19	11.75	12.30	12.84	13.37	13.90	14.43	14.95	15.47	15.99	16.50	17.02
25	h	613.8	625.0	635.8	646.5	657.1	667.7	678.2	688.8	699.4	710.1	720.8	731.6
(−7.96)	s	1.362	1.386	1.408	1.429	1.449	1.468	1.486	1.504	1.522	1.539	1.555	1.571
	v	9.25	9.731	10.20	10.65	11.10	11.55	11.99	12.43	12.87	13.30	13.73	14.16
30	h	611.9	623.5	634.6	645.5	656.2	666.9	677.5	688.2	698.8	709.6	720.3	731.1
(−0.57)	s	1.337	1.362	1.385	1.406	1.426	1.446	1.464	1.482	1.500	1.517	1.533	1.550
	v		8.287	8.695	9.093	9.484	9.869	10.25	10.63	11.00	11.38	11.75	12.12
35	h		622.0	633.4	644.4	655.3	666.1	676.8	687.6	698.3	709.1	719.9	730.7
(5.89)	s		1.341	1.365	1.386	1.407	1.427	1.445	1.464	1.481	1.498	1.515	1.531
	v		7.203	7.568	7.922	8.268	8.609	8.945	9.278	9.609	9.938	10.27	10.59
40	h		620.4	632.1	643.4	654.4	665.3	676.1	686.9	697.7	708.5	719.4	730.3
(11.66)	s		1.323	1.347	1.369	1.390	1.410	1.429	1.447	1.465	1.482	1.499	1.515
	v		6.213	6.538	6.851	7.157	7.457	7.753	8.045	8.335	8.623	8.909	9.194
45	h		618.5	630.5	642.1	653.3	664.4	675.3	686.2	697.1	707.9	718.8	729.8
(17.87)	s		1.304	1.328	1.351	1.372	1.392	1.411	1.430	1.448	1.465	1.482	1.498
	v			5.988	6.280	6.564	6.843	7.117	7.387	7.655	7.921	8.185	8.448
50	h			629.5	641.2	652.6	663.7	674.7	685.7	696.6	707.5	718.5	729.4
(21.67)	s			1.317	1.340	1.361	1.382	1.401	1.420	1.437	1.455	1.472	1.488
	v			4.933	5.184	5.428	5.665	5.897	6.126	6.352	6.576	6.798	7.019
60	h			626.8	639.0	650.7	662.1	673.3	684.4	695.5	706.5	717.5	738.6
(30.21)	s			1.2913	1.3152	1.3373	1.3581	1.3778	1.3966	1.4148	1.4323	1.4493	1.4658

		Temperature, °F											
		60	80	100	120	140	160	180	200	240	280	320	360
	v	4.401	4.615	4.822	5.025	5.224	5.420	5.615	5.807	6.187	6.563		
70	h	636.6	648.7	660.4	671.8	683.1	694.3	705.5	716.6	738.9	761.4		
(37.7)	s	1.294	1.317	1.338	1.358	1.377	1.395	1.413	1.430	1.463	1.494		
	v	3.812	4.005	4.190	4.371	4.548	4.722	4.893	5.063	5.398	5.73		
80	h	634.3	646.7	658.7	670.4	681.8	693.2	704.4	715.6	738.1	760.7		
(44.4)	s	1.275	1.298	1.320	1.340	1.360	1.378	1.396	1.414	1.447	1.478		
	v	3.353	3.529	3.698	3.862	4.021	4.178	4.332	4.484	4.785	5.081		
90	h	631.8	644.7	657.0	668.9	680.5	692.0	703.4	714.7	737.3	760.0		
(50.47)	s	1.257	1.281	1.304	1.325	1.344	1.363	1.381	1.400	1.432	1.464		
	v	2.985	3.149	3.304	3.454	3.600	3.743	3.883	4.021	4.294	4.562		
100	h	629.3	642.6	655.2	667.3	679.2	690.8	702.3	713.7	736.5	759.4		
(56.05)	s	1.241	1.266	1.289	1.310	1.331	1.349	1.368	1.385	1.419	1.451		
	v		2.166	2.288	2.404	2.515	2.622	2.727	2.830	3.030	3.227	3.420	
140	h		633.8	647.8	661.1	673.7	686.0	698.0	709.9	733.3	756.7	780.0	
(74.79)	s		1.214	1.240	1.263	1.284	1.305	1.324	1.342	1.376	1.409	1.440	
	v			1.720	1.818	1.910	1.999	2.084	2.167	2.328	2.484	2.637	
180	h			639.9	654.4	668.0	681.0	693.6	705.9	730.1	753.9	777.7	
(89.78)	s			1.199	1.225	1.248	1.269	1.289	1.308	1.344	1.377	1.408	
	v				1.443	1.525	1.601	1.675	1.745	1.881	2.012	2.140	2.265
220	h				647.3	662.0	675.8	689.1	701.9	726.8	751.1	775.3	799.5
(102.42)	s				1.192	1.217	1.239	1.260	1.280	1.317	1.351	1.383	1.413
	v				1.302	1.380	1.452	1.521	1.587	1.714	1.835	1.954	2.069
240	h				643.5	658.8	673.1	686.7	699.8	725.1	749.8	774.1	798.4
(108.09)	s				1.176	1.203	1.226	1.248	1.268	1.305	1.339	1.371	1.402
	v				1.182	1.257	1.326	1.391	1.453	1.572	1.686	1.796	1.904
260	h				639.5	655.6	670.4	684.4	697.7	723.4	748.4	772.9	797.4
(113.42)	s				1.162	1.189	1.213	1.235	1.256	1.294	1.329	1.361	1.391
	v				1.078	1.151	1.217	1.279	1.339	1.451	1.558	1.661	1.762
280	h				635.4	652.2	667.6	681.9	695.6	721.8	747.0	771.7	796.3
(118.45)	s				1.147	1.176	1.201	1.224	1.245	1.283	1.318	1.351	1.382

Source: Data from Tables of Thermodynamic Properties of Ammonia, *Natl. Bur. Std. (U.S.) Circ.* 142, 1945.

TABLE A-15

PROPERTIES OF FREON 12 (DICHLORODIFLUOROMETHANE): SATURATION TEMPERATURE TABLE (ENGLISH UNITS)

		Specific volume, ft³/lbm			Enthalpy, Btu/lbm			Entropy, Btu/lbm·°R		
Temp. t, °F	Abs. Press. p, psia	Sat. Liquid v_f	Evap. v_{fg}	Sat. Vapor v_g	Sat. Liquid h_f	Evap. h_{fg}	Sat. Vapor h_g	Sat. Liquid s_f	Evap. s_{fg}	Sat. Vapor s_g
−130	0.41224	0.009 736	70.7203	70.730	−18.609	81.577	62.968	−0.049 83	0.247 43	0.197 60
−120	0.64190	0.009 816	46.7312	46.741	−16.565	80.617	64.052	−0.043 72	0.237 31	0.193 59
−110	0.97034	0.009 899	31.7671	31.777	−14.518	79.663	65.145	−0.037 79	0.227 80	0.190 02
−100	1.4280	0.009 985	21.1541	22.164	−12.466	78.714	66.248	−0.032 00	0.218 83	0.186 83
−90	2.0509	0.010 073	15.8109	15.821	−10.409	77.764	67.355	−0.026 37	0.210 34	0.183 98
−80	2.8807	0.010 164	11.5228	11.533	−8.3451	76.812	68.467	−0.020 86	0.202 29	0.181 43
−70	3.9651	0.010 259	8.5584	8.5687	−6.2730	75.853	69.580	−0.015 48	0.194 64	0.179 16
−60	5.3575	0.010 357	6.4670	6.4774	−4.1919	74.885	70.693	−0.010 21	0.187 16	0.177 14
−50	7.1168	0.010 459	4.9637	4.9742	−2.1011	73.906	71.805	−0.005 06	0.180 38	0.175 33
−40	9.3076	0.010 564	3.8644	3.8750	0	72.913	72.913	0	0.173 73	0.173 73
−30	11.999	0.010 674	3.0478	3.0585	2.1120	71.903	74.015	0.004 96	0.167 33	0.172 29
−20	15.267	0.010 788	2.4321	2.4429	4.2357	70.874	75.110	0.009 83	0.161 19	0.171 02
−10	19.189	0.010 906	1.9628	1.9727	6.3716	69.824	76.196	0.014 62	0.155 27	0.169 89
0	23.849	0.011 030	1.5979	1.6089	8.5207	68.750	77.271	0.019 32	0.149 56	0.168 88
10	29.335	0.011 160	1.3129	1.3241	10.684	67.651	78.335	0.023 95	0.144 03	0.167 98
20	35.736	0.011 296	1.0875	1.0988	12.863	66.522	79.385	0.028 52	0.138 67	0.167 19
30	43.148	0.011 438	0.907 36	0.91880	15.058	65.361	80.419	0.033 01	0.133 47	0.166 48

40	51.667	0.011 588	0.761 98	0.77357	17.273	64.163	81.436	0.037 45	0.128 41	0.165 86
50	61.394	0.011 746	0.643 62	0.65537	19.507	62.926	82.433	0.041 84	0.123 46	0.165 30
60	72.433	0.011 913	0.546 48	0.55839	21.766	61.643	83.409	0.046 18	0.118 61	0.164 79
70	84.888	0.012 089	0.466 09	0.47818	24.050	60.309	84.359	0.050 48	0.113 86	0.164 34
80	98.870	0.012 277	0.399 07	0.411 35	26.365	58.917	85.282	0.054 75	0.109 17	0.163 92
90	114.49	0.012 478	0.342 81	0.355 29	28.713	57.461	86.174	0.059 00	0.104 53	0.163 53
100	131.86	0.012 693	0.295 25	0.307 94	31.100	55.929	87.029	0.063 23	0.099 92	0.163 15
110	151.11	0.012 924	0.255 77	0.267 69	33.531	54.313	87.844	0.067 45	0.095 34	0.162 79
120	172.35	0.013 174	0.220 19	0.233 26	36.013	52.597	88.610	0.071 68	0.090 73	0.162 41
130	195.71	0.013 447	0.190 19	0.203 64	38.553	50.768	89.321	0.075 83	0.086 09	0.162 02
140	221.32	0.013 746	0.164 24	0.177 99	41.162	48.805	89.967	0.080 21	0.081 38	0.161 59
150	249.31	0.014 078	0.141 56	0.155 64	43.850	46.684	90.534	0.084 53	0.076 57	0.161 10
160	279.82	0.014 449	0.121 59	0.136 04	46.633	44.373	91.006	0.088 93	0.072 60	0.160 53
170	313.00	0.014 871	0.103 86	0.118 73	49.529	41.830	91.359	0.093 42	0.066 43	0.159 85
180	349.00	0.015 360	0.087 94	0.103 30	52.562	38.999	91.561	0.098 04	0.060 96	0.159 00
190	387.98	0.015 942	0.073 476	0.089 418	55.769	35.792	91.561	0.102 84	0.055 11	0.157 93
200	430.09	0.016 659	0.060 069	0.076 728	59.203	32.075	91.278	0.107 89	0.048 62	0.156 51
210	475.52	0.017 601	0.047 242	0.064 843	62.959	27.599	90.558	0.113 32	0.039 21	0.154 53
220	524.43	0.018 986	0.035 154	0.053 140	67.246	21.790	89.036	0.119 43	0.032 06	0.151 49
230	577.03	0.021 854	0.017 581	0.039 435	72.893	12.229	85.122	0.127 39	0.017 73	0.145 12
233.6 (*critical*)	596.9	0.028 70	0	0.028 70	78.86	0	78.86	0.135 9	0	0.135 9

Source: E. I. du Pont de Nemours & Company, Inc., copyright 1955 and 1956.

TABLE A-16
PROPERTIES OF SUPERHEATED FREON 12 (ENGLISH UNITS)

Temp., °F	*v*	*h*	*s*	*v*	*h*	*s*	*v*	*h*	*s*
	5 psia			10 psia			15 psia		
0	8.0611	78.582	0196 63	3.9809	78.246	0.184 71	2.6201	77.902	0.177 51
20	8.4265	81.309	0.202 44	4.1691	81.014	0.190 61	2.7494	80.712	0.183 49
40	8.7903	84.090	0.208 12	4.3556	83.828	0.196 35	2.8770	83.561	0.189 31
60	9.1528	86.922	0.213 67	4.5408	86.689	0.201 97	3.0031	86.451	0.194 98
80	9.5142	89.806	0.219 12	4.7248	89.596	0.207 46	3.1281	89.383	0.200 51
100	9.8747	92.738	0.224 45	4.9079	92.548	0.212 83	3.2521	92.357	0.205 93
120	10.234	95.717	0.229 68	5.0903	95.546	0.218 09	3.3754	95.373	0.211 22
140	10.594	98.743	0.234 81	5.2720	98.586	0.223 25	3.4981	98.429	0.216 40
160	10.952	101.812	0.239 85	5.4533	101.669	0.228 30	3.6202	101.525	0.221 48
180	11.311	104.925	0.244 79	5.6341	104.793	0.233 26	3.7419	104.661	0.226 46
200	11.668	108.079	0.249 64	5.8145	107.957	0.238 13	3.8632	107.835	0.231 35
220	12.026	111.272	0.254 41	5.9946	111.159	0.242 91	3.9841	111.046	0.236 14
	20 psia			25 psia			30 psia		
20	2.0391	80.403	0.178 29	1.6125	80.088	0.174 14	1.3278	79.765	0.170 65
40	2.1373	83.289	0.184 19	1.6932	83.012	0.180 12	1.3969	82.730	0.176 71
60	2.2340	86.210	0.189 92	1.7723	85.965	0.185 91	1.4644	85.716	0.182 57
80	2.3295	89.168	0.195 50	1.8502	88.950	0.191 55	1.5306	88.729	0.188 26
100	2.4241	92.164	0.200 95	1.9271	91.968	0.197 04	1.5957	91.770	0.193 79
120	2.5179	95.198	0.206 28	2.0032	95.021	0.202 40	1.6600	94.843	0.199 18
140	2.6110	98.270	0.211 49	2.0786	98.110	0.207 63	1.7237	97.948	0.204 45
160	2.7036	101.380	0.216 59	2.1535	101.234	0.212 76	1.7868	101.086	0.209 60
180	2.7957	104.528	0.221 59	2.2279	104.393	0.217 78	1.8494	104.258	0.214 63
200	2.8874	107.712	0.226 49	2.3019	107.588	0.222 69	1.9116	107.464	0.219 57
220	2.9789	110.932	0.231 30	2.3756	110.817	0.227 52	1.9735	110.702	0.224 40
240	3.0700	114.186	0.236 02	2.4491	114.080	0.232 25	2.0351	113.973	0.229 15
	35 psia			40 psia			50 psia		
40	1.1850	82.442	0.173 75	1.0258	82.148	0.171 12	0.80248	81.540	0.166 55
60	1.2442	85.463	0.179 68	1.0789	85.206	0.177 12	0.84713	84.676	0.172 71
80	1.3021	88.504	0.185 42	1.1306	88.277	0.182 92	0.89025	87.811	0.178 62
100	1.3589	91.570	0.191 00	1.1812	91.367	0.188 54	0.93216	90.953	0.184 34
120	1.4148	94.663	0.196 43	1.2309	94.480	0.194 01	0.97313	94.110	0.189 88
140	1.4701	97.785	0.201 72	1.2798	97.620	0.199 33	1.0133	97.286	0.195 27
160	1.5248	100.938	0.206 89	1.3282	100.788	0.204 53	1.0529	100.485	0.200 51
180	1.5789	104.122	0.211 95	1.3761	103.985	0.209 61	1.0920	103.708	0.205 63
200	1.6327	107.338	0.216 90	1.4236	107.212	0.214 57	1.1307	106.958	0.210 64
220	1.6862	110.586	0.221 75	1.4707	110.469	0.219 44	1.1690	110.235	0.215 53
240	1.7394	113.865	0.226 51	1.5176	113.757	0.224 20	1.2070	113.539	0.220 32
260	1.7923	117.175	0.231 17	1.5642	117.074	0.228 88	1.2447	116.871	0.225 02

Source: E. I. du Pont de Nemours & Company, Inc., copyright 1955 and 1956.

TABLE A-16
(Continued)

Temp., °F	v	h	s	v	h	s	v	h	s
		60 psia			**70 psia**			**80 psia**	
60	0.692 10	84.126	0.168 92	0.580 88	83.552	0.165 56	—	—	—
80	0.729 64	87.330	0.174 97	0.614 58	86.832	0.171 75	0.527 95	86.316	0.168 85
100	0.765 88	90.528	0.180 79	0.646 85	90.091	0.177 68	0.557 34	89.640	0.174 89
120	0.801 10	93.731	0.186 41	0.678 03	93.343	0.183 39	0.585 56	92.945	0.180 70
140	0.835 51	96.945	0.191 86	0.708 36	96.597	0.188 91	0.612 86	96.242	0.186 29
160	0.869 28	100.776	0.197 16	0.738 00	99.862	0.194 27	0.639 43	99.542	0.191 70
180	0.902 52	103.427	0.202 33	0.767 08	103.141	0.199 48	0.665 43	102.851	0.196 96
200	0.935 31	106.700	0.207 36	0.795 71	106.439	0.204 55	0.690 95	106.174	0.202 07
220	0.967 75	109.997	0.212 29	0.823 97	109.756	0.209 51	0.716 09	109.513	0.207 06
240	0.999 88	113.319	0.217 10	0.851 91	113.096	0.214 35	0.740 90	112.872	0.211 93
260	1.0318	116.666	0.221 82	0.879 59	116.459	0.219 09	0.765 44	116.251	0.216 69
280	1.0634	120.039	0.226 44	0.907 05	119.846	0.223 73	0.789 75	119.652	0.221 35
		90 psia			**100 psia**			**125 psia**	
100	0.487 49	89.175	0.172 34	0.431 38	88.694	0.169 96	0.329 43	87.407	0.164 55
120	0.513 46	92.536	0.178 24	0.455 62	92.116	0.175 97	0.350 86	91.008	0.170 87
140	0.538 45	95.879	0.183 91	0.478 81	95.507	0.181 72	0.370 98	94.537	0.176 86
160	0.562 68	99.216	0.189 38	0.501 18	98.884	0.187 26	0.390 15	98.023	0.182 58
180	0.586 29	102.557	0.194 69	0.522 91	102.257	0.192 62	0.408 57	101.484	0.188 07
200	0.609 41	105.905	0.199 84	0.544 13	105.633	0.197 82	0.426 42	104.934	0.193 38
220	0.632 13	109.267	0.204 86	0.564 92	109.018	0.202 87	0.443 80	108.380	0.198 53
240	0.654 51	112.644	0.209 76	0.585 38	112.415	0.207 80	0.460 81	111.829	0.203 53
260	0.676 62	116.040	0.214 55	0.605 54	115.828	0.212 61	0.477 50	115.287	0.208 40
280	0.698 49	119.456	0.219 23	0.625 46	119.258	0.217 31	0.493 94	118.756	0.213 16
300	0.720 16	122.892	0.223 81	0.645 18	122.707	0.221 91	0.510 16	122.238	0.217 80
320	0.741 66	126.349	0.228 30	0.664 72	126.176	0.226 41	0.526 19	125.737	0.222 35
		150 psia			**175 psia**			**200 psia**	
120	0.280 07	89.800	0.166 29	—	—	—	—	—	—
140	0.298 45	93.498	0.172 56	0.245 95	92.373	0.168 59	0.205 79	91.137	0.164 80
160	0.315 66	97.112	0.178 49	0.261 98	96.142	0.174 78	0.221 21	95.100	0.171 30
180	0.332 00	100.675	0.184 15	0.276 97	99.823	0.180 62	0.235 35	98.921	0.177 37
200	0.347 69	104.206	0.189 58	0.291 20	103.447	0.186 20	0.248 60	102.652	0.183 11
220	0.362 85	107.720	0.194 83	0.304 85	107.036	0.191 56	0.261 17	106.325	0.188 60
240	0.377 61	111.226	0.199 92	0.318 04	110.605	0.196 74	0.273 23	109.962	0.193 87
260	0.392 03	114.732	0.204 85	0.330 87	114.162	0.201 75	0.284 89	113.576	0.198 96
280	0.406 17	118.242	0.209 67	0.343 39	117.717	0.206 62	0.296 23	117.178	0.203 90
300	0.420 08	121.761	0.214 36	0.355 67	121.273	0.211 37	0.307 30	120.775	0.208 70
320	0.433 79	125.290	0.218 94	0.367 73	124.835	0.215 99	0.318 15	124.373	0.213 37
340	0.447 33	128.833	0.223 43	0.379 63	128.407	0.220 52	0.328 81	127.974	0.217 93

TABLE A-16
(Continued)

		250 psia			300 psia			400 psia	
160	0.162 49	92.717	0.164 62	—	—	—	—	—	—
180	0.176 05	96.925	0.171 30	0.134 82	94.556	0.165 37	—	—	—
200	0.188 24	100.930	0.177 47	0.146 97	98.975	0.172 17	0.091 005	93.718	0.160 92
		250 psia			**300 psia**			**400 psia**	
220	0.199 52	104.809	0.183 26	0.157 74	103.136	0.178 38	0.103 16	99.046	0.168 88
240	0.210 14	108.607	0.188 77	0.167 61	107.140	0.184 19	0.113 00	103.735	0.175 68
260	0.220 27	112.351	0.194 04	0.176 85	111.043	0.189 69	0.121 63	108.105	0.181 83
280	0.230 01	116.060	0.199 13	0.185 62	114.879	0.194 95	0.129 49	112.286	0.187 56
300	0.239 44	119.747	0.204 05	0.194 02	118.670	0.200 00	0.136 80	116.343	0.192 98
320	0.248 62	123.420	0.208 82	0.202 14	122.430	0.204 89	0.143 72	120.318	0.198 14
340	0.257 59	127.088	0.213 46	0.210 02	126.171	0.209 63	0.150 32	124.235	0.203 10
360	0.266 39	130.754	0.217 99	0.217 70	129.900	0.214 23	0.156 68	128.112	0.207 89
380	0.275 04	134.423	0.222 41	0.225 22	133.624	0.218 72	0.162 85	131.961	0.212 53
		500 psia			**600 psia**				
220	0.064 207	92.397	0.156 83	—	—	—			
240	0.077 620	99.218	0.166 72	0.047 488	91.024	0.153 35			
260	0.087 054	104.526	0.174 21	0.061 922	99.741	0.165 66			
280	0.094 923	109.277	0.180 72	0.070 859	105.637	0.173 74			
300	0.101 90	113.729	0.186 66	0.078 059	110.729	0.180 53			
320	0.108 29	117.997	0.192 21	0.084 333	115.420	0.186 63			
340	0.114 26	122.143	0.197 46	0.090 017	119.871	0.192 27			
360	0.119 92	126.205	0.202 47	0.095 289	124.167	0.197 57			
380	0.125 33	130.207	0.207 30	0.100 25	128.355	0.202 62			
400	0.130 54	134.166	0.211 96	0.104 98	132.466	0.207 46			
420	0.135 59	138.096	0.216 48	0.109 52	136.523	0.212 13			
440	0.140 51	142.004	0.220 87	0.113 91	140.539	0.216 64			

TABLE A-17

THERMODYNAMIC PROPERTIES OF AIR AT LOW PRESSURE (ENGLISH UNITS)

Reference level: 0°R and 1 atm
Absolute entropies may be calculated from

$$s^\circ = \phi - R \ln (p/p_0) + 1.0 \text{ Btu/lbm·°R}$$

where p_0 is the reference pressure of 1 atm.

Temp., °R	h, Btu/lbm	P_r	u, Btu/lbm	v_r	ϕ, Btu/lbm·°R
200	47.67	0.043 20	33.96	1714.9	0.363 03
220	52.46	0.060 26	37.38	1352.5	0.385 84
240	57.25	0.081 65	40.80	1088.8	0.406 66
260	62.03	0.107 97	44.21	892.0	0.425 82
280	66.82	0.139 86	47.63	741.6	0.443 56
300	71.61	0.177 95	51.04	624.5	0.460 07
320	76.40	0.222 90	54.46	531.8	0.475 50
340	81.18	0.275 45	57.87	457.2	0.490 02
360	85.97	0.3363	61.29	396.6	0.503 69
380	90.75	0.4061	64.70	346.6	0.516 63
400	95.53	0.4858	68.11	305.0	0.528 90
420	100.32	0.5760	71.52	270.1	0.540 58
440	105.11	0.6776	74.93	240.6	0.551 72
460	109.90	0.7913	78.36	215.33	0.562 35
480	114.69	0.9182	81.77	193.65	0.572 55
500	119.48	1.0590	85.20	174.90	0.582 33
520	124.27	1.2147	88.62	158.58	0.591 73
540	129.06	1.3860	92.04	144.32	0.600 78
560	133.86	1.5742	95.47	131.78	0.609 50
580	138.66	1.7800	98.90	120.70	0.617 93
600	143.47	2.005	102.34	110.88	0.626 07
620	148.28	2.249	105.78	102.12	0.633 95
640	153.09	2.514	109.21	94.30	0.641 59
660	157.92	2.801	112.67	87.27	0.649 02
680	162.73	3.111	116.12	80.96	0.656 21
700	167.56	3.446	119.58	75.25	0.663 21
720	172.39	3.806	123.04	70.07	0.670 02
740	177.23	4.193	126.51	65.38	0.676 65
760	182.08	4.607	129.99	61.10	0.683 12
780	186.94	5.051	133.47	57.20	0.689 42
800	191.81	5.526	136.97	53.63	0.695 58
820	196.69	6.033	140.47	50.35	0.701 60
840	201.56	6.573	143.98	47.34	0.707 47
860	206.46	7.149	147.50	44.57	0.713 23
880	211.35	7.761	151.02	42.01	0.718 86
900	216.26	8.411	154.57	39.64	0.724 38
920	221.18	9.102	158.12	37.44	0.729 79
940	226.11	9.834	161.68	35.41	0.735 09
960	231.06	10.610	165.26	33.52	0.740 30

Source: Abridged from J. H. Keenan and J. Kaye, "Gas Tables," John Wiley & Sons, Inc., New York, 1948, by permission.

TABLE A-17
(Continued)

Temp., °R	h, Btu/lbm	P_r	u, Btu/lbm	v_r	ϕ, Btu/lbm·°R
980	236.02	11.430	168.83	31.76	0.745 40
1000	240.98	12.298	172.43	30.12	0.750 42
1020	245.97	13.215	176.04	28.59	0.755 36
1040	250.95	14.182	179.66	27.17	0.760 19
1060	255.96	15.203	183.29	25.82	0.764 96
1080	260.97	16.278	186.93	24.58	0.769 64
1100	265.99	17.413	190.58	23.40	0.774 26
1120	271.03	18.604	194.25	22.30	0.778 80
1140	276.08	19.858	197.94	21.27	0.783 26
1160	281.14	21.18	201.63	20.293	0.787 67
1180	286.21	22.56	205.33	19.377	0.792 01
1200	291.30	24.01	209.05	18.514	0.796 28
1220	296.41	25.53	212.78	17.700	0.800 50
1240	301.52	27.13	216.53	16.932	0.804 66
1260	306.65	28.80	220.28	16.205	0.808 76
1280	311.79	30.55	224.05	15.518	0.812 80
1300	316.94	32.39	227.83	14.868	0.816 80
1320	322.11	34.31	231.63	14.253	0.820 75
1340	327.29	36.31	235.43	13.670	0.824 64
1360	332.48	38.41	239.25	13.118	0.828 48
1380	337.68	40.59	243.08	12.593	0.832 29
1400	342.90	42.88	246.93	12.095	0.836 04
1420	348.14	45.26	250.79	11.622	0.839 75
1440	353.37	47.75	254.66	11.172	0.843 41
1460	358.63	50.34	258.54	10.743	0.847 04
1480	363.89	53.04	262.44	10.336	0.850 62
1500	369.17	55.86	266.34	9.948	0.854 16
1520	374.47	58.78	270.26	9.578	0.857 67
1540	379.77	61.83	274.20	9.226	0.861 13
1560	385.08	65.00	278.13	8.890	0.864 56
1580	390.40	68.30	282.09	8.569	0.867 94
1600	395.74	71.73	286.06	8.263	0.871 30
1620	401.09	75.29	290.04	7.971	0.874 62
1640	406.45	78.99	294.03	7.691	0.877 91
1660	411.82	82.83	298.02	7.424	0.881 16
1680	417.20	86.82	302.04	7.168	0.884 39
1700	422.59	90.95	306.06	6.924	0.887 58
1720	428.00	95.24	310.09	6.690	0.890 74
1740	433.41	99.69	314.13	6.465	0.893 87
1760	438.83	104.30	318.18	6.251	0.896 97
1780	444.26	109.08	322.24	6.045	0.900 03
1800	449.71	114.03	326.32	5.847	0.903 08
1820	455.17	119.16	330.40	5.658	0.906 09
1840	460.63	124.47	334.50	5.476	0.909 08
1860	466.12	129.95	338.61	5.302	0.912 03
1880	471.60	135.64	342.73	5.134	0.914 97
1900	477.09	141.51	346.85	4.974	0.917 88
1920	482.60	147.59	350.98	4.819	0.920 76

TABLE A-17
(Continued)

Temp., °R	h, Btu/lbm	P_r	u, Btu/lbm	v_r	ϕ, Btu/lbm·°R
1940	488.12	153.87	355.12	4.670	0.923 62
1960	493.64	160.37	359.28	4.527	0.926 45
1980	499.17	167.07	363.43	4.390	0.929 26
2000	504.71	174.00	367.61	4.258	0.932 05
2020	510.26	181.16	371.79	4.130	0.934 81
2040	515.82	188.54	375.98	4.008	0.937 56
2060	521.39	196.16	380.18	3.890	0.940 26
2080	526.97	204.02	384.39	3.777	0.942 96
2100	532.55	212.1	388.60	3.667	0.945 64
2120	538.15	220.5	392.83	3.561	0.948 29
2140	543.74	229.1	397.05	3.460	0.950 92
2160	549.35	238.0	401.29	3.362	0.953 52
2180	554.97	247.2	405.53	3.267	0.956 11
2200	560.59	256.6	409.78	3.176	0.958 68
2220	566.23	266.3	414.05	3.088	0.961 23
2240	571.86	276.3	418.31	3.003	0.963 76
2260	577.51	286.6	422.59	2.921	0.966 26
2280	583.16	297.2	426.87	2.841	0.968 76
2300	588.82	308.1	431.16	2.765	0.971 23
2320	594.49	319.4	435.46	2.691	0.973 69
2340	600.16	330.9	439.76	2.619	0.976 11
2360	605.84	342.8	444.07	2.550	0.978 53
2380	611.53	355.0	448.38	2.483	0.980 92
2400	617.22	367.6	452.70	2.419	0.983 31

TABLE A-17M

THERMODYNAMIC PROPERTIES OF AIR AT LOW PRESSURE (SI UNITS)

Reference level: 0 K and 1 atm
Absolute entropies may be calculated from

$$s^{\circ} = \phi - R \ln (p/p_0) + 4.1869 \text{ kJ/kg·K}$$

where p_0 is the reference pressure of 1 atm.

T, K	h, kJ/kg	P_r	u, kJ/kg	v_r	ϕ kJ/kg·K
100	99.76	0.029 90	71.06	2230	1.4143
110	109.77	0.041 71	78.20	1758.4	1.5098
120	119.79	0.056 52	85.34	1415.7	1.5971
130	129.81	0.074 74	92.51	1159.8	1.6773
140	139.84	0.096 81	99.67	964.2	1.7515
150	149.86	0.123 18	106.81	812.0	1.8206
160	159.87	0.154 31	113.95	691.4	1.8853
170	169.89	0.190 68	121.11	594.5	1.9461
180	179.92	0.232 79	128.28	515.6	2.0033
190	189.94	0.281 14	135.40	450.6	2.0575
200	199.96	0.3363	142.56	396.6	2.1088
210	209.97	0.3987	149.70	351.2	2.1577
220	219.99	0.4690	156.84	312.8	2.2043
230	230.01	0.5477	163.98	280.0	2.2489
240	240.03	0.6355	171.15	251.8	2.2915
250	250.05	0.7329	178.29	227.45	2.3325
260	260.09	0.8405	185.45	206.26	2.3717
270	270.12	0.9590	192.59	187.74	2.4096
280	280.14	1.0889	199.78	171.45	2.4461
290	290.17	1.2311	206.92	157.07	2.4813
300	300.19	1.3860	214.09	144.32	2.5153
310	310.24	1.5546	221.27	132.96	2.5483
320	320.29	1.7375	228.45	122.81	2.5802
330	330.34	1.9352	235.65	113.70	2.6111
340	340.43	2.149	242.86	105.51	2.6412
350	350.48	2.379	250.05	98.11	2.6704
360	360.58	2.626	257.23	91.40	2.6987
370	370.67	2.892	264.47	85.31	2.7264
380	380.77	3.176	271.72	79.77	2.7534
390	390.88	3.481	278.96	74.71	2.7796
400	400.98	3.806	286.19	70.07	2.8052
410	411.12	4.153	293.45	65.83	2.8302
420	421.26	4.522	300.73	61.93	2.8547
430	431.43	4.915	308.03	58.34	2.8786
440	441.61	5.332	315.34	55.02	2.9020

Source: Adapted to SI units from H. Keenan and J. Kaye, "Gas Tables," John Wiley & Sons, Inc., New York, 1948.

TABLE A-17M
(Continued)

T, K	h, kJ/kg	P_r	u, kJ/kg	v_r	ϕ kJ/kg·K
450	451.83	5.775	322.66	51.96	2.9249
460	462.01	6.245	329.99	49.11	2.9473
470	472.25	6.742	337.34	46.48	2.9693
480	482.48	7.268	344.74	44.04	2.9909
490	492.74	7.824	352.11	41.76	3.0120
500	503.02	8.411	359.53	39.64	3.0328
510	513.32	9.031	366.97	37.65	3.0532
520	523.63	9.684	374.39	35.80	3.0733
530	533.98	10.372	381.88	34.07	3.0930
540	544.35	11.097	389.40	32.45	3.1124
550	554.75	11.858	396.89	30.92	3.1314
560	565.17	12.659	404.44	29.50	3.1502
570	575.57	13.500	411.98	28.15	3.1686
580	586.04	14.382	419.56	26.89	3.1868
590	596.53	15.309	427.17	25.70	3.2047
600	607.02	16.278	434.80	24.58	3.2223
610	617.53	17.297	442.43	23.51	3.2397
620	628.07	18.360	450.13	22.52	3.2569
630	638.65	19.475	457.83	21.57	3.2738
640	649.21	20.64	465.55	20.674	3.2905
650	659.84	21.86	473.32	19.828	3.3069
660	670.47	23.13	481.06	19.026	3.3232
670	681.15	24.46	488.88	18.266	3.3392
680	691.82	25.85	496.65	17.543	3.3551
690	702.52	27.29	504.51	16.857	3.3707
700	713.27	28.80	512.37	16.205	3.3861
710	724.01	30.38	520.26	15.585	3.4014
720	734.20	31.92	527.72	15.027	3.4156
730	745.62	33.72	536.12	14.434	3.4314
740	756.44	35.50	544.05	13.900	3.4461
750	767.30	37.35	552.05	13.391	3.4607
760	778.21	39.27	560.08	12.905	3.4751
770	789.10	41.27	568.10	12.440	3.4894
780	800.03	43.35	576.15	11.998	3.5035
790	810.98	45.51	584.22	11.575	3.5174
800	821.94	47.75	592.34	11.172	3.5312
810	832.96	50.08	600.46	10.785	3.5449
820	843.97	52.49	608.62	10.416	3.5584
830	855.01	55.00	616.79	10.062	3.5718
840	866.09	57.60	624.97	9.724	3.5850
850	877.16	60.29	633.21	9.400	3.5981
860	888.28	63.09	641.44	9.090	3.6111
870	899.42	65.98	649.70	8.792	3.6240

TABLE A-17M
(Continued)

T, K	h, kJ/kg	P_r	u, kJ/kg	v_r	ϕ kJ/kg·K
880	910.56	68.98	658.00	8.507	3.6367
890	921.75	72.08	666.31	8.233	3.6493
900	932.94	75.29	674.63	7.971	3.6619
910	944.15	78.61	682.98	7.718	3.6743
920	955.38	82.05	691.33	7.476	3.6865
930	966.64	85.60	699.73	7.244	3.6987
940	977.92	89.28	708.13	7.020	3.7108
950	989.22	93.08	716.57	6.805	3.7227
960	1000.53	97.00	725.01	6.599	3.7346
970	1011.88	101.06	733.48	6.400	3.7463
980	1023.25	105.24	741.99	6.209	3.7580
990	1034.63	109.57	750.48	6.025	3.7695
1000	1046.03	114.03	759.02	5.847	3.7810
1020	1068.89	123.12	775.67	5.521	3.8030
1040	1091.85	133.34	793.35	5.201	3.8259
1060	1114.85	143.91	810.61	4.911	3.8478
1080	1137.93	155.15	827.94	4.641	3.8694
1100	1161.07	167.07	845.34	4.390	3.8906
1120	1184.28	179.71	862.85	4.156	3.9116
1140	1207.54	193.07	880.37	3.937	3.9322
1160	1230.90	207.24	897.98	3.732	3.9525
1180	1254.34	222.2	915.68	3.541	3.9725
1200	1277.79	238.0	933.40	3.362	3.9922
1220	1301.33	254.7	951.19	3.194	4.0117
1240	1324.89	272.3	969.01	3.037	4.0308
1260	1348.55	290.8	986.92	2.889	4.0497
1280	1372.25	310.4	1004.88	2.750	4.0684
1300	1395.97	330.9	1022.88	2.619	4.0868
1320	1419.77	352.5	1040.93	2.497	4.1049
1340	1443.61	375.3	1059.03	2.381	4.1229
1360	1467.50	399.1	1077.17	2.272	4.1406
1380	1491.43	424.2	1095.36	2.169	4.1580
1400	1515.41	450.5	1113.62	2.072	4.1753
1420	1539.44	478.0	1131.90	1.9808	4.1923
1440	1563.49	506.9	1150.23	1.8942	4.2092
1460	1587.61	537.1	1168.61	1.8124	4.2258
1480	1611.80	568.8	1187.03	1.7350	4.2422
1500	1635.99	601.9	1205.47	1.6617	4.2585

TABLE A-18

THERMODYNAMIC PROPERTIES OF GASES AT LOW PRESSURE (ENGLISH UNITS)

Reference level: 0°R and 1 atm; h in Btu/lbm mol; ϕ in Btu/lbm·mol·°R
Absolute entropies at other pressures may be calculated from

$$\bar{s}^\circ = \bar{\phi} - \mathcal{R} \ln(p/p_0) \text{ Btu/lbm·mol·°R}$$

where p_0 is the reference pressure of 1 atm.

Temp., °R	Products of Combustion, 400% theoretical air		Products of Combustion, 200% theoretical air		Nitrogen		Oxygen		Water Vapor		Carbon Dioxide		Hydrogen		Carbon Monoxide	
	$\bar{h}$	$\bar{\phi}$	$\bar{h}$	$\bar{\phi}$	$\bar{h}$	$\bar{\phi}$	$\bar{h}$	$\bar{\phi}$	$\bar{h}$	$\bar{\phi}$	$\bar{h}$	$\bar{\phi}$	$\bar{h}$	$\bar{\phi}$	$\bar{h}$	$\bar{\phi}$
537	3746.8	46.318	3774.9	46.300	3729.5	45.755	3725.1	48.986	4258.3	45.079	4030.2	51.032	3640.3	31.194	3729.5	47.272
600	4191.9	47.101	4226.3	47.094	4167.9	46.514	4168.3	49.762	4764.7	45.970	4600.9	52.038	4075.6	31.959	3168.0	48.044
700	4901.7	48.195	4947.7	48.207	4864.9	47.588	4879.3	50.858	5575.4	47.219	5552.0	53.503	4770.2	33.031	4866.0	49.120
800	5617.5	49.150	5676.3	49.179	5564.4	48.522	5602.0	51.821	6396.9	48.316	6552.9	54.839	5467.1	33.961	5568.2	50.058
900	6340.3	50.002	6413.0	50.047	6268.1	49.352	6337.9	52.688	7230.9	49.298	7597.6	56.070	6165.3	34.784	6276.4	50.892
1000	7072.1	50.773	7159.8	50.833	6977.9	50.099	7087.5	53.477	8078.9	50.191	8682.1	57.212	6864.5	35.520	6992.2	51.646
1100	7812.9	51.479	7916.4	51.555	7695.0	50.783	7850.4	54.204	8942.0	51.013	9802.6	58.281	7564.6	36.188	7716.8	52.337
1200	8563.4	52.132	8683.6	52.222	8420.0	51.413	8625.8	54.879	9820.4	51.777	10 955.3	59.283	8265.8	36.798	8450.8	52.976
1300	9324.1	52.741	9461.7	52.845	9153.9	52.001	9412.9	55.508	10 714.5	52.494	12 136.9	60.229	8968.7	37.360	9194.6	53.571
1400	10 095.0	53.312	10 250.7	53.430	9896.9	52.551	10 210.4	56.099	11 624.8	53.168	13 344.7	61.124	9673.8	37.883	9948.1	54.129
1500	10 875.6	53.851	11 050.2	53.981	10 648.9	53.071	11 017.1	56.656	12 551.4	53.808	14 576.0	61.974	10 381.5	38.372	10 711.1	54.655
1600	11 665.6	54.360	11 859.6	54.504	11 409.7	53.561	11 832.5	57.182	13 494.9	54.418	15 829.0	62.783	11 092.5	38.830	11 483.4	55.154
1700	12 464.3	54.844	12 678.6	55.000	12 178.9	54.028	12 655.6	57.680	14 455.4	54.999	17 101.4	63.555	11 807.4	39.264	12 264.3	55.628
1800	13 271.7	55.306	13 507.0	55.473	12 956.3	54.472	13 485.8	58.155	15 433.0	55.559	18 391.5	64.292	12 526.8	39.675	13 053.2	56.078
1900	14 087.2	55.747	14 344.1	55.926	13 741.6	54.896	14 322.1	58.607	16 427.5	56.097	19 697.8	64.999	13 250.9	40.067	13 849.8	56.509
2000	14 910.3	56.169	15 189.3	56.360	14 534.4	55.303	15 164.0	59.039	17 439.0	56.617	21 018.7	65.676	13 980.1	40.441	14 653.2	56.922
2100	15 740.5	56.574	16 042.4	56.777	15 334.0	55.694	16 010.9	59.451	18 466.9	57.119	22 352.7	66.327	14 714.5	40.799	15 463.3	57.317
2200	16 577.1	56.964	16 902.5	57.177	16 139.8	56.068	16 862.6	59.848	19 510.8	57.605	23 699.0	66.953	15 454.4	41.143	16 279.4	57.696
2300	17 419.8	57.338	17 769.3	57.562	16 951.2	56.429	17 718.8	60.228	20 570.6	58.077	25 056.3	67.557	16 199.8	41.475	17 101.0	58.062
2400	18 268.0	57.699	18 642.1	57.933	17 767.9	56.777	18 579.2	60.594	21 645.7	58.535	26 424.0	68.139	16 950.6	41.794	17 927.4	58.414
2500	19 121.4	58.048	19 520.7	58.292	18 589.5	57.112	19 443.4	60.946	22 735.4	58.980	27 801.2	68.702	17 707.3	42.104	18 758.8	58.754
2600	19 979.7	58.384	20 404.6	58.639	19 415.8	57.436	20 311.4	61.287	23 839.5	59.414	29 187.1	69.245	18 469.7	42.403	19 594.3	59.081
2700	20 842.8	58.710	21 293.8	58.974	20 246.4	57.750	21 182.9	61.616	24 957.2	59.837	30 581.2	69.771	19 237.8	42.692	20 434.0	59.398
2800	21 709.8	59.026	22 187.5	59.300	21 081.1	58.053	22 057.8	61.934	26 088.0	60.248	31 982.8	70.282	20 011.8	42.973	21 277.2	59.705
2900	22 581.4	59.331	23 086.0	59.615	21 919.5	58.348	22 936.1	62.242	27 231.2	60.650	33 391.5	70.776	20 791.5	43.247	22 123.8	60.002
3000	23 456.6	59.628	23 988.5	59.921	22 761.5	58.632	23 817.7	62.540	28 386.3	61.043	34 806.6	71.255	21 576.9	43.514	22 973.4	60.290
3100	24 335.5	59.916	24 895.3	60.218	23 606.8	58.910	24 702.5	62.831	29 552.8	61.426	36 227.9	71.722	22 367.7	43.773	23 826.0	60.569
3200	25 217.8	60.196	25 805.6	60.507	24 455.0	59.179	25 590.5	63.113	30 730.2	61.801	37 654.7	72.175	23 164.1	44.026	24 681.2	60.841
3300	26 102.9	60.469	26 719.2	60.789	25 306.0	59.442	26 418.6	63.386	31 918.2	62.167	39 086.7	72.616	23 965.5	44.273	25 539.0	61.105

TABLE A-18
(Continued)

Temp., °R	Products of Combustion, 400% theoretical air		Products of Combustion, 200% theoretical air		Nitrogen		Oxygen		Water Vapor		Carbon Dioxide		Hydrogen		Carbon Monoxide	
	$\bar{h}$	$\bar{\phi}$	$\bar{h}$	$\bar{\phi}$	$\bar{h}$	$\bar{\phi}$	$\bar{h}$	$\bar{\phi}$	$\bar{h}$	$\bar{\phi}$	$\bar{h}$	$\bar{\phi}$	$\bar{h}$	$\bar{\phi}$	$\bar{h}$	$\bar{\phi}$
3400	26 991.4	60.734	27 636.4	61.063	26 159.7	59.697	27 375.9	63.654	33 116.0	62.526	40 523.6	73.045	24 771.9	44.513	26 399.3	61.362
3500			28 556.8	61.329	27 015.9	59.944	28 273.3	63.914	34 323.5	62.876	41 965.2	73.462	25 582.9	44.748	27 261.8	61.612
3600			29 479.9	61.590	27 874.4	60.186	29 173.9	64.168	35 540.1	63.221	43 411.0	73.870	26 398.5	44.978	28 126.6	61.855
3700			30 406.0	61.843	28 735.1	60.422	30 077.5	64.415	36 765.4	63.557	44 860.6	74.267	27 218.5	45.203	28 993.5	62.093
3800			31 334.8	62.091	29 597.9	60.652	30 984.1	64.657	37 998.9	63.887	46 314.0	74.655	28 042.8	45.423	29 862.3	62.325
3900			32 266.2	62.333	30 462.8	60.877	31 893.6	64.893	39 240.2	64.210	47 771.0	75.033	28 871.1	45.638	30 732.9	62.551
4000					31 329.4	61.097	32 806.1	65.123	40 489.1	64.528	49 231.4	75.404	29 703.5	45.849	31 605.2	72.772
4100					32 198.0	61.310	33 721.6	65.350	41 745.4	64.839	50 695.1	75.765	30 539.8	46.056	32 479.1	62.988
4200					33 068.1	61.520	34 639.9	65.571	43 008.4	65.144	52 162.0	76.119	31 379.8	46.257	33 354.4	63.198
4300					33 939.9	61.726	35 561.1	65.788	44 278.0	65.444	53 632.1	76.464	32 223.5	46.456	34 231.2	63.405
4400					34 813.1	61.927	36 485.0	66.000	45 553.9	65.738	55 105.1	76.803	33 070.9	46.651	35 109.2	63.607
4500					35 687.8	62.123	37 411.8	66.208	46 835.9	66.028	56 581.0	77.135	33 921.6	46.842	35 988.6	63.805
4600					36 563.8	62.316	38 341.4	66.413	48 123.6	66.312	58 059.7	77.460	34 775.7	47.030	36 869.3	63.998
4700					37 441.1	62.504	39 273.6	66.613	49 416.9	66.591	59 541.1	77.779	35 633.0	47.215	37 751.0	64.188
4800					38 319.5	62.689	40 208.6	66.809	50 715.5	66.866	61 024.9	78.091	36 493.4	47.396	38 633.9	64.374
4900					39 199.1	62.870	41 146.1	67.003	52 019.0	67.135	62 511.3	78.398	37 356.9	47.574	39 517.8	64.556
5000					40 079.8	63.049	42 086.3	67.193	53 327.4	67.401	64 000.0	78.698	38 223.3	47.749	40 402.7	64.735
5100					40 961.6	63.223	43 029.1	67.380	54 640.3	67.662	65 490.9	78.994	39 092.8	47.921	41 288.6	64.910
5200					41 844.4	63.395	43 974.3	67.562	55 957.4	67.918	66 984.0	79.284	39 965.1	48.090	42 175.5	65.082
5300					42 728.3	63.563	44 922.2	67.743	57 278.7	68.172	68 479.1	79.569	40 840.2	48.257	43 063.2	65.252

Source: Abridged from J. H. Keenan and J. Kaye, "Gas Tables," John Wiley & Sons, Inc., New York, 1948.

TABLE A-18M

THERMODYNAMIC PROPERTIES OF GASES AT LOW PRESSURES (SI UNITS)

Reference levels: Enthalpy and internal energy at 298 K; entropy at 0 K and 0.1 MPa; $\bar{h}$ and $\bar{u}$ in kJ/kg mol; $\bar{s}°$ in kJ/kg mol·K

Absolute entropies at other pressures may be calculated from

$$\bar{s}° = \bar{s}°_{ref} - \mathcal{R} \ln(p/p_0) \text{ kJ/kg mol·K}$$

where p_0 is the reference pressure of 0.1 MPa and $\bar{s}°_{ref}$ is the tabular value.

	Water Vapor, H_2O			Carbon Dioxide, CO_2		
T, K	$\bar{h} - \bar{h}_{298}$	$\bar{u} - \bar{u}_{298}$	$\bar{s}°$	$\bar{h} - \bar{h}_{298}$	$\bar{u} - \bar{u}_{298}$	$\bar{s}°$
0	−9 904	−7 425	0.0	−9 364	−6 885	0.0
100	−6 615	−4 939	152.390	−6 456	−4 666	179.109
200	−3 280	−2 453	175.486	−3 414	−2 447	199.975
250	−1 610	−1 210	182.940	−1 737	−1 337	207.446
298	0.0	0.0	188.833	0.0	0.0	213.795
300	63	47	189.038	67	54	214.025
350	1 748	1 317	194.234	1 987	1 554	219.940
400	3 452	2 605	198.783	4 008	3 161	225.334
450	5 176	3 914	202.843	6 119	4 857	230.303
500	6 920	5 246	206.523	8 314	6 636	234.924
550	8 697	6 603	209.904	10 581	8 487	239.244
600	10 498	7 988	213.037	12 916	10 406	243.309
650	12 326	9 401	215.965	15 310	12 385	247.141
700	14 184	10 843	218.719	17 761	14 420	250.773
750	16 073	12 316	221.324	20 265	16 508	254.226
800	17 991	13 820	223.803	22 815	18 642	257.517
850	19 942	15 354	226.166	25 409	20 821	260.660
900	21 924	16 920	228.430	28 041	23 037	263.668
950	23 937	18 518	230.608	30 706	25 286	266.553
1000	25 978	20 143	232.706	33 405	27 570	269.325
1050	28 057	21 806	234.731	36 138	29 887	271.986
1100	30 167	23 500	236.694	38 894	32 227	274.555
1150	32 307	25 224	238.594	41 679	32 118	277.023
1200	34 476	26 978	240.443	44 484	36 986	279.417
1250	36 676	28 762	242.237	47 312	39 399	281.721
1300	38 903	30 575	243.986	50 158	41 828	283.956
1350	41 163	32 417	245.688	53 024	44 279	286.122
1400	43 447	34 286	247.350	55 907	46 746	288.216
1450	45 759	36 183	248.973	58 803	49 227	290.252
1500	48 095	38 103	250.560	61 714	51 721	292.224
1550	50 459	40 051	252.108	64 640	54 232	293.642
1600	52 844	42 020	253.622	67 580	56 856	296.010
1650	55 255	44 015	255.106	70 531	59 245	297.823
1700	57 685	46 030	256.559	73 492	61 836	299.592
1750	60 139	48 068	257.979	76 462	64 391	301.312
1800	62 609	50 122	259.371	79 442	66 955	302.993
1850	65 102	52 200	260.737	82 431	69 528	304.631
1900	67 613	54 295	262.078	85 429	72 111	306.232
1950	70 144	55 986	263.391	88 435	74 701	307.792
2000	72 689	58 540	264.681	91 450	77 300	309.320

TABLE A-18M
(Continued)

	Water Vapor, H_2O			Carbon Dioxide, CO_2		
T, K	$\bar{h} - \bar{h}_{298}$	$\bar{u} - \bar{u}_{298}$	$\bar{s}°$	$\bar{h} - \bar{h}_{298}$	$\bar{u} - \bar{u}_{298}$	$\bar{s}°$
2050	75 252	60 686	265.947	94 471	79 906	310.810
2100	77 831	62 850	267.191	97 500	82 519	312.269
2150	80 426	65 029	268.410	100 534	85 138	313.698
2200	83 036	67 224	269.609	103 575	87 763	315.098
2250	85 658	69 430	270.788	106 620	90 392	316.465
2300	88 295	71 651	271.948	109 671	93 027	317.805
2350	90 942	73 883	273.087	112 727	95 667	319.120
2400	93 604	76 128	274.207	115 788	98 312	320.411
2450	96 279	78 386	275.310	118 855	100 964	321.675
2500	98 964	80 657	276.396	121 926	103 619	322.918
2550	101 661	82 939	277.463	125 004	106 281	324.135
2600	104 370	85 231	278.517	128 085	108 947	325.332
2650	107 087	87 533	279.550	131 169	111 615	326.505
2700	109 813	89 844	280.571	134 256	114 287	327.658
2750	112 549	92 163	281.573	137 349	116 964	328.793
2800	115 294	94 492	282.563	140 444	119 643	329.909
2850	118 048	96 831	283.538	143 544	122 327	331.005
2900	120 813	99 180	284.500	146 645	125 013	332.085
2950	123 582	101 534	285.447	149 753	127 704	333.146
3000	126 361	103 896	286.383	152 862	130 398	334.193
3050	129 147	106 267	287.303	155 977	133 097	335.223
3100	131 942	108 647	288.211	159 092	135 796	336.235
3150	134 744	111 033	289.108	162 212	138 500	337.233
3200	137 553	113 426	289.994	165 331	141 204	338.218
3250	140 368	115 825	290.865	168 458	143 916	339.178
3300	143 197	118 224	291.715	171 588	146 628	340.123

TABLE A-18M
(Continued)

	Hydrogen, H_2			Carbon Monoxide, CO		
T, K	$\bar{h} - \bar{h}_{298}$	$\bar{u} - \bar{u}_{298}$	$\bar{s}°$	$\bar{h} - \bar{h}_{298}$	$\bar{u} - \bar{u}_{298}$	$\bar{s}°$
0	−8 468	−5 989	0.0	−8 669	−6 190	0.0
100	−5 293	−3 979	102.145	−5 770	−4 115	165.850
200	−2 770	−1 969	119.437	−2 858	−2 040	186.025
250	−1 357	−965	125.175	−1 403	−1 002	192.520
298	0.0	0.0	130.684	0.0	0.0	197.653
300	54	38	130.864	54	39	197.833
350	1 503	1 072	135.318	1 512	1 081	202.326
400	2 958	2 111	139.215	2 975	2 129	206.234
450	4 419	3 157	142.647	4 447	3 185	209.701
500	5 883	4 204	145.738	5 929	4 253	212.828
550	7 346	5 252	148.512	7 428	5 334	215.681
600	8 812	6 302	151.077	8 941	6 432	218.313
650	10 279	7 355	153.418	10 472	7 546	220.765
700	11 749	8 410	155.608	12 021	8 680	223.062
750	13 225	9 468	157.633	13 589	9 832	225.224
800	14 703	10 531	159.549	15 175	11 003	227.271
850	16 189	11 601	161.347	16 777	12 189	229.215
900	17 682	12 678	163.060	18 397	13 393	231.066
950	19 180	13 760	164.674	20 034	14 615	232.836
1000	20 686	14 850	166.223	21 686	15 851	234.531
1050	22 200	15 949	167.696	23 354	17 103	236.155
1100	23 723	17 057	169.118	25 033	18 367	237.719
1150	25 254	18 172	170.476	26 725	19 642	239.221
1200	26 794	19 295	171.792	28 426	20 928	240.673
1250	28 346	20 432	173.054	30 141	22 362	242.070
1300	29 907	21 578	174.281	31 865	23 535	243.426
1350	31 480	22 734	175.467	33 597	24 852	244.732
1400	33 062	23 900	176.620	35 338	26 177	245.999
1450	34 661	25 084	177.739	37 090	27 513	247.227
1500	36 267	26 277	178.833	38 848	28 856	248.421
1550	37 890	27 482	179.894	40 613	30 205	249.577
1600	39 522	28 698	180.929	42 384	31 560	250.702
1650	41 165	29 926	181.940	44 159	32 920	251.795
1700	42 815	31 161	182.929	45 940	34 284	252.861
1750	44 478	32 406	183.891	47 729	35 658	253.896
1800	46 150	33 663	184.833	49 522	37 035	254.907
1850	47 831	34 928	185.754	51 321	38 419	255.891
1900	49 522	36 204	186.657	53 124	39 807	256.852
1950	51 222	37 489	187.540	54 931	41 197	257.792
2000	52 932	38 782	188.406	56 739	42 590	258.710
2050	54 651	40 085	189.257	58 555	43 989	259.603
2100	56 379	41 397	190.088	60 375	45 394	260.480
2150	58 116	42 719	190.905	62 195	46 798	261.335
2200	59 860	44 048	191.707	64 019	48 206	262.174
2250	61 612	45 384	192.494	65 847	49 619	262.996
2300	63 371	46 727	193.268	67 676	51 032	263.802
2350	65 140	48 080	194.030	69 509	52 450	264.589

TABLE A-18M
(Continued)

	Hydrogen, H_2			Carbon Monoxide, CO		
T, K	$\bar{h} - \bar{h}_{298}$	$\bar{u} - \bar{u}_{298}$	$\bar{s}°$	$\bar{h} - \bar{h}_{298}$	$\bar{u} - \bar{u}_{298}$	$\bar{s}°$
2400	66 915	49 440	194.778	71 346	53 870	265.362
2450	68 700	50 809	195.512	73 183	55 292	266.121
2500	70 492	52 186	196.234	75 023	56 716	266.865
2550	72 287	53 565	196.946	76 868	58 145	267.594
2600	74 090	54 952	197.649	78 714	59 576	268.312
2650	75 900	56 346	198.338	80 561	61 007	269.014
2700	77 718	57 748	199.017	82 408	62 438	269.705
2750	79 540	59 155	199.684	84 261	63 876	270.394
2800	81 370	60 569	200.343	86 115	65 314	271.053
2850	83 203	61 987	200.994	87 970	66 755	271.711
2900	85 044	63 412	201.636	89 826	68 193	272.358
2950	86 890	64 842	202.266	91 683	69 635	272.993
3000	88 743	66 279	202.887	93 542	71 077	273.618
3050	90 597	67 718	203.500	95 404	72 525	274.232
3100	92 458	69 163	204.104	97 270	73 974	274.839
3150	94 325	70 615	204.701	99 133	75 422	275.435
3200	96 199	72 072	205.293	100 998	76 871	276.023
3250	98 077	73 534	205.874	102 865	78 323	276.603
3300	99 964	74 996	206.442	104 736	79 775	277.166

	Nitrogen, N_2			Oxygen, O_2		
T, K	$\bar{h} - \bar{h}_{298}$	$\bar{u} - \bar{u}_{298}$	$\bar{s}°$	$\bar{h} - \bar{h}_{298}$	$\bar{u} - \bar{u}_{298}$	$\bar{s}°$
0	−8 669	−6 190	0.0	−8 682	−6 203	0.0
100	−5 770	−4 115	159.813	−5 778	−4 124	173.306
200	−2 858	−2 040	179.988	−2 866	−2 045	193.486
250	−1 403	−1 002	186.479	−1 407	−1 006	199.994
298	0.0	0.0	191.611	0.0	0.0	205.142
300	54	39	191.791	54	39	205.322
350	1 511	1 080	196.282	1 531	1 100	209.874
400	2 971	2 124	200.180	3 029	2 181	213.874
450	4 436	3 173	203.632	4 546	3 284	217.451
500	5 912	4 233	206.739	6 088	4 411	220.698
550	7 395	5 302	209.570	7 656	5 562	223.685
600	8 891	6 384	212.175	9 247	6 737	226.455
650	10 406	7 481	214.598	10 862	7 937	229.041
700	11 937	8 594	216.865	12 502	9 161	231.272
750	13 480	9 723	218.998	14 162	10 404	233.758
800	15 046	10 871	221.016	15 841	11 669	235.924
850	16 623	12 034	222.931	17 536	12 947	237.973
900	18 221	13 217	224.756	19 246	14 242	239.936

TABLE A-18M
(Continued)

T, K	Nitrogen, N_2			Oxygen, O_2		
	$\bar{h} - \bar{h}_{298}$	$\bar{u} - \bar{u}_{298}$	$\bar{s}°$	$\bar{h} - \bar{h}_{298}$	$\bar{u} - \bar{u}_{298}$	$\bar{s}°$
950	19 832	14 413	226.498	20 970	15 551	241.798
1000	21 460	15 625	228.166	22 707	16 872	243.585
1050	23 103	16 852	229.768	24 458	18 207	245.288
1100	24 757	18 090	231.308	26 217	19 550	246.928
1150	26 426	19 344	232.791	27 987	20 905	248.499
1200	28 108	20 609	234.264	29 765	22 266	250.016
1250	29 799	21 885	235.604	31 554	23 640	251.475
1300	31 501	23 171	236.940	33 351	25 021	252.886
1350	33 214	24 469	238.231	35 155	26 410	254.245
1400	34 936	25 774	239.484	36 966	27 805	255.564
1450	36 666	27 089	240.698	38 785	29 208	256.836
1500	38 405	28 411	241.877	40 610	30 618	258.078
1550	40 150	29 741	243.022	42 441	32 033	259.274
1600	41 903	31 078	244.137	44 279	33 455	260.446
1650	43 664	32 424	245.219	46 121	34 882	261.575
1700	45 430	33 775	246.275	47 970	36 314	262.685
1750	47 203	35 132	247.302	49 826	37 755	263.757
1800	48 982	36 495	248.304	51 689	39 202	264.810
1850	50 764	37 861	249.281	53 559	40 657	265.832
1900	52 551	39 233	250.237	55 434	42 116	266.835
1950	54 349	40 610	251.167	57 315	43 581	267.807
2000	56 141	41 991	252.078	59 199	45 050	268.764
2050	57 943	43 377	252.967	61 090	46 524	269.697
2100	59 748	44 767	253.836	62 986	48 005	270.613
2150	61 557	46 161	254.687	64 891	49 494	271.508
2200	63 371	47 559	255.522	66 802	50 898	272.387
2250	65 187	48 959	256.336	68 715	52 487	273.245
2300	67 007	50 363	257.137	70 634	53 990	274.090
2350	68 827	51 768	257.919	72 561	55 501	274.918
2400	70 651	53 176	258.689	74 492	57 016	275.735
2450	72 480	54 589	259.441	76 430	58 539	276.533
2500	74 312	56 005	260.183	78 375	60 068	277.316
2550	76 145	57 423	260.908	80 322	61 599	278.088
2600	77 973	58 843	261.622	82 274	63 136	278.848
2650	79 819	60 265	262.322	84 234	64 680	279.594
2700	81 659	61 690	263.011	86 199	66 230	280.329
2750	83 502	63 116	263.686	88 170	67 784	281.051
2800	85 345	64 544	264.350	90 144	69 343	281.764
2850	87 190	65 973	265.004	92 126	70 909	282.466
2900	89 036	67 403	265.647	94 111	72 479	283.157
2950	90 887	68 838	266.279	96 103	74 055	283.837
3000	92 738	70 274	266.902	98 098	75 634	284.508
3050	94 591	71 712	267.513	100 096	77 216	285.169
3100	96 446	73 151	268.116	102 102	78 806	285.822
3150	98 303	74 592	268.710	104 113	80 398	286.464
3200	100 161	76 034	269.295	106 127	82 000	287.098
3250	102 021	77 478	269.872	108 145	83 601	287.723
3300	103 883	78 922	270.433	110 174	85 202	288.333

TABLE A-19
ONE-DIMENSIONAL ISENTROPIC COMPRESSIBLE-FLOW FUNCTIONS FOR AN IDEAL GAS WITH CONSTANT SPECIFIC HEAT AND MOLECULAR WEIGHT AND $\gamma = 1.4$

M	M^*	$\frac{A}{A^*}$	$\frac{p}{p_0}$	$\frac{\rho}{\rho_0}$	$\frac{T}{T_0}$
0	0	∞	1.000 00	1.000 00	1.000 00
0.10	0.109 43	5.8218	0.993 03	0.995 02	0.998 00
0.20	0.218 22	2.9635	0.972 50	0.980 27	0.992 06
0.30	0.325 72	2.0351	0.939 47	0.956 38	0.982 32
0.40	0.431 33	1.5901	0.895 62	0.924 28	0.968 99
0.50	0.534 52	1.3398	0.843 02	0.885 17	0.952 38
0.60	0.634 80	1.1882	0.784 00	0.840 45	0.932 84
0.70	0.731 79	1.094 37	0.720 92	0.791 58	0.910 75
0.80	0.825 14	1.038 23	0.656 02	0.740 00	0.886 52
0.90	0.914 60	1.008 86	0.591 26	0.687 04	0.860 58
1.00	1.000 00	1.000 00	0.528 28	0.633 94	0.833 33
1.10	1.081 24	1.007 93	0.468 35	0.581 69	0.805 15
1.20	1.1583	1.030 44	0.412 38	0.531 14	0.776 40
1.30	1.2311	1.066 31	0.360 92	0.482 91	0.747 38
1.40	1.2999	1.1149	0.314 24	0.437 42	0.718 39
1.50	1.3646	1.1762	0.272 40	0.394 98	0.689 65
1.60	1.4254	1.2502	0.235 27	0.355 73	0.661 38
1.70	1.4825	1.3376	0.202 59	0.319 69	0.633 72
1.80	1.5360	1.4390	0.174 04	0.286 82	0.606 80
1.90	1.5861	1.5552	0.149 24	0.256 99	0.580 72
2.00	1.6330	1.6875	0.127 80	0.230 05	0.555 56
2.10	1.6769	1.8369	0.109 35	0.205 80	0.531 35
2.20	1.7179	2.0050	0.093 52	0.184 05	0.508 13
2.30	1.7563	2.1931	0.079 97	0.164 58	0.485 91
2.40	1.7922	2.4031	0.068 40	0.147 20	0.464 68
2.50	1.8258	2.6367	0.058 53	0.131 69	0.444 44
2.60	1.8572	2.8960	0.050 12	0.117 87	0.425 17
2.70	1.8865	3.1830	0.042 95	0.105 57	0.406 84
2.80	1.9140	3.5001	0.036 85	0.094 62	0.389 41
2.90	1.9398	3.8498	0.031 65	0.084 89	0.372 86
3.00	1.9640	4.2346	0.027 22	0.076 23	0.357 14
3.50	2.0642	6.7896	0.013 11	0.045 23	0.289 86
4.00	2.1381	10.719	0.006 58	0.027 66	0.238 10
4.50	2.1936	16.562	0.003 46	0.017 45	0.198 02
5.00	2.2361	25.000	$189(10)^{-5}$	0.011 34	0.166 67
6.00	2.2953	53.180	$633(10)^{-6}$	0.005 19	0.121 95
7.00	2.3333	104.143	$242(10)^{-6}$	0.002 61	0.092 59
8.00	2.3591	190.109	$102(10)^{-6}$	0.001 41	0.072 46
9.00	2.3772	327.189	$474(10)^{-7}$	0.000 815	0.058 14
10.00	2.3904	535.938	$236(10)^{-7}$	0.000 495	0.047 62
∞	2.4495	∞	0	0	0

Source: Abridged from J. H. Keenan and J. Kaye, "Gas Tables," John Wiley & Sons, Inc., New York, 1948.

TABLE A-20
ONE-DIMENSIONAL NORMAL-SHOCK FUNCTIONS FOR AN IDEAL GAS WITH CONSTANT SPECIFIC HEAT AND MOLECULAR WEIGHT AND $\gamma = 1.4$

M_x	M_y	$\frac{p_y}{p_x}$	$\frac{\rho_y}{\rho_x}$	$\frac{T_y}{T_x}$	$\frac{p_{0y}}{p_{0x}}$	$\frac{p_{oy}}{p_x}$
1.00	1.000 00	1.000 00	1.000 00	1.000 00	1.000 00	1.8929
1.10	0.911 77	1.245 0	1.169 1	1.064 94	0.998 92	2.1328
1.20	0.842 17	1.513 3	1.341 6	1.128 0	0.992 80	2.4075
1.30	0.785 96	1.805 0	1.515 7	1.190 9	0.979 35	2.7135
1.40	0.739 71	2.120 0	1.689 6	1.254 7	0.958 19	3.0493
1.50	0.701 09	2.458 3	1.862 1	1.320 2	0.929 78	3.4133
1.60	0.668 44	2.820 1	2.031 7	1.388 0	0.895 20	3.8049
1.70	0.640 55	3.205 0	2.197 7	1.458 3	0.855 73	4.2238
1.80	0.616 50	3.613 3	2.359 2	1.531 6	0.812 68	4.6695
1.90	0.595 62	4.045 0	2.515 7	1.607 9	0.767 35	5.1417
2.00	0.577 35	4.500 0	2.666 6	1.687 5	0.720 88	5.6405
2.10	0.561 28	4.978 4	2.811 9	1.770 4	0.674 22	6.1655
2.20	0.547 06	5.480 0	2.951 2	1.856 9	0.628 12	6.7163
2.30	0.534 41	6.005 0	3.084 6	1.946 8	0.583 31	7.2937
2.40	0.523 12	6.553 3	3.211 9	2.040 3	0.540 15	7.8969
2.50	0.512 99	7.125 0	3.333 3	2.137 5	0.499 02	8.5262
2.60	0.503 87	7.720 0	3.448 9	2.238 3	0.460 12	9.1813
2.70	0.495 63	8.338 3	3.559 0	2.342 9	0.423 59	9.8625
2.80	0.488 17	8.980 0	3.663 5	2.451 2	0.389 46	10.569
2.90	0.481 38	9.645 0	3.762 9	2.563 2	0.357 73	11.302
3.00	0.475 19	10.333	3.857 1	2.679 0	0.328 34	12.061
4.00	0.434 96	18.500	4.571 4	4.046 9	0.138 76	21.068
5.00	0.415 23	29.000	5.000 0	5.800 0	0.061 72	32.654
10.00	0.387 57	116.50	5.714 3	20.388	0.003 04	129.217
∞	0.377 96	∞	6.000	∞	0	∞

Source: Abridged from J. H. Keenan and J. Kaye, "Gas Tables," John Wiley & Sons, Inc., New York, 1948.

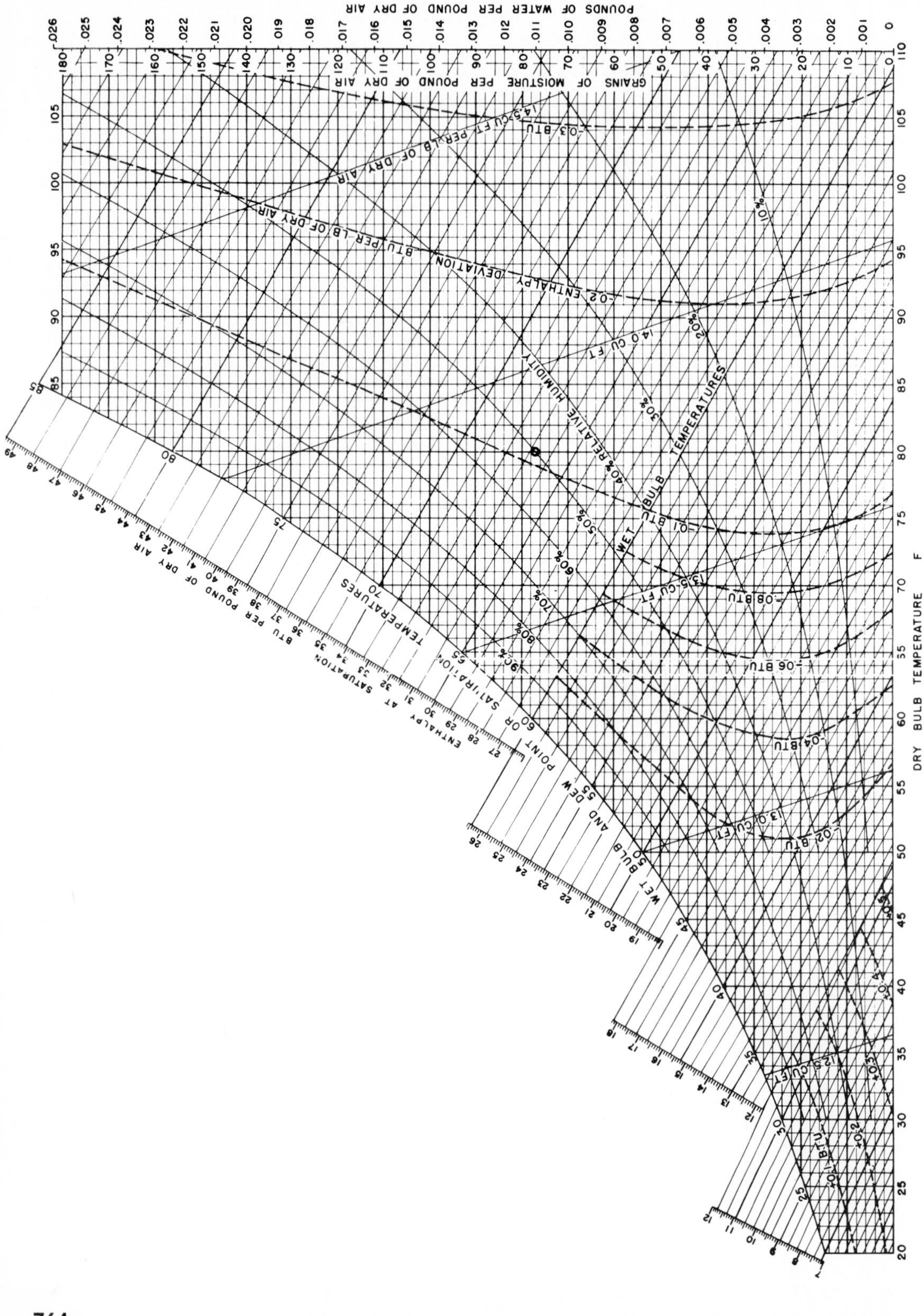

FIG. A-1 Psychrometric chart, normal temperatures, English units, $p = 1$ atm. (Carrier Corporation.)

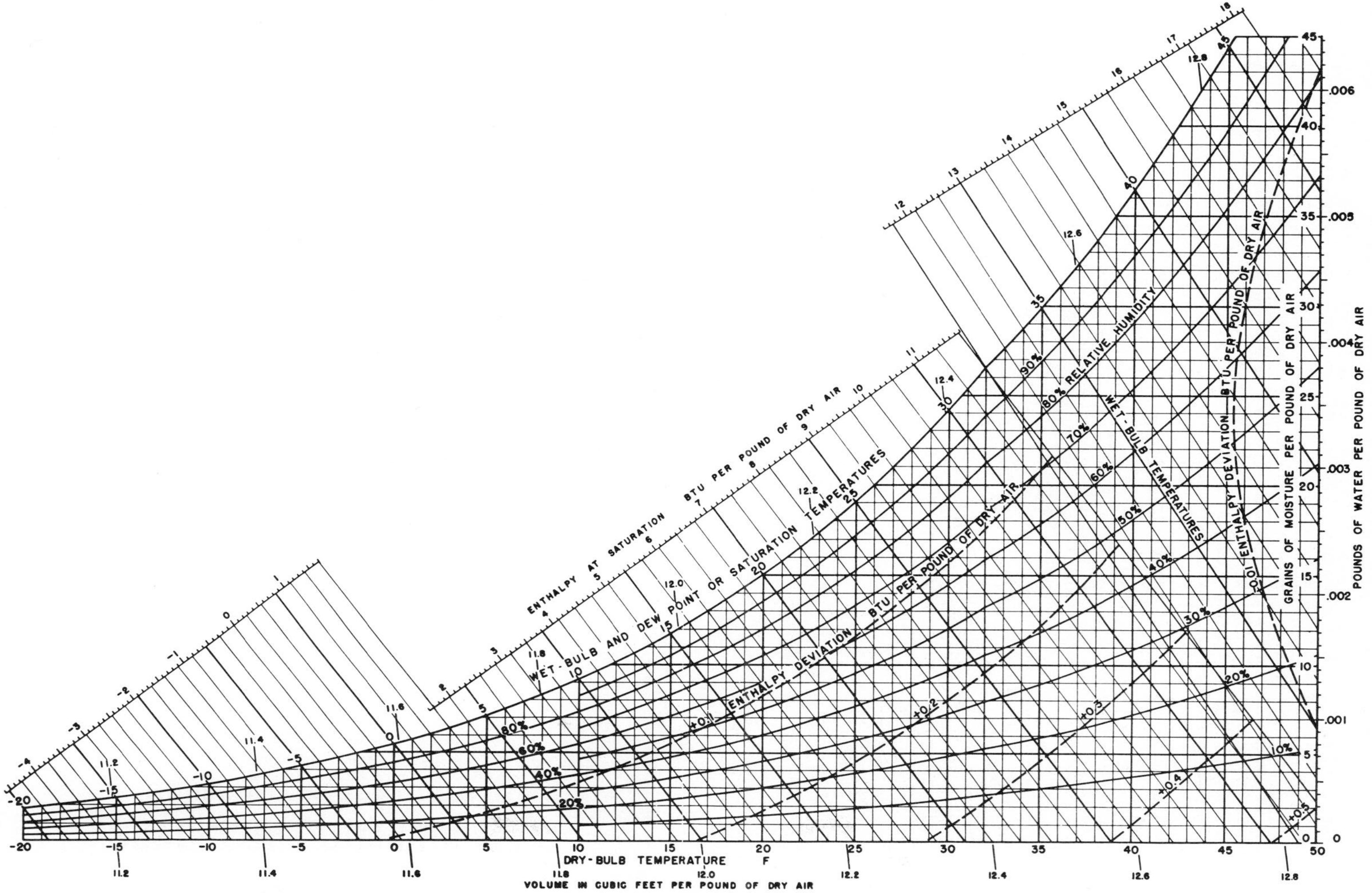

FIG. A-2 Psychrometric chart, low temperatures, English units, $p = 1$ atm. (Carrier Corporation.)

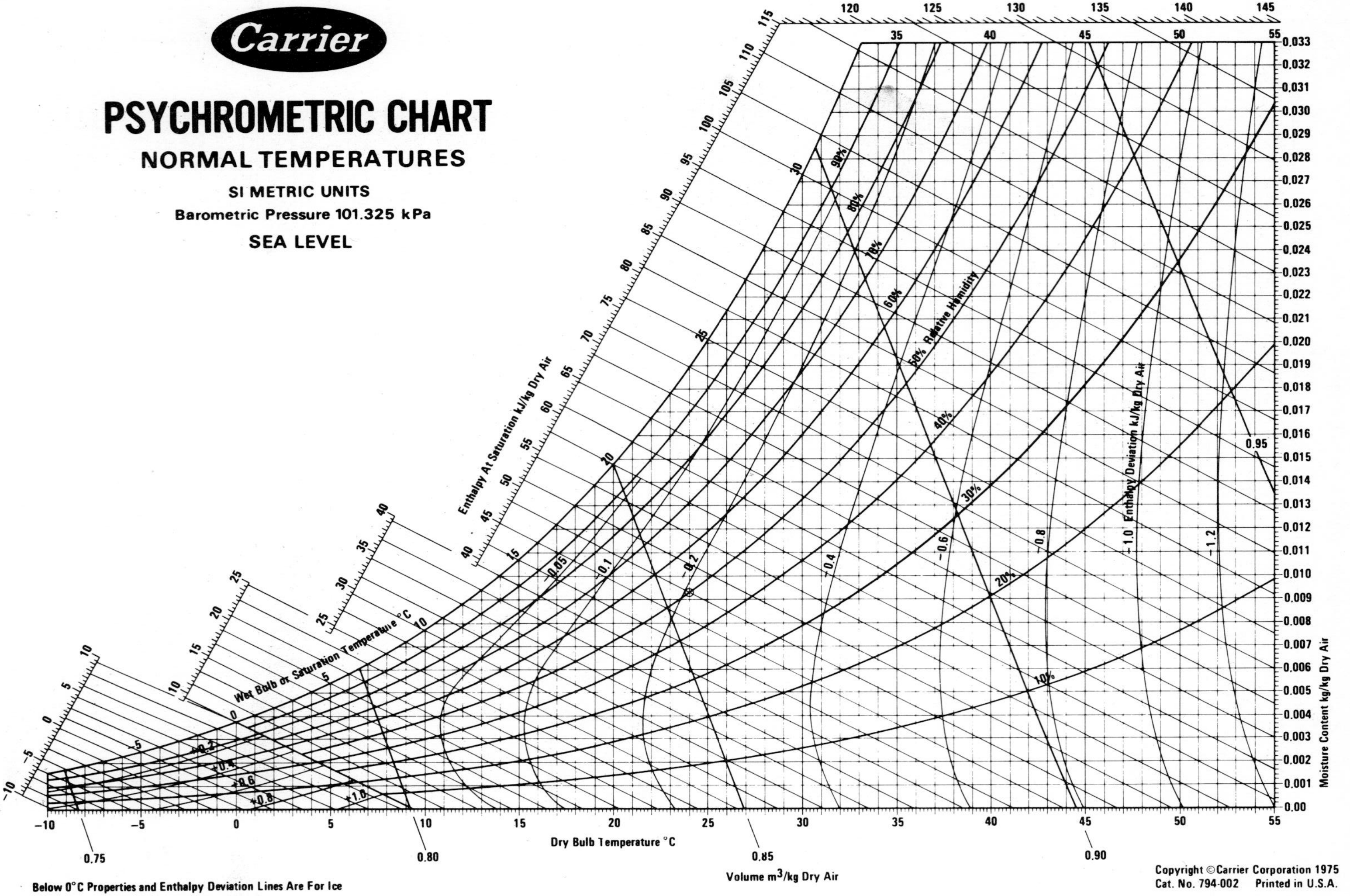

FIG. A-3 Psychrometric chart, normal temperatures, SI units, $p = 1$ atm. (Carrier Corporation.)

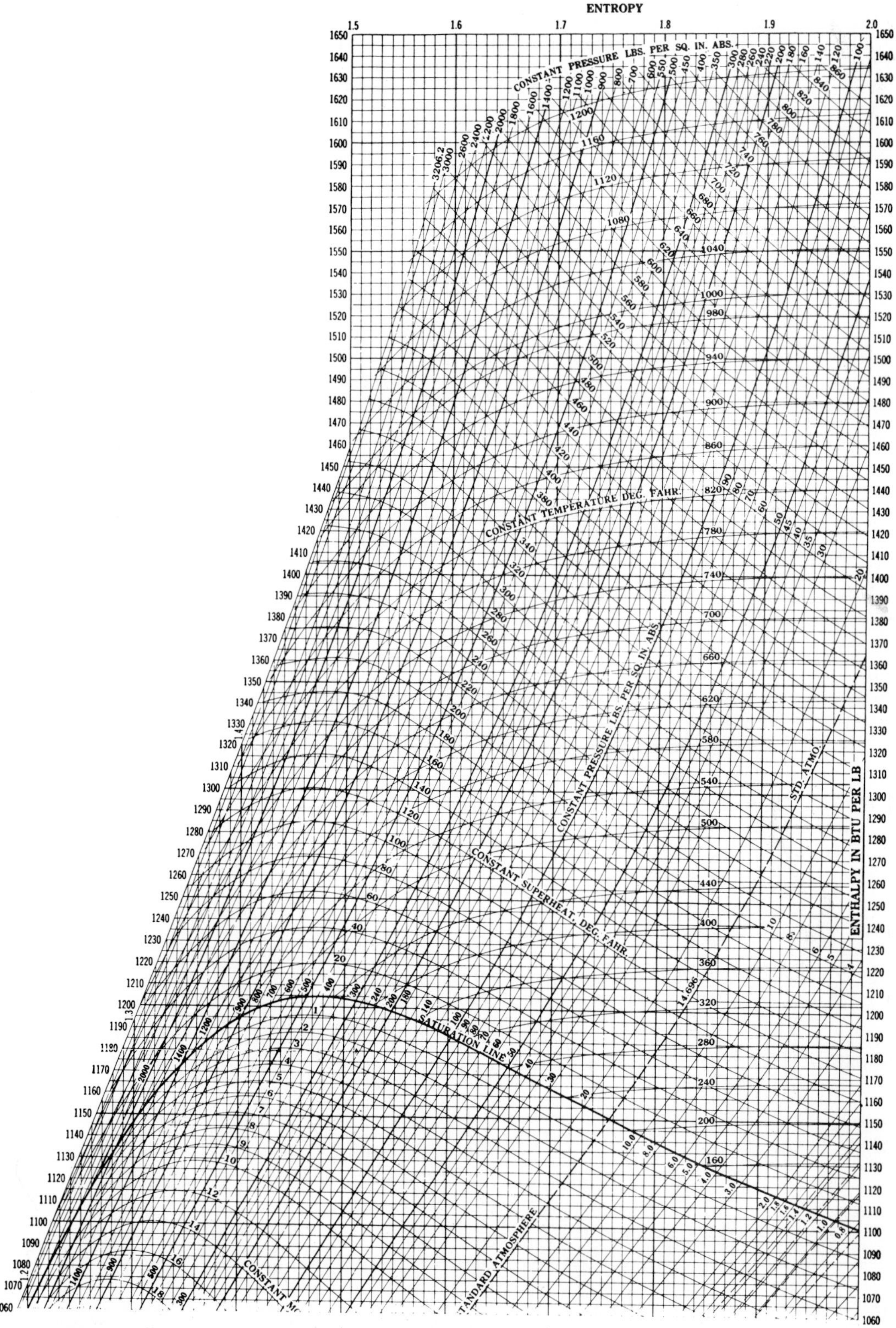

FIG. A-4 Mollier diagram for water vapor, English units.

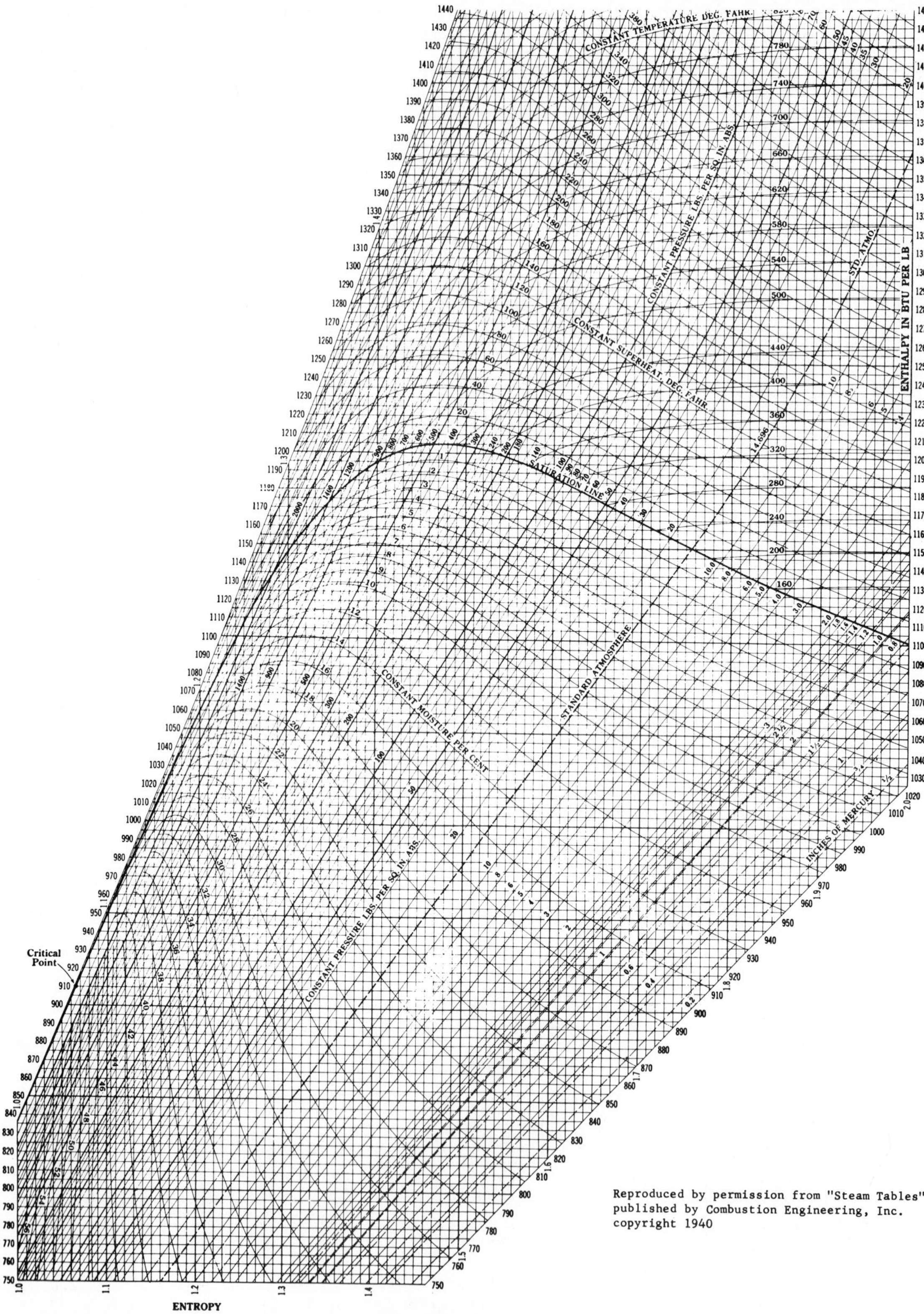
CONSTANT TEMPERATURE DEG. FAHR.
CONSTANT PRESSURE LBS. PER SQ. IN. ABS.
CONSTANT SUPERHEAT, DEG. FAHR.
STD. ATMO.
ENTHALPY IN BTU PER LB
SATURATION LINE
CONSTANT MOISTURE PER CENT
STANDARD ATMOSPHERE
INCHES OF MERCURY
Critical Point
ENTROPY

FIG. A-4 (cont.)

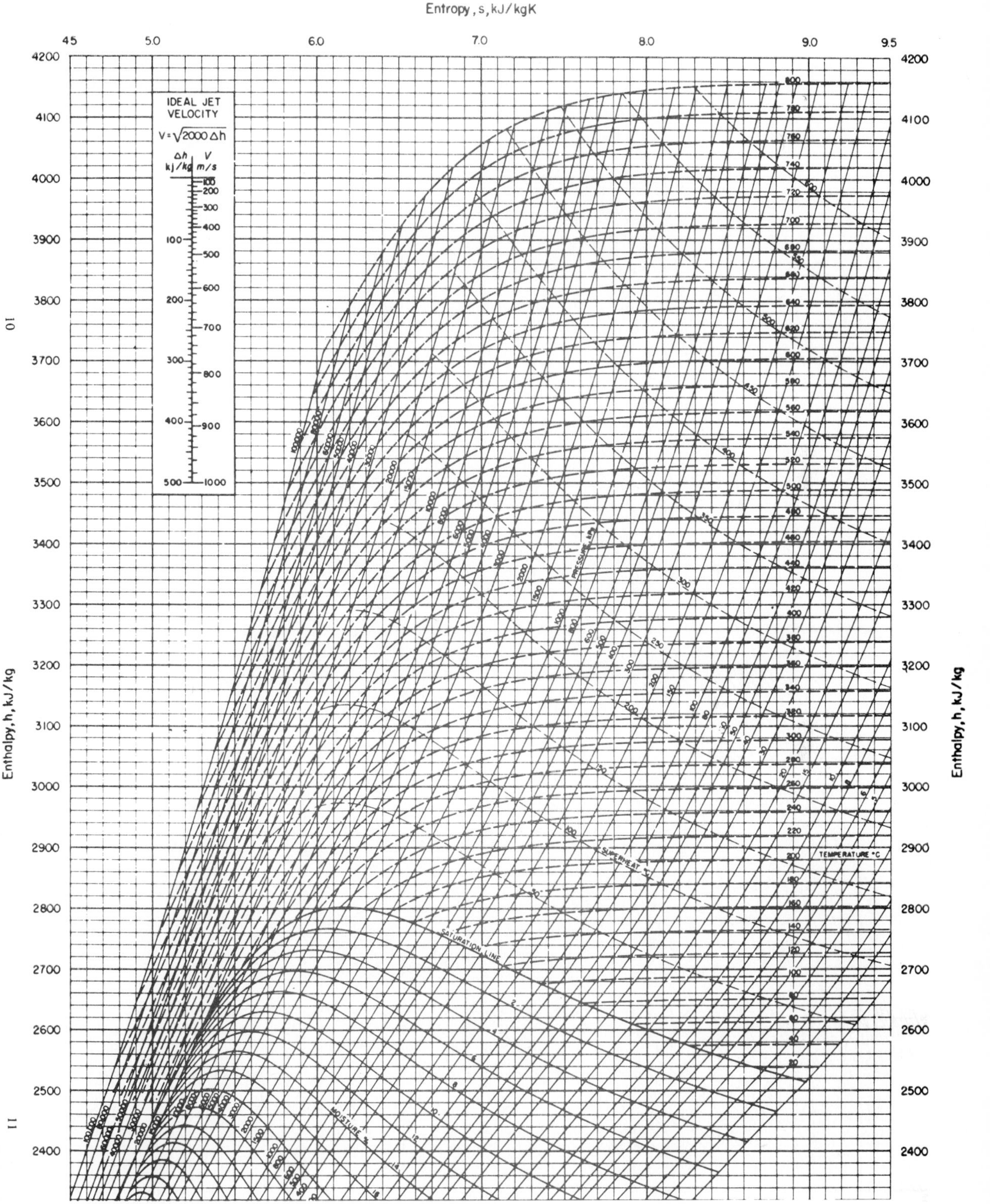

FIG. A-4M Mollier diagram for water vapor, SI units. (By permission of American Society of Mechanical Engineers.)

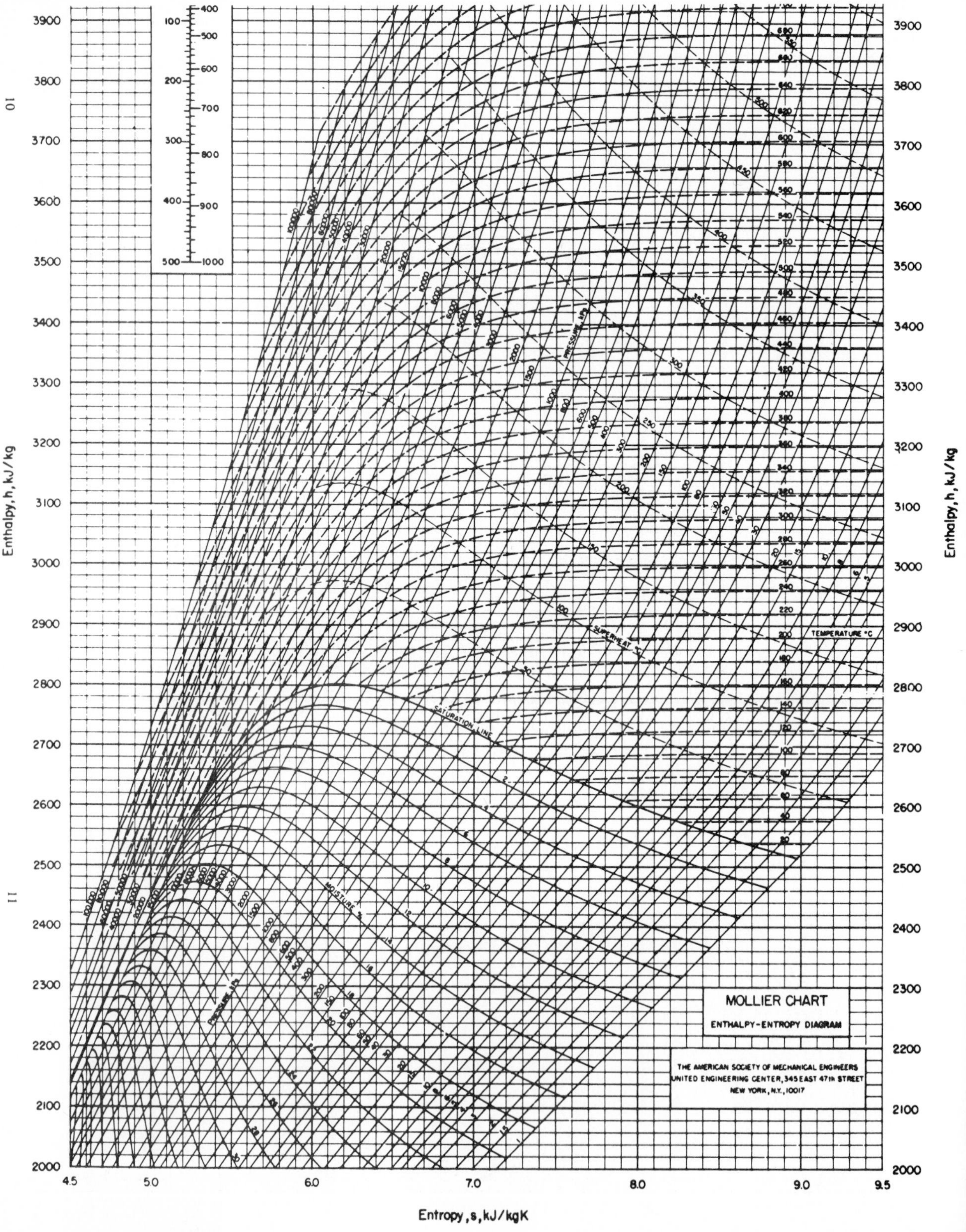
MOLLIER CHART
ENTHALPY-ENTROPY DIAGRAM
THE AMERICAN SOCIETY OF MECHANICAL ENGINEERS
UNITED ENGINEERING CENTER, 345 EAST 47th STREET
NEW YORK, N.Y., 10017
Enthalpy, h, kJ/kg
Entropy, s, kJ/kgK
SATURATION LINE
MOISTURE %
SUPERHEAT °C
PRESSURE, kPa
TEMPERATURE °C
4.5
5.0
6.0
7.0
8.0
9.0
9.5
2000
2100
2200
2300
2400
2500
2600
2700
2800
2900
3000
3100
3200
3300
3400
3500
3600
3700
3800
3900

FIG. A-4M (cont.)

B

PROPERTIES FOR HEAT-TRANSFER CALCULATIONS

TABLE B-1
PROPERTIES OF METALS

Metal	T °C	T °F	ρ, kg/m^3	c_p, kJ/kg·°C	k, W/m·°C
Aluminum, pure	20	68	2707	0.896	204
	200	392	. . .	. . .	215
	400	752	. . .	. . .	249
Lead	20	68	11373	0.130	35
	300	572	. . .	. . .	29.8
Iron					
Pure	20	68	7897	0.452	73
	300	572	. . .	. . .	55
	1000	1832	. . .	. . .	35
Wrought	20	68	7849	0.46	59
Carbon steel (max. 0.5% C)	20	68	7833	0.465	54
Carbon steel (1.5% C)	20	68	7753	0.486	36
	400	752	. . .	. . .	33
	1200	2192	. . .	. . .	29
Stainless steel	20	68	. . .	. . .	12–45
Copper					
Pure	20	68	8954	0.383	386
	300	572	. . .	. . .	369
	600	1112	. . .	. . .	353
Bronze (75% Cu, 25% Zn)	20	68	8666	0.343	26
Brass (70% Cu, 30% Zn)	20	68	8522	0.385	111
Silver, pure	20	68	10524	0.234	407
Tungsten	20	68	19350	0.134	163

TABLE B-2
THERMAL CONDUCTIVITY OF SOME NONMETALS

Substance	T, °C	k, W/m·K
Structural and heat-resistant materials		
Asphalt	20–55	0.74
Brick		
Building brick, common	20	0.69
Building brick, face	. . .	1.31
Diatomaceous earth,	200	0.24
molded and fired	870	0.31
Fireclay brick, burnt at 2426°F	500	1.04
	800	1.07
	1100	1.09
Cement, portland	. . .	0.29
Concrete, stone 1-2-4 mix	20	1.37
Glass, window	20	0.78
Plaster, gypsum	21	0.48
Plaster, metal lath	21	0.47
Plaster, wood lath	21	0.28
Stone:		
Granite	. . .	1.73–3.98
Limestone	100–300	1.26–1.33
Wood, across the grain		
Balsa, 8.8 lb/ft^3	30	0.055
Yellow pine	24	0.147
White pine	30	0.112
Insulating material		
Asbestos		
Asbestos-cement boards	20	0.744
Asbestos sheets	50	0.166
Asbestos cement	. . .	2.07
Asbestos, loosely packed	−45	0.148
	0	0.154
	100	0.160
Corkboard, 10 lb/ft^3	30	0.043
Diatomaceous earth (Sil-o-cel)	0	0.061
Fiber insulating board	21	0.048
Glass wool, 1.5 lb/ft^3	24	0.038
Kapok	30	0.035
Magnesia, 85%	38	0.067
	93	0.071
	150	0.074
	204	0.080
Sawdust	23	0.059
Rock wool, 10 lb/ft^3	32	0.040

TABLE B-3
PROPERTIES OF AIR AT ATMOSPHERIC PRESSURE

The values of μ, k, c_p, and Pr are not strongly pressure-dependent and may be used over a fairly wide range of pressures.

T, K	ρ, kg/m^3	c_p, kJ/kg·°C	μ, kg/m·s × 10^5	ν, m^2/s × 10^6	k, W/m·°C	α, m^2/s × 10^4	Pr
100	3.6010	1.0266	0.6924	1.923	0.009 246	0.025 01	0.770
150	2.3675	1.0099	1.0283	4.343	0.013 735	0.057 45	0.753
200	1.7684	1.0061	1.3289	7.490	0.01809	0.101 65	0.739
250	1.4128	1.0053	1.5990	11.31	0.02227	0.156 75	0.722
300	1.1774	1.0057	1.8462	15.69	0.02624	0.221 60	0.708
350	0.9980	1.0090	2.075	20.76	0.03003	0.2983	0.697
400	0.8826	1.0140	2.286	25.90	0.03365	0.3760	0.689
450	0.7833	1.0207	2.484	31.71	0.03707	0.4222	0.683
500	0.7048	1.0295	2.671	37.90	0.04038	0.5564	0.680
550	0.6423	1.0392	2.848	44.34	0.04360	0.6532	0.680
600	0.5879	1.0551	3.018	51.34	0.04659	0.7512	0.680
650	0.5430	1.0635	3.177	58.51	0.04953	0.8578	0.682
700	0.5030	1.0752	3.332	66.25	0.05230	0.9672	0.684
750	0.4709	1.0856	3.481	73.91	0.05509	1.0774	0.686
800	0.4405	1.0978	3.625	82.29	0.05779	1.1951	0.689
850	0.4149	1.1095	3.765	90.75	0.06028	1.3097	0.692
900	0.3925	1.1212	3.899	99.3	0.06279	1.4271	0.696
950	0.3716	1.1321	4.023	108.2	0.06525	1.5510	0.699
1000	0.3524	1.1417	4.152	117.8	0.06752	1.6779	0.702
1100	0.3204	1.160	4.44	138.6	0.0732	1.969	0.704
1200	0.2947	1.179	4.69	159.1	0.0782	2.251	0.707
1300	0.2707	1.197	4.93	182.1	0.0837	2.583	0.705
1400	0.2515	1.214	5.17	205.5	0.0891	2.920	0.705
1500	0.2355	1.230	5.40	229.1	0.0946	3.262	0.705
1600	0.2211	1.248	5.63	254.5	0.100	3.609	0.705
1700	0.2082	1.267	5.85	280.5	0.105	3.977	0.705
1800	0.1970	1.287	6.07	308.1	0.111	4.379	0.704
1900	0.1858	1.309	6.29	338.5	0.117	4.811	0.704
2000	0.1762	1.338	6.50	369.0	0.124	5.260	0.702
2100	0.1682	1.372	6.72	399.6	0.131	5.715	0.700
2200	0.1602	1.419	6.93	432.6	0.139	6.120	0.707
2300	0.1538	1.482	7.14	464.0	0.149	6.540	0.710
2400	0.1458	1.574	7.35	504.0	0.161	7.020	0.718
2500	0.1394	1.688	7.57	543.5	0.175	7.441	0.730

Source: From *Natl. Bur. Stand. (U.S.) Circ.* 564, 1955.

TABLE B-4
PROPERTIES OF WATER (SATURATED LIQUID)

°F	°C	c_p, kJ/kg·°C	ρ, kg/m³	μ, kg/m·s	k, W/m·°C	Pr	$\frac{g\beta\rho^2 c_p}{\mu k}$, 1/m³·°C
32	0	4.225	999.8	1.79×10^{-3}	0.566	13.25	
40	4.44	4.208	999.8	1.55	0.575	11.35	1.91×10^{9}
50	10	4.195	999.2	1.31	0.585	9.40	6.34×10^{9}
60	15.56	4.186	998.6	1.12	0.595	7.88	1.08×10^{10}
70	21.11	4.179	997.4	9.8×10^{-4}	0.604	6.78	1.46×10^{10}
80	26.67	4.179	995.8	8.6	0.614	5.85	1.91×10^{10}
90	32.22	4.174	994.9	7.65	0.623	5.12	2.48×10^{10}
100	37.78	4.174	993.0	6.82	0.630	4.53	3.3×10^{10}
110	43.33	4.174	990.6	6.16	0.637	4.04	4.19×10^{10}
120	48.89	4.174	988.8	5.62	0.644	3.64	4.89×10^{10}
130	54.44	4.179	985.7	5.13	0.649	3.30	5.66×10^{10}
140	60	4.179	983.3	4.71	0.654	3.01	6.48×10^{10}
150	65.55	4.183	980.3	4.3	0.659	2.73	7.62×10^{10}
160	71.11	4.186	977.3	4.01	0.665	2.53	8.84×10^{10}
170	76.67	4.191	973.7	3.72	0.668	2.33	9.85×10^{10}
180	82.22	4.195	970.2	3.47	0.673	2.16	1.09×10^{11}
190	87.78	4.199	966.7	3.27	0.675	2.03	
200	93.33	4.204	963.2	3.06	0.678	1.90	
220	104.4	4.216	955.1	2.67	0.684	1.66	
240	115.6	4.229	946.7	2.44	0.685	1.51	
260	126.7	4.250	937.2	2.19	0.685	1.36	
280	137.8	4.271	928.1	1.98	0.685	1.24	
300	148.9	4.296	918.0	1.86	0.684	1.17	
350	176.7	4.371	890.4	1.57	0.677	1.02	
400	204.4	4.467	859.4	1.36	0.665	1.00	
450	232.2	4.585	825.7	1.20	0.646	0.85	
500	260	4.731	785.2	1.07	0.616	0.83	
550	287.7	5.024	735.5	9.51×10^{-5}			
600	315.6	5.703	678.7	8.68			

Source: Adapted from A. I. Brown and S. M. Marco, "Introduction to Heat Transfer," 3d ed., McGraw-Hill Book Company, New York, 1958.

INDEX

An introductory MONEY-SAVING book offer. . .

For: Engineering Students Using Holman's THERMODYNAMICS, 4/e

From: McGraw-Hill's Professional and Reference Division

The textbook selected for use in your current course was published by the College Division of McGraw-Hill Book Company. We're proud of this division's deserved reputation as a premier name in quality educational publishing.

As you advance through the college program you're now pursuing, we want you to know about some of the renowned reference books, published by the Professional & Reference Division, that you can use right now (as supplemental course references) and also in the coming years as working professionals.

To introduce our division to tomorrow's practicing engineers, we have a Special Offer that will *save you 40%* on the purchase of a best-selling reference Handbook—the type of book that engineers turn to again and again in their day-to-day work.

Here's how the offer works. When you've completed this course, fill out the information at the bottom of this page and return your copy of THERMODYNAMICS, 4/e to us. As soon as we receive *your* book, we'll send you the perfect companion reference—HANDBOOK OF HEAT TRANSFER APPLICATIONS, 2/e for 15 days' FREE examination. What's more, if you decide to keep the book, *you'll save $33.00* off the regular price of $82.50 (see details below).

Edited by Warren M. Rohsenow, James P. Hartnett, and Ejup N. Ganić and written by a team of world-renowned experts, HANDBOOK OF HEAT TRANSFER APPLICATIONS, 2/e is *the* source for all major developments and data in heat transfer practice. Nine of the twelve sections in this second edition are brand-new, and nearly all the material in the other sections has been revised and updated. In 976 pages, with over 500 illustrations, you'll find complete coverage of mass transfer cooling, heat exchangers, heat pipes, thermal energy storage, cooling towers and ponds, geothermal heat transfer . . . and much more. This Handbook will prove an invaluable tool for anyone who needs a comprehensive reference on today's developments in *applied* heat transfer.

Check the box below, fill in your name and address, and mail this copy of Holman's THERMODYNAMICS, 4/e to:

Edward T. Matthews
McGraw-Hill Book Company
Professional and Reference Division
11 West 19th Street—4th Floor
New York, NY 10011

☐ Here's my copy of Holman's THERMODYNAMICS, 4/e. In return, I understand you'll send me a copy of HANDBOOK OF HEAT TRANSFER APPLICATIONS, 2/e (053553-1) for 15 days' FREE examination on approval. At the end of that time, I can either keep the book and remit $49.50, plus local tax, postage and handling—**a SAVINGS of $33.00**—or return the book with no further obligation.

Name *(Please Print)* ______________________________

Permanent Mailing Address ______________________________

City ____________________ State __________ Zip __________

All orders subject to acceptance by McGraw-Hill.